HumAnIty

Should We Be Frightened by AI Super-Robots?

> *"We can be overwhelmed by the complexity of the problems, or we can fall in love with the creativity and the possibility of the solution."*
>
> **Mary Heglar**

By

Howard Lack

Table of Contents

I dedicate this book to my wonderful children Oliver, Amelia and Bella.

Acknowledgements

This book has taken three years of research and writing and many days when the topic appeared insurmountable given its breadth and complexity. What allowed me to persevere was the support and encouragement of my wonderful children Oliver, Amelia and Bella, my twin brother Kelvin and my partner Deborah. Each in different ways has allowed me to reconnect with why I wrote the book. My children have all been taught to have curiosity and to embrace creativity and risk- taking and as they have grown into accomplished adults, I have seen that these qualities are central to who they have become. My brother Kelvin's insatiable thirst for enquiry and understanding has always inspired me. I am indebted to Deborah's love and patience in watching me spend thousands of hours at the computer whilst in the US and UK and tolerating my ups and downs in completing this project.

Many of the questions that I have posed in the book in the context of what it means to be human are those you encounter as a parent and in your relationships with others. Without those fundamental relationships and the fulfilment and forging of bonds that come with them, we might never ask ourselves the important questions about life and purpose and why we are as we are. Why do we make our choices, and decisions, have dreams and aspirations and search for something beyond the material self? My family and my friends are the reason I have asked myself those questions and are what made me want to capture what it means to be human. In the modern world with the inexorable march of AI and the rapidly increasing presence of AI robots in everyday life, our specialness as a species is under the spotlight and now is the time to seek answers.

Any misconceptions or false conclusions are mine alone as I have worked very much based on my research and I apologise to any of the great and superior minds out there if I have failed in my humble attempt to reflect their intellectual stances or the implications of their work. At the end of the day, we all face a battle between subjectivity and objectivity, of habit and change and experience and new learning but I hope I have maintained a healthy balance. Ultimately though it is you the reader who will bring your preconceptions, biases and experiences to these pages. In advance, I want to acknowledge those readers who have the curiosity, attention span, motivation and willingness to bring their attention and time to the topics that I have covered in the book. To me these topics are key to understanding ourselves as individuals, in our close relationships and as we intersect within communities and generally as a species. Anyone who shares that journey with me by reading my book has in advance my appreciation and thanks.

"I do not want to be human. I want to be myself. They think I'm a lion and that I will chase them. I will not deny that I have lions in me. I am the monster in the woods. I have wonders in my house of sugar. I have parts of myself I do not yet understand. I am not a Good Robot. To tell a story about a robot who wants to be human is a distraction. There is no difference. Alive is alive. There is only one verb that matters: to be."

Catherynne M Valente. American fiction writer, poet, and literary critic.

Chapter 1

Introduction

CHAPTER 1

Robots

In November 2023, China announced to the world its plan to mass-produce thousands of GR-1 humanoid robots that within two years will be as "disruptive as smartphones "and "reshape the world."[1] US-based Agility Robotics is opening a robot factory in Oregon to make robots that "mimic human movements like walking, crouching, and carrying packages." Elon Musk announced the release in 2024 of the generation of humanoid robots called Tesla Bots, which will be 5'8" tall, weigh around 125 pounds, and be designed to perform repetitive, boring or dangerous tasks with current specifications giving a top speed of 5 MPH, a lifting capacity of 45 pounds, dead lift of 150 pounds, and an arm-extended lift of 10 pounds. Tesla claims to have deliberately built-in limitations to provide the consumer with comfort that the robot will not try to control us, run amok or be able to outrun or overpower a human. The era of smart AI robots is upon us.

Professor of AI and robotics Ken Goldberg observes how looking into the eyes of robots should make us reflect on our human qualities, strengths and weaknesses, individually and collectively and make us question what it means to be human perhaps because they remind us of our perceived specialness. Much research on Artificial Intelligence (AI) and robotics over the last 15 years has been led by the Australian author and Panasonic Professor of Robotics at the MIT AI Laboratory, Rodney Brooks (1954–) who observed how "We have programmed ourselves to believe in our tribal specialness as proof of our uniqueness" but are essentially machines. He believes that the "body, this mass of biomolecules, is a machine that acts according to a set of specifiable rules. We are machines, as are our spouses, children, and dogs... I believe myself and my children to be mere machines. But this is not how I treat them. I treat them in a very special way, and I interact with them on an entirely different level. They have my unconditional love; the furthest one might be able to get from rational analysis. Like a religious scientist, I maintain two sets of inconsistent beliefs and act on each in different circumstances. I think this transcendence between belief systems will ultimately enable humanity to accept robots as emotional machines and, after that, start to empathize with them and attribute free will, respect, and rights to them... When our robots improve enough beyond their current limitations, and when we credit humans, we will also break our mental barrier, need, and desire to retain tribal specialness, differentiating ourselves from them."[2] Computer scientist Sebastian Thrun(1967–) defines the field of robotics as "The science of perceiving and manipulating the physical world through computer-controlled mechanical devices"[3] but over the last twenty years, android science has brought together robotics, the cognitive sciences, and AI with the purpose of making robots that physically resemble people in their appearance, movements, reactions, and perceptual abilities. The Latin root android means resembling a man while gynoids are gendered feminine but we will use android to generically represent both androids, gynoids or any other gender identity to avoid confusion or controversy. As of writing, several companies have developed androids, for example, Nadine, Jia Jia, Geminoid DK, and Sophie, etc, all of whom resemble humans in realistic ways with human-like eyes and skin, walking, talking, and interacting with humans, answering questions, and demonstrating a range of apparent emotions and human behaviours, including recognizing people, making eye contact, displaying micro facial expressions, and remembering conversations. The most advanced android to date is Sophie, designed to acquire social skills and is equipped with artificial neural networks and AI that allow her to appear to understand human gestures and emotions and achieve a realistic interface with people. Sophie's behaviour is yet to reflect genuine understanding rather than mimicking human behaviour, but the implications for whether this makes them more of or less of a threat to us will not be clear until we agree on what we want and expect from such robots. Driven by industrial and consumer applications, research into humanoid and android robots is proceeding rapidly, as is their ability to perform some specific tasks far better than humans.

The last few years have seen rapid progress in developing robots of all shapes and sizes, including nano-robots, biorobots, and medical robots using nanotechnology and biological and medical engineering and combining these with robot technology. Researchers are finding ever ingenious ways of increasing robotic intelligence, reducing the physical size of robots, making them lifelike and even harnessing quantum technology.

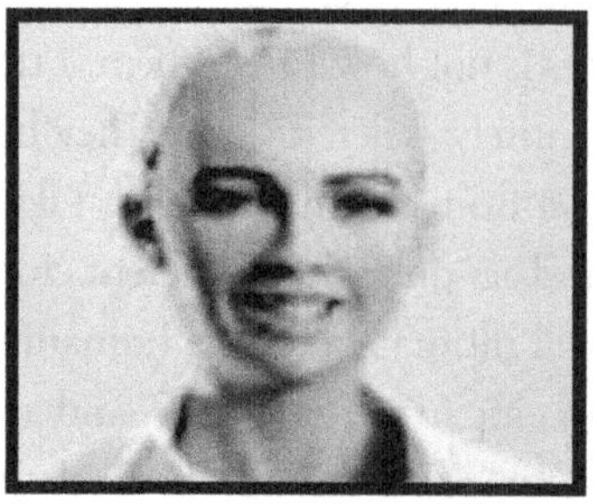

Sophie[4]

Geminoid DK[5]

The German online research firm Statista recently estimated that the global market for industrial robots will more than double from U.S. $81 billion in 2021 to over U.S. $165 billion in 2028.[6] Most robots currently operate in very controlled and programmed environments to make our lives easier but the goal is to make robots "autonomous agents" operating in real-world environments without constant reprogramming. Xabier Barandiaran, who focuses his research on the trans-disciplinary intersection between the cognitive sciences and complex systems, together with colleagues, set out three conditions that a system must meet to be considered a genuine AI agent. These are that a system must, "(1) Define its individuality (2) Be the active source of activity in its environment (Interactional asymmetry) and (3) Regulate this activity about certain norms." They define agency as "An autonomous organization that adaptively regulates its coupling with its environment and contributes to sustaining itself as a consequence."[7] We will use this definition of agency and the definition of autonomy provided by UNESCO which requires that autonomous, robots must possess "the ability to carry out tasks and achieve goals to some degree or in certain respects in novel contexts or environments, without complete external oversight or control" which involves making decisions, planning courses of action and setting proximate goals.

Friends or Foe?

Philosopher Brian Hutler and colleagues point out that "In addition to the technical challenges, designers need to determine an appropriate sphere of responsibility for robots and to specify which of various types of harms must be avoided or prevented."[8] In an open letter to AI magazine in 2015, several eminent scientists across a range of disciplines sought to establish the research priorities for "Robust and Beneficial AI" to ensure that "our AI systems must do what we want them to do."[9] They vigorously contest the characterization that they are anti-AI but are aiming to provide guidelines for the safe development of AI, particularly autonomous AI. They listed short and long-term research priorities aimed at ensuring that this goal is adhered to, observing that "Success in the quest for AI has the potential to bring unprecedented benefits to humanity, and it is, therefore, worthwhile to investigate how to maximise these benefits while avoiding potential pitfalls." Their short and long-term priorities are as follows, "Short-term to (1) Optimise the economic impact of AI (2) Address important ethical and legal questions (3) Ensure that AI systems can

satisfy certain desired formal requirements (verification) (4) Ensure that the system does not have unwanted behaviours and consequences (validity), the prevention of intentional manipulation by unauthorised parties (security) (5) Ensure the ability to control the system (control). Long-term to (1) Constantly reprise the themes of short-term research, to eliminate large classes of bugs and problems in general AI systems and (2) Focus on the types of unexpected generalisation that would be most problematic for very general and capable AI systems. This includes verification which embraces reasoning and decision under bounded computational resources, considering correlations between AI systems' behaviours and those of their environments or other agents, and how agents that are embedded in their environments should reason which includes uncertainty over logical consequences of beliefs or other deterministic computation to align the values of powerful AI systems with our values and preferences (3) Security which envisages as AI systems become more complex and networked together, intelligently managing their trust, which should motivate research into statistical behavioural trust establishment and (4) Control in preventing undesired consequences.[10]

In 2016, researchers at Stanford University started publishing a report every five years called, "A one hundred year study of AI" focusing on how AI is changing people's lives. This highlighted concerns that "We could one day lose control of AI systems via the rise of super intelligences that do not act in accordance with human wishes and that such powerful systems would threaten humanity."[11] Laurent Orseau and Stuart Armstrong at Deep Mind who are investigating ways to ensure that robots will not be able to learn how to prevent being interrupted by the environment or a human operator, suggested that in such a scenario a "Big red button" could serve as an off switch. Such concerns are reflected in the title of the 2021 report called Gathering Strength, Gathering Storms[12], an interdisciplinary view from insiders of the field of AI. President of the Future of Life Institute, Max Tegmark (1958–), famously observed that "Everything we love about civilization is a product of intelligence, so amplifying our human intelligence with AI has the potential of helping civilization flourish like never before – as long as we manage to keep the technology beneficial."

We must determine whether the technology is beneficial and remain alert to the cyber-armageddon warnings. The American author, scientist, inventor, and futurist Ray Kurzweil (1948–), postulated that in 2045, Robo-sapiens would supersede homo-sapiens and "to this extent, our independent biological evolution will come to an end, and we will merge with and be dependent upon the super intelligent and autonomous machine world."[13] The great American science fiction writer and biochemist Isaac Asimov (1920–1992) asked, "Can we help but wonder whether computers and robots may not eventually replace any human ability and render us obsolete? Whether artificial intelligence of our creation is fated to replace us as the dominant entities on the planet should make us all pause to consider the limitations and fallibility of human knowledge and caution against a hubristic and arrogant belief in ourselves as a superior and privileged species.

Swedish-born philosopher Nick Bostrom (1973–) who has a background in theoretical physics, computational neuroscience, logic, and AI, has raised significant ethical concerns about the future of machines that potentially possess Artificial Super Intelligence (ASI). Bostrom describes ASI as "not just another technology, another tool that will add incrementally to human capabilities" but as "any intellect that vastly outperforms the best human brains in practically every field, including scientific creativity, general wisdom, and social skills."[14] He warns that the transition from the human level to artificial general intelligence (AGI) to ASI may be much more sudden than we realize and that we should determine the goal of ASI in advance while recognizing the prospect that ASI robots may determine their own goals. Bostrom believes that the top goal should be friendliness. We currently have no real idea of whether AI will enhance or dilute what it means to possess knowledge and wisdom and may discover that compared to androids, humans are simply meat machines with no free will and all our achievements will ultimately be reduced to zero.

Anthropomorphising or Dehumanising

Because androids potentially look and behave like humans, we must avoid anthropomorphizing them, which means "looking for and attributing human characteristics or behaviour to a god, animal, or object, which is something we do with the supernatural, and other biological or mechanical entities."[15] Humans do this with animals, objects or concepts by imbuing them with voices, thoughts, feelings, faces, physical characteristics, morals, and values. For example, in 2015, the robotic company Boston Dynamics posted a video of their dog-like robot called Spot being kicked by people to show how stable it was but the images prompted many complaints from the public, even though it was clear that Spot was just a robot and experienced no pain.

We have anthropomorphised non-human entities since the Greek philosopher, theologian and poet Xenophanes described the similarities between religious followers and their gods.[16] We see anthropomorphised objects and animals in popular films like Finding Nemo, Shaun the Sheep, Toy Story, Walli, and R2D2, etc, that entertain children and adults alike. We all blame our computers, TVs, cars, etc, when they go wrong as though it's their fault and scientists have even observed that when we stick a face on a computer, we attribute personality traits to the computer. Strangers change their personas presenting themselves more positively to the computer or robot.[17] Scientists have even demonstrated that we use the same brain processes when we represent the actions of industrial robots as when we represent another person's actions, although do this more when objects or animals aren't simply repetitive in their movements but appear goal-directed.[18] We attribute sensory features to plants and natural phenomena and talk to them believing they can receive our communication but we must avoid attributing human qualities to them that don't exist.[19]

Scientists believe we anthropomorphise for three reasons based on our thinking and motivation. The first is when the object or animal evokes or activates knowledge about humans, particularly us. Hence, the more similar it is in speed and smoothness of movement or response, the more it will do this.[20] Secondly, we aim to look for social connections, like the American actor Tom Hanks with his friend Wilson the basketball when stranded on the island in the film Castaway.[21] Thirdly, humanizing makes them easier to understand, less unpredictable and potentially less frightening, which explains why we even name hurricanes. By anthropomorphizing, we may fail to recognize any inherent danger they pose to us because we have made them in our image.

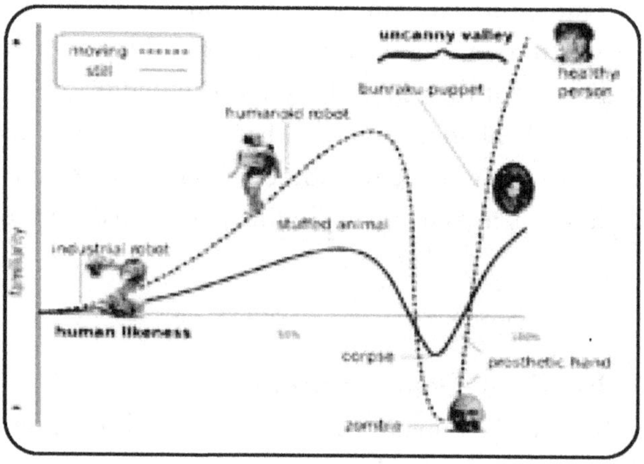

Based on his personal experience with robots, Japanese Professor Masahiro Mori (1927–) observed that when robots become too like humans, they become very creepy or eerie and called this the Uncanny Valley. This is considered an evolutionary feature that confers humans with sensitivity to threat, and the challenge for roboticists is how to cross this uncanny valley. Mori's uncanny valley diagram postulates a human reaction on a spectrum of negative to positive to different types of robots correlated with how realistically human they appear.[22] The reaction to androids given their similarity to humans is lodged deep in the floor of the uncanny valley alongside zombies and animated characters. There are three current theories for the uncanny valley (1) It occurs at the boundary between where transitions from one thing to another like, between non-humans and humans (2) When people think the robot possesses the ability to sense and experience things, but if it does not appear to possess a mind, it is not frightening [23] (3) There exists a mismatch between the robot's appearance and/or behaviour.[24]

Isaac Asimov wrote the three laws of robotics, which state that a robot (1) May not injure a human being or, through inaction, allow a human being to come to harm. (2) Must obey the orders given it by human beings except where such orders would conflict with the First Law, and (3) Must protect its existence if such protection does not conflict with the First or Second Law. We must bear these laws in mind as we consider the future of androids, although recognise the risk that robots may decide to make their laws, in which case we may be unable to enforce our law of robots.

The flip side of anthropomorphising robots is dehumanizing them because they are different from us. Will a point come when we regard them as no longer fulfilling our needs like the children in the film Toy Story who grow up and put their friends back into the cupboard or recycle them, or might we dispose of androids in a huge scale recycling program or extermination? Dehumanizing is particularly true when someone is not perceived as part of our social group or shares our values and beliefs. However, psychologists have observed that this exacerbates the less we interact with such people and that we tend to dehumanize others when we treat them as subservient to our goals and ends. This happens when we "stop focusing on their essential human qualities," particularly when there is a power dynamic in which we think we exercise power over another person.[25] This assumes we know what the essential human qualities are reminding us to keep sight of what we might be creating and what we want from robots.

What Do Humans Want from Robots?

In 2004, Japanese Professor Yoshio Tsukio, who focuses on information and digital technology, gave a lecture with the title, Living with Robots-Symbiosis of Robots and Human Beings, focusing on our having to live with and become increasingly dependent upon robots or these symbiotic artificial machines. This reflects how our increasingly symbiotic relationship with robots is becoming blurred as robotic devices surround us, whether it's shopping, controlling our media outlets, our sound systems, flying our planes, sliding doors, chat Bots on the computer, or soon autonomous vehicles, etc.

Roboticists David Vernon (1958–) and Markus Vincze, who work at the Vision for Robotics (V4R) laboratory at the Vienna University of Technology, reported the results of the RockEU2 survey of industrial developers of cognitive robots and identified 11 functional features which developers were looking for in robots which were "(1) Safe, reliable, transparent operation that generates trust in the robot (2) High-level instruction and context-aware task execution (3) Knowledge acquisition and generalization (4) Adaptive planning (5) Personalised interaction (6) Self-assessment (7) Learning from demonstration (8) Evaluating the safety of actions (9) Development and self-optimization (10) Knowledge transfer and (11) Communicating intentions and collaborative action."[26] These eleven desiderata largely envisage robots having limited autonomy to set intermediate goals when performing tasks set by human users, with the goals determined by high-level instructions tied to the robot's limited contextual knowledge and operating within well-defined parameters of normal behaviour. Shingo Shimoda and colleagues observed that "Robots need to perceive the environment, understand the surrounding circumstances, communicate with people, and move safely, sharing the same environment with humans."[27] They must display critical functions, for example, active vision and perception and social interactions, including predicting others' intentions and recognizing activities and gestures while understanding the relationship between speech and gaze dynamics. Users typically place a high value on the personalized nature of the robot-human interaction, though many developers want to be able to fluidly switch between full autonomy, partial autonomy, or manual control, which sounds like a parent not yet prepared to trust their child. Collaboration with others requires robots to be able to infer the intentions of others while communicating and explaining their intentions and actions. Vernon and Vincze observe that "much work is required on the scientific and technological tools and techniques to deliver all these various features into a single robot," and feel that humanoid robots are a long way off from being able to act as autonomous agents or behave like humans.

Biology and Robots

Not currently being able to use biological ingredients to build androids is no barrier to designing robots based on nature's biological processes. Scientists studying brain disorders have already succeeded in developing a three-dimensional organoid culture system called cerebral organoids that develop various discrete, albeit interdependent, brain regions using human pluripotent stem cells. They are cells that can self-renew by dividing and developing into the three primary groups of cells that comprise the human body, enabling researchers to track development and disease in the most complex human tissue[28]. What would happen if these cerebral organoids achieved the level of integration and differentiation that characterizes the human brain cells? Could super-intelligent robots harness cerebral organoids for their purposes like we are attempting with AI?

Scientists claimed in 2021 to have created the first living robots known as Xenobots from the stem cells of the African clawed frog, Xenopus Laevis. Xenobots can reproduce and although only a millimetre (0.04 inches) wide but can move, work together in groups, and self-heal. An AI program developed the optimal shape for Xenobot replication, which resembled a C, and sculpted the Xenbots in that shape, allowing them to replicate even more efficiently. They call them robots because, as Josh Bongard the lead author of the project says, "Most people think of robots as made of metals and ceramics, but it's not so much what a robot is made from, but what it does, which is act on its own on behalf of people." Making robots as human as possible requires the application of the latest knowledge about neuroscience, psychology, and AI and understanding how humans process information, acquire knowledge, intelligence, and understanding and whether by creating android brains we could encounter a new form of intelligence and knowledge not constrained as humans are by the processes of evolution.

Austrian-born Canadian scientist Hans Moravec (1948–), a pioneer in AI, made the following observation "Encoded in the large, highly evolved sensory and motor portions of the human brain is a billion years of experience about the nature of the world and how to survive in it. The deliberate process we call reasoning is, I believe, the thinnest veneer of human thought. It is effective only because it is supported by this much older and much more powerful, though usually unconscious, sensorimotor knowledge. We are all prodigious Olympians in perceptual and motor areas, so good that we make the difficult look easy. Abstract thought, though, is a new trick, perhaps less than 100 thousand years old. We have not yet mastered it. It is not intrinsically difficult; it seems so when we do it."[29] Can AI robots ever expect to perform at the level of humans when they lack the billion years of cumulative evolved experience about the nature of the world and how to survive in it? Over the last twenty years the research field of evolutionary robotics has emerged to address this problem.

Evolutionary Robotics

Evolutionary robotics involves designing the sensory apparatus, morphology, and control centre simultaneously because of their interdependence, which is how human biology works. The evolutionary component comes from the evolutionary algorithm, which tests an application for its fitness in terms of its physical and intellectual performance on assigned tasks. It translates this genotype into a robot phenotype but compresses the evolutionary life cycle to permit laboratory study. The robot must interact with its environment for some time to determine the fitness of phenotypes in terms of sensory performance, morphology or control. This fitness value is used for the next generation/iteration and repeated over generations until a quality threshold is met. Unlike biological life, which controls reproduction, humans need to employ costly resources and laboriously build robots but it is envisaged that evolutionary robots that pass the fitness test threshold will one day be built by other robots thus achieving autonomy over their reproduction.[30]

Biologist John Long uses robots to study animals and describes what happens when we let robots play the game of life.[31] He uses evolved robots that mimic selected animal features like fish swimming, which have given us great insight into our evolution, the potential for evolutionary robots and the future of technology.[32] Long asserts that these evolving biorobots can show us in real time what happens to evolving entities when faced with unexpected environmental challenges. Long creates robots that look and behave like extinct species and then subjects them to evolutionary pressures, such as competing for mates and resources and mutating their genes. They demonstrate evolution's power to solve difficult challenges without human input and that even simple robots can exhibit complex behaviour when interacting with the environment.

Most scientists in this field now believe that starting simple and progressively becoming more complex is the best approach consistent with the best practices of nature and science. John Lehman and Kenneth Stanley argue that there needs to be no pressure on the genotype to adapt and that evolvability increases because of a passive drifting process over genotypes, such that evolvability is inevitable.[33] Others argue that selective pressure on evolution is as important as the complexity of the encoding used in designing AI systems. This is based on the observation that evolution often converges prematurely onto a narrow range of designs because the fitness functions in AI are rewarded for having performed a designated task, which focuses on the short-term benefits and limits the exploration of new and innovative designs.

To overcome this convergence problem, several scientists have advocated that the focus should be on developing diverse behaviour, which has been shown to improve evolutionary progress rather than diverse genotypes.[34] Intriguingly, studies show that the more complex the morphology becomes, the more robust the robot's behaviours. Robotics expert Josh Bongard claims that embodied AI reflects how the "body and brain play equally important roles in the generation of adaptive behaviour" and observes how "robots that successfully evolved to accomplish all of their objectives, those composed of evolved rather than fixed morphologies generalized better to previously unseen environmental conditions."[35] This has the distinct advantage of improving the complex behavioural capacity of robots because the initial design need not be constrained by having to explain the workings based on an analogy with or biased by reference to human processes. It poses the potential for robot intelligence to surpass human intelligence by evolving new and creative ways of thinking that we don't possess and we need to keep in mind whether androids are constrained rather than enhanced by applying the psychology and neuroscience of human behaviour. This applies at all levels, and when you consider the many aspects of human behaviour that are problematic and self-destructive, must we seek to build robots that avoid these negative traits and failings or try to replicate the many negative and self-destructive features of humans, which have led to wars and the destruction of our environment or embrace the unexpected and innovative rather than mimic the human species?

Knowledge and Understanding in Humans v Robots

Human knowledge is no more than the manipulation of raw data and information taken from our environment, which allows us to interpret, predict, give meaning to, adapt and respond to our world. Acquiring intelligence through interacting with our environment has been termed embodied intelligence, which Professor of Machine Learning and Robotics Angelo Cangelosi and colleagues recently defined as "the computational approach to the design and understanding of intelligent behaviour in embodied and situated agents through the consideration of the strict coupling between the agent and its environment (situatedness), mediated by the constraints of the agent's own body, perceptual and motor system, and brain (embodiment)."[36] Conscious awareness, sustaining our primary drive to survive, and possessing the capacity for higher cognition and intelligence all depend on the information we extract from the data in our real-world environment. Cognitive comes from the Latin verb Cognosere, which means to know and includes knowing about us and how we see, hear, acquire information, remember, believe, understand, speak, think, solve problems, make decisions, etc. Will androids use such cognition and extract knowledge from their environment, in conjunction with the information we program into them, to carve out their ecological niche on the planet and maybe beyond?

In 1990, Allen Newell (1927–1992), an American researcher in computer science and cognitive psychology, laid out a general set of highly influential assumptions now known as the Unified Theories of Embodied Minds (UTEM). He introduced a new cognitive architecture called SOAR which envisages a unified level of cognition meeting the following criteria: (1) Behaving flexibly as a function of the environment. Exhibiting adaptive (rational, goal-oriented) behaviour (2) Operating in real-time (3) Operating in rich, complex, detailed environments (4) Using symbols and abstractions (5) Using language (6) Learning from the environment and experience (7) Acquiring capabilities through development (8) Operating autonomously, but within a social community (9) Being self-aware and have a sense of self (10) Being realizable as a neural system (11) Being constructible by an embryo-logical growth process and (12) Arising through evolution. Despite Newall's attempt to clarify the role of cognition in AI, there is currently no prevailing unified theory of cognition that cognitive architects universally adopt. Professor Paul Verschure (1962–) recently recalibrated Newall's model to avoid cognition being achieved through mimicry rather than representing genuine cognition and advanced the Unified Theories of Embodied Minds – Standard Model (UTEM-SM), which requires candidates to address functional and structural constraints, including biological validity and physical realisability.

Shukla Shubhendu and Jaiswal Vijay define AI as "The study of ideas to bring into being machines that respond to stimulation consistent with traditional responses from humans, given the human capacity for contemplation, judgment, and intention. Each machine should critically appraise and select differing opinions within itself. Produced by human skill and labour, these machines should conduct themselves in agreement with life, spirit, and sensitivity. In reality, they are imitations." They observe how AI "combines sophisticated hardware and software with elaborate databases and knowledge-based processing models to demonstrate the characteristics of effective human decision making" and how "Powerful AI software develops the high precision machine capabilities often freeing them from direct human control and vastly improving productivity. When a robot interacts with a richly populated and variable world, it uses its senses to gather data and information and then compares the sensate inputs with expectations embedded in its world model. The effectiveness of the robot is limited by the accuracy to which its programming models the real world."[37] They must be able and allowed to operate in the wider real world, not controlled environments like factories, dangerous situations, safety cages or merely performing programmed tasks as dumb robotic servants to humans.

These diverse approaches to AI and robots reflect a long-standing philosophical debate between rationalism and empiricism and whether human knowledge emerges from the use of logic and reason because of our unique human brains or by our active exploration of the environment through the senses. This requires us to explore the role of robotic senses in extracting information from the environment rather than by equipping them with a facsimile of our reasoning capabilities. SOARS involves "Knowledge grounded in the interaction between embodied agents and the physical and social environments in which they are situated, constrained by mechanisms of learning and memory and bootstrapped from minimal priors that emerged during evolution."[38] Knowledge and understanding require generating new learning and comparing it to prior knowledge to predict our world and constantly adapt and change is a powerful form of learning based on the theory that the learning process depends upon the memory and existing knowledge. Generative AI has rapidly emerged as the dominant AI learning method closely resembling human behaviour and requires knowledge sharing, which improves the efficiency and the rate of learning and abstract reasoning to generalize the knowledge to different environments and situations.

The eighteenth-century philosopher and statistician Thomas Bayes (1702–1761) established how we infer knowledge from our environment, and using prior knowledge, our brains act as sophisticated predictive mechanisms. AI must determine how much robots must be able to learn through interaction with their environment to assess what is probable, real, or not. Too much programming might make them less adaptable in real-world environments that are unpredictable, uncertain and messy with lots of risks, noise, and distractions. AI already far exceeds humanity in certain areas like chess, the game Go, or mathematical calculations. but is far less capable than even young children in making informed guesses about things or constructing mental models of the world and exploring and learning from the world. Alison Gopnik (1955–), who is an American Professor of psychology and philosophy, has been working with AI scientists to incorporate AI systems with what we understand about how children learn with little input or supervision from adults. Gopnik has observed, "Despite enormous strides in machine intelligence, even the most powerful computers still cannot learn as well as a five-year-old does."[39] Computer technology has made computers faster, more precise, and accelerated yet AI, can't yet compete with the human brain in dealing with novel, ambiguous, and highly complex environments.

The Explainability of the AI problem

Hans de Bruijn and colleagues observe, that we must "tackle the criticisms of the opaqueness of algorithmic decision-making with AI."[40] This has been dubbed the XAI problem, which means explaining AI needs to be addressed to inspire transparency and trust among users of robotic agents The problem is becoming more difficult as algorithms become increasingly autonomous and invisible and replace human decision-making.[41]

In 2021 Vaishak Belle and Ioannis Papantonis, provided four suggestions for addressing the XAI problem, including "Explanation by simplification, Describing the contribution of each feature to the decisions, Explaining an instance instead of in general, and Using graphical visualization methods for explanations."[42] AI systems are often described as expert systems because of their complexity. XAI issue differs from systems that use symbolic representation and languages (symbols), which humans can understand and use to verify reasoning and generative AI models. Generative models and systems are not directly programmed by humans but employ deep learning, which relies on complex processes and sophisticated statistical models that teach themselves how to get the correct answers from the input to the output. Hans de Bruijin and colleagues observe, "The internal representation of these non-symbolic AI systems does not contain a collection of human-readable rules but instead a collection of non-linear correlations."[43] This is further complicated because they are continuously updated as they learn, which requires updating explanations and interpretations.[44] This may explain why XAI has typically focused on interpreting whether the output is correct and currently "There are not yet widely available techniques that can be easily deployed to provide unambiguous explanations of the underlying AI models." Computer scientist Sule Anjomshoae and colleagues have observed how most studies attempting to peer into the black box deal with interpreting and explaining the data-driven algorithms, with very few explaining the inner workings of human-robot interaction HRI models.[45] The dual challenge facing AI developers is to allow humans to peer into the black box of AI and understand how and what robots themselves are thinking if they are thinking.

Can Robots have a soul?

Socrates famously asked, "What is it that, when present in a body, makes it live? A soul?" The Britannica dictionary describes the "Soul in religion and philosophy, as the immaterial aspect or essence of a human being, which confers individuality and humanity, and is often considered synonymous with the mind." By the end of the 5th century BC, the "soul was typically thought and spoken of as the distinguishing mark of living things, as something that is the subject of emotional states and that is responsible for planning and practical thinking, and also as the bearer of such virtues as courage and justice."[46] Many still believe that the soul distinguishes us from other animals and that the self doesn't die when our bodies die but leaves the body as the soul. This conflation of self and the soul was central to the French philosopher, scientist, and mathematician René Descartes's (1596–1650) dualistic distinction between the mind and the body. Descartes asserted that (1) The mind has no place in the spatio-temporal world. (2) It Is not in motion. (3) It Is not divisible and (4) Anything we cannot conceive of as capable of belonging to the body must be attributed to our soul. Because he didn't attribute thinking to the body, he believed the soul, joined to the body, must be thinking, and regarded the soul as the principle of thought, not of life. By separating the mind or soul from the body, he believed that he had incorporated into the non-material mind the non-material concept of God as the supremely perfect being who can do anything without limitations. Descartes's rationale for merging the soul and mind was that thought was a necessary and sufficient quality of the soul. He tried to reconcile this position and the problem of the exploratory gap between the indivisible mind and the divisible body by locating the seat of the soul and rational thought, imagination, and common sense in a tiny organ in the mid-line of the brain called the pineal gland.[47]

The Roman/Greek medical doctor and philosopher Aelius Galen (ca. 130-ca. 210 CE) thought the pineal gland served to support the blood vessels and tried to refute the belief that it controlled what was then thought to fill the ventricles of the brain, which was called "psychic pneuma," a fine, volatile, airy, or vaporous substance which he described as "the first instrument of the soul." Galen believed the worm-like appendage we now call the cerebellum, not the pineal gland, was responsible. Descartes chose the pineal gland because, unlike most brain structures, which have two versions, there is only one pineal gland in the human brain, full of animal spirits that inflate the brain's ventricles. We now know that the

ventricles are not full of animal spirits but brain cavities that produce and store Cerebrospinal fluid. The pineal gland remained a mystery until 1958 when American dermatologist Aaron Lerner (1920–2007) discovered a molecule produced exclusively by the pineal gland, which he isolated and called Melatonin, which helps regulate the body's internal clock. It produces several hormones that contribute to cardiovascular health and mood stability and is just one of many endocrine glands in the body. It is easy to see why Descartes's theory has been universally rejected.

A recent study examined how exposure to neuroscientific research impacts the still widely held belief in the existence of the soul. It suggests that those who find neuroscience to have established a strong mechanistic link between the mind and the brain are more likely to have little belief in a soul. In contrast, when neuroscience emphasizes the explanatory gap between the mind and the brain, some use this as evidence for a soul.[48] This raises a more general concern with scientism which the Oxford English Dictionary defines as the "Belief in or advocacy of science and the scientific method; specifically, the theory or belief that all phenomena can be scientifically explained" (OED), which is exactly what neuroscience and some would argue psychology appear to do. We will discuss the implications of this later. That's why the debate between science and faith and the evidence of God and a soul continues to rage and will continue to do so for some time. Psychologist Professor Paul Bloom (1963–) explores some of the things that make us peculiarly human, such as feelings and beliefs like humour, disgust, art, religion, and morality and argues that we all, even babies, see the world in terms of bodies and souls. He believes we are all "natural-born dualists" who "see the world as containing both physical things, which are governed by principles such as solidity and gravity, and immaterial minds, which are driven by emotions and goals."[49] Bloom explains that for some people, the "notion that our souls are flesh is profoundly troubling." Bloom argues that babies can distinguish between objects and people who hold beliefs (even false beliefs) as evidence of dualism suggests that dualism represents the difference between people and objects rather than the mind and the body although there is no scientific evidence that babies can make this distinction.

The soul the self and the non-material mind have been linked for centuries and despite neuroscientific discoveries across many areas to explain the neural underpinnings of human behaviour without a soul, the belief among many is enduring. There needs to be a universal consensus as to what the soul consists of and how it relates to the human body or the mind but there is no scientific evidence even to support the existence of a soul. For people who believe only humans have souls, can androids qualify as sufficiently human to possess a soul or would this diminish and be an anathema to us as humans in all our specialness. If the soul is a construct of the human mind, then maybe ASI robots will make up their minds about whether they have a soul, and what may constitute a soul, for a robot soul could look very different.

Endnotes

1 Centre for AI Safety. 2023.
2 Flesh and Machines (2002).
3 Probabilistic Robots (2005).
4 Image Source Hanson Robotics. Hong Kong.
5 Image Source Henrik Sorensen.
6 https://www.statista.com/statistics/728530/industrial-robot-market-size-worldwide/
7 Barandiaran X., et al., 2009.
8 Hutler, B, et al., 2023.
9 https://futureoflife.org/data/documents/research survey.pdf
10 Russell, S., et al., 2015.
11 https://ai100.stanford.edu/about/reflections-and-framing
12 https://www.researchgate.net/publication/364932697_Gathering_Strength_
13 The Singularity, is Near (2005).
14 Bostrom, N., 2003.
15 Barrett, J L, VanOrman, B., 1996.
16 Waytz A, et al., 2010.
17 Sproull, L et al., 1996.
18 Gazzola,V., et al., 2007.
19 Gray, HM et al., 2007.
20 Morewedge, CK., et al., 2007.
21 Epley, N., et al., 2008.
22 Source: AI and Society.
23 Gray K, Wegner DM., 2012.
24 Tinwell, A, Abdel N, John D, John C., 2013.
25 Gruenfeld, DH., et al., 2008.
26 Chrisley, R., et al (2017).
27 Shimoda, S., et al., 2021.
28 Lancaster, MA., et al., 2013.
29 Mind Children: The Future of Robot and Human Intelligence (1988).
30 Lipson, H., 2000.
31 Darwin's Devices (2012).
32 Long, J., 2012.
33 Lehman, J, Stanley, KO., 2013.
34 Mouret, J B, Doncieux, S., 2012.
35 Bongard, J C., 2010.
36 Cangelosi, A., et al., 2015.

[37] Shubhendu, S, Vijay, J., 2013.
[38] Verschure, P., 2012.
[39] Gopnik, A., 2017.
[40] de Bruijn, H, Warnier, M, Janssen M., 2022.
[41] Brauneis, R, Goodman EP., 2018.
[42] Belle, V, Papantonis I., 2021.
[43] de Bruijn H, Warnier M, Janssen M., 2022.
[44] Belle, V, Papantonis I., 2021.
[45] Anjomshoae, S., et al., 2019.
[46] Stanford's Encyclopaedia of Philosophy. https://plato.stanford.edu/
[47] The Passions of the Soul and Treatise on Man (1649).
[48] Preston JL, Ritter RS, Hepler J. 2013.
[49] Descartes' Baby: How Child Development Explains What Makes Us Human (2005),

"The true sign of intelligence is not knowledge but imagination."

Albert Einstein. German-born theoretical physicist who is widely held to be one of the greatest and most influential scientists of all time.

Chapter 2

Intelligence Learning & Understanding

CHAPTER 2

Human Intelligence

Intelligence Generally

THERE REMAINS NO CLEAR CONSENSUS on what constitutes intelligence, the underlying neural correlates or the interplay of the wide-ranging variables creating intelligence. Studies seeking to establish whether intelligence is linked to a single neural system have concluded that it comprises multiple independent components, with each behavioural component associated with a distinct functional brain network. In 2012 Dr. Adam Hampshire and colleagues observed that "Intelligence is an emergent property of anatomically distinct cognitive systems, each of which has its capacity."[1] In collaboration with Roger Highfield of the Science Museum Group, Hampshire and Adrian Owen conducted a landmark test on 110,000 applicants, out of which a representative group of 45,000 people was chosen to carry out a mass intelligence test. Participants had 30 minutes to complete 12 cognitive tests involving memory, reasoning, attention, and planning abilities and a questionnaire about their background and lifestyles. They concluded that "When a wide range of cognitive abilities are probed, the variations in performance can only be explained with at least three distinct components: short-term memory, reasoning and finally, a verbal component." Functional magnetic resonance imaging (fMRI) tests, albeit on only a small population of 16 study participants demonstrated these three different brain networks were involved in these cognitive functions and concluded that intelligence is composed of multiple factors with the various cognitive systems engaged depending upon the cognitive demands.

Clinicians, when assessing intelligence look for such things as verbal comprehension, working memory, perceptual reasoning, cognitive efficacy, etc. Intelligence at the neural level appears to be a fragmented concept rather than attributable to a single factor and an ill-defined construct upon which to base AI. For many years, psychologists have employed psychometric tests to assess how psychological constructs such as intelligence or mental states like depression etc can be related to observable tests, genetic profiles, and neuroscientific data etc. They examine complex attention, executive function, learning and memory, language, perceptual-motor control, and social cognition. All these domains need to be efficiently and smoothly integrated and operate and function smoothly for the brain to exhibit various levels of intelligence. The higher you score across these domains, the more likely you will be labelled intelligent. Problems in any domain can negatively impact general intelligence.

In the 1960s, the English psychiatrist and pioneer in cybernetics and systems theory, Ross Ashby (1903–1972), made a strong case for the connection between and adapt to continuously changing and unpredictable environments and intelligence.[2] He observed how evolutionary constraints, physiological reactions, sensory adjustment, and learning new behaviour, must be kept within well-defined boundaries. Ashby's formulation of adaptive intelligence has inspired much subsequent research into models of adaptation that embrace the body's capacity for self-regulation and adaptive change, such as in the fundamental mechanism of homoeostasis that maintains the organism's internal state within critical parameters. Adaptability appears to reflect a natural level and measure of intelligence allowing us to survive and thrive, and probably essential to autonomous intelligent robots.

General intelligence was first operationalized by the English psychologist Charles Spearman (1863–1945), who described intelligence as "The construct that the different abilities and skills measured on intelligence tests have in common."[3] Scientists since Spearman's time have defined the factors of general intelligence that allow someone to learn faster, which are (1) The capacity to think abstractly which includes acquiring knowledge (2) The ability to reason abstractly and the ability to adapt to novel situations and (3) Be able to learn through instruction and experience. Most intelligence tests involve various tasks assessing to what degree a person displays these capabilities and recognize that intelligence is multidimensional.[4, 5] The realization that intelligence is multifaceted was led by Harvard psychologist Howard Gardner (1943)[6, 7] who formulated the concept of "multi-intelligence" which he believes is composed of eight different types of intelligence. These are (1) The ability to speak and write properly (2) Using logic and maths to solve problems (3) Thinking and reason in three dimensions, (4) Performing and enjoying music (5) Displaying agility and athleticism (6) Interpersonal skills (7) Self-awareness (intra-personal skill) and (8) Naturalistic skills in terms of recognizing, animals, plants, or other natural organisms. In addition to general and specific intelligence, psychologists look at fluid intelligence, which is measured by problem-solving and speed of reasoning and processing of information, and crystallized intelligence, which refers to the level of knowledge that you have acquired over your lifetime, often measured by knowledge and vocabulary.

Intelligence at the neurological level, reflects unique neural patterns and neural pathways, with different levels of connectivity between brain regions and different capacities for integration and processing information. The combination of various functional domains has a genetic component in that people are born with a functional arrangement of the various brain structures but predominantly reflect different life experiences particularly early in life. This was all too evident in the tragic case of the Romanian orphans who were discovered living in appalling conditions following the removal of Romania's communist dictator Nicolae Ceausescu in 1989 and, due to neglect and abuse, have grown up with smaller (8.6%) than average brains. These very young children had been physically and psychologically deprived, some chained to their cots with little social contact or stimulus in terms of toys and playing or even interaction with others. A study led by Professor Edmund Sonuga-Barke (1962–) from Kings College University London followed the long-term plight of the orphans, who had lived in these appalling conditions of deprivation for between 4 weeks and four years.[8] Follow-up studies have shown many experiencing ongoing problems in adulthood ranging from ADHD, autism, and a condition known as disinhibited social engagement disorder, which means they don't display normal levels of caution or fear of strangers. The research showed the effect involved three main regions associated with the functions of (1) Organisation and integration of information (2) Motivation and (3) Memory. The damaging effects of early deprivation can't always be rectified, even if children have subsequently been looked after by loving and nurturing adoptive families.

Savants and Super Intelligence

Psychologists recognize that a person's general intelligence may not reflect intelligence about specific knowledge in certain domains or subjects of interest. Human intelligence can be very high in one but not across all the domains and often accompanied by major deficits in other domains, as evident in the fascinating condition of Savantism. Savant comes from the Latin word Sapere, for wise, and the French Savoir, to know and Sage, meaning very wise. Savants usually born with the condition appear to display super intelligence, and remarkable mental capability, with many studies showing that most suffer from severe mental disabilities and Autistic Spectrum Disorder (ASD).[9] Much of what we know now about savants was first developed by the American psychiatrist Darold Treffert (1933–2020), who pioneered studies into the epidemiology of ASD and savant syndrome and was fascinated by what he called "These islands of genius." Some savants have such prodigious talents that they are called mega-savants, and we marvel at how mundane most of our brains appear to be in comparison.

One of the most famous savants was Kim Peek (1951–2009)[10], affectionately known as "Kimputer," who inspired the character Raymond Babbitt, played by Dustin in the 2009 movie Rain Man. When he was 18 months old, Kim seen in the picture below could simultaneously read both pages of an open book, one with one eye and the other with the other eye, and remembered about 98% by reading just once. Although not on the autistic spectrum, brain scans revealed he was missing the corpus callosum, which are white matter fibres that connect the two brain hemispheres, and they think Kim developed the ability to

process language in both brain hemispheres when it is normally processed in just the left hemisphere.

Leslie Lemke (1952–) a blind American savant was born with severe defects necessitating that doctors remove both eyes and has ASD, but his adoptive parents bought him a piano at 7. At 14 years old, his mother woke up in the middle of one night to hear him flawlessly playing Tchaikovsky's Piano Concerto No. 1 despite having had no classical music training, having heard it just once earlier on the TV. He learned the chord organ, drums, and accordion even though he could not hold a knife and fork to eat. His unique brain allows him to remember and play any musical piece flawlessly, having heard it once. Englishman Daniel Tammet (1979–) suffered childhood epileptic seizures and is Aspergeric, which is on the autistic spectrum, yet appears normal, is a writer and novelist and has written an autobiography,[11] has exceptional mathematical and language abilities, and has recited from memory Pi to 22,514 decimal places, speaks seven languages, learned the Icelandic language in a week and is creating his language. Daniel has linguistic, numerical, and visual synaesthesia, which involves the brain routing sensory information through multiple but unrelated senses, allowing a person to experience more than one sense simultaneously. For example, tasting shapes, hearing colours, seeing letters as colours, etc. His perception of words, numbers, and colours are woven together, giving him a different way of perceiving and understanding the world. About 1 in 2000 people have the condition, which is more prevalent amongst creative people such as musicians or artists. Savants fascinate us because they display prodigious feats of intelligence or memory far beyond those of ordinary people like fictional characters with superpowers which in part is why the prospect of super-intelligent robots remains so compelling.

Although intelligence is typically associated with grey matter in the brain, this fails to capture the major role played by communication connectivity and integration between the different cortical and sub-cortical grey matter regions, related to various intelligence domains. These remarkable islands of genius are intriguing examples of how critical the extent and nature of connectivity and integration of brain regions or hemispheres are to specific forms of intelligence, explaining outstanding mental feats and cognitive and social deficits. The disabilities often accompanying savant skills are typically associated with problems in integrative cognitive processing, manifesting in deficits in social cognition, executive function, restricted interest, and compulsive behaviour, all of which are symptoms seen to varying degrees in people with ASD. Neuropsychological studies in the 1990s reported clinical cases of individuals suffering from the early stage of Frontotemporal Dementia (FTD) who acquired an artistic savant-like skill.[12] FTD is associated with problems in integrative cognitive processing. Different neurological models have been developed to explain ASD, and two relate to problems in connectivity and integrative cognitive processing. The first is based on neuro-imaging studies of ASD, which suggested that long-range connectivity between brain regions is disrupted, while local brain regions are intact or display enhanced connectivity in certain brain regions. Hence, it is called the Under Connectivity Theory, attributing the problems to a lack of inhibitory control exercised by the frontal lobe over other brain regions, coupled with possible abnormal organization of the cortical regions due to problems during development or damage arising from diseases such as FTD.[13] The other model developed by neuroscientists Keisuki Takahata and Motoichiro Kato from the Japanese National Institutes for Quantum Science and Technology attributes it to the "Altered balance of local/global connectivity patterns that contribute to an altered functional segregation/integration ratio." Our brain always balances chaos by keeping information separate against the need to integrate to make the brain function in response to the integrated and coherent information.[14]

Creativity

There is growing acknowledgment that the measures for general or specific intelligence fail to adequately capture creativity, reflecting someone's ability to think of new ideas and adapt to novel situations. Psychologists try to measure aspects of creativity in terms of the level of expertise in an area of interest, imaginative ways of thinking, the willingness to take risks, the level of intrinsic motivation, and their propensity to work in a creative environment with others. Professor of psychology Anna Abraham, who specializes in the study of human creativity, observes that we are only beginning to understand the role of creativity in the human brain and the relationship between creativity and heritability of brain damage, neuro-degenerative conditions, and mental illness all of which can impact and be related to creativity.[15]

Psychologists Andrea Andiliou and Karen Murphy were dissatisfied with the level of scientific creativity in high school students, which depends on the different teaching and learning approaches employed. This includes "Teaching thinking skills, technology-aid teaching and learning approach, experiment and practical based learning, inquiry-discovery based learning, cooperative based learning, and project-based learning."[16] Their in-depth research used teacher case studies, albeit very small samples, which showed that the teachers, the students, and the education setting influence creativity in schools. They observed that most teacher's beliefs "are often misaligned with the scientific theories of creativity." Abraham describes how, in psychological research, the most common distinction is made between scientific and artistic creativity, and creativity has been cited as a "Basic competency for being able to navigate the increasingly complex and unpredictable nature of life in a digital age."[17] Creativity requires the fundamental ingredients of originality and appropriateness of expression.

In 2009, American psychologist James Kaufman, known for his work on creativity in collaboration with Dr. Ronald Beghetto, an internationally recognized authority on human creativity in educational settings, sought to provide greater definitional clarity by proposing the four C's model of creativity. They intended to address what they described as the dichotomy of creativity that perpetuates the belief that everyday creativity called "Little C" can be found in nearly all people. In contrast, eminent creativity called "Big-C" is reserved for the great. Their 4 C model suggests that everyone has the potential to be creative, which has important implications for teaching and the classroom in nurturing creativity alongside knowledge and acquisition of skills. It also has implications for AI. The four Cs are "(1) Big-C or eminent creativity, exhibited by great artists or scientists (2) Pro-C or expert level creativity that is the creativity of professional-level creators who have not yet attained eminent status (3) Little-C or everyday creativity, which represents creativity involved in daily activities and experiences, and (4) Mini-C or subjective creativity which refers to the novel and personally meaningful insights that one encounters during learning and experience."[18]

Psychologist Gary Davis has spent his career researching creativity and revealed how creativity researchers have highlighted numerous personal features, including "cognitive abilities such as divergent, critical, analogical and reflective thinking, and personality characteristics, such as openness to experience, creative self-efficacy, task motivation, domain knowledge and expertise, risk-taking and resilience in the face of criticism that one requires to be creative."[19] Considering this confusion, building creativity into AI androids is challenging from a hardware and software perspective and AI developers must assess whether AI can be genuinely creative or mimic human creativity and dilute our appreciation of what human creativity can produce. We are already seeing famous works of art replicated and generated by AI, which are impossible to distinguish from those created by humans. A stunning example of this is the AI-generated 3D painting called The Next Rembrandt, which was created in 2016 by researchers including art historians, material researchers, and computer scientists from Delft University of Technology, Microsoft, and the Museum Het Rembrandthuis, following Rembrandt's style and using over 168,263 painting fragments from his works. The goal was to teach the AI machine to think and paint like Rembrandt, and the results are remarkable for the realism and likeness to his style. It has started a global discourse on where data and technology will take us and AI's relationship with creativity, and readers can view it on the Next Rembrandt website. This should make us question whether creativity has been exalted to a level that again seeks to specialize humans but ignores that machines can already provide creative output in the same way some humans can. It begs the question of whether this is genuine creativity or merely mimicking human examples.

Emotional Intelligence

A major challenge is whether A1 and AC technology will ever imbue a machine with emotions and feelings, motivation and drive, or personality traits. Do machines need these to function intelligently and autonomously or are they merely traits and qualities that would enhance the interface with humans, making them appear more real to us? Psychologists describe this as the difference between Hot and Cold cognition, with hot cognition describing how a person's thinking is influenced by their emotional and motivational states, whilst cold cognition is just the cognitive analysis and processing of information without emotion or feeling. They describe the process of executive decision-making and action planning formulated in human brains. Do robots need human-like capabilities such as emotional motivation and drive to provide the impetus to exercise their intelligence and cognitive decision-making and interact with the real world like humans do?

Practical Intelligence and Common Sense

People commonly talk about practical intelligence and how often have you heard the accusation levelled at a highly intelligent person, that they might be bright, but they appear to have no common or practical sense which is usually attributed to general life experience. American psychologist and professor of human development at Yale University, Robert Sternberg (1949–) defined practical intelligence as the "Ability that individuals use to find a more optimal fit between themselves and the demands of the environment through adapting, shaping, or selecting a new environment in the pursuit of personally valued goals." Despite this seemingly comprehensive definition and the common perceptions of and references to practical intelligence there appears to be no scientific evidence suggesting that practical intelligence constitutes a specific type of intelligence. Androids will need to display the type of practical intelligence or common sense envisaged by Sternberg in finding the optimal fit between themselves and the demands of the environment through adapting, shaping, or selecting a new environment. Will they ever be able to demonstrate common sense particularly when even the brightest of humans appear to struggle with this? We will look at the important role of common sense in AI robots and the challenges they face as currently AI fails miserably when confronted by problems that can be solved simply using common sense. The definition of common sense is "The basic ability to perceive, understand, and judge things that are shared by ('common to') nearly all people and can reasonably be expected of nearly all people without the need for debate." We use common sense to deal with unlikely or unforeseen situations when we draw upon all our implicit knowledge based upon observation, trial and error and learning what works and what doesn't to interpret and deal with these situations. The problem as we will see is that the reasoning behind common sense is rather obscure and although it is pervasive in human behaviour it is extremely difficult to describe in a way that it can be encoded into AI systems which are rule-based. Dr Howard Shrobe worked with Massachusetts Institute of Technology's (MIT) Computer Science and Artificial Intelligence Lab (CSAIL), where he led research into creating high-performance, reliable, and secure computing systems and technology and in his current role with the Defence Advanced Research Projects Agency (DARPA) he observes how "Machine common sense has long been a critical but missing component of AI. Recent advances in machine learning have created new AI capabilities, but machine reasoning across these applications remains narrow and highly specialised. Current machine learning systems must be carefully trained or programmed for every situation" and he goes on to acknowledge that "humans are usually not conscious of the vast sea of common-sense assumptions that underlie every statement or action. This shared, unstated background knowledge includes a general understanding of how the physical world works (i.e., intuitive physics), a basic understanding of human motives and behaviours (i.e., intuitive psychology), and a knowledge of the common facts that an average adult possesses". To address this problem Schrobe, with others, has developed the Machine Common Sense Program (MCSP) which purports to adopt two strategies which are to "(1) Create a service that learns from experience, like a child, to construct computational models that mimic the core domains of child cognition for objects (intuitive physics), agents (intentional actors), and places (spatial navigation) and (2) Develop a service that learns from reading the Web, like a research librarian, to construct a commonsense knowledge repository capable of answering natural language and image-based questions about common-sense phenomena". Using these strategies DARPA appear to have demonstrated significant improvements in the performance of robots during multiple experiments "ranging from understanding how to grasp objects and adapting to obstacles, to changing speed/gait for various goals." Common sense is an imporant if not vital requirement for adaptive survival and something that AI will have to address if machines are to be able to operate in the environment and be able to make decisions and choices that are approriate to the context and situation that they find themselves in.

Schrobe believes that "By focusing on common sense, we are creating the possibility for systems to have the flexibility of human learning and the breadth of human knowledge," and that "Fusing this knowledge with advanced robotics could result in highly capable, mission-critical systems that humans will want to have as partners."

Human Learning

The following section touches on some milestone theories establishing how humans learn subconsciously and consciously. The twentieth century saw a battle between different schools of thought as to how we learn, why we behave as we do, and the role of memory in learning. Some thought learning and behaviour were inexplicable and a vital element of an organism; others a reflexive mechanical process or largely learned through association.

Associative Learning

Russian physiologist Ivan Pavlov (1849–1936) discovered unconscious learning, now known as classical or learning by association. He was studying how dogs respond before delivering their food by salivating, an automatic, reflexive, instinctual and non-conditioned response. He observed that they salivated on seeing the white lab coats of the researchers who fed them which meant a neutral stimulus (the white coat) had become associated with food prompting an instinctual, unconditional response. He demonstrated that dogs could be taught to salivate in response to a bell, which initially is a neutral or unconditional stimulus until the dogs associate it with the food when it becomes a conditioned stimulus, provoking salivation. Pavlov identified this response as encoded and originated in the dog's cerebral cortex and this has now been observed in most vertebrates. Classical conditioning therapy modifies negative or harmful behaviour in many conditions, such as anxiety, depression, obsessive-compulsive disorder (OCD), substance abuse, and some phobias. For example, alcoholics may be given medication that makes them sick if they drink so they associate alcohol with nausea. This is useful with children if positive experiences can be used alongside something a child does not like so they associate the challenge with positive stimuli. It is linked to the placebo effect, in which people who have taken something that has no medicinal benefits have shown reduced activity in the brain regions linked to pain, suggesting pain reduction as though they were taking anti-pain medicine and works even when they know they are taking a placebo.[20]

Pavlov was very influential in the work of the behaviourist psychologists, led by John B Watson (1978–1958) and Burrhus Frederic Skinner (1904–1990) who argued that all behaviours are not innate or inherited but learned and remembered through interaction with the environment and involve either classical or operant learning. Watson maintained that "Psychology is a purely objective experimental branch of natural science" and that "the theoretical goal is prediction and control."[21] The behaviourists believed that everything we do, including our speech and emotions, are just examples of a stimulus and response dynamic and that all differences in individual behaviour were simply the product of different learning experiences. Watson boldly claimed, "Give me a dozen healthy infants, well-formed, and my own specified world to bring them up in, and I will guarantee to take anyone at random and train him to become any type of specialist I might select doctor, lawyer, artist, merchant-chief and, yes even beggar-man and thief, regardless of his talents, penchants, tendencies, abilities, vocations and the race of his ancestors."[22] They only employed observable stimulus-response behaviours that they could study systematically, arguing that emotions are internal events that require you to get inside the black box of people's heads and understand their thoughts, meanings, motives, or unconscious yearnings and desires. We could call this the explainability of the human intelligence problem. Behaviourists argued that learning in humans was like learning in animals thus animals, particularly rats and pigeons could be used to study human behaviour.

Watson became notorious for a highly unethical experiment in which he and an assistant used classical conditioning to show how a nine-month-old orphan called little Albert could learn how to fear stimuli by observing his response to various stimuli, which included including a white rat, burning newspapers and a hammer striking a four-foot steel bar just behind his head. Initially, only the hammer sound frightened him, so 11-month-old Albert was allowed to stroke and pet the white rat, but when he did so, they struck the steel bar behind him, which frightened him. Over seven weeks, they demonstrated that Albert associated the hammer noise with the rat. He became frightened of the rat which had become the cue to triggering his fear response, and this lasted for periods of weeks afterwards, though it became less marked. This reaction to a stimulus that acts as a cue has been evidenced in people with addiction problems, such as alcoholics, smoking, and drug addiction, who react strongly to certain cues that trigger the craving and physiological arousal they associate with their addiction. Psychologists now call this cue reactivity, and it has been demonstrated that people associate certain situations often with social gatherings, with, for example, the rewarding effects of smoking, which act as a cue to trigger the craving.

Skinner thought classical conditioning was too simple an explanation for human behaviour and established that the consequences of behaviour or our actions or responses determine the likelihood of that behaviour being repeated, which he called Operant conditioning. Skinner believed that people's minds are not amenable to scientific study but he could study how if pleasant consequences follow the behaviour, it is more likely to be repeated, or when followed by unpleasant consequences, it is less likely. Although obvious now, this was a highly influential milestone in our understanding of memory and learning by showing that reinforced behaviour, will be repeated. If you don't, that behaviour slowly dies away. Skinner placed an animal, usually a rat in the "Skinner Box", to observe their behaviour and response when either positively rewarded for pressing a lever or punished. The reward is the reinforcing stimulus and is a primary reward if it occurs naturally, like food or water, but secondary if associated with a primary reward. An example of a secondary reward is money that gives you access to the primary rewards. The negative reward can also remove something unpleasant, although punishment weakens or stops the behaviour, and the difference between negative reward and punishment can be blurred. Studies since Skinner have shown that punishment causes many problems because it can generate fear and merely teaches someone what not to do rather than what they should do and after removing the punishment, the behaviour usually returns.

Skinner established that different schedules of reward and punishment affected behaviour and the greatest impact of operant conditioning has been seen in the subsequent techniques employed to modify behaviour in children and adults, which include changing the environment in which behaviours are learned, shaping behaviour over time and the use of tokens to reward behaviour. Tokens are regarded as secondary rewards, such as badges, stickers, or toy money, which can be exchanged for primary rewards, such as food or privileges. This has shown to be very effective when dealing with patients in psychiatric care. However, they tend to become too attached to receiving the secondary rewards, making it difficult in normal society.

American psychologist Professor Charles Gallistel (1941–), an expert in the cognitive processes of learning, and psychologist and neuroscientist Professor Peter Balsam (1947–) argue that we should not assume that most learning involves association or pairing and what psychologists call temporal contiguity, which is that people learn quicker when related words and pictures are presented simultaneously. The premise is that they enter working memory simultaneously, so are processed together, facilitating greater integration. Gallistel and Balsam have observed how much learning is very rapid, doesn't require temporal contiguity (closeness in time) between events and repetition and people can rapidly learn associations between events even with a delay over long periods lasting days with the difference being in how that learning is expressed. This is important because much of the research has focused on identifying the neural changes that come about when contiguity and repetition occur. They argue that associative learning has no critical interval which has implications for robot learning.[23] Several different terms are used to describe the role of associative learning in AI, with Reinforcement Learning (RL) being the most common.

Non-Associative Learning

Non-associative learning results from a single stimulus without reinforcing behaviour, so you don't pair the behaviour with the stimulus. It involves three key methods of learning in response to stimuli, which are (1) Sensitization, (2) Habituation and (3) Imprinting which are forms of learning often regarded as a prerequisite for all types of learning. Sensitization occurs not through association but through repeated exposure to a stimulus, making a person increasingly sensitive to a stimulus that may not have previously provoked a response. An example is a child who has been repeatedly verbally abused may, over time, become very sensitive to a perceived threat from all children and become withdrawn. The opposite is when someone becomes accustomed to or habituated to a stimulus that ceases to have the same impact. This is crucial for allowing people to tune out non-essential sensory information. The extent to which we become habituated is affected by various factors, including (1) A person's age and temperament. (2) How strong the stimulus is, and the duration of the exposure. If strong, it may take longer to habituate, and the longer the exposure and the more often it is repeated, the more likely someone is to become habituated.

(3) New stimuli will take longer than familiar stimuli and previous exposure will make habituation faster, as will the context. If you don't expect a stimulus like a loud noise it will surprise you, but if you are in a loud environment, it won't. (4) If someone is motivated to attend to the stimuli, it may take longer to habituate. It is not clear what happens in our brains to cause habituation, with some thinking that our brains create a representation of how we should behave in response to the stimuli and others that our brains contain neural mechanisms to regulate the response to stimuli in our environment. Neurons may become less reactive to repeated stimuli, but this is an automatic process, whereas people have more conscious control over the habituation process. Habituation is often used to reduce psychological stress and anxiety triggered by exposure to certain stimuli. For example, repeated exposure therapy involving exposing someone to fear or anxiety-inducing stimuli over time can reduce the anxiety, and it is used to treat PTSD, OCD, and a range of phobias. These learning processes are central to how we develop habits in life that govern our behaviour and how we respond to the world. We will be discussing habitual behaviour later in both humans and AI.

Imprinting was first discovered by British naturalist Douglas A. Spalding (1840–1877), who noticed how newly hatched chicks would follow the first thing they see. The term imprinting is attributed to the Austrian comparative anatomist and animal psychologist Konrad Lorenz (1903–1989) who used to spend the summers at his family home in Altenburg, Germany. He called these his "Goose Summers" because he spent so much time observing the Greylag geese and observed that within a genetically short time frame of between 13–16 hours, the birds were able to accept the first thing they saw as an adoptive mother, even if it was an animal of a different species like Konrad. This is why there are many pictures of him being followed adoringly by a little gaggle of goslings. Lorenz and a Dutch colleague Nikolaas Tinbergen (1907–1988) and Karl Von Frisch (1886–1982) discovered that animal species have a genetic programming of certain survival behaviours that can be activated during a critical period. Lorenz had long attributed the spontaneous behaviour of young animals to the reflex theory, but realized that attachment was genetically innate and programmed, not mechanical or reflexive. This happens in the critical period of brain development in young animals, including children, when our brains are optimally placed to learn things essential to survival and at the highest level of plasticity before synaptic connections (see later) have become more stable and enduring. We can't recreate this period. A child not repeatedly exposed to stimuli during this period may be unable to develop the associated functions later as we saw earlier with the Romanian orphans. This

period of sensitivity to stimuli is a property of the neural circuits that represent information in an adaptive way for the child or young animal and been hypothesized to modify the architecture of neural circuits in a fundamental way "which causes certain patterns of connectivity to become highly stable and, therefore, energetically preferred." As Knudsen observes, this demonstrates the "importance of genetics and the environment in shaping the information processing capabilities of a neural circuit" and how this affects subsequent behaviour and learning.[24]

Although Lorenz and others believe that this is simply innate, others describe it as biological preparedness in that it is a more readily learned behaviour that reflects our disposition to fear things that might pose a threat to us or our survival, such as a fear of snakes or foods that will make us ill. Preparedness acts along a continuum with some stimulus activating an inherent biological readiness to learn at one end and at the other contra preparedness in which no biological preparedness exists, and no association is made because it is not inherent. In the middle is unprepared learning, in which, although we are not biologically disposed or prepared to make an association, we can learn to make an association.[25] Intriguingly, psychologists have observed people display what they now call the baby duck syndrome during computer use, which describes how when people are learning something, they often regard the first thing they learn as the best. Later, things measured against that are worse. This explains why people often resist and are reluctant to change to updated software versions, having learned how to interface with the existing software in a particular way. It reflects how people tend to be much more comfortable with the familiar in their environment, which has important implications for both how humans and robots learn when operating in uncertain and unfamiliar environments.

Social Learning Theory

Canadian American psychologist Albert Bandura (1925–2021) agreed with the behaviourists but did not believe that all learning involved an association between stimuli or reinforcement or a stimulus(input) and response (output) equation because humans exist in highly complex social environments. He developed the Social Learning Theory that children learn by observing, imitating, and modelling other's behaviour. Simulation has been theorized as a process whereby our brains build mental constructs encoded in neural patterns that constantly flow between sensory perception, planning, decision-making or choices, and adjusting, allowing us to execute appropriate motor actions or responses and is how our brains represent others behaviour.

Bandura observed preschool children watching adults verbally and physically abuse a Bobo clown-faced inflatable toy and copying that behaviour by treating the doll similarly. He argued that learning requires a degree of self-agency, which reflects a child's need to choose to pay attention to others and how they behave, levels of motivation and drive or engagement with their environment, attitudes in terms of traits, personality and values and beliefs, and emotional states or moods. Bandura described how observation could be done by watching a real person behaving, fictional characters in the visual media, or just reading or listening to descriptions and explanations of how people behave. Bandura's social learning theory consists of (1) Observation of others. (2) Inferring mental states to others and (3) Acknowledgment that even knowledge learned in this way need not change a result of a behaviour's behaviour. In 1977, he demonstrated how the greater an individual's belief in their capabilities, the more effort they will put into doing things or overcoming challenges to achieve their goals, which affects how they feel as they are doing it.[26] This Self-Efficacy Theory, reflects Bandura's realization that people play a major role in adapting and self-regulating their behaviour to achieve their goals or desires by addressing their shortcomings or failings and that everyone can do so. Bandura advocated the technique of mastery in which by setting goals and persisting, you achieve mastery of a skill or activity, and that is how you derive real satisfaction, which he compared to achieving success too easily and getting so used to it that you become too easily discouraged when you face difficulties.

Curiosity and Exploration

The need for children to be allowed to explore and be curious for them to learn has been forcefully argued by American educator, author, and politician Julie Lythcott-Haims (1967–).[27] A common observation is that the millennials, defined as born between 1981 and 1996, are the "Snowflake generation," which is a term that came into public use after the publishing in 1996 of Chuck Palahniuk's book, The Fight Club which was used as follows "You are not special. You're not a beautiful and unique snowflake. You're the same decaying organic matter as everything else." Palahniuk used this to describe how Millennials are "overly sensitive and think that the world revolves around them." The Financial Times included snowflake in their 2017 Year in a Word list, defining it as "a derogatory term for someone deemed too emotionally vulnerable to cope with views that challenge their own, particularly in universities and other forums once known for robust debate." Having snowplough parents who produce snowflakes might be strangely appealing as a wintry metaphor, but how does this translate into memory and learning?

Protecting children from risks has been coupled with attempting to constantly build up the children's self-esteem but linked to anxiety and mental health issues. Resilience requires memory based on experience and the knowledge to deal with challenges. Over the last twenty years, there has been a huge growth in the research into the neuroscience of resilience, specifically about stress and trauma. Snow Plough parenting has been linked to anxiety in children, and neuroscientists are starting to investigate the neurobiology of anxiety and self-efficacy. Yet as recently as 2020 it was observed that "no studies are highlighting the relationship between self-efficacy and neuro-circuitry in youth." This prompted Krystal Lewis and colleagues to measure parent, child, and clinician ratings of anxiety symptoms against child-reported self-efficacy in

an albeit very small sample of 51 children between 8- to 17-year-old. They found no evidence in the neurocircuitry or morphology of the brains, but concluded, "Self-efficacy is an important treatment target for anxious youth."[28] Neuroscientists operationalize resilience to study it by defining it as the ability to achieve a successful outcome in the face of adversity. This research has shown how biological, psychological, and social factors and developmental contexts contribute to or degrade resilience. Scientists have developed a new model called Life Course Health Development (LCHD), which harnesses research from biological, behavioural, and social science disciplines to understand how these affect well-being and mental health, including resilience's role.[29] This reflects the growing acknowledgement that stress and adversity in early life are significant risk factors for disease in later life. Studies on depression's impact on our brain identify stress as a major influence on brain structure and functionality and even on gene expression and neural connectivity within the brain.

Continual Learning

Humans and animals can learn a range of tasks sequentially or segregate one at a time without confusing the task requirements or experiencing mutual interference between tasks, preserving the learning from previous tasks with minimal forgetting. We often encounter significant problems when required to learn multiple tasks simultaneously.[30] Yifan Chang and colleagues point out in their review of the topic that continual learning' is not just about remembering or not forgetting but utilizing previous learning in future learning, which they break down into the three processes of (1) Information Retrospection (2) Information Prospection and (3) Information Transfer.[31] Incremental learning requires remembering learned information, the continuous inferring of new information, which means remembering to carry out intended actions in the future and transferring useful information from one task to related tasks. Many scientists use the phrase continual when they talk about learning which suggests that learning happens frequently but with intervals, whereas continuous more accurately describes the constant stream of perception from stimulus in the environment. The distinction assumes greater relevance when we look at learning in AI and the search for algorithms that can learn from a continuous stream of perception data from the real world. It can be unclear when we talk about reinforcement learning, which typically involves discrete rather than continual learning, but despite this nuance, we use them interchangeably for present purposes.

Humans are continuously learning and adapting to the environment, and through incremental learning, develop more complex knowledge and skills. It is thought that this relies on the mechanism of reactivation of neuronal activity patterns responsible for stabilizing new memories.[32] The complex biological mechanisms associated with continual learning remain unclear, but Geram Parisi observes that they are known to "contribute to our sensorimotor skills and long-term memory consolidation and retrieval."[33] Continuous learning has been attributed to the stability versus plasticity mechanism, which describes how our brains can adapt to input new information but be stable when faced with irrelevant information. Humans apply continual learning to various tasks, even with complex inputs whilst AI tends to learn using large batch data presented at once, which requires the "retraining of the network parameters using the entire dataset to adapt to changes in the data distribution." This is highly inefficient and prevents learning in real-time and Parisi and colleagues have laid out the "set of widely studied biological aspects of lifelong learning and their implications for the modelling of biologically motivated neural network architectures."[34] These include "(1) Structural plasticity which refers to diversity in the anatomical properties and neural thickening due to neural configurations in terms of, for example, size, location, and branching patterns (2) Memory replay (3) Curriculum and transfer learning which involves organizing the learning process in a way that gradually increases the complexity of tasks or training and avoids learning from scratch (4) Intrinsic motivation in terms of the performing a task for its inherent satisfaction rather than a separable or external consequence and (5) Multisensory integration which refers to the process by which the CNS integrates inputs (information) from two or more sensory modalities to form a stable and coherent percept of the world.[35]

Although the precise biological mechanism remains unclear, neuroscientist Dr Helen Barron, who focuses on the neural mechanisms for continual learning, memory, and decision-making, recently proposed the critical role in humans of neural inhibition which "regulates the stability of neural networks to gate cortical plasticity and memory retrieval" and "protects memories from interference to permit continual learning. " This is evidenced by the decreased concentrations of the inhibitory neurotransmitter GABA during learning, which provides homoeostatic control to restore network stability. She concludes that memories are dormant unless latent inhibitory connections are unmasked, and cortical inhibition protects overlapping memories from interference, thus permitting continual learning.[36] This illustrates how our brains rely on a dynamic and ongoing balance between excitation and inhibition and stability and adaptability to constantly respond to the environment, which is a major challenge to building autonomous androids.

The continual learning process seems to involve us factorizing or segregating our representation of tasks, reflecting the brain's tendency to represent the stimulus space in a well-structured way that optimizes our ability to segregate and learn new tasks. Guido van de Ven and Andreas Tolias who work on continual learning in AI and are seeking a more structured method for comparing continual learning in different AI systems, distinguish between three types of continual learning scenarios, which are "(1) Task-incremental, which involves learning a set of clearly distinguishable tasks (2) Domain-incremental, which involves learning the same kind of problem in different contexts and (3) Class-incremental learning; which involves learning to distinguish between an increasing growing number of objects or classes." They argue that when we look at continual learning in AI, each type presents different challenges.[37] Continual learning facilitates scalability of intelligence and optimizes efficiency in terms of the use of the brain resources to avoid reinventing the wheel with every new task or being faced with new information and where scenarios keep changing or more complex tasks are involved.

Human Understanding

What it means to understand something is a classical philosophical question that has been sharply brought into focus by the rise of AI and the growth of society's reliance on and belief in the validity of reductionism and scientism. Reductionism describes a complex phenomenon in terms of its simple or fundamental constituents and is believed by many scientists to provide a sufficient explanation for treating complex systems as the sum of their parts, for example, neuroscience seeks to deduce a system's collective behaviour from knowledge of its components. Many believe though that knowledge of all the brain's components can only guarantee a partial understanding of the system's behaviour's complexity. Biologist Gilles Laurent (1960–),while musing on the meaning of understanding in 2000 in Nature Neuroscience, observed that it needs to consider the "distributed interactions that underlie the global properties of systems." He argues that "neuroscience is (to me) a science of systems in which first-order and local explanatory schemata are needed but insufficient." In other words we may never reach a satisfactory level of understanding of the world around us, including ourselves, and that reductionism by itself will not do so, which begs the question as to what level of understanding AI would be able to achieve or in reality need.

Despite however much intelligence or knowledge they display based on their level of algorithmic competence, computational speed, access to data, and vast memory banks, will AI robots ever understand the world and themselves or their knowledge as humans do." This is vital if we accept that understanding is inherently more valuable than pure knowledge. We have all heard of people with prodigious memories who possess a vast repository of knowledge and information that they can readily retrieve, and with Google, this is potentially available to all of us based on a quick search, but what use is this if someone understands little of what they know. The American Philosopher Professor Linda Zagbeski (1946–) observes that the word understanding has been used to mean so many different things that the concept of understanding almost completely disappears.[38] American philosopher Jonathan Kvanvig argues that the value of all knowledge comes from realizing that understanding has its value. He questions what has been described as

the epistemology of understanding in the theory of knowledge, especially regarding its methods, validity, and scope and the distinction between justified belief and opinion.[39, 40] Epistemology comes from the Greek words Episteme- Knowledge or understanding or acquaintance" and Logos- "Account or argument or reason."[41] Kvanvig argues there are three types of understanding, which are (1) Propositional understanding, that can be applied to different subject matters, such as understanding morally why something is right or wrong.for example (2) Understanding the value of an object aesthetically and (3) Objectual understanding and the mathematical foundations of something or the scientific reasons why something happens, which can provide the motivational basis that can give people a sense of purpose and goals, rather than simply understanding what and how things happen.

Three conditions are individually or jointly required to possess such knowledge which is the belief condition, which means you accept what you know; the truth condition, which requires what you know to be factual; and the justification condition, which means that knowledge is not simply true belief or in other words the believer has adequate indication or evidence that it is true. right or wrong. Understanding why has been described as a cognitive achievement, although as philosopher Insa Lawler, who focuses on the epistemology (the theory of knowledge) of scientific inquiry, observes, "Understanding-why featuring rudimentary explanations and understanding-why concerning very simple causal connections are not cognitive achievements in a strong sense."[42] Lawler stresses the importance of different levels of understanding. For example, most people know that an atomic bomb relies for its power on the release of energy, so have knowledge of why, but most people will not understand the complex maths and physics underlying this superficial understanding, so do not have a full and complex understanding of why atom bombs are so destructive. Objectual understanding compares to the understanding of isolated propositions in propositional understanding because it involves understanding a topic, subject matter, or body of information that may contain a multitude of propositions (including mathematical sentences such as equations) that are true or false. Duncan Pritchard argues that understanding is a cognitive state, whether in a strong or weak sense[43] which Kvanvig and Pritchard observe has an intrinsic value.

If we accept the compelling proposition of understanding as a cognitive state, we need to reach some consensus on what level of understanding AI systems could achieve and what that cognitive state constitutes. What will be the minimum criteria or cognitive state that establishes that they do understand, and can they possess the cognitive state required, whether weak or strong or somewhere along a continuum of understanding? Yan Ufik and Karl Friston observe how the human nervous system "Underlies the development of intelligence, enabling a progression from the ability to sense the environment to the ability to understand it."[44] We will see how this perspective may allow us to determine whether robots can ever progress from simply sensing the environment and extracting data and information to understanding at any level, even potentially beyond that of humans.

Artificial Intelligence

AI was first conceived in 1956 by American computer scientist and cognitive scientist John McCarthy (1927–2011), Alan Turing (1912–19540, Marvin Minsky (1927–2016), Allen Newell (1927–1992), and Herbert A Simon (1916–2001). McCarthy proposed a 2-month conference on AI with the objective of "proceeding based on the conjecture that every aspect of learning or any other feature of intelligence, can in principle be so precisely described that a machine can be made to simulate it." In 1979, he wrote an article entitled Ascribing Mental Qualities to Machines, in which he claimed that "Machines as simple as thermostats can be said to have beliefs, and having beliefs seems to be a characteristic of most machines capable of problem-solving performance."[45] This controversial assertion was challenged by the American Philosopher John Searle(1932–), who vehemently argued that machines can't hold beliefs because they are incapable of consciousness. In 1980, to support his position, Searle formulated the Chinese Room argument which states that a computer cannot have a mind or understand things because they are executing a program, however much it appears to simulate human intelligence. Searle refuted the concept of strong AI and functionalism in which every mental state, like a belief, has a function and this is simply a product of the brain's biological organization and processes. This would mean that if computers can perform the appropriate functions based on input into the software program, which is their equivalent of the biological processes, the output of that function can create a mental state or function akin to that of humans. Functionalists believe that we don't need to understand the underlying physical medium, only the higher functions that arise from that medium, be it a computer or brain, and that if computers have the right programs, they can possess conscious minds and cognitive or thinking states. In contrast, Weak AI offers an alternative perspective asserting that computers merely simulate thought based on an IF calculation which checks whether a condition is met and returns one true value and another value if false, but can't and never will be able to understand anything as we humans genuinely do. Proponents of weak AI recognize their usefulness for studying the mind by simulating it.

Searle's Chinese Room argument requires us to imagine being locked in a room only being able to speak English, with two documents consisting of a script with Chinese symbols and a story with Chinese symbols, which you can identify solely by their shapes and one set of rules. You are given English instructions called the program for correlating the two documents and a third document with some Chinese symbols representing answers to the questions and more instructions in English to allow you to correlate parts of this third batch with the other Chinese script and story. You aren't told any of this and are just given the information. Still, after practicing, you become so good at following the instructions that you can work out how to give answers in a way that no one outside the room would know, that you can't speak or understand a word of Chinese. Searle compared your performance to a computer running a script applier mechanism used in Roger Schank's (1946–2023) and Robert Abelson's (1928–2005) program in the 1970s. The Schank theory states that all human behaviour consists largely of patterns or scripts and that memory is episodic, being based on episodes of personal experiences rather than semantic categories. Generalized episodes function like scripts with specific memories stored as pointers to scripts, and we retrieve these scripts to guide our behaviour. He argued that you don't understand Chinese or the answers, follow a program or script, and simulate understanding by following a syntactic process without semantic meaning. Syntactic means that you have worked out the rules that govern the sentence structure, whereas semantic involves understanding the sentence's meaning.

Others have challenged Searle with various replies and rejoinders to the Chinese Room thought experiment. The reply of immediate interest to us involves building a robot friend and goes as follows. The person in the room can't understand the Chinese meaning because it assumes that the person is disconnected or not grounded in the outside world or the environment in which the symbols are causally related to things. But if the person were a robot equipped with a computer that allowed it to perceive the world interactively, it would have access to this causal meaning and so would be able to understand. Searle countered by having himself as the mind of the robot in the room with access to the symbols via interacting through a camera installed in the robot and its movement but observed that it can still only manipulate the symbols and won't understand or have mental states that arise from manipulating those symbols. At the heart of the discussion is whether a computer program that relies on processing syntactics and the manipulation of symbols can possess the semantic mental content necessary to have an understanding mind and whether syntactic processing by itself can ever be sufficient to constitute semantics, meaning or understanding.

Part of Searle's reply in 1989, which he called "Derivation from axioms," was to observe that we need to decide whether computers, without a human-like brain, can achieve states of mind that will allow them to understand the world as humans do. In 1948, leading brain surgeon Geoffrey Jefferson (1886–1961) decried the notion of thinking machines that were touted as equivalent to humans, arguing that they were not conscious and that "Not until a machine can write a sonnet or compose a concerto because of thoughts and emotions felt, and not by the chance fall of symbols, could we agree that machine equals brain—that is, not only write it but know that it had written it." It is not just about doing things and performing tasks, however complicated and impressive or creative, but about understanding what it is doing and why it is doing it that is key to being human and key for robots if they are to be like humans? Some scientists argue that if a computer can simulate the actual sequence of neuron firings at the synapses of a Chinese speaker when he understands stories in Chinese and gives answers to them, then this must represent intelligence in the form of understanding. Searle argued that this only simulates the formal structure and needs to be revised to generate genuine human understanding, which requires intrinsic intentionality and not derived intentionality. Intentionality is not used in this context to mean someone does something deliberately, but as the German Philosopher Franz Brentano (1838–1917) conceptualized, it is a characteristic of all mental states. Jessica Pfeifer and Sahotra Sarkar described derived intentionality as "A thing that has derivative intentionality because "the fact that it represents such-and-such can be explained in terms of the intentionality of something else."[46] Robert Van Gulick proposes that "Intrinsic intentionality depends on a system understanding that the symbols it processes have the semantic content that they have," which constitutes genuine understanding.[47]

An intriguing rejoinder to Searle was the Connectionist stance proposed by Canadian neuro-philosophers Paul (1942–) and Patricia Churchland (1943–), who in 1990, envisaged a more realistic brain architecture than can even currently be achieved by imagining a central processing unit (see later) consisting of a serial architecture machine and the combined processing of all the various components, including the person in the room, which would acquire the content of mental states, which they argue is what our brains do. A real potential danger or strength that emerge from AI systems and robots comes not from the prowess of the individual systems or robots but the potential level of collective connectivity, sharing, learning, and collaboration between these autonomous robotic agents which would generate a collective intelligence and ensemble learning. Is that not how humanity has evolved over the millennia, and are we witnessing the dawn of the android epoch in the clumsy early movements of embryonic androids?

In 1995, British computer scientist Stuart Russell (1962–) and American computer scientist Peter Norvig (1957–) sought to clarify the field of AI and elevate the philosophical debate defining AI as "The study of intelligent agents that receive precepts from the environment and take action." They described how AI could be achieved in four different ways, which is that it could build systems that "(1) Think like humans (2) Think rationally (3) Act like humans or (4) Act rationally and the many cognitive architectures have explored all these various goals."[48] With machines that act like humans or act rationally, the systems are not focused necessarily on the thinking aspects, whereas thinking machines are either aimed at simulating human intelligence and thinking by including making human-like errors or producing the optimal rational thinking that consistently produces the correct answers across a range of tasks, which is not as we know how humans typically behave.

In 2007, Shane Legg and Marcus Hutter observed that "A fundamental problem in AI is that nobody knows what intelligence is" and sought to address this by compiling a comprehensive collection of definitions of intelligence, acknowledging that there is no standard definition of intelligence but that there are obvious and "strong similarities between many of the definitions."[49] They constructed from the most common and essential features of all the definitions a single definition of intelligence called Universal Intelligence for application to machine intelligence, given how inappropriate standard human intelligence tests are for machine systems. They identify the two common features as "(1) That intelligence is the property of an entity which is interacting with an external environment, problem or situation" and (2) An agent's intelligence is related to its ability to succeed in an environment, which they define as the goal". Their proposed definition of intelligence and the formal framework that they developed is that intelligence for AI should be the measure of "an agent's ability to achieve goals in a wide range of environments." They support this with the argument that what we are trying to achieve in machines is "a very general and flexible capacity to succeed when faced with a wide range of problems and situations."[50]

The Fundamental Hypothesis of AI

The fundamental hypothesis of AI has been that the human mind and intelligence are cognitive systems and computational entities that can be generated using a range of physical mechanisms and algorithms. More recently, there has been a paradigm shift to embrace a range of new AI theories, including Dynamic Systems Theory (DST) and Connectivity Theory (CT) that harness systems biology by exploring new biomimetic approaches beyond the pure computational perspective. DST studies the behaviour of complex dynamic systems with internal states that evolve (i.e., internal dynamics) and how these interact with exogenously applied inputs called perturbations. A 2017 study of 6–10-year-old children by Laura Golenia and colleagues even cited DST as a powerful tool for studying developmental change and explains such things as "Motor behaviour which emerges from interactions of multiple components entailed by the person, environment, task" and how "changes in all these components can potentially contribute to the emerging behaviour."[51] CT recognizes that learning and intelligence cannot simply be attributed to what occurs inside a person and must reflect connectivity to sources of information, others, and even technology. Spanish social scientist Ingrid Del Valle García Carreño has observed how "Technology has created and defined a new culture, reshaping the way we communicate, study, and learn. With this phenomenon, the need arises to create a new theory that can adapt to these changing circumstances. This emerging theory is the theory of connectivity." Carreño claims that this involves the "Integration of principles explored by chaos, network, and complexity and self-organization theories," and argues that "Connectivism is an epistemological approach grounded in the interactions within networks both inside the individual mind and outside to the world, rather than to the individual memory of what to do (behaviorism), what to think (cognitivism) or how to make meaning (constructivism)."[52]

DST and CT illustrate how AI research is increasingly seeking to reflect the dynamic nature of human learning, knowledge, and understanding through the connection with the real world, which can't be treated as just a computational or a pure algorithmic exercise. Biomimetics involves emulating nature's models, systems, and elements to solve complex human problems, which means the human body and brain, and includes neuromorphic computing and other biomimetic technologies, which we will discuss later. For example the new field of neuromorphic computing is inspired by the way our brains work with neuromorphic chips containing millions of artificial neurons and synapses to improve self-intelligence by being able to manipulate the data received through the various AI sensors. Neuromorphic chips are an example of how AI is beginning to attempt to resemble more closely the way the brain processes information, with billions of neurons and trillions of synapses responding to sensory inputs to produce intelligent output.

Types of Artificial Intelligence

There are two ways to classify AI machines, the first by reference to how alike they are to humans, and the second into the types of intelligence they exhibit, which are widely accepted as Artificial Narrow Intelligence (ANI), Artificial General Intelligence (AGI) and Artificial Super Intelligence (ASI).

In the first classification of how alike to humans they are, there are four types of AI machines which are (1) The oldest form of machines, which are purely reactive in that they respond to certain stimuli but have no memory, only respond to a limited range of inputs, and cannot learn. IBM's Deep Blue chess machine is an example (2) Machines with limited memory but can learn to improve and adapt their responses which describe most of the AI systems currently in use, such as Chatbot and virtual assistants (3) Machines that understand the intentions and needs of other intelligent entities so possess a theory of mind (TOM) which have yet to be developed and (4) Machines that possess human self-awareness and intelligence which are no more than currently a hypothesized type of machine. The problem with these various distinctions or classifications of types of AI machines is that they need to reflect the more granular nature of what AI developers seek to achieve in AI.

Acknowledging this problem and oversimplification, cognitive scientist Dr Ron Sun, who is investigating the fundamental nature of the human mind and made significant contributions to computational psychology, published in 2004 a comprehensive and highly influential review of the various competing AI projects and initiatives around the world revealing significant disagreement amongst developers of AI cognitive architectures as to what artificial intelligence consists of.[53] In a 2016 paper which is complementary to Sun's earlier review, David Vernon, Claes von Hofsten and Luciano Fadiga, motivated by studies in developmental psychology, cognitive neuroscience, and enactive cognitive science, set out the following ten desiderata which are "(1) Dealing with value systems & motives (2) Embodiment, (3) Sensorimotor contingencies (4) Perception (5) Attention (6) Prospective action (7) Memory (8) Learning (9)Internal simulation and (10)Constitutive autonomy." All ten desiderata focus on the "ultimate aspects of cognitive development, why a feature is necessary and whether it enables a machine to be intelligent".[54] We will look at all these desiderata as we go, but first, let's examine the more general technical classifications of the types of AI that are used by computer scientists, namely, Artificial Narrow Intelligence (ANI), Artificial General Intelligence (AGI) and Artificial Super Intelligence (ASI) bearing in mind that the ten domains set out in Sun's desiderata apply across these classifications.

Artificial Narrow Intelligence (ANI)

Many AI developers have focused on developing specific competencies and behaviours of AI machines rather than trying to simulate the complex goal of AGI. Ray Kurzweil describes this as focusing on Narrow AI (ANI), and as of the time of writing, all forms of AI and robotics are ANI.[55] Examples of ANI designed to perform specific tasks include image and facial recognition used in social media by Facebook and Google, Chatbots, and conversational assistants, which many of us are all too familiar with, such as the voice activation devices in our homes like Alexa, Google Assistant, and Siri. They respond to voice commands and are examples of ANI that have no real general intelligence or understanding of what we or they are saying or doing. Nor do autonomous self-drive vehicles or programs that make user recommendations, such as those employed by Netflix and Amazon.

ANI is used in predictive maintenance models in which data is collected over time to monitor the condition of the systems and predict when maintenance is required. It has applications in various industries, such as power outages, aircraft maintenance, refrigeration sensors, and the oil and gas industry. Some ANI systems like IBM Watson developed in IBM's Deep QA project are data analytics processors that use natural language processing to analyse syntax and meaning and answer questions posed in natural language much more rapidly than humans in fractions of seconds. One of the applications was to assist doctors when making data-driven decisions and improve healthcare by making it quicker and safer. Unfortunately, the system experienced major problems due to not using the necessary and correct data, for example, it was not trained with real patient data in cancer diagnosis. It used hypothetical cases from a small group of doctors in one hospital, so was error-prone, illustrating the challenges of building systems that can display intelligence in terms of understanding. ANI systems can pick up language patterns and, based on pre-programming, execute certain commands, giving us the impression that they are responsive to our needs and know what they are doing. The reality is that they don't know what we want or why and have no understanding of what we mean or what they mean when they respond to us.

A major drawback of ANI systems has been that they can only do what they have been programmed to do by humans and make all their decisions based on the training data. Because of the volume of data required, they are prone to bias inherent in that data and can't explain or understand the bias or know what they are saying. Despite these limitations, ANI is used widely to improve the quality of people's lives in specific areas and begin to understand the difference between what people say and appear to want and convert what we say into text language with much greater accuracy. Most importantly, they are starting to learn as they go, which has the potential to transform the nature and scope of AI fundamentally.

Artificial General Intelligence (AGI)

The goal of AGI is to equip machines and robots with the human ability to understand language and abstract concepts, learn by themselves, use logical thinking to plan and decide to act and solve, perform tasks, adapt and transfer knowledge to different contexts and environments and be creative. In other words, integrate perception, attention, memory, motor actions, social behaviour, planning, motivation, emotions, and self-awareness and perform any intellectual task that humans can. In 2012, computer scientist Sam Adams from IBM and colleagues from a wide range of institutions sought to bring clarity to the field of AI by trying to "map the landscape of AI," purporting to establish "An initial capability landscape …drawing on major themes from developmental psychology and illuminated by mathematical, physiological, and information-processing perspectives."[56] The objective was to provide an AGI road map with categories of intelligence largely reflecting the domains that psychologists use to measure human behaviour and cognition, which include perception, attention, memory motor actions, social behaviour, planning, motivation, emotions, and self-awareness which make people aware of and understand their actions and the consequences of their behaviour. This has led to the focus of AI research on specific psychological functions of intelligence as sub-approaches rather than representing a unified and integrated approach to creating the goal of AGI. One major problem is that different researchers construe AGI more broadly or narrowly than others with an ongoing lack of consensus or commitment as to what constitutes AGI. Ben Goertzel (1966–) in 2014 described the AGI community at that time as "a fuzzy set containing researchers with various interpretations and varying levels of commitment to the AGI concept."[57] Nearly ten years later, although AI can process data far faster than humans, it has yet to master the ability to think by integrating prior knowledge or harnessing its thoughts and memories to make informed decisions, think in abstract terms, strategize, or generate creative ideas entirely by itself. These are all qualities associated with sentience and consciousness in humans, and building robots with AGI without human intervention to reprogram or reconfigure the software and hardware has yet to be achieved.

Two major limitations on AI highly pertinent to the field of AGI are the volumes of input data required to train AI systems and getting AI systems to generalize their intelligence or transfer their knowledge to different applications and environments, particularly real-world environments. This is the difference between being taught to do things under supervision and generative knowledge which involves learning through experiencing and interacting with the real world. Autonomous agents must be able to adapt to the world, continuously learn from experience, and behave in a habitual and goal-orientated way without being instructed or supervised by humans using knowledge to learn and create more knowledge. This means dealing with situations and problems in ways their creators had yet to envisage and going beyond the solutions and approaches that humans might have come up with. Because of the complexity involved in graduating from ANI to AGI, which involves thinking of abstractions and interpretations and being self-aware, AGI is currently largely confined to research programs, although some scientists maintain that AGI has already arrived in NLP models like ChatGPT, Bard, LLaMA, and Claude, which although they have major flaws satisfy certain metrics for AGI. These metrics have been described as the ability to (1) Cover any topic that is put online by using hundreds of gigabytes of text from internet sources and audio, video, and other media (2) Ability to complete most tasks, whether it is answering questions, translating, composing stories and music, even explain why it decides to act or do things, etc (3) Handle any sensory or motor modality although integration of information from multi modalities remains a major challenge (4) Speak and employ most languages and (4) Respond to instructions or prompts when completing tasks. Based on these capabilities it may be decades before they are not plagued by problems.

Others observe that AGI is nowhere near the unrivalled capacities of the human brain, whilst some scientists argue that scaling up existing AI models will be enough to produce human-level AGI, although what scaling up constitutes needs to be clarified. Demis Hassabis, the head of Google's AI division Deep Mind, believes that AGI is just a few years away, maybe within the next decade and research director Nando de Freitas claims that DeepMind's Gato AI, a generalist agent, is already close to rivalling human intelligence. In 2023, Masayoshi Son, the CEO of SoftBank, in an interview with CNN Business, said that artificial intelligence that surpasses human intelligence in almost all areas and ten times more capable and powerful than human intelligence, will be realized within ten years. He observes, "It is wrong to say that AI cannot be smarter than humans as humans create it," as "AI is now self-learning, self-training, and self-inferencing, just like human beings."[58] He acknowledges that he thinks he is the only person who believes AGI will come within a decade although Sam Altman, the CEO of Open AI, declared in 2023 that," I think AGI will be the most powerful technology humanity has yet invented," particularly in democratizing access to information globally.[59]

Artificial Super Intelligence (ASI)

Nick Bostrom defines ASI as "Any intellect that greatly exceeds the cognitive performance of humans in virtually all domains of interest." ASI are software-based systems with intellectual powers beyond those of humans across a comprehensive range of categories and fields of endeavour. Other than in specific domains, general super intelligence has yet to be achieved, and there are currently no AI systems that are superior to humans across all accepted fields or domains of intelligence. It is the super intelligence that many people believe to be our greatest threat and, when harnessed in autonomous robots, is the stuff of potential nightmares for many people. Ray Kurzweil who postulated the singularity scenario, believes that we could coexist with such machines, that they could reinforce human abilities rather than take over, and that we could do this by treating them as "one synthetic neocortex in the cloud." The fact is that humans remain more generally intelligent than any existing AI systems or robots, which have yet to display real-world general intelligence and understanding comparable to, or superior to humans. AI displays ASI in certain tasks such as chess, but this is only in some domains.

Artificial Learning

The increasing number of researchers focusing on AGI use a variety of approaches to AI learning, including machine and deep learning and different types of ANNs. We often hear AI and machine learning used interchangeably, but they differ in several ways. AI is a very broad field in which machines and computers are being built to mimic human cognition and intelligence, but simply mimicking the output of humans is different to learning like humans. Conceptualized by American physicist Roger Osborne and the American educational psychologist Merlin Wittrock (1931–2007) in the 1980s, generative learning has been the focus of much research into AI learning. It requires actively constructing meaning, which involves "Building relationships between stimuli and their stored information such as knowledge and their experiences."[60, 61] The Generative Learning Theory (GLT) postulates that learning occurs when people actively organize and integrate new information into their existing knowledge structures. The problem is that AI isn't a specific computerized system but refers to many different technologies incorporated into different systems to allow them to reason, learn, and solve complex problems using various learning approaches, such as machine and deep learning, which are all subsets of AI. The different approaches and AI systems that have emerged are typically categorized as either (1) Symbolic (Logistic) (2) Emergentist or (3) Universalist (or Hybrid) with the major distinction between them being the degree of independent agency or autonomy of the AI system when learning, solving a problem or choosing an output or response.

Symbolic (Logistic) Reasoning

Symbolic or logistic reasoning reflects how humans encode knowledge and communicate in language mainly by logically using a constant and unchanging lexicon of combined signs, strings of characters, numbers, and symbols to represent features of the environment or us. This allows abstraction and knowledge transfer between situations and contexts or environments, such as using symbols for abstract categories like a table or chair. Humans naturally combine symbols such as letters or numbers to express the interrelation between symbols so that inductive or deductive reasoning can be used to convey information in thoughts through manipulating those symbols. The symbolic approach in AI involves a high-level, finite and unchanging set of symbolic representations of problems specified by programmers and employs inductive logic and search. A fundamental difference is that they don't permit interpretation, emerge naturally from, or are generated by the system itself.

Symbolic reasoning faces a major problem when a robot is required to operate in the real world because for symbols to have meaning, they must be grounded in some reality, and symbols aren't grounded in reality because they are just a human rule-based construct. A major obstacle to using symbolic reasoning is that humans use both symbolic and sub-symbolic processing as an integral part of general intelligence. Ben Goertzel and colleagues have intuitively described sub-symbolic as "Referring to things like pattern recognition in high-dimensional quantitative sensory data and real-time coordination of multiple actuators utilizing multidimensional control signals, while symbolic referring to things like natural language grammar and (certain or uncertain) logical reasoning, that are naturally modelled in terms of manipulation of symbolic tokens in terms of particular (perhaps experientially learned) rules."[62]

In evolutionary terms, symbolic language and communication are believed to have evolved after sub-symbolic processing with symbolic reasoning employed to communicate and teach in a didactic sense. Although symbolic reasoning was popular in the early years of AI research, it is now called Good Old-Fashioned AI (GOFAI). Humans create the rules governing the symbols as we do in speaking different languages governed by different rules but to exhibit genuine autonomy, algorithms need to learn the rules that govern the correlation between inputs and outputs not rely on the human input of symbol rules. Most AI developers are moving away from GOFAI towards emergentist approaches.

Emergentist approach

Emergentist approaches assert that complex behaviour and abstract complex symbols emerge from the sub-symbolic processes, such as meaningful patterns in high-dimensional quantitative sensory data, without applying symbolic rules. The challenge has been the limited ability of emergentist AI systems to construct complex meanings such as value notions, including beliefs or abstract concepts and our ability to represent these distinguishes us from the rest of the animal kingdom. The emergentist approach is popular in deep learning and specific industrial applications and is considered the most feasible for robots because it does not involve higher levels of abstraction but relies on vast data to mimic general human intelligence even if not displaying genuine understanding and real human intelligence. Although deep learning employing the emergentist approach has led to much rapid progress in AI, it has many dissenters who complain that it is not possible to interpret the model because it has no explanatory power and requires vast amounts of data. Although it can be used for unsupervised learning, it is a supervised learning method and has yet to form an adequate platform for genuinely autonomous AGI. The approach gaining the most traction in the AI community is the Universalist or Hybrid approach.

Universalist or Hybrid approach.

The Universalist or Hybrid approach involves AI mapping highly compressed representations of a task and environment to maximize the task's rewards. It utilizes symbolic reasoning and the emergentist approach by employing deep learning to select the optimal program and responses based on a combination of probabilistic correlations and abstract higher-order concepts, allowing for greater data compression. Various initiatives like MIT, DeepMind, and IBM are now combining symbolic reasoning with emergent ANN as a hybrid or universalist approach, which partly reflects the dissatisfaction with deep learning. It solves the problem of being unable to interpret or explain the process and the need to deal with more complex abstractions and value notions such as beliefs. Using a highly compressed representation of the environment and task with all behaviours explained by a reward system makes learning complex behaviours highly challenging because associating complex behaviour with specific rewards or punishments is highly complex. All the various systems and approaches continue to have their advantages and limitations, which we will discuss later.

One of the challenges when comparing AI to neuroscience and psychology is that the same word may have different meanings in the various disciplines and the word learning in the context of ANNs is one such word. AI learning is used as a precise technical description of the "process of extracting structure statistical regularities from input data and encoding that structure into the parameters of the network in terms of the weightings and thresholds."[63] There are different types of data learning in AI with the main data science and data mining approaches used in developing AI machine learning systems generally categorized as (1) Supervised learning (2) Unsupervised learning and (3) Reinforcement Learning (RL). Supervised learning involves a much greater level of human intervention, whilst unsupervised learning, or RL requires little or far less human supervision during training and learning. RL is based on rewarding desired behaviours and/or punishing undesired ones and learning by trial and error.

ANI has typically used a bottom-up rule-based type of learning in which AI uses rules encoded in if-then-else statements, which allow the machine to follow data programs. If-then-less statements are known as decision-making statements because the AI system decides based on a given condition or expression. For example, an IF statement will check a Boolean or truth value or condition such as whether is it greater than equal to, or less than, and if it is true, it will execute a piece of code and the value or condition is false or null it will skip the statement or code and execute another block of code. ANI uses machine learning to identify in given data or input, features, trends, patterns, and anomalies that humans would probably be unable to detect, which allows the machines to send commands or notifications when certain thresholds are reached. The consensus in the neuroscientific world appears to be that the human brain is algorithmic in the manner of processing raw data, although some scientists continue to hotly contest this. This algorithmic perspective could be said to have its origins in the seminal work of Warren McCulloch and Walter Pitts in 1943[64] which highly influenced the field of cybernetics. The state of any neural network be it BNN or ANN is deemed to be the current 1 or 0 of all its neurons with the logical Boolean function governing its behaviour. This assumes the definite next 1 or 0 value for each neuron, so at each moment, the brain passes from one state to another state. Because of the finite number of states which is 2 to the Nth power it repeats around a state cycle, which is the attractor that we discussed earlier in the context of neural networks and those with neural loops have many attractors that appear to provide the stability required for memories and the process of classification that the brain performs. Dr Joe Tsien is a neuroscientist who focuses on cognitive and systems neurobiology and argues that our brains employ basic algorithmic calculations in a form of simplistic logic which he maintains, like DNA, is a fundamental design feature of the brain. Tsien published his Theory of Connectivity (TOC) in 2015 and with a colleague, purports to have identified algorithmic calculations at work in seven different brain regions suggesting that it is at work generally throughout most of the brain. His work is supported by neuroscientist Dr Thomas C. Südhof, who studies synapse formation and function and won the 2013 Nobel Prize in Physiology or Medicine. TOC maintains that groups of similar neurons form different cliques and are prewired during development to handle each basic calculation like the classification of objects and these cliques cluster into Functional Connectivity Motifs (FCMs) with a correlation between the complexity and the number of cliques involved. Tsien acknowledges that there are exceptions in the brain including the brain's reward circuits which are governed by the activity of the dopamine neurons with cells tending to be more binary in judging whether, for example, something is good or bad. At the centre of the TOC theory is the algorithm, n=2. -1, which defines how many cliques are needed for an FCM and which has even enabled the scientists to predict for example when selecting food, the number of cliques needed to recognize food options.

Machine learning tries to mimic the way humans make decisions based on previous knowledge by using previous computer data but the extent of human involvement in inputting data and monitoring the output is fundamental to the extent of robot autonomy, so let's look briefly at what these three types of learning means for AI and autonomous robot design. We will also need to look at some of the more common algorithms used in AI machine learning to primarily understand the principles behind the methods employed in the various types of AI learning to classify or categorize data and make accurate predictions and decisions based on the data. An algorithm is though just a mathematical procedure for solving problems or performing computations that provides a precise series of step-by-step instructions to be followed and guides the system in performing calculations or operations, each with different goals and methods. We employ algorithms in our everyday lives for example our morning, bedtime, or exercise routines, driving somewhere, deciding what to eat typically, or tying our shoes all involve replicating a series of steps that we follow.

Computer algorithms can appear highly complex but simply stipulate the steps to be taken by the system when processing data to extract information and we encounter them all the time without being aware of them in facial recognition, auto-correct, search engines, or social media algorithms, etc. Many machine learning algorithms use training data supplied by a programmer or obtained from the environment and use that training data whether it is categorical, numerical, image, video data, or anything else to ideally learn on its own although many require human intervention. Algorithms are either classification or regression and if classifiers they produce a classification of something as either one or another class or they may classify the data into defined categories or types so use multi-classification. Regression algorithms produce real numbers that predict the probability of something to make estimations.

Supervised Data Learning

Supervised data learning is a form of machine learning that uses labelled input and output data training sets to train or supervise the algorithms that process and assign the test data to categories or classes. Using these categories or class systems can more accurately predict outcomes and make decisions when using other data. The accuracy of the model is directly proportional to the quality of the training data because the data consists of an input item such as an image and its label to find network parameters i.e. weights and thresholds, that generate the correct tags or labels for novel pairs. When developing a machine learning model, labelling the data or data annotation is part of the pre-processing stage which involves identifying the raw data (i.e., images, text files, videos) and adding one or more meaningful tags or labels to specify its context which allows the machine learning model to make more accurate classification or categorization, predictions, and decisions. The label forms a representation of the class of or category objects the data belongs to and machine learning trains the machine to learn to identify that class of objects when encountered in data without a label. For example, whether a photo contains a dog or a horse a brain X-ray contains cancerous growth, or identifying the words used in a recording. Labelling is used in computer vision, natural language processing, and speech recognition and although data labelling may appear simple it can be very complicated with multiple factors affecting the choice of what labelling method to use, each with pros and cons.

Much progress in developing ANNs has focused on the tools required for supervised learning but a major challenge is the ability of the supervised learning systems to generalize learning which means adapting to and classifying and forecasting new previously unseen data that was not part of the training data, but drawn from the same distribution as the one used to create the model. If a model performs well or accurately when presented with new data, it is termed generalizable although this typically does not refer to all data types but to data with similar domains or datasets. Supervised learning is regarded as more accurate than unsupervised learning but usually requires a significant level of Human-In-The Loop (HITL) input to label the data appropriately, with the specific goal of predicting what types of results are to be expected, rather than leaving the machine to work it out by learning itself.

Data annotation can be performed manually or automatically using advanced machine learning algorithms. Supervised machine learning involves HITL participation to guide the labelling process and typically involves humans making judgments about a piece of unlabelled data which can be a very simple yes or no it contains a dog or horse or at the pixel level whether those pixels are associated with a particular object and the labels are then used to create, train, or refine and test the learning models. Labelled data is more expensive and time-consuming to generate and store and common labelling methods include employing internal data science experts, generating new project data from pre-existing datasets called synthetic labelling, automatic programming labelling using scripts, outsourcing the process and even crowd-sourcing. Properly labelled training sets are often referred to as Ground Truth with the accuracy depending on the labelling in the ground truth data. Much machine learning currently uses labelling algorithms that match an input to an output, but variance errors can result from sensitivity to small fluctuations in the dataset used for training and can cause algorithms to model and extract meaning from irrelevant data or noise in the dataset, instead of the intended outputs called signals. This is called overfitting the data which can result from coding and manual entry errors etc, making HITL involvement in quality control essential.

Data labelling is most used in certain applications such as computer vision, natural language processing (NLP) applications, spam detection, machine translation, speech recognition, text summarization, virtual assistants, chatbots, voice-operated GPS systems, etc. Labelling allows the trained AI system employing iterative learning to use categorical data to classify observations, events, or inputs. It does this mainly by classification with labels if the data is categorical or if the data is numerical by regression to predict

a quantity. Categorical data is a statistical type of data that is qualitative because it describes an event using a pattern of words rather than numbers although it can be calculated in numerical values as well. By sorting the data into categories or classes such as distinguishing between types of objects or things, classification and prediction algorithms are powerful tools for decision-making and strategy formulation. Regression algorithms detect relationships between dependent and independent numerical variables. By comparison to supervised machine learning in AI, human children learn many complicated skills such as speaking a language with apparent ease not based not on supervised learning but simply early exposure to and imitation of the spoken language. The challenge with supervised learning in AI is that the greater the level of HITL involvement the less autonomous the AI system or android will be.

In supervised machine learning many classification tasks can be performed though each problem requires a different algorithm which involves choosing the optimal classification algorithms. Classification involves recognizing, understanding, and grouping ideas and objects into predetermined categories or sub-populations, and classification algorithms are used in machine learning, to predict the likelihood that subsequent data will fall into one of the predetermined categories, so is a form of pattern recognition in which the machine uses algorithms to apply to training data and predict similar patterns in other data. Some of the most widely employed classification algorithms, used in supervised learning are (a) Linear Regression (b) Logistic Regression (c) Support Vector Machines (SVM) (d) Kernelized Support Vector Machines (KSVM) (e) Naïve Bayes (f) Decision Trees/Forest Trees and (g) K- Nearest Support.

Supervised learning has many advantages in that it makes training easier and clarifies the exact classes in the data and how many there are which makes it possible to be very specific about the definition of classes, so decision boundaries are very clear which allows accurate distinction between different classes. It is also easier to understand in that we know how the machine is learning. Supervised learning though has significant limitations and can't handle some of the more complex tasks in machine learning such as extracting unknown information from the data training clustering or classifying data by discovering features on its own and is heavily reliant on the accuracy of the input and samples used in the training data. It takes a significant amount of time and data to train systems and ultimately there is a limit to how much information can be given which poses an obvious problem for achieving complete autonomy which ultimately requires a system to be able to learn by itself. This is where unsupervised data learning comes in.

Unsupervised Data Learning

Unsupervised data learning resembles more closely the way the human brain works but is far more complex and challenging than supervised AI machine learning. It uses different learning algorithms on unlabelled data or unstructured data sets to sort the data points into predefined clusters, detect features and patterns in the data, and generalize that to similar though not necessarily identical data. The goal of unsupervised learning is to identify statistical regularities in the data without guidance as to which regularities to search for. Machine learning can do this with significantly less HITL involvement and unlike supervised learning can determine what is interesting or different about the data. If you want to group or cluster some data that you know little about, unsupervised learning is useful, although typically requires huge amounts of unclassified data and training sets. It tends to be used in data mining for grouping unlabelled data based on their similarities or differences, to find relationships or associations between variables in each dataset to reduce the number of data inputs to a manageable size whilst preserving data integrity. Unsupervised machine learning can produce wildly inaccurate results and usually needs some HITL intervention in the form of a data analyst to validate the output variables which sounds like some humans we all know and have probably worked with. The most common unsupervised learning algorithms are (a) K-means (b) Hierarchical Cluster Analysis (HCA) (c) Expectation Maximisation (d) Locally Linear Embedding (LLE) (e) Principal Component Analysis (PCA) (f) Kernel PCA (KPCA) (g) t-distributed Stochastic Neighbour Embedding (t-SNE) (h) Apriori and (i) blat. The details of which are beyond the remit of this book.

Semi-Supervised Learning.

To overcome the drawbacks of supervised and unsupervised learning, some systems are using semi-supervised methods that utilize both labelled and unlabelled data and tend to be used when it is difficult to extract relevant features from a high volume of data. They use both supervised and unsupervised algorithms and are the most used learning approach in ANNs because semi-supervised can be used for both classification and pattern recognition processing.

Machine Learning (ML)

Machine Learning has been described as the "automatic discovery of regularities in data through the use of computer algorithms...".[65] ML can learn using algorithms and statistical models to analyse the data and imitate the way humans learn by gradually improving the accuracy of the analysis and can involve any of the four different types of learning supervised, unsupervised, semi-supervised, and reinforcement learning). Some of the most common applications are natural language processing, facial recognition, data mining, product recommendations so improved marketing and advertising, email automation and spam filtering, financial accuracy, fraud detection, healthcare, and predictive analysis. ML relies largely on HITL intervention to learn, with humans collecting the training data and preparing the data which includes determining the hierarchy of features to distinguish between data inputs which typically requires structured data, so is more supervised. Structured data has a standardized format for easier access by software and humans. A model must be chosen and trained using the data which is typically in table form with rows and columns and clearly defined with labelled data attributes that are quantitative like names, telephone numbers, etc more easily processed to extract information but heavily reliant on the accuracy of the information contained in the data. The parameters often need tuning before predictions can be made and there are several major limitations with ML with a major challenge being how difficult they are to interpret or to understand how they reach a judgment or decision. They are often called black boxes because of the lack of transparency and pose the XAI problem.

Another problem with the large datasets required to train ML models is the possibility of bias and discrimination in the data. An example of this was the problems that some facial recognition systems have in detecting darker skin types with the result that they produce lots of false negatives and positives and are known to display gender biases. A 2018 collaborative study performed by MIT and Stanford University, co-authored by Joy Buolamwini and Timnit Gebru, a scientist at Microsoft Research, evaluated three of the largest facial recognition programs from Microsoft, IBM, and Megvii of China. They were assessing how well they could classify gender with different skin tones. It emerged that one company that had claimed 97% accuracy employed a data set to evaluate the performance that used more than 77 per cent male and more than 83 percent white participants. What it demonstrates is that an AI machine learning system is only as clever as the data that is used to train it. If it involves many more white men than black women, it will probably be poorer at identifying black women.[66] In 2015 Google was forced to issue an apology when its facial recognition App identified two African Americans as gorillas, but their highly unsatisfactory solution at the time was to simply block the algorithms from recognising gorillas. Joy Buolamwini has gone on to become a leading advocate of algorithm accountability and given a TED talk that has been viewed nearly a million times, which illustrates people's concerns about the impact that ML can have on lives. Another major problem with ML is that they are frequently either over-fit or under-fit data. With over-fitting it has been almost trained too well on the learning data or if under-fit it has been trained too simplistically and has poor performance on both the training and new data. Another problem is the lack of sufficient data due to sometimes restricted availability with the computational costs of training on large data sets making some machine learning very expensive and not available to everyone. With machine learning, it is also almost impossible to determine whether its predictions relate to an identifiable cause of the predicted behaviour. Finally, many other major social and ethical considerations arise using ML that are employed to make decisions about things that affect people's lives, like recruitment, healthcare, law enforcement etc.

Model Parameters

We have seen how for any AI model to learn it must have certain values or variables i.e. parameters that it can change or adjust independently as it learns to make decisions and predictions and effectively encode the model's knowledge and skills. The largest models now possess billions of parameters which has rapidly increased the size of models, although some systems prefer to increase the data set rather than the model size through increasing the number of parameters. Parameters in neural networks as we have seen are given weights and biases of probabilities which have a critical impact on the performance of the model in allowing it to improve its learning and these are produced during training. In linear regression, the parameters are the m (slope) and c (intercept). There are two main types of parameters in ML models, which are the Model parameters and Hyper-parameters. Model parameters are not set manually but learned or chosen by the model from the data whilst training. The parameters are coefficients which in mathematical terms are numerical or constant quantities placed before and multiplying the variable in an algebraic expression. For example, in the expression: 3X2 + 5, X is the variable and 3 is the coefficient whilst 5 is the constant. If the coefficients are variables, they are often called parameters and are usually a number or sometimes a symbol (letter) that represents a constant value which is multiplied by the variable of a single term or the terms of a polynomial. Polynomials are simply expressions consisting of such variables and constants and the operations performed such as addition, subtraction, and multiplication which represent the relationship between the variables and constants. The final parameters after training ultimately determine the model's performance on unseen data. Hyper-parameters are fixed instructions for the learning process and are the values set manually before the training so cannot be learned by the model. They are needed to estimate and optimize the model parameters and have a major impact on the accuracy and efficiency of the model. The model size in ML in AI is typically measured by the number of free parameters that must be fit to the data and the amount of data required to train the model. Data fitting is the process of fitting models to data and analysing the accuracy of the fit with observed data being fixed while the model is mutable because the parameters are estimated which means that the model is made to fit the data not the other way around.

The more parameters there are in a model the greater its capacity to handle multi-modal data sources like images and text as input and to produce for example text as the output. They can execute much more complex tasks, generate coherent texts, and potentially exhibit human-like intelligence. The dataset size (although not confirmed by Open AI) of ChatGPT-4 is 45TB and 1.7 billion parameters which dictate how it generates text. Bites are used to measure storage size or capacity and most people are familiar with the common Megabyte and Gigabyte. For example, a photo taken on an iPhone 11 is approximately 3MB. 1024GB = 1TB and about 1 trillion bytes. This means that Chat GPT is 1000 times larger than ChatGPT-2 and approximately 1000 times larger than ChatGPT-3, with 1.5 billion and 175 billion parameters respectively. A 2022 review of the growth in model sizes called Machine Learning Model Sizes and the Parameter Gap found that "From 1950 to 2018, model size in language models increased steadily by seven orders of magnitude. The trend then accelerated, with model size increasing by another five orders of magnitude in just 4 years from 2018 to 2022. Vision models grew at a more constant pace, totalling 7 orders of magnitude of growth between 1950 and 2022."[67]

Deep Learning (DL)

In 2006, computer scientists Geoffrey Hinton, Simon Osindero, and Yee-Whye Teh published a fast-learning algorithm for Deep Learning (DL), describing their innovation as follows "Using complementary priors, we derive a fast, greedy algorithm that can learn deep, directed belief network one layer at a time, provided the top two layers form an undirected associative memory (also called an auto-associative memory or auto-encoders and Recurrent Neural Networks (RNN),". Hinton, Osindero, and Teh's fast-learning algorithm has been described as marking the dawn of the deep learning era of AI.[68] You will remember that an auto-associative memory or RNN is one in which the output is fed back into the input, which enables the network to learn and remember the patterns it was trained on. Auto-associative means to associate something like a pattern with itself, not with something else or another pattern like hetero-associative processes.

Many probability distributions in sample data contain unknown parameters and Bayesian statistics is used to estimate and determine the probability of an unknown parameter being correct. Bayesian assessment of probability typically reflects our prior beliefs or uncertainties that we bring to the data, so we compare it for example to patterns etc, that we already know hence the term auto association. The assessment of likelihood is not the same as the probability but is proportional to the probability and is key to assessing the probability of unknown parameters because it describes how well the sample data fits a particular statistical model. This is because it determines the probability of observing the data given specific model parameters. The likelihood function tells us how well our data summarises these unknown parameters. A prior that is complementary is chosen so that the product of the prior and the likelihood function is constant because the prior probability is a uniform and constant function. This approach recognises that all sample data is subject to random errors which reflect the fact the incomplete nature of most raw data which is what the human brain must constantly deal with.

When assessing the accuracy of the data a consistent estimator is sought which means that there will be insignificant errors or variations as the sample size gets larger. With even more data collected a consistent estimator means that it will be close to the real population parameter. Although it can use labelled data sets, to learn to recognize features and patterns in the data DL models typically use unstructured raw data which is more difficult to analyse, and the learning is more unsupervised because the machine learns as it goes.[69] DL is effectively a subset of ML that automates a lot of the feature extraction thus reducing the level of HITL intervention, which means that it can be used on much larger unstructured data sets. DL can learn appropriate low-level representations such as visual features from the data it is trained on rather than having them explicitly programmed in. DL is evolving rapidly due to increased chip processing abilities like GPUs, the significantly reduced cost of computing hardware, and major advances in machine learning algorithms. Both ML and DL use ANNs, but DL consists of more than three neuron layers having multiple hidden neural layers as part of DL algorithms with each layer having its activation functions. This includes input and output layers and uses back propagation which we discussed earlier. All the outputs from the hidden layers are used to provide inputs to the output layer to allow the output of the neural network to be calculated. Unlike linear regression in which you can change single inputs without affecting the other inputs, in DNNs the act of changing one input can have a cascade effect on the whole network but the use of many layers enhances the system's capacity for learning good internal representations which potentially adds to the accuracy of the final output. Although DL has revolutionised AI there remains a long way to go if DL is to compete with the human brain. This is reflected in the words of Japanese mathematical neuroscientist, Professor Shunichi Amari's observation of how there is currently "no theory of deep learning except for individual minor ones. So far these individual theories are not sufficient to achieve a comprehensive understanding of our brain."[70]

Reinforcement Learning (RL) and Deep Reinforcement Learning (DRL).

One of the most promising methods of artificial learning for AI generally and robots is Reinforcement Learning (RL) and increasingly Deep Reinforcement Learning (DRL) which is modelled on the learning methods employed by humans that we discussed earlier.

RL is the third main type of learning in AI and a form of unsupervised learning. It is arguably the most relevant to us as it originates from autonomous robotic agents learning through trial-and-error through search and exploration and using feedback in terms of what to do or not to do, which may be reinforced by a reward or punishment process. RL is based on the work of Edward Thorndike (1874–1949) in his 1911 Law of Effect and Pavlov's work with dogs that elucidated selection and associative learning that we discussed earlier. Computationally RL relies heavily on dynamic programming and the Markov Chain and Markov decision processes which help a system gradually reach the correct solution or answer by making successive guesses as to the likelihood of the solution being correct in classification and prediction tasks.[71] Remember that the Markov Chain assesses how probable it is to transfer from one state into another based only on the previous state.

In 1953 the American applied mathematician Richard Bellman (1920–1984) introduced dynamic programming, from probability statistics as a basis for what is known as the Optimal Control Theory. Optimal control reflected in the Bellman equation is often employed in RL algorithms and directed at maximizing the returns from and minimizing the costs of the operation of physical, social, and economic processes which is partly why statistical analysis is considered such a core element of AI. RL has led to far greater accuracy in robotic system performance, allowing robotic devices to perform arduous tasks in cost-effective ways and often dangerous situations. It has enabled automated robots to perform NLP and mimic and more accurately predict how people speak to each other. In 2016, a Stanford University team in collaboration with Microsoft used RL in NLP with an AI system designed to generate accurate and realistic dialogue with the system rewarded based on the metrics of coherence, informativity and ease of answering. The novelty came from the fact that the system was focusing on predicting the future outcome of the dialogue. We are seeing RL in AI being applied across a wide range of situations including marketing and advertising, image processing, trading and finance, systems for recommending likes for consumers based on previous buying patterns, gaming like AlphaGo Zero, energy conservation, traffic control, healthcare treatment options and automated vehicles which uses deep RL to deal with speed limits, parking and avoiding collisions and in robotic manipulation and handling of objects etc.

Despite the promise of RL and DRL some like Google research scientist Alex Irpan citing papers from Berkeley, Google Brain, DeepMind, and Open AI publicly observed in 2018 that DRL doesn't work well yet.[72] He illustrated this claim by reference to the Arcade Learning Environment (ALE) and learning the game Atari which is the most well-known benchmark for DRL and used as an evaluation platform for empirically assessing the generality of agent's performance across dozens of Atari 2600 games. Deep Q-Networks (DQN)[73] was the first algorithm to achieve human-level control in many games[74] but Irpan points out that it takes the computer using DRL about 18 million frames which corresponds to about 83 hours of play experience, plus however long it takes to train the model compared to a few minutes for most humans to pick up. The second most popular benchmark is the multi-joint dynamics with contact (MuJoCo) system involving a set of tasks using the MuJoCo physics simulator. MuJoCo is a free and open-source physics engine now owned by Deep Mind which facilitates research and development in robotics, bio-mechanics, graphics, and animation by allowing developers to access much more cheaply, huge amounts of data required to train robots compared to real-time environments which are expensive, and robots break, etc. Irpan observes that even using MhJoCo which is a relatively simple environment, depending on the task the benchmarks tests take between 105105 to 107107 steps to learn.

Irpan concludes that "Reinforcement learning can theoretically work for anything, including environments where a model of the world isn't known. However, this generality comes at a price, namely that it's hard to exploit any problem-specific information that could help with learning, which forces you to use tons of samples to learn things that could have been hard coded. The rule of thumb is that except in rare cases, domain-specific algorithms work faster and better than reinforcement learning. He also observes how RL assumes the existence of a reward function throughout the course of learning and that "for RL to do the right thing, your reward function must capture exactly what you want. And I mean exactly. RL has an annoying tendency to over fit to your reward, leading to things you didn't expect." Some RLs will perform well if the reward consists of a simple or single goal like getting good at one game like Atari, but it doesn't generalize to other tasks or games. Iran claims that "the generalization capabilities of deep RL are not strong enough to handle a diverse set of tasks yet" and he explains how "it's hard to find cases where deep RL has created practical real-world value." Dirk Heinemann and colleagues in a February 2022 paper titled Quantum Deep Reinforcement Learning for Robot Navigation Tasks observed how "Classical techniques for deep reinforcement learning for robotic behaviours have produced impressive results but they suffer from enormous sample complexity, limited generalization capabilities and limited applicability for flexible long-term autonomy." We will look later at how Quantum DL methods may be a way to overcome the current limitations of RL and DRL but the reality is that RL and DRL remain subject to considerable challenges and limitations.[75]

Continual Learning

We discussed the crucial role of continual learning in humans and how it confers significant benefits, primarily obtained from learning over a lifetime by adding incremental skills and by applying or transferring and generalizing learning to diverse applications which are regarded as a vital attribute of human intelligence.[76] The problem is that it poses significant challenges in AI and AL. AI systems adopt a variety of learning methods based on consideration of the major features of data which include the variety of data, the volume of data, the speed or velocity at which the data arrives, and the need for various types of data processing. Learning can depend on the nature of the transition between tasks and whether it is smooth or discrete, the length or duration of the task, how many times it is repeated, and the type of learning in terms of whether it is supervised, unsupervised, or RL.

Raia Hadsell who is a Senior Director of Research and Robotics at Google DeepMind pursues new solutions for AGI and has observed how the most powerful modern ML methods perform well only when presented with data that is "carefully shuffled, balanced, and homogenized. Not only do these models under-perform when presented with changing or incremental data regimes, but in some cases, they fail or suffer from rapid performance degradation on earlier learned tasks" for as they observe "the world is not stationary, but machine learning has typically involved fixed datasets and stationary environments."[77] We have seen how humans and other animals can "incrementally learn new skills without compromising those that were already learned."[78] Traditional ML algorithms can access all the training data at the same time whereas, in continual learning models, the data is supplied in sequence or incremental steps with the underlying data distribution changing over time. Incremental learning of new information from a dynamic data stream is complicated by what Gido van de Ven and colleagues describe as three fundamental types or scenarios which are whether a model needs to (1) Adapt to incrementally learn a new set of distinguishable tasks (2) New domains or (3) The same problem but in a different context or learn to distinguish between an increasing number of new objects or classes. They claim that continual learning research "lacks a shared framework making systematic comparison between continual learning algorithms challenging."[79]

At the heart of the challenge facing ANNs, ML, and DL is the extent to which they can process data dynamically by harnessing present or new data with current and past data to capture changes in the raw data available from the environment, to make accurate predictions and decisions and continually learn. Continual learning has important implications in allowing machine models to deal with evolving data, to allow them to operate in real-time, and make updates to the model and significantly improves the efficiency of the system because it learns more and more about the underlying patterns and relationships in data as it is updated. ML has in recent years, as we have seen, been equipped through developments in central processing technology with highly improved parallel processing capabilities which enables the machine to train and consider vast amounts of data simultaneously which has facilitated rapid scaling in their function in speech, vision, and language applications.

A major problem facing AI systems and ANNs is that despite the increase in parallel processing they have been unable to learn continually without erasing previous learning, with this degradation of earlier learned tasks termed by cognitive scientists, Michael McCloskey, and Neal Cohen in 1989 as "Catastrophic Forgetting."[80] Catastrophic forgetting limits the ability of AI systems to learn multiple tasks sequentially which interferes with the continual learning process and inevitably the autonomy of robotic agents. ML systems require all the task data to be available, whereas humans can efficiently and robustly learn incrementally from sequential experience even in a highly dynamic and changing environment, although as we saw they perform more poorly when required to train simultaneously on multiple tasks.

Batch learning, also called offline learning, unlike continual learning requires training AI models with the entire training data set before they can function as designed. Because of this, they are static when trained and their performance will not improve until a new model gets retrained. This leads to the performance of these models deteriorating over time, which is known as model rot or data drift because the system is not learning from changes in its environment. Online training or learning is by contrast incremental and involves continuously feeding data in small batches with performance reflecting its learning rate. Van de Ven GM and colleagues point out that the challenge with task-incremental learning, "is not or should not be to simply prevent catastrophic forgetting, but rather to find effective ways to share learned representations across tasks, to optimize the trade-off between performance and computational complexity and to use information learned in one task to improve performance on other tasks (that is, to achieve positive forward or even backward transfer between tasks)."[81] To address the continual learning problem in ML computer scientists have been seeking to emulate the human capacity for continual learning in a dynamic environment utilizing a stream of correlated data and replicating the capacity of humans to learn over an entire lifetime, which involves adapting to change whilst retaining old knowledge. This would appear vital for ML with robots needing to expand their skill sets to new tasks and environments rather than be constrained to do what they were initially trained to do.

We have seen how the goal of any ANN machine learning is to have parameters that minimize the error between true and predicted outputs and thus minimize the loss function of the network. Hadsell has even set out the desiderata of continual ML required for real-world situations which are as follows "(1) Minimal access to previous tasks (2) Minimal increase in model capacity and computation (3) Minimising catastrophic forgetting and interference (4) Fast adaptation and recovery (5) Maximising forward and backward transfer and (6) Task-agnostic learning.[82] Continuous learning involves some clear steps which are the initial training of the model, the updating of the model based on new data whether it is in the form of a continuous flow of real-time data or regular batches of new data, the use of a learning algorithm that can update the model's parameters without the need to retrain the model and regular the evaluation of the new adjusted model to ensure that it is improving over time and finally continually repeating this cycle. It is known to significantly improve the predictive capacity of AI systems. Intriguingly in 2022 a team at the University of Catania, Italy, looking to solve the problem of catastrophic forgetting developed a new method of training AI called Wake-Sleep Consolidated Learning (WSCL).[83] This harnesses the process whereby people shuffle short-term memories during sleep into long-term memories by training the machine to have periods in which they sleep, during which they parse a sample of awake data, and a highlight reel from previous training. Several methods to achieve continual AL have been explored and reflect the fact that continuous learning in humans and biological systems is not the result of a single biological mechanism but involves multiple systems which include the memory system and neural plasticity.

Scientists have been seeking to understand whether continual learning in humans is based on an algorithmic foundation and how this could be applied to ML as autonomous systems and robots need the capacity for continual learning. Continual learning in humans has been attributed to the way the PFC in the brain actively maintains relevant task information whilst suppressing irrelevant task information over time, a process that neuroscientists call context-dependent gating. It has been hypothesized that the storing of learned tasks in different and independent neural representations, prevents the forgetting of tasks. Timothee Resort whose research focuses on continual learning for robots, together with colleagues recently laid out a Framework, Learning Strategies, Opportunities, and Challenges in Continual Learning for Robotics, which postulates a mathematical formulation for continuous learning algorithms.[84] They observe that an important challenge for ML is "not necessarily finding solutions that work in the real world but rather finding stable algorithms that can learn in the real world" and the "ideal approach would be tackling the real world in an embodied platform, an autonomous agent." They argue that "most recent papers on continual learning only use experimental approaches in simulations or with static datasets" but "the ability to forget what is not important and retain what matters for the future are the main issues that continual learning targets focus on."

This essential lifelong learning component in the developmental robotics model can only be achieved if the AI system can learn cumulative skills and progressively improve the complexity and diversity of tasks handled.[85] Without the capacity for continual learning, robots could not operate in the real world with complete autonomy. Timo Flesch and colleagues recently set out to "identify the basis-set of algorithmic motifs required to mimic human-like continual learning." They proposed a novel training method for ANNs using an algorithmic approach that involves "task neurons that carry information over successive trials, and an update step that links those to other neurons that encode task-relevant information" and they claim that this allowed them to model human continual task performance in an ANN.[86] A system used by many researchers to assess the capacity of a system for continual learning, known as Permuted MNIST consists of 70,000 images of handwritten digits from 0 to 9, where 60,000 images are used for training, and 10,000 images for testing. They introduce a random image permutation which means the images are very noisy and differ significantly from the original image to the extent they can't typically be recognised by humans. The learning performance of the model is evaluated based on its ability to recall data learned in prior training. Although widely adopted, it has been criticized by computer scientists Sebastian Farquhar and Yarin Gal and others as insufficiently robust. They argue that "instead of assessing performance on challenging and representative experiment designs, recent research has focused on increased dataset difficulty, while still using flawed experiment set-ups." They propose fundamental desiderata for future evaluations of continual learning in machines, which they list as follows "(1) Input data from later tasks must resemble old tasks enough such that they at least sometimes result in confident predictions of old classes, early in training (2) If each task is given a different output vector (which are all the output values from artificial neurons during training and testing, it must be explicitly and prominently mentioned (3) Acknowledgment that sometimes-continual learning is not appropriate if two tasks are better learned separately (4) Succeeding at two-task transfer does not guarantee good performance on more tasks so more tasks are required. They conclude that unless" experiments better reflect the continual learning problem, recent leading approaches will continue to fail even on simple datasets like MNIST."[87]

This brief discussion illustrates not just the importance of continual learning for AI systems particularly autonomous androids, the significant challenges to equipping robots with the ability to learn continually as well as the problems associated with evaluating the capabilities of AI systems in this respect.

Social Learning

We saw how Albert Bandura refused to believe that all learning was brought about by association as a simple stimulus (input) and response (output) equation or through reinforcement using reward and punishment because humans exist in highly complex social environments. Bandura's Social Learning Theory (SLT) was based on how when children observe aggressive behaviour in adults, they copy and simulate that behaviour. Human intelligence generally is remarkably dependent on our being able to acquire knowledge efficiently from other humans through simulation which has been theorized as a process

used by our brains to build mental constructs. These are encoded in neural patterns that constantly flow between sensory perception, planning, decision-making or choices, and adjusting which allows us to execute the appropriate motor action or response and is a feature of how our brains represent the behaviour of others. Professor of Psychology Cecilia Hayes who is a Senior Research Fellow in Theoretical Life Sciences and Professor of Psychology at All Souls

College, University of Oxford, corroborates Bandura's work.[88] She does so though not by reference to associative or reinforcement learning but by refuting the widely held view that humans have cognitive instincts that involve us being born "with complicated mechanisms for reasoning about causation, reading the minds of others, copying behaviours, and using language." She argues instead that these "Cognitive gadgets" are not the product of genetic evolution but cultural evolution being constructed by our malleable minds during childhood through social interaction. Heyes argues that social learning couldn't possibly be associative and provides seven objections to the American psychologist Andrew Meltzoff's (1950–) widely applied Associative Sequence Learning (ASL) model which suggests that human infants learn to imitate using mechanisms that they share with other animals.[89] Although ASL is predicated on associative mechanisms it is not behaviourist in ignoring the black box of the mind because it attributes the behaviour to internal representations percepts, memories, and motor programs and postulates that learning constitutes changes in a person's internal representations which can't always be inferred from, behaviour. Heys does this by (1) Providing evidence that born babies do not imitate (2) Arguing that infants receive a plentiful supply of the kind of experience necessary for learning to imitate (3) Suggesting that neither infants nor adults can imitate elementally novel actions (4) Explaining why nonhuman animals have a limited capacity for imitation (5) Discussing the goal-directedness of imitation (6) Presenting evidence that improvement in imitation depends on visual feedback and (7) Reflecting on the view that associative theories steal 'the soul of imitation." She believes that although humans are great imitators, we are not made that way but become so through learning and culture.

Social learning is also known as Cultural Transmission and in 2023 Avishkar Bhoopchand and colleagues were exploring cultural transmission in the context of AI, describing it as a "Domain-general social skill that allows agents to acquire and use information from each other in real-time with high fidelity and recall" and a foundational social skill in artificial agents. They set out to measure "an agent's capacity to improve performance based on expert demonstrations, emphasizing the crucial distinction between authentic learning and mere memorization." They recognized the "Integral role played by memory, expert demonstrations, and attention mechanisms" and designed a method for generating cultural transmission in AI agents, involving "Few-shot imitation" claiming that their agents succeeded "at real-time imitation of a human in novel contexts without using any pre-collected human data."[90] They focused on observational learning using particular training tasks delineated into what they describe as diverse but realistic environments of world, game, and co-players as a form of cultural transmission with efficient knowledge acquisition reliant on knowledge transfer between individuals, in this case, expert demonstrations. The goal was to chart the robot's evolution through social learning and imitation to independent adaptive behaviour, which equipped the agent with a memory-based policy regarded as essential for autonomous task-solving. Their Reinforcement

Deep Learning model called MEDAL-ADR contains a (1) Memory module which employs a recurrent neural network (RNN) with a Long Short-Term Memory (LSTM) core that encodes observations and constructs a belief-state representation frame-by-frame (2) An Attention Loss (AL)module which directs attention towards co-players (3) An Expert Co-player (E) that intermittently appears (D) to facilitate learning and (4) The Automatic Domain Randomization (ADR) which "diversifies the task distribution, fostering cultural transmission in a broader task space," which allows the model to generalization the learned capabilities beyond the training tasks. Their research appears to highlight the critical role of social learning in shaping AI systems toward autonomous, adaptive behaviours. Together with continual learning, these have the potential to make robotic systems much more akin to the way humans learn.

Artificial Understanding

We have seen the different types and levels of understanding and their intrinsic value over knowledge, which extends to understanding the emotions and motivation of others and why people behave as they do. Whether AI systems and robots will be able to understand the world around them in the way humans can and, if so, at what level has important implications for trusting robots. The more robots interact with humans, the more both will need to understand the other in what Professor of Haptic human-robot interaction David Abbink (1977–) describes as "Human-robot symbiosis." Trust is much more challenging if we don't believe that another person or a robot understands what they are doing, how they are doing it, or why they are doing it. We will look later when we discuss the human brain in comparison to the robot brain at the neurobiology related to understanding, so this brief discussion relates to the concept of artificial understanding (AU), not the underlying neural basis.

There is a classical pyramidal approach and theoretical construct to understanding and wisdom called DIK (W), representing data, information, knowledge, and wisdom. We live in a world based on access to vast amounts of information knowledge, which ascribes greater knowledge to machines than people. Despite all the progress, wisdom and understanding seem outside the reach of AI machines. A 2021 study performed by Jarkko Hautula at Valmet Automotive considered the most robotised factory in Finland, with 269 people asked whether the robots possessed knowledge, and half thought they did while the other half didn't. Those who did were much more inclined to trust the robots, though, no one was asked whether they thought the robots understood what they were doing or why. Hautala points out that the "topic of knowledge is lacking in robot-related research,"[91] and we might add that the topic of understanding is also lacking.

One study into workplace competencies about human employees and robots observed that companies place "high emphasis on technical skill and knowledge, soft competencies are considered important, as well as competencies of continuous learning, flexibility, creativity, problem-solving, critical, and analytical thinking."[92] These soft competencies could be regarded as ingredients for proper understanding and not a necessary or inevitable outcome of possessing knowledge. This may be why managerial, professional, and artistic jobs requiring high-level reasoning, empathy, and creativity are not currently threatened by robots in the job market in the way that some of the more mundane, dirt, routine, and dangerous jobs are. This has important implications for the division of labour and potential collaboration between humans and robots, for as Hautala observes, "the better robots can process data, information, and knowledge, the more complex tasks they can perform."[93] Gerald Mathews and colleagues argue that trust towards autonomous robots is achieved on two levels, which are the (1) Level of transparency of the machine's data-analytic capabilities and (2) the Humanlike nature of its social functioning as a team member.[94] Although this references data analytical capabilities, it equally applies to competence assessment based on the degree of understanding and trusting a human colleague who needs help understanding the job, which, as we all know, is extremely difficult.

The challenge facing AU becomes immediately apparent when we look at whether robots can even understand some basic things that we take for granted, such as understanding the words we use in language communication, which mostly have abstract rather than concrete meanings because robots currently struggle to understand abstract concepts or words. This requires us to define what we mean by understanding a sentence. The dictionary definitions are unhelpful in that they commonly refer to the ability to understand something or comprehend, know, or extract the meaning of the words. In practice, understanding a sentence requires thinking about not just the literal meaning of individual words but also context, abstract concepts vis, realizations comparisons, etc, all of which constitute the act of thinking, and there remains no consensus in the field of AI whether machines can at any level think in this way. People typically speak in imprecise ways, and getting robots to understand natural speech or instructions without programming step by step remains a problem.

One innovative approach to overcome this problem has been pioneered at Cornell University, where their project called "Tell Me Dave" involves equipping the robots with the capacity to associate words and objects with capabilities and functions, which means that the robots can relate words and objects to actions in the real world. Words grounded in actions allow the robot to give meaning to and understand the word and allow the robot to use visual and non-visual data to understand objects and learn from watching people using objects like a mug or kettle to hold and pour water. Covariant was established in 2017 by Pieter Abbeel, Peter Chen, Rocky Duan, and Tianhao Zhang to find ways to apply their cutting-edge AI research to solve real-world problems. Peter Chen, CEO and founder distinguishes between mechanical calculations and more cognitive or thinking tasks, such as understanding what an object is and what it does and why, which require a degree of understanding, reasoning, or decision-making. Without understanding and the capacity to perform cognitive tasks, robots will continue to lack the flexibility to perform tasks in different environments and adapt to change. They developed the Covariant Brain, an AI platform that enables robots to see, think, act, and understand their surroundings, make informed decisions, and adapt their actions to changing circumstances. According to Chen, they do this by possessing cognitive abilities and dynamically perceiving their environment, allowing them to deal with novel tasks and multiple scenarios. Chen stresses that the knowledge and intelligence exhibited by the Covariant Brain are driven by the data rather than from human programming, and the understanding based on learning from the data comes from learning of all the connected robots. It is the only robot that was able to compete with other robots to solve 26 complex piece-picking challenges.

Another Google-led technology called PaLM, short for Pathways Language Model, capitalizes on the increased ability of robots to understand language to apply the learning to open-ended, complex problems in the real world. According to the developers, the technology, called PaLM-SayCan, can "interpret a vague human command and string together a sequence of actions to respond," reflecting increased abstraction abilities such as relating colours metaphorically to landscapes. For example, blue is the ocean, green is the jungle, and yellow is the desert. To train PaLM-SayCan, Google has "used a huge 6,144-processor machine vast multilingual collection of web documents, books, Wikipedia articles, conversations and programming code found on Microsoft's GitHub site." This training and learning from such vast quantities of data has enabled the robot to marry its language skills to its range of over 100 skill sets, increasingly allowing the robots to operate outside the world.

Researchers Tony Prescott and Stuart Wilson at Sheffield University have said that for robots to achieve cognition and the ability to think as a prerequisite for proper understanding requires a level of biological architecture in the robots and their being embodied in the environment and interacting with the real world. Higher brain functions like cognition and understanding depend on the integrated activity of an entire nervous system, reflecting the biological system's morphology, dynamic nature, and interaction with the environment. They expressly recommend that "Future neurorobotic models integrate multiple neurobiological constraints and be hybrid."[95] Animal and computational models have proved limited in generalizing the findings and robotics has provided an invaluable platform for modelling using neurorobotic models based on being embodied in the environment.

Endnotes

1 Hampshire, A., et al., 2012.
2 Design for a Brain, the Origin of Adaptive Behaviour (1960).
3 General Intelligence, Objectively Determined and Measured (1904).
4 Sternberg, R, J., 2003.
5 Gottfredson, LS., 1997.
6 Frames of Mind (1983).
7 Intelligence Reframed: Multiple Intelligences for the 21st Century.
8 Sonuga-Barke, EJS., et al., 2017.
9 Takahata, K, Kato, M., 2008.
10 Source Neatorama.
11 Born on a Blue Day (2006).
12 Miller, BL., et al., 1998.
13 Just, MA., et al., 2004.
14 Takahata, K, Kato, M., 2008.
15 The Neuroscience of Creativity (2018).
16 Andiliou, A, Murphy, K., 2010.
17 Craft, A, 2011.
18 Kaufman, JC., Beghetto, RA., 2009.
19 Davis, G A., 2004.
20 Lee, IS., 2021.
21 Watson, J.B., 1913.
22 Watson J.B., 1924.
23 Gallistel, CR, Balsam, PD., 2014.
24 Knudsen, EI., 2004.
25 Davis, B., Francis, K., 2022.
26 Bandura, A., 1977.
27 How to Raise an Adult: Break Free of the Overparenting Trap and Prepare Your Kid for Success (2015).
28 Lewis, KM, et al., 2020.
29 Halfon, N., et al., 2014.
30 Kudithipudi, D, et al., 2022.
31 Chang, Y., et al., 2021.
32 van de Ven, G.M., et al., 2020.
33 Parisi, G I, et al., 2019.
34 Parisi, G I, et al., 2019.
35 Yau., et al., 2015.

[36] Barron, HC., 2021.
[37] van de Ven GM, Tolias A, et al., 2022.
[38] Zagzebski, L., 2009.
[39] Kvanvig, J., 2003.
[40] KvanvigJ., 2009.
[41] Stanford Encyclopaedia of Philosophy 2005.
[42] Lawler, I., 2019.
[43] Knowledge and Understanding (2010).
[44] Yufik, YM, Friston, K., 2016.
[45] Ascribing Mental Qualities to Machines 1979.
[46] Philosophy of Science: An Encyclopaedia, ed. Pfeifer J and S. Sarkar (Routledge).
[47] Van Gulick, R., 2012.
[48] Russell S, Norvig, P., 1995.
[49] Legg, S, Hutter, M., 2007.
[50] Legg, S, Hutter, M., 2006.
[51] Golenia, L, et al., 2017.
[52] Carreño, IDVG., 2014.
[53] Sun, R., 2004.
[54] Vernon, D, von Hofsten, C, Fadiga L., 2016.
[55] When Humans Transcend Biology (2005).
[56] Adams, S., et al., 2012.
[57] Goertzel, B., 2014.
[58] https://edition.cnn.com/2023/10/04/tech/japan-softbank-ai-hnk-intl/index.html
[59] https://time.com/6344160/a-year-in-time-ceo-interview-sam-altman/
[60] Wittrock, M., 1990.
[61] Osborne, R, Wittrock, MC., 1983.
[62] Goertzel,B., et al., 2010a.
[63] Zador, AM., 2019.
[64] The Logical Calculus of Ideas Immanent in the Mind 1943.
[65] Bishop, CM., 2006.
[66] Buolamwini, J, Gebru,T., 2018.
[67] Villalobos P, et al., 2022.
[68] Hinton, GE, Osindero S, Teh, YW., 2006.
[69] Schmidhuber, J., 2015.
[70] AIBS 2020.
[71] Sutton, RS, Barto, AG., 2018.
[72] Irpan, A., 2018.
[73] Mnih, V., et al., 2015.
[74] Bellemare, M. et al., 2013.

75 Heimann,, D., et al., 2022.
76 Hassabis, D, et al., 2017.
77 Hadsell, R., et al., 2020.
78 Kudithipudi, D., et al., 2022.
79 Van de Ven GM, Tuytelaars, T, Tolias, AS., 2022.
80 McCloskey, Cohen, N J., 1989.
81 Van de Ven GM, Tuytelaars T, Tolias, AS., 2022.
82 Hadsell, R,et al., 2020.
83 Golden R, et al., 2022.
84 Lesort, T, et al., 2019.
85 Lungarella, M., et al., 2003.
86 Lungarella, M., et al., 2003.
87 Farquhar, S Gal,Y., 2019.
88 Cognitive Gadgets The Cultural Evolution of Thinking, 2018.
89 Meltzoff AN. 1988.
90 Bhoopchand, A, et al., 2023.
91 Hautala, J., 2021.
92 Elprama, S., et al., 2016.
93 Hautala, J., 2021.
94 Matthews, G., et al., 2020.
95 Prescott TJ, Wilson, SP., 2023.

"The human brain is by far the most complex physical object known to us in the cosmos."

Owen Gingerich, Professor Emeritus of Astronomy and the History of science at Harvard University

Chapter 3

The Brain

CHAPTER 3

The Human Brain

HUMANS DON'T LIVE IN TOTAL isolation but are constantly and dynamically changing and adapting our responses to the people we are with and our environment. Everything we do involves attaching salience and relevance to typically incomplete and insufficient raw sensory data from our environment and extracting meaningful information to guide our actions and behaviours. This is believed to involve constructing internal representations of the world, comparing them with prior knowledge or experiences, calculating the balance between reward and punishment, approaching or withdrawing from stimuli, and performing cognitive tasks to initiate appropriate decisions, choices, responses, and behaviour. We do not process information as though it already exists out there and American philosopher and psychologist Anthony Chemero argues that "Perception and cognition should be viewed as the result of our interaction and engagement with our environment, not merely from a representational perspective."[1]

Understanding ourselves from the perspective of adaptivity and change and interaction with the environment, rather than computational mechanisms, captures our individuality and subjective uniqueness and reflects how we and other organisms and animalds survive and thrive. Neuroscientists know that it is impossible to calculate the effect our experiences have on the information transmitted by any bit of information, given the number of possible connections between neurons containing the information and the complexity of neural patterns in the human brain. The theoretical limit has been roughly calculated as follows. Suppose we have 100 billion neurons, each with a possible 250,000 connections, which is about 25,000 trillion or 25 quadrillion connections. This means that we have 100 million more connections in our brains than there are estimated to be stars in the Milky Way. Autonomous androids must be able to perceive and think as the result of their interaction with their environment, not just because they have been programmed and trained by humans using huge volumes of static or stationary data or symbolic rules developed by us. This goal has given birth to the emerging field of probabilistic robots, with probabilistic algorithms dominating robotic software design because of the implications for dynamic perception and control of the robots' action. The German American

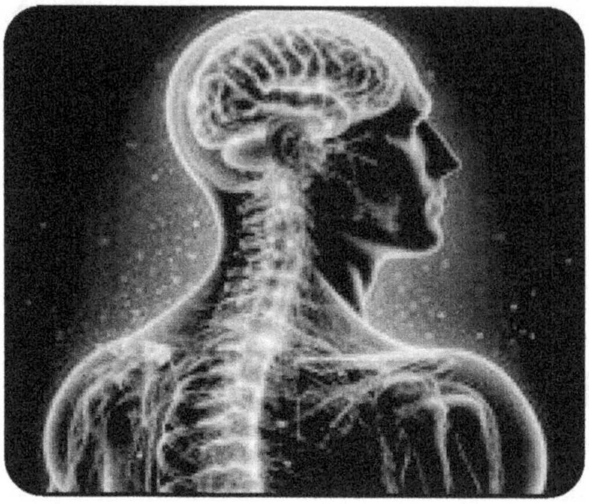

entrepreneur and roboticist roboticist Sebastian Thrun(1967–) with Professors Wolfram Burgard (1961–) and Dieter Fox, describe how "Robots have to be able to accommodate the enormous uncertainty that exists in the world." Many factors contribute to this uncertainty because real-world environments are dynamic, uncertain and unpredictable. Robotic sensors are as we will see, currently limited in their capacity to roam freely and vulnerable to damage, as are the mechanisms for actuation or movement called actuators, which can break and suffer wear and tear. Current robot software, at best, approximate abstractions of the real world and remains relatively crude. They observe that 'managing uncertainty is possibly the most important step towards robust real-world robot systems" and the key is to "represent uncertainty explicitly using the calculus of probability." Dealing with uncertainty and the ability to predict our environment is a major issue in building robust robots and the probabilistic approach seems to be the way to make them more robust in the "face of sensory limitations and model limitations."[2] Let's look at how the human brain accomplishes this and will start with some of the new technology giving us remarkable insights into how our brains operate.

New Technology for Exploring the Brain

One of the most valuable tools that has greatly enhanced our understanding of neural pathways and neural networks is called Optogenetics, which uses a combination of genetic manipulation and optics to modulate and reveal the behaviour of excitable neurons in tissue. Initially suggested as an idea by the English molecular biologist, biophysicist, and neuroscientist Francis Crick (1916–2004) it was first used in 2005 when scientists harnessed the properties of highly light sensitive, microbial proteins, known as Type 1 Opsins, that are proteins found in prokaryotes, algae, and fungi. They act as light-sensitive channels or pumps to allow the movement of ions across the neuron membrane and can be activated by light. Type 11 Opsins can be found in some animals but are called rhodopsin. Using viral vectors, scientists introduce the microbial Opsin gene into the neuron, and its sensitivity to light, allows them to induce electrical signalling or inhibit signalling in a very precise manner. Thus they can manipulate the neurons without damaging them and monitor the activity of neural circuits. Scientists can now perform millisecond-scale control of neural circuits, which gives extremely high temporal (when activity happens) and spatial (where activity happens) resolution to monitor and manipulate neuron cell activity, with vast applications in neuroscience and the potential to elucidate the role of functional networks and specific neurons. This can be used in live mobile animals with over 800 laboratories worldwide involved in optogenetic research with exciting implications for health research. Optogenetics is allowing researchers to precisely control the timing and spatial activity of neurons in specific neural pathways, and by using mice, they have even demonstrated it in vivo (which means in a living organism) as a tool for changing behaviour by controlling which neural pathways are activated.[3, 4] Optogenetics is proving invaluable for exploring how deficits in certain types of neurons due to disease or brain damage impact sensations, feelings, thoughts, and movements, which could lead to ways to correct dysfunctional neural circuits in humans that create problems with cognition.

In 2017, a research team designed, fabricated, and tested a new silicon probe, Neuropixel, which they claim is "compatible with use in unrestrained animals and has combined high spatiotemporal resolution with large volume coverage."[5] The dense recording sites and high channel count yield well-isolated spiking activity from hundreds of neurons implanted in mice and rats per probe. Using two probes, more than 700 well-isolated single neurons were recorded simultaneously from five brain structures in an awake mouse, and the team believes that the neuropixel probe will "open a new path towards the recording of brain-wide neural activity during behaviour."

Another new technology called Cell-based Neurotransmitter Fluorescent Engineered Reporters or CNiFERs (pronounced sniffers) is getting us closer to observing how neurons behave in live tissue, communicate with each other and pass on information. CNiFERs combine molecular biology with in vivo optical imaging techniques, such as two-photon microscopies (2PEF), to allow imaging of living tissue up to about one millimetre in thickness to optically detect the release of neurotransmitters from a cell.[6, 7] CNiFERs are biological cells designed to change colour when a particular neurotransmitter binds to a specific receptor on the neuron, which allows scientists to monitor the overall increase or decrease in the amount of neurotransmitter. Neuroscientist David Kleinfeld, professor of physics and behaviour, who founded the CNiFER technology, describes how it gives us access to "A world of signalling between cells that we were blind to before." Kleinfeld and his team have used it to demonstrate how a popular class of drugs called atypical neuroleptics is used to treat schizophrenia by blocking the release of the neurotransmitter acetylcholine. In one study, CNiFERs were implanted in the frontal cortex of mice, and scientists could watch the release of dopamine in real-time as the mice were being taught to respond to something. They could watch how the dopamine was initially released when the mouse was rewarded and then shifted to when they learned to associate a cue with the reward.[8]

iGluSnFR (Intensity-based Glutamate-Sensing Fluorescent Reporter), known as Glue sniffers, is even more precise than CNiFERs in that they can detect the signalling molecule glutamate in the synapse between single neurons such as pyramidal neurons (see below).[9] The major limitation in humans is that these sniffer technologies require injecting chemically modified cells, which would be highly unethical. Still, they are revolutionizing our understanding of the behaviour of neurons in live brain tissue using animals. Let's now turn to the unique neurobiological features of the human brain that need to be considered when comparing human and robot brains and the diverse initiatives being pursued to imbue AI and robots with the capabilities possessed by the human brain.

Neurobiologically Unique Features

To make sense of how all the sensory information is transmitted through the neural connections and patterns in our brains, psychologists have traditionally focused on key domains, including emotions, cognition, personality traits, memory and learning, motivation and beliefs. Behaviour scientists seek to identify the underlying and integrated neural correlates that support these domains and the mind states that arise and have traditionally attributed specific functions to certain brain regions to explain how our brains make sense of the world. Current science is demonstrating that the brain's complexity comes not from functional segregation in which one part of our brain deals with emotions and the other thinking etc, but from our brain's ability to rapidly communicate and integrate all the sensory information between regions. This is what allows our brain to plan and formulate appropriate responses to the sensory stimuli in its environment.

In a study of primary consciousness Professor of Clinical Psychiatry and Neurology Todd Feinberg (1952–) and Jon Mallett, argued that the first complex brains accompanied by consciousness emerged about 520 to 560 million years ago, in the great Cambrian explosion of animal diversity and postulated that human consciousness is key to our survival as a species. They observe how consciousness results from "neurobiologically unique features of complex nervous systems, especially of complex brains, that all together create consciousness." They maintain these neurobiologically unique features are fundamental to how our brain gives rise to our conscious self and makes sense of our world and identify them as follows " (1) The diversity in function and morphology and the sheer number of brain cells (neurons) (2) The neural patterns and maps in the cortex and particularly the patterns that relate to prior knowledge and experience of the world and how prior knowledge influences our perception and interpretation of new information and experiences (3) The extensive local and long-range connections and capacity for communication between brain regions, and integration of information related to emotions, thinking and motor responses (movement) and social cognition (4) The greatly enhanced long-term and episodic or autobiographical memory capacity of the human brain (5) The reward and motivation systems that drive human behaviour and influence our emotions (6) The distinct roles of the left versus right brain hemisphere in creating the self and how we perceive our world (7) The capacity of our brains to ruminate and over-think and worry and (8) How our brain appears to coordinate and process information related to the self."[10]

Connectivity and integration appear essential to converting all the raw data from uncertain environments into a continuous yet ever-changing stream of coherent representations and information that form the basis of all our continual learning, knowledge, intelligence, and understanding of our world. By focusing on these special neurobiological features rather than attributing distinct functions to different brain regions, the emphasis is on how we perceive the world through the dynamic, highly complex, and largely unconscious and subconscious process of encoding and storing memories, learning, and adaptation in response to the raw data from our environment.

A Brief Tour

It is necessary to refer to different brain areas and structures, which can seem confusing but is much easier using a few of the basic rules that brain scientists use to orient themselves around the brain. Neuroscientists use spatial references to accurately identify where parts or regions of the brain are or where a cluster of neurons are about the middle or outer regions of the brain and other internal regions. This allows them to locate, recognize, and understand the significance of the important neural connections and pathways that communicate and integrate information between the various brain regions. Scientists typically slice the brain in three ways to expose different brain regions from different perspectives and access regions that we can't see when the brain is intact or alive. These are the Sagittal, Coronal, and Horizontal (Axial) perspectives shown below.[11]

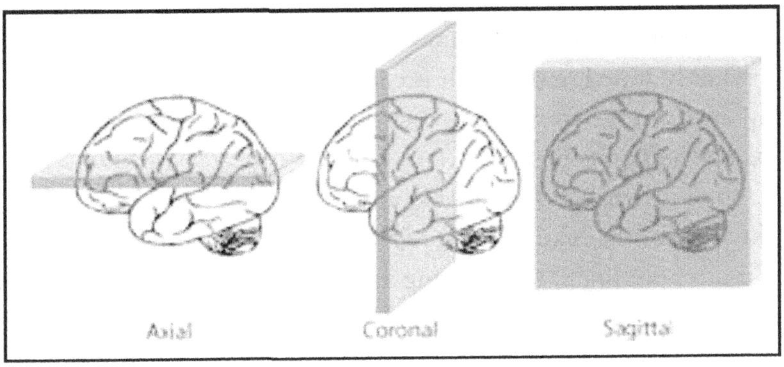

The human brain, encased in its bony skull, sits at the top of the spinal column, to which it is joined by the brain stem but is set at an angle to the spine. We can't describe brain regions as up or down in relation to other regions because that wouldn't be accurate enough, so scientists distinguish between the longitudinal axis of the forebrain at the front and the longitudinal axis of the brain-stem and spinal cord. We also can't just say above or below because we need to know whether we are referring to parts of the brain or the spine and brain stem, so scientists use the phrases Ventral and Dorsal by reference to these two longitudinal axes. Ventral means underside even if it isn't pointing strictly downwards, but in relation to its position to the longitudinal axes. Dorsal refers to the top or back of something in relation to one of the longitudinal axes. Rostral refers to near the head and caudal near the tail of a structure concerning one of the longitudinal axes. Anterior means towards the front of the brain but may refer to the front portion of a brain structure, not actually at the front of the brain. The posterior is to the back of the brain with the same caveat. Superior means above or towards the top, and inferior means towards the bottom of the whole brain or concerning a specific part of the brain. Lateral means towards the edge or side of a structure in the brain rather than the middle. The brain's midline is the line down the middle along the sagittal (imaginary) plain, which divides the two hemispheres, which gives you a medial view of the brain as though it has been split and pulled apart, and you look at each hemisphere's inner side. Medial means closer to the midline of the CNS and associated subcortical areas.

Although each brain region has specific functions, many are interchangeable and cross and overlap and we should heed the advice of Professor of Neuroscience and Psychology Joseph LeDoux (1949–), who cautioned us to "Be suspicious of any statement that says a brain area is a centre responsible for some function. The notion of functions being products of brain areas or centres is left over when most evidence about brain function was based on the effects of brain lesions localized to specific areas. Today, we think of functions as products of (neural) systems (networks) rather than of areas, with neurons in areas contributing because they are part of a system. The amygdala, for example, contributes to threat detection because it is part of a threat detection system, but just because the amygdala contributes to threat detection does not mean that threat detection is the only function to which it contributes. Amygdala neurons, for example, are also components of systems that process the significance of stimuli related to eating, drinking, sex, and addictive drugs."[12]

Raw data from internal and external stimuli is processed and moves through the cortex toward the front of the brain. This is reflected in neural pathways and neural patterns towards the front of our brain, representing a higher level of conscious awareness, perception, and the capacity for reflective thought. Lower levels of awareness do not constitute perception of our world, even though people often conflate them. For example, when under anaesthetic, our brain can still be aware of music but doesn't perceive or recognize it as music even if the normal brain areas become active. Although aware of the auditory signals, the brain seemingly cannot attach meaning to the information because the various brain regions are not in communication or sufficiently connected to integrate the information into a coherent representation to perform cognitive tasks or further subconscious processing. Awareness is a level of consciousness whereby a person is aware of the existence of sensory data. Perception is "The process or result of becoming aware of objects, relationships, and events using the senses, which includes such activities as recognizing, observing, and discriminating."[13] To appreciate perception, we need to understand how our brain processes, communicates, and integrates raw sensory data across the different brain regions, allowing our brain to recognize and discriminate between data at the different scales of biological organization.

We are constantly engaged in actively sensing our environment, processing sensory raw data that the brain and body are made aware of, and using that data to form our perceptions of the world. Based on the unconscious, subconscious, and conscious processing of the information our brains make predictions about the nature of that information to guide our decisions and choices, perform cognitive tasks and carry out actions and behaviours.

The Cortex

The cerebrum or cerebral cortex is responsible through its neural connections for receiving and attributing meaning to sensory information, particularly sight, and hearing, cognition, generation of motor commands, spatial reasoning, and language.

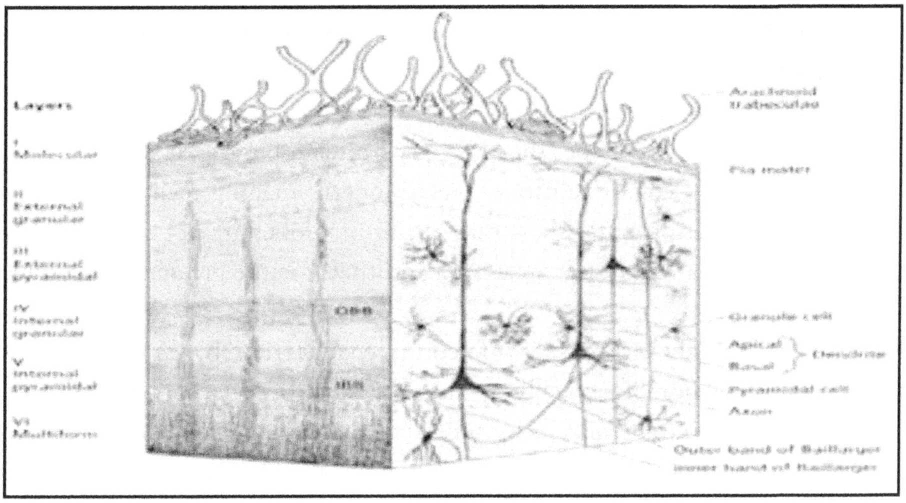

The Six Layers Of the Cortex[14]

The cortex has been identified as the part of the brain that endows humans with the capability to know about and be cognisant about our world and ourselves, generate consciousness, free will, intelligence, and wisdom or insight. It consists of a sheet of grey matter that varies in thickness. It is called grey matter because it mostly consists of neuron bodies, compared to white matter which consists of communicating fibres or axons that are sheathed in white fatty myelin to facilitate rapid signalling. The neocortex is arranged in vertical columns and six horizontal layers, which connect with different brain regions and are equipped with excitatory and inhibitory brain cells called neurons. It is the most complex and divergent part of the mammalian brain with intricate connectivity patterns between neurons and multiple layers that play a vital role in generating the awareness and perception associated with consciousness and self. "Expressed as a percentage of total brain volume, cortical grey matter is about 25% for insectivores (animals that feed on insects, worms, and other invertebrates) and 50% for humans."[15]

The other part of the cerebral cortex is the allocortex (or heterogeneous cortex), which covers the boundaries of the cortex but takes up only around 10% with three or four layers which is why it's also called heterogenic. Nevertheless, it has an important role in constructing the sense of self as it covers the hippocampus (associated with memory), which is embedded deep in the temporal lobes at the sides of the brain, the olfactory (smell) regions, and plays an important role in perception, emotions, awareness, memory, attention, thought, language, and consciousness. The very large surface area of the human cortex, which is about 2000 cm2, can only be accommodated in the typical human size skull because the surface of the cerebrum is characterized by many bumps and grooves known as gyri (bumps) and sulci (groves or fissures) with it folded in a very precise way and approximately two-thirds of the volume hidden within the folds. Some animals like rodents and squirrels don't have grooves and folds, while dolphins have more than humans. An odd feature is that the front cortex has far more grooves than the back, which may be because this area processes more abstract information. It was thought humans and the great apes share a larger frontal lobe, which is often cited as the reason for our superior cognitive capabilities. Modern imaging techniques and recent comparisons with other primates show that our frontal lobes are not disproportionately larger than other primates and the great apes and that it is the greater interconnectivity that matters.[16]

The cell layers contain various types of neurons characterized by varying shapes, densities, sizes, and organization of nerve fibres, glial cells which are vital support cells, one layer of axon endings, and apical dendrites from cortical neurons. The location of the neurons in the cortex is largely determined during development in the womb, with cortical connections demonstrating a bias towards specific layers. The densities of cell receptors, which are proteins located either inside a neuron or on its surface, that receive sensations in the form of chemical signals at the synapse, are not uniform across cortical areas and cluster into definable specific areas.

The process of generating our conscious perception of our world starts with the cell receptors in sensory neurons detecting a sensory stimulus and generating a strong enough electrical potential to cause the sensory neuron to produce an action potential that is transmitted into the Central Nervous System (CNS). In the CNS with highly connected yet differentiated neural networks, the generated signals are integrated with other sensory information and undergo the brain's subconscious and higher cognitive and emotional processing. The ability to rapidly integrate multi-modal sensory information in such a highly connected network of synchronized neuronal activity presents the main challenge to all AI systems, even those employing ANN and deep learning. There are many types of cell receptors, but a major role of sensory receptors is to support awareness, perception, and learning about our internal and external environment.

It has taken a long time for scientists to understand the role of electricity in transmitting information in the human brain and after many years of seeking to understand how neurons communicate with tissue in the body, it was observed that neurons generate electrical activity. Only when scientists developed the measuring technology did they realize that the transmission of signals between neurons and tissue is mediated by "Action Potentials." Action potentials are sudden, fast, transitory, and propagating changes in the neuron's resting membrane potential, generating electrical signals. Only neurons and muscle cells can generate an action potential and send electrical signals to other cells in an excitatory state. In contrast, inhibited neurons will not transmit information through electrical signals. Hungarian Professor of Neuroscience Gyorgy Buzsáki observes "how 60–80% of energy is devoted to communication within the brain, but far less is understood about the metabolic cost of inhibition, and the brain is largely inhibited with so much of our behaviour depending on inhibition."[17]

Buzsaki describes how "if only excitatory cells were present in the brain, neurons could not create form or order or secure some autonomy for themselves" and observes how neural "networks built from excitatory and inhibitory neurons can self-organize and generate complex properties" and it is the balance between excitation and inhibition that gives the neurons that diversity.[18] This finely orchestrated balance between the excitation or inhibition of the electrical activity of neurons is central to human behaviour because which neurons are activated, inhibited, or modulated in response to stimuli determines which neural networks are engaged or kept quiet and how they behave. We have and maintain a coherent sense of self because the processing of sensory information in our brain is largely inhibited, which means that our brain can be selective as to what it processes. Continual learning, memories and stability depend on inhibition and the diversity of cortical functions a critical feature of how our brains inhibit neural activity. The rapid transfer of information between neurons is coupled with vital inhibitory, excitatory, and synchronization processes, allowing a stable sense of self to emerge from our brains' electrical and chemical chaos. Hallucinogenic psychedelics are a potent example of how when you remove this inhibition across most of the brain using these drugs, it is possible to unleash unconstrained connectivity and hyperactivity. Experiencing a psychedelic trip all the time might sound like great fun to some, but would unsurprisingly render you unable to function normally. Inhibition of neuronal activity is vital to a stable functioning self and when the brain is unable to balance inhibition with excitation, we experience major problems controlling our motor movements, emotions, thinking, and moods and, in extreme cases, psychosis.

The Number and Diversity of Neurons

The electric signals and neurotransmitters that neurons generate, send and receive are vital to communicating sensory information, which allows us to be aware of, perceive the outside world, have emotions, feelings, and thoughts, move, behave, and plan in response to that information. There are three main functional types of neurons in the CNS which are (1) Sensory neurons, which receive sensory input from the external environment (2) Motor neurons that transmit motor commands to muscles for movement and are the most common and (3) Inter-neurons which transform and relay electric signals and information at stages between neurons of the information received by the cell through the cell receptors. They all have a core structure consisting of the cell body (Soma), which contains the nucleus, which stores the cell's genes, axons that transmit the electrical signals, and the chemical neurotransmitters that communicate and modulate the signal between the sending neuron's axon, and the dendrites of the receiving neuron. Dendrites typically branch out like bushy trees from the cell and receive and process the information, sometimes passing it on to other neurons in the neural networks. Plasma membranes separating the interior of the neuron from the outside environment are in very close contact with other neuron membranes, separated only by a very narrow space called a synaptic gap or cleft which the electric signal between neurons cannot cross, but is transmitted across this gap by chemical modulators called neurotransmitters. The synaptic cleft requires a vast array of proteins to function properly, and any mutation or interference causes problems with synaptic connectivity.

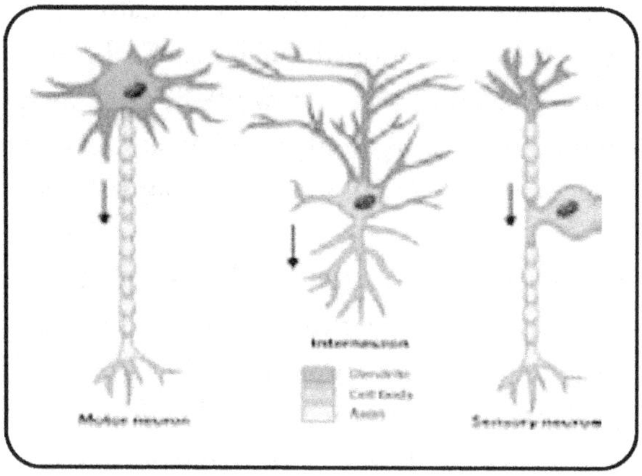

Neurons can perform a range of functions which include receiving information from the internal and external environment, transmitting signals to other neurons and to effector organs, processing and integrating information and determining or modifying the differentiation of sensory receptor cells and effector cells. In 1952 Alan Hodgkin (1914–1998) and Andrew Huxley (1917–2012) following their study of the giant axons of the squid, published the Hodgkin-Huxley (H-H) theory of how the action potential in neurons is dynamically initiated and propagated and established the biophysical characteristics of cell membranes and ion channels and the excitability of single cells.[19] Their mathematical equations underlie our understanding of current flow through ion selective channels, potassium (k) and sodium (Na) in the cell membrane and electrical spiking and information transmission in the brain and at its most simple, demonstrate that the cell membrane behaves like an electrical circuit. Their work continues to form the basis of much computational neuroscience and is acknowledged as one of the great scientific achievements of the twentieth century applying to a wide range of phenomena, which is why they won the 1963 Nobel Prize in Physiology or Medicine for their work.

If the electrical signal is strong enough to generate an action potential, a neuron communicates a signal through the axons to other neurons which receives the information through their dendrites. Axodendritic connections are formed between the axon of one neuron and the dendrite of another and tend to be excitatory synapses, while Axosomatic involve a direct connection between the axon of one neuron to the cell body of another neuron and tend to be inhibitory synapses. Neurons are activated or described as lighting up (although they don't light up or we would all resemble light bulbs) when your brain is processing information and thinking and when a sensory stimulus is centrally processed by neurons, which happens whether it produces a motor action in response to that stimulus. The synchronised oscillations of electrical signals between connected neurons form the neural pathways and patterns that encode all the vastly different information arriving through our senses. They are unique to everyone's brain structure and function with around 95% of what we consciously become aware of or perceive existing already in neural patterns stored in our long-term memory and based on existing synaptic connections.

Computer scientist Colin Ware when describing visual thinking in the context of design describes somewhat tongue in cheek, how "The reason we have the impression of perceiving a rich and complex environment is that we have rich and complex networks of meaning stored in our brains. This meaning becomes activated by the quite small amount of information that makes it through the (largely) visual processing system to visual working memory. This is why, many husbands do not notice when their wives have new hairstyles and wives do not notice when their husbands shave off beards or moustaches. What is perceived of our wives and husbands is mostly what we have previously learned about them."[20]

When we are young our brains undergo a process of synaptic pruning to reduce the excess numbers of synapses, many of which aren't used during development. Pruning peaks in the teenage and early adult years but intriguingly excessive synaptic pruning has been associated with the condition of schizophrenia whilst too little synaptic pruning with autism, with obvious and potentially serious implications for how the brain operates as a child develops.[21] Yet another illustration of how the level of synaptic connectivity which is constantly being modified through learning is a finely balanced mechanism that is fundamental to the effective differentiation and integration of information in our brain.

The traditional view of perception was that the brain receives and becomes aware of information from the senses, processes the basic raw data and then in a bottom-up manner processes that information upwards until it involves higher-level skills like reasoning and judgement. Current science suggests that this process happens both ways simultaneously so there is a top-down and bottom-up flow, with perception a reciprocal process by which the cortex and the various senses are constantly communicating with information mediated and modulated by the perception processes rather than it being one-way traffic upwards. Much human learning utilises prior knowledge stored in our brain and enhanced by default levels of connectivity and neural architectures common to all normal human brains, although we still undergo extensive training and learning to be highly proficient in specific areas. Later we shall see how the attention we pay to stimuli in the environment because they are relevant to or salient to us plays a vital role in shaping our neural patterns and activating existing neural patterns. These patterns represent long-term memories stored throughout the cortex which determine how we respond to the stimuli from our environment.[22]

The human CNS is a dynamic, constantly moving system, which has evolved to extract information from the stimuli that we attend to, become aware of and attribute meaning to, some of which we encode, actively integrate and store in neural patterns in the brain as prior knowledge and memories. Attribution of meaning to stimulus constitutes our perception of and knowledge of our world and others and is a vital element of the higher conscious experience because simply being aware of raw data and not being able to attribute meaning to it would not be conducive to our survival. The perception of the meaning of information about our sense of self and our relationship with the world makes us complex and subjectively unique. Perception knowledge and understanding are the outcome of higher cognitive processing and actions on that information and depend upon the capacity of neurons to encode data as single neurons and perform many functions individually and collectively in both local and global networks across the brain.

Most information processing takes place outside our fleetingly conscious theatre of awareness, in the multitude of subconscious and unconscious processes in the various distributed brain regions. This saves valuable and limited conscious processing capacity for which competition is high, given that we can only process 50 bits per second, but this means that our subconscious processes the majority of the 11 million bits of sensory raw data per second that we receive from the environment. Our brains learn by integrating new information with existing schemata or units of understanding. The word schema comes from the Greek word σχήμα (skhēma), which means shape, or more generally plan. The plural σχήματα (skhēmata) and schemata refer to our existing mental frameworks which for example in reading have been observed to consist of linguistic, formal, and content, schemata all of which allow us to comprehend the written words and detect relationships between ideas and words.[23] The brain's complexity and ability to communicate and integrate vast amounts of information, stems from the huge number and variety of neurons, with studies estimating that the CNS contains around 86 billion neurons and a hundred or more different types. Scientists are still discovering new variations, classified by morphology (shape and function) and you can visit the fascinating and free online library of neurons at Neuro Morpho.Org[24], which is a curated inventory of digitally reconstructed neurons and associated glial cells which support the neurons. At the time of writing, it has contributions from over 800 laboratories worldwide and peer-reviewed publications and is continuously being updated as new morphological reconstructions are collected, published, and shared.

Examples of Key Neurons

The amazingly diverse functionality of neurons belies their basic structure because they come in many shapes and sizes, some performing specialised functions, and others various roles with some operating locally, whilst others operate over long distances between different brain regions.[25] The diversity in function and morphology and sheer number of neurons is key to the rapid processing, communication, and integration of sensory information in the brain. Neuroscientists at the Max Planck Institute in Frankfurt, Germany recently identified 96 specific mutations that changed the structure of brain proteins in the brains of ancestors a few hundred thousand years ago, which appears to have accelerated the production of more neurons in the frontal lobe of our brain, which is responsible for abstract thought. One gene called the TKTL1 found in animal studies was claimed to be responsible for the generation of new neurons in the frontal lobe although this finding has been disputed by other scientists who believe that the "special cognitive abilities attributed to a frontal advantage may be due to differences in individual cortical areas and to a richer interconnectivity, none of which required an increase in the overall relative size of the frontal lobe during hominid evolution."[26] Other research is finding mutations that alter the number of connections that can be made and the pace at which the brain cells divide, and further research will potentially throw light on what made our brain so different from our ancestors that it gave us the specific capacity for consciousness and self-awareness. Neurons are usually named, based on their resemblance to some object or function and we will briefly look at some specific and key neurons which are known to play a critical role in the communication and integration of information within our brain.

Pyramidal Neurons

Pyramidal neurons are by far the most common neuron in the human brain, forming 70–80% of the neurons in the cortex. Named after the very large pyramid-shaped body or soma they are found in five of the six layers of the cortex concentrated in the internal (the third) and external pyramidal layers (the fifth). They are multi-polar neurons which means that they have a cell body from which emerges a single long axon, as well as a crown of many shorter branching dendrites and projection neurons sending signal messages over long distances to and between other brain regions. They are found extensively in the neural maps in the primary motor cortex, with many found in much thicker layers in the pre-frontal cortex, the cortico-spinal tract, and sub-cortical structures in the temporal lobes,

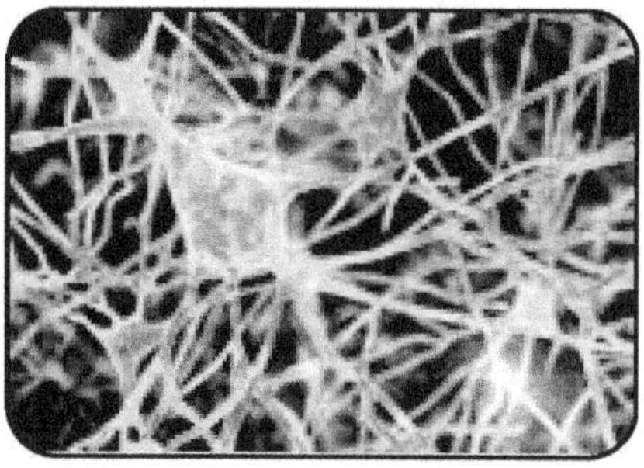

like the hippocampus and the amygdala, which are associated with memories and emotions. These long-distance connections predominate in the front of our brain whilst the back which is involved in visual processing and precise movements tends to rely on local shorter connections. Cognitive neuroscientists believe that cortical pyramidal neurons and their dendritic trees play a central role in the mechanisms of consciousness with single pyramidal neurons involved in the separation of forward and backward flow of information in the cortex and critical to movement, thinking, executive control, emotions, memory, and orientation.

Estonian neuroscientist and science author Jaan Aru who focuses on the neuronal basis of visual awareness and states of consciousness, is researching with others into the modelling of learning and memory processes, to empirically narrow down the neural processes associated with the subjective experience or the Neural Correlates of Consciousness (NCC). They regard the problem of consciousness as one of the biggest challenges in modern science. There are two approaches to exploring NCC, which are firstly studying the content based for example in comparing the NCC for say faces to cars when studying overall conscious states and secondly the contrast between dreamless sleep vs the awake state known as the state of consciousness. When studying the conscious state, scientists tend to focus on the neural activity going back and forth between different cortical regions but when studying the content, they focus on the neural activity going back and forth between the cortex and the thalamus (see below). Aru observes that the "Hallmark of conscious processing is the flexible integration of bottom-up and top-down data streams at the cellular level" and that the subjective characteristics of consciousness are achieved by "Neurobiological mechanisms, facilitated by pyramidal neurons."[27]

The state and content consciousness can never in reality be disassociated from each other and are always integrated so studying one inevitably involves taking account of the other. Aru and a team of neuroscientists have even claimed to have discovered the exact type of pyramidal neurons that link the state of consciousness with the content of consciousness. This is the cortical layer 5 pyramidal L5p neurons which appear to couple in a "Unique way the neural loop that governs the transmission of information between regions of the cortex so cortico-cortical transmission and the neural loop that transmits information between the thalamus and cortex, called the Thalamocortical loop."[28]

Another remarkable feature of pyramidal neurons is their role in coincidence detection which appears to allow the neuron to encode information based on distinct but concurrent signals converging on that neuron. One example is how our brain integrates sound signals from both the left and right ear, which activates the target neuron and localises the sound through the auditory neural maps we saw earlier. Two neuronal pathways converging on a target neuron increases connections between neurons or synaptic plasticity which is crucial to learning by enabling our brain to change and adapt to new information and experiences by strengthening or weakening synapses (junctures between neurons) over time due to increased or decreased activity.[29] Recent fMRI studies have shown that external stimulation like a painful stimulus still activates primary sensory cortices in some vegetative patients, showing that parts of the cortex still function but are functionally disconnected from the higher-order associative areas needed to perceive the world.[30] Pyramidal neurons are vital to connecting the brain's primary brain regions, those regions where information is integrated such as the associative regions of the cortex and the sub-cortical regions.

A note of caution is required whenever we think we are beginning to understand how our brain works and NCC. Professor Jakob Howhy from the Monash Centre for Consciousness Studies argues that examining neural activity about the content a person consciously perceived compared to when it is absent, a process called contrastive analysis, often confuses the prerequisites for conscious processing of content, with the consequences of the processing which makes it very difficult to isolate the NCC, related to specific content.[31] He observes that these approaches are failing to identify the neural correlates and believes that we need new experimental paradigms if we are to find them. He is a powerful advocate for the modern theory in neuroscience that the brain acts like a hypothesis-testing mechanism which is constantly seeking to minimise the error of its predictions about the sensory input it receives from the world. To do this it employs probabilistic and statistical calculations to represent and sometimes misrepresent the world which explains many aspects of the human mind including attention, the unity of consciousness, introspection, the self, and the private nature of our mental worlds.

Spindle Neurons

Another key neuron discovered in 1925 which forms at around 4 months of age and as far as we know found only in humans and other primates is the Spindle neuron which appears to be restricted to layer V of the front insular cortex, and the anterior cingulate cortex of the brain. They are very large stick-rod-like or spindle- bipolar corkscrew-shaped structures and are called Spindle neurons because of their large spindle shaped body or von-Economo neurons (VEN) after their discoverer the Austrian Psychiatrist Constantin von-Economo (1876–1931).[32] Though far less numerous than pyramidal neurons they play a critical role in information processing and generating self-awareness and complex emotions by refracting or deflecting waves of neural signals from different brain regions, which facilitates high-speed channel signalling from deep within the cortex. This connects brain regions and appears essential for forming quick emotional and intuitive judgments and has been described as the air traffic controller for emotions. Their location at vital neural hubs means they appear to be involved in generating self-awareness and complex emotions. They also play an important role in the development of intelligence and adaptive responses to changing conditions as well as cognitive dissonance, which happens when thoughts contradict each other, which may include ideas, beliefs, or the knowledge of your behaviour which seem contradictory. Damage to these neurons is implicated in frontopolar dementia a degenerative brain disease that causes deterioration of social and emotional self-awareness, moral reasoning, empathy, and theory of mind.[33] A possible cause for developmental delay in conditions such as autism may be due to impairment in the vital role of spindle neurons in mediating social sensitivity in that autism often manifests in difficulty making eye contact, recognizing faces, responding to others' emotional cues, and reciprocating another person's intentions.

Rose Hip Neurons

In 2018 in their search for what makes the brain special at a cellular level, scientists from the Tamas Lab at the University of Szegedin in Hungary and the Allen Institute for Brain Science in Seattle independently discovered a new bushy-shaped neuron, which they believe to be unique to humans, although this may prove not to be the case, as has happened all too often in the past. They called it the Rose Hip neuron because the axonal boutons and surrounding dense bundle of nerve fibres resemble the shape of a rose after the petals have fallen off. Boutons are club-shaped endings by which axons make synaptic contacts with usually the dendrites of other nerve cells or with effector cells (muscle or gland cells). They are inhibitory neurons and make up about 10–15% of all the inhibitory (GABAergic) neurons in Layer 1 of the neocortex and appear to largely target pyramidal neurons in Layer 3 by attaching only to specific parts of the pyramidal neurons, thus potentially manipulating the information flow by quieting or inhibiting the activity of the pyramidal neuron. GABAergic neurons produce (gamma-aminobutyric acid) which is an inhibitory neurotransmitter which reduces neuron's ability to send and receive chemical messages throughout the CNS. Low levels of GABA are linked to anxiety, autism, Parkinson's disease, and other conditions. Although in its early stages this research reinforces how animal models may work for studying the human body but are much less appropriate when modelling the human brain with its remarkable capacity and output and such unique features as Rose Hip Neurons.

Basket Cell Neurons

First described in 1911 again by Ramon y Cajal Basket cell neurons are the most common inhibitory inter-neurons and found in the neocortex, cerebellum and and hippocampus and form baskets of axonal arborizations onto and around the body of target neurons. Inter-neurons come in two types, local, which have short axons and form small local circuits with other neurons to transmit small amounts of information and relay inter-neurons which have long axons and connect circuits of neurons between brain regions. Their axons can connect to dendrites but specifically target the soma or body of the neuron and envelop pyramidal cell bodies in shapes resembling baskets. By inhibiting excitatory pyramidal or other neurons they serve to modulate the activation of neural patterns.[34, 35]

Engram Neurons

The name Engram Neurons was first used in 1904 by the German zoologist, explorer, evolutionary biologist and memory researcher Richard Semon (1859–1918) when describing their role as the neural substrate for storing memories. A neural substrate is a neuroscientific term used to indicate a part of the CNS that underlies a specific behaviour or psychological state with the substrate being an underlying substance or layer. Semon proposed that when we experience information from our environment, this produces physical and chemical changes in populations of neurons that make them become engram neurons. When engrams that were present in the initial experience are reactivated by cues, the brain can retrieve that information in the form of a memory stored in those engram neurons. Engram thus refers not to a particular type of neuron but to the function of the neurons that populate the neural assemblies involved in retaining specific memories. New technology is allowing scientists to explore whether the enduring changes that characterise synaptic plasticity are linked to specific memory retention in these engram neurons. Recent research has found that the more excitable the neuron the more likely it is to be allocated to this functional assembly of engram neurons. When they studied the engram neurons for changes, they found "increased synaptic strength and spine density in these neurons as well as preferential connectivity to other downstream engram cells."[36] They now know that intrinsic excitability and synaptic plasticity combine to form engrams which are involved in the processes of memory consolidation and retrieval. Being an engram neuron as part of the engram neural assembly appears to give them a vital role in memory formation and retrieval and engrams are found in different brain regions supporting different types of memory. Scientists have even demonstrated that it is possible to manipulate memory encoding and retrieval by activating these engram neurons to generate entirely false memories and create memories without any sensory experience. They have also discovered that by reactivating silent engram neurons in amnesiac mice they can induce the retrieval of memories, which they couldn't do using natural cues, which suggests that they can exist in different states of activity or dormancy. The latest hypothesis is that memory consists of connected engram ensembles dispersed across multiple brain regions, which form an engram complex of overall memory composed of these components.

Mirror Neurons

Although scientists are learning about neurons all the time, with new technology such as optogenetics etc, we must avoid attributing functions to them which are not supported by scientific evidence or acknowledge when new evidence points to a different function. This is illustrated by the controversial mirror neurons, discovered over 25 years ago, in the motor cortex of Macaque monkeys, which appear to activate in response to the execution and observation by others of goal-directed motor acts.[37, 38] Any interaction between people involves the perception of ourselves and others, and the action and perceived outcome is known as the Perception Action Loop (PAL) which is one of the ways humans learn. The Simulation Theory asserts that to make sense of other individuals, neurons are activated and create mental processes including emotions (including empathy) and intentional behaviour or actions resulting in mirroring behaviour. Simulation has been theorised to explain how we build mental constructs encoded in neural patterns, which constantly flow between sensory perception, planning, decision-making or choices and adjusting which allows us to execute the appropriate motor action or response. Mirroring appears critical to this PAL, but it remains unclear how certain neurons acquire this mirroring property.

From a computational perspective, predictive assessment relying on past experiences or existing neural patterns and Bayesian inference suggests that prior experience or knowledge is a significant factor in mirroring.[39] Bayesian inference involves updating or adjusting the probability of something happening or an outcome based on more information, with prior experiences updated when new data is acquired. Scientists first associated mirroring with a particular neuron found in the premotor region of the cortex and the parietal lobe, which they unsurprisingly named the mirror neuron. Many scientists including the highly renowned Indian American neuroscientist, VS Ramachandran (1951) and cognitive and computer scientist Marvin Minsky who as we saw was one of the founding fathers of AI, believe that mirror neurons play a key role in self-awareness.Ramachandran maintains that mirror neurons were the driving factor behind the "Great leap forward in human evolution" and that the discovery of "mirror neurons will do for psychology what DNA did for biology." The mirror neuron has even been employed in the broken mirror hypothesis for autism, a condition associated with difficulty in reading the perspective of others. Composer, music educator and researcher of neuroscience, Odeh Turjman (1961–) has hypothesised that "The self-experience is sustained by two mirror neuron systems involved in the simulation of the self-face and the self-voice, both of which are implicated in self-recognition and self-concept."[40]

Despite these bold claims even the existence of mirror neurons is disputed and the hyped and ubiquitous notion that we feel empathy through mirror neurons, remains highly controversial, given they do not even appear necessary to experience empathy. It may be that rather than causing our response, mirror neurons are active when we use others' actions to make our own choice of how to act or respond, so they are a consequence not a cause of the response. Professor of cognitive science Gregory Hickok "remains highly cynical about the evidence underlying the bold claims for mirror neurons and his research has revealed that they may involve a broader class of sensorimotor neurons with alternative explanations.[41] The ongoing dispute illustrates how much we are learning about the role of neurons in constructing our world and how much we still don't know.

Glial Cells

We can't discuss neurons without mentioning their vital supporters, the glial cells (pronounced, "glee-uhl") or neuroglia which outnumber them by 10–50 times. Glial cells are found in the CNS and peripheral nervous system and amongst other functions, provide physical and metabolic support to neurons. This includes insulation, communication (they form the myelin sheath around axons) and nutrient and waste support which keeps the nerve cells in place and helps them work properly. The most relevant glial cell in terms of consciousness and self is the Astrocyte cell (from the Ancient Greek "star" plus "cavity") because they are star-shaped and come in fibrous or protoplasmic form. Protoplasm is the transparent liquid inside all living cells. Protoplasmic astrocytes occur in the grey matter, whilst fibrous astrocytes are prevalent in myelinated nerve fibres in the white matter. The traditional perspective that astrocytes are simply there to support neurons has over the last twenty years been radically challenged with new unknown functions being discovered including the regulated release of gliotransmitters like glutamate and their vital neuro-modulatory role. Astrocytes are excitable and message other cells in a process known as glial transmission.

Andrea Volteri and Jacopo Meldolesi in a 2005 study, concluded that astrocytes should be "Seen as local communication elements of the brain that can generate various regulatory signals and bridge structures (from neuronal to vascular) and networks that are otherwise disconnected from each other."[42] Their excitability is chemically encoded and involves "Assays of intracellular Calcium (Ca2+) concentration, transients and oscillations." Volterra and Meldolesi claim that astrocytes "Exert a range of non-stereotyped feedback and/or feed-forward effects on neighbouring neurons, glial cells, and blood vessels. In neuronal circuits they can fine-tune the balance between excitation and inhibition and synchronise the activity of contiguous neurons.

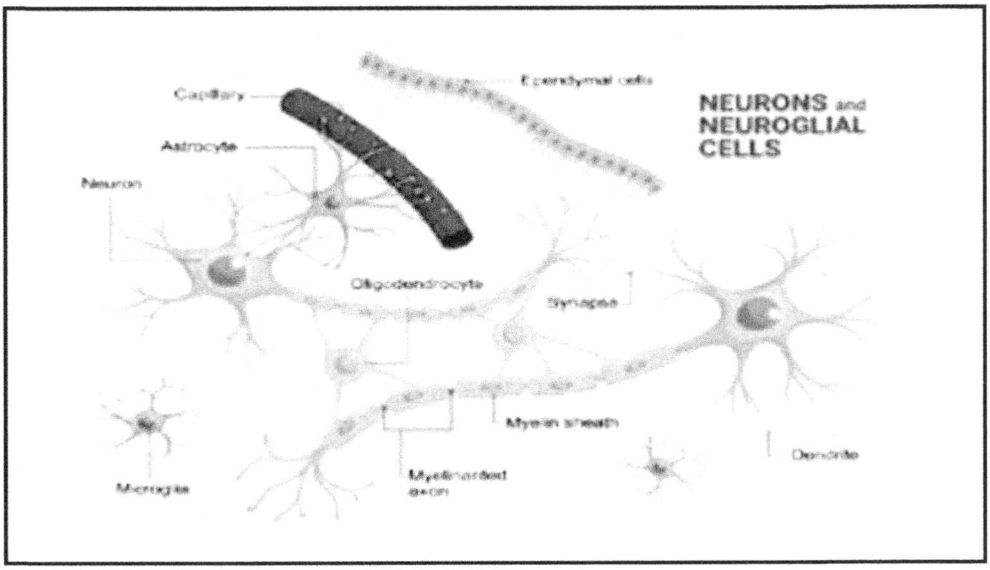

They can also control blood flow by inducing local vasoconstriction or vasodilation responses." When the CNS is injured, astrocyte cells divide and fill the spaces left by injured neurons take up the inhibitory neurotransmitter GABA and modulate and terminate synaptic transmission, as well as acting as a neurotransmitter storage system and synchronising the activity of neighbouring axons. What is increasingly evident is their major and complex role in the healthy functioning of the human brain beyond merely supporting neurons. Research at the National Institute of Neurological Disorders and Stroke in the United Kingdom has shown that the support cells, in this case oligodendrocytes which form the white fatty matter surrounding the axons, boost local energy production allowing axons to maintain sufficient energy production for their vital work in information transmission. This has obvious implications for Parkinson's and Huntington's disease and Amyotrophic Lateral Sclerosis (ALS) which have been linked to disruptions in axonal energy supply. Oligodendrocytes appear to release an enzyme SIRT2, which is not naturally present in neurons, that modifies proteins specifically those linked to ATP production and mitochondrial activity along often long axons and thus increases energy production in the axons. Enzymes are naturally produced substances that in living organisms act as catalysts to bring about biochemical reactions. When researchers genetically turned on SIRT2 in neurons, they observed significant increases in the levels of ATP production.[43] Damage to the various glial cells such as Oligodendrocytes and Astrocytes may result in several neurodegenerative diseases and impairment in learning and memory all related to self. Abnormalities in astrocytic Ca2+ signalling are also implicated in neurodevelopmental disorders and neurodegenerative diseases.[44]

Neural Plasticity

It is the cumulative effect of the neural pathways and neural patterns formed by all the connected neurons that integrate and contain all the sensory information related to the self and our environment, which means that we are essentially the sum of our neural pathways and neural patterns. The neural pathways and patterns when activated, enable us to predict things in our environment and reflect values, prejudices, dispositional traits, and beliefs. For many years scientists believed that the brain stopped evolving and changing after childhood, which would imply that neural patterns remain unchanging, but we now know that the brain possesses a remarkable capacity for neuroplasticity, which is defined as "The ability of the brain to continue growing and evolve in response to life's experiences by forming and reorganising synaptic connections, especially in response to learning or experience or following injury" (OED). The first mention of plasticity in the nervous system was made by William James (1842–1910) in 1890, although the term neural plasticity, was first used in 1948, by the Polish neurophysiologist, Jerzy Konorski (1903–1973) and popularised by the Canadian psychologist Donald Hebb (1904–1985), when he coined the now famous phrase "Neurons that fire together wire together."[45] Hebb, often considered the father of neuropsychology because of his ability to merge the psychological and neuroscience worlds, worked closely with the famous behaviourist and memory researcher, Karl Lashley (1890–1958) and Wilder Penfield with much of his research focused on the impact of surgeries and head injuries on specific aspects of brain functioning, rather than just general intelligence.

The Hebbian theory of learning and memory explains how neural pathways develop based on experiences and that when connections are used more frequently, they become stronger and faster. Hebb described how "When an axon of cell A is near enough to excite a cell B and repeatedly or persistently takes part in firing it, some growth process or metabolic change takes place in one or both cells such that A's efficiency, as one of the cells firing B, is increased." Scientists have since learned much about how neuroplasticity results in adaptive changes to the structure and function of the brain and although we are still learning about the complex process of plasticity, it has major relevance in clinical situations following injury. This includes stroke or traumatic brain injury and the degree of recovery in brain function based on neuro plasticity and in general learning and behavioural change. In clinical situations, scientists have observed three largely distinct phases of neuroplasticity, which depend on the nature of the damage to the brain. In the first two days the brain tries to use alternative brain networks to compensate for the loss of whatever neurons have been damaged or killed off. In the weeks that follow, the brain recruits support cells like the glial cells that forge new, excitatory, rather than inhibitory pathways and thus establish new connections between neurons to restore communication and integration of information in the brain. It's as though microscopic engineers are being mobilised to restore the energy network when it goes down and for many months afterwards, the brain attempts to remodel structural changes, through regeneration of neurons, which involves the sprouting and growth of new axons and dendrites. Nature seeks to rapidly restore brain function and structure to its optimal state. This has major implications not just for sensorimotor functioning, often affected by strokes, but across the various domains of cognition involving complex attention, executive function, learning and memory, language, perceptual-motor control and social cognition. That is why, the appropriate therapy in terms of stimuli, is so vital in these stages of recovery and can include exercise, exposure to stimuli in the environment, repetition of tasks, motivation, and medications.[46] Long-term plasticity is believed to happen at the neuronal level by the growth of more neuron receptors which "Lowers the level at which a neuron can be excited and thus transmit electrical information."[47]

Current research into the role of plasticity involves understanding why some neurons are strengthened, and why certain networks are restored whilst others are not, as well as the mechanisms, including why only certain types of glial cells help restore the functions of neurons.[48] In terms of functional reorganisation rather than structural restoration, scientists have observed certain key processes and developed certain theories. The first is how, if one side of the brain is damaged, the opposite side will compensate by taking over certain roles, which involves regions on the opposite side taking over or reorganisation of brain areas to assume functions that they were not designed to perform. Another theory is that a person can lose function in an undamaged region, because of the loss of connected pathways.[49] Modern brain imaging techniques have demonstrated that even in extreme cases where one-half of the brain is removed to treat certain types of epilepsy the brain appears able to compensate for this.[50] Neuroplasticity demonstrates that even in the absence of clinical issues, it is usually possible throughout our life, to change dysfunctional neural patterns that are dictating our emotional responses and habitual ways of thinking and behaving and by doing so develop new mindsets, new memories, new skills, and new abilities and reshape our self.

We must always exercise caution when we talk about the brain and our understanding of the scope of brain plasticity in regeneration, as shown by recent collaborative research at the University of Cambridge and John Hopkins University. In a 2023 article in eLife, Professors Tamar Makin (Cambridge) and Director of the Centre for the Study of Motor Learning and Brain Repair at Johns Hopkins University, John Krakauer (Johns Hopkins) reviewed ten seminal studies purporting to demonstrate that the brain possesses the ability to reorganise. They concluded that "the notion that the brain, in response to injury or deficit, can reorganise itself and re-purpose regions for new functions is fundamentally flawed, despite being commonly cited in scientific textbooks. Instead, they argue that what is occurring is merely the brain being trained to utilise already existing, but latent, abilities". Krakauer said "The idea that our brain has an amazing ability to rewire and reorganise itself is an appealing one. It gives us hope and fascination, especially when we hear extraordinary stories of blind individuals developing almost superhuman echolocation abilities, for example, or stroke survivors miraculously regaining motor abilities they thought they'd lost. This idea goes beyond simple adaptation, or plasticity, it implies a wholesale repurposing of brain regions. But while these stories may well be true, the explanation of what is happening is, in fact, wrong." They argue that we should not invoke an "Extreme or special form of plasticity" and that what happens when the brain appears to recover is that it is being trained to utilise existing but latent abilities and not to create new ones.[51]

Neuron Coding

The study of neural coding focuses on understanding how neurons process and transform the raw data received from the environment and interpret the neural activity to explain computational and cognitive functions. Neuroscientist Zhengwei Wu and colleagues recently described the challenge as how can we "Use behaviour and neural activity to understand this internal model and its dynamic latent variables?[52] Scientists are still trying to understand how the coordinated firing of individual neurons results in the functioning of larger neural regions or groups of neurons. Populations of interconnected neurons execute a wide range of linear and non-linear processes in response to sensory inputs with neurons fluctuating enormously in their firing rates which may constitute random noise or contain and transmit information between neurons. These are inevitably constrained by the context of the stimuli, the experience already represented by existing neural encoding in our brains and the unique anatomical features of each of our brains.

Neuroscientist Barry Richmond who works in the field of AI and neuronal coding and computation explains how "There is a sequence of action potentials containing information about the membrane potential at the neuronal cell body, and it is the information carried by the train of action potentials that provides information at the projection targets for the neurons." "This train of action potentials (the spike train) can be considered as elements of a neural code." The challenge he observes is to determine "What is being encoded, what is the code used to transmit the information, how reliable (noisy or not) is the code, and how the information is utilized or decoded."[53] Neural codes are activated in response to new sensory information to match up the best fit to determine which existing codes the new information most closely resembles. Since the 1990s there has been an ongoing, almost quasi-political debate within the neuroscience community as to how neurons encode information. The orthodox view called Rate Coding suggests that coding and transfer of information relates to the rate of firing of the neurons and the average number of spikes per unit of time. The non-orthodox view called Temporal Coding is that coding depends upon the precise timing of single spikes in the electrical signalling from neurons.[54] It is believed that in rate coding, information related to a stimulus is contained in the firing rate of the neurons and because the sequence of action potentials generated by a given stimulus varies, neuronal responses are usually determined statistically or probabilistically.

Neurons respond to stimulus or task features and when these are present or salient fire spikes in generally have a Stochastic rate or a Poisson rate that is proportionally related to the feature's intensity or salience which means how noticeable or prominent it is. Stochastic describes the fact that they fire in a random probability distribution or pattern that can be analysed statistically but cannot be predicted precisely. Named after the French mathematician and physicist Simeon Denis Poisson (1781–1840) the Poisson rate involves a discrete probability distribution, which determines the probability of a given number of events happening several times within a given or fixed interval of time or space. The occurrence of one event does not effect on the probability of another event occurring, and the mean or average rate is constant, whilst events cannot occur simultaneously. This model works well with stimuli varying slowly over time because the spikes need to be integrated over a time window and appears consistent with the discovery of Hubel, Wiesel, and Sperry that some neurons only respond to information coming from one eye, which they called ocular dominance and cluster together in columns in the brain's visual cortex.[55] They demonstrated that certain neurons respond to visual features such as a line's orientation or specific patterns which suggests a coordinated communication of information based is based on neuron firing.

A major criticism of the rate coding theory was advanced by Bruno Olshausen and David Field in a 2004 paper when they demonstrated how little we still understand the visual cortex in humans of which V1 is a region of neurons and maintained that "we still do not have a decent picture of how V1 operates i.e. how a population of its neurons represents natural scenes under realistic viewing conditions." They identified five problems with the current view "stemming largely from biases in the design and execution of experiments, in addition to the contributions of non-linearities in the cortex that are not well understood." They argued that the Hubel, Wiesel Sperry theory only explains what around 15% of the cells in the visual cortex are doing and rate coding doesn't make metabolic sense in that it consumes too much energy.[56]

The Temporal coding theory was originally inspired by digital computation and the all-or-none nature of the action potential so postulated neurons encoded as bitstreams of timed spike (...010110011...) like computers. With the spike rate, the temporal average requires counting the number of spikes in a period and dividing this by the duration. There are several versions of temporal coding which reflect the flexibility of the concept of timing, but the major problem is that the brain does not have any central control over the synchronisation process or the ability to chunk time which appears necessary to enable temporal coding to operate, which is why it remains controversial.

Neither the rate coding nor temporal coding theories appear to provide a definitive explanation of how neural activity relates to the coding and transmission of raw data and information and although new technology like optogenetics may well start to resolve the issue, the evidence remains inconclusive. A major challenge facing scientists is trying to identify how much encoding or processing of information takes place in single neurons and to understand their specific functions. A widely held theory is that neurons use a process known as sparse coding.

Sparse Coding

Since the 1980s, there has been a widely held assumption among neuroscientists that our brains must maintain a minimum level of neural activity called the Sparse Coding Theory (SCT). SCT is often attributed to HB Barlow (1921–2020), the great-grandson of Charles Darwin, who in 1972 wrote an article in which he purported to explain the subjective experience by reference to the activity of neurons.[57] Barlow asserted that the higher our brain information processing gets, the less the number of neurons involved and for each stimulus a subset of neurons is used so neural coding gets sparser. This supposedly reflects how raw data is represented in our brain by the activation of single neurons and the pattern of activation of groups of neurons forming neural codes. It is believed that the new raw data that needs to be processed determines the appropriate response and is represented by the strong simultaneous activation of a relatively small set of neurons at that time. The rationale appears to be that the fewer neurons that are active the lower the metabolic cost to the brain, thus simplifying the process and saving energy by reducing the number of signals required to convey information. This is evidenced by only a relatively small number of neurons potentially less than 10 % appearing to spike at any moment in time, although it has been calculated that an average neuron in the human brain transmits a spike of about 0.1–2 HZ per second. The problem for researchers trying to record neural activity is that they can rarely take account of the neurons that are silent and not firing and can overestimate the number of neurons that are firing, and many reports of firing levels lack clear evidence.[58]

The brain's metabolic fragility has been postulated as a reason for limiting brain size in primates and a fundamental influence on determining the architecture of our brain's neural circuitry, neural activity patterns and neural coding. Our brains have evolved to optimise the energy efficient encoding of information transmission, evidenced for example by the fact that spiking neurons can encode precise information with a high level of reliability at about 2–5 bits of information per spike. Indian theoretical physicist Vijay Balasubramanian (1969–) observes that they appear naturally highly efficient in that they only use about "half the total entropy available in their spike trains."[59] The visual system has been shown to use a lower number of neurons early in the visual processing of information which suggests efficient coding is achieved through discrete bursts of spiking activity followed by long quiet intervals. The all-or-nothing nature of action potential spiking means that it acts as a binary source of information with zero information being transmitted in non-spiking phases. Each firing encodes discrete information based on the number of spikes in the firing event, so it is possible to measure energy consumption related to the firing event and determine the energy cost of the information contained. Energy is mainly consumed by neurons in the operation of the pumps that transport ions across the cell membrane which create the depolarisation necessary for action potentials.[60, 61] At the neuron level this allows you to calculate the energy cost by adding the baseline metabolic cost to the metabolic cost of spiking which is the ratio of the energy cost of spiking to the baseline metabolic cost over a specified period varies for different neurons. An alternative method is to radioactively label metabolic precursors like glucose and measure the uptake during spiking activity. At the tissue level, the metabolic cost can be assessed using oxygen consumption or the amount of heat produced. Balasubramanian and colleagues have studied the costs and constraints of biological computation involving sensory processing to assess the value of information theory as a tool for predicting the structure of neural codes in optimising the amount and kind of information transmitted. This has potentially important implications for the capacity of artificial neurons in AI systems to convey information.[62]

Claude Elwood Shannon (1916–2001), known as the father of information theory and founder of the digital revolution, developed the Mutual Information theory (MIT).

MIT stemmed from Shannon's cryptography research into the problem of how best to encode the message a sender wants to transmit so that only the intended recipient can read it. Warren Weaver popularised Shannon's concept of information which is that entropy is a measure of the amount of information content and is determined by the level of reduction in uncertainty that the information results in.[63] MIT purports to measure the entropy and therefore the amount of information contained in a random variable which reduces the level of uncertainty. Randomness in information theory reflects the lack of predictability or pattern in information, knowing the probability distribution makes it possible to predict the probable frequency of different outcomes over time and repeated events in an event space, as we just saw with for example stochastic or Poisson rate analysis etc.[64] Mathematicians typically assign a numerical value to that probability to reflect the statistical likelihood of a possible outcome. If there are two random variables, mutual information measures the mutual dependence between the two variables in terms of the amount of information in one random variable by observing the other random variable. With two variables the joint entropy is the measure of the reduction of uncertainty associated with both and so it is a measure of the level of information generated. It is the amount of information or reduction in uncertainty you can get about B from information gained from A because you have reduced the uncertainty and information gained from A.

Mutual information is used to measure the minimum energy required to transmit information at a certain rate, which in organisms can directly determine their survival in certain environments when they need to transmit information quickly and reliably although numerous factors influence the value of the information, which comes at a metabolic cost. This involves studying how codes are structured to determine the energy required to transmit the information. The amount of energy required is determined by the amount of information required from the environment and reflects the level of exploration and associated energy cost at the neuronal or tissue level. It suggests that early in sensory information processing neurons independently transform the raw data inputs into an efficient representation to optimise the amount of data that can be extracted given the finite neural and energy resources of our brain. The problem with the assumption that sparse coding conserves energy is that neurons aren't independent and neural pathways are very noisy communication channels, so sharing patterns of activation would make the transmission of information more efficient rather than simply saving energy which is partly why sparse coding is still regarded as controversial.

The Grandmother Cell Theory

The most extreme example of sparse coding is postulated as encoding by a single neuron, which in the Grandmother Cell Theory (GCT) has been dubbed the Grandmother Neuron which competes with the rather more mundanely titled Distributed Representation Theory

(DRT). GCT maintains that because single neurons can contain a large amount of information of around 250,000 bits, one neuron can recognise a complex thing like an object, a word or even a celebrity like Jennifer Aniston, so a single neuron could perform the function of recognising your grandmother, although why your grandmother was chosen no one seems to know. GCT suggests that neurons are incredibly selective but there is a problem with the theory because scientists have established that in humans, every neuron responds to more than just one thing, with one study by Stephen Waydo and colleagues for example, finding that single neurons are activated in response to between 50–150 faces.[65] If single neurons were so highly selective it would also mean that if one grandmother neuron dies our brain would need backups or redundancy or you would lose the ability to recognise your granny and try to explain to your granny, why you can't recognise her might be awkward.[66] Jeffrey Bowers who is a major advocate for GCT stresses that it only applies to familiar not unfamiliar faces, but there remains considerable scepticism about the largely discredited GCT.[67] Sparse coding generally, not just GCT, has recently been further challenged by researchers Anton Spanne and Henrik Jörntell who maintain that sparse coding is "simply wrong", arguing that the brain has many neural connections involved in processing sensory information.[68]

The Distributed Representation Theory of Coding

The Distributed Representation Theory (DRT) seeks to address the criticisms of GCT by maintaining that single neurons aren't that selective and perform many roles and representation of objects is distributed across many neurons in the form of neural patterns.[69] The information is represented by a pattern of activation over a group of neurons with each concept represented by multiple neuron activation and each neuron participating in the representation of multiple concepts.[70] The DRT postulates that the brain stores knowledge in patterns of activation distributed across large numbers of neurons. It highlights the fact that for our brain to generalise the sensory information based on existing codes and experience, neural coding can't stay local or be sparse and must be disseminated to other brain networks. Proponents observe that it is unclear whether sparse coding is asserted to be confined to motor movement or applies to all sensory systems and whether it applies to complicated learning and memory. Their argument is further supported by the observation that many neural connections are maintained in a state of readiness to be activated and if they weren't the brain would be unable to learn new information rapidly and in a reasonable time. Anton Spanne and colleagues point to the vital role of inhibitory neurons in the role of memory and observe that "Silent synapses and inhibitory inter-neurons can permit learning speed and memory capacity, not only sparse coding. Combining these properties, without the exaggerated sparse coding, improves the capacity for generalisation and neural patterns represent the sum or aggregate at any moment in time of the electrical and chemical, signals of neurons acting in networks which contain the information and facilitate learning of models of a complex and high-dimensional reality."[71] This could have important implications for designing ANNs for AI systems.

The debate about sparse coding, rate and temporal coding and distributed representation remains unresolved but highlights how much we still need to learn about the role of single neurons and groups of neurons in terms of how they communicate with each other to process sensory information and where and how in the brain knowledge is stored and represented. The problem is further complicated because neurons don't just communicate through electrical signalling but employ chemical intermediaries known as neurotransmitters which are vital to the neuron's ability to communicate with each other and regulate and modulate the activity of other neurons.

Neurotransmitters

Although neurons communicate through electric signals generated in the axons and dendrites they rely on chemicals called neurotransmitters to transmit the information represented by the electric signal between neurons at the synaptic gap between the axons and dendrites. Generated in response to these electrical charges they are critical to neuron communication and because neurotransmitters either excite or inhibit other neurons they play a vital role in determining the level of activity and information processing. The most abundant excitatory neurotransmitter in vertebrates is Glutamate which is concentrated in the CNS and has been dubbed the "Master Neurotransmitter."[72] Glutamate is a non-essential amino acid which means that our body typically produces enough by itself, and it is the most common amino acid found in baby's milk. You can consume it from foods, and you may have heard of Glutamates in the form of the food additive monosodium glutamate, but this is not quite the same as that produced in your body or glutamate supplements. Being excitatory means that Glutamate stimulates the receiving neuron and increases the likelihood that the information in the form of the signal will be passed on to other neurons. It is produced and recycled by glial cells to make GABA which paradoxically is the main inhibitory neurotransmitter and a naturally occurring amino acid which blocks nerve impulses and reduces nervous activity.

Although naturally occurring GABBA is widely used as a supplement, only minimal amounts of GABA can cross the Blood-Brain Barrier (BBB), so such supplements have little if any effect on our brains. The BBB is made up of tightly packed specialised cells called endothelium that have a lipid-based outer membrane and line the inner surface of our brain capillaries or blood vessels which acts as a very effective filter to prevent large or water-soluble molecules passing into our brain. Many medicines and alcohol can pass the BBB and inflammation and many conditions can weaken it. The prescription drug,

Valium, is classified as a Benzodiazepine which works in the CNS to increase GABA levels, and because of its inhibitory function it is used to treat severe anxiety, alcohol withdrawal syndrome, seizure disorders, and muscle spasms. Other excitatory neurotransmitters, modulate neuron behaviour and include Dopamine (DA), which is critical to the reward system and motivation, Serotonin to mood, Norepinephrine which is made from dopamine and important in the flight or fight mechanism of the sympathetic nervous system and Acetylcholine which is involved in generating or inhibiting nerve impulses.

Russian neurologist Eugene Izhikevich (1967–) has been seeking to address what is known as the distal reward problem or how our brains can associate a reward with an earlier action when the neural activity and firing patterns related to the action no longer exist. It is further complicated by neurons and synapses that are active whilst waiting for the reward. Izhikevich has observed how spike-timing-dependent plasticity (STDP) associated with an action is modulated by DA because "the kinetics of the subsequent synaptic plasticity is sensitive to changes in the extracellular dopamine concentration during the critical period of a few seconds."[73] STDP is a phenomenon in which the order and precise temporal interval between presynaptic and postsynaptic electrical spikes affect the sign and magnitude of changes in synaptic strength. Professor of Neurobiology Daniel Feldman who specialises in how neural circuits in the brain's cortex mediate sensation, memory, higher functions, and the implications in neurological disease, observes that "Spike timing is just one of several factors (including firing rate, synaptic co-operativity, and depolarization) that govern plasticity induction, and its relative importance varies across synapses and activity regimes."[74] Izhikevich argues that the STDP that generated the action is insensitive to ongoing neural activity and that the dopamine concentration selectively influences the correct synapses at the right time.

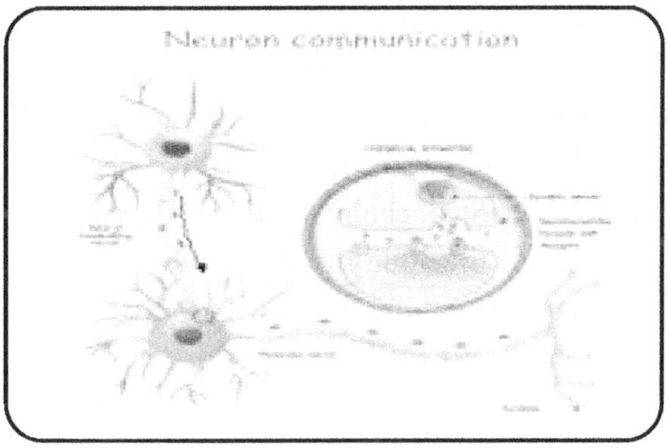

This avoids other theories that seek to explain the distal reward problem by assuming that the neural network goes quiet during the waiting period or that neural patterns are preserved during this period, despite there being no evidence of either happening. It illustrates the exquisite role of neurotransmitters in modulating synaptic activity in important learning mechanisms associated with the action reward and reinforcement dynamic which as we will see are governed by DA. Another chemical modulator of the activity of neurons that further complicates the process of understanding the processing and communication of information in the brain is the body's chemical messengers called hormones.

Hormones

Hormones like Adrenaline, Testosterone and Oestrogen are chemical signalling molecules in multicellular organisms that are constantly being released into the bloodstream and travelling to distant tissues and organs where they work slowly over time to regulate physiology and behaviour. Even tiny amounts can influence multiple processes such as growth and development, metabolism, reproduction, sexual function, and mood.

Hormones are produced in special groups of cells in the endocrine glands in the Endocrine system (ENS) with the main hormones being the Pituitary, Pineal, Thymus, Thyroid, Adrenal glands, and Pancreas. The sexual/gonadal hormones androgen, testosterone, oestrogen, and progesterone are continuously produced by the male testes and ovaries in women and regulated throughout adolescence and adulthood. Signs of problems with the ENS include mood swings, tiredness, muscular weakness, fluctuations in weight and blood glucose and cholesterol levels.

Hormones are believed to affect neuronal networking and the mechanisms of cognitive function including thinking, spatial tasks, memory, cognitive flexibility, problem solving and the ability to learn new things. Particularly the stress hormone Cortisol because stress hormones can cross the BBB and influence cognition and mental health.[75] All hormones can have both a positive and negative impact depending on the balance and whether levels are too high or too low.[76] Negative hormonal effects can include cognitive deficits including language problems, difficulty with attention, conceptual reasoning, memory difficulties and impairment in the speed of processing information. Oestrogen is believed to act as a neuro-protector and selectively benefits certain cognitive tasks such as forming new memories, which is why it has been explored as a potential therapy in post-menopausal women at risk of ageing-related risk brain atrophy, the frontal lobe and hippocampal region.[77] We have an inadequate understanding of the neural pathways affected by the hormones, but it seems they play an important regulatory role and influence cognition.

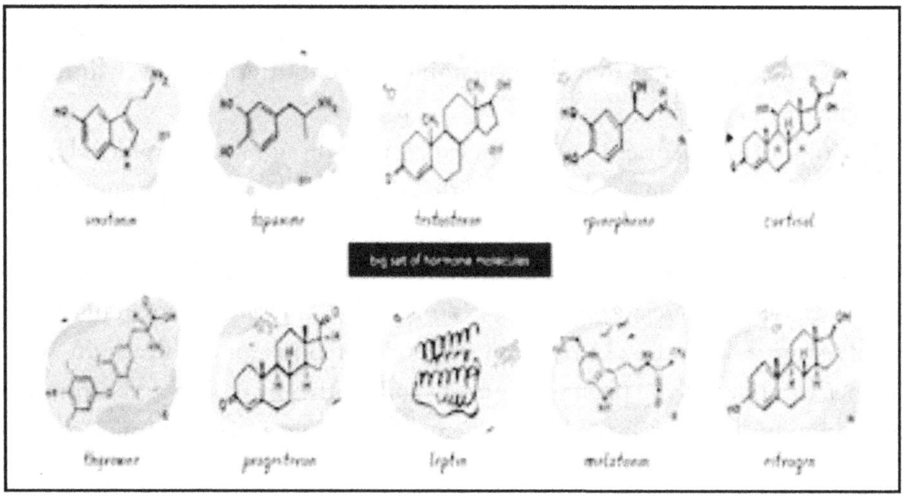

Connectivity

Connectivity in the human brain is characterised as either Structural or Functional with structural connectivity referring to "The anatomical organisation of the brain by means of fibre tracts" and functional connectivity referring to "Statistical dependence between time series of electro-physiological activity and (de) oxygenated blood levels in distinct regions of the brain."[78] In 2002 Richard Passingham and colleagues observed how "The functions of a cortical area are determined by its extrinsic connections and intrinsic properties. Each cortical area has a unique pattern of cortico-cortical connections, a 'connectional fingerprint" They went on to claim that "No two areas share identical connection patterns. Instead, there are families of areas that share a resemblance in their connections" and "Cortical areas also have their own 'functional fingerprints': the proportions of cells that fire in association with different tasks or task events differ between areas. The connectional fingerprint seems to underlie such cell-firing differences."[79]

The Connectome

Using massive brain maps called the connectome, neuroscientists are striving to trace all the structural and functional connections in the hope of deciphering how through these neural connections and patterns our brains work, and why based on our perceptions of our world we think and behave as we do. We have seen a growing interest in the role of neural networks in the brain and the field of Connectomics, a term coined in 2005 by Professor of Radiology, Dr Patric Hagmann and Professor in Psychological and Brain Sciences Dr Olaf Spons describes the organization of connections throughout the CNS whilst Ome stands for the whole class of or complete set of something.[80] A major challenge facing the field of connectomics arises from the fact that natural behaviours involve interactions between us and our environment and the multiple complex neural networks or circuits which change with that behaviour often overlap in the process of integrating the information. Attempts to decipher this have been compared to recording a snapshot of conversations at a busy cocktail party because it cannot adequately or accurately reflect all the connections and dynamics underlying the operation of our brain and body. It requires studying animals including humans in their natural environment to reflect the involvement of a vast variety of neurons, the multiple nature of the connections, whether they act locally or across long distances in our brain and which complex networks are involved. Correlation between neurons and neural network activity has been shown to change depending on the complexity of a task, the sensory richness of the environment and whether an animal is resting, moving freely, or exhibiting different behaviours. We have already seen how scientists have established that single neurons can perform specific or many different roles, which adds to the brain's flexibility but also its complexity.

Despite the challenges posed by this complexity, gradual progress is being made. Scientists have been studying the connectome of a transparent microscopic roundworm

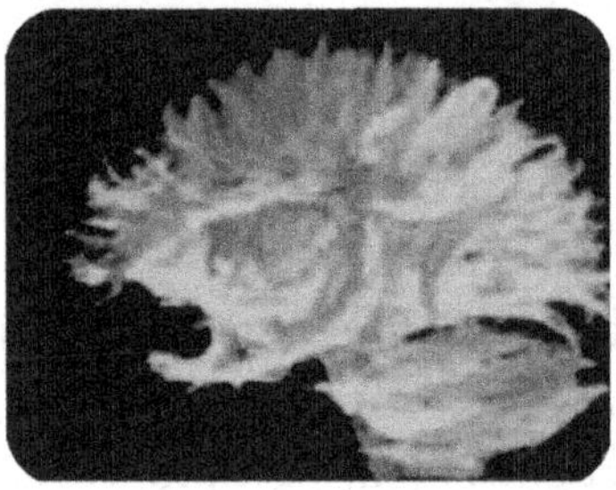

called C. elegans, which has 19,000 genes, compared to around 25,000 genes in humans. Around 40% are very similar to other organisms including humans and its genome has been fully mapped, making it a popular research subject. With only 302 neurons it is much easier to study than our brain and by superimposing a map of activated neural networks over the worm's connectome map, they have even developed an algorithm which can predict the worm's mating behaviour. They observed that each neuron is functionally unique with many neurons consistently activated or inhibited by different combinations of behaviour. In one study, neuroscientist Vladislav Susoy and colleagues found that when they disrupted a single neuron, the mating behaviour broke down and concluded that it is possible to show how diverse brain-wide dynamics emerge from the integration of sensory perception and motor actions in a natural context, thus allowing us to predict behaviour, albeit so far only in a tiny worm.[81] Studies of the brain region used by fruit flies to navigate, recently mapped over 25,000 neurons and 20 million connections, are also providing rich data on how they manage the complex task of navigating their environment.[82]

When the synchronised macro-oscillations that form the neural patterns and neural pathways that represent our memories are impaired as in many neurological diseases like dementias including Alzheimer's or the result of traumatic brain injury or neurological conditions like epilepsy, we see the devastating implications for cognition and behaviour and our sense of self and identity. This is why epilepsy, one of the most common neurological disorders with an estimated 68 million people affected worldwide, is so devastating to the sufferers and challenging to clinicians.[83] Epilepsy is "characterised by "spontaneous and recurrent seizures that are mainly generated in sub-cortical areas of the hippocampus (associated with memory) or the cerebral cortex"[84] and if the epilepsy is so bad and the hippocampus is involved, the epileptogenic hippocampus, may be surgically removed. Epileptogenic has been described as the convergence of neural circuit dysfunction and is defined as "The process by which the previously normal brain is functionally altered and biased towards the generation of the abnormal electrical activity" by reference to the brain state which "supports recurrent, unprovoked seizures."[85]

Recent research into epilepsy is opening the potential for new therapies based on a clearer understanding of why recurrent seizures characterising various types of epilepsy happen. This has been gleaned through an examination of samples of the cerebral cortex taken during surgery from patients with epilepsy and autopsies of affected individuals. Out-of-body experiences are predominantly observed in individuals suffering from focal epilepsy or schizophrenia in which seizures start in and affect one region of the brain. They are usually preceded by an aura, which is the term for the sensory changes that happen before a seizure or for some people before they experience a migraine. A team at Harvard

University led by American neuroscientist Jeff Lichtman (1951–) removed a tiny piece of brain from a 45-year-old woman with drug-resistant epilepsy. Another team from Google then used the brain slices to form a three-dimensional volume and from this tiny sample which included around 50,000 cells made the most detailed map ever seen of brain connections and observed that these cells are connected by hundreds of millions of dendritic tendrils forming 130 million synaptic connections. The data set measuring 1.4 petabytes, is around 700 times the storage capacity of the average modern computer. They revealed an enormous discrepancy in the number of connections between neurons, with stronger connections speculated to be involved in fast network communication and learning. They also found new neuron pairs deep in the cortex, which for no obvious reason were facing in exactly opposite directions. Interested readers can access this brain map freely online.

Recent studies focusing more on functional connectivity have observed a dysfunction of specific subtypes of neurons generating the inhibitory neurotransmitter GABA (discussed earlier) which is used by inter-neurons to inhibit brain wave activity.[86] You can imagine how removing the ability to inhibit some neuron signalling causes chaos in the brain and remember that cortical inhibition has been hypothesised as protecting overlapping memories from interference thus permitting continual learning.[87] We will discuss structural connectivity first and look later at functional connectivity, although they are for obvious reasons typically assessed together.

Structural Connectivity

Neuroscientists and psychologists traditionally distinguish between the biological (motor actions and bodily sensations), emotional, cognitive, and social selves. In the past, scientists believed that motor actions and bodily sensations, emotions, and thinking were hierarchically processed in the brain, with motor control and bodily sensations at the bottom of the hierarchy, followed by emotions and, finally, the higher powers of cognition and thinking at the top. This belief perpetuated the now largely discredited concept of the Tribunal or three-in-one brain theory developed in the 1960s by the American physician and neuroscientist Paul MacLean(1913–2007). MacLean postulated three evolutionary and supposedly independent but connected regions of the brain called the Reptilian brain associated with movement (the Basal Ganglia), the Paleo-mammalian brain, the centre for emotions such as happiness, surprise, disgust, fear, and anger, etc. (the Limbic system) and the Neo-mammalian brain responsible for rational thinking (the Neocortex meaning new). Maclean's hypothesis involved the evolving neocortex layering itself over the limbic or emotional feeling mid-brain just above the instinctual, automatic reptilian brain stem, which houses the primary functions for keeping you alive, including movement and breathing. By the 1990s, scientists had rejected the idea that evolution added layers to the brain or that these three regions are part of a functional hierarchy. Popular scientific literature continues to disseminate theories such as emotional intelligence, reflecting the belief that the more recently evolved neocortex regulates the emotions generated in the more primitive brain. Modern brain imaging studies show a different story, with multiple brain regions becoming active when responding emotionally and thinking.

Canadian neuroscientist, psychologist, and author Professor Lisa Feldman Barrett (1963–), who focuses on the affective (emotional) sciences, observes that "Brains aren't fundamentally different from reptiles, rats or even fish" and describes how "Human brains did not emerge from reptile brains by evolving extra parts for emotion and rationality. Instead, something more interesting happened." Barrett goes on to state: "… your misnamed neocortex is not a new part; … Anything you read or hear that proclaims the human neocortex, cerebral cortex, or pre-frontal cortex to be the root of rationality, or says that the frontal lobe regulates so-called emotional brain areas to keep irrational behaviour in check, is simply outdated or woefully incomplete."[88] The triune brain's conceptualization of the epic battle between emotion, instinct, and rationality appears to be a modern myth.

A non-controversial neurobiological feature of the complex human brain is the high level of structural connectivity that enables rapid communication and integration of sensory information between cortical and sub-cortical brain areas, which includes information related to internal bodily data and homoeostasis, emotions, thinking, and motor responses related to external stimuli. Structural connectivity depends on the white matter axons and dendrites responsible for carrying and receiving all this information between the neurons. Extending from the neuron body are axons, which carry electrical impulse signals to other neurons and appear whiter because of the white myelin/fatty sheath around the fibre, facilitating faster transmission of the electrical signals, although not all axons are myelinated. Dendrites are short, unmyelinated, branch-like extensions that dominate the cortical neurons and receive, integrate, and communicate sensory information to the neuron's body. Incoming electric impulses from axons at the synaptic junctures enter through the tens of thousands of receptors on the surface of the dendrites. Dendrites produce sustained oscillations at specific frequencies, facilitating information processing in the neural circuits connected to the dendrites which enables trillions of brain connections and the rapid exchange of complex information about internal and external stimuli. Axons and dendrites allow the scaling up of neural assemblies and networks and the synchronization that creates the neural pathways and patterns that form the basis of our emotions, feelings, thoughts, and memories and enable our brain to communicate with the body and vice versa.

Axonal Pathways

A new technology called Diffusor Tensor Imaging (DTI) has facilitated a much greater understanding of axonal microstructure, and the connectome has revealed that white matter consists of three main tracts or bundles of fibres, which are (1) Projection fibres

(2) Association fibres and (3) Commissural fibres. Projection fibres connect the cortical areas with the sub-cortical nuclei, brainstem, cerebellum, and spinal cord, or vice versa. Afferent neurons carry information from sensory receptors of the skin and other organs to the CNS. In contrast, efferent neurons carry motor information away from the CNS to the muscles and glands of the body. Association fibres connect different cortical areas within the same brain hemisphere, while commissural fibres connect similar cortical areas in the two hemispheres, including the Corpus Callosum and the Anterior Commissure.

It has traditionally been thought that humans have bigger brains than most animals because of their size and that bigger means more grey matter or neurons, which makes us capable of higher intellect. A wealth of modern imaging technologies and research methodologies have revealed that it is not brain size that matters but the level of connectivity locally and between cortical and sub-cortical brain regions, which gives our brains the capacity to integrate large amounts of sensory information. Dr Michael Hoffman observes how the development of the cortex harnessed folding and connectivity to produce smaller, faster brains but reached a critical point at which the brain grows beyond a critical processing capacity in terms of size; it will limit any gains in cognitive power. Hoffman argues that "with the evolution of the human brain, we have nearly reached the limits of biological intelligence."[89] Hoffman's point is that our cognitive capabilities are not limitless, and it is not the size of our brain but the processing speed, integration of sensory information, and complex neural circuitry of the cortex that determines the nature and adequacy of our adaptive survival response to our environment.

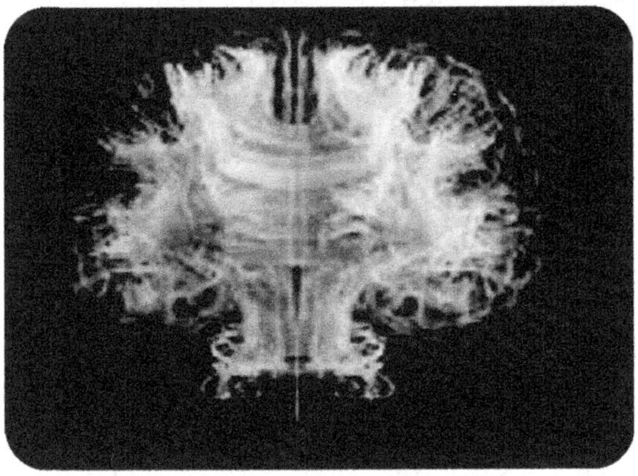

Some studies have suggested that the tension along axons or fibre bundles in the underlying white matter is the primary driving force for cortical folding. By keeping the aggregate axonal and dendritic wiring length low, tension contributes to the compactness of neural cortex circuitry, which has allowed us to have such a complex brain in a relatively small skull. The white matter volume increases disproportionately with brain size. In contrast, grey matter contains neurons connected by dendrites and unmyelinated axons, and volume increases as a linear function of the brain volume. It is not just the number and diversity of cortical neurons (grey matter) that drives intelligence and cognition but the progressive expansion of axonal mass that facilitates global communication and integration of information across the brain. Consistent with this are other studies showing that human brains appear to have a greater proportion of their cortical surface allocated to the higher-order association cortex, which processes and integrates information in neural patterns.[90] We will look at the role of genes later. Still, given the brain's complexity, you can begin to appreciate the difficulty in identifying the underlying neurobiology of our emotions, thoughts, behaviour, and other domains of self and the challenge facing designers of AI architecture and autonomous androids when using the brain as the blueprint for design.

There are many types of simple and highly complex neural pathways, some of the most important ones are the reflexive pathway, the visual pathway, the pain pathway, the dopamine pathway, and the corpus callosum. They are usually named according to their function, location, specific characteristics, and whether they converge, diverge or reverberate. Pathways converge when signals from several neurons converge on one neuron, enhancing excitation or inhibition. In contrast, with diverging pathways, signals from one neuron split and are sent to several neurons. In reverberating pathways, neurons connect with earlier neurons, which means the signal will be sent through the pathway again. We will look briefly at the Corpus Callosum and the Thalamus as two major and critical pathways for integrating information.

The Corpus Callosum and Asymmetry in the Two Hemispheres of the Brain

A unique biological feature of the human brain is the pronounced asymmetry in the brain, often cited as reflecting distinct roles of the left versus right brain hemispheres and implicated in how we perceive our world. A major commissural fibre pathway connects the two hemispheres called the Corpus Callosum (Latin for the tough body), the largest bundle of nerve fibres in the human body, containing about 200 million nerve fibres and passing information back and forth between the hemispheres. It sits just below the cortex and consists of the Rostrum and Genu which connect the brain's left and right frontal lobes and the body, and the Splenium, which connects the Temporal and Occipital lobes. It is only about 10 cm long, shaped like a C, and doesn't stop growing until a child is 12 years old. Many studies have demonstrated functional and anatomic lateralization in the brain with functional asymmetries observed about a range of cognitive functions, such as expression and understanding of language on the left.[91, 92] Scientists exploring hemispheric asymmetry have long sought to identify and elucidate the distinct roles of the left and right hemispheres, evidenced by the dominance of one side over the other. This is most obviously reflected in people tending to be either left or right-handed, even though nearly 90% of people and primates are right-handed. However, the neural correlates for this almost universal tendency remain a mystery. We know that genetics influences asymmetry, but twin studies have shown that genes can only account for 25% variation in the difference in handedness in twins[93] with the rest attribute to non-genetic factors, including epigenetics and neural mechanisms.[94]

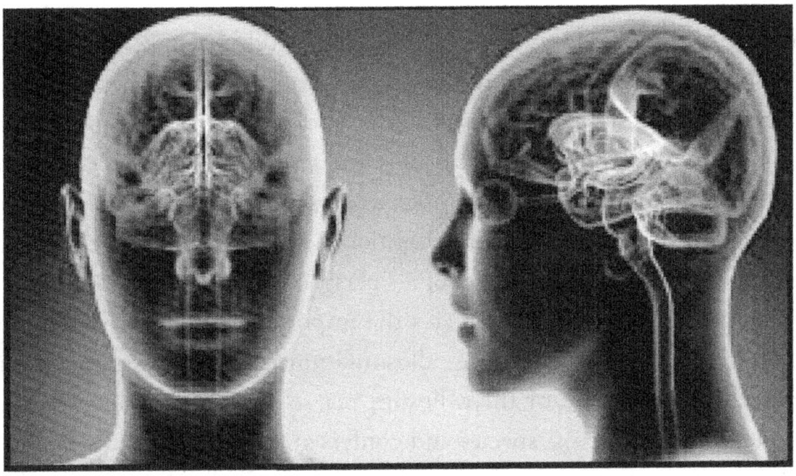

The Corpus Callosum

In 1865, the French physician Pierre Paul Broca (1824–1880), while working at the hospital Bicètre in Paris, treated a patient called Leborgne, often referred to as Tan because he could only speak a few words and constantly repeated the word Tan, but was able to understand questions when asked. After Leborgne's death, Broca found a lesion in the left hemisphere, and based on this and a further five years of work on other patients with speech defects, Broca announced to the scientific world that "Nous parlons avec l'hémisphère gauche" (We speak with the left hemisphere). At that moment, Broca seemingly ended the scientific notion of a symmetric or holistic brain while appearing to confirm the uniqueness of humans given that asymmetry relates to language and we are the only species equipped with language. Since the 19th century, many comparative studies on brain asymmetry have been performed but have yet to gain impetus largely due to the highly flawed assumption that this asymmetry was confined to humans. We now know that brain asymmetry is prevalent in other invertebrates and vertebrates and has been identified in over 200 species, including worms. In the 1950s and 1960s, neurobiologist, cognitive scientist, and Nobel Prize winner Roger W. Sperry (1913–1994) and his student Ronald Myers discovered that when removing the corpus callosum, the two hemispheres continued to operate as two independent brains and removal did not seriously impair the patient's faculties, although were unable to detect visual stimuli coming from the opposite side of the body. Sperry explored how the callosum integrates information between the brain hemispheres and sought to elucidate its role in coordination and complex problem-solving.

Each hemisphere controls movement and feeling on the opposite side of the body. The hemispheres process language information, so sensory information coming into one half of the brain must be transmitted to and integrated with the other hemisphere. The callosum is involved in eye movements and eyesight by integrating visual inputs from each hemisphere and helping focus improve attentive behaviour and bodily balance. Rarely in a condition called Agenesis, a child like the savant Kim Peek, whom we met earlier, may not develop a callosum, which has given scientists insight into its role because they suffered from a range of deficits, including reading facial expressions and voice tones, difficulty in problem-solving, performing complex tasks, assessing risks, understanding abstract concepts, and reading emotions. They might have trouble with attention, a lack of self-awareness, and an inability to read social cues or understand other people's perspectives, and it can manifest in obsessive-compulsive behaviour. All of this illustrates the important role of information integration between the brain hemispheres and the callosum's function. Brain asymmetry, rather than the existence or non-existence of the callosum per se appears to enhance sensory, cognitive, and motor efficiency across species and confers an evolutionary advantage. It has been evidenced by a variety of examples, including pigeons with visual asymmetry finding more grain,[95] chimps with stronger handedness catch more termites when using sticks to dig them out,[96] and humans with higher IQs correlated with pronounced language asymmetry.[97] Such advantages have been attributed to three distinct mechanisms, namely (1) Animals that favour one side will experience life-long perceptual or motor learning, which increases the lateralization (2) Increased use of one side shortens reaction times; and (3) Parallel computing of complementary functions reduces cognitive redundancy and increases overall efficacy.[98]

A landmark study in 1994 hypothesized that due to the slow conductance of signals through the corpus callosum and the large size of the human brain, the two hemispheres were prohibited from communication and interaction during simultaneous information processing but asymmetry in animals with much smaller brains suggests this is incorrect and not due to brain size.[99] Asymmetry does not appear to depend on the presence or otherwise of the callosum, with studies showing that cerebral asymmetries are "Only subtly associated with callosal thickness."[100] Asymmetry enhances parallel computing processing in the two hemispheres of complementary information. It allows the simultaneous performance of two tasks in parallel by utilizing both hemispheres for different uses.[101] The dominance of the right hemisphere in emotions is seen in all the primates studied and suggests a long evolutionary history over millions of years.[102] Asymmetry has also been implicated in specialization in emotion and perception in the right hemisphere and is cited as relevant to maintaining psychological balance or equilibrium.[103] Atypical or abnormal lateralization has been associated with such conditions as autism and schizophrenia.[104, 105, 106]

Since Myer's and Sperry's work, there has evolved a widespread view, which some might say amounts to a dogma that asymmetry means that left-brained people are more logical, analytical, mathematical, detail, and fact-orientated. In contrast, right-brained people are more creative, intuitive, and visual and tend to see the gestalt or the whole picture rather than focus on the detail. Hemispheric lateralization has been widely used to suggest that personality, behaviour, and cognitive style or ways of thinking are dominated by the dominance of one side of the brain or the other. This perspective ascribes different information processes to the two hemispheres with the left processing information sequentially and in order, while the right processes the whole and in parallel. As early as 1992 it had been observed that in the context of learning and education, "Instructional methods in today's classroom are sometimes based on the notion of cerebral dominance" and that teachers assessed individual differences in cerebral dominance using a paper-and-pencil survey called Your Style of Learning and Thinking (SOLAT)" However Zalewski and colleagues evaluating this widely used metric failed to confirm the "notion of hemisphere at least not as measured by the SOLAT."[107] Despite such findings, the belief in the dominance of hemispheric styles has become part of a widespread public impression of how our brains work.

This hypothesis has a long history since Arthur Ladbroke Wigan (1785–1847) in 1844 attributed mental disease to the independent hemispheres of the brain opposing each other rather than working together[108] which are believed to have inspired Robert Louis Stevenson's Jekyll and Hyde story. A leading figure in the modern field of brain asymmetry is philosopher, neuroscientific researcher, and former psychiatrist Dr Iain McGilchrist (1953–), who has rightly established an international reputation as a polymath with profound insights into the roles of the two hemispheres and life generally. McGilchrist's overarching message, elegantly argued in his most recent book, The Matter with Things (2021), is reflected in the enigmatic title, which seeks to draw attention to how we have lapsed into a reliance on scientism, defined as "an exaggerated trust in the efficiency of the methods of natural science applied to all areas of investigation."[109] Science's reductionist approach is driven by understanding how matter and things work, whilst Scientism rejects any knowledge not based on scientific evidence, resulting in the application of reductionist scientific methods to areas that don't necessarily lend themselves to science. One effect of scientism is to create the artificial distinction that we discussed earlier between intelligence and understanding when, as McGilchrist rightly points out, intelligence means understanding.[110] McGilchrist maintains that problems arise if intelligence is approached through the lens of scientism, which is attributed to a predominance of left-brain thinking, which leaves little place for human understanding that arises from insights to be gained from such fields as philosophy, music, dreams literature, emotions, and spirituality, etc. which are attributed to right brain thinking. As a rigorous scientist, McGilchrist goes out of his way to stress that he does not reject science but rejects scientism as the holy grail and makes a compelling argument for us not to focus on things in a reductionist sense but to appreciate that true understanding and intelligence comes not just from scientific validation, but from insights gleaned by being part of the gestalt or the whole flow of life and the universe of which we are all part and to which we all ultimately return. These do not lend themselves to reductionism and he asserts that the over reliance, particularly in the Western world, on left-hemisphere thinking with its rational and logical perspective has led to the neglect of right-hemisphere thinking which gives us greater insight and understanding from a non-scientism or left-brain perspective.

Whatever the science supporting the lateralization of thinking styles, McGilchrist's profound observations on human understanding and intelligence should have enormous resonance in the modern world and increasingly in the role and future of AI. The problem with the reliance on the general polarisation and lateralization of the hemispheres to support broad phenotypes, behaviour, and thinking styles is that recent studies need to provide evidence of this, and suggest that thinking styles differ within each hemisphere. A 2003 meta-analysis of 65 neuro-imaging studies found "no support for the hypothesis of overall right-lateralization of emotional function." It acknowledged that "lateralization of emotional activity is more complex and region-specific than predicted by previous theories of emotion and the brain."[111] A 2013 study evaluating the left versus the right brain hypothesis by studying over 1000 people using MRI found that although the hemispheres do function differently, the brain does not appear to favour one side or the other. Input is received from both sides when carrying out tasks. They concluded that "Lateralization of brain connections appears to be a local rather than a global property of brain networks, and that the "data are not consistent with a whole-brain phenotype of greater "left-brained" or greater "right-brained" network strength across individuals."[112] Despite the widespread and popular belief in the different styles of thinking and seeing the world in the two hemispheres, the prevailing scientific view appears that although there are some functional asymmetries, the two brain hemispheres work together in every cognitive task and not in isolation. This has relevance to AI, as we will see, because of the vital role of integrating information between the brain hemispheres and the connectivity function of the callosum, which allows integration to take place.

The Thalamus

A crucial region of our brain, involved in extensive communication between brain regions, is a group of densely packed oval-shaped structures of neurons that sit just above the brainstem, deep in the brain centre called the Thalamus.

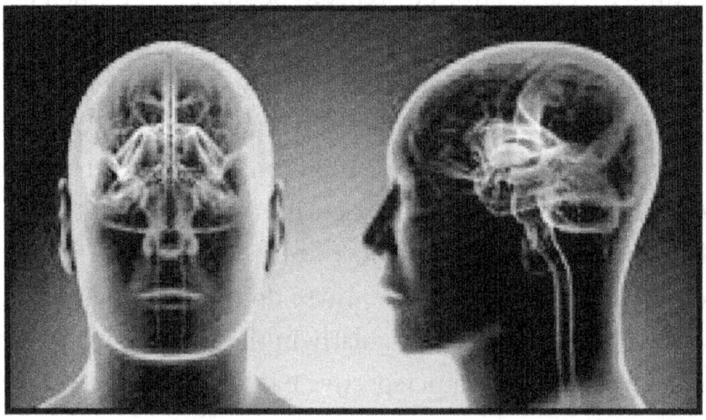

The Thalamus

They are two large oval-shaped clusters of distinct regions of nuclei, connected by inter-thalamic adhesion. The Thalamus has traditionally been described as the gateway to the cortex because it relays auditory, visual, somatosensory, and gut information, though not smell (Olfaction), between the cortex and other brain regions and plays a critical role in integrating emotions, memory, and motor functions. The Thalamic Reticular Nucleus (TRN) forms a capsule around the Thalamus that contains GABA neurons, which, constrain or inhibit information going through the Thalamus.

The TRN regulates interactions between the Thalamus and the cortex through the thalamocortical pathway, which is critical for sensory processing, attention, and cognition. The other thalamic neurons are dopamine-producing and excitatory. A recent study of the TRN using multi-scale single-cell analyses in mice revealed two important neural net works regulated by the TRN.[113] When quiet, the TRN participates in sleep rhythm generation, stability, and memory consolidation, when active, the TRN filters sensory information, which is crucial in attention. These thalamic neurons are either inter-neurons projecting within the Thalamus or are connected reciprocally to the cortex on the same side of the brain.

Information constantly flows around the brain, fed back and forth through the various pathways, with the Thalamus acting as the control centre with the inhibitory and excitatory thalamic neurons key to this. The traditional view of the Thalamus as a passive control centre or gateway is being revised, given its enormous metabolic cost to the brain regarding energy consumption, and some scientists question its role purely as a relay station. More and more evidence suggests that its contribution is far greater and more complex than previously thought. Neuroscientists Nima Dehghani and Ralf Wimmer have described it as acting like a "Blackboard or global workspace in the brain"[114] on which, as Robert Worden and colleagues claim, "Computations of different cortical modules are composed, coordinated, and integrated."[115] They believe the following features are consistent with this role, namely its "lack of olfactory relay function, its lack of internal excitatory connections, its regular and conserved shape, its inhibitory inter-neurons, triadic synapses, and diffuse cortical connectivity." They concluded that it has an important role in information processing, which includes "Spatial cognition, planning of movements, and risk-reward trade-off decisions" and that it "instantiates a generative model of objects in a three-dimensional peri-personal space (PPS) (i.e., around the person or body)." Worden and colleagues link its role to Karl Friston's notion of free energy that we discuss later and point to how the brain data mines its sensory inputs because to be adaptive the brain must minimize the long-term average of surprises to resist the tendency to disorder in terms of uncertainty and information and does so by reducing uncertainty.[116, 117] Neuroscientists Mathieu Woolf and Seralynne Vann describe it as the "Gateway to Mental representation," which plays key integrative roles in cognition, ranging from learning and memory to flexible adaptation."[118] The human ability to exhibit flexibility and switch between behavioural options has been consistently linked to the Thalamus.

In a 2017 study, Ian Schmitt and colleagues concluded that there is an unknown principle in neuroscience, which is the thalamic control of functional cortical connectivity, and that "this function, which is dissociable from categorical information relay, indicates that the thalamus has a much broader role in cognition than previously thought."[119] A 2021 study conducted by MIT's McGovern Institute for Brain Research discovered that certain brain circuits allow the brain to deal with ambiguity and uncertainty, whether causality or making inferences.[120] The fMRI research revealed that when people are asked to study a scene but aren't sure what to pay attention to, the mediodorsal Thalamus becomes active, and they observed that "the dorsomedial thalamus helps the PFC tune in to the right signals during decision-making, adjusting signalling as needed when circumstances change." Inhibitory neurons in this region appear to dampen downstream signalling and stop the frontal cortex using unreliable information. Studies show that the less guidance the brain is given regarding ambiguity and uncertainty, the harder the mediodorsal Thalamus must work. This has very important implications, for example, in individuals with the condition of schizophrenia who exhibit less activity in these circuits, which may be implicated in the condition. It illustrates the exquisitely balanced mechanism involved in the brain's connectivity and communication of information between brain regions when processing information. This brief discussion illustrates the vital role played by the Thalamus in constructing and representing information and relaying information to the cortex for further integration, inference, and prediction.

Integration of Information in the Brain

Another special neurobiological feature of the brain is how prior knowledge influences our perception and interpretation of new information and experiences and how this information is integrated within existing neurons, neural pathways, and neural networks. Integration is essential to generative learning and acquiring new knowledge, experiences, and beliefs. Integration of all the information passing through thalamic pathways occurs not in the cortex's primary regions, even though these are heavily connected, but in the association areas of the cortex. In and between these association areas, the vital neural pathways, patterns, and localized neural networks are established during neural development, shaping our cognitive capacity and the emotions and feelings representing the self and who we are. That is why, while a child is developing a coherent and stable sense of self, the nature of the stimuli and experiences that that child is exposed to are so vital. Laying down neural pathways, patterns, and networks to facilitate effective communication, connectivity, and integration between brain regions and networks is critical and influences a child's subsequent levels of intelligence, understanding of the world, and behaviour. As we saw with the Romanian orphans, early deprivation that prevents this can have long-term and tragic consequences for cognitive development. Although cognition is developed in parallel with and is highly influenced by sensory motor skills, the latter typically develops much faster, partly because the neural maps for sensory motor function are already largely installed. In contrast, cognition relies on the communication between brain regions and integration, which depends on the development of its neural maps, patterns and networks, which continues to take place into early adulthood and constantly changes and adapts throughout our lives in response to experience.

Neural Maps and Neural Patterns

The activation pattern of neural populations forms a neural code that defines what pattern of neural activity corresponds to representing types of information. Integration of information involves encoding and storing the prior knowledge and experience of the world in neural patterns and maps in the cortex, and these patterns or schemata influence our perception and interpretation of new information and experiences. When our brains process sensory information, the balance between neural excitation or inhibition dictates and controls our perception of and response to that information. Perception of information happens when sensory stimuli produce a processed sensation and either create a new neural pathway or are incorporated into existing neural pathways. Conscious perception of stimuli and representation of the information in our theatre of consciousness or working memory (see later) may produce a decision or action as a motor response. Neural patterns formed by multiple pathways rely on the potentially excitable nature of most neurons, which respond to changes in the body's environment and send signals throughout the brain's network of neurons. A vital function of the cortex's grid and columnar structural layout is to generate sensory pathways and patterns between neurons, with evidence pointing to the fourth layer as the primary location. To encode the many features of information related to events or episodes, sensory information must be linked during the encoding process and kept separate from other memories to allow for subsequent retrieval or recall. We saw the importance of this when we discussed continual learning earlier, and this vital neurobiological process is called binding. It explains how, when subject to multiple sensory inputs, our brains subconsciously combine objects, the environmental context in terms of colour, location, and abstract emotional and thinking processes into a unified experience and create coherent and unified memories of those experiences. We will look at binding further when we discuss the role of the visual pathways and the subconscious in our perception of the world and its role in AI. Consciousness involves the cortex, where most sensory neural patterns form and, when aggregated through integration, create our autobiographical memories and higher consciousness. Although neural patterns are also generated in the brain stem, these tend to be of lower resolution than those in the cortex.

A ground-breaking discovery made in the 1930s by the American Canadian neurosurgeon Wilder Graves Penfield (1891–1976) revolutionized brain surgery methods and techniques giving us a remarkable insight into how our brain creates the self. Focusing mostly on mental processes, he pioneered mapping the functions of various brain regions while exploring the scientific evidence for the soul and how our brain generates that feeling of déjà vu. As part of the surgical procedure on patients with epilepsy, Penfield would insert an electric probe into the patient's brain to identify the least disruptive area to operate on. His key discovery was the existence in the cortex of stable neural patterns in which bodily parts are represented topographically, with separate cortical patterns for the Motor, Somatosensory, Auditory, Visual, and Olfactory senses. The cerebrum is where we process emotions and thinking and is the largest part of the brain, containing the cerebral cortex and sub-cortical regions and involved in sensory processing, like vision, touch, hearing, and movement. The cerebrum contains two (left and right) hemispheres, each with a Frontal, Parietal, Temporal, and Occipital lobe, divided into primary and association areas.

Primary and Association Regions of the Brain

The location and configuration of these neural maps are predetermined from birth, with each of the four main lobes having a primary area containing topographical maps. Although metaphorically called neural maps because they map the body parts, they consist of dynamic signalling between neurons and are not static, two-dimensional physical maps. Professor Nicholas G. Hatsopoulos has focused much of his neuroscientific research on the neural basis of motor control and learning, investigating what information features of motor behaviour are encoded and represented in the collective activity of neuronal ensembles in the motor cortex. Hatsopoulos and his team have explored how AI machines might harness brain mechanisms to learn and realized the importance of how neurons are spatially laid out in the cortex.

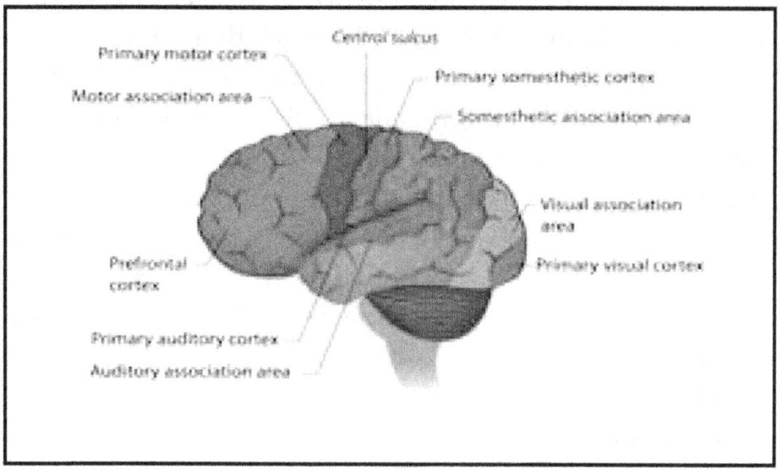

Primary and Association regions of the Four Brain Lobes.

They have challenged the traditional assumption that these patterns merely reflect the difference between wakefulness and sleep and studied neural patterns to understand how the brain organizes information. This involved measuring the brain activity in Macaque monkeys while they were pushing a joystick, looking specifically at high-frequency neural wave patterns in the motor region of the cortex. They found that these neural waves consisted of the propagating activity of hundreds of neurons travelling in different directions, which told them exactly what the animal was doing.[121] They concluded that the spatially organized patterning in the cortex contains and carries the sensory information with neural patterns reflecting how the neurons are organized. The connected pathways between neurons create the patterns that encode information from our environment as memories. The electric waves generated in these neural pathways and neural patterns represent the dynamic transfer of information between brain regions. Signs are generated in response to stimuli propagating through these pathways, which, if repeated or strong, will form new connections and memories. These patterns constitute the basis of our thoughts.

Computer scientists Alex Huth, Jerry Tang and Shailee Jain have even been able to use a non-invasive AI system called a Semantic Decoder to convert the private thoughts generated when a person is listening to or imagining a story into a stream of words and text. The algorithm is trained on brain patterns while a person thinks about certain things, and it can recognize patterns when they are generated during subsequent brain activity. The major limitation is that it currently involves participants lying in an fMRI scanner, but future iterations could involve a transplant directly into the brain. The Vocoder algorithm has been used to synthesize speech once trained using people's voices and can do this because our thoughts generate tell-tale patterns in the brain that the system can detect.

It can decode complex language using thoughts for long periods and when tested across the cortex has established that continuous language can be separately decoded from multiple regions.[122] Although the prospects of an AI system reading our thoughts might sound scary, the reality is that it currently requires significant collaboration from the individual in training to apply the system and learn their unique brain patterns and activity. The information encoded in neurons, neural patterns, and neural maps is key to how humans perform sensory movement and perception.

Sensorimotor Perception and Movement

In the frontal lobe, we find the somatosensory and primary motor cortex maps in which a specific body part is associated with a distinct location in the cortex and organized topographically concerning somatosensory and motor activity. Soma means the body as distinct from the soul or mind. This means that the location of body parts, about other body parts, is mirrored in neural patterns in the somatosensory cortex, so fingers are near thumbs, ankles near toes, etc.

The number of cortical neurons allocated to different body parts reflects their level of sensitivity, so the pattern is distorted to reflect sensitivity rather than the relative size of body parts. Locomotion in humans and many animals involves many cortical, sub-cortical, and spinal neuron centres, and focusing on spinal features needs to reflect the complex mechanisms involved in how humans move, grasp, avoid obstacles, walk, run, and navigate their environment, etc. New approaches seek to reflect how the sensory information is processed, for example, through the intrinsic biomechanical properties of the hand and muscles that provide sensorimotor control and automatic reflexive actions (discussed earlier). The spinal cord uses a feed-forward and feedback processing mechanism, which facilitates rapid dexterous movement involving adjusting movements or regulating more complex actions related to events.

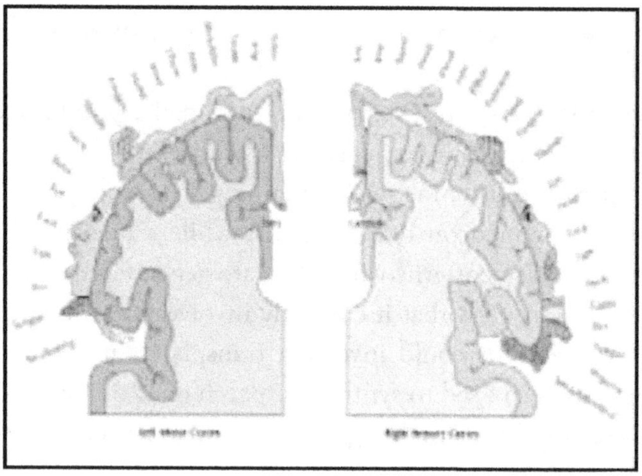

Sensorimotor Cortex.

All this feeding forward and feeding back of information is processed in the brain-stem and Thalamus to plan and execute movements about tasks and goals. At the cortical level, perception of an object involves integration with motivation and impetus and prior knowledge or experience to achieve desired goals like touching or grasping the object, etc and this hierarchical control over movement is evident in the simple decision to pick up and read a book. The goal of grasping the book arises in higher cortical structures, then through sub-cortical structures involved in motivation reward and movement, through the brain-stem to the spinal cord, which generates the motor plan and reflexive adjustments to the movements necessary to grasp the book. When we move muscles, the motor system generates an out-flowing neural signal (efferent) and an internal copy. Efferent means "Conducted or conducting outwards or away from something (for nerves, the CNS; for blood vessels, the organ supplied."[123] This copy is collated with a reafferent sensory input generated by the movement. It allows comparison between the actual and desired movement, which enables our brain to predict the effects of the movement. Studies have shown that an efferent copy is generated even when silent inner speech is produced.[124, 125] Given the complexity of the sensorimotor mechanism in humans, the challenge is to equip AI robots with sensory motor processes that allow them to move and plan movements, grasp objects, and navigate in real environments, which are often noisy, messy, and uncertain and typically involve partial or incomplete information.

Functional Connectivity

The Oscillating Behaviour of Neurons

Understanding functional connectivity at the neuronal level is vital to understanding how our brains operate. Sending electric signals depends upon neurons' ability to use charged particles (ions) to build up a charge across the cell membrane, and neurons have many electrically charged channels in the cell membrane. Neuronal signals involve a rapid increase of electricity along the membrane, with the electric charge in each neuron going up if stimulated and down if inhibited, which is how electric signals are propagated in our brain. Oscillating behaviour is a ubiquitous feature of the natural world, existing at all levels of organization from the subcellular, cellular, tissue, organs, and the organism, each representing greater organizational complexity. In response to outside stimuli or spontaneously, for example during sleep, neurons oscillate, representing rhythmic or repetitive patterns of periodic to-and-fro motion of the cell membrane from its mean position. Measured in Hertz (Hz), oscillations at the micro-scale are driven by internal neuron mechanisms and at the macro scale by interactions between neurons. The macroscopic oscillation between groups or clusters of neurons arises from the feedback connections between neurons, which synchronize their electrical firing patterns. Synchronized oscillatory behaviour has been posited in the communication through coherence theory (CTC) as a key mechanism for information sharing and binding together of perceived stimuli. CTC suggests that oscillations of a certain type of brain wave frequency called Gamma waves (see later) passing between different brain regions allow for the selective processing of a single item one at a time. An alternative Theta-Gamma Neural code theory (TGNC) proposes that a sequence of items is processed simultaneously during one oscillatory cycle. Although complicated and difficult to reconcile, these theories purport to explain how the brain can process several items of information at the same time through synchronized oscillations.

In a 2016 study, Douglas McLelland, Rufin VanRullen, and Louisa Lavergne claim to have demonstrated that both types of processing are plausible and concluded that they "coexist in the brain (and)... enables dynamic switching between exploratory and selective modes of attention."[126] The synchronized macro-oscillations of millions or even billions of neurons form stable neural pathways connecting different brain regions, and the neural patterns encode the brain's existing information upon which it bases its perception of our world. Neural patterns represent different levels of data extraction by our brain when processing sensory information and develop in response to experiencing our environment Yann Le Cu and colleagues described them as "more multisensory, unified and complex than anything AI can currently achieve."[127] and Portuguese-American neuroscientist Antonio Damasio (1944–) believes that they should be used as a marker for primary consciousness.[128] New sensory information potentially, though not always, generates new neural pathways and patterns and activates existing neural patterns and pathways. Each neural pathway and pattern created by synchronized oscillations represents an encoded experience with multiple types of pathways. However, some are better formed because they have been reinforced through repetition or practice of whatever activity or experience created them.

Based on extensive empirical data, brain research has for years treated physical brain function and mental states as complementary rather than contradictory. The most obvious example is how various brain areas oscillate at different frequencies, creating various neuronal assemblies. Through coupling, these are coordinated in space and time to form active neural networks that constitute the mental state of thinking, attending to stimuli, remembering, planning, etc. This evidences microscopic level activity and self-organizing at the macroscopic level to produce what American medical doctor and scholar Gerhard Werner (1921–2012), who researched pharmacology, psychiatry, and cognitive neuroscience, described as a "more abstract rendition of the microscopic level."[129] Relying on the Critical State Transitions theory and using classic examples such as the transition from water to ice, Werner argued that brain activity is a large collection of microscopic constituents that experience state transition to macro states that display qualitatively novel features and properties. The macro state's novel properties have no referent at the microscopic level and require new descriptors. This describes how conscious lower-level neurobiological processes cause states, and as Werner observes, we, as "Cognitive agents, must be viewed as coupled with the world in reciprocal interaction."[130] According to Werner's perspective, for robots to be cognitive autonomous agents they also need to be coupled with their environment to automatically achieve what the human brain does, which maintains a level of internal coherence essential to interpreting and utilizing the raw data that comes through the senses.

Coherent States in the Human Brain

In 1989, American neuroscientist and Professor of Complex Systems and Brain Sciences J A. Scott Kelso (1947–) postulated his theory of Coordinated Dynamics (CD) in which he described the brain as "A pattern forming system that can switch flexibly from one coherent state to another" and "describes, explains and predicts how patterns of coordination form, adapt, persist and change in living things."[131] Kelso postulates a regime of coordinated dynamics called Metastability in the Metastable Brain Mind (MBM) theory which says, "Rather than teetering between order and randomness, seizure and inactivity, the brain (or more properly, the brain-environment system) lives in a sea of metastability."[132] Kelso describes how "(1)The functioning brain has a vast repertoire of coexisting tendencies/ dispositions for regions of the brain to integrate and segregate simultaneously. This assures the brain a vital mix of both flexibility and stability (2) The degree to which one or the other tendency may dominate (or one might say 'the balance' between these tendencies) will depend on several factors, including age, dementia, task, memory, functional connectivity, propagation delays, circulating neurohormones, etc and (3) There are a near infinite, uncountable number of metastable tendencies exist between order (stable) and disorder (uncoupled). This happens when all the states have disappeared. The outcome is that rather than teetering between order and randomness, seizure and inactivity, the brain (or more properly, the brain environment system) lives in a sea of metastability." As the poet Robert Frost (1874–1976) said in his 1942 poem, The Secret Sits, "We dance around in a ring, the secret sits in the middle and knows." Kelso observes that the middle is more of a niche for the brain than a 'sweet spot or a boundary between order and randomness favoured by the critical brain hypothesis." To date no AI system has come near to achieving Kelso's sweet spot, but Kelso hypotheses that what if instead of zeroes and ones, computations were based on collective coordination variables having the characteristic of a phase "based on known properties of coordination or collective behaviour of neurons in the brain, not simply the on-off threshold-like elements/ neurons of current computational neuroscience."[133] Kelso claims that this would usher in a biological dynamical neuroscience with far greater potential computational capabilities and far greater biological realism. According to CD theory, realism complements the sweet spot theory of brain criticality with a brain mind that functions naturally in an immense sea of metastability. Kelso's notion of coherence and the dynamic shifting between coherent states appear central to the challenge facing AI systems and robotic brains if conscious autonomous robots ever become a reality. This begs the question of how our brains make sense of our world.

How Do Our Brains Make Sense of Our World?

The Human Brain and Information

Everything humans do is an adaptive response to our environment, involving attaching salience and relevance to sensory raw data from our environment, attributing meaning to the information, comparing it, and assimilating it with familiar information/prior knowledge or experiences. To make sense of how we process all the sensory information, psychologists focus on what they consider to be the key domains of the human self, which are

Emotions, Cognition, Personality, Memory, Learning, Motivation, Beliefs, Values, and Behaviour, all of which influence how we process raw data and the information we extract from it. The brain's complexity comes not as was once thought from functional segregation in which one part of our brain deals with emotions and the other thinking, etc, but from the rapid differentiation, communication, and integration of sensory information into a coherent mental representation which allows our brains to plan and formulate appropriate responses to the sensory stimuli in its environment. Our perception of the world is the outcome of a lifelong, dynamic, highly complex, and largely unconscious and subconscious process of memory learning, adaption and change in response to the raw data and information from our environment, that we process through our body, brain, and mind.

Raw data and information are different, and we must first appreciate and understand the meaningful difference between them. Based on the then prevailing laws of physics, scientists during the early 1940s were becoming aware that information is nature's way of reducing uncertainty because it increases the probability of something being true or false, existing or not existing or happening or not happening, etc and they developed the field of information theory. The simplest way of measuring uncertainty is to look at probability as a series of binary or yes or no states, which means that you can halve the level of uncertainty because there is a 50/50 chance that it will exist in one of these states. The term bit was used as a shortcut for binary, reflecting the probability of information as and if it exists is a yes or a no if it doesn't exist calculation. If we regard humans as merely complex sensory information channels, then every unit of information that flows through us can be treated as binary in nature existing in a yes probability state, so the amount of information is measured by the number of yes bits transmitted through the system per second. Using words like processing and bits suggests that our brains act like computers in that we operate on symbolic representations of our world. It is inaccurate given that we are born with little, if any essential computer software such as data and rules, algorithms, pre-set programs, memories, or binary formatting that allows computers to represent patterns. Some scientists believe that our brains harness the basic algorithms of nature to function,

although this remains controversial. AI has in recent years moved away from the early reliance on symbolic representation in which objects stand for or suggest an idea, belief, action, or material entity, to the concept of complex systems generating mental neural representations and patterns of integrated information based on the embodied experience to make predictions about the environment. We make these representations to understand incomplete and noisy information coming through our senses and use the data to update our internal models or representations. This conceptualization reflects the modern understanding of how the human brain makes sense of the world more accurately rather than processing complete information that already exists in the world.

The computer analogy has been a misleading perspective from which to research and understand the brain, and AI expert George Zarkadakis describes it as "just another metaphor used to explain our intelligence."[134] The computer metaphor was given significant academic credence in the 1950s with the publication of George Miller's book Language and Communication (1951) and mathematician John von Neumann's Computer and the Brain (1958), in which he deemed our brains to be digital processing machines. The physicist Stephen Hawking (1942–2018) even stated, "I think the brain is essentially a computer and consciousness is like a computer program. It will cease to run when the computer is turned off. Theoretically, it could be re-created on a neural network, but that would be very difficult, as it would require all one's memories." The computer metaphor or the notion of information being stored in our brain like a computer stores information is something we must try to resist when we refer to information processing in the brain. It has posed one of the major challenges facing AI in developing generative computer model simulations that get close to capturing the dynamic complexity of the whole brain, including the background neuronal activity and conscious and subconscious processes involved.

In 2008, computational neuroscientists Eugene Izhikevich and American biologist Gerald Edelman (1929–2014), who shared the 1972 Nobel Prize in Physiology or Medicine with the British biochemist Rodney Robert Porter (1917–1985) for their work on the human immune system, constructed a large-scale computer model of a mammalian brain which included three anatomical scales of white matter connectivity, cortical microcircuitry and 22 basic types of neurons. They observed that this large-scale computer model "exhibited behavioural regimes of normal brain activity that were not explicitly built-in but emerged spontaneously as the result of interactions among anatomical and dynamic processes. This included "spontaneous activity, sensitivity to changes in individual neurons, the emergence of waves and rhythms, and functional connectivity on different scales."[135] Such models represent the early stages of our deeper understanding of how a complex biophysical entity like our brain appears to operate and seem to demonstrate that focusing on multi-scale dynamics is potentially the most productive way to understand how observable data and phenomena emerge from largely unobservable latent states like those characterizing the brain. Izhikevich explains the relationship between electrophysiology, non-linear dynamics, and the computational properties of neurons and maintains that to model neuronal behaviour in the brain and interpret the results, neuroscientists must call upon methods of non-linear dynamics which means understanding the role and nature of chaos in the human brain.[136]

Chaos in the Brain

Many people are frightened that the modern technological world personified by AI is teetering on the edge of chaos brought about by the ever-increasing complexity. Not, however, Professor Douglas Hofstadter (1945–),an American scholar of cognitive science, physics, and comparative literature who focuses on consciousness, complex systems, and how people think and perceive the world. In 1996, he wrote a collection of essays called Meta Magical Themes (1996), which blends mathematics with philosophy and psychology to integrate many of the remarkable features of this complex world.[137] Hofstadter recognizes that humans have survived, adapted, and thrived because our brains have evolved to be the most complex and dynamic system in the known universe and that our natural curiosity should drive us to understand and marvel at this complexity. Nature by itself does a wonderful job of creating order out of complexity and chaos, and Hofstadter and other scientists and mathematicians are driven by the need to understand how order and apparent predictability emerge from the chaos of complex dynamic systems like our brains and how, for example, coherent thoughts arise from such chaos.

German theoretical neuroscientist Gregor Schöner (1958–) is a pioneer of dynamic field theory (DFT), which seeks to explain how our brain gives rise to behaviour through the coordination of the activity of populations of neurons. Schöner hypothesized that our motor actions,such as running, walking, brushing our teeth etc, reflect a dynamic balance between stability and flexibility. This happens because the MBBN has multiple components or dynamic systems that process the complex information or input from the senses. Such dynamic systems are non-linear because small changes or inputs can lead to major changes in the system's behaviour and changes in the system's output are not proportional to changes to the input. You may have heard of this as the butterfly effect which recognizes that although the output is unpredictable, it is still based on the basic laws of physics, which do not depend on chance. They are called deterministic because the output is determined solely by the input and initial conditions, which if maintained, always return the same results; hence, if you know the input and initial conditions, you should, in theory, be able to determine the output. Our body and brain comprise a non-linear dynamic system because any change of output, in terms of movement, is not proportional to the change in information input into the body and brain. Non-linear systems like our body and brains change from millisecond to millisecond, which, in terms of the movements or motor actions of the body, accounts for the tiniest single action becoming a multitude of actions over time. Most real physical systems in nature are inherently non-linear, making them of vital interest to scientists.[138] Although this applies to motor movements, Schöner and others believe that because cognition and emotions are embodied, they must always be intimately linked to the sensory-motor systems and are part of the same complex non-linear dynamic system. Not all non-linear systems are chaotic, but all chaotic systems are non-linear and most dynamical systems exhibit non-linear behaviour involving complicated motion and the potential to result in chaos. What do we mean when we talk about the chaos in the context of the non-linear dynamic system we call the brain?

The laws of physics, particularly the laws of thermodynamics, tell us that the universe is heading towards increasing levels of Entropy (Greek En-divergent) and Tropos-trend) or growing randomness, lack of predictability, and disorder. Entropy always happens when heat passes to another system, meaning that all natural processes reliant upon energy in heat form trend towards a state of greater disorder or divergence. We witness this all around us in, for example, bonfires, melting, salt or sugar dissolving, making popcorn or boiling water, and in ourselves as we get older. Fortunately, the natural world, including humans, is built upon the existence of order, not chaos, which means that over time, we are rarely made aware of the relentless march of entropy toward disorder. Everything naturally appears stable and ordered; otherwise, nothing would exist as we know it, but it is nevertheless subject to the relentless march of entropy. As complex, stable organisms, humans operate far from a state of maximum entropy or disorder but only do so by constantly inputting energy in the form of oxygen, food, water, etc.

It may not just be entropy at play for Ulisse di Corpo (1959–) and Antonella Vannini argue that our brains and all living matter also benefit from the law of Syntropy (Greek Syn- which means converging and Tropos- for tendency), which is the tendency to converge into order as opposed to chaos.[139] The theory of syntropy emerged from Einstein's laws of general relativity and the energy / momentum /mass equation, E = mc2, which predicts the existence of three types of time. Theoretical biologist and professor of biophysics Robert Rosen (1934–1998) describes how we typically describe the brain as predictive. Syntropy is not a predictive model and asserts that life depends on future causes and "should be explained using backwards in time causation" as biological systems anticipate the environment.[140] It is so counterintuitive to imagine and use the time working backwards that Rosen attempts to steer the subject away from the "metaphysical, magical, or paranormal" by observing that the three types of time in Einstein's equation envisage are "(1) Causal time which characterizes diverging systems, for example, our expanding universe where causes precede effects, entropy prevails, and time moves forwards (2) Retro Causal time which characterizes converging systems, for example, black holes where effects precede causes and syntropy prevails and time moves backward and (3) Super Causal time which characterizes systems which are balanced between diverging and converging forces, such as atoms and causality and retro-causality coexist and time is unitary: past, present and future are present together." Corpo and Vannini based a lot of their work on the pioneering theories of Luigi Fantappiè (1901–1955), who asserted that the laws of syntropy govern all living systems.[141] Fantappiè was never able to experimentally verify this in his lifetime because it envisages a world where time flows backward, cause and effect are reversed, and so are retro-causal. In the macro classic world, we only see and experience entropy, with time always appearing to move forward and cause preceding the effect. But in the quantum world, we can see syntropy at play with time flowing backward and effect preceding cause. Since Fantappiè, experimental data supporting syntropy has been obtained using electronic devices called Random Event Generators (REG) which use quantum tunnelling to generate completely random outputs, such as sequences of zeros and ones. Quantum tunnelling is a quantum mechanical phenomenon in which a particle can penetrate through a potential energy barrier higher in energy than the particle's kinetic energy. Based on REG experiments, it appears that in the quantum world, syntropy can prevail, which means that living systems can anticipate future stimuli using advanced waves that move back in time, with some evidence seemingly suggesting that this also happens at the macroscopic level. Neuroscientist mathematician and AI researcher Guillem Collell, and psychiatrist and clinical neuroscientist Jordi Fauquet point out that our brains are both thermodynamic and information processors, which means that the brain is subject to the constraints of both entropy and syntropy.

Most brain models are based on the laws of thermodynamics but they observe that while thermodynamics allows us to model brain activity by applying the law of entropy, information theory allows us to model cognition using syntropy and mind operation.[142] Anyone who uses a computer is intuitively aware of how energy is used to convert data into useful digital information, which is binary, either a zero that produces no response or a response. This resembles the binary nature of all-or-nothing action potentials in neurons. They either happen at a certain threshold or they don't, so either transmit an electrical signal or don't, which may explain why so many scientists treat the brain like a computer. Collell and Fauquet stress the "existence of deep theoretical connections between thermodynamics and information theory" and that "unlike in physics or chemistry, a formalization of the relationship between information and energy is currently lacking in neuroscience but is necessary to provide insight into the "formal relationship between cognition and neural activity."[143] They argue that both models can be used to "predict the change in a system in a certain direction" and cite deep theoretical connections. For readers interested in going into more detail, these are the Landauer limit, the energetic cost of processing information, the Gibbs and Boltzmann formulas, and the concepts of entropy and neg-entropy.

Entropy measures the state of uncertainty in a system and reflects the physical qualities of that system, with higher entropy reflecting higher physical disorder. First developed in 2014, the Entropic Brain hypothesis emerged from the focus on the psychedelic state in the brain and asserts that within upper and lower limits beyond which consciousness may be lost, entropy is associated with spontaneous brain activity and reflects the informational richness of the conscious state.[144] Professor Robin Carhart-Harris even maintains that the psychedelic state is regarded as a primitive brain state that is an elevated state of entropy and the "Primary state of consciousness that preceded the development of modern, adult, human, normal waking consciousness."[145] It has been cited as the reason why elevated brain entropy is reflected in a "repertoire of functional connectivity motifs that form and fragment across time" during the psychedelic state. The primary state appears to operate at the critical point of transition between order and disorder, and only during wakeful consciousness is entropy or uncertainty suppressed.[146] Collell and Fauquet argue that we have evolved an artificial distinction between cognition as information processing and brain activity as a thermodynamic physical process and recommend a more integrated approach to understanding the "fundamental relationship between cognition and physical brain activity."[147, 148]

If AI is ever to succeed in the attempt to replicate consciousness and sentience in robots, it may be necessary to reconcile the theories of syntropy and the entropic brain at the microscopic level of neuronal activity to integrate our knowledge of information processing and the laws of thermodynamics. In January 2024, Patrick Cole and colleagues announced that they had built a prototype thermodynamic computer that takes advantage of the random noise of the thermodynamic environment. Cole referred to "themselves as making lemons out of lemons." It employs a stochastic processing unit or SPU with eight connected circuits using electrical oscillations to store energy and take its inputs from the environment. It uses the energy of whichever components are hottest to perform computations which appears to more closely resembles the way the brain operates.

A constant dynamic balance between disorder/chaos and order is evident all around us in weather patterns, fluid flow, the motion of the planets in the solar system, the spread of epidemics, changing populations of animals, birds, and insects, and even the motion of electrons in atoms. The transition into chaos is most evident in extreme weather events, epidemics, earthquakes, or avalanches, which remind us that nature is always teetering on the edge of chaos. Gregor Schöner and others recognize that thoughts and other stable features are constituted by the synchronized neural patterns that emerge as ordered from our brains' nonlinear electric and neurochemical chaos. Schöner argues that characterizing non-linear systems should be "sufficient for understanding the behavioural or neural system and how cognition is embodied by systems whose structure reflects functional adaptation to an environment and a specific developmental and evolutionary history."[149] Scientists have recently established evidence that our brains, as natural organisms, exist on the edge of chaos and that this is essential to our survival. It explains why, despite their non linearity and complexity, our brains exhibit order yet constantly adapt to our environment and that random thoughts arise because of the unpredictable nature of the chaos in our brain and represent random unsynchronised neuronal activity that is unpredictable. John Beggs, whose research interests cover biophysics, computational neuroscience, statistical physics, and neural networks, while investigating the spontaneous firing of neurons in rat brains found that each random firing of a single neuron triggers the firing of only one another neuron and so on and that at any moment in time, the smallest perturbation will cause an avalanche of random neural firing.[150] Our brain appears to operate like a tightrope walker, functioning precariously on the edge of chaos yet achieving stability by constantly adjusting and adapting to new information in its environment. They achieve this by utilizing random inputs drawn from the chaos of randomly firing neurons. Like the motor movements of our bodies on the tightrope, these adjustments and random inputs keep our brains on that tightrope or close to the critical point.

To prove that our brains are chaotic, neuroscientists Henri Korn and Phillipe Faure examined and reviewed all the experimental data and computer simulations using formal brain chaos models.[151] They acknowledged that their review relied largely on the studies utilizing computer simulations because of the difficulty in conducting experimental research on the non-linear properties of large cortical networks and higher brain functions in real brains. Nevertheless, they claim to have confirmed clear evidence of the "existence of chaos at all levels from the simplest to the most complex forms of organization of the nervous system," which they called "Neuro-Chaos." Given the neuro-chaos of our brains, they advocate that new models are needed to describe how the external world is represented in our brains. They have challenged the traditional computational theory of mind (CTM) first proposed in 1943 by the American neurophysiologist, Warren Sturgis McCulloch (1898–1969) and computational scientist Walter Harry Pitts (1923–1969). McCulloch and Pitts developed the first artificial neuron using the Threshold Logic Unit (TLU), or Linear Threshold Unit(LTU), which was a computational model of what they described as the "Nerve Net" in the brain. CTM's assertion that neural activity in our brains is computational has led to a widely held perception that the human mind is like an information processing system that is deterministic and that cognition and consciousness are a form of computation. Korn, Faure, and many other scientists increasingly argue that because our brains are non-linear, dynamic and chaotic systems, computational comparisons can only adequately reflect how our brains work if they contain the random stochastic noise that chaos involves.

Random events are not subject to the laws governing determinism and modelling the brain accurately requires incorporating the random electrical firing of neurons called random or stochastic noise, which causes those unpredictable random thoughts to pop into our heads. Schöner suggests that only by understanding the characteristic features of non-linear dynamic systems containing such random noise can we understand how stability emerges from the chaos of our brains and how our brains operate. However, we should acknowledge neuroscientist Paul Miller's observation in his review of this topic, that "Despite the tremendous quantity of electrophysiological data, room for debate remains as to which type of dynamical system is best used to model and understand brain function," the challenge is that the "relevant continuous variables necessary for a full description of the behaviour of a functioning neural circuit are typically hidden from us," and issues the following warning about our extent of knowledge, which we should heed. Namely that "Any one of these types of a dynamical system (or any other not mentioned) could provide the most accurate basis for understanding a particular neurological function, another may be compatible with the observed data and more useful for explaining features of the neural behaviour, while others may at heart be incompatible with the known circuit properties. Therefore, it behoves us all to avoid entrenchment in our favourite paradigm and to improve our understanding of the variety of dynamical systems when attempting to understand neural circuit function."[152] Miller's note of caution should remind us that when it comes to our brain and nature, we must never think we have all the answers or explanations and that AI will always run into the limitations of our knowledge about the human brain. A feature ubiquitously exhibited by all non-linear dynamic systems in nature, including our brains, is that of being dynamically attracted toward Self-Organized Critical States (SOCS). Let's see what this means in the context of how our brains work and, later, the implications for AI systems and robots and how, for example, our brains are able to understand the world.

Self-Organized Critical States. (SOCS)

Regarded as one of the most important contributions to science, the principle of Self-Organized Critical States (SOCS) was first published in a 1987 seminal paper, co-authored by Danish theoretical physicist Per Bak (1947–2002), Tang Chao (1958–) and Kurt Weisenfeld (1958–). Bak worked with the Niels Bohr Institute in Copenhagen and was internationally renowned for his contributions to the physics of complex systems and the application of the principle of SOCS to a wide range of naturally occurring phenomena. These encompass such diverse events as avalanches, prices on the commodities market, traffic jams, biological evolution, the distribution of galaxies in the universe, and the brain etc and as Bak explained are critical states that apply to every branch of science.[153] A classic and simple but elegant example of SOCS cited by Janine Hesse and Thilo Gross, first visualized by Bak[154] and experimentally replicated by Vidar Frette and colleagues in 1996[155] is the dynamics of sand piles. When the number of grains added to the sand pile equals that lost (on average) the sand pile has self-organised into a critical state and adding one more grain may or may not cause an avalanche, but one grain eventually will. They exhibit a feature better known by the popular idiom of the straw that broke the camel's back or the domino effect. In other words, the sand pile though organised and ordered is always perched precariously on the edge of chaos and the potential for avalanches triggered by one grain of sand being added. A considerable body of experimental research over the last twenty years has shown that the human brain "with its diversity in cell types, interconnectivity, and neural plasticity exhibits self-organized critical states in the form of biological neural networks."[156]

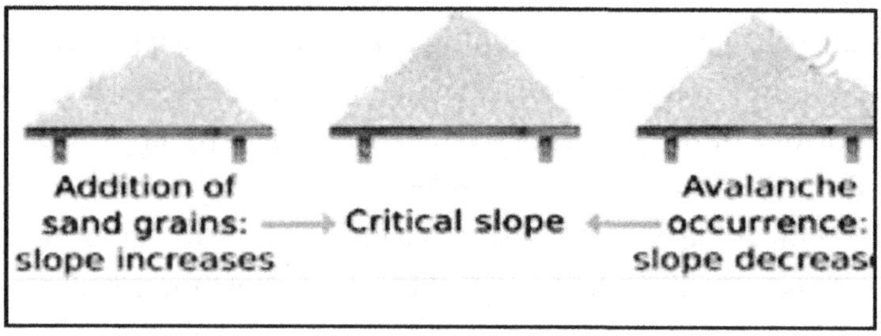

The Sandpile Model.[157]

Criticality denotes the transition between ordered and disordered states, and it has been argued that critical states enable optimal processing and properties in biological neural networks that "include information transmission, information storage, dynamic range, metastable states, and computational power."[158] Given our understanding of the non-linear dynamic and chaotic nature of our brains, let's look at biological neural networks (BNNs) and the exquisite balance between stability and disorder.

Biological Neural Networks (BNNs)

All networks are "Groups formed from parts that are connected"[159] and the type of network typically determines how resilient that network or system might be to disruption or attack or how flexible and adaptive it is. Stable states in a network or system are conventionally represented by nodes or in the brain by neurons or groups of connected neurons called hubs. Interactions or connections between nodes or hubs are represented by edges, which in the brain are the axons that transmit information between neurons or hubs which represent transitions between brain states. BNNs consist of neural networks formed by the synchronized oscillations of these groups of neurons and hubs that we discussed earlier, which represent the functional connectivity within and between various brain regions. They are another remarkable biological feature of the way our brains organize and communicate information within the brain and are called scale-free networks, which refers to the fact that they possess the virtue of being able to rapidly propagate information between the nodes. How do we know that our brains use scale-free networks? Scale-free networks follow a power law of distribution, which is a universal organizing principle underlying the formation of diverse networks in nature including our brains.

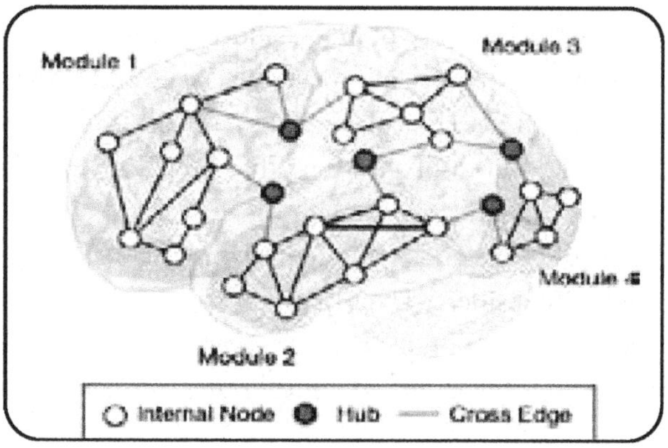

Networks can be extremely complex and exhibit highly complicated patterns, although you can sometimes identify the type of network that it is by looking at it or by using what mathematicians call the degree of distribution. This is a method of understanding the characteristics of a network by focusing on explaining certain elements more simply. This involves assessing the number of connections of a node which is called the degree of the node and counting how many degrees a node has makes it possible to calculate the degree of distribution of nodes. In real-world networks, most nodes have a relatively small degree of distribution, but a few will follow the power law and have a very large degree of distribution so are highly connected. In the power law of distribution, a small number of occurrences are clustered at the top of a distribution (or at the bottom) and take up almost all the resources, with small numbers of occurrences more common, and larger occurrences rarer. There are many examples in life, for example in the distribution of income with very few billionaires and the bulk of the population having little capital, the magnitude of earthquakes, the size of cities according to population, and the size of corporations with few being large and most small, trading volumes on the stock market and word frequencies. BNNs in the human brain are scale-free and follow the power law because a small percentage of neurons cluster together in hubs that are highly connected to many other neurons and each other and a much higher percentage of neurons that are not in hubs with far fewer connections. They are called small world networks because they are arranged in hubs that have the most connections and are connected so have a very large degree of distribution. This hub arrangement makes it faster and easier for information to move around the network with more intense cycles than other networks.

Network analysis of connectivity in the human brain as a dynamic system has identified that neural hubs are vital for efficient neuronal signalling, communication, and integration of information between connected regions, which is the basis of most cognitive functions and necessary for learning, thinking, reasoning, remembering, problem-solving, decision making, and attention. Dr John Beggs who explores how neural networks as dynamic systems exhibit metastable states, has with others been experimenting with new technology that allows thousands to millions of interconnected neurons to be grown on what are called micro-fabricated arrays, which can be kept alive for weeks. Beggs describes how they can measure the spontaneous electrical activity of these brains in a dish and have observed that they spontaneously organize into metastable states. He maintains that the micro-fabricated arrays simulations establish that "Living neural networks do organize themselves so that many metastable states exist. In addition, these networks appear to operate at the critical point, producing distributions of event sizes that can be described by a power law." He observes that this "indicates that metastable states can be used to store information and that the critical point involves optimizing information transmission while preserving network stability."[160]

If we were to remove the neural hubs the number of connections between all the neurons would start to become more equal so it would not follow the power law but become a random network because it could be created randomly simply by connecting neurons without having to prioritize the hubs. Such a random configuration would overwhelm the brain with the random transmission of huge volumes of information. We saw this earlier when in epilepsy electrical activity overwhelms the brain, and all order and stability disappear in such a way that the brain transitions into random chaos or 'primitive state of consciousness." Being a small-scale network that follows the power law of distribution vitally also means that our three-pound brains, with their enormous capacity for complexity, communication, and integration of information can fit inside our relatively small skulls.

Our brain's ability to achieve critical states through synchronized networks involves the converging of information to invariant states. This means that the state remains unchanged after operations or transformations of a certain type are applied to it and is thus stable. In chaotic systems, a set of invariant states toward which a system tends to evolve for a wide variety of starting conditions of the systems are called attractors. The attractors in our brains to which the system has evolved are the distinct networks of neurons with excitatory interconnections that can settle into a stable and synchronized pattern of firing found in local regional hubs in the cortex. They do so through a process known as phase locking which means that information or signalling in the form of input volleys arrive at the peaks of neural excitability of the receiving neural hub to communicate effectively.[161] Attractors provide essential stability for the processing of information and enable order to emerge from the chaos of our brains.

Computational neuroscientist Professor Edmund T Rolls has observed how "Attractor networks in the cerebral cortex and separate states permit high memory capacity across the whole neocortex and are important for long-term memory, short-term memory, attention, and decision making" because of their stability. Rolls maintains that "The random firing of neurons can influence the stability of these (attractor) networks by introducing stochastic noise and that these effects are involved in probabilistic decision making but also implicated in some disorders of cortical function such as poor short-term memory and attention, schizophrenia, and obsessive-compulsive disorder." He describes how "Each memory pattern stored in an attractor network by associative synaptic modification consists of a subset of the neurons firing. These patterns could correspond to memories, perceptual representations, or thoughts, and our thoughts are generated by the neural patterns that emerge from the stability of attractor states."[162] Synchronized neural activity in cell assemblies is modulated by sensory input with neural activity rapidly converging towards transient stability corresponding to the input pattern of the information that the brain is analysing or processing. Synchronization in neural networks provides essential stability but means that our brains like the sand piles need to stay close to the critical state. It is known that motor action reflects a dynamic balance between reliance on this combination of stability and flexibility which happens because the system (our body) has multiple components that process the information coming through the senses (the input) and scientists now believe that cognition operates in the same way.

Cognition or thinking requires the brain to be flexible enough to incorporate new information into the system which means it must always be dynamically balanced between stability and flexibility. New information comes not from synchronized neural networks but from the random or spontaneous unsynchronised firings (stochastic noise) or instabilities called bifurcations or neuronal avalanches which represent a change in the qualitative behaviour of the network system. This randomness is ubiquitous in all neural systems and our brain's neuronal avalanches arise from the unpredictable and random transition of a neuron at rest to periodic spiking activity in superficial layers of the cortex. Beggs and Dietmar Plenz revealed that the equations that were being used to explain local field potentials which are the aggregate transient signalling of small populations of neurons that reflect the dynamic flow of information across neural networks, could be described by the same equations that govern neuronal avalanches.[163] They hypothesized that although they only last for a few milliseconds, these spontaneous avalanches represent the very rapid propagation of local synchrony with a "rich spatio-temporal diversity" which "recur over several hours" and are therefore "optimal for information transmission."[164, 165] They also concluded that the " long-term stability, diversity and temporal precision of these neuronal avalanches means that these spatiotemporal patterns may be involved in information storage and act as a substrate for memory."[166] The flexibility produced by the inherent instability of this unsynchronised stochastic noise means that synchronized neural networks can either incorporate or suppress input from other competing stimuli depending on the relevance or salience of the stimuli. It is believed to be the basis upon which change involving working memory and decision-making operate and how instability improves information processing in non-linear systems like the brain, although it is acknowledged that much more work is required to establish this for certain.[167] Wang and his team have shown that in the critical state, even very small perturbations or instabilities (like the grain of sand added to the sand pile) can result in chain reactions or neural avalanches which may result in changes to any elements within the system.[168]

The property of critical states and its relevance to the brain's capacity to process new information and incorporate it into the dynamic system is even being harnessed in clinical situations and exploited in AI and machine learning. Learning involves ordered information which, like synchronized networks, decreases the tendency towards chaos due to the reduction in the uncertainty of the information in the system, hence maintaining the critical state and delaying the level of disorder or entropy. Learning in the brain appears to involve integrating random information from the stochastic noise into our synchronized adaptive (attractor) neural networks whilst staying near the critical state. Neural patterns are created within the various neural networks whilst brain waves (discussed below) represent distinct mental states and levels of alertness with us constantly switching between mental or brain wave states. Going from being relaxed and daydreaming to performing a task involves a transition between mental states that require cognitive resources or effort. Whether our brain is operating near the critical state is a measure of how hard our brains find it to transition between brain states and if they don't or can't transition there is usually a major problem. Using the sandpile example, the further away from the critical state the brain is, the greater the required energy resources required to affect the transition so you would need more energy. A team from Cambridge University demonstrated experimentally not just that our brains exist on the edge of chaos, but as co-author Manfred Kitzbichler observed, this enables our brains to "respond very rapidly and extensively to minor changes in their inputs" and "allows us to switch quickly between mental states to respond to changing environmental conditions." They used brain imaging technology to measure dynamic changes in the synchronization of activity between different regions of functional networks. Kitzbichler believes that future research should seek to establish how "measures of critical dynamics relate to cognitive performance or neuropsychiatric disorders and their treatments."[169] Critical states describe a dynamic system like our brain at points of transition or change and research has demonstrated that operating close to the critical state is very important for the healthy functioning of the brain and the ability to use stochastic noise is vital to dynamic learning and adaption, whilst also explaining those pesky random thoughts.

Clinicians use a constant called the Hurst constant (H) as a proxy for how close a brain is to a critical state which reflects how optimal the neural network is in terms of the energy employed in neural activity. A constant in maths is used to represent a function whose output is the same regardless of the input value and the H constant exploits a feature of neuroimaging data such as fMRI and EEG, which is that the information they extract contains fractal patterns which exhibit scale invariance. Scale invariance simply means that the qualities of something don't change because the scale, energy, or other variables change and therefore they represent universal properties of that system. An example of this that we see throughout nature is in fractals which are infinitely complex patterns that are self-similar across different scales. Shapes don't have to be the same but must display inherent and repeating similarities. Fractals is a word coined by the American French-Polish mathematician and polymath Benoit Mandelbrot (1924–2010) from the Latin word fractus which means broken or fractured. Common examples in nature include branches of trees, animal circulatory systems, lightning and electricity, plants and leaves, geographic terrain and river systems, clouds, and crystals. Two beautiful examples commonly seen are ferns or snowflakes which are made up of the same general structure repeated over and over.

For brains with states near the critical state, they appear to have smooth temporal brain waves and are therefore near an H of 1, which means they are exhibiting scale-free invariance as evidenced by the scale invariance of the fractal components of the brain wave images. You would expect to see an H near 1 in a brain state near the critical state. Brains further from the critical state have an H nearer 0.5 and exhibit brain wave fluctuations that look like random noise in the brain and do not exhibit the same level of scale invariance in the images reflected in the H constant score.[170] An important feature of the H measure is that brains that are closer to the critical state, find it easier to transition to task-relevant states which require cognitive effort /energy, and this may even enhance the performance of the task. Clinicians regard a lower H constant to be associated with multiple mental disorders in children and adults such as learning in particular working memory, experiencing task difficulty, and in the ageing brain, and has been identified as a useful potential measure of certain psychopathologies including depression and ADHD, with the implication that these individuals are expending greater cognitive effort in switching brain states and impulsivity and poor performance appears to correlate with lower H levels. This has been attributed to an imbalance in resource allocation between brain regions involved in performing tasks and is known to be a clinical feature of depression. Lower H levels have even been shown to correlate with those who don't benefit from practising a task when compared to those who do.

Xiao Wang and Henry Kennedy in a short review in 2016 focused on the relationship between form (structure) and function (dynamics) in the brain have observed that structural connectivity within the brain is "essential for understanding the constraints of brain function, (but) alone is insufficient to predict the dynamical behaviour of neural circuits."[171] They challenged the widely accepted hypothesis that phased synchronization is the mechanism for coordinating interactions between brain areas and gating information transmission according to cognitive demands and necessary for behaviour and cognition. Using the example of two identical neurons interconnected by mutual inhibition they argue that dynamical behaviour often cannot be straightforwardly deduced from anatomy, pointing to how experimentation and theory have uncovered multiple possibilities when trying to predict the network's behaviour from this structural connectivity. These multiple possibilities include both neurons remaining inactive or silent, acting as a switch with only one neuron active at a time, or they could take turns to be active leading to a "half centre" oscillator, which is the core of rhythmic central pattern generators in which two (groups of) neurons generate alternating oscillatory activity by reciprocally inhibiting each other." Lastly, they could be perfectly synchronized in that they spike at the same time thus they mutually inhibit each other but spike again after a short period when the inhibition has decayed. They observe how local connections account for about 80% of the total cortical brain connections so they "constitute a large part of the cortical machine" and these local connections "exhibit non-random configurations with a wide range of synaptic strengths." They describe how they found that "strong reciprocal connections are mostly found between pairs of neurons with highly correlated responses whereas neurons with uncorrelated response are infrequently and weakly connected." They hypothesized therefore that "local circuit neurons may form sub-networks that could provide an important infrastructure supporting global and local integration." They further observed that the strongly interconnected neighbouring neurons in the supra-granular layers of the cortex constitute cell assemblies which in a sensory area encode not just for sensory input but integrate contextual information such as reward, attention, and expectation.

Supra-granular layers as a cytoarchitectural term describe two layers of the cerebral cortex, which are the external granular layer (II) and the external pyramidal layer (III), that are categorized as superficial to the internal granular layer (IV). An earlier study involving Francis Song as well as Wang and Kennedy had provided substantial amounts of data on areal connectivity underlying distributed processing in the cortex but that the "fundamental principles that govern the large-scale organization of cortex remain unknown." Areal connectivity simply means connectivity between areas or regions of the brain. They concluded that "functional similarity between areas as defined by the pattern of shared (neural) inputs or (neural) outputs is a key to understanding the areal network of the cortex."[172] They also observed that spatially embedded random networks are not accounted for in existing models and believe that spatially embedded random networks give rise to "a systematic relation in… the cortex between the occurrence of connections from one area to another and their similarity distance rooted in the wiring distance dependence of connections in the brain." They developed what they call their "weighted, spatially embedded random network model" which they claim allows connections to "emerge from an underlying distribution of spatially embedded axons thereby integrating the two scales of cortical connectivity (which are) individual axons and interareal (white matter) pathways into a common geometric framework". According to Song, Kennedy, and Wang, this model has the potential to be a generative model that can explain how areal connectivity in the brain arises and better explains why "unlike the large and sparsely connected architecture of the neuronal network of the brain, cortical areal networks are relatively small and densely connected." They claim that the properties that define conventional complex large and sparsely connected networks such as the degree distribution (number of areas connected to an area), the average path length (smallest number of connected steps between a pair of areas), and clustering (density of connections among areas connected to the same area) are not very informative when it comes to explain these small and densely connected cortical areas. This has potentially important implications for the development of AI and robotics.

Hidden Latent Brain States

One recent study observed how "dynamic transitions between hidden latent brain states allows flexible reconfiguration of functional brain circuits" and that "failure to engage these states promptly, is associated with poorer task performance and weaker decision-making dynamics." The authors of the study argue that a "central unaddressed challenge is that human cognition is influenced not only by (known) external task demands but also by (unknown) latent mental processes that change with time."[173] Their work focused on dynamic brain states and circuits associated with fronto-parietal cortical regions involved in working memory which is key to cognition and if correct potentially has significant implications for AI systems. They observe that "changes in brain states and connectivity can be induced by both external stimuli and latent factors such as motivation, alertness, fatigue, and momentary lapse in attention, which can dramatically impact behaviour." The question is how this would impact machine and deep learning if robots were to mimic or replicate human brain processes. They used the Bayesian Switching Dynamical Systems (BSDS) which is an unsupervised Bayesian learning algorithm that estimates hidden latent brain states and dynamic brain connectivity[174] and concluded that these "hidden brain states are only weakly aligned with external task conditions."[175] They applied these findings to use the BSDS to implement an unsupervised learning algorithm to detect hidden and latent brain states from fMRI data and examine the dynamic interconnections between brain regions associated with salience, central executive, and default mode networks which plays a critical role in working memory and cognition. Although this AI model is not being used in a robot to detect brain activity from brain imaging data it illustrates how the dynamic nature of the brain's interconnectivity plays a key role in cognition, performance and decision-making. Each latent state appears to be associated with a distinct pattern of inter-regional connectivity and the role of optimal brain states in working memory performance.

The brain is a highly dynamic and non-linear system that operates on the edge of chaos order and stability, which is what allows it to transmit and store information and formulate thoughts, whilst hidden latent brain states influence cognition and decision-making. We briefly discussed the importance of understanding our environment rather than the simple processing of information and the acquisition of knowledge so we need to look at how understanding rather than simply knowledge emerges from self-organising nervous systems like the brain. Given that coherence is required for us to be able to think, let's first look at what constitutes a thought.

What is a Thought?

At any moment in time, neural patterns represent the aggregate of the electrical and chemical signals of neurons acting in networks that contain information. How does the electrical signalling mediated and modulated by neurotransmitters that form these neural patterns create a coherent and transient thought, when neural patterns are highly dynamic? Psychology defines a thought as a representation of information through the association of characteristics that allows comparisons to be made between different information. Neural patterns represent the organization and association of sensory information to other sensory information received from the environment. The synchronized firing of neuron assemblies and associated neurotransmitter activity form distinct neural patterns that act like symbols that represent or stand for something else. For example, when a child meets its first dog, a neural pattern is formed which will be used to represent the dog, even if the child is too young to know it's a dog or has the language to identify a dog. Eventually, many patterns are formed to recognize many types of dogs and distinguish between them.

IBM physicist Rolf Landauer (1927–1999) described information "not as an abstract entity, but existing only through a physical representation, thus tying it to all the restrictions and possibilities of our real physical universe." Landauer observed that 'information is always tied to a physical representation. It is represented by engraving on a stone tablet, a spin, a charge of elementary particles such as electrons, a hole in a punched card, a mark on paper, or some other equivalent."[176] The information contained physically in each neuron aggregates with some of the information contained in the neural assemblies to constitute the physical substrate of the information contained in a neural pattern which can constitute a thought. Abstract thoughts are therefore merely representations of the sensory information which if triggered by external stimuli starts with the first physical perception of the stimuli from the environment which as this is processed becomes representations layered upon representations that correspond to our perception of our environment and sometimes materialize as conscious thoughts, although more often in subconscious thought processes of which we are unaware. We will see the parallels with deep learning models in AI which as computer scientist Yann le Cunn explains "are composed of multiple processing layers to learn representations of data with multiple levels of abstraction."[177]

Thoughts reflect the information contained in the neurons and neural patterns which correspond to the raw data and information that we have perceived and processed through our senses and may represent an emotion or learned action. Many patterns may be fleeting but if they elicit a much stronger response or are consistently repeated, they may be consolidated as long-term memories. For example, if you saw and heard something as a child that frightened you, like an aggressive dog, the visual and auditory sensory information would have been processed in the various brain regions to form a widely distributed network of neural activity. This includes neural patterns in the cluster of neurons we call the amygdala (see later) which are associated with emotions and if strong enough, cause changes in the expression of proteins in the neuron membranes which form more connections, thus strengthening the signalling between neurons.

Thoughts become encoded in your brain as long-term memories and if they represent unpleasant data are probably stored as an unpleasant memory and a representation of that aggressive dog's behaviour. Some neural pathways and patterns may be more fragile and cease to be used so the connections or pathways are lost over time. Neural patterns are stored around the cortex encoded in the neurons in the form of long-term memories which can be accessed and recalled later through activation of these patterns. These patterns are like fingerprints, unique to each of us because they reflect our highly personal and subjective experiences and prior knowledge of the world. The conscious experience arising from the correlation of oscillations on the macro scale corresponds to specific mental states because the integrated information reflects our moods, depression, happiness anxiety, shyness, courage, risk-taking, dispositional traits, rumination, dreaming, cognitive biases, defence mechanisms, prejudices, values, etc. Recursive introspection or reflection on one's own subjective experience through time which allows us to perceive continuity in one's identity from the past to the present and the future, involves accessing the neural patterns that form our memories, without which we would not be able to reflect. Drugs and many neurological conditions including the various types of devastating dementias or traumatic brain injury, that disrupt these neural patterns, may disastrously impact the ability of the brain to process, integrate, or store new sensory information or activate existing neural patterns.

When measuring brain signals to relate neural activity to behaviour, the most common methods for recording electrical brain activity on a large scale use meso or macroscale field potentials. The technologies employed include electroencephalograms (EEGs) which involve small sensors being attached to the scalp to pick up the electrical signals produced by the brain, magnetoencephalography which records magnetic fields measured outside the head, produced by electrical activity within the brain, and electrocorticography (ECoG) which involves placing electrodes in direct contact with the cerebral cortex or surface of the brain. Local field potentials and local field potentials (LFPs) unlike the ECoG involve inserting a small size metal or glass electrodes or silicon probe into neural tissue within the brain. Hence LFPs are known as micro-EEG, depth-EEG, or intra-cranial EEG and because they contain action potentials and membrane potential-derived fluctuations in a small neuronal volume they yield the most informative signal for studying where cortical signals are generated which is known as electrogenesis. Using micro-EEG, the LFPs amount to a summation signal of excitatory and inhibitory dendritic potentials from many neurons in the neighbourhood of the recording site. This contrasts with the spiking activity from a single neuron which is usually measured at the tip of whatever type of electrode is used to measure the electrical signals. Micro-EEG allows you to measure local electrical activity in specific brain regions.[178] Bijan Pesaran and colleagues explain in their review of this topic that the interpretation of the data yielded typically involves using "the forward model which describes how field potentials are generated by the activity of populations of neurons or the inverse model which describes how to infer the activity of populations of neurons from field potential recordings."[179] The technologies in this field are being developed and rapidly extending the potential for research and the range of clinical applications and even beginning to allow us to study thoughts as they arise in the brain.[180]

Although difficult to define, we experience thoughts all the time which are mental activities involving reasoning, remembering, imagining, and problem-solving. Psychologists regard thoughts as the brain's way of processing, organizing, and interpreting information and employing past experiences to predict or plan. When accompanied by language we use thoughts to communicate with ourselves, through an ongoing internal dialogue, and with others to describe feelings and create marvelous art and music, etc. From a neurological basis, thoughts are generated through the complex interaction of many neural networks, involved in planning decision-making and memory and inevitably entwined with emotions. Given this complexity, it has been challenging for neuroscientists to agree not just on what a simple thought constitutes but on how and where thoughts are generated in our brains. If thoughts are the abstract mental activity containing stored information in our brains and are generated by external sensory stimuli or internally, why do some thoughts appear random, appearing unexpectedly out of nowhere and for no obvious reason? Thoughts require and involve the passing of information in our brain between neurons and neuronal networks but efforts to measure this information transmission have to some extent, been hindered by the limits on EEG and brain imaging technology, which for different reasons provide suboptimal spatial or temporal resolution of precisely what is happening in the brain at a given moment in time or whilst a thought is being generated and processed.

The problem with EEGs is that it is difficult to establish the firing patterns of the contributing individual neurons because the soft and hard tissues (skull) distort and attenuate the signals, although as Paul Nunez and Ramesh Srinivasan in their review of the electrical fields of the brain acknowledge, high-density EEG recordings, in combination with source-modeling that can account for the gyri and sulci (as inferred from structural MRI imaging) of the subject, have substantially improved the spatial resolution of EEG.[181] Studies using ECoG, have involved placing hundreds of tiny electrodes directly against the exposed cortex of patients with epilepsy. researchers from the University of California, Berkeley, have been able to monitor, track, and measure neural activity much more precisely when a person is engaged in simple or even more complex tasks. Using ECoG in healthy patients is unethical but not in those who suffer from epilepsy that impairs their quality of life and for whom this knowledge could be life-changing.ECoG provides much greater spatial detail than EEG when used for measuring electrical activity in the brain and far better resolution in time or temporal detail than fMRI and can discover epileptogenic zones which as we saw earlier are cortical regions that generate epileptic seizures. They discovered that the PreFrontal Cortex (PFC) is active almost throughout the processing of all sensory information and appears to orchestrate all the neural activity and is where all the information is integrated to generate a final output such as a thought. They observed that the more complex the tasks the more brain areas are involved, so it is not simply a case of neurons working harder but reflects the level of connectivity required. They observed how several brain regions are activated quite early in the process as though they are preparing the PFC and the team leader Dr Avgusta

Shestyuk tentatively suggested that this might explain why sometimes people say things before they think. This research builds upon the work of Patricia Shoer Goldman-Rakic (1937–2003), who pioneered research into the neural basis of working memory located in the PFC which it appears is crucial to being able to generate abstract thoughts. She was the first to identify spatially tuned, persistent neural signalling, which allows precise information to be held in mind, and observed how this involves recurrent excitation within glutamatergic pyramidal cell circuits in layer III of the cortex, whilst tuning and modulation the cell signalling performed by GABAergic lateral inhibition of that firing.

This provides another marvellous example of how our brain utilizes both excitation and inhibition of neural activity when processing sensory information and by slowly revealing the many ways that neural activity is implicated in our mental health and contributes to human behaviour, science is gradually bridging the gap between the mind and the brain. This means recognizing that we simply can't control everything, even the random thoughts that keep popping into our heads, although as we know not all thoughts are random or spontaneous but are triggered by certain events or memories and may even be generated without any external sensory stimulus. It raises fundamental questions as to what at the biological level constitutes understanding, what benefits understanding confers on the human CNS and what at the biological level is the difference between learning, knowledge, and understanding, all of which are highly relevant to AI systems and robots. Let's start with the neurobiology of understanding.

The Neurobiology of Understanding

A major challenge for neuroscience is posed by the multi-scale nature of the way the brain is organized, with "networks existing at the microscale level of cells and synapses, the mesoscale level of cell populations and axons, and the macroscale level of brain regions and white fibre tracts."[182] Scientists believe that cognition results from interactions among functionally specialised but widely distributed brain regions which means that trying to explain how understanding emerges in the brain necessitates consideration of Structural, Functional, and Effective connectivity. Effective connectivity is the union of functional and structural connectivity. Neuroscientist Barry Horwitz when explaining the elusive concept of connectivity in the brain observes that "Neurons and neural populations do not function as islands onto themselves. Rather, they interact with other such elements through their afferent and efferent connections in an orchestrated manner, to enable different sensorimotor and cognitive tasks to be performed."[183] Eshel Ben-Jacob describes how, albeit concerning biotic self-organization at the bacterial level, self-organized biological systems are imbued with "internal degrees of freedom, informatic capabilities, and freedom to respond by altering itself and others via emission of signals in a self–regulated manner."[184] When scientists talk about internal degrees of freedom, they are referring to statistics as any of the statistically independent values of a sample that are used to determine a property of the sample, such as the mean or variance. In physics, it refers to any of the minimum number of statistically independent values of a sample that are used to determine a property of the sample, or the state of a physical system measured as the mean or variance.

British neuroscientist Karl Friston and physicist and cognitive scientist Yan Yufik who is an internationally renowned authority on brain imaging, sought in a 2016 paper to frame a theory of understanding (TOU) at the level of biotic self-organising systems. The premise at the heart of TOU is that understanding in humans as they describe it is "the desire to resolve uncertainty which is not fully discharged until a complete understanding is attained." Whilst this accords with many philosophical perspectives including even Aristotle's views expressed in his Metaphysics as early as 350 BC, it fails to explain the science or physics of the mind behind understanding. Uncertainty is measured by the level of entropy, but to give it greater granularity Friston observes that from the scientific perspective, the various cognitive models of understanding in neuroscience involve "(1) Active inference or volitional sampling and sensing of the world (2) Structural analogues of the world and representational models that allow simulation of processes and events in the world (3) Satisfaction of the systems needs and optimization of resources, unification (4) Explanations predicated on partitioning heterogeneous into homogeneous groups or subclasses and (5) Prediction."[185]

Friston points out that the purpose of understanding is the "resolution of surprise and uncertainty about the causal structure and relationships in the environment and in particular the relationship of self to the environment (and others)." He describes how the success of phenotypes depends on a system successfully anticipating and avoiding surprises in the environment which requires a deep understanding or representational neural modelling of the environment and describe this as a generalized form of homoeostasis (see later). With this in mind they define understanding in TOU as "denoting the ability to compose and manipulate mental models representing persistent stimuli groupings, or objects, their behaviour under varying conditions, and different forms of behaviour coordination (i.e., relations between objects)." Understanding as a process "overcomes the inertia of prior learning and enables the construction of adequate responses under novel and unfamiliar circumstances." TOU purports to provide a neurobiological basis of understanding through analysis of the role of brain connectivity at the micro-scale, mesoscale, and macroscale.

Functional connectivity is related to the statistical relationship between specific physiological signals in time (i.e., the electrical signalling between neurons). This statistical relationship is reflected in the oscillatory neuronal activity because it is dynamic and thus provides a mechanism for dynamic network coordination. Functional connectivity is based on the notion that synchronization between two brain structures or regions (either at rest or during the performance of a specific task) constitutes communication between these regions. Functional and structural connectivity are not necessarily correlated and there may be functional connectivity without structural or anatomical connectivity. Yufik and Friston hypothesized that the science of human understanding starts with Markov blankets which "operate in the nervous system and underlie the development of intelligence, enabling a progression from the ability to sense the environment to the ability to understand it."[186]

Understanding starts with input into the senses from the environment and the internal process of modelling that input. Markov blankets were first conceived of by the Israeli American computer scientist Judea Pearl (1936–) who has been described as one of the giants in the field of AI. She named them after the Russian mathematician Andrey Markov (1856–1922) who became famous for his work on stochastic or random processes,the Markov model, Markov chains, and Markov processes that are widely used in AI to model randomly changing systems. The basic Markov chain used in machine learning is a stochastic model which means it has a random probability distribution that makes a probabilistic assessment of a random sequence of possible events or the series of transitions between states, which is known as the Markov property in which the probability depends only on the current state attained, not the past states or how they arose from their predecessors. It is memoryless and does not need to remember past states and future states, depending only on current states which can be used mathematically to predict the probability of the next event or future state occurring. A good common example is the ability to predict the next word in a sentence or a game of snakes and ladders. Markov chains are commonly used in sequential data because it simplifies predictions involving complex transitions. On a graph, this involves a random walk along the chain, with after a fixed time the probability function calculates the distance from the point from the origin and is excellent in predicting the most likely future events or states that a system will produce.

Markov decision processes are used for modelling decision-making problems where the outcomes are partly random and partly controllable and can solve most reinforcement learning problems by following three steps which are (1) Establishing the state space with the data needed to make decisions, determine rewards and guide transitions (2) An action state with a set containing all the feasible actions that can be taken and (3) A reward function that is the reward when action (a) is taken in the space (s). Hence the agent or system moves to a new state driven by the reward. The brain's neural networks comprise SOCS in the form of sets of neurons connected by certain mechanisms or processes and methods of communicating and these SOCS constitute Markov blankets. They do so because at the micro-scale, mesoscale, and macroscale they are "bounded subsets (of the overall system/ the brain) that maintain a degree of statistical independence from the rest of the network" and their internal states are separated from the surrounding states.

As Friston observes, they represent "flexible but stable biophysical structures capable of withstanding entropic erosion." This is explained by the associative networks undergoing self-partitioning (folding) into neural packets in which internal nodes (neurons) are connected more strongly to each other than to external nodes. This process of self-partitioning into packets produces what Friston describes as a "coarse representation of statistical regularities in the environment." Friston argues that mental modelling by the brain involves "self-directed (deliberative, attentive) composition of these neural packets into groups (mental models) such that mutual constraints in the packets' responses can be explored in search of the best fit between implicit models of stimuli and attaining a good enough fit (which) underlies the experience of reaching, grasping or understanding." By doing this they reduce the uncertainty or entropy with cognition predicated on the active censoring of these associative neuronal groups, assemblies, or bounded structures (neuronal packets). Because Markov blankets maintain a degree of statistical independence from each other they "capture regularities in the Sensorium which are the parts of the brain or mind concerned with the reception and interpretation of sensory stimuli which gives rise to the perception of objects namely, the external causes of sensations which when combined with the compositional mental (generative) models reflect the coordinated dynamics of objects in the world that cause sensory inputs."

Ines Hipólito believes that Markov blankets "highlight how open systems exchange matter, energy or information with their surroundings" and represent partitioning in "neural systems at multiple scales from "single neurons, brain regions, and brain-wide networks."[187] Partitioning separates the internal state from its surrounding states and at the highest level acts as a boundary between an agent like a human or potentially a robot and its environment. This is essential to maintaining the ordered structure of our bodies. Michael Kirchhoff and colleagues observe that a "collective of Markov blankets can self-assemble into a global system that itself has a Markov blanket; thereby illustrating how autonomous systems can be understood as having layers of nested and self-sustaining boundaries."[188] Markov blankets give SOCS-like neural assemblies their stability enabling them to maintain their structure under changing conditions and allowing them to be involved in the perception of objects with groups of neurons conserved over many representations of objects over time.

When represented as a graph the condition of a bounded structure or SOCS is represented by the node or hub and a transition between states is the edge connecting the nodes, with all the groups of states bound by transitions forming a dynamic system. Some structures or states are more likely than others, but the total of all possible states is called the State Space, with the complexity increasing with the scale of the state space. In probability theory, dependent events influence the probability of other events, or their probability of occurring is affected by other events, so they are conditional upon each other, but independent events do not affect one another and do not increase or decrease the probability of another event happening. Markov blankets rely on the concept of conditional dependency and independence and are "a set of states that separates two other sets in a statistical sense and implies that any state either is or is not coupled to another. This involves a partition into external states and internal states which are only conditionally dependent upon each other through the third states which are the Sensory or Active states. External states need to be inferred by the brain but can only be seen "vicariously by the internal states, through the Markov blanket". Inference is conditional upon the sensory states and changes in internal states can only take place via sensory states. The internal states can only influence external states via the active states which refers to the fact that the brain as a system intentionally constructs mental representations or models of the world. We influence external states through these mental models or active states. In simple terms, our internal states require our senses to be influenced by the stimuli from the environment, and to influence the environment we use active states to build the representational models.

Friston whose work has explored the variational free energy theory (VFET) of the brain has light-heartedly observed that "Life is an inevitable emergent property of any (ergodic) random dynamical system that possesses a Markov blanket. Don't leave without it!" Ergodic in this context simply means "that the time average of any measurable function of the system converges (almost surely) over a sufficient amount of time" which allows you to infer that the average amount of time a state exists as the probability of the system being in that state when observed at random. Friston argues that "biological self-organization is not as remarkable as one might think and is (almost) inevitable, given local interactions between the states of coupled dynamical systems". He maintains that "the events that 'take place within the spatial boundary of a living organism' may arise from the very existence of a boundary or blanket, which itself is inevitable in a physically lawful world." He claims that this arises as a natural feature of organisms maintaining their sensory states within physiological bounds and that the free energy theory explains how the brain reduces uncertainty by using internal models or representations to make predictions and updates these models using sensory input.[189] This means that perception is an inevitable consequence of active exchange with the environment and utilizes the free energy principle predicated on aligning brain functions with conservation laws and the laws governing neuronal energy.

Friston defines free energy as "an information theory measure that bounds or is greater than the surprise on sampling some data, given a generative model" with data from sensory inputs modelled by the brain which means the free energy can "be evaluated because it is a function of sensory data and brain states". Biological systems resist entropy and the tendency to disorder by minimizing the uncertainty or prediction error in their sensory states. Our brains don't like surprises or uncertainty which is a measure of salience based on the divergence of the information in the data that can be recognized and improbable information that represents a surprise in that data. Prior in this sense means the probability distribution or density of the causes of data that encode beliefs about those causes before observing the data and sensory data is sampled consistent with the current representation of that data using the Markov chain principle. Variational in VFET reflects how a system that minimizes the entropy in the system whilst maximizing the evidence for prediction does so by employing bounded rationality or evidence bounds. Bounded rationality in humans reflects the fact that we are usually confronted by imperfect information such as that coming from the raw data received through our senses and rather than seeking the optimal solution or prediction when deciding, our brains make a trade-off by accepting a good enough estimate and in that sense, the brain is also bounded by the resources available to it. It manifests not just in the neuronal processing of raw data but in the psychology of our behaviour which we see when we make decisions based on a false sense of rationality because we have insufficient information available to make an optimal or rational decision.

The concept of conditional dependence is a crucial element in the increased use of graph models in AI because they provide a visual representation of probability distributions. In neural networks or assemblies, each node or neuron has a probability distribution over its possible states which in mathematical terms means a map or function p that assigns a number (positive or zero) not necessarily between 0 and 1 to every possible value of X which enables us to estimate the probability that an event may occur or estimate the variability of its occurrence. Connected nodes or neurons are dependent on each other in that information about one node will change the probability distribution of the other node. The number of connections of a node is the degree of the node and by counting how many nodes have a degree we can calculate the degree of distribution of nodes. The degree of distribution in the brain is the probability distribution of these degrees over the entire network which gives us important insights into the type of network it may be. If nodes are not connected, they are independent although nodes can also be conditionally dependent such that even though they are not connected if another event happens, they become dependent upon and influence each other.

Neural assemblies at the various Micro, Meso and Macro levels are always updating their predictions based on information reflecting conditional dependency and independence and Markov blankets provide them with the stability and structure to allow them to do this in a constantly changing environment. At the macro level of the brain the conditional dependency of the internal state of the brain is mediated by the active state and the conditional dependency of the external state is mediated by the sensory state. Internal states model and engage in Bayesian inferencing, which reflects the growing awareness of the way humans and animals gather sensory data from the environment as an active sensorimotor process and make inferences about the (hidden or latent) causes of that sensory data.[190]

Charles Schroder and colleagues observe that most sensory processing is "largely determined by motor/attentional sampling routines" because we "actively locate and acquire most sensory input using a motor sampling routine" which is modulated by motor cortical oscillatory rhythms in particular delta, theta, and beta brain wave bands (see below). Although it is recognized that not all the senses rely on these motor rhythms and vary as to the extent, they do.[191] Internal states can only influence external states via active states, while external states can only influence internal states via the sensory states, but how does this conditional dependency which underpins Markov blankets translate into our brain's understanding of the world? Yufik and Friston define understanding as a form or component of intelligence "involving the self-directed (deliberative, attentive) composition of packets into groups (mental models) and manipulation of the ensuing "mental models" constituted by (these) neuronal packets, whose dynamics capture relationships among objects. The neuronal packet hypothesis associates understanding with the "ability to detect and generate coordinated rotation of population vectors in the neuronal packet in the associative cortex and other regions in the brain."[192] In neuroscience, a population vector is the sum of the stimulus orientation or preferred directions of neuron populations, weighted by the respective spike counts. This was first hypothesized in 1968 by Apostolos Georgopoulos when they detected how the very specific and controlled movements of a monkey's arm were predicted by the action of a population of motor cortical neurons. They observed how "when individual cells were represented as vectors that make weighted contributions along the axis of their preferred direction the resulting vector sum of all cell vectors (population vector) was in a direction congruent with the direction of movement."[193] The research on Markov blankets and understanding reflects the notion of effective connectivity which combines structural and functional connectivity to characterize functional integration which describes how brain regions work together to process information and determine responses. Klaas Enno Stephan and Karl Friston describe how "functional connectivity describes statistical dependencies between data, whereas effective connectivity rests on a mechanistic model of the causal effects that generated the data."[194] It reflects the influence that a neural assembly has over another one and establishes the statistical causation from their signals and a direction for their interaction.

Friston had many years before defined effective connectivity as the "causal influence that neural units exert over another" reflected in the degree distribution and the probability distribution.[195] Friston argues that factorization of probabilistic representations (which in physics is known as mean-field assumption under the mean field theory (MFT) provides an incredibly efficient process to infer states of the world (and respond adaptively) and believes that many aspects of the brain can therefore be understood in terms of MFT. In physics and probability theory, MFT facilitates the study of the behaviour of high-dimensional random (stochastic) models by using a simpler model that approximates the original by averaging over degrees of freedom. The effect of any neural unit on any other neural unit is approximated by a single averaged effect which replaces all the interactions to one neural unit with the average or effective interaction. It provides a way of inferring brain states based on the interactions of neural units using MFT as a heuristic by replacing the many interactions between neurons with one effective interaction, predicated on the assumption that if a field or particle exhibits many random interactions in the original system, they tend to cancel each other so the mean effective interaction is more accurate. MFT provides a crude approximation for understanding behaviour in interacting systems but can explain the principle of functional segregation, through the dynamic and context-sensitive maintenance of neuronal packets, groups, or cell assemblies. The assumption is that each weight for each neuron in the neural network is partitioned and independent of all the others.

The conclusion reached is that learning without understanding confines performance to situational envelopes narrowly constrained by past exposures. Understanding expands the envelope indefinitely, enabling counterfactual (what if) modelling, simulation of the future, and an implicit ability to anticipate the consequences of action. In statistical mechanics, a partition describes how n particles are distributed among k energy levels and a partition function is a mathematical way of counting the particles of a system when the particles are distributed over the available energy levels in accord with the Boltzmann distribution. Boltzmann distribution is a probability function that is used to describe the distribution of particles, such as atoms or molecules, and gives the probability that a system will be in a certain state as a function of that state's energy and temperature. Computing the partition function of a complex system is extremely difficult because it is attempting to identify how many micro-states are accessible to a system in a specific ensemble. We rely on raw data in the environment and neural processes to generate information and knowledge and TOU may provide a biological explanation for how humans understand this information as a process of searching for a best fit between implicit models of stimuli and attaining a good enough fit (which) underlies the experience of reaching, grasping, or understanding. We know that information is transmitted and integrated through a combination of electrical and chemical activity, but how do the electrical brain waves generated by this activity affect the way our brains operate and function and our behaviour and mental states?

Brain Waves

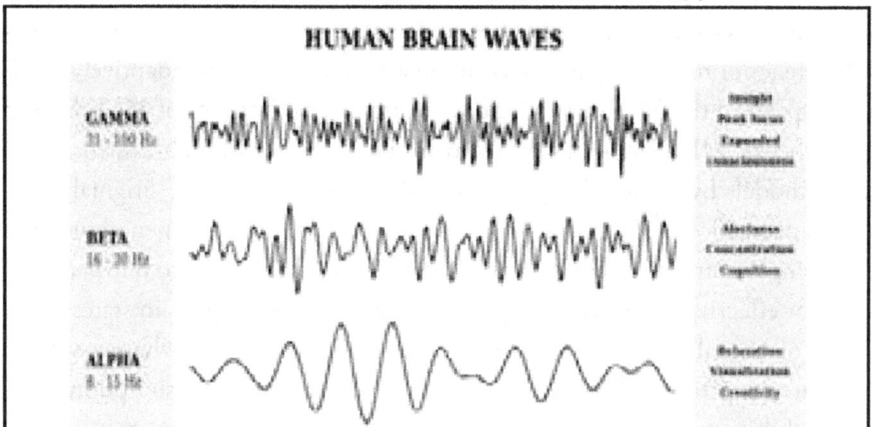

Brain waves are oscillating electrical voltages measuring just a few millionths of a volt. The most studied are Alpha, Beta, Delta, Gamma, and Theta waves, each possessing different frequencies, and oscillatory patterns reflecting different levels of awareness and mind states.The amplitude of a brain wave refers to the maximum disturbance from its undisturbed position and in the brain is measured in millimetres whilst wave frequency is the number of waves produced by a source each second or the number of waves that pass a certain point each second or cycles per second measured in Hertz (Hz). Scientists use an EEG machine, which involves the painless attaching of around 20–64 small sensors or metal discs called electrodes to the scalp, to pick up electrical signals received at the brain's surface. Invented in 1924 (to study telepathy, which he didn't find) by a German psychiatrist called Hans Berger (1873–1941), EEG gave us the first technology with which to study the fully functioning brain as opposed to relying on post-mortem samples. EEG has been thought to only measure the brain's spontaneous electrical activity. in the outer few millimetres of the cortex but not deeper in the brain. This is largely because the "source distributions are inherently four-dimensional (three in space and one in time) and few measurements are made outside the head"[196] making it difficult to identify the source particularly when it is sub-cortical neural activity. However, a study recognizing the importance of sub-cortical neuronal activity for mediating communication in large-scale brain networks measured the activity using EEG in conjunction with intracranial electrodes which are used in deep brain stimulation for clinical purposes (DBS).[197] 2019 They were able to record sub-cortical activity simultaneously with high-density (256 channel) scalp EEG and concluded that EEG can also sense sub-cortical signals which seems to confirm earlier findings.[198]

Although an EEG reveals all the brain waves present in our brain, depending on our state of consciousness, typically one brain wave will be dominant and reflect our state of mind at that moment. This dominance is essential when you need to sleep and theta or delta waves dominate, but problematic when we are trying to concentrate on a task and have slow waves like alpha or theta dominating rather than the beta waves associated with alertness and attention. In a deep restorative sleep, the EEG shows low-frequency, large-amplitude oscillations, but when awake and attentive, it shows primarily fast frequency, small amplitude rhythms. Transitioning between brain waves and brain states is essential to sleep, performing tasks, and managing stress, with our brain waves playing an important role in the optimal functioning of our brains. and problems with transitioning have been associated with many psychopathologies, like ADHD, depression, anxiety, etc. Brain waves are not only responsible for and crucial to our levels of awareness and mind states, but the major brain rhythms serve a largely inhibitory and constraining function on our brain. This was observed by Chris Timmerman who led the study on the potent effects of psychedelics and observed "What we have seen is that (the psychedelic compound) DMT(N-Dimethyltryptamine) breaks down the basic networks of the brain, causing them to become less distinct from each other. We see the major rhythms of the brain that serve a largely inhibitory, constraining function break down and in concert brain activity becomes more entropy or information-rich." Timmerman is alluding to the vital role that inhibition of neural activity plays in the healthy functioning of the brain and why psychedelics impact the self in such a dramatic way. By removing the constraint on information flow between brain regions, the brain becomes deluged with information and neural activity hence the overwhelming and often distressing or surreal nature of conditions like epilepsy. Although connectivity is vital it must be controlled, with the brain needing to maintain a constant balance between appropriate levels of connectivity and the overconnectivity produced by psychedelics and epilepsy.

Gamma waves

Gamma waves (30–80 Hz) are the fastest waves produced in short bursts, with the 40Hz level of oscillation associated with normal brain activity and important for learning, memory information, and cognitive processing. When asleep our brains cycle between two very distinct electrical and biochemical states, and we move through four cycles of sleep. Each lasts about 90 to 120 minutes with each cycle consisting of one period of rapid eye movement (REM) sleep and three periods of non-REM sleep. The discovery of REM sleep in 1953 by an American PhD student Eugene Aserinsky (1921–1998) established how during REM sleep our brains remain highly active, our eyes move rapidly behind closed eyelids, our heart rate gets faster, and breathing is irregular, compared to the steady slower breathing in non-REM sleep and when we dream. Gamma waves in REM sleep, reflect our brains becoming highly active and very similar to wakefulness which is why we can wake up more easily during REM sleep. Aserinsky became a pioneer in sleep research who fundamentally changed the way we think about sleep and dreaming.

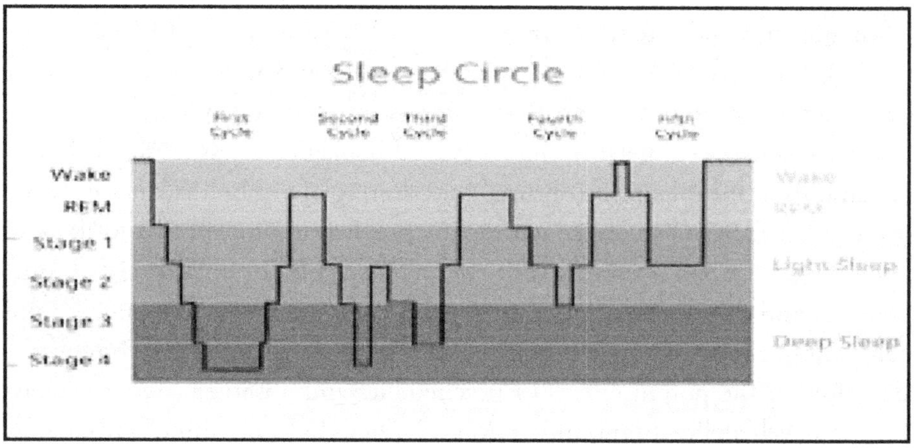

Non-REM sleep consists of two periods of light sleep, followed by heavy or deep sleep, slow delta waves, and then REM sleep again. With each new cycle, we spend more time in REM sleep with most REM sleep happening during the second half of the night. Researchers were so surprised by this when they first discovered it that they called it the active or paradoxical sleep phase.

Your brain waves become more variable, and you temporarily lose muscle tone, which scientists think is nature's way of protecting us against potentially acting out our dreams and hurting ourselves. Another common brain myth is that you only dream during REM sleep because we can dream at any time, but dreams during REM sleep tend to be more vivid. REM sleep involves brain development, the generation of new cells, which is why babies need to sleep so much and the processing of emotions which explains why our dreams may be more vivid. During REM sleep our brains are processing and consolidating new learning and motor skills that we experience during the day, storing some in memory, maintaining others, and deleting some.[199, 200] Babies and young children need most sleep because their brains and neural pathways are developing which requires energy consumption.[201] By the time we are adults we need about 2 hours of REM sleep a night, though this varies nightly reflecting our individual bodily needs. Studies on students appear to show though that those who nap score higher accuracy on tests, and the more REM sleep they experience, the higher their accuracy.[202] Being very fast, Gamma waves allow us to process information very quickly and focus attention which may explain why they increase during meditation. Too many Gamma waves are associated with anxiety, high arousal, and stress, and too little with ADHD, depression, and learning disabilities, and altered gamma levels are seen in Alzheimer's, schizophrenia, and epilepsy. Because of their speed, they are very important in linking distinct brain regions and are associated with binding the information from the different senses across brain areas.[203] This is known by scientists as the binding problem mentioned earlier which is how our brain combines objects, the background, and abstract or emotional features, particularly when there are several involved, into a unified and coherent single experience.

Two Princeton University researchers, Margaret Livingstone (1950–) and David Hubel (1926–2013) a founder of modern neuroscience, in a now famous paper in 1988, provided what has become a widely accepted model for how the brain achieves this binding concerning visual stimuli.[204] Studying monkeys, they observed that different visual features like colour, motion, orientation, and retinal disparity are analysed in anatomically separate processing streams. Retinal disparity is the result of the left and right fields of vision producing slightly different visual images when focusing on the same object. Different visual features are processed in parallel but in different parts of our brain, starting in different cells located near the inner surface (the ganglion cell layer) of the retina of the eye and processed in separate layers in the thalamus. The brain subconsciously dismantles the sensory information and needs to reassemble and reintegrate it in a way that makes the process seamless and provides a unified experience. This is crucial to attention and distinguishing between information that needs to be encoded and remembered although the mechanism by which information is integrated and bound remains a mystery. Neuroscientists have long known that damage to different brain areas can disrupt the perception of features like motion, colour, shape, size, or location without disrupting others because different features are represented by different neural activities in different brain areas. Psychologists use the illusory conjunction to illustrate the binding problem and how the brain normally generates a coherent impression of a whole experience. The illusory conjunction occurs when attention is spread over several stimuli and the characteristics of one stimulus such as colour or sound are mistakenly combined with another stimulus to form one object. With vision and touch you may attribute textures to something you are looking at to something that you are feeling. It typically happens when we see something quickly and is more likely the closer two objects are together. Research has shown though that the illusory conjunction theory may be a shadow.[205] This illustrates how critical attentional focus is in accurately representing our world and the vulnerability to misrepresenting information during integration in our brains, even with the features of objects. We will look at the role of attention later, but one explanation for how binding is achieved has been advanced by Professor of neuroscience Wolf Singer who observes that the dynamic neural activity of neurons that represent specific features are bound into functional neural assemblies which "represent the constellation of features defining a particular perceptual object."[206]

These are assembled into higher-order assemblies that constitute conscious awareness and perception of the experience and as we saw are implicated in our understanding of the information. The synchronized neural activity in cell assemblies is modulated by sensory input and the neural activity rapidly converges towards transient stability in that activity, which corresponds to the input pattern of the information being analysed or processed. It happens so rapidly and dynamically that it creates the sensation of stable processing.

Research over the last 40 years has revealed the remarkable organizational complexity of the brain and cortical areas are being subdivided into ever-smaller regions, each characterized by distinct response properties and connections to other brain structures. Representation of sensory information is typically allocated to distinct neuronal assemblies which are selective in being activated in response to a particular feature of an object and these features are summed to give the overall visual representation of an object. The problem arises from the fact that most realistic scenes comprise many objects and the brain must distinguish between them with fast Gamma waves appearing central to this.

Beta waves

Beta waves (12–30Hz) are fast and very common whilst awake and involved in conscious thinking, socializing writing, reading, problem-solving and focusing on tasks. Higher frequencies are associated with over-arousal and observed when you have stimulants like coffee, excess levels cause anxiety and stress, which is why too much coffee can be bad, whilst too little beta waves are associated with ADHD, rumination, depression, and impaired cognition.

Alpha waves

Alpha waves (8–12 Hz) predominate when we are awake and attentive and calm us down, acting as the bridge between the theta waves involved in sleep and our conscious and subconscious minds with people tending to fall asleep when they start generating large amounts of theta waves. If we become over-aroused, it can be a sign of too great a dominance of beta waves which can block out the alpha waves which causes difficulty relaxing, producing stress, anxiety, and insomnia, whilst too much alpha can make us too relaxed, daydream, and unable to focus. This can be induced by marijuana, relaxants, some antidepressants, and alcohol. A recent study led by Professor of Experimental Psychology, Zoe Kourtzi, albeit only involving 80 participants, exploited the fact that every person has their own unique natural brain rhythm generated by neural oscillations. Focusing on alpha waves every person has a distinct peak wave during a cycle of -125 milliseconds. The scientists used an optical pulse consisting of a white square pulsing on a dark background aligned to that person's alpha wave to get the brain waves working at the natural rhythm in a process known as entrainment. They were then given a difficult cognitive visual task to perform over 800 times with slight variations, with the optical pulse either given at the trough of the brain wave, the peak, or randomly. They found that those who received the pulse at the trough, which is when neurons are at the most excitable state, were most receptive to new information, learned three times quicker, and maintained that performance over time. This suggests that the information we obtain from our environment is almost like a snapshot but the neurons signal to each other and connect information making it appear seamless.[207]

Another study led by neuroscientist Victoria Long using a technique called hyperscanning discovered that babies when solving a puzzle together with their mothers, learn from their mothers when their brain waves are in sync and that the peaks and troughs of each person's oscillations consistently track that of the other person. This has been called a mega network exhibited by mothers who show a happy positive state of mind, maintain eye contact, and use infant-directed language like the higher-pitched, sing-song cadences people naturally use with babies.[208] This has enormous implications for communication in that when people express positive emotions it influences the neural connectivity between people in terms of the efficient flow and sharing of information. Long observed how when the neural connection between mothers and babies is strong, babies are more receptive and ready to learn from their mothers. Tuning into a person's brain waves appears to significantly improve that person's ability to share and learn new information with our emotions affecting how we share information. This appears to extend to relationships like couples, close friends, and siblings, where people are more highly attuned to and trust each other.

Brain wave synchronization is part of a phenomenon known as bio-behavioural synchrony and scientists believe it exists at four levels, the other three being (1) Synchronisation of behaviour in terms of eye gaze and speech (2) Physiologically in terms of heartbeat and breathing rates and (3) Endocrine secretions of hormones like Oxytocin which is produced in the hypothalamus and released during sexual activity and giving birth, hence it is known as the love hormone. Traditional theories about the mating process tend to have focused on fixed metrics such as physical attractiveness and supposed fertility of the potential partner, but as Zeevi and colleagues found when looking at speed dating, bio-behavioural synchrony is a much more dynamic process that correlates with the outcome of romantic dates and pair bonding.[209]

Theta waves

Theta waves (4–8Hz) are as we saw linked to sleep and daydreaming and can produce an unexpected, unpredictable, dreamlike state with very vivid images (hypnagogic images) accompanied by intense memories. They are involved during our experiencing raw emotions, and linked to creativity, emotional connection, and intuition and appear to predominate when the senses are processing internal information, and the individual is self-absorbed. They are known to be very important during learning and memory and are produced during the transition between wakefulness and sleep.

Delta waves.

Delta waves are the slowest waves (0–4Hz) typically dominating in babies and young children and present in unconscious bodily functions such as regulating heartbeat and digestion and important for the immune system, natural healing, and restorative/deep sleep. Too much Delta can cause major problems such as severe ADHD, learning difficulties, and cognitive deficits or as with some brain injuries difficulty maintaining conscious awareness, whilst too little can cause poor sleep and insomnia and difficulty revitalizing the brain or body.

A recent development in understanding how our brains communicate through electrical signalling and brain waves has emerged from studies that have identified that our brains may not communicate all information via synaptic transmissions and gap junctions or according to the Hodgkin–Huxley (HH) theory of neural spiking. This is based on observing how some neural spiking travels far too slowly to constitute conventional electrical signalling under the HH theory and Chen Qiu and colleagues claim that very low amplitude brain waves are being transmitted by a weak electrical field, which does not involve synaptic transmission but excites and activates neighbouring neurons which have been detected in mice. They observed how "field effects alone can indeed mediate propagation across layers of neurons." They believe that this "phenomenon could be important to explain the slow propagation of epileptic activity and other normal propagations at similar speeds."[210]

The Robot Brain

A Brief Tour of The Robot Brain.

We have discussed the various psychological theories of learning such as associative learning processes, reinforcement learning, social learning, and continual learning. We have looked at how the human brain functions, particularly electrical and chemical communication between neurons, neural networks, structural and functional connectivity between and within brain regions essential to the integration and processing of raw data and information, the role of brain waves in our mental states and how understanding arising from this relates to learning. We have talked about neural plasticity, the synchronized oscillatory behaviour of neurons and self-organization of networks, latent brain states, the theory of FEP, and Markov blankets in understanding etc but how are computer scientists and designers of cognitive architectures in AI systems and robots translating this complex yet vital knowledge into developing the AI robot brain?

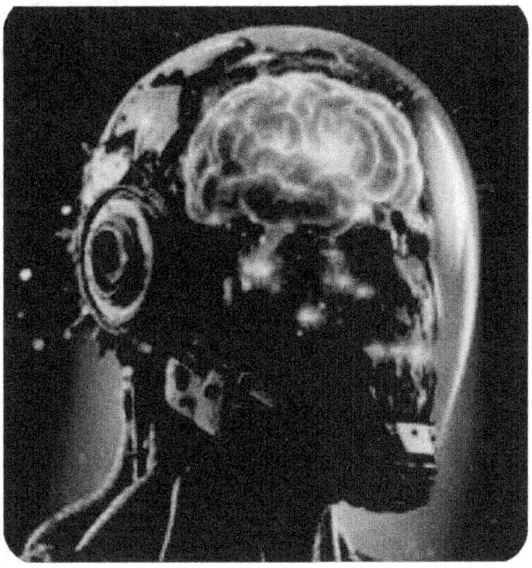

There are different schools of thought reflecting the extent to which computer scientists are seeking to emulate these features of the human brain in AI. Some are attempting to reverse engineer the specifics of neural circuits in the human brain whilst others like Dr. Demis Hassabis and a team of neuroscientists in their review of neuroscience-inspired AI, point out that "from a practical standpoint of building an AI system, we need not slavishly enforce adherence to biological plausibility. From an engineering perspective, what works is ultimately all that matters. For our purposes, biological plausibility is a guide, not a strict requirement. What we are interested in is a systems neuroscience level understanding of the brain, namely the algorithms, architectures, functions, and representations it utilizes."[211] Whether it is necessary or not to slavishly adhere to the blueprint offered by the human brain, neuroscience in its broadest terms including all the research fields associated with human behaviour and cognition including psychology continues to inspire and validate developments in AI and robotics. Biotechnical engineer and lifelong neuroscience researcher Dr Frederic Narcross in a 2021 study of Artificial Nervous Systems (ANS) describes how three AI methodologies stand out for their diversity and adherents and how they "increasingly incorporate biological inspiration for performance improvement" and claims that these initiatives could expedite the "common path toward biological authenticity in the field of AI and that this needs to be encapsulated in a common biological framework."[212] The three methodologies are (1) The various types of AI learning supported by the different ANNs (2) The trend to represent nervous system temporal connectivity pioneered by Jeff Hawkins in the Thousand Brain Project and (3) Neuromorphic computing. Although AI robots come in many different versions, they tend to share many common features which include (1) Sensors that gather information about the environment, (2) Effectors and actuators which are the parts of the robot, or the tools built in that are designed to perform the work and physical tasks and (3) Control systems which determine the behaviour of the robot acting like its brain. The combination of these features determines the capabilities of the robot, but the more autonomous the robot, the greater the complexity of the control system and the requirement for sensors that can extract and integrate real-time raw data available from the environment.

Robots including self-driving cars for example use many different types of AI/Machine Learning (ML) models that involve different types of networks, and the field is increasingly complex in terms of the specific processes employed by the different AI models. We have seen that human intelligence, learning, and understanding involve a highly complex but highly efficient biological process including stochastic (random) parallel distributed processing (PDP) of raw data to extract information and meaning. Replication of such processes and features in the early days of AI appears to have often been overlooked due to the emphasis on the output of the systems rather than the processes although PDP has been a goal of much AI research. A 2015 review by Hassabis and colleagues of neuroscience-inspired AI described how the PDP movement acknowledges that "human cognition and behaviour emerge from dynamic, distributed interactions within networks of simple neuron-like processing units, interactions tuned by learning procedures that adjust system parameters (see below) to minimize error or maximize reward." They observe how the conceptualization of PDP has "introduced a diverse collection of ideas that have had a sustained influence on AI research."[213] PDP has been a key factor in the advances made in AI and robotic systems and has resulted in the development of vital processing unit technology essential to most modern AI. The success or otherwise of all AI systems relies largely on the quality of the data training and the quantity and quality of data available to input into the system, without which AI can produce inaccurate and useless results and little information or knowledge. Most AI still requires humans to define the problem, identify and select the appropriate AI technology, train the technology with the appropriate quality and quantity of data, validate the results, monitor its performance, and ensure that errors are corrected. There are uncanny parallels with training humans, though not yet in terms of the biological functions that deliver the output and the level of autonomy in converting new data into useful information, applying creativity and new ideas employing abstract thinking that is so important to humans. Let's begin our exploration of what constitutes a brain in the world of AI and robotics, by looking at the workhorses of AI systems which are the Processing Units. These are defined as "The part of a computer that performs logical and arithmetical operations on the data as specified in the instructions."[214] The types of processing units used increasingly reflect the requirement for fast and high levels of computational power with the trend being away from simply programming of AI systems and sequential processing to an emphasis on PDP and generative learning.

Processing Units

The Central Processing Unit (CPU)

CPUs have huge instruction sets and manage all the computer inputs and outputs, but although they are fast, individual CPUs don't process data in parallel, which means that they can't compete with systems that do employ parallel processing units like the increasingly widely used Graphical Process Units (GPUs) discussed below. That said, although CPUs are designed to handle one task at a time, many systems now contain multiple CPUs with each containing 4–8 cores that are simply small CPUs built into a bigger CPU which significantly enhances the computer performance and enables multi-tasking. Each core performs like a very small version of a brain which receives instructions and performs calculations or operations, so the cores can work on different tasks at the same time. In recent years, companies like AMD and Intel have been regularly increasing the number of cores used in their systems. The control unit with its clock sends timing signals which precisely control the timing and synchronisation of systems operations and take input from the registers, which typically include the Status registers, which for example include the Flags and the Instruction register. Status registers test for various conditions in an operation such as what is the result of the operation e.g. is negative or positive. Instruction registers store the instructions which are currently being executed. Flags determine the outcome of the preceding ALU operation and the exact status of the processor which acknowledges and interrupts signals and control signals to and from the Control Bus (BUS) which is used by the CPU to communicate with other devices within the computer.

Control Units

Control Units usually have a software or hardware cache which temporarily stores frequently or often accessed information providing a fast albeit costlier memory to enhance the performance.

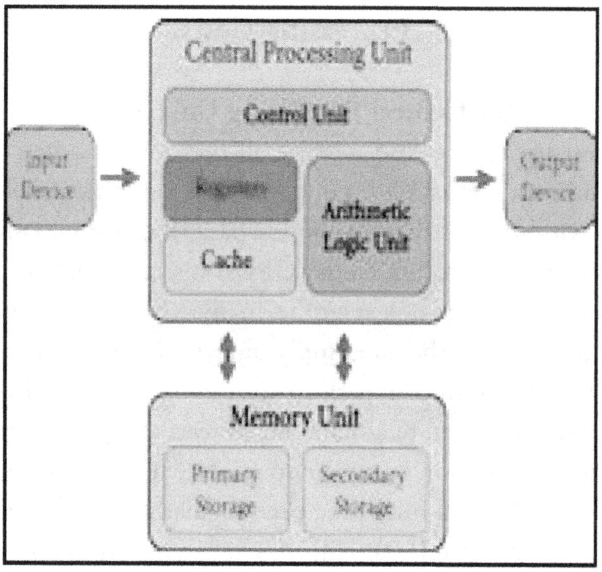

Arithmetic Logic Units (ALUs)

The Arithmetic Logic Unit (ALU) performs the arithmetic and logic operations on the operands, for example, X or Y in the instructions. Operands like X and Y are simply the objects of the mathematical operation and are used in conjunction with operators such as +, or, = etc, to create equations that produce results or values, based on how the operators and operands are positioned within the equations. Robotics like a lot of AI involves the application of many mathematical operations to detect patterns, the relationship size and speed of objects etc, in the environment, to navigate and move using coordinates. This includes calculus, geometry, advanced linear equations, ordinary differential equations, and complex numerical analysis.

The Memory Unit (MU).

The main memory unit (MU) is a small device which stores the program information, data and instructions of the CPU and consists of a non-volatile unit which means that the memory is not lost when the power is turned off and a volatile unit in which the memory is lost when the power is turned off. It is called other names like main memory, internal memory, or primary memory. Primary memory consists of Random Access Memory (RAM) which is volatile memory in which information can be temporarily stored which can be read and changed in any order and Read Only Memory (ROM) which is a non volatile memory but can't be modified. Secondary memory consists of hard drives, CDs and Pen Drives which are portable Universal Serial Bus (USB) flash memory devices for storing and transferring audio, video, and data files etc.

The CPU consists of complex electronic circuitry designed to operate the machine's system and all the various applications whilst decoding, interpreting, processing, and executing instructions, from hardware and software programs. CPUs are general purpose processors used with most general-purpose computers based on Von Newman architecture and work with almost any kind of software and multiple applications. In Von Newman architecture both the data and instructions are stored as binary digits in primary (memory) storage with instructions fetched serially i.e., one at a time and in order from memory. A CPU loads the values accessed from the MU and based on those values the ALU performs calculation and for every calculation stores the results in the MU. The biggest problem using this architecture is known as the Von Neumann bottleneck which arises because the MU and CPU are separate, which requires the CPU for every operation to fetch data from memory, which means that memory access is relatively slow when compared to the calculation speed of the ALU and inevitably limits the total system throughput. Using the ALUs the CPU employs a combination of arithmetic, logic and multiple different scalar, vector, matrix, and other mathematical operations fundamental to the functions and performance of the computer brain. Scalars are single numbers or numerical values which only have a magnitude or amount such as length, area, volume, speed, mass, density, pressure, temperature, energy, entropy, work power etc. For example, 5 miles, or 5Kg are scalar measures which must be non-negative numbers and you only compare scalars using the magnitude or amount. Scalar operations include basic things like addition, subtraction, multiplication, or division to transform numerical data into usable information. Vector operations have both a magnitude and a direction, typically indicated by an arrow, which includes, for example, displacement, velocity, acceleration, momentum, force, lift, drag, thrust, weight etc. Magnitudes are the maximum extent of size and the direction of an object, for example, 5 miles north is a vector measure. Vector operations essentially extend the laws of algebra used in scalar calculations to vectors which in mathematical terms enables the representation of a quantity in 3-D space that has both magnitude (norm or length) and direction but not position. It is commonly represented by a directed line segment whose length is the magnitude with an arrow indicating the direction in space and vectors can be used to perform operations like algebra. Two of the most widely used vectors in AI are velocity and force. Matrix operations involve matrices which are sets of numbers arranged in rows and columns to form rectangular arrays which are used to represent an object or its properties. Matrices are multi-dimensional arrays of numbers often known as black boxes because of the difficulty in analysing patterns in them and they typically store data for mathematical computations, image processing and record management. They can be used to represent multiple same-type data items using a single name and implement other data structures.

CPUs are the workhorse of computers, an AI performing multiple arithmetic, logic and multiple different scalar, vector, matrix, and other mathematical operations and transferring data between the memory and I/O devices and executing programs stored in the computer's MU CPUs have huge instruction sets and manage all the computer inputs and outputs, but although they are fast, individual CPUs don't process data in parallel, which means that they can't compete with systems that do employ parallel processing units like the increasingly widely used Graphical Process Units (GPUs) discussed below. Although CPUs are designed to handle one task at a time, many systems now contain multiple CPUs with each containing 4–8 cores which are simply small CPUs built into a bigger CPU which significantly enhances the computer performance and enables multitasking. Each core performs like a very small version of a brain which receives instructions and performs calculations or operations, so the cores can work on different tasks at the same time. In recent years, companies like AMD and Intel have been regularly increasing the number of cores used in their systems. A technology which was originally designed in the 1990s called FinFET (Fin field-effect transistor) involves a special kind of metal-oxide semiconductor transistor (MOSFET) which has revolutionised the nature of chip production and processing units. Since 2014, the almost universal adoption of FinFETs by major chip manufacturers like Samsung, TSMC, Intel, and Global Foundries has allowed the production of smaller and smaller components. A new transistor technology the Gate-All-Around Transistors (GAAFET) or Nanosheet is expected to replace FinFET to stack the ends of the transistor vertically rather than just sideways which will enable much higher performance for future processors and memories. This may also involve 3D Integrated circuits composed of multiple chips stacked vertically rather than monolithic chips. Currently, the most widely adopted and standard CPU is called NVIDIA, which is produced by the Nvidia Corporation, an American multinational technology company, a leading supplier of AI hardware and software.

Despite advances in technology, there are major limitations with CPUs including limited processing speeds because they can only process a certain amount of data at a time which slows down performance when dealing with large data. They are high in energy consumption, generate lots of heat, and have a limited memory capacity. Because of these limitations, many different types of processing units have been developed to meet the ever-changing requirements and needs of computer users and to overcome the Von Newman bottleneck problem and lack of parallel processing. The main ones we will briefly look at are the (1) Graphical Process Unit (GPU) (2) Intelligence Processing Unit (IPU) (3) Associative Processing Unit (APU) (4) Vision Processing Unit (VPU) (5) Tensor Processing Unit (TPU) (6) Field Programmable Gate Array (FPGA) and (7) Quantum Processing Unit (QPU).

(1) Graphical Process Unit (GPUs)

Since 1999 AI has seen the widespread adoption of Graphical Process Units (GPUs) which thanks to hundreds of smaller cores means they can break complex problems into thousands or millions of separate tasks and simultaneously perform multiple calculations across streams of data to work them out all at once. This makes them more flexible and scalable than CPUs and although originally developed for accelerating graphics image and visual processing they have since been used to dramatically increase the speed of computational processes for AI learning and general research. Machine vision often converts images into digital signals for image processing and because of its capacity for parallel processing, the GPUs have become the most widely used image processor because they provide the bandwidth required to accommodate large datasets and distribute the training processes. GPUs can optimise linear algebra operations such as vector and matrix operations which are the backbone of most AI systems. NVIDIA is the world's fastest GPU and is credited with driving the growth of the gaming industry, revolutionizing computer graphics, the metaverse, and generative AI and it is increasingly common for developers to use multiple GPUs to improve performance.

In 2007, NVIDIA launched the Compute Unified Device Architecture (CUDA) which is a parallel computing architecture and API (Application Programming Interface) model and programming language that effectively allows software to use certain GPUs for general-purpose processing such as processing matrices and other linear algebra operations. It is useful for applications that require significant amounts of parallel processing like processing matrices with potentially enormous implications for enhancing the compute capability of systems. Its great advantage is speeding up computer-intensive applications and it currently dominates the field of deep learning in AI by providing approximately 30–100x speed-up over other microprocessors for some applications. This is because although others like AMD offer competition and Open CL which is a standard interface for parallel computing using task and data-based parallelism, NVIDIA uses its GPUs in conjunction with CUDA. OpenCL was launched in 2009 to level the playing field in terms of computing because CUDA is a closed program, but it has not performed as well as CUDA and is not supported on many deep learning systems. AI has also seen rapid growth in the use of Intelligence Process Units (IPUs) over the last few years.

(2) Intelligence Process Units (IPUs)

Intelligence processing Units (IPUs) are totally different from CPUs and GPUs and designed specifically for machine learning and AI applications rather than office applications like CPUs or graphics like GPUs. They are programmable networking devices that offer efficient and massive compute parallelism combined with huge memory bandwidth and are predicted to have a major impact on the AI market by providing almost supercomputer performance levels. Graphcore is a British semiconductor company that develops accelerators for AI and machine learning with the declared aim of developing massively parallel intelligence and is one of the leading manufacturers of IPU chips. The company has claimed that in benchmark tests its chips using Poplar™ which is a graph-programming framework for the Graphcore chip are 16x faster than Nvidia's. Virtually all the major manufacturers are now producing IPU technology. Graphcore in their IPU user manual which is freely available online explains how IPU-based systems for example IPU-POD™ or a Bow™ Pod can be connected to a host computer which can then execute code on more than one IPU with a focus on machine learning algorithms. IPUs have a unique memory architecture that gives them the capacity for In-Processor-Memory™ within the IPU is made up of what are called SRAM (organized as a set of smaller independent distributed memory units) and a set of attached DRAM chips which can transfer to the In-Processor-Memory via explicit copies within the software. You can use the same IPU for training, inference, and prediction which means that they can process many times faster than even GPUs. The improvement has been brought about by switching from the manipulating of 2D matrices in GPUs which relies heavily on large data batches to employing what is known as the graph model.

Nigel Toon, the CEO of Graphcore in an interview in 2017 observed how 'the knowledge models that we are trying to build and manipulate in machine intelligence systems are most naturally expressed as graphs" and all the major machine learning systems have now adopted graphs as the fundamental data structure. Graph models refer to graph databases consisting of nodes and edges that connect the nodes and essentially describe their relationship to each other, and a node-edge-node combination contains a piece of information, and when many nodes are connected by edges it produces a web of interconnected points which is the graph. Graph modelling more accurately represents the connected nature of the world that we live in, the BNNs of the brain that we discussed earlier, and also makes the model more flexible and expressive. When comparing the processing units which are a core element of the robotic brain, CPUs, GPUs, and IPUs have progressed from scalar to vector to graph models with a resulting improvement in performance and parallel processing capabilities.

(3) Associative Processing Units (APUs)

Associative Processing Units (APU) involve associative memories also called Content Addressable Memory (CAM) in which associative memory is used to pair things it wants to compare (comparators) with using data from its memory. APUs have been around for a long time but their use is limited due to size and operational constraints. Comparators allow the system to simultaneously make parallel comparisons by using input from memory rather than sequentially processing the data which, like associative memory in humans, involves the ability to remember relationships between unrelated items. In AI systems it constitutes a special type of memory that optimizes data search by storing the set of patterns in memory and when presented with a pattern, retrieves the stored pattern that most closely compares or correlates to it. Data is stored with additional tags or meta-data with meta-data providing information about and contextualizing other data though not the content. For example, the author, date created, date modified, and file size which makes it easier to find, understand, use, and manage that data. Web pages often include meta-data using meta tags which are Hyper Text Markup Language (HTML) tags used to provide additional information about a page to search engines and other users. The associative memory compares the search task with all the stored data tags and retrieves the data that most closely matches the task. APUs are highly flexible and can handle bit-wise operations, masking, and bit-masking operations which involve keeping, changing, or removing parts of the information. They can also work with variable-length words and comparisons. Using parallel processing facilities, APUs can rapidly find matching data even when using incomplete or imprecise information. Bit-wise operation involves working with individual bits each with a single binary value of 0 or 1. Bit masking uses the binary representation of numbers unique to each number to solve difficult problems. The bits of the number 11 are "1011", because in base 2: $11 = 8 + 2 + 1 = 1*2^3 + 0*2^2 + 1*2^1 + 1*2^0$ = "1011." The processors operate on bit strings, bit arrays, or binary numerals (considered as a bit string) at the level of individual bits.

The main disadvantage of APUs is that they are usually more expensive than RAM and require each cell to have storage capability and logical circuits for matching content. In RAM the user supplies the memory address with RAM returning the data word from that address. The two types of associative memory are Auto and Hetero-associative memory. Auto-associative memory involves the output being fed back into the input, which enables the network to learn and remember the patterns it was trained on. It is often used on input data that may be incomplete or noisy such as speech and image recognition. Hetero-associative memory networks learn and remember the associations between different input and output patterns and are often used in data compression and data retrieval.

(4) Vision Processing Units (VPUs)

A Vision Processing Unit (VPU) is another type of emerging microprocessor specifically designed to accelerate vision in machine learning and run machine vision algorithms with robotics, one of the target markets, particularly where deep learning, and Convolutional Neural Networks (CNNs) which we discuss below are used. They can provide more accurate results with higher precision and lower error rates. They are small and low cost and in 2017 Intel even released a USB stick called the Neural Stick Compute 2, which allows the computer to transfer computations using machine vision algorithms to the stick. They can operate many processes in parallel and Intel advertised the chip as being able to run 6 neural networks which included pedestrian detection, age estimation, gender classification, face detection, body orientation, and mood estimation (Edge AI: The Ultimate Guide to Vision Processing Units (VPUs) 2022). Most devices now have VPUs installed but one of the biggest challenges with VPUs is that because they must process large amounts of visual data, they consume a lot of energy and so are problematic for autonomous or battery-powered devices.

(5) Tensor Processing Units (TPUs)

Tensor Processing Units (TPUs) are Google's custom-developed application-specific integrated circuits (ASICs) chips which are intended for tensor operations rather than chips that have a general purpose or use. They were originally designed to accelerate machine learning workloads. Tensor operations have been described by software developer Joel Tovar in his online comparison of GPUs and TPUs as the "Building blocks of neural network computations."[215] The word tensor does not mean the same in physics, maths, or computer sciences. In computer science, tensor refers to the Data Structures, which typically consist of multidimensional arrays or matrices that store a specific type of value or information and functions that can be performed on objects that use the data to operate.

The more complex the data structure the more multi-dimensional the arrays or matrix etc, which often distinguish between classes of data and objects but are hard to visualize. All deep learning systems rely on tensor operations like arrays that consist of a collection of elements (values such as numbers etc, or variables each identified by at least one array index or key). Unlike physical objects such as cameras or robots, data structures have no attributed behaviours, and data (or State) is assigned certain properties (also called variables or variable values) and put into classes of data that contain members of that class. For example, the data structure for a robot could include the robot's name, colour, weight, etc. Objects such as cameras or robots are constructs that, unlike data structures, do things that have functions. If an object is included in the data class it brings the properties and the functions together, so that they represent something, someone, or some object like a particular camera or robot, etc, with each object having different properties or variables and different functions. For example, one robot with the property name Tom may have the functions of walking, talking, or going to sleep whilst another may have the property name Jessica with the functions of dancing or running, etc. When incorporated into an object such as a camera or robot, the properties are called instance variables or attributes which are specific to a particular or instantiated object such as one version of a robot or camera, and that version or object will have methods, procedures or processes it uses to operate on the data structures. The class with its members of the class, effectively provides a blueprint for making the object like the camera or robot, because you need to know what properties and functions the object possesses, but a class will not refer to a particular object until it is assigned values in terms of the variables or properties and functions etc. An object like a camera or robot that has private data (state) but public observable functions, is described as encapsulating boundaries around the data or state. This means creating self-contained modules or classes that hide the data that classifies, instantiates, or represents the physical object like the camera or robot. The object if it's a camera or robot exposes its behaviour by using the processing functions that act on that data member in that class but the data in the data structure and class is kept private. Each class typically sets out rules for what can be publicly accessible and what if any modifications can be made to the data. If for example a robot has a name (a property or instance variables or attributes) and possesses the function of being able to introduce itself the process will allow that robot from the class of robots to use the function to give you, its name. By using the class of data, you can create the object by giving it both properties and functions.

Although announced by Google in 2016 TPUs have been installed in their systems for many years and facilitated high volume low precision computing primarily aimed at the embedded and robotics market. Like GPUs, they are capable of parallel processing and contain thousands of small cores optimized for handling vector and matrix operation. Unlike GPUs which are flexible and general purpose, TPUs are purpose-built for machine learning tasks and accelerating the tensor operations and tend to have higher memory than GPUs, faster inference times, and consume less power than GPUs. The major constraint is that there are currently fewer software and support tools available for users of TPUs and these are primarily accessible through the Google Cloud platform which may not suit all users.

(6) Field Programmable Gate Arrays (FPGA)

Field Programmable Gate Arrays (FPGA) are integrated circuits that contain configurable arrays or programmable logic blocks (CLBs) and programmable interconnects and are described as field programmable because the configurable blocks can be connected in various configurations, to meet the specific needs of different users. Their great advantage is their genuinely parallel processing with each section of the chip functioning entirely independently without affecting the performance of the other blocks and their flexibility, speed, and reliability.

(7) Quantum Processing Units (QPUs)

Quantum Processing Units (QPUs) are the undisputed brains of quantum computing and harness the quantum behaviour of particles like electrons or photons to make calculations much faster than processors in conventional computing. Rather than classical physics they use behaviours like superposition in which a particle can be in many states simultaneously and use qubits that can be more than 0 and 1 and represent many different quantum states or possibilities, rather than the binary bits 0 or 1 that simply represent the on-off state of electrical currents. Nvidia in an online review describes how one of the ways the computer scientists are producing these, is called superconducting qubits in which tiny metallic sandwiches called Josephson junctions allow electrons to tunnel through a non superconducting insulating layer between two layers of superconducting materials.[216] The Josephson effect is named after the Welsh theoretical physicist Brian Josephson (1940–), who in 1962 postulated that pairs of superconducting electrons could tunnel through a non-superconducting energy barrier to cross from one superconductor to another. Superconductors are materials that at low temperatures lack electrical resistance. It capitalises on the quantum feature that matter is both wave and particle so even though in classical physics it should not have the energy to pass through the barrier, the quantum wave-function can penetrate the barrier. This means that atoms in very close proximity can connect by tunnelling through. In March 2023 it was announced that physicists had for the first time been able to watch molecules forming using quantum tunnelling.[217] Robert Wild, an experimental physicist and his colleagues supercooled negative deuterium ions to a temperature that rendered them close to a standstill and then introduced a gas made of hydrogen molecules and watched as they bonded.

Artificial Neural Networks (ANNs)

There are major differences between BNNs in the human brain and the ANNs used in AI, and there is currently no agreed definition of what constitutes an ANN. Synapses in the BNNs in our brains carry out highly complex non-linear, rapid parallel processing of multiple types of information including perceptual, pattern recognition and motor control etc. The human brain learns its own rules through development and experience with neural plasticity creating new synaptic connections or modifying existing synapses allowing it to use continual learning to adapt and learn. By comparison, ANNs have traditionally carried out sequential processing with each artificial neuron trying to replicate the neural coding and transmission processes employed in the human brain. This is further complicated given that coding in human neurons remains vigorously disputed as to whether rate or temporal coding is involved and the controversy over sparse coding. The use of multiple layers of neurons and neuron stacking in ANNs also means that non-linearity is distributed throughout the ANN rather than in BNNs where the output is non linear at the neuron level.

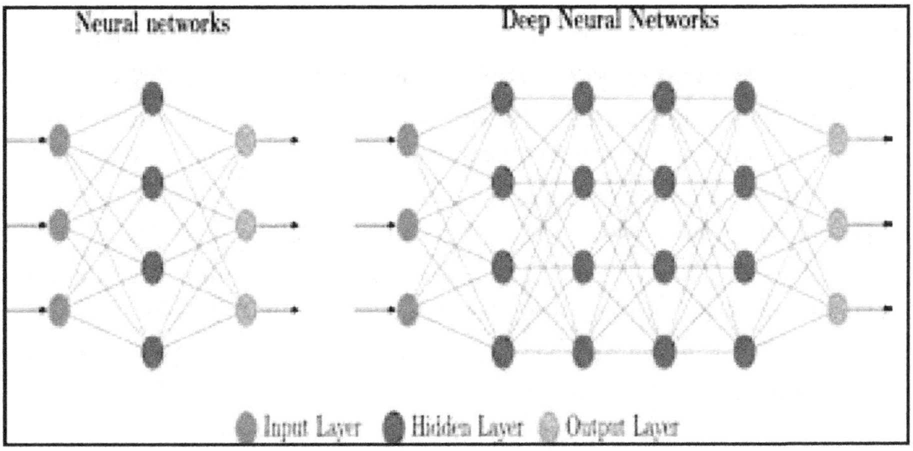

The complexity and ability of BNN are attributable in part to the ability to integrate large amounts of information due to the diversity in the types of neurons, their functions and how they receive input through multiple dendrites which gives rise to a mind-boggling level of connectivity between brain regions and the critical role of inhibition and excitation modulated by the neurotransmitters. Neurons in BNNS are typically significantly slower than the transmission speeds in ANNs but this is compensated for by the staggering number of synaptic connections in the human brain. The human brain is a highly efficient energy structure requiring far less energy than most supercomputers. We use around 20 watts, compared for example to the Fugaku supercomputer which needs 28 megawatts, so our brain needs about 0.00007% of Fugaku's power supply and unlike supercomputers that need elaborate cooling systems, the brain is kept at a constant 37°C within its bony skull encasing.

The configurations of artificial neurons in ANNs outperform many conventional algorithms in terms of speed and accuracy. Each artificial neuron or node is a computational component with the initial step in training a multi-layered ANN involving the assigning of initial values and random weights and biases to the individual neurons, and then training the model to perform sequences of calculations. This seeks to mimic the way the brain organises and differentiates between information and involves information being propagated forward to provide a meaningful output. Weights set the standards for the neuron's signal strength and the biases are simply neural network parameters that are adjusted based on the difference between the expected values and the calculated values which gives an error amount by which to correct the parameters. Artificial neurons sum up all the different weighted values for each input, which is known as the transfer function which is just a mathematical function that models the system's output for each possible input. The neuron multiplies the initial value by the weight and sums up the other values that are approaching that neuron, which is further regulated by applying the bias, which is additional units with only one per layer, involved in sending data to the correct end unit.

To fire like a real neuron when it generates an action potential all artificial neurons must be activated and pass on information to the next neuron, which requires a neuron or node to pass a certain threshold, with the activation function determining the output of that neuron given its input or set of inputs. Activation functions are crucial as they determine the information that is processed through the ANN. Weights are probably the most important factor in converting input to output at each layer because by multiplying the input by a weight they act as the parameters that will determine the extent to which that neuron or input affects the other neurons and the output from the layer which will determine whether the neurons in the next layer are activated. An important development in ANNs has been the use of Sigmoid neurons.

Neural networks must not be over-trained which means that given an unmanageable number of weights that clutter neuron signals, a weight regularisation method is often used which keeps connection weights small by performing learning algorithm updates. Most ANNs currently rely on Sigmoid neurons and variations of activation functions which have evolved from the artificial neuron called a perceptron that has been widely used for many years. In early ANNs, the perceptron processed weighted real value inputs to produce binary results with positives being 1 (Yes) and negative 0 (No) classifications or outputs decided by a binary threshold. The weights that indicate the importance of the input in the decision-making process are updated to correct for any misclassification, hence it learns which are the best weights. The model only fires or is activated once the threshold or bias is reached and if it fires, that neuron will send data to the next level of neurons in the network and continue to be repeated until a single decision or output is reached. This is very useful if you want to classify something in a binary way between 0 and 1. The problem with the perceptron was that a small change to any input could cause the output to flip between binary states in a step-like transition and this is not how real-life decisions are made. The Sigmoid neuron provides a much smoother activation and output function by employing a real value between 0 and 1 which also in ANNSs addresses the risk of exponential growth in output calculation by always limiting it to 1.

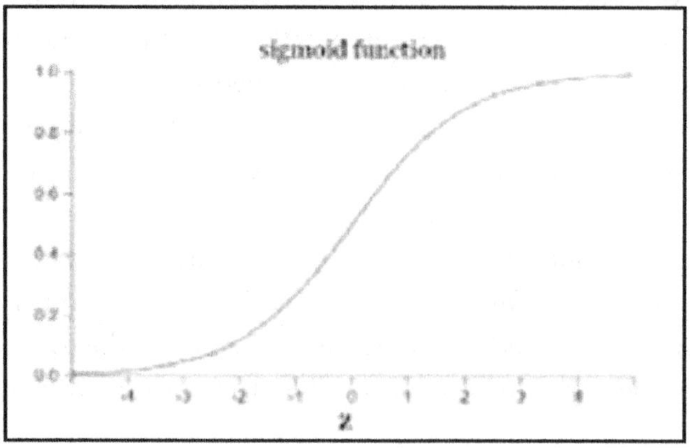

Rather than acting like a step function with a harsh binary transition between either 0 or 1 or Yes or No the Sigmoid neuron treats the output as a probability of existing between these states which is represented by a real number between 0 to 1. The type of sigmoid function most used is the logistic function which maps predicted values to probabilities and acts as an activation function that maps any real value into another value between 0 and 1 to model the probability of binary events happening. The value will determine whether it is classed by the system as falling within the 0 class or the 1 class, which more accurately represents how we make decisions in real life and allows for changes in weighting and biasing without creating major changes to or distorting the output. AI commonly uses other activation functions which have evolved since the original binary function used by the perceptron and are variations of the Logistic or Sigmoid function. They include Tanh, ArcTan which is just the inverse of the Tanh function Rectified Linear Unit (ReLU) which is very popular in deep learning and normal neural networks because whenever the function is less than zero it gives you zero, Leaky ReLU which performs a similar function to ReLU but rather than making negative input zero reduces the magnitude of it. SoftMax imparts probabilities so several outputs can be passed through to give a probability distribution of those outputs. These are all very important and widely used activating functions that enable the system to perform non-linear functions which are vital to determining the accuracy of models.

Back Propagation

AI learning has also been revolutionised by using a method known as Back Propagation or Backprop of errors algorithms. Invented in the 1970s, backprop was widely adopted after 1986, following the publishing of a paper by David Rumelhart, Geoffrey Hinton and Ronald Williams.[218] When the weight and bias assigned to an artificial neuron have caused that node to activate and pass on the information to the next layer, that node is described as having learned a feature. The features are how well the system has learned something because if subsequently, the features are present in the input into that layer, the node will be activated and pass that feature on until the output of the system constitutes the accurate or not-so-accurate internal representation of the information such as the image of a face or a cat or dog. The overall representation of the information containing the features is built in a hierarchical fashion using backprop to correct errors as it goes which achieves far greater accuracy in identifying information. Backprop works as a multi-step process that involves calculating the gradient of the error function concerning the neural network's weights and simultaneously updating the network parameters. An error gradient is the direction and magnitude calculated when training the network and employed in updating network weights by the correct amount. During updates, error gradients can accumulate resulting in extremely large error gradients. Unlike forward propagation, backprop starts at the final output layer of weights by comparing the predicted output with the actual output and proceeds backwards until the error gradient of the last layer of weights at the input layer.

Calculating the gradient of the error function in neural networks is crucial to optimising the performance because it signifies the rate of change with reference to the networks parameters and iteratively adjusts the parameters in a way that reduces the errors. Gradients are calculated by multiplying the previously calculated activation gradients with the corresponding inputs to each layer. Calculating the gradient of an error function for a neural network though is a computationally intensive task, especially for complex architectures. Backprop in deep learning allows the system to learn features using error correction, thus reducing the need for humans to hand engineer the features into the system from the outset and is a much more efficient algorithm that allows the systems to be applied to a much wider range of problems and situations.

Backprop is based on the realisation that artificial neurons in ANNs are like the neurons in our brain which learn features or specific information and are the building blocks of a larger information set that we use to recognise things. For example, when we see a familiar object, all the neurons that were previously activated and connected by signalling to each other over the many times you might have seen that object will be reactivated because the object contains the features that activate the neurons and allow you to recognise it. There are two major problems with error gradient calculation that plague AI which are the Vanishing gradient and Exploding gradient in which the weights become unstable which we will discuss later but as we saw, have been increasingly addressed using activation functions like ReLu.

ANNs are designed to compute predictions or outputs based on inputs, for example, a given vector X input/also called Domain produces a vector Y prediction output/also called Codomain. In mathematics, the product of multiplying two scalars when it gives an ordinary or scalar number such as the length, area, volume, speed, mass, density, pressure, temperature, energy, entropy, work power etc, is called a dot product. The product of multiplying two vectors indicates magnitude and direction but is called a cross product. A dot product (scalar) or cross product (vector) provides the input for the next neuron layer upon which further mathematical computations are performed. The important thing for us is not the details, but the fact that mathematical processes are conducted on the data, be it scalar, vector, matrix, or other mathematical calculations at various stages to produce outputs which form the input to the next layer and so on until producing the final output.

Whilst seeking to emulate the human brain most cognitive architects aspire to avoid the constraints imposed on BNN by evolution and many ANN designs bear no obvious relation to biological reality. Designing an ANN involves many choices such as the size of each input and output layer, where and when to apply batch normalisation layers and dropout layers (see below), and the choice of activation functions etc. When training a deep learning machine with many layers, they can be extremely sensitive to the initial random weights and the learning algorithm configuration which involves the automated search for the best parameter configuration (which are the settings that affect database behaviour) of a parameterized algorithm. In deep learning the model is updated, working as we have seen backwards from the output layer-by-layer which involves an estimate of the error which assumes that the prior layers have used fixed weights and the parameters do not change. The problem is that all layers are usually updated simultaneously which means that the input from prior layers can change after weight updates, making the model difficult to train. Computer scientists call this Internal Covariate Shift and to overcome this problem, Batch Normalisation Layers are used as a means of coordinating the update of multiple layers. They do this by standardisation of the activations of the prior layers which means that the subsequent layer will not change the assumptions it makes about the spread and distribution of inputs during the weight update. Standardisation involves calculating the mean and standard deviation of each input variable to a layer per mini-batch and using these statistics. Dropout layers are also often employed and involved during model training nullifying some randomly selected neurons as a method to stop over-fitting the model. It was first introduced by Nitish Srivastava in 2014[219] and means that other neurons must handle the representation required to make predictions for the nullified neurons leading to the learning of multiple independent internal representations making the model better able to generalise the data and less likely to overfit the training data.

The overriding goal of the machine and deep learning is not to recreate a human brain but to enable an AI system to do what the brain does so efficiently, which is to process information in real-time, recognising that it is currently only possible to do this on a much smaller and simpler scale than nature has organised for the human brain. A BNN learns its own rules through development and experience, with neural plasticity creating new synaptic connections or modifying existing synapses allowing it to adapt and learn. The neural synapses in a BNN carry out highly complex, non-linear, parallel processing, including perceptual processing, pattern recognition, motor control etc, and rapidly process multiple types of information. Deep learning using DNNs allows artificial nodes with their level of connectivity determined to be stacked in multiple neuronal layers which allows them to process hierarchical extraction of features recognise patterns in data and even generalise information or pattern recognition to other new data. ANNs have traditionally carried out sequential processing which can include linear and non-linear processing and new hardware devices such as the processor units that we discussed means that ANNs are increasingly able to perform parallel computations. Processing in AI systems is centralised through the CPUs or other types of processing units whereas BNN have no central control over the process and our brains as we have seen comprise both small-scale networks and larger networks, reflecting various levels of anatomical organisation from the neuronal to the network to the whole CNS and body. In ANN the system's unique functional memory is in one location whereas in BNNs the memory is distributed throughout the neural network in the neural patterns and neural connections. BNNs are much more fault tolerant and robust, particularly given synaptic plasticity in BNNs and their ability to perform much more complex and abstract tasks than ANNs.

Humans process some sensory stimuli, though by no means all that we become aware of because we tend to attend to and focus only on stimuli which are salient or relevant to us and attribute meaning to the information that is selected. If we did not select only some of the information of which we are aware in this way, we would easily become overwhelmed by the sheer volume of information coming through our senses. This highlights a major difference between the human brain and AI which typically requires vast amounts of data to learn and requires varying degrees of human involvement in data classification or supervision of learning and validation of the output. Although humans are exposed to vast amounts of information in the environment, they use far less information to learn than AI systems currently require and can build abstract mental representations and concepts without much effort. AI systems typically compute far faster than humans and are making remarkable progress in reinforcement and unsupervised learning, but still currently come nowhere near to being as adaptable as humans in their ability to perform a wide variety of different tasks and functions.

In their 2013, comparative review of BNNs and ANNs, O.S. Eluyodel and Dipo Theophilus Akomolafe observed how over the last 70 years computing has evolved through the processing of Numerical data (1950–1970), Text data (1971–1980), Image and Graphic data (1981–1990) and Audio and Video (1991-to date).[220] The metrics that they used to compare BNNs and ANNs are ones that we should be aware of because technical advances and progress within each of these vary, with obvious implications for how fast android robots are being developed and their future potential. They are "Structures, layers, size of neurons, functional capabilities of neurons, their learning capabilities, the style of computation, processing elements, processing speed, connections, strength, information storage, information transmission communication media selection, signal transduction and fault tolerance."[221] AI is proceeding at a pace, with much of the recent AI research and development employing ANNs, machine learning and deep learning, accelerating the rate at which AI appears to be getting closer to the goal of AGI and potentially even ASI. However, the complexity of BNNs substantially exceeds the current complexity of any of the current ANNs, even Deep Neural Networks (DNNs).

Cognitive Architectures in AI

Cognitive architectures are the computational instantiation of the theory of how the human mind is structured but unlike intelligent software, seek to model human behaviour and the underlying cognitive processes and are distinct from intelligent software systems that most AI systems use to solve specific types of tasks, such as playing chess or games, etc. All intelligent software systems in AI have memory storage, control components, data representation, and input/ output devices. These do not include a cognitive architecture mimicking the human mind and complex behaviour and American computer scientist John E Laird, one of the developers of the cognitive architecture called Soar, observed how good cognitive architecture must be able to "change through development and efficiently use knowledge to perform new tasks." Building even a limited range of abilities into any cognitive architecture is not easy, and modelling the human mind and creating AGI with the potential for ASI remains challenging and not a goal that all cognitive architects are pursuing. Around a quarter of AI's current cognitive architecture applications are employed in robots, predominantly focusing on sensory-motor challenges in navigation and obstacle avoidance in autonomous vehicles. Not all AGI research uses the same approach to cognitive architectures, using the analogous form of natural intelligence, computational digital intelligence, and biologically inspired (biomimetic) cognitive architectures. Biomimetic architectures typically focus on the developmental aspect of physical and sensorimotor functions and include some robots like iCub, Dav Sai,l, and Isaac, which mimic human functions such as grasping objects and vision-guided navigation that are attempting to make these robots more human.

Many cognitive architectures employ the Standard Model of Mind (SMM), which, although not a cognitive architecture, is a theoretical model for cognitive architectures. Neuroscientists John Laird, Christian Lebiere, and Paul Rosenbloom, in a 2017 paper claimed that over the last 40 years, cognitive architectures have converged onto a collection of shared core assumptions related to the structure of the human brain, in terms of memory, learning, perception and motor functions which govern the architecture of the human mind and which could form the basis of a machine-based mind. They attribute the supremacy of SMM to the convergence of a "likely number of factors, most notably, the cumulative lessons learned from the design of intelligent systems in AI and robotics, continual progress in cognitive psychology, and advancements in neuroscience." They argue that SMM "reflects a common architecture for intelligent, human-like behaviour, which can be implemented in a biological or artificial system." This has been supported by fMRI studies of humans when compared to a variety of other cognitive architectures not using SMM, which concluded that "the SMM significantly outperforms the other models, suggesting that the SMM best captures the functional requirements of brain dynamics in fMRI data among these alternatives" and they compare SMM to the Standard Model in physics. We must also consider other theories of mind employed in AI, like the Global Workspace Theory and the Predictive Theory of Mind, and the attempts to reconcile these with SMM.

The cognitive architecture ACT-R, developed by John Robert Anderson and Christian Lebiere at Carnegie Mellon University, is the most advanced form of a unified theory of cognition, described as playing "a central role in many research and applied efforts and now serves as a central repository of cognitive theory."[222] ACT-R is emergentist, using both symbolic and sub-symbolic representations and procedural and declarative memory as the theoretical basis for simulating and understanding human cognition. Procedural memory refers to the learned information necessary to perform certain skills like tying your shoelaces whilst declarative memory relates to personal or episodic memory like life events and general semantic information and knowledge about the world, which must be consciously retrieved. ACT-R uses three different spatial representations of how humans focus on objects, grasp objects, and navigate environments. ACT-R also seeks to predict cognitive, perceptual, and motor behaviour and, by comparing this behaviour quantitatively to human behaviour in performing the same tasks, explore how the human mind works. ACT-R has demonstrated that a robot appearing to use a limited type of theory of mind (TOM) or mentalization (which we will discuss later) can, when realizing that an instruction has changed and that a human is failing to act by the change, infer that the human has forgotten and will instruct them of the change.[223]

Possessing intelligence in certain tasks does not achieve the broader goal of building cognitive architectures into machines that can simulate human behaviour and thinking and AGI. To achieve AGI, computer programs must allow a machine "To perceive the world, solve problems by generalizing their knowledge across a wide range of domains, be able to develop insights, learn from experience, adapt to new situations and be able to reflect on themselves, which arguably confers them with mind states and a self-concept."[224] According to Adams, it should include fundamental features such as "Perception, memory, attention, actuation (what makes us act as we do), social interaction, planning, motivation, emotion, etc," which would appear to be a necessary part of any AGI landscape.[225] Goertzel described the need for AGI researchers to come to terms with fundamental cognitive processes and interactions, including "Working and long-term memory, deliberative and reactive processing, perception, action and reinforcement learning, metacognition."[226]

Over the last forty years, nearly 300 different cognitive architectures have been developed, including Soar, ACT-R, SiMA, Soar Sigma, and Cog Prime, which seek to achieve AGI and potentially ASI, while others such as CLIPS, CLARION, ICARUS, EPIC, and LIDA aren't expressly aiming for AGI, although ACT-R and CLARION appear to be making substantial progress toward integrating the various key functions and domains of the brain and mind, such as action, skill, learning, memory, concept, reasoning, motivation, emotions, personality, and meta-cognition. CLARION employs novel processes to integrate and connect various mind functions and holds significant promise in the potential to create a robot with a mind. Memory, in particular, plays a crucial role in the construction and maintenance of a coherent self, as we can all see when and tragically, an individual's memory is lost to the ravages of dementia and the devastating impact this can have on the capacity to maintain a coherent sense of self over time and others perception of that self.

The major advance in modern machine intelligence comes from machine learning, particularly deep learning, in that some AI machines can constantly build up knowledge by learning from the environment, with obvious parallels to human learning. To date, deep learning has largely focused on solving specific problems and decision-making rather than developing AGI, with some cognitive architectures, as we will see, designed to recognize patterns. Others use decision trees to improve the quality of decision-making but pattern recognition has become the most powerful tool in big data analysis and extracting meaningful information from vast amounts of data. It uses special algorithms to distinguish and segment data using regularities in the data that can be sorted into set criteria or common elements and used in learning because patterns can be encoded, stored, remembered, and constitute prior knowledge. It can perform pattern recognition on any data, including text data, images, etc., explore the data for regularities, and then use a descriptive process to categorize it and render it meaningful.

No AI cognitive architecture currently allows the processing of sensory data in a way humans do when attributing meaning, knowledge, and understanding to that information but develops capability at different levels for each sensory modality of vision, smell, hearing, taste, and touch. Nicholas Cassimatis believes this may be due to the immense challenge faced by AI, of needing to integrate multiple representations and inference techniques about perception, cognition, and action, which is something our brains do naturally all the time. Humans don't live in isolation but are embedded or embodied in an environment as robots will have to be, and we can't just rely on a high level of reasoning to give meaning to all the raw input of external stimuli from our environment and the inner thoughts to generate appropriate behaviour and actions. Unless AI robots are embodied and learn from their experience of interacting with the world, the big challenge will be how they will even begin to be able to perceive the world as we do or give meaning to the world, particularly as we will see much of this processing relates to information that is associated with the self and the preservation of the self.

Different Types of ANNs

There are various types of ANN used for different data types with different applications, different approaches to processing the data and producing output and learning and different architectures. The main ones are (1) Feed-forward Networks (FFN), (2) Recurrent Neural Networks (RNN) (3) Convolutional Neural Networks (CNNs) (4) General Adversarial Networks (GANs) (5) Self-Organised Neural Networks (SONN) and (6) Spiking Neural Networks. We will look at these at a relatively superficial level because the point is to try and understand how AI is mimicking the BNNs to process information and learn and identify some of the major differences and current shortcomings of ANNs when compared to the human brain and BNNs.

(1) Feed-forward Neural Networks (FFNNs)

Feed Forward Networks contain an Input layer, Hidden layer (s) and Output layer, with the required task, for neurons and perform a mathematical function to calculate and produce an output based on the weights of individual inputs. FFNNs are known as bottom-up networks typically employed, in classification and regression analysis when the data is not sequential nor time dependent. Modern FFNNs pass data through multiple layers of neurons with hidden layers between the input and output layers and it is the hidden layers which are the computational engine. They mimic the way our brains assign groups, often millions of neurons to specific tasks or functions and simultaneously assign other neurons to different tasks or functions.

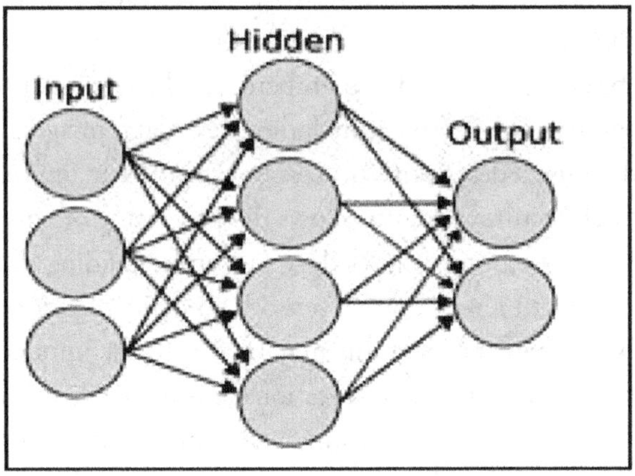

Simple Feed Forward Neural Network

Data only flows in one direction which is forward, so FFNNs don't form a continuous cycle or have a feedback loop allowing them to combine the previous input and output to generate new information or insights which means that the output of any layer does not affect that same layer.FFNNs accept multiple inputs and multiply each input by a weight which are then all added together, and an activation function determines whether a neuron is activated or not and the system generates an output. Because they only consider the current output and don't memorise past output, simple FFNNs have significant and obvious limitations and can't handle sequential data. They effectively take a snapshot of data but can't process the data dynamically or provide insights arising from dynamic changes in the data. The reason FFNNs can't handle sequential data becomes apparent when you weather data using a time series of data on humidity and temperature. If you feed the data points into the input layer, the output will use that data to classify the weather as sunny or otherwise. The problem is that FFNNs do not remember the previous data so it can only take one classification based on that data, not the sequence of data.

Traditional FFNNs assume that data points are independent and non-sequential so all the inputs are processed in isolation which means they cannot pick up any dependencies in the data like the impact of the previous day's weather on the subsequent days. This means that highly relevant or useful past information is ignored, whereas humans would intuitively and logically utilise such information. It is often vital to understand changes in data patterns over time and whether they appear sequentially over time in an order or identify a particular dynamic phenomenon such as the change in weather or stock prices. However, they can handle high dimensional and non-linear data so are useful for example in data compression, pattern and image recognition, speech recognition, natural language processing, and financial forecasting. High dimensional data is where the number of features or dimensions often exceeds the number of observable features. One example is in healthcare, where a person's data may contain an enormous range of features like blood pressure, resting heart rate, immune system status, surgery history, height, weight, existing conditions etc or financial information such as with stocks/shares etc and information like PE Ratio, Market Cap, Trading Volume, Dividend Rate, etc.

Non-linear data is data in which there is no clear relationship between the variables which refers to an object, event, person, place, thing, idea, feeling, period, or any other type of category or phenomenon that you are trying to measure in some way, and the data concerning the variables cannot be represented on a line graph, so it can't be separated or analysed using linear classifiers that we discussed earlier. Linear data is much easier to predict and model but non-linear data can often give greater insights into complex data and systems. The major limitation of FFNNs includes the loss of information, and the number of parameters involved which increase the computations required making the system potentially unstable and preventing continual learning. They are also not translationally invariant which simply means that the system does not produce the same response (output) regardless of how its input is translated. Although easy to train, they require huge amounts of data to function properly, are computationally very power-hungry and suffer from the vanishing gradient and exploding problem mentioned earlier, which negatively impacts the accuracy of the predictions and outputs. The vanishing gradient and exploding problem faces all ANNs.

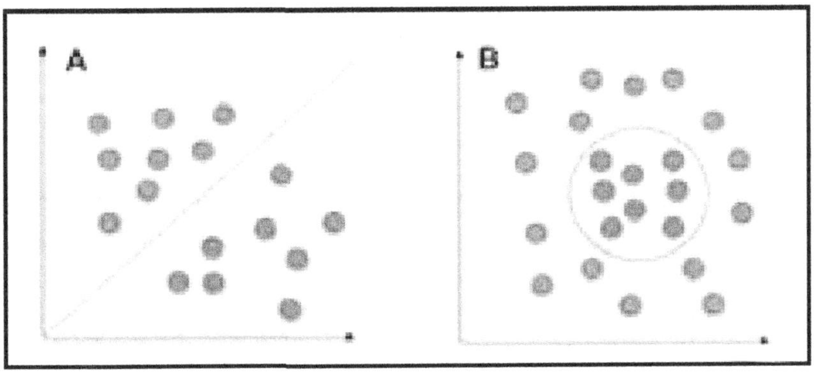

Linear data. **Non-Linear data**

(2) Recurrent Neural Networks (RNNs)

First developed in the 1980s, Recurrent Neural Networks (RNNs) also known as Feedback Networks and Interactive Networks evolved from FFNNs. They comprise three main types which are (a) Simple RNNs, (b) Long Short-Term Memory Neural Network (LSTMN) and (c) Gated Recurrent Unit Neural Network. The increase in computational power, the vast amounts of data now available and the development of the LSTMN have begun to realise the potential of RNNs in deep learning and Apple's Siri and Google search use RNNs. Unlike FFNNs, they are specialised to use sequential, or time series data so are highly popular algorithms for sequence models. A common example of sequential data is stock prices where a dot product represents an observation of the stock price at a particular point in time, or gene sequences and weather data.

(a) Simple RNNs

RNNs have hidden layers between the input and output layers, using signals travelling in both directions in the network which is achieved using self-feedback links or recurring loops to propagate the outputs of some of the neurons in a layer, back to some neurons in a preceding layer which allows them to perform repeated computations. They essentially recur new states to themselves with the inputs all related to each other which means they can become extremely complicated. They can use prior or past outputs from neurons with links that receive feedback that influences the current input. A major advantage over standard neural networks is that they share the same features or weight parameters across each network layer, with information cycling through a recurrent loop to hidden layers so that the network can remember previous inputs and past computations.[227]

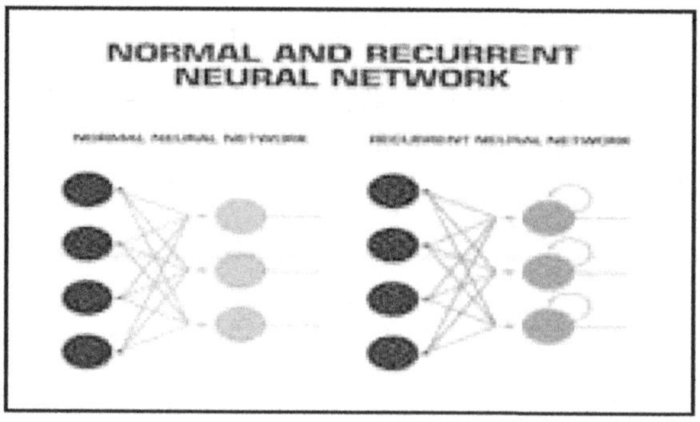

RNNs can pick up dynamic changes in sequential data and patterns over time which means that their internal state or memory changes continuously until they reach an equilibrium point. The equilibrium point is when the system achieves a stable-input stable-output function which with a single neuron is when it has learned a specific input-output relationship and the state need no longer change. They remain at the equilibrium point until the input changes and a new equilibrium is needed. Because RNNs can use the memory to process arbitrary input and output lengths and sequences of inputs, they can perform dynamic analysis of data such as predicting what word comes next in a sequence of words in a sentence, which is why they have been widely used in deep learning. They effectively calculate the current state using both the current and previous input which is fed through the feedback loop to the current input so like humans, it has two inputs the present and the past. A classic example of an early RNN is the Hopfield network which can store patterns in its associative memory, and was mainly used in pattern recognition and optimization. With the feature of being dynamic comes the problem of internal stability and the importance of analysing the equilibrium point in RNNs.[228] Chinese computer scientists Xiaobing Nie and Jinde Coa, describe the challenge as that "small changes in this model may shift a network from a relatively quiet state to some other state with highly complex dynamics" which makes the need for global exponential stability in the network as it changes vital.[229]

RNNs can possess continuous attractors which as we have seen are implicated in the stability of dynamic neural networks such as the human brain and memory and learning.[230] Remember that attractors are "a set of states toward which a system tends to evolve, for a wide variety of starting conditions of the system." For example, we will see when we discuss memory how when we retrieve memory of events, we retrieve not only the content but the temporal structure and context and how experiences or events unfolded which allows us to relive parts of the experience. We don't simply retrieve a snapshot or static memory and nor do we when we imagine the future. Davis Spalla and colleagues have developed a continuous attractor model which reflects how neurons communicate with each other to encode and store memories which maintains that the brain represents the memories as continuous attractors. Their AI model "spontaneously breaks the equilibrium and stability of the stored memory configurations and produces dynamic retrieval" of the appropriate memories.[231] We will discuss this in more detail later, but for present purposes it illustrates the importance of having a dynamic feature provided by RNNs and the need for global stability in the system. This dynamic function means that they are mainly used to detect patterns in a sequence of data and time series data so they are widely used for text streams, audio and video clips, natural language processing, speech recognition and image capturing in devices like Siri and Google translate. For example, if the system is given an audio clip as input the RNN will generate the text script so both the input and output are sequence data. In sentiment classification, they categorise the input which is a sequence of words into a sentiment. This means that sometimes the input and outputs represent sequential data whilst at other times either the input or output will be non-sequential. They are commonly used in NLP and translation, music generation, name recognition and sentiment classification. RNNs have though some major limitations or disadvantages which include the training procedures required which are highly complex and slow, the fact that they cannot be stacked up on each other which as we will see is important, difficulty dealing with long sequences and like FFNNs experience the vanishing or exploding gradient problem.

(b) Long-Term Short-Term Memory (LSTM) Networks

LSTM networks are a type of RNN but much better at dealing with sequential data where order matters because it can memorise or hold the required data sequence and dispose of unnecessary or unused data, which enhances data classification. For this reason, LSTMs are widely used in learning sequential data prediction problems. Some sequential data has a time series or sequence of measurements of some quantity taken at several discrete times such as the share price of a company over the day measured each minute. For other sequential data, time is not important, and the time stamp does not matter, for example, it may be a DNA sequence or text data in which the order matters but not time. Time series and sequential data are often used interchangeably even though they are not the same. A simple version of an LSTM network consists of an input, output and forget gate using the Sigmoid function with the forget gate ignoring information that tends towards Zero rather than one.

(c) Gated recurrent Unit Neural Networks Memory (GRUN)

Gated Recurrent Unit Networks (GRUN) were introduced in 2014 by Kyunghyun Cho and colleagues to solve the vanishing gradient problem that plagued RNNs and is a variation of the LSTM networks. They use update gate and reset gates which are two vectors which select the information to be sent to output and can retain past information so they can store and filter the information using the update gate and reset gate. The update gate allows the model to determine what past information from previous time steps should be used whilst the reset gate allows the model to decide how much past information to forget.

(3) Convolutional Neural Networks (CNN)

Convolutional Neural Networks ("Conv Nets" or "CNNs") involve multiple layers of nodes or neurons, learn directly from the data, and are often utilised for classification and computer vision tasks by applying learned filters to input images and create feature maps that summarise those features. A major problem with images or feature and activation maps is that minor changes in feature positions due to re-cropping, rotation, shifting etc, of an image, can result in different features or activation maps. It is called both a feature or activation map because it maps the activation of different parts of the image and where a feature is found in the image, but both refer to the activation of the hidden neurons output from the layer. CNNs can have tens to hundreds of layers, each of which learns and detects features of an object or image so they can be trained to classify scenes and detect objects.

In CNNs, not all input nodes are connected to all output nodes which significantly reduces the total model size and parameters required when compared to a fully connected DNN. Layers close to the input level learn low-level features such as lines and edges and deeper layers learn more abstract features, like shapes or specific objects. CNNs rely on certain key principles such as (1) Local Receptive Fields which means that only a small region of input layer neurons called local receptive fields are connected to neurons in the hidden layer to create a feature or activation map (2) Shared Weights and Biases which mean that the weights and biases are the same for all the hidden neurons in a layer so they all detect the same feature such as an edge or line in the image, so can more easily translate features across images (3) Activation functions which uses activation functions on the outputs, with the most common being Sigmoid & Tanh, Rectified Linear Activation Function (ReLu), Leaky ReLu and Parametric ReLu, Exponential Linear Units, Scaled Exponential Units and Swish Activation Functions. For example, the commonly employed ReLu function maps the output of a neuron to the highest positive value or zero. Because the derivative used is 1 for all values larger than zero, the ReLu function overcomes the vanishing gradient problem in machine learning. The loss function in machine learning represents how accurate or poor the model's prediction is and generates a loss penalty associated with this. It does this by computing the distance between the current and expected output of the algorithm evaluating how well the algorithm models the data. The weights and biases are designed to reduce the loss function but in a network with what is known as a vanishing gradient, it can't update the weights, which means that the network can't learn and therefore the performance of the system deteriorates.

The vanishing gradient phenomenon occurs during the training of DNN and involves the error gradient discussed earlier that is used to teach or update the system vanishing during backpropagation from the outer to inner layers. Early layer gradients are derived by multiplying with later layers gradients but if they are less than one the multiplication rapidly vanishes. There are many ways to overcome the vanishing gradient problem and ReLu is just one example. Others that are employed but we won't go into our Weight initialization, ResNets, Multi- level Hierarchy, Batch normalisation, Long Short-Term Memory (LSTM) discussed earlier, Gradient clipping, Skip connections, Gated architectures, Reduced Network Depth and Faster Hardware and Rectified feature maps which are simply feature maps created using ReLu (4) Pooling which involves a pooling operation or filter which condenses small region of neurons into a single output, because the size of the pooling operation is smaller than the size of the feature map. For example, the pooling layer will usually always reduce the size or patch of each feature map by a factor of 2, so each dimension is halved, reducing the number of pixels or values in each feature map to one-quarter the size. A patch is just a subsection of an input image and feature detectors (kernels) process one patch at a time to detect features such as edges etc, so the patch is the input to the kernel.

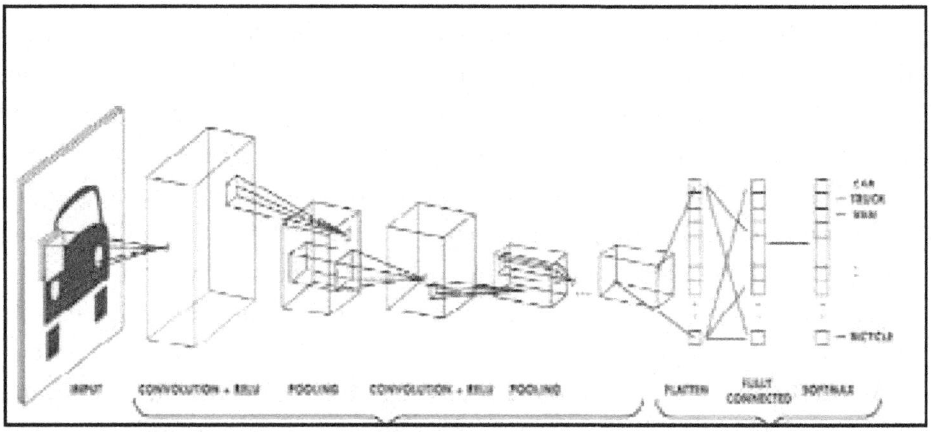

A Convolutional Neural Network

Average pooling calculates the average value for each patch on the map and Max Pooling calculates the maximum value for each patch. Current state-of-the-art CNNs are increasingly attempting to incorporate many fundamental features of neural computation, often described as canonical neural computations. Canonical simply means authoritative and accepted rules, because they are "hypothesised to apply the same mathematical operations in a variety of contexts" and "are capable of explaining and unifying seemingly unrelated neural and perceptual phenomena."[232] Examples are Non-linear Transduction, Divisive Normalisation and Max Pooling. Non-linear Transduction means converting something especially energy or a message into another form. In biological cells for example transduction is governed by the interaction of interacting proteins with signalling molecules binding to cell receptors and inducing stochastic transitions between different conformational state signalling requires multiple binding events so the rates of reaction have a non-linear dependence on the amount of signalling molecule. This non-linearity "allows robust switch-like properties critical to their biological functioning" but what is remarkable is that non-linear networks exhibit bi-stability, which is where the network activity can switch between states that correspond to a low and high activity level.[233] Divisive Normalisation (DN) was first proposed to explain many neurophysiological and behavioural aspects of how the human senses integrate multi-sensory information and cognition. It means that the output of a group of neurons does not just reflect the stimulus input but also the "contextual activity of other neurons" and the modulating factor is the ratio between these two components.[234]

This is particularly important in the visual system which must represent concurrent stimuli and involves normalisation which captures the neural spike responses and local field potential responses of nearby neurons in reaction to stimuli. The result is the balancing of "a neuron's net excitation with inhibition reflecting the overall activity of the neuronal population." Max Burg, in a study on the primary visual cortex, describes how a "neuron's driving input is normalised divisively by a weighted sum over nearby neurons' responses."[235] For example,neurons in the Primary Visual Cortex (VI) of the Occipital lobe are extremely sensitive to orientation of the stimulus and shapes and increase or inhibit signal activity for specific degrees of orientation.[236] This balancing of neuron's net excitation with inhibition has been cited as important for healthy brain function and alterations to the DN process have been implicated in autism and even hypothesised to "contribute to a wide array of the disorder's behavioural consequences". This is attributed to an "increased ratio of neural excitation to inhibition (E/I) in the disorder" and described by Rosenberg as reflecting a less social neuronal population.[237]

Max Burg and his team have exploited the feature of DN by building a machine learning system based on how the primary visual cortex (VI) works in humans. Their stated goal is to " build a model that predicts neural responses to arbitrary stimuli with high accuracy" and this involves allowing their model to have one layer of subunits and pooling of a group of neurons with Learned Orientation Specific DN, so in effect, they learn how to employ normalisation in the way the brain uses this process. Burg reports how the model outperforms linear-nonlinear and wavelet-based feature representations and makes a significant step towards the performance of state-of-the-art CNNs.[238] Max Pooling of inputs involves a pooling operation or filter which condenses a small region of neurons into a single output because the size of the pooling operation is smaller than the size of the feature map and max pooling simply calculates the maximum value for each patch. Dense layers are often used in CNNs and involve a linear operation, with every output formed by the function based on every input. The dense layer does not employ a filter like the convolution layer in which the output is formed by just a small size of inputs depending on the size of the filter and the weights. Dense layers are used when there is no prior information about the data or the relationship to the prior data, so the deep layer must learn the relationship.

Variational Auto-encoders (VAEs) which are widely used in CNNs are generative models that have emerged from traditional auto-encoders and useful in creating art, music, text and even fake human faces because they can alter or explore variations in existing data in very specific ways. In computer science a traditional auto-encoder comprises two connected networks with an Encoder and Decoder. The encoder receives the input and compresses it into a more dense and small representation whilst the decoder converts it back into the input. They are used in CNN as convolutional layers to convert a large image into a smaller representation which is then used to classify an image with the auto-encoders relying on the encoder to intelligently discard irrelevant information. In practice traditional auto-encoders have had limited applications because they do not sample data continuously or randomly from a generative continuous space and cannot deal with gaps between data clusters, because they only use the encoded vectors they are trained on. Deep learning networks including CNNs are vulnerable to subtle modifications to the input data for example pixel perturbation or 3D scene transformations which can result in incorrect predictions.

By contrast, VAEs are designed to sample randomly from a continuous space distribution area by making the encoder employ random sampling and standard deviation from the mean to generate input stochastically which involves each encoding generating a new variation. They employ the Kullback–Leibler(KL) divergence which measures how much two probability distributions diverge from each other. Using KL divergence, the decoder learns not to just decode specific single encodings but the ones that vary slightly as well so it is exposed to and encodes a range of variations and produces smoother continuous sampling. Sampling aims to encode all data which though distinct is very similar and to quantify and ideally minimise how much useful information is lost during encoding. The objective is to quantify the amount of information or entropy or the minimum number of bits in the raw data required to encode the information. In discontinuous sampling, more information will be lost so continuous sampling considering variations is much more effective. All neural networks involve approximations that try to minimise the loss and VAEs are a very efficient way of approximating the information in any given data set allowing much more efficient inferences to be made. Research scientist Carl Doersch who works at Deep Mind observed how only three years after their introduction VAEs had become a very popular approach to unsupervised machine learning.

(4) General Adversarial Networks (GANs)

General adversarial networks (GAN) are an unsupervised deep-learning-based generative model. GAN automatically discovers and learns pattern regularities in input data which it uses to produce new examples of the data. They approach the problem by framing the unsupervised training as a supervised problem involving two models. The first is the generator model that creates the examples whilst the second is the discriminator model that in a zero-sum, adversarial way decides whether the input is real or fake and if it appears real at least 50 % of the time it is regarded as the most plausible representations.

(5) Self-Organised Neural Networks (SONN)

Some ANN developers have employed an unsupervised learning model called, Self-Organised Neural Networks (SONN) often also known as Auto-Organised Networks or Kohonen maps. First introduced in 1988, SONN differs from other ANNs because instead of error-correction learning methods like backpropagation with gradient descent, they involve competitive learning for the model training. All the nodes on the map grid are connected directly to the input vector, but not to one another so they don't know the value of the nodes around them and learn competitively rather than employing backprop error correction.

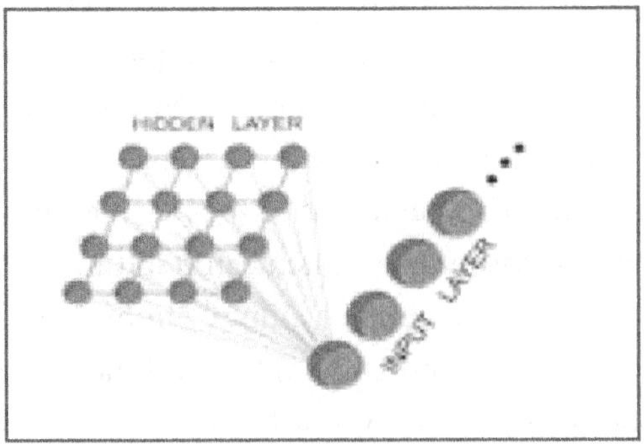

Kohonen Maps.

In SONN if neurons are inactive or don't respond to a specific input pattern, they will shift the weights of these neurons to active neurons which have won the competition to respond to the input. The chosen node called the Best Matching Unit (BMU) is selected according to the similarity, between the current input values and all the nodes in the grid. SONN also have another self-organising feature that resembles the way biological neurons work which is that they can create 2D feature maps, known as Kohonen maps after Finnish professor and researcher Dr Teuvo Kohonen, who pioneered them in 1982. They involve topology conserving and unlike ANNs involve only two neuron layers with a fully connected input layer and an output layer which is the map or Kohonen layer that is used to classify higher dimensional data in a lower dimension or topology. The learning phase involves the construction of the maps and in the prediction phase,new data is classified with a specific location provided on the converged map.

There are major limitations with SONN which include the fact that Kohonen networks are less effective on categorical data which you remember is a type of qualitative data that can be grouped into categories instead of being measured numerically. They perform very poorly with mixed types of data that can't be used in generative models as the data cannot be developed or trained against slowly evolving data and the learning phase of the model when training is very time-consuming.

In a study published in the November 2023 edition of Nature Machine Intelligence, Jascha Achterberg, and co-lead Dr Danyal Akarca and colleagues created an artificial system intended to model a very simplified version of the brain and applied physical constraints. Achterberg observes how "Not only is the brain great at solving complex problems it does so, while using very little energy". Akara observed how "This stems from a broad principle, which is that biological systems commonly evolve to make the most of what energetic resources they have available to them. The solutions they come to are often very elegant and reflect the trade-offs between various forces imposed on them." They discovered that their system developed key characteristics and methods like those found and employed in human brains. They applied a physical constraint on their AI brain by allocating each artificial node to a specific location in a virtual space, and the further the distance between two nodes the more difficult it was for them to communicate, which resembles how human brain neurons are organised. The system had to navigate a simple maze which required it to maintain information like where to start and end and intermediate steps, which made it possible for the researchers to track the nodes that were active during the different task stages. The physical constraints meant that the distant nodes found it more difficult to build a connection between the two nodes in response to the feedback which reflects how in the human brain, connections involving a long distance are expensive to establish and maintain. They observed that the system employed some of the human brain tricks to solve the task. by starting to develop hubs which as we have seen are highly connected nodes that pass information across the network. What was remarkable was that the response profiles of individual nodes changed so instead of each node coding for one particular property of the maze task, like goal location or next choice, the nodes developed a flexible coding scheme. At different moments nodes might fire for a mix of the maze properties. For example, the same node might encode multiple locations of a maze, rather than specialised nodes encoding specific locations, which as we have seen is a feature of complex brains.[239]

(6) Spiking Neural Networks (SNNs)

ANNs are based on the physical theories of how our brains communicate internally via electrical signals in neural networks. Galinsky and Frank observe that a "wide range of ANNs has been based on the mechanism of memory and learning using the Hodgkin–Huxley (HH) theory and ad-hoc circuit analogies, but the theory is too complicated to explain even relatively small neural networks." Hence in ANNS they use a "much-reduced model of a Leaky Integrate-and-Fire (LIF) neuron which replaces multiple gates, currents, channels and thresholds with just a single threshold and time constant." This has led to the use of standard algorithms and the simplistic LIF neuron as the basis of AI machine learning and deep learning with the majority being Spiking Neural Network(SNN) models.[240]

SNNs involve neurons only being connected to the neurons near them which are only activated when the membrane potential reaches the activation potential, so like human neurons, they continuously communicate with each other using discrete spiking signals. It recognises that when compared to BNNs the ANNs and DNNs used in AI are very different structurally and computationally and employ different learning rules. As Kashu Yamazaki and others observe in a recent review of SNNs, "DNNs are very resource-intensive in terms of energy consumption, data requirements, and high computational costs. With the recent increasing need for the autonomy of machines in the real world, e.g., self-driving vehicles, drones, and collaborative robots, the exploitation of deep neural networks in those applications has been actively investigated. In those applications, energy and computational efficiencies are especially important because of the need for real-time responses and the limited energy supply". They acknowledge how despite all the progress in deep learning "ANNs still lag behind the biological neural networks in terms of energy efficiency and abilities for online learning" and that a promising solution often cited as the third generation ANNs are the biologically plausible SNNs which purport to "bridge the gap between neuroscience and machine learning."[241]

Nikola Kasabov explains how "most important information of an input pattern is contained in earlier arriving spikes. It establishes a priority of inputs based on the order of the spike arrival on the input synapses for a particular pattern. This is a phenomenon observed in biological systems as well as an important information processing concept for spatio-temporal problems, such as computer vision and control which allows them to process parts of an input separately. Rank order learning (RO) makes use of the information contained in the order of the input spikes (events)."[242] This results in fast learning and asynchronous, data-driven processing, real-time adaptation and an open evolving structure based on RO learning. Using a variety of algorithms this has been extended to generate populations of Artificial Spike Trains (AST) which seek to achieve realistic or actual patterns of neuron inputs seen in BNNs such as a group of neurons converging onto target neurons, thereby closely mimicking the non-linear and input specific dependence of synaptic integration in BNNs.[243]

Although these various types of ANNs employing artificial neuron arrangements are attempting to capture how our brains process data and turn it into information through neural networks, none yet come close to achieving the level of diversity and functions of human neurons. We have seen how our brains use and identify features from the raw data of sensory stimuli to compile an accurate internal representation of the information contained in that raw data, which allows us to act upon it by performing cognitive processes and making decisions or generating behaviour etc. For example, all the diverse data features of a familiar human face will have been stored in the neurons and neural networks in our brain such that when we see that face again, the features that we rapidly extract from the raw visual data of the face are processed to build information based on an internal representation which allows us to recognise or not recognise the face. An additional challenge to AI is that the human brain does not just recognise a face or features, but imbues it with emotions, memories, biases and prejudices, abstract concepts like friend, lover, stranger, foe etc, and even a dynamic narrative so that our reaction to every face that we recognise is unique and highly subjective to us as an individual. To get AI to achieve facial recognition is one thing but to allow it to bring the richness of interpretation and response that the human face evokes is a major challenge. With AI systems it is a difficult balance between building the data features in from the outset or allowing the system to learn to extract features and by learning the features, use them to build representations of the information by itself. Being able to do so by learning rather than pre-programming makes the system more flexible in its ability to operate in uncertain and unfamiliar environments by generalising those features across environments and enabling it to use the data to gain information and knowledge of its environment. This will typically depend on the type of learning algorithm employed and within AI there are the different subsets of machine learning and deep learning which have important differences in the way they operate and their potential application in robotic autonomy. One of the main reasons that AI technology seems to be progressing rapidly is the trend for open-sourcing commercial tools from legacy and emerging tech companies. This has been described as the democratisation of AI and could be why major advances in AI technology are on the horizon and offer a perfect example of collective human intelligence in operation.

What this brief review of these separate initiatives shows us is that rapid progress is being made in ANN and much research has been applied in applying our understanding of the functional capabilities of neurons to the artificial neurons used in the various ANNs.

Key to this is the technical development in the artificial neurons that comprise the ANNs, so we need to look more closely at these artificial neurons.

Artificial Neurons

Developers of ANNs have typically attempted to replicate the role of the neuron in the human brain by using artificial neurons and other vital brain features when selecting and processing vast amounts of data and information. The first artificial neuron model was proposed in 1943 by McCulloch and Pitts (McCulloch and Pitts, 1943) to address one of the major challenges facing robotic AI, which is replicating the human neuron with its remarkable properties and diversity. ANNs using artificial neurons have been in use since the 1960s with modern ANNs using complex configurations of the CPUs or new types of processing units.

Artificial neurons or nodes are connected by wires that mimic the role of the axons and dendrites and the different levels of connectivity between these artificial nodes are usually characterized by an assigned number and differ between different types of neural networks. Enormous progress is being made in recreating the functionality of biological neurons in artificial neurons and we will look at a few of the recent innovations. Because the field is moving rapidly, it is impossible to capture all the developments and those selected represent just a few examples of the progress being made and the implications for building autonomous androids. ANNs are essentially mathematical models driven by algorithms and comprise many processing components and although ANNs using artificial neurons and machine and deep learning algorithms etc, attempt to mimic our brain, they by no means accurately replicate the function and structure of the human brain and there are many major fundamental differences.

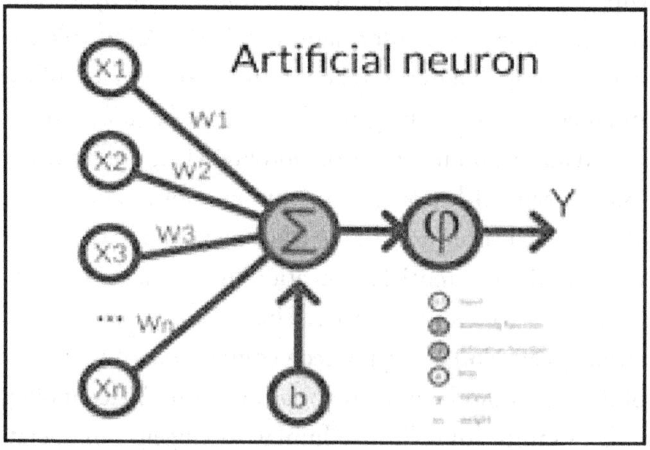

Unlike the biological neurons, artificial neurons are required to store and process information separately and AI has struggled to replicate the synaptic plasticity exhibited by biological neurons or their remarkable functional diversity and morphologies. The architectures in AI systems are physically separated, which has important implications with Sterling and Laughlin pointing out that "Biological nervous systems significantly outperform ANNs based on conventional silicon hardware in terms of energy/area efficiency."[244] Although the biological functions of the neuron and in particular the synapse where most of the activity takes place are highly complex and not fully understood, computer scientists have typically reduced their main functionality to multiple-input and single-output devices where algorithms describe the input-output transfer relationship. Over most distances in the human brain communication of electrical signals is digital but as we saw at the synapse it involves chemical neurotransmitters which are a much more efficient analogue technique. By contrast, computers have traditionally relied entirely on digital communication and processing, although we will see later how computer scientists are beginning to mimic the role of modulatory chemicals like neurotransmitters and hormones in ANNs.

Coding and Artificial Neurons

No one AI cognitive architecture or ANN allows for the processing of multi-modal sensory data in a way that humans do when attributing meaning, knowledge, and understanding to that information, although computer scientists are rapidly developing capability at different levels for each of the sensory modalities of vision, smell, hearing, taste, and touch. Nicholas Cassimatis and colleagues observe the immense challenge faced by AI of "needing to integrate multiple representations and inference techniques about perception, cognition, and action, which is something our brains do naturally all the time."[245] Unless AI robots are embodied and learn from their experience of interacting with the real world, will they ever be able to perceive the world as we do, particularly as we will see much of this processing relates to information associated with the self and preservation of the self? AI learning algorithms must extract structure from the data and encode that into the network parameters which are the weights and thresholds of the artificial neurons. Most AI systems are to varying levels extremely sensitive to the data sets that they use, with changes to external stimuli, noise, missing values etc, all capable of distorting their ability to operate the necessary algorithmic functions potentially resulting in them being unable to extract meaningful information from the data or stimuli in the environment.

ANNs have developed a wide variety of supervised and unsupervised algorithms for a diverse range of applications, including classification into categories, pattern recognition clustering, optimization, prediction, forecasting function approximation etc. Although computers using algorithms can efficiently execute arithmetic and logic operations, the human brain is remarkably efficient in executing operations that involve pattern recognition and matching and AI has long strived to match this pattern recognition prowess.[246] Humans naturally learn to infer invariances from the environment which means that the information contained by invariant features is "constant, unchanging and specifically unchanged by specified mathematical or physical operations or transformations."[247] An example is translational invariance, which means that the system produces the same response, regardless of how its input changes so such invariances can form the basis for generalising and inferred learning across different and multiple environments. In theory, any ANN must be able to recognize such invariances and extract essential parameters from complex high-dimensional data.

Some programmers believe that to learn, all AI systems require enormous amounts of prior data and sufficient installed or pre-programmed invariants for the system to be able to make accurate predictions and generalize patterns. Taken to extremes this requires very detailed generative data models describing how the real world works including many pre-programmed invariants, which in theory would enable it to deal with all potential situations. Unless such models are used only in tightly controlled environments this would require almost impossibly complex models making the system far less flexible or capable of dealing with unexpected and unpredictable situations or uncertain environments, which are inevitable in the real world. Some designers prefer to build in larger ANNs with more nodes and connectivity and rely on the greater flexibility and learning capacity of the program to come up with the optimal output and deal with uncertainty. At the heart of the different approaches is the level of connectivity of the neural network and the level of parameters that need to be programmed into the AI system.

Parameters

Parameters are " a set of facts or a fixed limit that establishes or limits how something can or must happen or be done."[248] The human brain is constantly adapting to its environment and making predictions and self-correcting without supervision. This involves the construction of useful representations or mental models, the synthesizing of novel patterns, combining and integrating ideas from different sources and grouping these to present and interpret new patterns. On a hierarchical basis biological neurons can predict the state of neighbouring neurons and adjust their synaptic parameters based on how different their predictions are from the observed signal and are constantly inhibiting or activating other neurons to process information contextually. This generative human learning process at the neural level does not involve enormous numbers of labelled examples or supervision and has recently been explained not by reference to clever learning algorithms in humans but by being encoded in the human genome. Neuroscientist Anthony Zadar who studies how brain circuitry gives rise to complex behaviour argues that since animals are born with "highly structured brain connectivity which enables them to learn rapidly, that this high level of connectivity is specified in the genome" and reflects how we rely heavily on a combination of innate mechanisms and learning. He maintains that "the wiring diagram is far too complex to be specified explicitly and must therefore be compressed through a genomic bottleneck which might suggest a path toward ANNs capable of rapid learning." Zador and colleagues are currently developing new methods for determining the complete wiring instructions of the mouse brain at single-neuron resolution. He soberingly observed that "We cannot build a machine capable of building a nest, or stalking prey, or loading a dishwasher. In many ways, AI is far from achieving the intelligence of a dog, a mouse, or even of a spider, and it does not appear that merely scaling up current approaches will achieve these goals."[249]

Google researchers recently asserted that unshared weights, allow an AI system to learn more invariances and that by not having to build in so many invariances, the learning system is more flexible. Google's rationale relies on using biological sparse coding in machine learning like the grandmother neuron principle that we discussed earlier below when we looked at the types of neurons in the brain. Many current AI systems use labelled data for things like face recognition, but this is a problem when there are few labelled data sets. Using work by Olshausen and Field from the 1990s, albeit that this work only captured low-level concepts and simple, not complex invariances, Google postulated that a system using unlabelled data and unsupervised deep learning could use sparse coding to self-learn face recognition.[250] Initially, it was only able to self-learn low-level features such as an edge or a blob and even this required a long learning time, but Google is seeking to address this by scaling up the datasets, the model, and the computational resources. What does sparse coding mean in the context of AI?

Sparse Coding in AI.

Some large-scale AI unsupervised learning systems harness the biological feature of sparse coding in high-level feature recognition. It provides a powerful machine learning technique that permits compact representations in complex data and is increasingly used in brain imaging, natural language processing, and image and sound processing. Sparse coding involves finding a sparse representation of given data which means that most of the vector values in the new representation will be zeros. Any single item such as an image in the dataset or stimulus is represented by the stronger connection of a comparatively small number or set of neurons and fewer features from a dictionary of learned features (factors) and the sparse representations can be transformed into original ones. Artificial neuron technology is advancing rapidly with the potential to transform the capabilities of AI inspired as we will see by the features of biological neurons. Let's look at some of the most interesting developments.

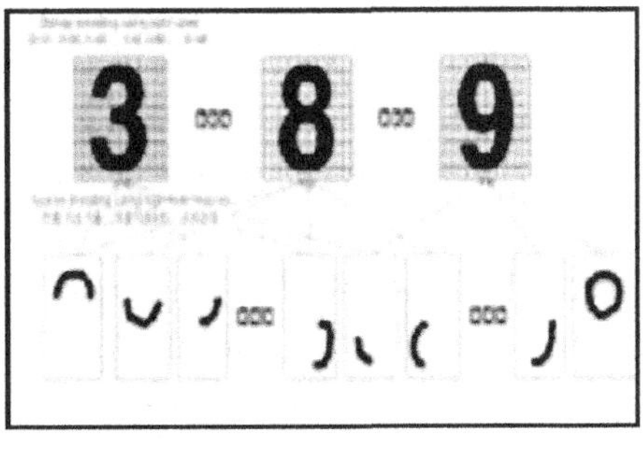

Developments in Artificial Neuron Technology

Biotechnology engineer and neuroscientific researcher, Frederic Narcross who has focused on modelling nervous systems observes how three dissimilar methodologies are being pursued in AI towards a "common path towards biological authenticity" and believes that progress towards biological realism would be expedited by using Artificial Nervous Systems (ANS).[251] The three are (1) Machine Learning which utilizes a variety of networks that we have discussed (2) Jeff Hawkin's company Numenta, which uses research into neuroscience to build computer models and algorithms that represent brain functions and (3) Neuromorphic computing which supports the trend towards biological realism in AI. Narcoss cites the following observations made in a 2020 paper observing that "Despite the recent progress in deep learning methodologies and their success in various fields, such as computer vision, speech technologies, natural language processing, medicine, and the like, it is obvious that current models are still unable to compete with biological intelligence. It is, therefore, natural to believe that the state of the art in this area can be further improved if bio-inspired concepts are integrated into deep learning models."[252] Narcross advocates that we need to integrate bioinspired concepts into deep learning models and like many others feels that real progress in AI will be made at the intersection of biological and machine intelligence.

The effectiveness of artificial neurons appears to depend on how well they can replicate the remarkable properties of human neurons and the growing interest in the performance of artificial neurons that mimic real neurons, has focused on developing what are known as bio-inspired microelectronics and neuromorphic computing. These include (1) Memristors which are some of the earliest examples of bio-inspired artificial neurons (2) Organic Electrochemical Transistor (OECTs) (3) Conductance-Based Organic Electrochemical Neurons (c-OECNs) (4) Neuromorphic chips (e) Dendritic neurons and (5) Astrocytes.Let's briefly look at these.

(1) Memristors

In 1947 the Bell laboratory pioneered transistors which replaced the large and fragile vacuum tube technology used in electronic devices, were much smaller, required less power operated at a much cooler level, and revolutionised technologies such as radios, calculators, and telephones. The gradual integration of silicon transistor technology has allowed them to be produced at ever-increasing smaller and smaller scales. Transistors operate in a binary fashion in that electrical current is either switched on or off and conventional computing saves Boolean-type data (0 or 1) in a single unit. This compares to the synaptic plasticity of BNNs which save multiple states in a single synapse by adjusting the synaptic weight. To do this in artificial neurons requires one memory cell to gradually change the information and a type of resistor called a memristor has been designed to do precisely this.[253] First developed in 1971 by Leon Ong Chua, memristors, a combination of the words 'Memory' and 'Resistor', are electrical components in which the conductance in two-terminal electronic devices can be precisely modulated by the charge or flux through them.[254] Unlike an electrical resistor and transistors that have a fixed resistance, memristors possess a voltage-dependent resistance making the material's electric properties paramount although unlike electrical capacitors, memristors don't store energy. Memristors employ various switching mechanisms which means that in a larger ANN with multiple paths, the system can control or switch data to transmit it between specific points on a network and select the best route for data transmission, which is why memristors have been adopted to artificially perform the function of an artificial synapse. They can be produced down to 2–10 nanometres, are stable in state and non-volatile which means that they retain memory even without power or using stored power, and have recently achieved high endurance levels at 120 billion cycles and memory duration of 10 years. Their great advantage is that they are analogue rather than digital and work like the human brain which is analog in terms of the chemical transmission at the synaptic level. Combining memory and computational elements reduces component size and energy consumption. Mario Lanza who specializes in material science and engineering believes that memristors will eventually replace traditional transistors as the new switching technology standard, and vastly improve the speed and operational efficiency of AI systems.[255] They can currently be made in four different materials with four different applications and Lanza believes that there are many other potential material combinations with memristors made of two-dimensional layered materials and perovskite crystals which will rapidly improve their performance. Seung Ju Kim recently reviewed the main six competing memristor technologies which are (1) Ionic migration, (2) Phase change, (3) Spin, (4) Ferroelectricity, (5) Intercalation, and (6) Ionic gating and observed how memristors have the potential to address the fact that "conventional computing systems are prone to volatility, high power consumption and even delay between the processor and memory, due to the von Neumann bottleneck."[256] This bottleneck is the result of the processor and memory regions being physically separated, and data being transferred via bridged buses. Mogahid Zidan believes that memristors can overcome this problem and observes that they have three major applications in AI namely (1) On-chip memory and storage (2) Biologically inspired computing like neuromorphic computing and (3) General-purpose in-memory computing.[257] One study by John Hermiz and colleagues demonstrated that "using memristors as synapses in neuromorphic circuits (see below) can potentially offer both high connectivity and high density required for efficient computing."[258] Resistance can be programmed and stored in memory so they can limit or regulate electrical current in a circuit and remember the amount of charge that has previously flowed through the circuit. They can be scaled in arrays and made in natural biomaterials like proteins. Another study demonstrated that parallel memristor circuits could even emulate high-level biological behaviours such as decay and interference in short-term human memory which "implies the existence of a capacity limit beyond which memory cannot be securely formed and retained."[259]

In 2023 technology giant IBM launched a new brain-like chip based on memristors with a hugely increased level of efficiency-driven component design along the lines of the neural connections in the brain. Thanos Vasilopoulos, who is based at IBM's research lab in Switzerland, observed how memristors could allow "large and more complex workloads. (to be) be executed in low power or battery-constrained environments."[260] Prof Ferrante Neri from the University of Surrey describes how a memristor could "remember its electric history, in a similar way to a synapse in a biological system" and that "Interconnected memristors can form a network resembling a biological brain," He expresses cautious optimism that "These advancements suggest that we may be on the cusp of witnessing the emergence of brain-like chips shortly."[261]

Despite their obvious potential and IBM's recent announcement, there remain challenges to the widespread adoption of memristors. These include limited commercial availability, difficulty in designing new circuits with memristors, the fact that they currently exhibit much lower speeds than dynamic random-access memory (DRAM) which is a specific type of RAM, non-linearity in the write process, energy-costly switching, and the lack of biocompatibility for tissue interfacing. Hence their applications have to date been limited but they may well start to play a central role in the development of ANN and have even been described as "Shining in the field of neuromorphic computing."[262, 263] To address some of these problems of memristors a transistor called an Organic Electrochemical Transistor (OECT) is being developed.

(2) Organic Electrochemical Transistors" (OECTs).

Recent research led by Xudong Ji and Jonathan Rivnay is focusing on an Organic Electrochemical Transistor (OECT) which has been described as one of the most versatile bioelectronic devices available, capable of sensing, memory, and processing. OECTs can interface with living systems through organic electronic materials and can convert and amplify ionic and biological signals into electronic output.[264] The mixed (ionic and electronic charge) conduction properties of the material in OECTs act like a rapid switch in controlling the current flow through a circuit because the electrochemical state and the ability to impede the flow of the electrode enables high amplification of signals. Ji and the team observe concerning biological neurons that "a synapse relies on the action potential-driven transport of neurotransmitters, while an OECT relies on ions being transported similarly and they can therefore be used in biosensors, neuromorphic computing and studies that involve the mixed (ionic and electronic) conduction properties of materials."

A team of scientists has fabricated OECTs on conformable substrates which were implanted on the cortex and mapped electrocorticography signals over a large surface, achieving a higher spatial resolution and Signal-to-Noise Ratio (SNR) than conventional electrodes.[265] OECTs have the potential to be implanted with the human nervous system and if integrated with neuromorphic circuits could in robots allow ANN to come close to resembling how neural synapses work in the human brain by simultaneously sensing, processing, and storing data from the environment. OECTs are easy to fabricate and aqueous electrolytes making them compatible with biological substances and operate in a very similar way to a neural synapse in using transported ions. Shijie Wang and colleagues used OECTs to explicitly address the problem that in traditional AI systems sensing, information processing, and memory are physically separated and involve frequent data shuttling between hardware systems. This reflects the fact that traditional AI systems using ANNs compare unfavourably with BNNs in which sensing, information processing, and memory are co-located due to the need to support complex behaviours. They observed that OECTs have the potential to integrate "sensing, memory and processing functionality" by using a "vertical traverse architecture and electrode process, together with a crystalline–amorphous channel that can be selectively doped by ions to enable reconfigurable operation as a volatile receptor or a non-volatile synapse."[266] The vertical traverse architecture allowed them to keep the ions in the channel and the channel microstructure and electrode process allowed them to develop reconfigurable multi-modal sensing and non-volatile analogue memory capabilities. Another highly innovative technology called Conductance-Based Organic Electrochemical Neurons (c-OECNs) is also being developed.

(3) Conductance-Based Organic Electrochemical Neurons (c-OECNs).
In 2023, a team of computer scientists in Sweden made a technological breakthrough in producing synthetic artificial neurons called Conductance-Based Organic Electrochemical Neurons (c-OECNs), which they claim to mimic 15 characteristics of biological neurons and even stimulate natural nerves. They use n-type conducting polymers which are materials that can conduct negative charges and thus can incorporate ion modulation based on activation/inactivation of their sodium and potassium channels. c-OECNs can be inserted on a thin plastic foil enabling thousands of transistors to be produced on a flexible substrate to create artificial neurons which can be used to print complementary organic electrochemical circuits. The c-OECNs have even been connected to the Vagal nerve of a mouse and successfully used to stimulate the nerve cells to increase heart activity. Lead author Padinhare Harikesh attributes this to the "unique material which is an organic polymer semiconductor that is biocompatible, soft, and malleable and allows increase and decrease in the current a way that resembles the activation and inactivation of sodium ion channels found in nature". Their potential arises from the fact that they have exceptional sensing capabilities for biological and physical plus chemical signals, are biocompatible, switch speeds readily, and operate at low voltages and they believe that these combined features are impossible to achieve using previous technologies.[267]

(4) Neuromorphic Chips

One of the main drivers in edge computing is autonomous cars which can't afford the latency associated with relying on the cloud for information processing. Conventional computer chips are inadequate because even if the algorithms mimic the brain the chips don't and can't for example exhibit neural plasticity, whereas the neuromorphic chip is designed to form new neural connections and new physical structures just like the human brain. Existing machines and deep learning don't mimic biological neurons because all the layers are effectively connected, unlike in the human brain. The field of neuromorphic computing has led to developments in bio-inspired microelectronics and evolved from attempts to design computers along the principles underlying the structure and characteristics at work in our brains and in particular the neurons.[268]

Two notable projects developing neuromorphic chips are being pursued by Intel. The first, called Loihi, has 130,000 neurons and 139 million synapses and uses Intel's 14nm process technology The second supported by IBM is project True North which has 46 billion synaptic operations, with one chip capable of one million programmable neurons and 256 million synapses. They are employing SNNs and because only active chips fire whilst others are silent, they consume far less energy, are much faster, and learn quickly from much smaller data sets. Neuromorphic chips have the potential to generate the working memory more biologically and all the neurological processes which could accelerate the path towards autonomous robots.

Neuromorphic computing has become a major area of interest with computer elements and physical artificial neurons like memristors, OECTs, and c-OECNs, etc modelled after the human brain and CNS. As a subset of electronic computing, neuromorphic computing has been made possible by the ability to integrate or embed billions of transistors on a single silicon semiconductor microchip in a technology called Very Large-Scale Integration (VLSI). VLSI has been with us since the 1970s and is widely used for microchip processors, integrated circuits (IC), and component designing. Microscopic chips and transistors contain carbon and silicon molecules, which are perfectly aligned to move electricity through a circuit faster and the development of microchip technology has followed Moore's law, which is more of a goal than a law. It is attributed to Gordon Moore, the co-founder, and former CEO of Intel in 1965 when he observed that the number of transistors in an integrated circuit (IC) doubles about every two years and that the growth in microprocessors is exponential.[269] In the International System of Units, the prefix nano means one-billionth or 10–9, therefore one nanometre (nm) is one-billionth of a meter. Intel has produced a 7 nm chip which to put into perspective is one billionth of a meter smaller than the wavelength of visible light and atom diameters range from about 0.1 to 0.5 nm, a human hair is approximately 80,000–100,000 nm wide, and an inch is 25,400,000 nm.[270] There is a consensus amongst experts that we are nearing the limit of Moore's law because of the dual problem of power consumption and heat reduction. High temperatures of transistors eventually make it impossible to create smaller circuits because cooling the transistors takes more energy than already passes through the transistors. This appears to be leading to greater specialization in chip design with the aim of more efficient algorithm processing rather than custom design or standardized high-volume products. A major centre for neuromorphic engineering is the Institute for Neuroinformatic (INI) established in 1995 at the University of Zurich and ETH Zurich by Rodney Douglas and Kevan Martin. Their stated mission "is to discover the key principles by which brains work and to implement these in artificial systems that interact intelligently with the real world" and Douglas has explained that although there are many places to begin, they did so with the cerebral cortex. INI has developed neuromorphic vision sensors, silicon cochlea which play a vital role in hearing and auditory transduction, and medium-scale neuromorphic processors such as the Reconfigurable On-Line Learning Spiking (ROLLS) and cx Quad chips.

English computer scientist Steve Furber, in his 2016 review of Neuromorphic computing, explains how the cxQuad chips are used to implement a hierarchical CNN with each chip incorporating 1024 neurons and 65, 536 digital synapses. The ROLLS chip implements the classification layer for the DNN and incorporates 256 neurons and 128,000 analog synapses.Furber observes how all the latest large-scale neuromorphic systems face trade-offs between "energy-efficiency, integration density, flexibility, and configurability, analogue versus digital algorithms, hardware versus software."[271] There are currently several large-scale systems being developed each with advantages and disadvantages, which include, the IBM TrueNorth chip, the Stanford Neuro grid, the Heidelberg Brain Scales system, and the Manchester SpiNNaker machine. Furber describes how each of the large-scale systems described here has its strengths. True North offers a platform for very highly integrated and energy-efficient application delivery, SpiNNaker offers maximum flexibility for researching different neural models and plasticity rules, Brain Scales brings high acceleration for long-term learning and Neuro grid offers very good energy-efficiency with models that are closest to the physics at work in the biology. Although neuromorphic computing has enormous potential in AI and robotics and is evolving rapidly Furber concludes that "we are still some ways from identifying an optimal approach that is as general-purpose in this domain as is the general-purpose programmable processor in the conventional computing domain."[272]

(5) Dendritic Neurons

Biophysical models of neurons with a detailed morphology involve increased complexity in terms of overly complex equations with numerous free parameters, enormous demands being made on the CPU/GPU, and larger run times, all of which can negatively impact computational efficiency. Free parameters are those not pre-defined by the model but chosen or estimated based on theories or experimental data. To date, artificial neurons have not possessed the wide range of complex non-linear integrative properties displayed by human neurons, such as pyramidal neurons and the many diverse morphologies of many others.[273] Pyramidal neurons are known to integrate inputs at multiple dendritic segments whereas artificial point neurons can only integrate at a single point or zone and pyramidal neurons contain extensive dendritic arbours and many thousands of synapses, so a realistic model would in theory need to include these types of dendritic structure. In a 2021 paper computer scientists, Karan Grewal and colleagues observed how "dendrites impart fundamental computational benefits beyond a simple increase in complexity or computational power." They recognized that "the majority of ANNs ignore biological neurons' structural complexity and instead use simplified point neurons". To address this shortcoming, they incorporated into their ANN several properties of BNNs such as "active dendrites, local inhibition and sparsity, and complex synapses." To assess their impacts in continual learning tasks which traditionally unlike batch learning have posed a problem in an ANN, they introduced the Active Dendrite Neuron (ADN) and trained the systems using various continuous learning scenarios because they recognized it as a "core capability central to intelligent systems and that incorporating dendritic models significantly improved the level of retained learning or past information."[274, 275] Most dendritic synapses are not close (proximal) to the soma or body of the neuron but distal which means away from the centre and has minimal impact on the cell itself. But they do "process distal groups of synapses locally in a non-linear fashion and are thus known as active dendrites" which perform their computations and have been described as acting as subunits or like separate layers in a neural network.[276, 277] Single-point artificial neurons assume only a single input source and do not reflect the role of such distal dendrites in conjunction with proximal dendrites. It is known that the separate dendritic branches on a neuron can independently detect patterns and have a vital modulatory impact on the cell by keeping it depolarised for an extended time and thus nearer its firing threshold, in what has been described as a prepared state.[278] The ADN has been designed to receive two sources of input, analogous to the proximal and distal inputs and "the feed-forward inputs are processed just as in a point neuron, by computing a linear weighted sum of inputs. In addition, a set of dendritic segments on each neuron processes a context (distal) input, and the subsequent dendritic output modulates the neuron's response. The result is a neuron where the magnitude of the response to a given stimulus is highly context-dependent." They concluded that "augmenting point neurons with biological properties such as active dendrites, complex synapses, and sparse representations significantly improve a network's ability to learn continually", which we have seen is critical for autonomous agents. They maintain that a "3-layer ADN with SI can achieve greater than 90% accuracy when learning 100 permuted MNIST tasks in sequence."[279] Their results were applied using permuted MNIST so we must consider some of the earlier reservations about this as an evaluation tool.

In a 2023 study, Michalis Pagkalos and colleagues observed how many AI systems use SNNs that "typically comprise point, integrate-and-fire neurons that replicate basic biological features such as specific connectivity (network) motifs, excitation-inhibition dynamics and learning via synaptic plasticity rules." Network motifs are "over-represented and statistically significant patterns that are found in higher frequency in comparison to randomized networks."[280] Pagkalos claims like Grewal and others that SNNs ignore the role of dendrites and their "compelling computational capabilities" which "significantly influence both neuronal and circuit function which cannot be captured by point-neuron SNNs." They maintain that, although relatively efficient, "current SNNs are usually quite simplistic, overlooking essential dendritic properties." They observe that "circuit models with morphologically detailed neuron models are computationally costly, thus impractical for large-network simulations." To address the problem, they introduced Dendrify which is a free open-source Python package based on the Brian 2 Simulator which is expressly designed to capture only the essential electrophysiological characteristics of real neurons. Brian 2 is an open-source and intuitive model that enables scientists to efficiently and simply simulate SNN models. They claim that through simple commands Dendrify automatically generates reduced compartmental neuron models with simplified yet biologically relevant dendritic and synaptic integrative properties. They believe that "such models strike a good balance between flexibility, performance, and biological accuracy, allowing us to explore dendritic contributions to network-level functions while paving the way for developing more powerful neuromorphic systems."[281] They list the advantages that dendritic modelling provides including, (1) Improved associative learning (2) Better input discrimination (pattern separation) (3) Efficient binding/linking of information and (4) Increased memory storage and recall capacity." They appear to increase the continuous learning ability of the system, are realistic, highly accurate, lightweight, easy to develop, have far fewer parameters, and can be employed in larger networks.

(6) Astrocytes

Neuromorphic computing has inspired computer scientists to explore in collaboration with neuroscientists the latest understanding of the complex roles of the neurons, the morphology and function of dendrites and glial cells, and the interactions between them. This approach has given birth to the field of computational science which is slowly leading to modifications in the standard neural network models. Narcross how "over the last 30+ years neuromorphic chips have progressed from modelling several handfuls of "neurons" to modelling hundreds of thousands of artificial spiking neurons and astrocytes."[282] He stresses how "the inclusion of astrocytes in any AI system is a significant step in the direction of biological representation" because "Astrocytes sub-serve critical nervous system functions, outnumber neurons by a factor of 4:1, are presented as the gatekeeper of synaptic information transfer,[283] and an indispensable partner of synaptic plasticity (memory)." This reflects what Maurizio De Pittà describes as "the growing excitement around glial cells as compelling evidence points to new previously unimaginable roles for these cells in information processing of the brain, with the potential to affect behaviour and higher cognitive functions."[284] Only by understanding their role can the computational features that they bring to brain function be incorporated into biologically realistic ANNs. This includes for example the astrocytic contribution to brain functions with hundreds of computational models developed in recent years.[285] Tang and colleagues claim that "Astrocytes have largely been ignored in computational studies on the biological principles of intelligence" yet they 'sub-serve critical nervous system functions, outnumber neurons by a factor of 4:1, are presented as the gatekeeper of synaptic information transfer, and are an indispensable partner of synaptic plasticity (memory)."[286] They therefore introduced a piking Neuronal-Astrocytic Network (SNAN) into an Intel Loihi neuromorphic chip and claim that the inclusion of astrocytes in any AI system is a significant if not essential step in the direction of biological representation This is in fact "the basis of the Loihi module.. (which is) an end-to-end biophysically plausible compartmental model of an astrocyte that simulates the intracellular activity in response to the synaptic activity in space and time."[287] They maintain that the SNAN can "sense and induce activity-dependent neuronal synchronisation, switch on and off spike-time-dependent plasticity (STDP) to introduce single-shot learning, and monitor the transition between ordered and chaotic activity at the synaptic space."[288] Neuromorphic computing has even been able to model hundreds of thousands of artificial spiking neurons and astrocytes which has enabled them to explore and replicate what they describe as "the distinct computational roles that astrocytes have in forming biological intelligence."[289]

Neuromorphic Computing

We have seen how DL architects have been exploring ways of enhancing real-time operation, energy efficiency, and scalability by employing neuromorphic hardware such as neuromorphic chips. This reflects the growing need for faster computers driven by the widespread adoption of AI applications with AI algorithms consuming vast computing resources and the demand rather than relying on the cloud, to be able to use processes on individual devices using Edge computing.[290] One such initiative is the development of Artificial Nervous Systems (ANS) or artificial sensory neural systems which as Frederic Narcross argues would expedite the "Common path toward biological authenticity" in the field of AI.[291] Narcross has been researching neuroanatomy and neurophysiology for more than 30 years and programming computer systems since working at NASA in 1980. He claims that by using ANS, "any methodology, which wanted to reproduce human intelligence using mimicking central nervous systems, would have no limit or impediments to the progress which could be made. No further technologies would be necessary to achieve AI because the very inspiration to achieve the goal, nervous systems, would be fully accessible." In a 2023 article in Forbes Magazine Vitaly Kleban who is CTO at Everynet described how the major hurdle to the widespread impact of the digital transformation has been data starvation. He vigorously argues that "The world needs a large-scale artificial sensory neural system to enable such a link, stop data starvation and make a technology breakthrough possible in connecting AI to the real world for practical application beyond the confines of the data domain."[292] He observes how separate parts are already being built such as low-cost ML sensors and large-scale networks dedicated to low-power, low-cost sensor connectivity. These are necessary to equip AI systems with a body and as Narcross asserts any "NN in any form could be defined with realistic neurons, plasticity, and inhibition or excitation and connected in any biological or non-biological configuration. Multiple networks could be coupled naturally to analyse multiple data types simultaneously, and both input and output could be configured within the same tool set."

There are many more advantages and Narcross sets out what he describes as a non-comprehensive laundry list (1) Architecture that is developed both top-down and bottom-up (2) Astrocyte networks which can alert upstream astrocyte neighbours of local changes in activity and metabolic resource demand (3) Tripartite synapses giving astrocytes a key role in managing resources for their neuron partners by "listening" to the neurons' activity (4) Hormones and different neurotransmitters can communicate programmatic needs over long distances and persist over time to keep the environment balanced and optimized. This also obviates the need for direct neuron-to-neuron communications in many functions (5) Dynamic myelination to improve the efficiency of long-range neural connections when activity demands are sufficient, another example of optimization (6) Spike-timing-dependent plasticity (STDP) based upon clustering of dendritic spines, each cluster having the capacity for spiking and contributing to a soma spike. This includes both long-term and short-term plasticity and both synaptic growth and pruning (7) Plasticity based upon the initial tagging of what should be modified and then allowed to proceed based on past cytoskeleton modifications to the same synaptic site and on current resource availability (8) Global operating principles of optimization of resources and homoeostatic balance, including the peripheral nervous system (9) Customizable neuron characteristics sufficient for the six most abundant neuron types as well as the hundreds of other known neuron types (10) Cortical columns that can be customized to accommodate the five primary varieties of the cortex, as well as the transition areas between them and (11)Support for the creation of nervous system structures beyond the cerebral cortex, such as nuclei of the diencephalon and mid-brain, and the spinal cord and ganglia.[293]

Narcross envisages that "All ANS items should be customizable by the user and permit new application programming interfaces (APIs) beyond the basics provided, with the resultant code stored on GitHub or the like so that new APIs and modifications could be widely shared on sites such as Git Hub or other similar platforms. GitHub is a platform that since 2008 has provided a very important cloud-based service for software development and version control using Git. It allows developers to store and manage their code which is used by over 100 million developers who purport to shape the future of software, together. Microsoft acquired the platform in 2018 for $7.5 billion. Narcross tells us how "the tools do exist today that includes all the "what would it look like? features that can reproduce any user-definable nervous system and have a resultant nervous system model that is downloadable onto FPGA microchips for parallelism or otherwise operates on conventional computers.[294]

They support not only ML, Neuromorphic computing, and the work of Numenta on the Thousand Brains Theory of Intelligence (TBTI) (see below) but also the "fields of robotics, Internet of Things (IoT), cognitive computing, and biologically rigorous fields of medical diagnostics and remediation, such as for epilepsy and Alzheimer's disease." Narcross calls for wider collaboration to "improve the user-friendliness and usefulness of the tools and set ANS on a path into the public sector and fully elaborate the new paradigm." Event-based neuromorphic systems have been inspired by the remarkable efficiency and rapid response times facilitated by the human brain's use of SNNs for communication, to develop more power-efficient electronic sensors and processors. Computer scientist Shih-Chii Liu and colleagues observe how "Circuit building blocks are combined in architectures to construct complete systems. These include vision and auditory sensors as well as neuronal processing and learning circuits that implement models of nervous systems."[295] They recognize that the challenge is to enable the systems to deal with massive amounts of noisy multisensory data in the way the human brain does. Several initiatives in the field of neuromorphic computing illustrate these challenges but also the rapid progress that is being made and include (a) TrueNorth (b) SyNAPSE (c) Neurogrid and (d) CAVIAR the details of which are outside the remit of this book. What is clear is that the convergence of hardware technology and brain-inspired architectures is changing the face of AI with the potential for robots to harness some of the remarkable neurobiological features that make the brain so powerful.

Distributive Adaptive Control (DAC)

In 1991 scientists introduced the biologically inspired Distributive Adaptive Control (DAC) which is a biomimetic dynamic robot-based "Whole Brain" neural architecture that they claimed instantiates the theory of the MBBN and bridges the exploratory gap between the mind and brain. Originally designed as a model for operant and classical conditioning it has been applied in a range of AI adaptive behavioural systems and is grounded in the neuroscience of invertebrate and vertebrate biological systems. It claims to integrate and unify rather than negate the dominant paradigms of the 19th and 20th centuries across psychology, artificial intelligence, cognitive science, and neuroscience. It purports to synthesize intelligence and explicitly declares its ambition to explain the conscious, embodied MBBN.

DAC is an ANN that is organized like the human cortex in columns that process the

Environment, the Self-Action, or the Interaction between the environment and the Self and employs three layers called the (1) Reactive (2) Adaptive, and (3) Contextual layers.[296] In the Reactive layer, the processing is akin to allostatic control and the emergent non-linear output that we see in the human body and brain arising from the interconnectivity and integration of neuronal regions in the cortex. The Adaptive layer is intended to mimic learning in humans using the classical conditioning of reward and behaviour, so the neural substrates would be the Amygdala, the Neocortex, and Cerebellum, which govern motivation and movement and episodic memory mechanisms involving the hippocampus. The Contextual layer is the equivalent of the neural substrates of the Pre-frontal Cortex in terms of planning and executing decisions etc. Paul Verschure observes in his 2012 review of DAC over the previous twenty years that "It has built up a track record of explaining core aspects of mind, brain, and behaviour, has made testable and verified predictions at the level of behaviour, physiology, and anatomy, can control complex real-world artefacts and has been successfully applied to brain repair and neurorehabilitation. He acknowledges that it has yet to capture consciousness.[297, 298]

The Thousand Brains Theory of Intelligence

The Thousand Brains Theory of Intelligence (TBTI) approach has been pioneered by Jeff Hawkins, Donna Dubinsky, and Dileep George who set up Numenta in 2002 with the explicit goal of "applying twenty years of neuroscience breakthrough discoveries (and "neocortical principles") to dramatically increase performance and unlock new capabilities of AI." TBTI is derived from neuroanatomy and neurophysiology and is based upon specially designed network hardware architecture, data structures, and algorithms. The network hardware architecture reflects specific biophysical properties which they describe as including (1) Sparse connectivity (2) Neurons with dendrites and (3) The way cortical layers learn sequences and cortical columns learn to model objects through movement and common sensorimotor circuitry. The data structures have sparse contextual activations, and 3D reference frames and represent uncertainty.

The concept of sparse contextual activations in computer science has recently been developed to overcome some of the limitations of the point neuron that has been employed in computational studies of the neuron since it was first proposed in 1907, by the legendary doyen of French neurophysiology Louis Lapicque (1866–1952). Based on a simple linear integrate and fire mechanism it formed the basis of the Perceptron model developed by Rosenblatt and the early ANNs. The algorithms in TBTI employ a k-winner takes-all strategy, sensorimotor learning, and distributed model learning, which they claim allows the algorithms to intelligently act on data and adapt as the nature of the problems changes. A k-winner takes-all strategy reflects how in the brain rather than a single neuron becoming active there are k neurons that become active for a fixed number of k and involves a dynamic competitive learning in RNN in which non-linear inhibition is used to select the maximum neuron input and suppresses activity in all the other nodes of the network. TBTI was first validated in 2020 and commercialised in 2022.

Numenta's approach questions the traditional perspective of how the human neocortex operates, which is that information is processed step–by–step, from simple features to complex features to complex objects, as it passes from one region of the neocortex to the other. They instead use the theory first proposed by American neurophysiologist, Vernon Mountcastle (1918–2015) in 1978,[299] who discovered the columns in the somatosensory cortex of which there are around 150,000 which argues that information is processed in the cortical columns, which as we saw are the basic functional units tightly replicated across the cortex. These cortical columns are tiny and vary from 300 μm to 600 μm in diameter[300] with diameters estimated between 30–60 μm.[301] Numenta claims that "all cortical columns, even in low-level sensory regions, are capable of learning and recognizing complete objects through sensory inputs and movement. and they build tens of thousands of models for everything we know." Jeff Hawkins on the Numenta website describes it as "As if your brain is thousands of brains working simultaneously." Each column creates an internal representation or model of what it is sensing so thousands of models of the same data are created based on sensory information from different sense models according to visual, touch data, etc. They argue that the long-range excitatory lateral connections in the brains allow communication between these various columnar models which in "essence vote to determine which singular perception or representation should prevail." This is the k-winner takes all model in which "each cortical column builds predictive models of objects and connections between the output neurons shows competition between them with columns either 'ON' which means it is the winner and with others 'OFF' being the losers. The "columns integrate changing input over time to learn complete predictive models of observed objects."

Numenta employs a new technology called Hierarchical Temporal Memory (HTM) which represents the nervous systems' temporal connectivity which they describe as "biologically constrained rather than bio-inspired." The reason as explained in 2019 by researcher Vincenzo Lomonaco, is the need for "the biological details of the neocortex to be compatible with the theory, and that the theory can't rely on principles that can't possibly be implemented in biological tissue." Lomonaco illustrates this by reference to the dendritic tree-like structure of the pyramidal neurons that we discussed earlier which connect thousands of synapses and actively process elements which enables the neuron to "recognize hundreds of unique patterns with the pyramidal neuron doing this in multiple integration zones. Lomonaco stresses that "HTM neurons don't attempt to model all of the (detailed) aspects of biological neurons, only those that are essential for the (principles of) information-theoretic aspects of the neocortex."[302]

If ANNs are to replicate BNN they must incorporate such biological features, though as Lomonaco observes, "one day AI may be so advanced that it has departed from its biological origin." HTM reflects Mountcastle's observation that neurons don't have specific algorithms to process the separate senses but do so because they receive the data like visual data from certain senses like the eyes. This arises from connectivity within and between the various brain regions, which over time have become more homogenous and use universal algorithms that can process any type of sensor data and behaviour. Lomonaco asserts that the development of universal algorithms is what allowed the human brain to evolve relatively quickly.

HTM uses Sparse Distributed Representations (SDR) which are vectors with thousands of bits but only a small percentage are 1's rather than O's and is predicated on the belief that all intelligent systems must employ SDR. The more the SDR shares a 1 bit in the same location the greater the similarity in representations which means that neurons can simultaneously generate multiple representations and values and make multiple predictions. HTM uses the equivalent of human senses in the form of encoders that turn or convert data into SDRs and analyse it by using learning algorithms. This means that encoders can be created for any type of data even with no biological equivalent and embedded in a sensory-motor or behavioural framework akin to a body though the human body must be able to deal with constantly changing data which requires what the developers call Temporal Memory (TM). TM memorises sequences and transitions in data streams which allows the system to make inferences about the data and generate motor actions. Lomonaco claims that TM is the substrate upon which all neocortical functions are built and "the biggest difference between HTM and most other ANNs in assuming that everything the neocortex does is based on memory and recall of sequences of patterns." Encoding in AI systems in the human brain is central to how information is processed, transmitted, and stored and attempts to harness our understanding of this neuron coding are inspiring new initiatives in coding in artificial neurons.

The Thalamus in AI

We have seen that generative models are an important and growing form of unsupervised ML and DL. Generative models typically involve significantly smaller numbers of parameters than the amount of data they are trained on, so rather than just rely on memory they are forced to compress the data and discover and efficiently internalize the data's essence to generate features to learn and describe how the world looks. Recognising the crucial role of the thalamus AI developers have sought to develop brain-inspired algorithms which according to computational neuroscientist Ali Hummos from MIT use "optimization at inference time to dynamically generate internal representations of the current task."[303] In DL inference is the phase in development when the trained DNNs make inferences or predictions using new unseen data which involves forward propagation that gives an input generated and output. Knowing how long it will take for forward propagation to infer or predict something to get the output can obviously improve the learning model and thus computer scientists try to optimize the inference time. To measure this inference time computer scientists, calculate the total number of computations the model must perform. They use something called a Floating-Point Operation (FLOP) which is any mathematical operation (such as +, -, *, /) or assignment that involves floating-point numbers that have fractional points usually with a decimal point. Floating-point Numbers have decimal points, and the precision is the number of significant digits following the decimal point. FLOPS are the floating-point operations per second which tell an operator how efficient the hardware is because the more FLOPS the faster the operation and inference time will be. Designers typically aim for low enough FLOPS in the model to provide complexity and a higher level of FLOPS in the hardware. Optimization involves reducing the number of operations in the model and an example is the use of pooling which we discussed earlier which involves sub-sampling layers that reduce the data amounts or parameters passed from one layer to another.

Hummo's work focuses on expanding computations in the weight matrix of a neural network using the concept of top-down feedback control in biological brains. Hummos' brain-inspired algorithm has been developed with the goal of "biologically plausible continual learning and disentangled representations' to overcome one of the major problems facing AI robots which struggle to deal with constantly changing environmental demands at different time scales. He observes that AI systems "display prominent interference between old and new learning, limiting most training paradigms to using shuffled data" which is a problem when determining causality in events and achieving well-factorized or structured representations of information. A factored representation is one in which states are defined by a set of features and factored models have more internal structure.

Many neural networks struggle as we have seen to achieve continual learning and suffer from forgetting things when the environment is changing, and many different methods have been employed to try and address this. Hummos's answer is to employ an algorithm that optimizes inference time to "generate internal representations of the current task dynamically." This has important implications for AI generative learning models in terms of accuracy, rapid adoption, recovery, and backwards and forwards transfer of knowledge and transferability across tasks and the ability to change behaviour dependent upon context. Hummos explains that "the algorithm alternates between updating the model weights and a latent task embedding, allowing the agent (system) to parse the stream of temporal experience into discrete events and organize learning about them.

On a continual learning benchmark, it achieves competitive end average accuracy by mitigating forgetting, but importantly, by requiring the model to adapt through latent updates, it organizes knowledge into flexible structures with a cognitive interface to control them." He claims that the "Thalamus is a relatively simple algorithm that can label ongoing temporal experience and then adapt to future similar events by retrieving those internally generated labels." He goes on to assert that "another possible extension is once the gradients of the latent embedding are approximated with a second neural network, the algorithm can be deployed in hardware, allowing it to show adaptive behaviour and 'learn' without requiring backpropagation or weight updates."[304]

Neurotransmitters and Hormones in AI

Borrowing ideas from the concept of embodied intelligence, biologically inspired AI research has recently focused on mimicking the role of chemical neurotransmitters and hormones. Over the years, several different approaches to mimicking the role of chemicals in traditional electrical connections between artificial neurons and artificial synaptic signalling. This has included modifying the weights in ANN depending on a performance criterion for each iteration[305] or coordinating pre-determined behaviours with systems designed to switch between them.[306] In the 1990s Philip Husband developed a system called GasNets which sought to combine electrical and chemical signalling into a single ANN.[307] Joseph Gally and others started to explore chemical signalling by using gases to mimic neurotransmitters with the aim of modulating behaviour and controlling autonomous robots when performing a range of tasks.[308] Their Nitric Oxide (NO) hypothesis based on the nervous system of the human body postulates "that the release of NO and its subsequent diffusion may be critically related to three aspects of nervous system function (i) synaptic plasticity and long-term potentiation in certain regions of the adult nervous system (ii) the control of cerebral blood flow in such regions and (iii) the establishment and activity-dependent refinement of axonal projections during the later stages of development." Gally believed that "voltage-dependent postsynaptic release of this rapidly diffusing substance links the activities of neurons in a local volume of tissue, regardless of whether the neurons are directly connected by synapses."[309] This means that chemical signalling using gases could potentially play the role of neurotransmitters in ANN. Almost all GasNet parameters and variables are under evolutionary control which means that they are designed to evolve to perform a range of tasks. It wasn't until nearly 20 years ago though that significant progress was made in mimicking the role of the internal body state in modulating the human brain. In 2004 an Artificial Endocrine System (AES) was developed by Jon Timmis, and Mark Neal in 2004 to develop an Artificial Immune System (AIS) and ultimately an Artificial Homoeostatic System (AHS). The AES was "capable of autonomously interacting with an unknown and changing environment while maintaining its internal state (e.g., energy level and integrity)."[310]

In 2005 Vargas and colleagues explored the incorporation of the role of chemical hormones into ANNs by employing a Hormone Production Controller (HPC) that based on the internal state triggers a specific artificial hormone to activate target artificial neurons, which would affect the input weight of those neurons. Using a feedback mechanism, hormone production ceases upon task completion Vargas designed the NSGasNet model so that all emitted gases spread freely among the neurons which more closely resembles how the human CNS works. They describe how the nodes are in a non-Euclidean geometry and the RNN model could be fully or partially connected with a fixed or variable number of nodes. The gaseous connections have sensitivity limits between [0,1] which impose a filter on each network node to regulate the level of gas modulation.[311] They have tested the robots across various tasks including negotiating a T-Maze and turning based on remembering the position of a light and the results indicate that spatial constraints and a spatially embedded diffusion process are not necessary to explain the success of GasNet models.

In 2008 a team from the Centre for Computational Neuroscience and Robotics (CCNR), Department of Informatics, University of Sussex developed the Evolutionary Artificial Homoeostatic System (EAHS).[312] We will talk about homoeostasis and allostasis later in connection with consciousness, but EAHS extends the earlier work of Vargas and Moioli and is designed to cope with what the developers describe as "One of the most challenging issues when developing autonomous systems that can adapt to unforeseen situations and disruptions, is how to coordinate conflicting behaviours and to operate and remain adapted within a viability zone." Their work is derived from how organisms self-regulate to keep essential variables within a limited range and their architecture comprises two modules. The first is an Artificial Endocrine System (AES) which is synthesized using evolution (under two different performance criteria)' and the second is three evolved spatially unconstrained GasNet models, named Non-Spatial GasNets (NSGasNets). NSGas Nets are a new type of network model that complements the traditional electric point-to-point communication by employing diffusible chemical modulators that enable chemical neuro-modulation of behaviour as well as electrical synaptic interaction. The initial tests revealed it to be successful in controlling a robot in reacting to something but has yet to be applied to a delayed non-reactive task. This is evidence of a growing appreciation of the need to have a biologically inspired system that addresses the conflicting internal signals within the body and brain together and the choices made but also the requirement for the robotic system to maintain internal viability.

A Layered Reference Model of the Brain (LRMB)

Yingxu Wang and colleagues developed a Layered Reference Model Brain (LRMB) that is both hierarchical and integrated and seeks to explain "the functional mechanisms and cognitive processes (CPs) of natural intelligence."[313] It consists of 37 CPs cognitive processes (CPS) and from the bottom-up six layers which are (1) Sensation (2) Memory (3) Perception (4) Action (5) Meta-cognitive and (6) Higher cognitive layers. It facilitates the study of the inherited and acquired life functions and between the long and short-term memory and sensory and action buffers, with the hierarchical feature being divided between the subconscious and conscious life functions. The subconscious layers are described as inherited, fixed, and relatively mature and neither directly controllable nor intentionally accessible by the conscious life function layers. They concluded that "Complex and instant intelligent behaviours of the brain and mind could be reduced to the combinations of the six-layer CPs." Given all these diverse developments in the types of artificial neurons and biomimetic initiatives, we need to assess whether there is any indication that AI systems or robots can or will ever be able to understand and give meaning to the world as humans do.

Can AI Understand?

When we discussed the neurobiology of human understanding and how we attribute meaning to our world we looked at how biological systems self-organise, and structures emerge that establish boundaries between them and their environment. This occurs at nested levels from the microscopic to the macroscopic with Markov blankets hypothesized as essential to keeping us and all organisms alive by conferring the capacity to act as autonomous agents. It is these operationally closed mutually enabling constituent processes that organize systems and provide autonomy so if they break down, autonomy and agency are threatened or destroyed. An important part of this process is adaptivity to the environment with the combination of activity and autonomy being cited as what imbues an entity with meaning or understanding and what Tom Froese and Tom Ziemke describe as its "own view on the world."[314] It is what allows us to understand the world by reference to ourselves as distinct entities within that world and is predicated on the notion of active inferencing of the world which in humans underpins perception, action, planning, decision-making, and learning and theoretically in autonomous artificial agents as well.

Lancelot De Costa and colleagues in their review of this topic observe how it allows "continuous sensations to inform discrete representations which is a fundamental building block of generative models."[315] When we talk about the self in humans and robots, we will see it has proved immensely challenging to create artificial agents that are self-individuating in terms of possessing a clear sense of self that is separate from others and the environment. Active inference as we saw when we looked at human understanding is postulated as based as based on the Free Energy Principle (FEP) directed at the goal of minimizing free energy and employs the Bayesian brain hypothesis by using an internal generative model to predict and infer the causes of the sensory input. This allows us to constantly make decisions that involve recognizing features from stimuli, considering the options in terms of responses, choosing, and implementing the option, assimilating the outcome, and adjusting our behaviour appropriately.

Active inference and decision-making predicated upon self-organization have already been used in computer simulation of the brain to simulate a wide range of complex behaviours such as how people make decisions. Different approaches broadly described as Evidence Accumulation Models (EAMs) have been used in neuroscience and psychology that assume minimizing of free energy by accumulating evidence to predict sensory data. In EAMs the parameters determine how fast the evidence is accumulated with a programmed bias to select one response over another so if you change the parameters, you change the behaviour of the model's decision making which influences the speed and accuracy of the process, but once a threshold is reached a decision or choice is made. They include Drift Diffusion Models (DDMs) and Dynamic Programming Models (DPMs) and their great strength is identifying how latent cognitive processes such as decision making are linked to different brain substrates. Latent simply refers to the fact that they cannot be directly observed but only inferred based on the observed behaviour of a person. Standard DDMs tend to break the process into four distinct elements which are (1) How long it takes to detect and encode the stimulus (2) The time to decide and how long it takes to choose a motor response. More elaborate models have more free parameters that can be changed, but most include the four key free parameters of (a) non-decision time (Ter) in which the encoding and response time are combined into a single parameter, representing non-decision time[316] (b) boundary separation which describes the evidence required for a decision threshold to be reached which is adjusted by the speed–accuracy trade-off. Wider decision boundaries produce slower but more accurate decisions, whereas narrower boundaries result in faster but more error-prone decisions[317] (3) A response bias (z), reflects how when people are biased to prefer certain responses over alternative responses when deciding, they make the preferred response more quickly[318] and(4)At least one drift rate (d), which is the rate of accumulation of evidence which differs with different stimuli based on the difficulty of interpreting the stimuli or information.[319] These parameters combine to reflect the level of understanding of and meaning derived from the stimuli and the level of confidence in the potential accuracy of the underlying prediction process. In humans, this involves, "noisy sensory inputs, stochastic variation in the firing rate of neurons in the decision-making centres of the brain, and even momentary fluctuations in attention."[320]

Julian Kiverstein, Michael Kirchhoff, and Tom Froese have recently observed how enactive cognitive scientists have been exploring the design of artificial agents that possess sensorimotor autonomy and how "stable, self-sustaining patterns of sensorimotor interaction can ground values, norms, and goals necessary for encountering a meaningful environment" which they believe to be essential for autonomy.[321] Enactive Cognitive Science (ECS) asserts that cognition emerges from or is constituted by sensorimotor activity and the dynamic interaction between an acting organism and its environment and has emerged as part of a growing challenge to the Theory of Mind (TOM) that postulates that all human cognitive processes involve computation of internal representational states which entail the FEP. Julian Kiverstein and colleagues do not think that FEP and sensorimotor autonomy are incompatible and that the FEP may prove to be used to model autonomy and adaptivity, observing how FEP can be applied to many systems that do not satisfy the conditions for sensorimotor autonomy, citing swinging pendulums and Watt governors.

Free energy was after all originally conceived as a thermodynamic principle "denoting a measure of energy available to a system to do useful work e.g., maintaining a particular speed in a Watt governor or photosynthesis in plants."[322] Miguel Aguilera and colleagues argue that the FEP requires "the Markov blanket condition (i.e., a statistical boundary precluding direct coupling between internal and external states) and stringent restrictions on its solenoidal flows (i.e., tendencies driving a system out of equilibrium)." They point out that these are "are only valid for a very narrow space of parameters. Suitable systems require an absence of perception-action asymmetries that is highly unusual for living systems interacting with an environment." They conclude that FEP, as it stands, is not straightforwardly applicable to simple linear systems .... and suggest that more development is needed before the theory can be applied to the kind of complex systems that describe living and cognitive processes.[323]

The debate goes to the heart of whether robots designed based upon the FEP can be autonomous agents or whether this requires sensory motor autonomy, a combination of both, or maybe other features that have yet to be conceptualized or discovered. It highlights the clear challenges in designing autonomous agents that can attribute meaning and understanding to their world which are required for our definition of agency as a system "as an autonomous organization that adaptively regulates its coupling with its environment and contributes to sustaining itself as a consequence." This you may remember is derived from the assumption that it must (a) Define its individuality (b) Be the active source of activity in its environment (interactional asymmetry), and (c) Regulate this activity about certain norms.[324] Based on the various conflicting viewpoints and the fact that no robot has currently been built that satisfies this definition of agency or even a more intuitive sense of what constitutes agency, this remains unresolved. It highlights the vital nature of the role of sensorimotor perception and movement in both humans and robots, so let's look at this in robots.

Robots and Sensorimotor Perception and Movement

We have seen the challenges of providing robots with reflexive capabilities, getting them to walk like humans, plan motion, and move like humans with flexibility, fluidity, and adaptability. This is increasingly important as more and more robots are required to navigate among pedestrians in crowded unconstrained environments and need to employ collision avoidance algorithms, particularly with the increased deployment of unmanned aerial such as parcel delivery robots produced by tech giants like JDin China and Amazon and autonomous road vehicles. Mobile robots are used in an increasingly wide range of applications including agriculture, military, space, medicine, mining, rescue, etc. AI scientists have explored a wide range of algorithms for path planning in robots to find the optimal route between two points and Changmin Zhou observes how "Research of motion planning is going through a flourishing period, due to development and popularity of Deep Learning and

Reinforcement Learning that have better performance in coping with non-linear problems with complexity." They attribute this to the investment in autonomous vehicles by tech giants like Google and Tesla and the multinational Chinese company Baidu which is China's largest and most widely used search engine and the use of deep learning algorithms or integrating traditional motion planning algorithms with advanced machine learning algorithms.

Humans spend most of their time planning and moving without needing to give it much thought and motion planning in robots has become a growing and vital topic in the search to make robots more autonomous. Although planning and motion are often used interchangeably there are some important differences. For example, path planning aims to find the path between the origin and destination using the shortest distance or time. Motion by contrast, involves robots interacting with dynamic environments, which requires consideration of the kinetic aspects such as velocities and the robot's movements and nearby dynamic objects and obstacles, etc. This requires short-term optimal or sub-optimal reactive strategies to equip the robot with the ability to make an instant or reactive response. Path or trajectory planning in robotic navigation depends on the level of known and unknown information about the environment and identification of where the robot is, which is known as the Configuration space or "C"-space of the robot.

This is determined by its 2D coordinates which are the X and Y coordinates represented in a coordinate plane or Cartesian plane and 3D coordinates represented by the X, Y, and Z coordinates concerning all points on the robot which determines its footprint in the environment or C-space and its environment which is why all those points are called the configuration.

For path planning and motion, robots currently use a diverse range of multi-modal sensors but typically three types of movement or motion through their actuators which cause the links or limbs to move. These involve rotary or linear control based on robotic hardware like motors or servos. From the perspective of robotic and control engineering, this necessitates (1) Joint or non-linear, (2) Linear, or straight line, and (3) Arc/arch or circular motion. Using this range and types of movement, robots are programmed to be able to move and react using.

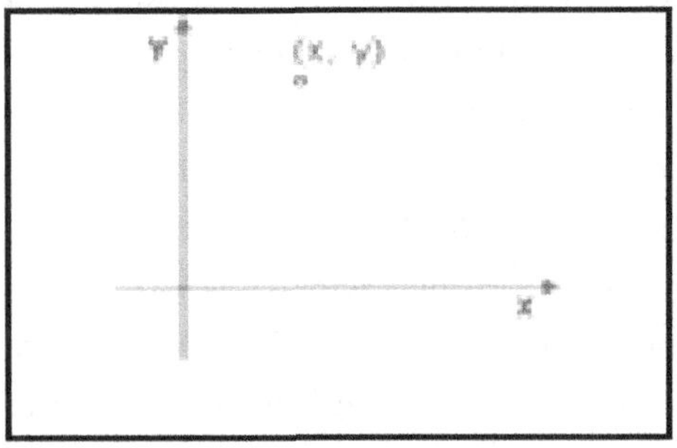

2D

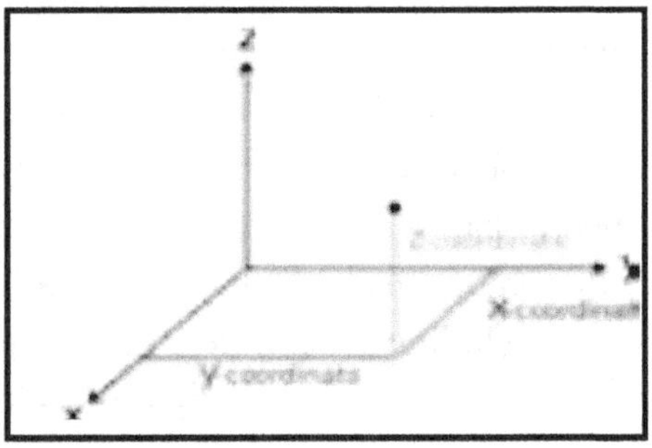

3D

Cartesian coordinate systems with two x and y-axes and a z-axis occupying the vertical space to make a 3D C-space of the environment they are operating in and any objects and obstacles, etc, which constitutes the robots' frames of reference. Hence, they will have more than one coordinate system. This may include a tool or tool centre point which indicates how the robot must move to operate the tool. If the robot is grasping an object the coordinate systems describe the location of the robot's hand in c-space and the distance of the coordinates of the object which requires the robot to know the position of its hand about its body and calculate the correct orientation to grasp the object. It will also require the robot to calculate the coordination points on the object. To perform actions requires the robot to change its coordinates and to grasp the object will require the robot to determine the position and orientation of the object relative to the robot's hand. It often will be used as a transformation matrix to do this which we will not discuss here as it is beyond the remit of this book.

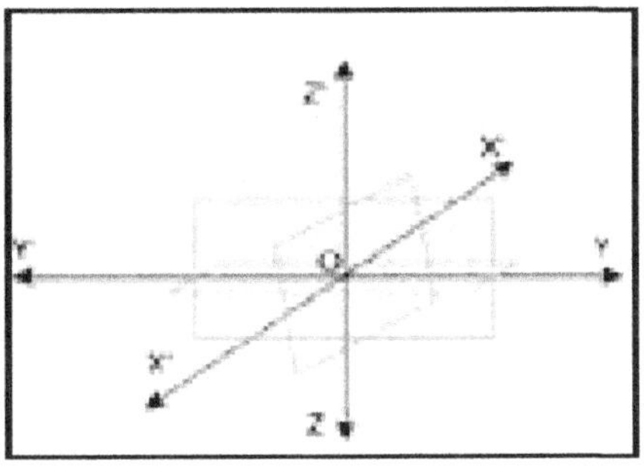

The C-space of a robot has a topology, which in mathematical terms is the invariant properties of the space (don't change) under any continuous deformation. For example, you can deform a square into a circle without deforming it, but you can't do that with figure 8. A torus is the surface of revolution created when you revolve a circle in a 3D space which allows you to deform it, but you cannot break or cut the space or object. It is also known as rubber-sheet geometry because you can stretch and contract the objects or space like rubber, provided you don't break or cut the space or object. The robot's topology in the c-space is a torus or doughnut shape or dimension, so for every configuration of the robot which can be represented by the circular motion associated with the coordinates, there is a unique 2D position on the torus. Any shape that can, without cutting or glueing assume the same shape as the other, share a topology, for example, a coffee mug can be transformed into a torus or doughnut shape without cutting or gluing it. They are regarded as topologically equivalent shapes. The C-space is the space of all the possible positions the robot may attain so is the precise specification of all the robot's degrees of freedom (DOFs) and a key element of motion planning is to represent any obstacle in the robot's environment as a set of points in C-space where the robot might collide with an obstacle. This means that a workspace obstacle is a set of points or regions of space that correspond to robot configurations that are not possible or not allowed. The robot needs to make sure, at every point in its computed path, that its footprint does not intersect the C-space of any obstacles.

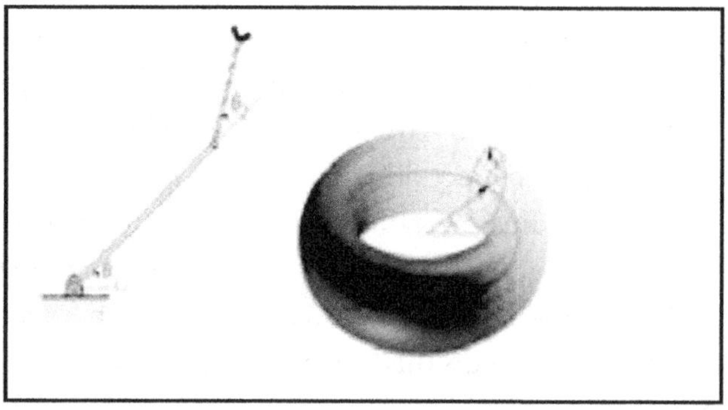

Coffee mug reduced to Torus shape.

This requires modelling real-world objects into the C-space of the robot and path planning involves calculating paths in which the robots use a detection collision module that uses computations to avoid obstacles as it moves through the C-space. Path planning is typically broken down into (1) Global which is about getting from the start to the goal and typically requires prior information about the environment, obstacles, and goals and involves path planning and (2) Local which is dealing with the dynamics of the environment and the relational position between the robot, things and objects so is more reactive and involves the motion of the robot. It requires real-time obstacle avoidance.

Traditionally robots used to be designed hierarchically to (1) Sense or acquisition and perception (2) Plan or Decision and (3) Act or Actuation-control, but although allowing the robots to generate long-term path plans, this was inevitably very slow and failed to allow them to react sufficiently quickly to situations they might confront in dynamic environments and when confronted by unforeseen objects and obstacles. This necessitates real-time updating of navigation paths, motion, and goal plans in a non-linear and complex real-world model when coping with uncertainty, ambiguity, incompleteness, and noise in the environment. Motion planning is far more complicated with autonomous robots that rather than rely on intensive programming and training, need to be able to decide on their motion plan, navigate and respond in real-time, and on occasions coordinate with other robots or people. Traditionally robots relied on path planning and motion software that consists of a decision tree algorithm approach which requires intensive upfront human programming and huge amounts of data and programming of key assumptions about the behaviour of other agents like people and other robots but as Michael Everett and colleagues observe, these assumptions often "deviate from reality as the number of agents in the environment increases."[325]

Some companies like Real Time Robotics, have used a combination of proprietary software and hardware to find the best and most highly efficient motion plan and claim to have reduced the programming requirement by nearly 90%. They can train the robot in just a few hours, but the robot is still entirely dependent upon human programming to plan and react rather than learning from and adapting to the environment. Path planning thus involves establishing a configuration space which is the set of all configurations or positions that a robot can attain and configuration space obstacles in which the robots could collide with objects. Zhou and colleagues in their 2022 review of the path planning algorithms used in AI systems observe that AI uses a combination of algorithms which include (1) Traditional Planning algorithms (2) Classical Machine Learning algorithms (3) Optimal Value Reinforcement Learning algorithms and (4) Policy gradient reinforcement learning algorithms. Let's look at some of the most used path-planning algorithms. This recognizes that these usually must be combined with motion planning algorithms for obstacle avoidance and real-time reactivity, which can be categorized as (1) Traditional algorithms and (2) ML-based algorithms.[326]

Oscillatory Behaviours of Artificial Neurons in AI

Spontaneous synchronous and asynchronous oscillatory behaviours and electrical signalling are ubiquitous with studies showing a significant relationship between cognitive processes and neural rhythms and this area has been the subject of much intense research in AI in recent years.[327] Developed in 1974 by Professor Emeritus of physics, Yoshiki Kuramoto, a leading model is the Kuramoto model which was initially motivated by the spontaneous synchronization among male fireflies which as we saw gather in their thousands and flash on and off in unison. The firefly algorithm is employed for the optimization of movement options related specifically to the strange feature of synchronization of behaviour. The Kuramoto model is a non-linear dynamic system that has initially random natural frequencies and phases which with strong enough coupling between oscillators evolves to a system in which all the oscillators are in phase and thus synchronized. A more generalized Kuramoto model has been developed to factor in inertia and noise present in naturally synchronizing oscillatory behaviour and become a widely used mathematical model for studying the synchronization of behaviour in populations of coupled oscillators.

Rennan Mioli and colleagues have applied the Kuramoto model to a simulated robotic agent engaged in a minimally cognitive benchmark task by employing a set of fully connected oscillators, some connecting to the robot's seven symmetrically placed sensor rays.[328] The purpose was to explore temporal dynamics in neural activity in generating embodied cognitive behaviours. The task required the robot to distinguish between circular and square objects by moving toward the circles and avoiding the squares. Their rationale is to reflect in AI systems the "temporal relations among groups of artificial neurons, either by external influences or sustained by internal mechanisms." To add flexibility Mioli's model incorporated a mechanism inspired by the Hebbian theory of synaptic plasticity which allows neuronal oscillators "with different natural frequencies to recruit and compete for coherence with other oscillators in the neural network".They concluded that the study demonstrated how in a robot "the external stimulus modulates the ongoing activity of individual neurons which in turn affect the strength of their synaptic connections, restructuring their mutual interaction and the phase dynamics which control the agent. Internally generated brain activity, regardless of environmental inputs, is stressed as one of the key elements in cognitive processes."[329]

In a 2000 paper British neuroscientist, Karl Friston made the following observation, "The distinction between linear and non-linear coupling has fundamental implications for the analysis and characterization of neuronal interactions, most of which are predicated on linear (synchronous) coupling (e.g., cross-correlograms and coherence)." Cross-correlograms are a function that indicates the firing rate of one neuron (the target neuron) versus another (the reference neuron), whilst neuronal coherence as we saw earlier is a fundamental mechanism of brain communication where synchronized field potentials coordinate synaptic and spiking events which supports plasticity and learning. Friston advocates that by viewing "the brain as an ensemble of connected dynamic systems one can show that a sufficient description of neuronal dynamics comprises neuronal activity at a particular time and its recent history." He goes on to observe how "This history constitutes a neuronal transient. As such, transients represent a fundamental metric of neuronal interactions and, implicitly, a code employed in the functional integration of brain systems. The nature of transients, expressed conjointly in distinct neuronal populations, reflects the underlying coupling among populations. This coupling may be synchronous (and possibly oscillatory) or asynchronous." Friston points out that neuromagnetic data has shown that nonlinear (asynchronous) coupling is more abundant and can be more significant than synchronous coupling.[330] The neural dynamics and the coupling of neurons are highly complex and if autonomous robots are to be able to mimic the human brain, it appears that AI systems employing ANNs may need to replicate the brain's use of both linear synchronous and non-linear asynchronous oscillatory behaviour.

Neural Plasticity in Robots

Computer scientists Andrea Soltoggo and Frank de Velde from Loughborough University point out in their 2015 paper, that building neural plasticity into robots poses a major challenge. They observe how "Models of adaptation and neural plasticity are often demonstrated in robotic scenarios with heavily pre-processed and regulated information streams to provide learning algorithms with appropriate, well timed, and meaningful data to match the assumptions of learning rules."[331] They note how "natural scenarios are often rich in raw, asynchronous, overlapping and uncertain inputs and outputs whose relationships and meaning are progressively acquired, disambiguated, and used for further learning."

Soltoggo and de Velde describe how realistic models of plasticity are increasingly exploring "the progressive acquisition of knowledge by neural systems through experience in environments that may be affected by ambiguities, uncertain signals, delays, or novel features."

The program director of AI at IBM Data and co-founder of Continual AI, German Parisi similarly observes that "While state-of-the-art models of action perception show remarkable results in large-scale action datasets, they mostly lack the flexibility, robustness, and scalability needed to operate in natural HRI scenarios which require the continuous acquisition of sensory information as well as the classification or assessment of human body patterns in real-time."[332] As we have seen and Parisi observes "Input-driven self-organization is a crucial component of cortical processing which shapes topographic maps based on visual experience" and architectures have been designed to capture the dynamics of Hebbian learning and structural plasticity. This extends to HRI situations where a robot is for a variety of reasons unable to extract and process facial expressions or speech features from a human to recognize affective states so needs to use body motions. Parisi argues that although CNNs have demonstrated high visual accuracy in recognizing action from videos "the majority of these models are computationally expensive to train and lack the flexibility and robustness to operate in HRI scenarios" explaining how" real-world environments "can be particularly noisy and susceptible to self-occlusion." For example, a person with whom the robot is interacting is not facing or close enough to the robot cameras or sensors. Various approaches have been employed to dynamic learning in terms of robots interacting and collaborating with humans which as you can imagine is essential in the healthcare environment and the role of neural plasticity when confronted by rich information flows. State-of-the-art models use a static batch of body patterns rather than dynamic body motion patterns on which to train robots but are plagued by the continual learning problem and unable to "acquire and fine-tune internal representations and behaviour in a continual learning fashion." An approach, proposed by Parisi and colleagues is the use of self-organizing networks to "learn action sequences and real-time inference." To address this, they employed a specific type of self-organizing network called a Growing When Required (GWR) network during the training phase which "Dynamically change their topological structure to better match the input space."[333] Parisi explains how GWR can "grow and remove neurons in response to a time-varying input distribution."

We saw earlier how the Neurobot CARL with its touching sensors was equipped with neurobiologically inspired plasticity through integrating sensory information. Jeroen Bumrs and colleagues demonstrated the possibility of modelling Hebbian plasticity to train robots in motor control, describing how they "apply the learning rule to optimize the connection weights of recurrent neural networks with different topologies and for various tasks and claim that they have demonstrated the universal applicability of reward modulated Hebbian learning. They go as far as to observe that "compliant robots should or can be seen as computational units, instead of dumb hardware that needs a complex controller."[334] Both oscillatory behaviour and neural plasticity remain problems that are being confronted by the designers of cognitive architectures in robots.

AI and Brain Waves

Although the relationship between EEG signals and brain activity remains poorly understood, a major question facing AI researchers is the potential role of human brain waves when communicating with AI robots in what is called the Brain-Computer Interface (BCI). This has raised the strange implication that AI robots might themselves begin to exhibit brain wave activity. DL has accelerated our ability to analyse pattern recognition in brain waves and decode brain wave activity, but what is unclear is whether robotic brains require the computer equivalent of brain wave activity. ANNs are based on physical theories of how our brains communicate internally via electrical signals in neural networks. Galinsky and Frank observe that a wide range of ANNs has been based on the mechanism of memory and learning using the Hodgkin–Huxley (HH) theory and ad-hoc circuit analogies, but recognize that the HH theory is far too complicated to explain even relatively small neural networks, This has led to the use of standard algorithms and the simplistic LIF neuron that we discussed earlier as the basis of AI ML and DL with the majority being Spiking Neural Network (SNN) models.

Galinsky and Frank have though developed an alternative physical model called Weakly Evanescent Brain Wave Propagation (WETCOW) which they claim can perform "much more sophisticated tasks of efficient coherent synchronization/ desynchronization guided by the collective influence of propagating non-linear near critical brain waves rather than SNN." This has potentially important implications for how ANN in the future attempts to replicate the behaviour of BNNs because it no longer treats the neurons as possessing "fixed properties that isotopically (i.e., exhibits the same properties or behaviour in all directions) collects input and fires when enough information has been collected." As we saw earlier, brain wave dynamics more accurately describe how our brains learn and remember which can potentially provide the "basis for a novel class of engineering analogues for both software and hardware learning systems that can operate with the extreme energy and data efficiency characteristics of biological systems that facilitate adaptive resilience in dynamic environments." Networks will not be formed from fixed neurons or nodes but use "flexible pathways encompassing propagating waves, or wave packets, or wave modes." This illustrates how we are only just beginning to understand the mechanism by which our brains process, learn, and remember information and that the current AI systems largely fail to reflect such biological intricacies as brain waves.

Can AI systems exhibit similar sorts of brain wave activity to that of the human brain? Intuitively you would think that this is a biological feature that a machine robot could not exhibit, but scientists at the University of California, Berkeley have in a study recently measured brain waves in both human and AI system participants. They used EEG to measure brain waves generated from the brain stem when participants listened to the word "Bah", and measured the signals generated by an unsupervised ANN, an AI system, trained in English that could interpret sound. The study lead, assistant professor of computational linguistics Gasper Begus observed that based on the raw data which did not undergo any transformation they found very similar signals emitted from humans and the AI system. The problem is that the XAI problems make it difficult if not impossible to know what is happening in the black box of the AI system to understand what exactly is going on. Begus is convinced though that language learning and the similarities between humans and AI as evidenced by these apparent AI brain wave patterns will give us greater insight into what is happening in AI systems.[335] If AI systems are generating brain waves this also has intriguing implications for determining consciousness and levels of awareness in AI machines which we talk about shortly.

Scientist Daniela Rus and other researchers from MIT have made a breakthrough in the way we can potentially communicate with robots using brain waves emitted by our brains i.e., BCI rather than operative systems or voice commands. Designed by Rethink Robots, the robot called Baxter allows a human wearing a mind-reading EEG helmet and muscle sensors attached to his arm to operate the robot with the robot sensors detecting the brain waves associated with awareness of an error in a process. The goal behind such research is to develop affordable and useful prosthetic devices but has wider implications for how we will be able to communicate with robots. Another study at the National University of Singapore, led by doctoral student Zijiao Chen, demonstrated the ability of a robot to construct images from what someone is thinking, by decoding human brain waves. This has potentially enormous benefits for enabling people who are unable to speak or communicate properly to let people know what they are thinking. It may also help us understand how the human imagination works and how brain waves respond to stimuli. It poses obvious ethical risks in terms of invasion of thoughts and privacy even if the technology is very much in its infancy. These human-robot interface studies are not an indication that robots exhibit brain wave activity but currently appear to demonstrate the capacity of AI to decode information contained in brain waves which is something that initiatives like WETCOW are exploiting. The ability of robots with the ability to read human brain waves would be remarkable though is currently confined to technology that can be attached to the human rather than mere proximity with the AI agent.

Edge, Fog, and Cloud Computing

Cloud Computing

Cloud computing has been defined as the delivery of computing services, including servers, storage, databases, networking, software, analytics, and intelligence over the Internet ("the

Cloud") to offer faster innovation, flexible resources, and economies of scale. Autonomous robotics is a very important part, with implications for such devices as self-driving vehicles, etc. Information technology has traditionally applied to almost entirely cyber-based environments like the Cloud whereas operational technology such as robotics almost inevitably requires and involves a cyber/physical environment interface.

Until relatively recently, robots were largely involved in industrial functions in which humans were kept apart from them, mainly for safety reasons but they are increasingly designed to work closely with humans. As robots become more and more autonomous and part of everyday life, the challenge has been to create individual robots without reinventing the wheel for each robot by allowing them to share allocated computer resources and robotic knowledge. This aspiration was reflected in the early field of networked robotics started in the 1990s with each robot regarded as a node in the network and "sensing data and information shared among nodes, the operators can transmit command data remotely and receive measurement feedback." This has allowed them to be used in long-distance medical surgery, disaster relief and other specialized cases[336] and evolved into what in 2010, American roboticist James Kuffner named "Cloud Based Robotics" (CBR).[337] The National Institute of Standards and Technology (NIST) defines CBR as "a model for enabling ubiquitous, convenient, on-demand network access to a shared pool of configurable resources (e.g., servers, storage, networks, applications, and services) that can be rapidly provisioned and released with minimal management effort or service provider interaction." CBR systems have the advantage of data and code being stored in the cloud using remote server farms with shared processors and memory and thus computational power and data offloading. The problem as Ben Kehoe and a team of computer scientists observe, is that they inevitably limit the potential autonomy of robots because they rely on data or code from a network, and not all sensing, computation, and memory are integrated into a standalone robotic system.[338]

Between 2010 and 2014, the EEC funded an initiative called Robo Earth which had the stated aim of allowing robots to "store and share information, offload computation and collaborate." The Robo Earth website describes how Robo Earth's World-Wide-Web style database stores knowledge generated by humans and robots in a machine-readable format, including software components, maps for navigation (e.g., object locations, world models), task knowledge (e.g., action recipes, manipulation strategies), and object recognition models (e.g., images, object models). In their review of Robo Earth as an initiative, Markus Waibel a Swiss entrepreneur, co-founder, director, and COO of Verity AG, whose research focus has been on multi-robot systems and cloud computing, described how "Humans can use the Internet to share knowledge and to help each other accomplish complex tasks. Until now, robots have not taken advantage of this opportunity. Sharing knowledge between robots requires methods to effectively encode, exchange, and reuse data and Robo Earth represents the design and first implementation of a system for sharing knowledge between robots."[339]

In 2016, computer scientists Kalpana Gilda and Sharwarwi Dixit published an overview of the CBR sector in which they envisage cloud robotics based on the basic sequence of robotic actions which involves data generation and capture, data processing, the application of the processed data and feedback to the computer server. They envisage a cloud-based system to address what they describe as the "effectiveness of the robotic network (being) limited by each robot's resources, including onboard computers, memories, and storage space. Physically, these on-board computing devices are restricted by the robots' size, shape, power supply, motion mode, and working environment." Gilda and Dixit identify many advantages of cloud computing for robots which they list as follows "(1) Big Data: A global library of images, maps, and object data, often with geometry and mechanical properties, expert system, knowledge base (i.e., semantic web, data centres) (2) Cloud Computing: Massively parallel computation on demand for sample-based statistical modelling and motion planning, task planning, multi-robot collaboration, scheduling, and coordination of system (3) Collective Learning: Robot sharing of outcomes, trajectories, and dynamic control policies and robot learning support (4) Human Computation: Human sharing of "open source" code, data, and designs for programming, experimentation, and hardware construction (5) On-demand human guidance and assistance for evaluation, learning, and error recovery and (6) Augmented human–robot interaction through various ways."

Their robot cloud architecture envisages two complimentary layers which are (1) Machine to Machine (M2M) communication which they describe as forming "a collaborative computing fabric (i.e., an ad-hoc cloud) and (2) Machine to Cloud communication (M2C) which provides a "pool of shared computation and storage resources that can be allocated elastically for real-time demand."[340] Effectively the robots communicate with each other and the cloud which according to them "allows the group of networked robots to offload computation-intensive tasks for remote execution." They recognise though that the robot network faces challenges with robots becoming unavailable or failing and the network becomes highly dynamic when the robots are highly mobile, which makes the design of effective routing protocols difficult. They employed the Robot Operating System (ROS) which is an open-source collection of software libraries and tools to assist in building robot applications, including drivers to state-of-the-art algorithms, and other powerful developer tools, The ROS is described on their website as a "loosely affiliated collection of engineers and hobbyists from around the globe with a shared interest in robotics and open-source software." They acknowledge that despite the apparent advantages of Cloud Robotics there are many challenges in "balancing between real-time demands and processing performance and the core issue of resource allocation". These include data interaction between the robot and the cloud, the requirement for new algorithms to deal with network latency and parallel processing, and the difficulty in choosing which resources to allocate to the robots and the cloud. They openly admit that this approach is not a solution for all robots and cite the following limitations: Controlling a robot's motion which relies heavily on sensors and feedback of controllers won't benefit much from the cloud. Cloud-based applications can get slow or unavailable due to high-latency responses or network hitch. If a robot relies too much on the cloud, a fault in the network could leave it brainless. Tasks that involve real-time execution require on-board processing and offloading large data to the cloud is taking time, just as on-board computation does.

Edge and Fog Computing

Despite the advantages of CBR, robotic network systems are regarded by some as less than ideal in terms of rapid real-time interaction and functionality, which explains why there has been a move to employ both Edge and Fog computing. Edge computing refers to a range of networks and devices processing data closer to where it's being generated thus at the edge. This enables greater speeds and volumes, reduces latency in response time, and real-time action-driven results for example in smart speakers, watches, and phones which use a combination of centralised technology in the form of the cloud technology and distributed or local edge technology. Edge reduces the need for centralized computer power and has the potential to lead a revolution in automation in open uncontrolled environments rather than controlled environments like factories. This is going to be critical if like humans our robots are going to be able to rapidly analyse data and make split-second decisions as humans do and not be dependent upon the internet to function. Fog computing is a computing infrastructure between the edge and the cloud that currently offers greater computer power than edge but less than the cloud and is where the data is processed and analysed, whereas in edge the data is generated and collected. Memristors, OECTs, and the range of new computer chip technologies that we discussed earlier have important implications in the field of Edge Computing of which operational technology like autonomous robots are central.

Although edge, fog, and cloud computing are all employed in the field of robotics, robots are currently limited in terms of computational capacity, memory, and storage. In a recent review of the future of edge computing for robots, Milan Groshev who focuses on networked robotics and colleagues observe how "the centralized nature of cloud computing is not well-suited for a multitude of Operational Technologies (OT) nowadays used in robotics systems that require strict real-time guarantees and security." They concur with earlier reviews that "Cloud facilities usually reside far away from the robots (i.e., separated by one or more uncontrolled transit networks), making it hard to achieve high-bandwidth, low-latency, and bounded jitter. Edge computing allows some of the more time-sensitive applications to be allocated on the robot which they cite as including "deep object recognition and grasp planning, visual odometry which means being able to assess its position and orientation, real-time control and object detection". In their proposed design the robots act only as sensors and actuators. They have embedded within them only the minimal components in charge of, reading data from the sensors (e.g., odometry, camera, lidar) sending them to the robot brain; and executing navigation instructions received from the robot brain."

Endnotes

[1] Radical Embodied Cognitive Science (2009).

[2] Probabalistic Robots.

[3] Bernstein, JG, Boyden, ES., 2011.

[4] Inoue, Ki, Takada, M, Matsumoto, M., 2015.

[5] un-JJ., et al., 2017.

[6] Nguyen, QT, Schroeder, L., Mank, M., et al., 2010.

[7] Lacin, E, Muller A., et al., 2016.

[8] Muller A., et al 2014.

[9] Marvin, J, Borghuis,B, Tian, L., et al., 2013.

[10] Feinberg, TE, Mallett, J., 2016.

[11] Source of Images: https://users.fmrib.ox.ac.uk/~stuart/thesis/chapter_3/section3_2.html

[12] LeDoux, J E., 2015.

[13] APA Dictionary of Psychology https://dictionary.apa.org/

[14] Source of Image: https://quizlet.com/138156185/cerebral-cortex-flash-cards/

[15] Hofman, M. A., 1988.

[16] Semendeferi, K., 2002.

[17] Buzsáki, G, Kaila, K, Raichle M., 2007.

[18] Buzsaki, G., 2006.

[19] Hodgkin, AL, Huxley, A.F., 1952.

[20] Colin, W., 2022.

[21] Tang, G., et al., 2014.

[22] Kenyon, GN, Sen, KC., 2014.

[23] An, Shuying., 2013.

[24] https://neuromorpho.org/

[25] Azevedo, FA., et al., 2009) (Herculano -Houzel, S., 2009.

[26] Semendeferi, K, Lu, A, Schenker, N., 2002.

[27] Aru, J., et al., 2011.

[28] Aru, J., et al., 2019.

[29] Citri, A, Malenka, RC., 2008.

[30] Laureys, S., 2005.

[31] The Predictive Mind (2013).

[32] Allman, M., et al., 2010.

[33] Seeley, W., et al., 2006.

[34] Czeiger, D, White, E L., 1997.

[35] Jones, BE, Hendry, SH., 1984.

[36] Josselyn, SA., et al., 2020.
[37] di Peelegrino, G., et al., 1992.
[38] Gallese,V., et al., 1996.
[39] Kilner, J.M, Friston, KJ, Frith C.D., 2007.
[40] Turjman, O., 2016.
[41] The Myth of Mirror Neurons (2014).
[42] Volterra, A, Meldolesi, J., 2005.
[43] Chamberlain, K.A. et al., 2021.
[44] Finsterwald, C., et al., 2015.
[45] The Organization of Behaviour: A Neuropsychological Theory (1949).
[46] Johnston, MV., 2009.
[47] Puderbaugh, M, Emmady, PD., 2022.
[48] Mateos-Aparicio, P, Rodríguez-Moreno, A., 2019.
[49] Finger, S., 2010.
[50] Graveline, CJ., 1998.
[51] Makin, TR, Krakauer, J W., 2023.
[52] Wu, Z, et al., 2020.
[53] Richmond, B.J., 2009.
[54] Martignon, L., et al., 2000.
[55] Berlucchi,G., 2006.
[56] Olshausen, B, Field, D., 2004.
[57] Barlow, HB., 1972.
[58] Fiorillo, CD, et al., 2014.
[59] Balasubramanian, V., et al., 2001.
[60] Siesjo, B.K., 1978.
[61] Laughlin, SB., et al., 1998.
[62] Balasubramanian, V, et al., 2001.
[63] The Mathematical Theory of Communication (1948).
[64] Cover, TM, Thomas, J.A., 1991.
[65] Waydo, S., 2006.
[66] Chang, L, Tsao, DY., 2017.
[67] Bowers, J. S., 2009.
[68] Spanne, A, Jörntell, H., 2015.
[69] Plaut, DC., et al., 2010.
[70] Plate, T., 2003.
[71] Spanne, A, Jörntell, H., 2015.
[72] Pal, MM., 2021.
[73] Izhikevich, EM., 2007.
[74] Feldman, D E., 2012.

[75] Lupien, SJ., et al., 2007.
[76] Ali, SA, BegumT, Reza F., 2018.
[77] Barth C, et al., 2015.
[78] Babaeeghazvini, P., et al., 2021.
[79] Passingham, R., Stephan, K, Kötter, RT., 2002.
[80] Hagmann, P, et al., 2012.
[81] Susoy, V., et al., 2021.
[82] Hulse, B., 2021.
[83] Ngugi, A. K., et al., 2010.
[84] Avanzini, G, Franceschetti, S., 2003.
[85] Goldberg, EM, Coulter, DA., 2013.
[86] Pfisterer, U., et al., 2020.
[87] Barron HC. 2021.
[88] Seven and a Half Lessons About the Brain (2020).
[89] Hoffman, M., 2014.
[90] Van Essen, DC, Dierker, D., 2007.
[91] Corballis, MC., 2009.
[92] Hugdahl, K, Davidson RJ., 2003.
[93] Medland, SE., et al., 2009.
[94] Conchla, ML, et al., 2012.
[95] Güntürkün, O, et al., 2020.
[96] McGrew, WC, Marchant, LF., 1999.
[97] Everts, R, et al., 2009.
[98] O, et al., 2020.
[99] Ringo, JL, et al., 1994.
[100] Cherbuin, N, et al., 2013.
[101] Rogers, LJ, et al., 2004.
[102] Lindell, AK., 2013.
[103] Harrington, A., 1987.
[104] Herbert MR, et al., 2002.
[105] Chance, SA, et al., 2008.
[106] Oertel-Knöchel, V, Linden DE., 2011.
[107] Zalewski, L J., et al., 1992.
[108] A New View of Insanity: Duality of the Mind.
[109] Merriam-Webster.com
[110] McGilchrist, I., 2021.
[111] Wager, TD, et al., 2003.
[112] Nielsen, JA, et al., 2013.
[113] Li, Y., et al., 2020.

[114] Dehghani, N, Wimmer, RD., 2018.
[115] Worden R, Bennett, MS, Neacsu, V., 2021.
[116] Friston K., 2003.
[117] Worden, R, Bennett, M S, Neacsu V., 2021.
[118] Wolff, M, Vann, SD., 2019.
[119] Schmitt, LI., et al., 2017.
[120] Mukherjee, A., et al., 2021.
[121] Hatsopoulos, N G., et al., 2023.
[122] Tang, J, LeBel, A, Jain S., et al., 2023.
[123] Oxford English Dictionary (OED).
[124] Whitford, T J., et al., 2017.
[125] Seminara, L., et al., 2023.
[126] McClelland, D., et al., 2016.
[127] LeCun, et al., 2015.
[128] Damasio, A., 2010.
[129] Werner, G., 2009.
[130] Werner, G., 2009.
[131] Werner, G., 2009.
[132] Kelso, JA., 1989.
[133] Kelso, S., 2013.
[134] Zarkadakis, G., 2015.
[135] Izhikevich, EM, Edelman, GM., 2008.
[136] Izhikevich, E M., 2010.
[137] Hofstadter, D., 1996.
[138] Krstic, M., et al., 1995.
[139] An Introduction to Syntropy (2017).
[140] Anticipatory Systems (1985).
[141] The Balancing Role of Entropy and Syntropy (2014).
[142] Collell, G, Fauquet, J., 2014.
[143] Collell, G, Fauquet J., 2015.
[144] Carhart-Harris, RL., et al., 2014.
[145] Carhart-Harris, RL., 2018.
[146] Carhart-Harris, RL., 2018.
[147] Collell, G, Fauquet, J., 2014.
[148] Cerra, P., 2003.
[149] Schöner, G., 2001.
[150] Beggs, J M.., 2006.
[151] Korn, H, Faure, P., 2003.
[152] Miller, P., 2016.

[153] How Nature Works (1996).
[154] Bak, P., et al., 1988.
[155] Frette, V., et al., 1996.
[156] Plenz, D., 2021.
[157] Hesse and Grosse 2014.
[158] Hoffmann, H, Payton, DW., 2018.
[159] Oxford English Dictionary (OED).
[160] Beggs, JM, Timme, N., 2012.
[161] Reyner-Parra, D, Huguet, G., 2022.
[162] Rolls, ET., 2010.
[163] Beggs JM, Plenz D., 2004.
[164] Plenz, D, Thiagarajan, TC., 2007.
[165] Beggs, JM, Plenz, D., 2003.
[166] Beggs, JM, Plenz, D., 2003.
[167] McDonnell, M., 2011.
[168] Wang, S J., et al., 2016.
[169] Kitzbichler, M., 2009.
[170] Witton, C., et al., 2019.
[171] Wang, XJ, Kennedy, H., 2016.
[172] Song, HF, Kennedy H, Wang XJ., 2014
[173] Taghia, J, et al., 2018.
[174] Lee, B, et al., 2022.
[175] Taghia J, et al., 2018.
[177] LeCun, Y., et al., 2015.
[178] Souza, A, Ribeiro S., 2018.
[179] Pesaran. B, et al., 2018.
[180] Kipke, DR, et al., 2008.
[181] Nunez, P Srinivasan, R., 2006.
[182] Fundamentals of Brain Network 2016.
[183] Horwitz, B., 2003.
[184] Jacob- Eshel, B., 2003.
[185] Yufik, YM, Friston K., 2016.
[186] Yufik, YM, Friston K., 2016.
[187] Hipólito, I., et al., 2021.
[188] Kirchhoff, M, et al., 2018.
[189] Friston, K., 2013.
[190] O'Reilly, RC, Frank MJ., 2006.
[191] Schroeder, CE, et al., 2010.
[192] Kirchhoff, M, et al., 2018.

[193] Georgopoulos, AP., et al., 1986.
[194] Stephan, KE, Friston K., 2010.
[195] Friston, KJ., 1994.
[196] Srikantan N, et al., 2016.
[197] Seeber, M., et al., 2019.
[198] Krishnaswamy, P. et al. 2017.
[199] Gaarder, K., 1966.
[200] Hennevin, E, Leconte P., 1971.
[201] Graven, S., 2006.
[202] Lau, EY, et al., 2015.
[203] Jia, X, Kohn A., 2011.
[204] Livingstone, M, Hubel, D., 1988.
[205] Henderson, CM, McClelland, JL., 2020.
[206] Singer, W., 2001.
[207] Michael, E, et al., 2022.
[208] Santamaria, L., et al., 2019.
[209] Zeevi L et al., 2022.
[210] Qiu, C., et al., 2015.
[211] Hassabis, D, et al., 2017.
[212] Narcross, F., 2021.
[213] Hassabis, D., et al., 2015.
[214] Collins Online Dictionary. 2023.
[215] Tovar J., 2023.
[216] Nvidia 2022.
[217] McRae, M., 2023.
[218] Rumelhart, D.E, Hinton, G.E, & Williams, R.J., 1986.
[219] Srivastava, N, et al., 2014.
[220] Eluyode1, OS, Akomolafe., DT, 2013.
[221] Akomolafe, DT., 2013.
[222] Salvucci, D, et al., 2013.
[223] Trafton, JG., et al., 2013.
[224] Ye, P, Wang, T, Wang, FY., 2020.
[225] Adams, S., et al., 2012.
[226] Goertzel, B., 2014.
[227] Schmidt, R., 2019.
[228] Xu, F, Liu, L, Xiao, J., 2017.
[229] Nie, X, Cao, J., 2012.
[230] Yu J, Yi Z, Zhang L., 2009.
[231] Spalla, D, Cornacchia, I M, Treves A., 2021.

[232] Aqil, M., et al., 2021.
[233] Hernandez-Hernandez, G., et al., 2017.
[234] Aqil, M., et al., 2021.
[235] Burg, MF., et al., 2021.
[236] Priebe, NJ., 2016.
[237] Rosenberg, A., et al., 2015.
[238] Burg, MF., et al., 2021.
[239] Achterberg, J, Akarca, D., et al., 2023.
[240] Galinsky, V.L, Frank, L.R., 2023.
[241] Yamazaki, K, Vo-Ho, VK, Bulsara D, Le N., 2022.
[242] Kasabov, N., 2019.
[243] Abbasi, S, Maran, S, Jaeger D., 2020.
[244] Sterling, P, Laughlin, S., 2015.
[245] Cassimatis, N., et al., 2004.
[246] Chen, H., et al., 1986.
[247] Merriam-Webster Dictionary.2023.
[248] Cambridge Dictionary 2023.
[249] Zador, AM., 2019.
[250] Olshausen, B, Field, D., 1996.
[251] Narcross, F., 2021.
[252] IEEE Access Trends and Advances in Biol.-Inspired Image-Based Deep Learning Methodologies and Applications. IEEE Access. 2020.
[253] Burr, GW., et al., 2017.
[254] Chua, L., 2017.
[255] Lanza, M., et al., 2022.
[256] Seung Ju, Kim., et al., 2021.
[257] Zidan, M.A., et al., 2018.
[258] Nano, Lett., 2010.
[259] Hermiz, J., et al., 2013.
[260] Vasilopoulos, T., 2023.
[261] Neri, F., 2023.
[262] Ji, X., et al., 2021.
[263] Groshev, M., et al., 2023.
[264] Wang, S., Chen, X, Zhao, C., et al., 2023.
[265] Lee, W., et al., 2017.
[266] Wang, S., et al., 2023.
[267] Harikesh, P., et al., 2023.
[268] Wesley, A., 1989.
[269] Moore, GE., 1965.

[270] https://www.nano.gov/nanotech-101.
[271] Furber, S., 2016.
[272] Furber, S., 2016.
[273] Spruston, N., 2008.
[274] Grewal, K.., et al., 2021.
[275] Hawkins, J, Ahmad, S., 2016.
[276] Beniaguev, D, Segev, I., 2021.
[277] Major, G., et al., 2013.
[278] Antic, SD., et al., 2018.
[279] Grewal, K., et al., 2021.
[280] Redhu, N, Thakur Z., 2022.
[281] Pagkalos, M., et al., 2023.
[282] Narcross, F., 2021.
[283] Volman,V., et al., 2007.
[284] De Pittà, M., Berry, H., 2019.
[285] Manninen, T., et al., 2023.
[286] Tang, G., et al., 2019.
[287] Homo Deus: A Brief History of Tomorrow (2016).
[288] Cambridge Dictionary.
[289] Bridgewater, A., 2018.
[290] Rao, A., et al., 2020.
[291] Narcross, F., 2021.
[292] Keban, V., 2023.
[293] Narcross, F., 2021.
[294] Narcross, F., 2021.
[295] Liu, S, et al., 2015.
[296] Reitter, D, Ritter, FE., 2016.
[297] Verschure, P., 2012.
[298] Verschure, P., 2018.
[299] Mountcastle, VB., 1978.
[300] Mountcastle, VB., 1997.
[301] Buxhoeveden, DP., 2002.
[302] Poirazi, P, Mel, BW., 2001.
[303] Hummos, A., 2023.
[304] Hummos, A., 2023.
[305] Di Paolo, Ezequiel., 2000.
[306] Dorigo, M, Colombetti, M., 1998.
[307] Husbands, P, et al., 1998.
[308] Gally, JA., et al., 1990.

[309] Gally JA., et al., 1990.
[310] Timmis, J, Neal, M., 2004.
[311] Vargas, P., et al., 2007.
[312] Moioli, RC., et al., 2008.
[313] Wang, Y., et al., 2006.
[314] Froese T, Ziemke, T., 2009.
[315] Da Costa, L, et al., 2020.
[316] Myers, CE., et al., 2022.
[317] Pedersen, ML., et al., 2017.
[318] Starns, JJ., Ma, Q., 2018.
[319] Ratcliff, R, McKoon, G., 2008.
[320] Myers, CE., et al., 2022.
[321] Kiverstein J, Kirchhoff, MD, Froese,T., 2022.
[322] Baltieri, M, Buckley, CL, Bruineberg, J., 2020.
[323] Aguilera, M, et al., 2022.
[324] Barandiaran, X., et al., 2009.
[325] Everett, M, Chen, F, How, J., 2018.
[326] Zhou, C., et al., 2022.
[327] Buzsaki, G., 2006.
[328] Moioli, R.C, Vargas, P.A,Husbands, P., 2010.
[329] Moioli, R.C., Vargas, P.A., Husbands, P., 2010.
[330] Friston, KJ., 2000.
[331] Soltoggo A, de Velde F Front., 2015.
[332] Parisi, GI., 2020.
[333] Parisi, G I, Weber, C, Wermter, S., 2015.
[334] Burms, J, Caluwaerts, K, Dambre, J., 2015.
[335] Beguš, G., et al., 2023.
[336] Wan, J., et al., 2016.
[337] Kuffner, J., 2010.
[338] Kehoe, B., et al., 2015.
[339] Waibel, M., et al., 2011.
[340] Gilda, K, Dixit, S., 2016.

"All our knowledge begins with the senses, proceeds then to the understanding, and ends with reason. There is nothing higher than reason".

Immanuel Kant was a German philosopher and one of the central Enlightenment thinkers.

Chapter 4

The Senses

CHAPTER 4

Human Senses

SINCE ARISTOTLE (384–322 BC) THE convention has been that humans possess the big 5 senses of touch, taste, smell, sight, and hearing. Neurologists believe that there are possibly between 9 and 21 senses, largely because senses are now defined as a group of sensory cells that respond to a specific physical phenomenon and correspond to a particular region of the brain, where the signals are received and interpreted. This includes a sense of Heat (Thermoception), Pain (Nociception), Balance (Equilibrioception), and Body Awareness (Proprioception). It has been estimated that vision, hearing, smell, taste, and touch alone receive approximately 11 million bits of information per second (bps), broken down as approximately 10m bps through the eyes, 1m bps the skin,100,000 bps the ears, 100,000 bps smell, and 1000 bps taste. Scientists estimate that the conscious mind only seems able to process around 50 bits per second, an astounding level of compression in the total volume of sensory information. Each neuron can store an estimated 250,000 bits of information which can be activated when new bits are processed. Our sensory systems are always between 50 to 100 milliseconds behind what happens and to compensate our brains rely predominantly as we have seen on prediction. For example, our brains predict the path of motion and location of a cricket or baseball ball in space and time before our senses have processed it, which means processing amounts to error correction to the extent the prediction is inaccurate. Some scientists argue that this applies not just to motion but to most of our complex conscious experiences making the challenge of simulating this level of sensory data to allow a robot to use the information from its sensors to adapt, plan, and execute action in the environment highly complex. The ability to integrate sensory information, including information generated from additional modalities, to achieve higher levels of cognition, consciousness, and self-awareness is key to whether AI robots will be able to learn, adapt, and be autonomous. It requires us to view perception and cognition in autonomous robots as the result of their interaction with the sensory information from the environment, not just the result of human programming. Let's start with vision which has also been the field most researched in both humans and AI systems and robots.

Vision

Vision requires the fusion of visual perception with motor control and decision-making to integrate movement planning and navigation grasping etc both rapid and more considered. More than 50% of the human cortex is devoted to vision with MIT estimating that

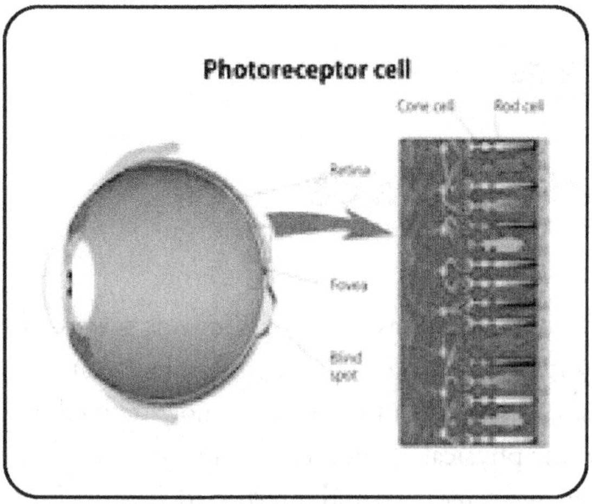

90% of information coming into the brain is visual and 80% of what we learn is obtained through visual processing of information involving roughly 70% of the brain's electrical activity. The human vision system comprises the eyeball, the optic nerve, the Lateral Geniculate body, and the visual cortex of the brain.[1] The human eye can receive light particles or photons in the range of wavelengths between 370 and 730 nanometres. The multiple directions and energies of photons are transformed by our brain into the shapes, and colours, brightness that constitute images. Below this visible light spectrum, we can't see infrared and radio waves, and above we can't see ultraviolet bands, then x-rays and gamma rays. Some people with a condition called Aphakia in which the lens of one or both eyes is missing due to congenital defects or surgical removal for cataracts, can though see ultraviolet light.

The eye contains cones and rods and our perception of the world relies on the cones to generate high-resolution colour vision. The visual colour spectrum is Red, Orange, Yellow, Green, Blue, Indigo, and Violet. The cones are called S (short), M (medium), and L (long) reflecting the short medium or long wavelengths that have peak sensitivity and response. The shorter wavelengths are bluer, the longer redder, so shorter wavelengths include blue and purple, medium yellow and green, and longer contain red and orange. There are about, 6–7 million cone cells located in the small central area of the eye called the fovea, which play a vital role in attention. There are about 100–120 million rod cells in the retina with light entering the eye through the coloured pupil which is the opening in the Iris and the lens focuses the colour and light onto the retina where cones turn it into signals for different colours. The rods detect light and dark and the cones either red, blue, or green shades called trichromacy, and all three must function to allow us to see colours. Photo-pigments are made of a protein called Opsin and a molecule known as 11-cis Retinalin which is sensitive to light and in the cones gives you the ability to perceive the full colour spectrum.

Each cone can register about 100 different colour shades and scientists claim that humans can distinguish around a million colour shades. Some people, only women called Tetrachromats, have a rare genetic mutation that gives them four types of cones and are often described as having supervision because they can see approximately 100 million colour

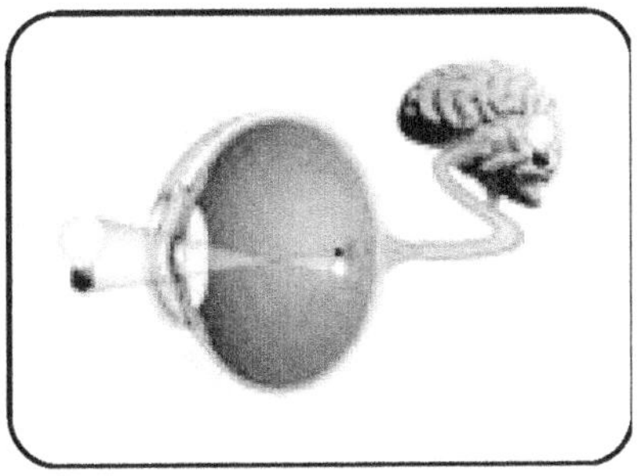

shades. The reason for anomalous Trichromacy or Tetrachromacy is that the gene for our red and green cone types is on the X chromosome (women have two X chromosomes, while men have one X chromosome and one Y chromosome). Very occasionally, women can receive two different versions of the gene and it is estimated that around 12% of women have an extra cone although they may still not have supervision which remains a very rare phenomenon.[2] If a man has a genetic mutation on the X chromosome, he is colour blind which happens when one type of cone is not functioning because typically, they overlap too much so that person may only see 10,000 shades. Some people can detect changes in the colour of a person's skin indicating illness even before doctors might detect something wrong. In some species, Tetrachromacy is an important survival tool playing a role in mate selection and even the ability of insects to pollinate certain flowers.

The visual system consists, of the Optic pathway and the Pupillary Reflex pathway with the Pupillary reflex involving the parasympathetic system constricting the pupil when light intensity increases to an uncomfortable level by decreasing the amount of light entering the eye. The optic pathway begins in the retina with a complex structure comprising ten different layers, each with a distinct function and containing the rods and cones. When information in the form of different nerve impulses or signals leaves the eye, the retinal vision cells (rods and cones) are projected via the optic nerve onto the visual cortex in the occipital lobe at the back of the brain. The total is about 140 million neurons of which half receive projections from the fovea (in the far back), with the other half receiving projections from the rest of the retina, mostly rods. Only a small percentage is sent to the occipital lobe, and scientists believe that mapping occurs in the retina before it is sent. Human vision involves a retinotopic map which is not like a camera in that we can construct 3-D images from 2-D information and organize them so that there is a close relationship between the position of an object in a person's visual field and the object's position represented on the cortex.

Each visual area has a retinotopic map of space which is a pattern of stimulation on the retina that involves visual information received from the retina being mapped on adjacent neurons along the cortical surface which receive input from adjacent points along the surface of the retina. In the 1950s and '60s, David Hubel and Torsten Wiesel in their study of cat vision discovered that specific neurons in our visual system correspond to specific pixel-like locations in the retina.[3] Humans visually perceive the world by hierarchical processing of the impoverished 2D image of the world on the retina to make a probabilistic calculation to interpret the image and decide what it is using partly prior knowledge and experience which reflects an individual's internal model of the world.

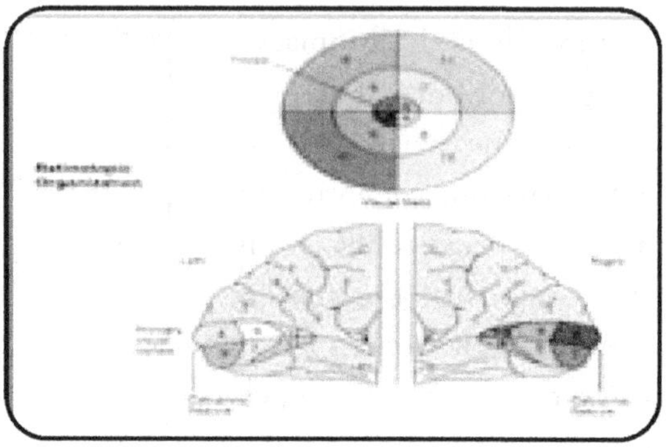

Retinotopic maps

It is not necessarily the eyes's design or capacity but the remarkably rapid processing by the human brain of the information that is received through the eye, that makes human vision so powerful and so complex. Neuroscientist Tai Sing Lee observes how "the number of possible images that can be expressed in a grey-level image patch of 30 by 30 pixels is 9002^6 practically infinite" and that "it is almost impossible to encode prior knowledge of such a scale in the brain even with its billions of neurons."[4] This rapid processing is evident in the fact that light appears continuous and steady to us because of what is known as the flicker-fusion threshold or critical flicker frequency (CFF) which is the point at which we cease to perceive a flickering light source as a series of flashes. This is why when TV screens and monitors are continually updating, we don't perceive the change giving us the illusion of a stable picture or image, and has important implications as a measure of cortical arousal alertness, and visual processing.[5] Most people cannot detect flickering above 50 to 90 Hz and it is thought that a frame rate of 72 Hz for computer displays is sufficient to avoid a perceived flicker, although it depends on the intensity and contrast, but some people appear to be able to distinguish between modulated and steady light at up to 500 Hz. This has been attributed to the unconscious rapid eye movements or saccades which "allow simple modulated LEDs to be observed at very high rates."[6] It is these rapid eye movements that cause the phantom array effect which turns a flickering light source or a lit-up object into a series of dots or lines or repeated images across your visual field.

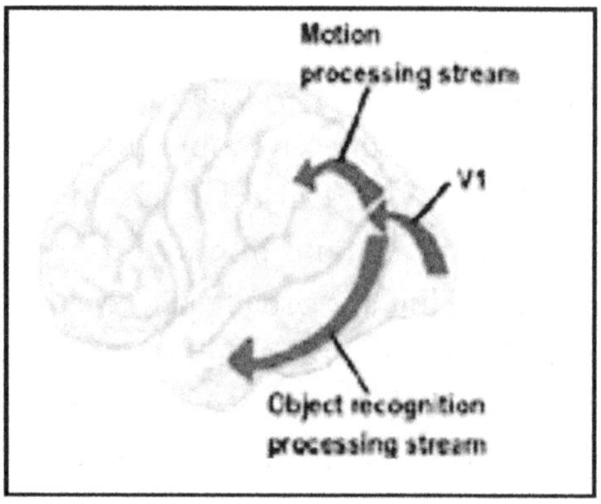

Processing of visual information received through the eyes starts in the visual cortex in the occipital region at the back of the brain. Where the information flows from the primary visual cortex to other association brain regions for integration, depends on the type of information. Over half of the caudal region of the cortex is devoted to visual processing and when stained has a unique layered appearance so is often called the Striate cortex. Striate simply means marked with long thin marks or stripes of colour. The Dorsal stream involves the association areas discussed earlier, in which the "Where is it?" Information is integrated and processes information related to spatial orientation, fusion of binocular vision and depth perception, and the location movement direction and velocity of objects in space. The Ventral stream involves association areas in which the "What is it?" information is integrated and processes information to recognize objects and colours, read the text, and learn and remember visual objects.

David Marr describes how humans process visual data in stages.[7] This involves parallel processing in terms of identifying what and where something is, then recognizing the object based on prior knowledge, predictive processes and error correction, and the assignment of salience relevance, and meaning. This is closely linked to the process by which we attend to our environment and the emotional and reward risk assessment of whatever it is that we are looking at. The gestalt function (see below) fills in the gaps based on prior experience and knowledge. For a long time, it was thought that the processing of visual signals is a sequential and passive pipeline process, but we now know that it is a dynamic process that involves humans constantly predicting what they are seeing, and perception is an active process mediated by prediction. In other words, our brains, based on experience, are constantly guessing what is there and filling in what they cannot see. For example, we all have a blind spot in the middle of the retina, but we don't notice it because the brain fills in what we don't see. As visual information passes through layers of cortical neurons, details about edges, textures, and colours combine to form abstract representations of what is being seen. Predictions appear to be generated in higher cognitive levels and processed backwards through our sensory processing areas in time to meet the sensory data they are predicting. There is no surface or canvas in the visual cortex, but a network of billions of neurons exchanging signals from signal flows containing information about the world, for example, the edge of an object, its colour, and moving features, and all these signals are organized into dynamic structural models allowing us to know what is out there.

One MIT study led by Mary Potter study in 2014 found that the human brain can identify images seen for as little as 13 milliseconds which is an incredible level of near-instantaneous conceptual activation that has been thought to be an evolutionary survival function that allows for immediate action when necessary. Potter observed that "The visual system is exquisitely adapted to the task of extracting conceptual information from visual input with every new eye fixation, three or four times a second."[8] The MIT team concluded that the process was "consistent with the feed-forward model, in which an initial wave of neural activity through the ventral stream is sufficient to allow identification of a complex visual stimulus in a single forward pass" and "Our ability to understand the information keeps pace with the rapid flow of information." They suggested that this may reflect how "the units that process a visual stimulus are hierarchically arranged: Units representing small regions of space (receptive fields) in the retina converge represent larger and larger receptive fields and increasingly abstract information along a series of pathways from V1 to the inferotemporal cortex, and higher to the pre-frontal cortex." This is at odds with the widely accepted theory that perception arises from feed-forward and feedback connections, with the initial feed-forward process generating possible interpretations which are then fed back and compared with lower levels of processing for confirmation. They concluded that "feed-forward processing alone is capable of decoding complex information in a novel image and sufficient for the conceptual comprehension of pictures" and that "a top-down attentional set is not required to rapidly extract and at least briefly retain conceptual content." Even periods as short as 13 milliseconds are sufficient to "drive conscious detection, identification, and immediate recognition memory."[9]

The pre-attentive process arises very early in the processing of sensory stimuli and is typically very weak and operates totally in our subconscious. Subconscious perception and thoughts influence everything we do from our vision to movement, attention, emotions, and our complex cognitive processes in terms of social judgments about other people or situations. Studies on visual perception show that we rely on our subconscious to fill in the gaps to provide us with what we think is a stable and consistent version of our real world, which includes key features such as invariances that we discussed earlier.

In 1910, Max Wertheimer (1881–1943), Wolfgang Köhler (1887–1967), and Kurt Koffka (1886–1941) published a paper explaining how, even though we perceive motion in rapidly flickering static images, we automatically group different objects into a single coherent whole or in groups, when presented with separate elements arranged together in a particular way. This Gestalt system has furthered our understanding of how we subconsciously organize sensory information. Gestalt meaning form or pattern reflects the idea that the whole is different from the sum of its parts and refers to the way that human perception is based on recognizing patterns and simplifying complexity in our brain's constant search for reducing uncertainty in the environment.

Pattern recognition is as we have seen a critical part of ML and DL in AI. Gestalt reflects our brain's natural inclination to find order from disorder and gain meaningful perceptions from the chaotic stimuli around them and involves our brain filling in the gaps. The key gestalt principles are (1) Similarity where similar elements are perceived to be more related than dissimilar elements (2) Continuation in which we are more likely to see continuous and smooth flowing lines rather than (3) Closure where prefer complete shapes to automatically fill in the gaps (4) Proximity where we group closer elements and separate them from those further away (5) Figure/ground which because we dislike uncertainty makes us look for solid stable items and (6) Symmetry & order which is when people perceive symmetrical elements as part of a unified group. We will briefly look at the main gestalt ways that our visual perception tries to fill in the gaps, which are, Multi-stability, Emergence, Reification, and Invariance.

Multi-Stability

Multi-stability is the tendency when faced with ambiguous perceptual experiences, to go back and forth, unstable between two or more alternative interpretations, which means that we can see two different things in a single image and have different experiences simultaneously. However, the two interpretations cannot be seen both at once so this is an illusion, and eventually, our brains tend to choose a dominant interpretation, and the longer you look the more difficult it becomes to juggle the two interpretations. You will see when you look at the Necker cube and Rubin's Figure/Vase illusion below.

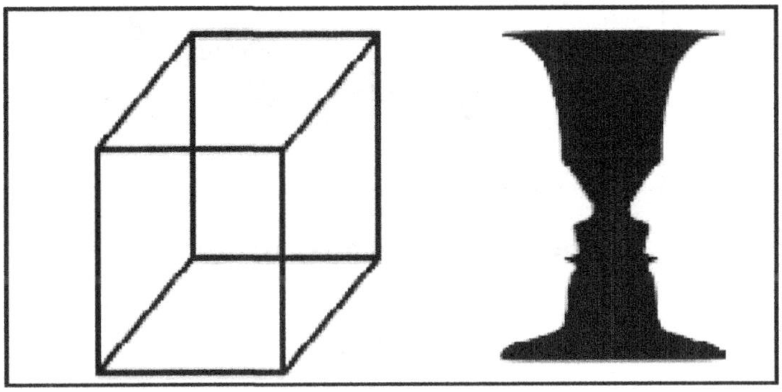

Necker cube, and Rubin's Figure/Vase illusion.

Emergence.

Emergence is when we form complex patterns, based on simple rules to identify something not from the parts, but by perceiving the whole object all at once. For example, the perception of the Dog picture below. Cognitive scientists Peter Tse and Howard Hughes explain how "many observers find that they first recognize one part of the dog, say the head which then makes the whole dog's shape (suddenly) apparent."

The Dog Perception Picture.

Reification

Reification, also called closure, is the tendency to consider an object in its entirety before it perceives the object as the sum of individual parts. This involves the human brain often perceiving more visual information than what it sees and filling in the gaps to complete an image, like the image below in which most people will see a triangle and a snake.

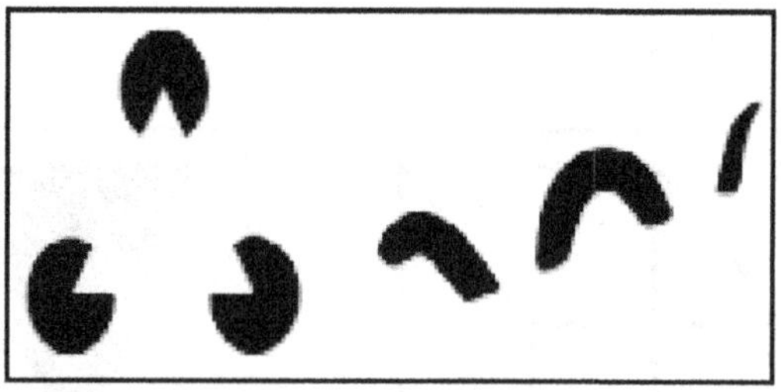

The Snake and Kanizsa Triangle.

Invariance

Invariance which we visited earlier describes how the mind can recognize certain objects independent of factors like rotation, scale, or distortion, so for example in the image below you will see the same cube despite the different perspectives and distortions.

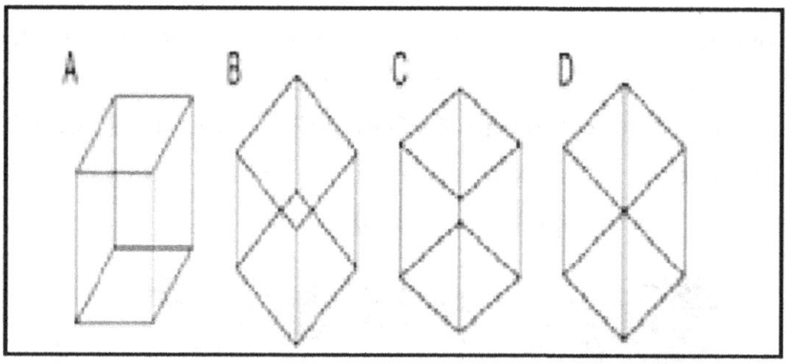

The gestalt processes by which our visual system completes the incomplete sensory information from our world to provide us with a seamless visual version of the world that we believe to be our reality is so automatic and subtle that it is entirely hidden from our consciousness control.

Hearing

The brain maps auditory information (Tonotopy) in the Auditory cortex within the temporal lobe, which is called frequency-to-place mapping, based on a mapping of the physical frequency in the cochlea, which is 10 millimetres (mm) wide hollow snail- or spiral-shaped structure, one of two main structures that form the inner ear.

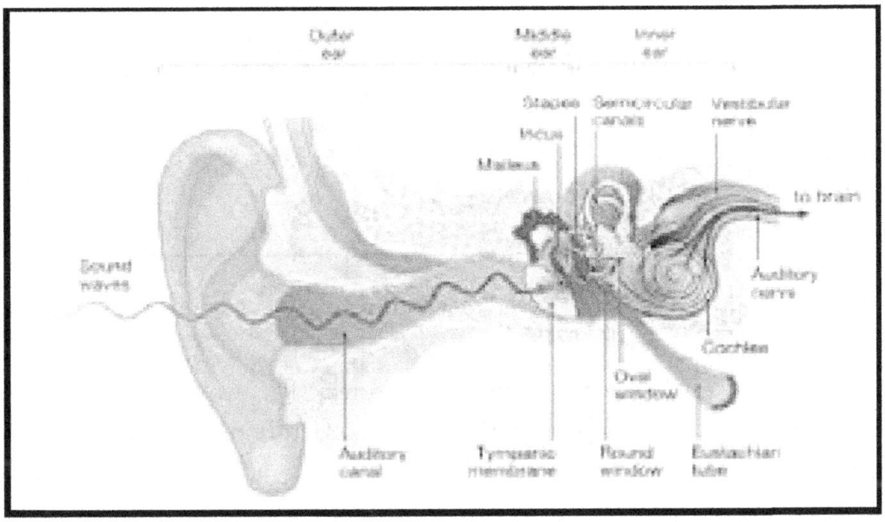

The Chochlea

The inner ear is located behind the eardrum and deep within the middle ear and has chambers filled with fluid which senses pressure changes and activates hair cells which results in the change in the balance of electrolytes in the fluid and the polarisation. This results in signals with varying frequencies being sent along the auditory nerve to the brain where they are interpreted.[10]

Smell

The smell is a primal and important arousal system calling attention to environmental events, air quality, and food. It gets processed in a very different way to visual information and so the systems designed to mimic visual input cannot simply be used for olfactory information. Olfaction research only really began in earnest in the 1990s, with the discovery by Linda Buck and Richard Axel, both then at Columbia University of the genes for odour receptors. It is now extremely well characterized, and we know there are over 1000 human smell receptors, compared to only four light receptors and four touch receptors, and like other animals, humans can store odour memories. We smell odours through our olfactory system (nose and the nasal cavities) when we inhale airborne molecules that bind to protein receptors located on over 100 million olfactory receptor cells (ORs) clustered in the olfactory bulb in the back of the nasal cavity just above the nose at the bottom of the front of the brain to form the olfactory epithelium.

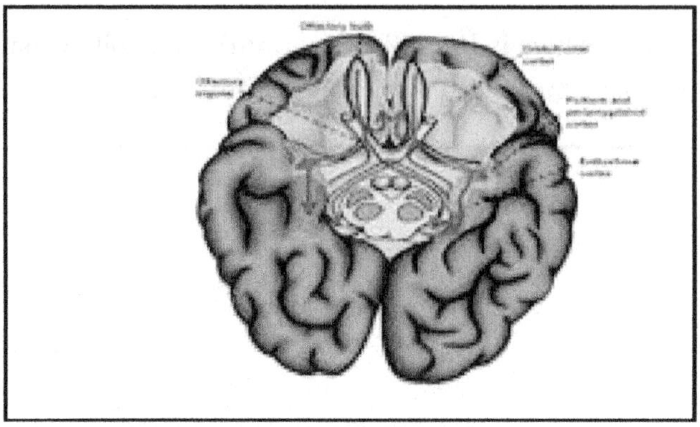

Olfactory Pathway

Recently scientists discovered the Piriform Cortex, (PC) which is the largest cortical recipient of direct olfactory bulb projections.[11] The olfactory receptor neurons expressing the same odourant receptors converge to form a receptotopic map which means they are the only sense that bypasses the thalamus. Smell is closely linked to many areas of the brain associated with emotions and reward-motivated behaviour which is why smell plays such an important role in detecting food and influencing sexual and social behaviour and is linked to subjective hedonic (relating to or considered in terms of pleasant or unpleasant sensations) processing. Many people have heard of Pheromones (Greek Pheran – to transfer – and Horman – to excite) which are non-odorous molecules secreted in body fluids like urine, sweat semen, and baby milk. They are widespread in the animal kingdom acting as undetectable or hidden forms of communication between individuals of the same species which can trigger hormonal and behavioural responses. Evidence of their existence in humans is weak though and this may be yet another brain myth that will be debunked in the future.[12, 13]

Touch.

Despite its crucial importance to us, touch is the sense about which scientists know the least. From an evolutionary perspective, the diversity in mammalian body shapes and sizes has led to each species developing sensitivity in specific areas, with mammals typically using the Glabrous (non-hairy) skin of their hands and feet to navigate and manipulate their environment.[14] In humans, the sense of touch, Haptics or as neuroscientists call it, Somatosensation, involves mechanosensory neurons in the body's largest sensory organ, the skin, that transform touch stimuli at the skin surface into electrical impulses that relay signals from the body's periphery to the CNS and particularly the cortex. Touching provides us with information we can't perceive through any other sense, such as the temperature, its texture, weight, and even the state of something. Haptic perception (Greek: haptis "palpable", haptikis "suitable for touch") is the ability to grasp something. Perception involves the active exploration of surfaces and objects when a person is moving to explore unknown objects or recognize familiar objects, rather than passive contact and tactile perception when someone is motionless.

Tactile feedback allows us to gather information about the world around us including people and objects and is known to play a vital role in cognitive development during the early years. It involves feeling texture, temperature, pressure, pain, or vibration and it has been shown that if you lose your sense of touch, you cannot sit get up, or walk. Over half of our brain is dedicated to processing the sensory experience of which touch is a fundamental part. The sensory cortex (see earlier) interprets the information that it receives from touch, and you will remember that the location of the body parts about other body parts, is mirrored in neural patterns in the somatosensory cortex, or somatosensory homunculus so fingers are near thumbs, ankles near toes etc. Homunculus is Latin for "little man", or "little person" and the number of cortical neurons allocated to different body parts reflects their level of sensitivity so the pattern is distorted to reflect sensitivity, rather than the relative size of the body parts. Some parts like hands and lips are more sensitive to the touch and our brain devotes a lot of space to sensitive and fine discriminative touch with the over-representation of sensitive skin surfaces in the brain emerging during adolescence. The traditional view was that this is achieved through over-concentration of sensory cells in certain skin surface regions, however, there is growing evidence that is challenging this. Professor of Biological Sciences, Alison Barth from Carnegie Mellon University is part of a group of international researchers who used transgenic fos-GFP mice to study activity in live neurons and small ensembles of neurons that are activated together in response to stimuli and discovered that certain neurons are very responsive to a specific stimulus, such as a puff of air on the whiskers. The fos-GFP mice express a fusion gene consisting of the murine FBJ osteosarcoma oncogene (Fos) and enhanced green fluorescence protein (EGFP) which allows scientists to visualize Fos-expression in neurons that are stimulated, which enables them to study the neurons level of excitability and synaptic efficacy. Barth and colleagues found that the neurons in question had "direct synaptic input from the posteromedial nucleus of the thalamus" and formed a "unique sub-network of neighbouring cortical excitatory neurons that display significant heterogeneity with communicate between each other amplifying the signal."[15] A Harvard study in 2011 led by Professor of Neurobiology David Ginty used mice to explore why the brain appears to allocate so much additional space to the touch senses. They discovered surprisingly that a relatively small number of neurons innervate the sensitive skin compared to what they had expected. They observed far greater connectivity between thalamic neurons that relay information from sensory neurons to more sophisticated, higher-order brain regions about sensitive skin surfaces that represent salient features of the physical world, than those with non-sensitive regions.

Robot Senses.

Robot developers must combine the latest developments in hardware which gathers the sensory data and the software which analyses and interprets the feed from those sensors. Machine sensors are an engineering and computer science discipline with a primary goal being to make robots safer and better equipped to operate outside safety cages, in highly unstructured environments and close collaboration with humans. We have seen that integrating raw data from multi-sensory modalities is going to be critical to achieving fully autonomous robots and applies across all the sensory modalities, particularly in the field of vision which has been the focus of much recent research. But let's first understand what we mean by data and information in AI.

Raw Data and Information in AI

We have seen how the processing of raw sensory input into meaningful information by our brains involves functions that are highly non-linear and dynamic and the forming by humans of mental content or percepts based on perception. A percept is simply an object or something that has been perceived by the brain, with lower levels of abstraction tied to precepts, but the higher the level of abstraction, the less connected they are with specific percepts so need to be built up using intermediate percepts. The human brain decomposes problems into sub-problems and multiple levels of representation, for example, when visually trying to identify an image, the brain identifies edges and shapes and more complex local shapes, to form abstract categories and classify or categorize the image. To function like the human brain, AI systems must have the capacity to construct multiple levels of abstraction from the lower to high levels, which requires a vast library of concepts, or the raw data cannot be perceived by the system and the computer equivalent of mental concepts can't be constructed. Perception requires an AI system to either be programmed with and trained on the raw data or to have the capacity to learn by processing raw data available in the environment like humans or a combination of both.

Data has various potential meanings relevant to the distinction between the way humans and robots process data. Israeli Professor Yuval Noah Harari explores where mankind is heading and amongst other highly insightful observations and warnings, describes the new philosophy of DATAISM which regards humans as no more than walking algorithms.[287] Noah Harari observes that like religion, freedom of information in the modern world is king with utopia achieved through greater connectivity and access to information for all. But the almost inevitable conclusion is that in a world governed by DATAISM, human experiences, and personal beliefs and values are subordinated to and trusted less than the presumed superiority of computer programs, the power of algorithms, and the data and information they process.Computer Data is defined as "information, especially facts or numbers, collected to be examined and considered and used to help decision-making or information in an electronic form that can be stored and used by a computer."[288] In computational terms, data is a unit of raw material that does not per se contain information but has been collected, analysed, translated, and organized into a form that is efficient for processing and by processing can be interpreted to give it meaning and provide information. In ordinary language and conversation, the concept of raw data is often casually conflated with the information it potentially contains once processed, which can detract from our understanding of the processes required by humans and AI systems to convert raw data into meaningful information, intelligent output, and understanding. The ability to convert raw sensory data from our environment into information which gives the data context and meaning goes to the heart of what it means to be autonomous for both humans and robotic systems.

There are as many as 32 types of classes or types of data that computer software developers can use as the source of different types of information, each having important implications in terms of the process required to convert the data into information and the type of information that can be extracted from the data. The type of data can significantly impact the amount of computer memory required, the time required to load certain features, and the program functions that can be performed, so programming languages tend to use a combination of different data types. Common examples of data include integers which are whole numbers, characters denoted by alphabet letters either as single characters or a string of letters, dates, floating points, or fractional numbers(FLOPS), long data types which are both positive and negative whole numbers that have many place values, Boolean which provides a logic code typically true or false, strings which are a combination of figures that includes letters and punctuation and many other data types which we won't discuss here, etc.

Data types vary enormously based on size, length, and their use depending on the coding language. Adrian Bridgwater described 13 different types of data which he observes as by no means exhaustive, but which form the building blocks of AI and incorporate many of the types of programming language data we have just discussed. These are (1) Big data (2) Structured and unstructured data (3) Timestamped data (4) Machine data (5) Spatiotemporal data (6) Open data (7) Dark data (8) Real-time data (9) Genomics data (10) operational data (11) High-dimensional data (12) Unverified outdated data and (13) Translytic data[289] (Translytic being a portmanteau of transaction and analysis with all the data processed in one layer and regarded as a single unified database). The type of data that is available and used has significant implications for how AI processes such data and poses a variety of challenges in allowing autonomous robotic agents to deal with the many different types of data sets with which it is programmed whilst potentially being required to extract data and meaningful information from their environment. The different types of data and the role of statistical analysis impact how AI systems are designed to learn from and convert data inputs into useful outputs in terms of information upon which it makes decisions and directs its output or behaviour. Humans do this automatically by using the raw data coming through our senses and storing the information extracted from that data in the form of neural patterns that represent prior knowledge and experience of the environment.

Information comes from the Latin word- informare which means to shape or to form or the act of informing and has been described as falling into seven categories, are (1) Factual information which is information that can be proven to be true using evidence and is universal and true for all cases and all people so can't be subjective (2) Conceptual information which is arrived at through concept formation which is regarded by psychologists as "the process by which we discover the feature or features which are 'common' to many objects and associate these with a symbol which thereafter may be applied to other similar objects." Conceptualization refers to how information forms the basis of concepts through a basis of abstraction, generalization, experience, and analysis. Experience means directly encountering something. Abstraction from the Latin abs, meaning away from and Trahere which means to draw, involves taking away or removing characteristics from something to reduce it to a set of essential characteristics and typically reduce complexity (3) Procedural information, which is about actions, conditions for actions which results from actions (4) Emotional information which is information that conveys a state of mind (5) Behavioural information which is about how and why people behave as they do (6) Attitudinal information which is about attitudes and beliefs and (7) Motivational information which is about people's motivations and drive for doing things.

Our ability to use the raw data from the environment to provide information, knowledge, and understanding is essential to our survival and allows us to be creative and inventive. We must recognize that in the future not all knowledge, information, intelligence, or understanding may necessarily be confined to current human knowledge or data and that AI may reveal and develop non-human data and knowledge, which is far superior to ours in every domain and at a speed we can't even currently imagine, especially with the potential for quantum computing which may use different sources of data. Much AI has originated from the computer sciences with statistics playing a central role in the complex algorithms employed in machine learning. Professor Sarah Friedrich and colleagues argue that statistics, as an interdisciplinary scientific field, plays a substantial role both in the theoretical and practical understanding of AI and for its future development and when applied to data analysis might even be "considered a core element of AI." The main types of data used in statistical analysis are typically classified as (1) Numerical (Quantitative) which involves a finite count of numbers or values such as how much, how often, how many (2) Categorical (Qualitative)which is not numerical but fits into the categories and involves categorical variables and features such as gender, cat, dog, etc (3) Ordinal which is a type of data which follows a natural order but the difference between the data values is not determined such as socio-economic status (low income, middle income, high income) (4) Continuous which can be calculated and can take any value and has an infinite number of probable values that can be selected within a given specific range such as temperature height, weight and length and (5) Discrete which can take only discrete values such as the number of students in a class or employees in a company and contains only a finite number of possible fixed values involving integers that cannot be subdivided into parts. Bearing this in mind let's look at the field of machine and robot vision.

Robot Vision.

Seeing in AI is called Machine Vision and with robots is largely used in industrial applications for guidance in picking, sorting, placing, and cataloguing items but increasingly important in quality assurance, object detection, material handling, navigation, and mapping. Robotic vision requires the AI system to generate real-time outputs that the robotic agent can use to perform physical tasks which involves the tight coupling between perception and the actions and behaviour of the robot. Although we compare machine vision to human vision, the reality is that nature did not confer on humans the most powerful eye in the animal kingdom and lots of animals such as eagles and octopuses have far superior eyes. We saw how in human visual processing the dorsal and ventral streams work in parallel with the integration of the information taking place in the frontal region of the brain, but parallel processing and integration of visual stimuli is currently lacking in machine vision. Our brains provide a continuous and unitary interpretation of our environment even though sensory information is changing all the time. The brain is always balancing the opposing processing needs for temporal integration of sensory information into a single event which occurs in milliseconds and the perception of separate events or entities by temporally segregating the information. Temporal segregation and integration are necessary to allow us to perceive objects as single or separate entities, but we also need to be sensitive to change. Temporal integration refers to our brain's capacity to self-organise information retention and involves persisting neural activity after the source of the input information is no longer available. Andreas Wutz and colleagues hypothesised that "the visual system might resolve this challenge by aligning integration windows to the onset of newly arriving sensory samples."[16] Their research shows that we still do not fully understand how the human brain maintains a seamless and continuous flow of integrated sensory information in vision processing, whilst constantly segregating that information.

Computer vision and image processing are often combined with machine vision which involves pattern recognition. This poses engineering challenges such as kinematics in terms of the motion of objects, reference frame calibration which defines the position of an object concerning the robot's position and orientation and the robot's ability to physically impact the environment it is operating in. The hardware equivalent of the human eye comes in various forms but replicating even the humble human eye with its level of high-resolution imaging and relative complexity is extremely difficult. Most

machine systems don't seek the same high level of clarity and focus provided by the human eye because this requires a very high level of computing power and battery life. The biggest challenge is developing the software and computer processor that allows the robot in real time to process and, interpret the visual information and decide how to act upon the information it receives through the sensors. AI systems like humans need to deal with temporal integration and segregation to allow the system to perceive information and act upon it. The complexity is illustrated by the simple act of making sandwiches which poses no problems for humans. Robots developed at Loughborough University in a project led by Dr Baihua Lu at the school of science, in conjunction with Millitec Food Systems to equip robots with computer vision designed to allow them to assemble and pack sandwiches which requires them to recognise a variety of items and pick up the correct ingredients in the correct amount and order. They are using Millitec's Delta robots whilst A U.S.-based company called Grote Company Inc has recently developed a Staubli TX2–60 SIX-axis robot with a target of producing 60 sandwiches. However, it consists of a conveyor belt delivering the sandwiches as two halves with the bottom slice, already buttered and filled, and the robot picks up the top slice and places it on the filling on the bottom slice. It is some way from making a sandwich itself let alone understanding the purpose of a sandwich or desiring one.

The modern field of computer vision has its origins in the seminal work of David Marr in 1982, who formalised the crucial abstract details of visual processing and broke the challenge down into problem specification, algorithms, and biological implementation.[17] Dana Ballard and Ruohasn Zhang recently observed how developers have since tended to approach the algorithmic problem at different levels of abstraction, distinguishing between the "Neural level, the Embodiment level and the Awareness level."[18] Ballard argues that "computational tools are essential for understanding brain function" and that the brain's hierarchical organisation has many parallels with the hierarchical organisation of computing with both being dramatically simplified by factoring in these different levels of abstraction. Let's therefore look at the three abstraction levels.

The Neural Level of Abstraction.

In 1976 the seminal work of David Marr and Tommaso Poggio developed an algorithm which successfully extracted information from random-dot stereograms which are computer-generated images that give a 3D representation of a solid object or surface.[19] Ballard and Zhang describe how "this ushered in the acceptance of an abstract algorithmic approach to account for biological observations."[20] The neural level reflects Hubel and Wiesel's edge orientation retinotopic maps that we discussed earlier and subsequent findings about how the human visual system extracts other basic image features from data and integrates these features into a coherent whole. In robotic vision there are two types of algorithms, the first is rule-based with the computer or AI systems programmed to identify an object using its edges and patterns in the sensory data, whilst the second involves DL with the robot trained to learn what it is seeing. DL has allowed developers to equip some robots with sophisticated search and corrective movement skills, such as eliminating distortions, overlaps or misalignments in the data. Quality assurance, object detection, material handling, navigation, and mapping all involve signal data processing. Machine vision in robots though is a far more complicated challenge than computer vision or image processing in fixed-platform computers which are predominantly designed to extract information from images to understand the visual information. The most widely used DL algorithm for vision which has revolutionised the field in the last decade is CNN.

CNN as we saw involves a convolutional layer which extracts the initial features, a pooling layer which extracts the main features, and a fully connected layer which summarises and classifies features to provide for prediction and recognition. Its great strength is the ability to automatically learn features from raw pixel data. One of the earliest CNNs that promoted DL, although no longer widely used, was Lenet-5 proposed by Yann LeCun and others in 1998 which has two sets of convolutional and pooling layers, two fully connected layers, and an RBD classifier as an output layer. In 2012 Geoffrey E. Hinton adopted a novel approach to machine vision when he published a paper with Alex Krizhevsky and Ilya Sutskever in which they launched a deep CNN model named AlexNet. They describe how AlexNet had "60 million parameters and 650,000 neurons, consisting of five convolutional layers, some of which are followed by max-pooling layers, and three fully connected pooling layers with a final 1000-way SoftMax classifier output layer." Alex Net vastly improved the level of accuracy in image recognition and is widely employed. The other main types of DCNN that we should be aware of are (a) R-CNN, (b) Fast R-CNN (c) GoogLeNet, (d) VGG, (e) ResNet and (f) Densely Connected Convolutional Networks (Dense Net).

In their 2022 review of robotic vision, Hongxin Zhang and Suan Lee despite all the advances that have been made in recent years observed that "bionic vision equipment is not as dexterous and intelligent as the human eye" and how "currently, robot vision can only imitate one or a few physiological functions of the human eye and cannot gather, process, and evaluate images in complex and changing settings as the human eye does."[21] This reflects different problems in the field of machine vision including at the embodiment and awareness levels of abstraction.

The Embodiment Level of Abstraction.

The embodiment problem acknowledges the need to embody algorithmic extraction of information into active perception by robots which if autonomous need to be exploratory, probing and searching their environment. Legendary American engineer and computer scientist Ruzena Bajcsy (1933–) over a lifetime in the field of robotics, recognised that active perception does not just introduce complexity but involves constraints such as perception and attentional focus when moving the robotic head which involves a relationship between the depth and flow of optical information. She won the 2009 Benjamin Franklin medal for her innovative work in robotics and artificial vision and founded the General Robotics and Active Sensory Perception Laboratory (GRASP) at Pennsylvania University. Bajcsy's major innovation was to introduce the idea of active perception in robotics as opposed to static cameras and her binocular computer-controlled moving camera system was the first example of embedding the capacity for active vision into an animated moving machine required to perceive information in its surroundings.[22] Subsequent research has led to more refined hardware developments in robotic heads and the eye movements displayed by humans. This includes incorporating saccadic rapid, ballistic movements of the eyes that abruptly change and reorientate the point of fixation or centre of gaze from one part of the visual field to another and vergence which is the simultaneous movement of both eyes towards or away from each other when focusing and pursuit which is the slower tracking movements of the eyes designed to keep a moving stimulus on the fovea.

The Awareness Level of Abstraction.

The awareness level in robotic vision reflects the visual processing required for real-time response to instructions and situations and visual, motor and language coordination and synchronisation. A major challenge with robot vision systems and algorithms embodied in a moving platform is to operate in real-time, integrate the sensory information and accommodate a major current problem of robotic platforms which is limited computational resources. Robot vision needs to integrate gaze and motor actions into cognitive models which allows the robot to be aware of and understand what it is seeing in real time. Particularly if they are to interact naturally with humans, which remains enormously challenging and why most robots still rely on verbal instructions rather than visual processing.

Visual sensors.

Robotic vision relies on the growing availability of hardware solutions in the form of low-cost sensors built into the robot. Much research has focused on various sensory modalities enabling robots to navigate, avoid obstacles and search for, identify or distinguish between objects and objects and backgrounds and between people and objects. This is achieved with varying degrees of reliability and accuracy involving either sensory devices integrated into the robot or visual simulation techniques. Sensory devices that construct 3D images include cameras, non-visual sensors such as sonar and ultrasonics, gyroscopes and compasses that provide a level of proprioceptive awareness, whilst structured light or laser sensors that project light onto an object can analyse the dimensions by reading the light or laser that bounces off. One widely used sensor is Lidar which stands for Light Detection and Ranging and uses infrared which as we know is invisible to the human eye to detect objects. Sonar which bats use involves echolocation and is probably the most used sensor for detecting objects. Sonar works well for short distances but not so well in noisy environments, whereas Lidar is highly accurate at long distances but involves enormous point clouds which are sets of a set of data points in a 3D coordinate system with each point measuring a coordinate space on the surface of the object, so the cloud represents the entire surface of the object. This is partly why companies are moving away from Lidar and are increasingly using arrays of camera sensors although these typically need good ambient light.

Simulated vision.

Many AI systems do not employ cameras but simulate vision and ANNs can achieve remarkable levels of accuracy, but the complexity makes it very difficult to understand how computers arrive at the most accurate prediction for what it has in front of it. Some biologically inspired systems use visual algorithms and saliency maps which highlight the most important parts of the input image which is usually what the human eye typically focuses on first, so find the most prominent or noticeable regions or features which then guide the attentional focus. AI correlates the brightness of the pixels to the level of saliency and spatial distribution of saliency is used in various visual attention models and often in conjunction with deep CNNS (DCNN) and are the ANN most frequently used to analyse structured arrays of data in visual imagery. Though increasingly used in natural language processing, DCNNs are still used mainly in visual AI systems, in object detection, image classification, and recommendation systems and are state of the art in terms of being able to identify patterns in images and lines, gradients, circles, or even eyes and faces.

They typically consist of feed-forward networks with 20–30 layers of activation, convolutional and pooling layers. They use fewer weights than normal feed-forward ANN in which every node in the input is connected to every node in the first layer, etc. The convolutional layers mimic the structure used by vision in the human cortex, in that each layer is more capable of progressively processing the image to recognise more complex features and sophisticated shapes, which is achieved by using, either convolutional kernels or filters. Kernels are 2D arrays of many small square templates, or weight matrices that slide across the image to search for patterns. When an image matches the kernel's pattern, a large positive value is returned and if it is no match, a zero or a smaller value. Filters are applied for 3D structures and perform the mathematical process called dot or scalar products which multiplies the weights with different input values and sums the total inputs to provide a unique value for each filter position. Once it passes through a convoluted layer it then passes through an activation layer which adds non-linearity to the function. They often use Rectified Linear Unit (ReLu) which replaces negative numbers of the filtered images with zeros and deals with the vanishing gradient problem. The pooling layer gradually reduces the image size by keeping only the important information. After potentially thousands of iterations of convolution and pooling layers, the system has created a multi-layered and fully connected neural network.

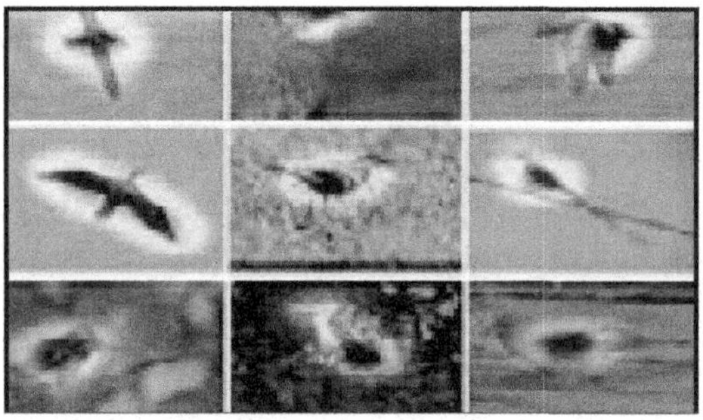

Example of a Saliency Map

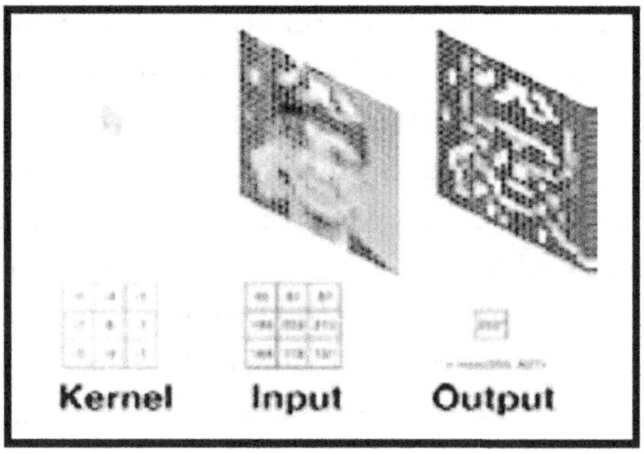

Convolutional Kernel

Humans see the world in 3-D, using stereo vision by having eyes placed at affixed distance from each other which gives the perception of depth. When the robots are required to pick up and place objects, they often use fixed-mounted In-Sight Vision sensors and there are many robots with 2-D and 3-D vision in the industry which have depth perception and can locate objects, handle material and to some extent can collaborate with their human operators based on their machine vision. The most complex technology currently involves visual feedback while the robot is moving around and requires iterative Visual Serving (VS) or Positional Feedback and correction in virtually real-time. VS techniques are well established and use visual feedback to control the robot's motion relying on a combination of image processing, computer vision, and control theory. Recently developers have been using Application-Specific Integrated Circuit (ASIC) chips customised for a particular use, rather than general-purpose, because they have higher processing speed and lower power consumption. Google has developed an ASIC-based TPU for accelerating the computing power of DNN and these tend to be more streamlined than CPUs with multiple cores that conduct multiple tasks simultaneously.

Some architecture designers treat vision as an engineering problem, requiring going through the various stages of visual processing and employing architectures inspired by robotic control biology. This involves vision algorithms or biometrics in terms of body measurements and calculations related to physical or behavioural human characteristics that harness predictive learning to authenticate an object or person. As we saw rule-based algorithms look for patterns and edges, whilst DL systems seek to interpret the visual data from the sensors like our brains do with the information coming through our eyes, which are no more than highly sophisticated sensors. Behavioural identifiers such as identifying patterns of behaviour are more difficult to rely on in terms of accuracy though good at distinguishing between robots and humans.

One technology that is a sign of the future visual capabilities of robots has been developed by the Intelligent Systems Lab at Rensselaer Polytechnic Institute and is a semi-humanoid robot called Pepper. Launched in 2014, Pepper was designed to form empathetic connections with people, serving for example as an airport concierge or hospital assistant. It has been equipped with enhanced vision capabilities that allow it to detect and acknowledge basic non-verbal cues such as body gestures, facial expressions, and body posture, which they claim allows it to display visual understanding. They mapped the human face and body and machine learning then allowed the robot to visually learn to perceive human actions and emotions. Pepper's capabilities include being able to maintain eye contact during an interaction with humans, recognise facial expressions, search for an individual in a crowded room and even estimate a person's age. It gathers data by learning about people's tastes, traits, preferences, and habits whilst conversing with them.

Pipitone and Chella demonstrated that allowing a robot to articulate what it is doing makes us trust them more and that inner speech is important to self, especially when like Pepper it appears to have a sense of humour. They tested Pepper with and without inner speech and found that Pepper had "a higher task-completion rate when engaging in self-dialogue" combined with vision, so thinking out aloud about what it is seeing has positive benefits for both the robot and those it is collaborating with.[23] Pepper was equipped with state-of-the-art sensors and devices such as NVIDIA's Jetson Nano Developer Kit and Intel's RealSenseTM Depth Camera and uses DL algorithms. Unfortunately the production of Pepper was stopped in 2021 because it appears that the developers faced challenges with limited functionality and unreliability and struggled to find a global customer base, only producing 27,000 Peppers.[24]

Some AI systems like the Ego Sphere attempt to mimic the memory system of our hippocampus. Ego Sphere involves the generation of a virtual sphere around the robot which allows it to create an appropriate representation of objects to remember and recognise them.[25] To do this it needs to be able to take the raw sensory data in the form for example of pixels and process it by reducing the data to an image with many fewer features, into the concept of an object using the low-level bottom-up attentional system used by humans which we will discuss later and integrate the information by mapping salient information within the ego sphere. It purports to act as a bridge or interface between the inadequate raw data from the sensors and the ability to represent the object as meaningful, although to date it has not been able to achieve this in a biologically plausible way.[26, 27]

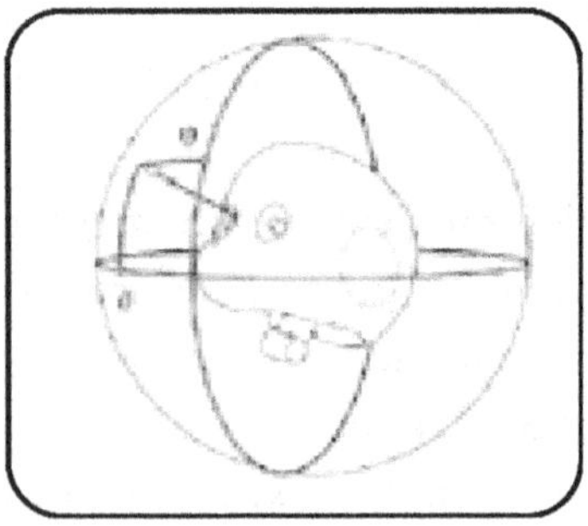

Other biomimicry systems, for example, the Leabra Vision System (LV) is one of several systems that attempt to mimic the visual pathways of the brain more precisely using a neural network simulation system, that uses a category filtering process involving reciprocal connections between layers and inhibitory dynamics, to recognise objects against cluttered backgrounds and purports to be robust against a wide range of invariant situations. This is something that in humans is enhanced by our binocular vision, but LV's virtual robot called Emer after the name Emergent for their neural network system, achieves stereo vision using two cameras offset on the robot's head.[28] This is used in navigational tasks etc, but so far has limited capability when compared to human vision which not only has the visual processing capacity but the ability to integrate data with a vast body of experience to represent and navigate our world and identify people and objects. This may be because it has been estimated that up to 50% of our brain is involved in visual processing.

The systems that use visual simulation tend to be employed in the later stages of visual processing. Systems like Soar employ simulation of vision which typically involves implicit DL and like many of the simulation systems provides visual data about the environment, objects and their features and spatial relation to each other. Most are limited to a small range of objects and are restricted to being used in controlled environments. In 2020, Sarah Wells and colleagues announced that they had built robot eyes using artificial retina light-sensitive nanowires that have a density about 6 times higher than that in a human retina and perform better than human eyes by embedding light-sensitive receptors directly into the surface pores of a 3D hemispherical aluminium oxide membrane. This enables them to be quicker in reacting to light and image resolution than humans and eliminates the blind spot we have. It can detect 86 light photons per second and see light waves in the 200 nm ~ 800 nanometres (nm) wavelength range, beyond the 400 to 700 nm of the human eye and recover from the light change in only 40 seconds compared to humans which is 150 seconds. The researchers claim that it has even greater potential with the main limit being the back contact electrode size.[29]

Comparing Human and AI Vision.

New vision technologies have resulted in interesting comparisons between human and robot vision. Humans, can process 10 to 12 images per second whereas AI systems can process hundreds to thousands of parts every minute, in high-resolution and magnified. Human vision is currently superior in interpreting unstructured and complex information, whereas AI is best for structured information and following programs very precisely. Some machines can function at infrared (IR), ultraviolet (UV), or X-ray wavelengths and have the major advantage that vision is not impaired by tiredness, boredom, or distraction so can operate efficiently 24/7. Despite these exciting developments state-of-the-art CNNs, DNNs and other mechanisms, current robots do not see the world as humans do and are effectively being programmed to see the things that humans think they will need to see. Machine vision can perform equal to and often exceed human performance with greater speed, continuity, and reliability over longer periods, but only as well as the rules that allow it to see. No machine can yet identify visual objects with the reliability and flexibility of humans and as Ognibene and colleagues have observed the research is often "seeking to identify and replicate the underlying design principles which enable humans to adaptively find and select relevant information."[30] The main area of progress in machine vision is thought to lie in the field of creating a robot with the capacity to learn how to see by mimicking the neural networks of humans but allowing the computer using machine learning to develop its algorithm for vision and share that learning so that each robot's visual capacity improves exponentially based on the learning of others. This process described as Hive learning has enormous implications for all the other domains that we will look at and the potential to create a population of robots that learn from each other through a process of distributed learning which unconstrained by the process of evolution may allow them to potentially have far superior vision to ours.

Hearing.

AI has developed the auditory capabilities of machines which almost match human performance in tasks such as automatic speech recognition and NLP.[31] They typically involve audio sensors or microphones which can also be used to navigate employing sonar or echolocation, or the detection of environmental pressure differences. Unlike visual systems, AI has not typically sought biological inspiration from the mechanism of the human ear and auditory systems but relied on "modelling the statistical structure of the temporal sequences in speech and language via hidden Markov models."[32] Markov models as we have seen are a stochastic method for randomly changing systems that possess the Markov property, which is that mathematical calculations can predict the probability of the next state or sequence of events based only on the current most recent state and are thus independent of past states. It means that it can predict a random variable based entirely upon the cue performance of machine hearing and processing of auditory information, this bears little if any resemblance to the human mechanism for hearing. It has been argued that without this AI systems will not fulfil the potential of robotic auditory capabilities and Nicolas Lesica and colleagues in a recent review of this field about clinical applications for improving human hearing using AI observed how innovation has stalled and "research suffers from a lack of computational tools with the capacity to match the complexities of auditory processing."[33]

Smell.

AI is now developing the machine sense of smell using a variety of technologies, inspired even by the olfactory circuits of flies and other insects. Some researchers regard smell,

rather than vision as the optimal way to understand how complex brain regions operate because when organisms process chemical information, they use coding strategies that could be employed in AI and robotics. Olfaction is much easier than vision to study in animals and insects etc, and could potentially address some of the computational challenges facing AI when mimicking the human body.

Unlike visual information, which is processed in a hierarchical fashion to use the different levels of abstraction to construct the representational image, smell is far less structured with no edges or spatial reference frame but consists of a huge number of variables in terms of composite molecules with varying compositions and concentrations. Human smell involves a sparse neural representation of the molecular information with many neurons being excited by a single odour and every neuron representing a range of odours. Employing such a mechanism in AI means that the molecules in the odour can generate a response with the initial neuronal layer requiring a minimal number of neurons to be able to pass the information on randomly to a greater number of neurons so they don't need to sample specific regions hierarchically like visual systems. They use an Anti-Map in which the smell information is distributed throughout the system and generates a sparse representation of the information in a higher dimensional space. At the later stage, a small percentage of neurons in this larger layer will be the most active in response to the information, effectively tagging or identifying the odour, whilst the other neurons are silent. This process mimics the role of the Piriform Cortex (PC) in human smell which as we saw is the largest cortical recipient of the odour information randomly projected by the neurons in the direct olfactory bulb and magnifies the amount of information from which to identify the molecules.[34]

It reflects the process used by fruit flies, which have 50 projection neurons that receive input from receptors with each sensitive to different molecules, so represents a 50-dimensional space which is just a mess of information. This is projected onto 2,000 Kenyon cells (KCs) which are the main type of cell in the mushroom body of the fly. KCs are the primary structure that receives sensory input from the antennae are involved in innate and associative learning behavioural responses and are dedicated to olfactory memory formation and retrieval. Unlike learning in the visual system in AI, this does not involve changing the strength of the connections between neurons and can be much quicker, requiring far fewer layers, because it seems like humans' basic smells are almost hard-wired into the system requiring little instruction or learning. Hence it can detect which smells are like previously encountered odours. Professor of Informatics Thomas Nowotny who has a theoretical and mathematical physics background, has compared this to the Support Vector Machines (SVMs) that use simple lines for 2D or hyperplanes for 3D spatially represented patterns to separate data points to distinguish classes of data through pattern recognition. They can use randomly non-linear organised information and dimensionality expansion to represent and analyse complex data efficiently and SVM learning produces significant accuracy with less computation power, which opens a new approach to encoded learning in AI which could be quicker and require less data training than in visual systems.

Touch.

Being able to touch and sense the environment and objects is a major step in robotics. Despite often having many sensors robots do not possess the dense network of sensory receptors which humans have in their skin but unlike vision which has been extensively researched, tactile sensors remain in their infancy and come nowhere near rivalling the human skin. A further challenge recognised by Sascha Fleer, and colleagues is that the interaction with an object changes its orientation and sometimes its shape, which needs the new coordinates to be accommodated through the robot's sensors. This makes it much more complex to build a machine-learning database which allows them to interact safely with objects and people.[35] Research is seeking to increase the tactile sensing and feedback of robots when interacting with objects, which is an important part of how humans move their bodies and their dexterity in response to feedback through for example the fingers and hands. The first requirement is hardware sensors with sufficiently high resolution which ideally replicate the sensitivity of the human fingers and skin in identifying information from touch, such as surface features, contact forces, and other object properties whilst withstanding wear and tear.

In 2020 an MIT spin-off, Meta AI and Gel Sight Inc. with unique elastomeric digital tactile-sensing technology and products, released a full open-source design of a computer tactile sensor called DIGIT which they are now commercially producing and can be used in conjunction with machine learning and PY Touch. On its website Gel Sight claims that DIGIT conforms to the topography of any surface, providing instant 3d Visualization and data output compatible with ROS/ ROS2 and Py Touch open-source library for touch processing. It purports to provide "spatial resolution well beyond that of human touch" and is advertised as "an easy-to-use, reliable, low-cost, compact, high-resolution, tactile sensor designed for robotic in-hand manipulation."

In 2022 scientists at Tokyo University led by Bio-hybrid engineer, Professor Shoji Takeuchi, mixed collagen, a protein found in human skin and human dermal fibroblasts to create a realistic living skin solution equivalent to human skin that generates the look and touch of real human skin, repels water, and repairs minor wounds and abrasions, much like living organisms. Called bio-hybrids they cover a robotic finger to create the appearance of a real human finger and can currently simulate the look and movement of a real finger and use the antagonistic mechanism of human muscles in which as one muscle contracts another stretches, thus forming antagonistic muscle pairs. They grow rat muscle cells in layered hydrogel sheets filled with myoblasts on a robotic skeleton and stimulate the muscle with an electric current forcing it to contract. Because they use live tissue, the robots can only operate in water and need to be constantly immersed in a soup of nutrients. They acknowledge that there is a long way to go before a robot could wear the skin outside and several significant problems to address including rapid degradation and dehydration of the tissue and problems with joint movements due to friction. They are exploring tissue containing motor neurons and incorporating hair follicles, nails, and sweat glands to give the skin a more human-like appearance. The major potential of using living skin is that the nerve cells allow the robot to sense its environment through the skin in the way humans do.

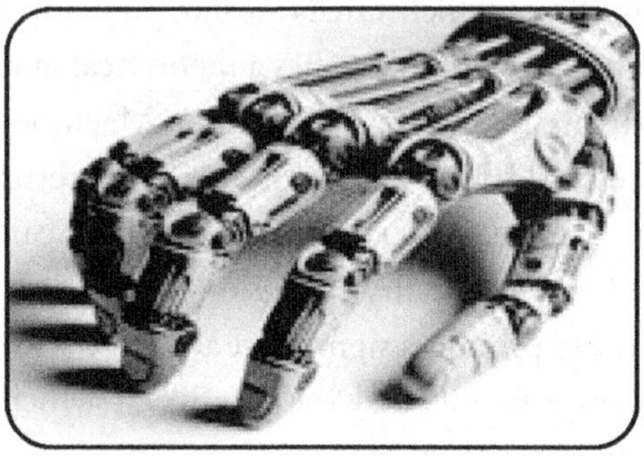

Reliable, wide range, and highly sensitive joint movement monitoring is essential for training activities, human behaviour analysis, and human-machine interfaces and scientists have developed a Conductive Porous MXene for Bionic, Wearable, and Precise Gesture Motion Sensors. This capitalises on how scorpions sense small vibrations by allowing for compression strain conversion from external mechanical vibrations through crack-shaped slit sensilla. Scorpions and spiders like many nocturnal arthropods possess micro-vibrational perception mechanisms which rely on remarkably sensitive micro-vibrational receptors, with the crack-shaped slits called Basitarsal Compound Slit Sensilla

(BCSS) amplifying the tiny mechanical signal from the receptors. This enables them to efficiently detect tiny biotic vibrational signals in noise environments. They claim to achieve this in robots through "precise and stable flexible hybrid electronic systems that integrate conformal physical sensors with miniaturised silicon-based rigid microcircuit chips."[36] Another remarkable example of nature inspiring robotic development.

Some research is using bioinspired touch to enable robots employing ML to discriminate between unfamiliar objects and shapes based on a sequence of actively controlled haptic contacts between the robotic sensors and the surface of the object. Scientists are seeking to model the sensory capabilities of the human CNS, which processes sensory-motor information in entwined parallel neural networks and integrates the information at multiple hierarchical levels. One approach used by Chou and colleagues is a neuro robot called CARL-SJR, which has a full-body tactile sensory area designed for petting. CARL encourages people to touch and interact with it by giving feedback in the form of displaying bright colours on its surface in a way that replicates the neurobiologically plausible model of fast and slow tactile sensing in humans. The developers observe that CARL-SJR is capable of learning associations between conditioned stimuli (CS), a colour pattern on its surface) and unconditioned stimuli (US) a preferred touch pattern by applying a spiking neural network (SNN) with neurobiologically inspired plasticity. They modelled and applied dopamine-modulated Spike Timing Dependent Plasticity (STDP) to the "primary somatosensory cortex, the pre-frontal cortex, the striatum, and the insular cortex, which are important for hedonic touch, to process noisy data generated directly from CARL-SJR's tactile sensory area." The SNN was "tuned to produce travelling waves of activity that carried spatiotemporal information."[37]

We have seen the challenges of incorporating sensors into robots that allow them to move, see, hear, smell and touch like humans such that they can use that sensory information to learn from and adapt to that information to make decisions and choices and act in real-time. This requires real-time integration of all this sensory information and a feature as we have seen that the human brain exhibits called neural plasticity which is central to the way the human brain learns and incorporates new information and adjusts prior knowledge. Currently, robots struggle to achieve this real-time integration of multi-modal sensory information which serves to highlight how remarkable the human brain and body are in maintaining this coherent sensory stream which allows us to use the sensory information from the environment to drive our cognition and behaviour.

Endnotes

1 Joukal, Marek 2017.
2 Jordan,G., et al., 2010.
3 Hubel D, Wiesel, T., 1962.
4 Lee, TS., 2015.
5 Mankowska, ND., et al., 2021.
6 Davis, J., et al., 2015.
7 A Computational Investigation into the Human Representation and Processing of Visual Information (2010).
8 Potter, M.C, et al., 2014.
9 Potter, M.C., et al., 2014.
10 Allen, EJ., et al., 2022.
11 Allen, EJ., et al., 2022.
12 Engen, T., 1988.
13 Taymour M, et al., 2012.
14 Lehnert, BP., et al., 2021.
15 Jouhanneau JS., et al., 2014.
16 Wutz, A., et al., 2016.
17 Marr, D., 1982.
18 Ballard, DH, Zhang, R., 2021.
19 Marr, D, Poggio, T., 1976.
20 Ballard, DH, Zhang, R., 2021.
21 Dosovitskiy, A., et al., 2020.
22 Bajcsy, R., 1988.
23 Pipitone A, Chella, A., 2021.
24 Reuters 2021.
25 Bodiroza, S,. 2011.
26 Peters, R.A., et al., 2009.
27 Peters RA, et al., 2012.
28 B. Mingus et al. STACS 98, 2011.
29 Wells S., 2020.
30 Ognibene D., et al., 2022.
31 Hinton, G., et al., 2012.
32 Lesica, N A., et al., 2021.
33 Lesica, N A., et al., 2021.

[34] Gottfried, JA., 2010.
[35] Fleer S, 2020.
[36] Duan S et al., 2021.
[37] Chou, TS., et al., 2015.

"Tell me what you pay attention to and I will tell you who you are."

Jose Ortega y Gasset.

Chapter 5

Attention

CHAPTER 5

Human Attention

WITHOUT UNDERSTANDING WHAT AND WHY our attention is subconsciously and consciously directed towards something or situations, we can never fully understand ourselves or others. Our ability to pay attention plays a major role in relationships, school, work and every aspect of life, determining the nature and quality of our interaction with our environment and forming our subjective experience of the world. For many years, attention was typically studied using introspection, with early psychological research regarded attention as a battle between our attention to the outside world and our subjective inner attention.

A common misconception is that attention is a distinct thing when it involves a complex process with our brains using information to form, retain, act upon, or transmit a representation of our world.[1] In 1890 in his book, the Principles of Psychology William James described attention as the "taking possession by the mind, in clear, and vivid form, of one out of what seems several simultaneously possible objects or trains of thought."[2] A range of neural processes control both covert and overt attention and switch attention between tasks or events. As Canadian computer scientist John Tsotsos (1952–) who focuses on computer vision and AI observes, attention in humans is "really an integrated set of mechanisms, too messy to cleanly and parsimoniously express with a single principle."[3] Neurophysiological studies have shown how "attention increases the mean firing rate of and variability and correlation across neurons."[4] Neuroscience has revealed various levels and types of attention[5] prompting computer scientist and former Dean of Yale College, Marvin Chun to observe how attention is "so ubiquitous that it is unwieldy to study."[6] Bernhard Hommel and colleagues claim that "No one knows what attention is" and even advocate "dropping the term attention as a label for a specific functional or neural system and instead focusing on behaviourally relevant selection processes and the many systems that implement them."[7] Chun proposes a taxonomy as the basis for providing a framework for understanding the neural mechanisms that at the most general level distinguish between external and internal attention, based on the types of information that attention is targeting. A taxonomy names and classifies things into groups within a larger system, according to their similarities and differences while high-level taxonomy distinguishes between external and internal attention. External attention involves alertness, vigilance, and attention to sensory information in the environment, the attention required to perform tasks all of which involve the selection and modulation of sensory information. Internal attention involves the "selection, modulation, and maintenance of internally generated information, such as task rules, responses, working memory executive control and decision making and attention involved in the storage and retrieval of long-term memory and learning."[8]

We each have a different capacity to pay attention or concentrate for many different reasons, but what you direct your attention to determines your current awareness and perception of your environment and experience and forms your highly subjective field of awareness and perception. Levels of attention naturally reflect our body's homoeostatic and circadian rhythms and our brain waves in terms of the level of wakefulness and the time of day. Most people have lower attention levels late at night and early morning and higher attention levels during early afternoon and in the early evening, with a similar rhythm seen in working memory and executive functioning like decision making etc. These can be different, depending on your body's natural inclination or chromotype which is why some people are night owls compared to the early birds.

One of the main ways that we measure a child's rate of development, including language, is the ability to attend to, focus on and learn through interactions with people and stimuli in the child's environment. Many studies have shown that the risk of behavioural and cognitive problems correlates with a baby or child's ability to learn, which it does by directing its attention and shifting between and focusing on events, objects and tasks, in the outside world. Listening and attention are critical to shaping our personalities and skill sets and a clinical attentional model developed in the 1970s by Mckay Moore Sohlberg, an internationally renowned expert in the field of traumatic brain injury rehabilitation and clinical neuropsychologist and author Catherine Mateer, demonstrated that attention consists of six levels of attentional sub-processes, which children normally acquire over time as the brain matures and neural pathways and circuits are strengthened. Their landmark work laid out important therapeutic interventions for a range of cognitive impairments, including deficits in attention, memory, executive functions, and communication. The attentional sub-processes are all important in enabling a child to learn and build memories effectively and often correlate with a child's levels of intelligence and are (1) Arousal which is the level of activation alertness and whether we are tired or energised (2) Focused Attention being the ability to focus on a stimulus or activity (3) Sustained Attention in being able to attend to a stimulus or activity over a long period (4) Selective Attention which requires attending to a specific stimulus or activity despite distractions (5) Alternating Attention which allows you to change the focus of attention between two or more stimuli and (6) Dividing Attention which is the ability to attend different stimuli or attention at the same time.

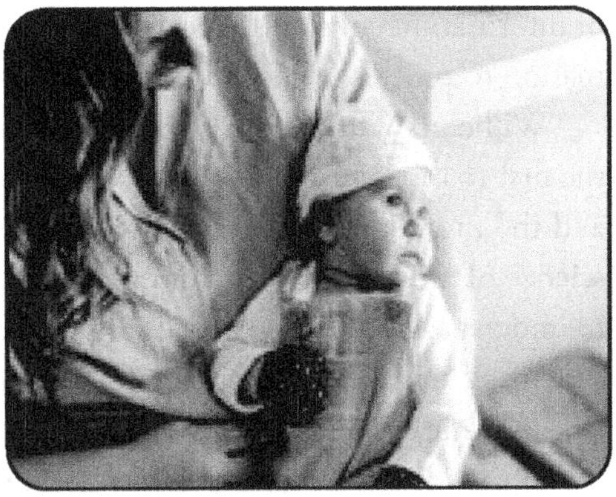

Attention in children typically develops as follows. Babies concentrate for very short amounts of time and their attention is involuntary and flits between objects, persons etc, but they increase concentration between the age of 1–2 years. By 3–4 years they can start to show signs of controlling attention after the age of five years, typically having learned how to control and pay attention, whilst being able to communicate and reflect on their motivations, desires, and interests. At this stage, they can normally exhibit a level of sustained, selective, and divided attention evidenced by their ability to continue to listen whilst performing a task. From the moment we are born, we are exposed to information about who we are and slowly how we can influence our environment, be it through interaction with objects, people, events, or situations. Babies transition from simply recognising themselves in the mirror so being able to pass the mirror test (see later) to using self-referential language and the use of pronouns such as I, me, you he, she etc. This is followed by more complex self-conscious emotions such as feelings of pride, guilt, or shame. They start to describe themselves, so develop a concept of self with attributes such as being good or bad and during late infancy self-esteem and body image develop. The development of the sense of the I or the Knower self and the Me or Known self which we will discuss shortly, means that the attention a child pays to stimuli in its environment, starts to be filtered through their subjective perception of the world and as they get older impacts their assessment of how it affects their Me self, combined with the growing cognitive capacity to reflect on an experience through the I self and indulge in introspection. This evolution of the different facets of attention has enormous implications for the development of healthy cognition and emotional systems in a growing child and determines the neural patterns and associated memories laid down in response to early experiences, which have a major influence on and govern an individual's future perceptions of and response to their world.

For our brains to recognise and process the staggering amount of sensory information, measured in bits, that as we have seen enters our big 5 senses alone, requires us to select what information to pay attention to and exclude competing stimuli or potential information. Our attention to specific and salient stimuli determines the nature and scope of the raw data that we use as individuals for processing this information. Only a small subset of information in the stimuli will be employed in guiding our decisions, choices and behaviour which makes what we select and process crucial. It remains unclear how stimuli that we do not select or consciously attend to affect us subconsciously or how divided attention affects our behaviour and performance. We cannot immediately process the sheer volume of information constantly received from our environment and must prioritise what information to process immediately, later, or not at all.

Wilhelm Wundt (1832–1920) a pioneering figure in modern psychology was one of the first to distinguish between our general field of awareness, which he called "Blickfield" and the more focused field of our attention he called "Blickpunt." Much of the modern science of attention stems from the 1940s and World War 11 when it became important to understand how soldiers could remain alert when in battle and whilst monitoring radar systems. This gave birth to the information theory emphasising attention as the processing of sensory information requiring effort to focus on and complete a goal or task. Scientists initially studied the capacity and content of our attention in terms of its intensity and how it affects our perception of stimuli, but in the late twentieth century, psychologists shifted the emphasis to understanding how deficits in attention affects our behaviour, which is most evident in such conditions, as, ADHD, ASD and Schizophrenia. Neuroscientists have focused on the neural processes or brain circuitry involved in the attentional process, recognising that many clinical and sub-clinical problems are associated with interference in how we process information in our brain, with serious implications for a person's ability to attend to their environment and extract the information vital to normal functionality and learning and memory.

One of the challenges facing researchers is that there is little consensus on what attention consists of.[11] Neuroscientist Stewart Shipp describes attention as a "matter of organising multiple brain centres to act in concert on the task at hand."[12] This neuroscientific conceptualisation more accurately captures how attention goes beyond simply focusing on or selecting something to attend to by highlighting how the brain processes and gives perceived meaning to the information by assessing how it relates to performing a task or to us and the implications for our well-being etc. We have seen how our brains process sensory information in different brain regions, and we will see shortly how information related to self is integrated into the brain and allows us to plan behaviour in response to the information, but this all depends on the information we attend to generated both internally and externally.

In 1990, Michael Posner (1936–) and Steven Peterson formulated an attention model, now known as the Posner and Peterson Model (PPM), which identified three different and discrete anatomical attentional systems called the (1) Alert system (2) Orientation system and (3) Execution system.[13] The alert system or Reticular Activating System is a pencil-shaped collection of interconnected nuclei which forms the core of the brain stem and controls arousal and sustained attention, which is important for detecting threats and salient information in the environment. Tonic alertness and sustained vigilance seem to be heavily lateralized to the right hemisphere.Tonic alertness describes the intrinsic arousal that fluctuates on the order of minutes to hours, whilst sustained attention is the state in which attention must be maintained over time.[14] Posner and Petersen have since revised the physiological and chemical understanding of the alert system because of the role of the neuromodulator nor-epinephrine (NE) which is produced in the tiny cluster of neurons called the Locus Coeruleus (LC) about 15mm large, in the brain stem which is connected by long-range projections throughout the cortex thus allowing them to modulate attention generally. The LC is now known to be highly active in acting as a warning system and implicated in pathways to the frontal and parietal cortex.

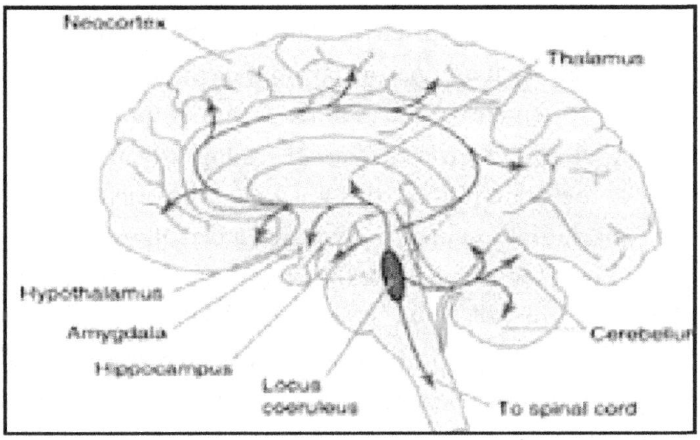

The Locus Coeruleus (LC)

The Orientation system or Posterior Attention System controls focused and selective attention so that it prioritises sensory input and comprises the Posterior Parietal Cortex, the Thalamus (The Pulvinar nuclei) and the Superior Colliculus. More recent evidence has shown that frontal areas including the frontal eye fields and specifically visually responsive neurons which correspond to what Kirk Thompson and colleagues describe as the "mental spotlight of attention via modulation of ongoing visual processing" are involved in converting visual information into saccadic commands about the orientation system.[15]

The Execution System also named the Target Detection System or the Anterior Attentional system is responsible for selective attention, sustained attention, and divided attention and consists of the Pre-frontal Dorsolateral Cortex, the Orbitofrontal Cortex, the Anterior Cingulate Cortex, the Supplementary Motor Area, and the Striatum in the Basal Ganglia.

Posner and Peterson have since revised their model and proposed two additional networks to reflect advances in neuro-imaging techniques such as PET, which have highlighted important changes that occur between infancy and childhood in the control systems and the role of genetic variation in the efficiency of attentional networks.[16] Attention is a limited resource because it requires the allocation and dedication of perceptual, cognitive and motor resources and various attentional systems, It is more likely to be allocated to tasks or situations that are novel or hold the prospect of reward than routine or unrewarding tasks.

Sustained attention is typically voluntary and associated with higher cognitive performance but is not automatic so requires effort and is something to be allocated carefully. There are three key components to the processing and perception of information and whether it is salient or relevant to us which are (1) You the perceiver (2) Your level of engagement and (3) Your dispositional trait and state of mind, the amount and nature of the information you obtain by paying attention, and the situation or context including the timing.

Effective attention is a fundamental survival feature which requires us to prioritise external sensory information from our environment or internally generated information both bodily and cognitively such as a particular train of thought. Babies instinctively display attention when they orientate towards a noise, sound or touch, in a reflexive non-conscious way. Discovered in 1863 by the Russian physiologist Ivan Sechenov and described in his book Reflexes of the Brain (1863), this was called the orienting response by Pavlov, it is an exploratory behavioural mechanism by which a person pays attention to what they become aware of as a novel (even if expected) or significant change in their environment, which doesn't trigger a startle reflex. We now know that with repeated familiarisation a person will stop responding to the stimulus, which as we saw earlier is called habituation and makes sense in that it would be exhausting and impossible to try and pay attention to everything with the same level of intensity and duration regardless of its novelty of level of familiarity to you. If there is a change to that stimulus a person may again redirect their attention towards it which is called dishabituation.

Most research into attention looks at how and what activates or engages a person's attention, what makes people select what to attend to and how much control a person has over attention and its intensity. The cognitive processes associated with attention have been studied extensively in relation to levels of awareness or arousal, the nature of vigilance in terms of being able to sustain attention, the nature of saliency, which is from the Latin verb "Salire" which means to leap, so why certain information leaps out at you, the nature of executive control over attention and the role of attention in learning and memory. Psychologists use common tests to measure attention processes related to the activation of attention, the selection of what to attend to and the level of executive control over attention. To assess how someone activates attention they may ask a person to respond to a target simulation following a cue or warning that it is about to be presented to the person or for a more prolonged test they might ask a person to notice when a clock hand jumps twice so they need to be more vigilant. When evaluating how a person selects between competing information or stimuli, they might ask them to identify a target that doesn't share basic features with distracting stimuli so it should pop out at you or look for cues that a change is about to happen. For more controlled or bottom-down attention, they might ask a person to find a target sharing a basic feature with a distracting stimulus or interpret a cue designed to orient them towards the target. When testing for control over attention, they look for the ability to inhibit or resolve conflicting attention and may test how quickly someone cannot respond to or inhibit a response to a stimulus which is presented frequently and rapidly amongst other similar stimuli. They may see how a person attends to competing stimuli or responds to a stimulus surrounded by distractors which assesses the ability to resolve conflicting attention between stimuli. Much research has focused predominantly on the visual process of attention given that most of what we attend to is generated in our visual fields.

Visual Attention.

Maurizio Corbetta and Gordon Shulman maintain that the scientific evidence points to visual attention consisting of two partially segregated networks which are (1) Goal Directed (top-down) and (2) Stimulus Driven neural networks performing different attentional functions, although interacting during normal vision.[9] Parts of the Intraparietal Cortex and Superior Frontal Cortex are involved in preparing and applying top-down selection for stimuli and responses. The other is the Ventral Frontoparietal Network which includes the Temporoparietal Cortex and Inferior Frontal Cortex which is mainly lateralized to the right hemisphere and specialises in the detection of behaviourally relevant stimuli, particularly when salient or unexpected. Visual-spatial and feature attention involves the automatic bottom-up process of identifying image patterns through "oriented edges, spatial frequency, colour contrast, intensity, or motion."[10] and is automatic because it is driven by external factors such as certain stimuli which might be inherently salient in a particular context or against a certain background. By contrast, Top-down attention involves the voluntary or deliberate allocation of attention to a target of interest be it an object or region in space driven largely by prior knowledge, intentional plans, and current goals. Neural representations as we have seen encode statistical information about the sensory environment which has been attended to with visual neurons appearing to code exclusively for features or objects including complex high-dimensional objects like faces. Our understanding of visual attention owes a great deal to the English cognitive psychologist, Anne Marie Treisman (1935–2018), who in the 1980s Triesman developed the Feature Integration Theory (FIT) which is a two-stage model by which people perceive things and integrate the different features of an object into a coherent whole. The first stage is General or Distributed attention during which a person visually scans their environment, requires no effort and is largely automatic, whilst the second Selective attention requires effort and a level of control. Tonic alertness is general or distributed attention, and sustained attention or phasic alertness is Triesman's selective attention. When something catches your attention, the neuronal activity typically moves from the tonic to the phasic level and back. This is because neuron firing in response to stimuli occurs against a background neural activity characterized by a steady action potential firing at a constant frequency and the phasic response consists of short bursts of phasic action potentials with the ability to accelerate, enhance and orientate attention towards a cued stimulus. One of Triesman's greatest contributions was the concept of a sensitivity threshold which contradicted previous assumptions that stimuli were filtered out and some ignored. She observed that these stimuli were not ignored but were attenuated so don't reach the threshold required to become selected for them to be perceived. This threshold prioritises certain stimuli in terms of salience and valence, for example, the threshold for a mother responding to her baby's cry or for you hearing your name spoken and orienting towards the speaker is set very low. The importance in terms of salience and valence, a person places on the stimulus and conflicting stimuli, determines the intensity and duration of the person's orientation and attention towards that stimulus, which explains the response of the mother to her baby's cry. This can also be triggered by something important or salient being removed or missing from the environment.

Humans are constantly engaged in a dynamic process of actively sensing our environment and processing sensory data to form our perceptions of the world and we constantly make predictions based on that data. The intensity of our attention reflects our current level of arousal or alertness in relation to the stimuli, because the attribution of a weighting to it in terms of how it relates to or potentially satisfies a need or goal, determines its salience and valence and whether it meets the sensitivity threshold. When we scan the environment, we use pre-attentive processing to process "a stimulus that occurs subconsciously, before our conscious attention has focused on this particular stimulus, from among the array of those present in a given environment and thought to identify basic stimulus features in parallel, with high capacity."[17] It consists of the eye and the brain in less than 500 milliseconds, subconsciously processing in spatial memory the properties of any stimulus. When our attention is directed towards a stimulus, patterns and features in the form and shape are assessed by the hierarchical visual processing of orientation, edges, colour contrast, intensity, or motion etc, which is largely an unconscious visual process.

All information is processed pre-attentively, filtered and processed further to determine what it is and how important it is in terms of relevance to what an individual is thinking about or feeling at the time or salient to the extent it stands out amongst competing stimuli. Triesman showed that when people visually scan their environment, they look quickly for features of an object such as its colour, which is the pre-attentive stage and then they look at the object which is the focused attention stage. It is during the focused attention stage that we associate an object, event or person with our existing experiences and memories to identify the whole and understand the information. Researchers exploring visual pre-attentive attention, often track saccadic movements that are made several times per second and reflect the brain visually processing image patterns. Our brain decides whether to allocate the fovea to that stimulus which is as we have seen where the centre of the field of vision is focused and the retinal cones are particularly concentrated, so it provides the highest visual resolution on the retina. The type of response and the reaction time reflect variations in a person's alertness or readiness to respond or in the intention to perform a task, which psychologists call the Preparatory Set (PS) or how prepared they are which recognises that we go rapidly through phases. These are the 'initial reflexive noticing and orienting to the stimulus, preparation, and execution of response" and this preparation "involves a coordination of many aspects of the organism: muscle tone, posture, breathing, autonomic functions, motivational/emotional state, attentional orientation, and expectations."[18] This has been studied extensively in the context of the preparatory adjustments involved in switching between tasks, which is an important indication of cognitive control over attention. Studies have shown that advanced task preparation can influence response times and error rates when switching between tasks.[19] Peter Payne and Mardi Crane-Godreau observed that PS provides a more nuanced and flexible perspective on the stress response than current theories of stress. Their premise is that a maladaptive PS is implicated in stress and trauma and that "meditative movement, meditation, somatic education, and the body-oriented psychotherapies are approaches that should be used as interventions on the PS to remedy stress and trauma."[20] This illustrates the importance of attention in creating stress, how stress can be highly subjective and involves a physiological and neurochemical response and sometimes but not always an external environmental stressor.

Anti-saccade tasks involve people making a voluntary or intentional saccadic eye movement or shifting their gaze away from a stimulus so inhibiting the response to look at something and disengaging. They usually take longer with people tending to make more errors when required to do this in attention tests, as evidenced by a rapid eye movement towards the stimulus and then a rapid correction to look away, which means that they couldn't inhibit the gaze. Studies evidence significant differences in the level of preparatory neuronal activity of single neurons in the frontal eye field of the PFC between pro-saccade and anti-saccade trials. People with damage to the PFC and with conditions like schizophrenia experience great difficulty, often failing to suppress the saccadic movement towards a stimulus and displaying delayed reaction times.[21] One study suggested this may relate to "an imbalance in favour of preparation for motor movement over inhibition of motor activity."[22] Young children struggle to perform anti-saccade tasks probably because they have difficulty in suppressing the automatic pro-saccade which is attributed to the fact their frontal lobes are not fully developed. People with ADHD struggle to suppress the automatic saccadic movements, even though they usually have normal reaction times. Because of these differences psychologists often use anti-saccadic tests to determine the level of control a person has over their attentional behaviour during this pre-attentive processing phase. Research using ocular-motor and manual-motor tasks has found evidence of genetic overlap in schizophrenia, ADHD, and autism spectrum disorder (ASD) and common deficits of being able to inhibit eye and motor movements.[23]

Auditory Attention.

Both visual and auditory attention can be covert which means that we can shift our attention without moving our eyes, so there are no related saccadic eye movements and with auditory stimuli we can listen to another conversation going on at the same time as your conversation, and both will have an impact on how something is perceived. Our understanding of auditory attention owes much to the early work of the British electronic engineer and cognitive scientist Edward Colin Cherry (1914–1979), known as Colin Cherry, who was the first to work on what has been called the "Cocktail party effect" which is our ability to follow a single conversation amongst many other conversations around us, which comes from our ability to focus our attention on whoever is talking.[24, 25] Cherry asked subjects to listen to two different messages from a single loudspeaker at the same time and to separate them and repeat only one of the conversations. He called this the shadowing task and discovered that our ability to separate sounds from background noise is affected by many variables, such as the speaker's gender, the direction and pitch of the sound and the rate of speech and when these were similar, people couldn't complete the task successfully.

Covert visual and auditory attention makes studying visual and auditory attention more difficult, but we know that orientation to an external stimulus usually involves interrupting attention or actions and a movement towards that stimulus. This typically involves physiological changes such as a general change in muscle tone, perhaps in preparation for further movement, a change in heart rate and respiration, skin conductance, and vasoconstriction in the limbs. Vasodilation in the brain with higher-frequency brain waves is indicative of increased cortical arousal or wakefulnes. The degree of orientation and arousal will reflect the level of novelty, level of familiarity and previous exposure or conditioning to that stimulus or conflict in attention. Initial attention requires activation of the attentional process, by some trigger or cue in the environment, which may be because something happened suddenly like someone waving at you which is an exogenous bottom-up process because it is driven by an external stimulus or because something is relevant to a goal or task you are undertaking which is called top-down attention. The trigger or cue may also be endogenous in that it is driven by an internal stimulus like a bodily sensation or a train of thought rather than something in your environment.

If information reaches the sensitivity threshold and is selected and attended to, it shifts into Triesman's focused or selective and sustained attention stage. The neural network related to working memory is in the PFC which is a key area of the Cortical Midline Structure (CMS) where control over our attention is processed and involves the information being held in short-term storage in our working memory, which typically can last up to 30 seconds. The greater the task demands the greater the attention required, which means the capacity for peripheral attention is reduced, reflecting the deployment of mental effort, involving both conscious and significant unconscious processing. Once the importance of the stimuli is established the top-down attentional process is involved. Attention to relevant and important visual stimuli is essential to perform daily tasks, and scientists have studied the different orienting mechanisms of both engagement and disengagement of attention. Selective attention is typically, voluntary, intentional, and effortful, rather than a learned automatic action which does not require attention like riding a bike, although voluntary attention can sometimes require the suppression of automatic behaviour to control our thoughts and actions when we need to use attention to achieve a particular goal.

Some scientists believe our visual working memory and selective attention involve the same processes and representational content, with a vital interaction between working memory and attention which influences focusing our attention on stimuli that match or are relevant to that held online in our working memory.[26] Sustained attention can be very tiring because it relies on diverse areas of the brain, and we all know how difficult it can be when we are performing a task to maintain concentration and attention for a long period. Sustained attention involves three distinct brain processes which are the (1) Monitoring and control of attention (2) Activation of task-relevant processes which includes enhanced visual processing and (3) Inhibition of processes that are not relevant to a task, such as peripheral noise or other activity. These reflect the brain constantly adapting the cortical processes to meet the required activity which is achieved through temporary strengthening or weakening of synaptic connections, in what scientists describe as synaptic competition. The long-term strengthening of connections typically lasting beyond 3 hours, requires the expression of new proteins and can result in strengthening certain functional clusters of neurons.[27] It has long been believed that this term memory formation involves and requires macromolecular synthesis which is crucial to how we learn and involves the binding of small chemical structures called monomers to form larger long chains of molecules called polymers,in this case proteins.[28]

Spontaneous Cortical Fluctuations.

Neuroscientists know that our brains exhibit a level of weak spontaneous electrical activity producing spontaneous cortical patterns, even in the absence of any sensory stimuli during rest or dreaming. For many years this was regarded as just background noise associated with slow alpha brain wave sleep states. The traditional view was that the asleep brain operates in a "synchronised state, characterised by strong low-frequency fluctuations in cortical activity"[29] but during wakefulness and REM sleep, our brain is highly desynchronized, and the low -frequency spontaneous fluctuations are suppressed. New research has shown that spontaneous low-frequency fluctuations occur even whilst awake and correspond with ongoing behaviour. Low-level spontaneous fluctuations have been found in the cortex the thalamus, the hippocampus, the basal ganglia, and the cerebellum.

For a while, scientists believed that synchronised cortical oscillations played an important role in higher cognitive activity during sustained attention and have assessed how this relates to fluctuations and deterioration in sustained attention over time. Neuroscientist Dario Ringach observes that the prevailing view is that the brain "operates on a continuum of states between synchronised and desynchronized states, with varying levels of spontaneous fluctuations in neural population activity along this continuum."[30] Ringach and others observe that this spontaneous electrical activity is highly structured in both space and time,linked to the connectivity between different brain regions and interacts with incoming sensory stimuli to generate responses. It has even been associated with visual hallucinations when people are deprived of external sensory input, whether through natural causes like blindness or deafness or the somatosensory sensations of touch and other input as experienced in sensory deprivation experiments or chambers. A common condition called Charles Bonnet syndrome happens when people lose some or all their vision and experience visual hallucinations, which is the brain making up images to compensate for the loss of visual input. They often occur when waking up and include landscapes, imaginary creatures, repeating patterns and shapes or lines and dots, both static or moving and can be fleeting or last for hours. Similar hallucinations can be attributed to medication, certain mental health conditions or neurological conditions which need to be discounted by clinicians when assessing people for Bonnet syndrome, but people suffering from Bonnet's know that the hallucinations are not real so don't act upon them.

Giovanni Pezzulo and colleagues describe the spontaneous neural activity of our brains as "The secret life of predictive brains that underlies top-down dynamics of generative models" with brains even when at rest being proactive and predictive. Pezullo and colleagues observe a "Rich mental life" and even "in the absence of any task, sensory stimulus and even consciousness, the brain can generate spontaneously complex scenes and manipulate them in cognitively sophisticated ways, prompting a reconsideration of the richness of the brain's "secret life."[31] The brain is always generating top-down predictive electrical signals and spontaneously emerging patterns of activity across cortical and sub-cortical regions. We saw how the brain consumes 20% of the body's energy and that 60 – 75% of this energy budget relates to the generation of electrochemical signals.[32] Our brains appear constantly ready and poised to react to our world and the weak spontaneous electrical activity producing spontaneous cortical patterns has a valuable though unclear function. Most neuronal activity involves a less metabolically costly infra-slow activity (ISA) frequency range below Delta brain wave frequency (<0.1 Hz) which is a distinct neurophysiological process detectable in fMRI. As Anish Mitra and colleagues observed in a recent study, this ISA has a "Unique spatiotemporal dynamics and laminar structure" which means that ISA activity travels slowly through specific cortical layers following a different trajectory to higher frequencies.[33] A major challenge for neuroscientists though has been to understand how weak spontaneous electrical activity or ISA is connected to observable behaviour like sustained attention during a task. Measuring blood flow in the brain using imaging techniques is too slow to capture the rapid neural dynamics involved in sustaining attention and many studies have attempted to use EEG to capture the electrical activity, in particular the relation between brain waves specifically theta and alpha waves.

Attention and Emotions.

Researchers have been fascinated by the link between arousal levels and attention as determined by the task demands requiring intention balanced against internal states, in particular the impact that cognition and emotions can have on perception, attention, memory, and decision-making.[34] External sensory information involves a staged brain process of orientation, attention, selection, processing, encoding consolidation and storage of some of the information if it is to be stored in long-term memory. Cognitive and emotional dispositions appear to play a vital role in influencing or biasing orientation with research showing for example that anxiety and depression can significantly impact an individual's concentration levels but also bias towards certain information, typically negative. An emotion is an event-focused process, based on the salience of the stimuli which results in generating a motivational response in terms of excitement or stress associated with an unpleasant or pleasant stimulus and the need to decide whether to withdraw or approach or engage or disengage from a stimulus. This involves physiological changes such as heart rate, skin conductance, changes in bodily, facial and voice expressions etc, subjective feelings generated through introspection and finally appraisal of how salient it is to the individual in terms of relevance to that individual's needs, goals, values, and overall well-being. Attention processing is entwined with emotional appraisal and stress levels or anxiety impact cognitive appraisal of information and decision making. The whole attention process is highly subjective and individuals vary enormously in the attention they allocate to and appraise information. The brain prioritises emotionally relevant information through pre-attentive processing, orientation, working memory and appraisal, long-term encoding and storage guided by the subjective bodily states that alert the body and brain to the information's importance. Many studies show that most emotional disorders are characterised by an attentional bias towards emotional stimuli. with anxiety about a perceived threat, compromising attention and constricting the visual focus of attention which means that stimuli in the peripheral vision are not attended to.[35] Anxious people differ in various ways in terms of the content of their attention, their capacity to attend, the level of distractibility and the selective nature of their attention.

Research in 2001 by psychologist Elaine Fox and colleagues challenged our understanding of how anxiety affects attention, showing that anxious people find it more difficult to be distracted by other things but also to disengage from information perceived to be a threat, even a social threat.[36] This suggests an anxious person's attention is not just drawn to threatening stimuli but they have trouble disengaging their attention from these

stimuli, which applies to anxious or negative thoughts as well. An individual's ability to engage and disengage attention has been linked to the onset and maintenance of anxiety disorders, because of the bias towards negative information. Anxiety is considered either trait or state and involves different neural patterns in the brain, with the more transient state anxiety associated with an adverse event or situation, whilst a trait suggestive of an anxious personality tends to be more enduring with a tendency to feel anxious in many situations. Recent research has revealed that chronically trait anxious individuals, even when focused on neutral rather than threatening stimuli appear less able to expand the scope of their attention and are thus seemingly more rigid in their attentional focus and have greater difficulty switching their attention away.[37] Coupled with previous research this shows anxious people concentrating on stimuli perceived to be threatening rather than competing stimuli. One study focusing on eye tracking research including a meta-analysis of studies in anxious individuals and depressed individuals showed that many emotional disorders are characterised by a bias towards emotional stimuli.

Thomas Armstrong and Bunmi Olatunji performed a meta-analytical review of the eye-tracking research on anxiety and depression and concluded "Relative to controls, anxious individuals showed increased vigilance for threat during free viewing and visual search and showed difficulty disengaging from threat in visual search tasks, but not during free viewing. In contrast, depressed individuals were not characterised by vigilance for threat during free viewing, but were characterised by reduced orienting to positive stimuli, as well as reduced maintenance of gaze on positive stimuli and increased maintenance of gaze on dysphoric stimuli." Dysphoria means feeling unhappy, uneasy, or dissatisfied. It appears vigilance to threat characterises anxiety but not depression. People with depression seem to have a blunted "orienting response to pleasant stimuli and will maintain their focus on negative stimuli.[38] " In contrast studies have shown that a positive emotional state result in the broadening of attention and according to Barbara Fredrickson the "experience of positive emotions broadens people's momentary thought–action repertoires, which in turn serves to build their enduring personal resources, ranging from physical and intellectual resources to social and psychological resources."[39]

Emotion shapes the content of our perception and awareness because we categorise the information to allow for a rapid emotional response where appropriate. A meta-analysis of 172 studies, showed an attentional bias towards perceived threat exhibited by anxious individuals, which is present not just during the cognitive appraisal of stimuli, but even about stimuli outside a person's conscious awareness.[40, 41] It affects a wide variety of clinically and non-clinically anxious adults and children and is known to be not just a cause of anxiety but responsible for maintaining anxiety. This has been attributed to an over-activation of a brain circuitry which reacts rapidly to threat[42] and anxious children are observed to display an attentional bias towards the faces of angry relatives when compared to neutral faces.[43] It is not clear whether this is the threat appraisal process in their attentional appraisal of stimuli or due to problems with emotional regulation though studies have found a bias in the initial orientation towards angry faces but no bias in sustained attention towards those faces.[44]

A study recently observed that attentional bias is linked to "Problematic internet use as well as other addictive disorders"[45] whilst another study of young people found that vigilance to external threats tends to result in social anxiety and school phobia, regardless of age and gender.[46] Research has demonstrated that modifying and reducing this attentional bias can reduce anxiety symptoms and new treatments involving Cognitive Bias Modification (CBM) are being employed.[47] The biased nature of attention determines what we are conscious of reflecting our prior experiences and knowledge which is unsurprising given the important neuromodulator role of the neurotransmitters, nor-epinephrine, acetylcholine and dopamine, in generating alertness, and the orienting of attention to important salient information, and the involvement of executive control of attention in the PFC.[48] Most of the nor-epinephrine that increases alertness and potentially attention, is produced in the LC in which neurons are connected by long-range projections throughout the cortex allowing them to modulate attention generally. Recent research has developed a new model explaining the LC's vital role in allowing us to rapidly switch between attention and inattention. When we are inattentive, the dominant brain wave is alpha with its synchronised oscillation at around 10Hz, which suppresses and filters the active processing of sensory information. Previous animal research had linked increased pupil size to higher levels of nor-epinephrine, and they found that by releasing nor-epinephrine, the LC made the alpha oscillations disappear which resulted in enhanced attention and focus on a task. One study concentrating on the thalamus, which acts as a pacemaker for the alpha brain waves, found transient LC activity stopped the thalamus from performing its role as the pacemaker, thus the LC appears to "dynamically sculpt neural synchronisation in thalamocortical networks to meet current attentional demands."[49] They also observed that as the brain gets older neural toxins and Alzheimer's-related tau precursors make the LC susceptible to neurodegeneration which potentially interferes with its role in synchronising the brain waves to generate attention.

Despite variations in alertness and attention levels due to energy and bodily states, a new encompassing theory on individual attention has recently been advanced by neuroscientists Moshe Bar, Noa Herz and Shira Baror which they call the changing states of minds theory (SoMs). SoMs postulate a holistic approach to understanding attention and attention bias and argue that we do so along a dimension from narrow to global attention which is related to, but not the same as our personality traits. SoM reflects a dynamic and transitory disposition when attending to information and combines with our personality traits to determine how we perceive our world, the nature and scope of their attention, emotional states, cognition, and behaviour. All interrelate and are connected such that Noa Herz and colleagues observed they "exert an all-encompassing and coordinating effect on attention."[50] This acknowledges that a person's current state of mind is a dynamic consisting of dispositions and tendencies and operates along a dimension. At one end is the narrow-minded state which tends to have a narrow perception of the world based on predicting the nature of the information, local attention, narrow thought, more selfish disposition, and negative emotions whereas at the other end of the scale is more global receptivity, open-minded state open to new sensory information, and new experiences, tends to be more curious and exploratory and involve more positive emotions. At times we may be more confident, and uninhibited, whilst at other times we may feel anxious or insecure, inadequate, inhibited or depressed or moody. This has parallels with people's style of thinking in that some people who think narrowly tend to "focus on selective content using rules, whereas broad thinking is unlimited in scope and is not governed by any constraining rule other than coherence."[51]

Cognitive neuroscientist Noa Herz and colleagues used ruminative thinking to illustrate an extremely narrow thinking style in which the individual repetitively reminisces about a specific event, fixates on surrounding topics, and exhibits difficulty disengaging from it even when this rumination is disadvantageous to that person. They concluded that the correlation between activity in the brain and a person's attentional behaviour depends largely on the state of mind at the time. This might sound obvious, but most people are typically unaware of how the way they attend to the world negatively or positively, significantly impacts their experiences in life and their perception of the world. A shift in the various dimensions of attention from broad to narrow attention and thinking tends to happen in tandem, leading to an overall bias in attention, perception and thinking which is often exacerbated by stress and anxiety. These tend to make people behave in a narrower pre-planned way, with over-reliance on prior expectations rather than exploring peripheral information and options or alternatives. This is evident in depression and rumination when the narrow focus on the negative can lead to a depressive bias and a state of mind that is difficult to break free from.

Multitasking and Attention.

We have been led to believe that we are all capable of multitasking and 51% of people surveyed thought that in a work context, switching between tasks made them more efficient and found it to be more satisfactory than focusing on a single task. Multitasking means trying to perform two or more tasks concurrently, which in practice involves repeatedly switching between tasks or leaving one task unfinished to do another. The human brain cannot though do several tasks simultaneously, although many people continue

to try and multi-task but research shows that there are cognitive and mental health issues that arise from constantly trying to multi-task. It takes us more time rather than saving time, it results in mistakes and errors and consumes far more energy. Multitasking is an executive function which is controlled by the pre-frontal cortex and requires the brain to dedicate separate regions of the brain in the frontal cortex to each task. The two halves of the frontal cortex operate independently and so the brain feels under strain if it tries to handle more than two complex tasks at the same time. It might appear that the brain is doing different tasks at the same time, but it is always switching between them, which is effortful and thus absorbs far more energy. Despite what people seem to report, completing one task at a time can even be far more rewarding as it appears to generate a sense of deep engagement and concentration with higher productivity, less distraction, and less stress. It leads to greater dopamine production in the reward mechanisms of the brain so can be highly addictive, by comparison, multitasking can be a sign of anxiety and stress and a signal that someone should slow down rather than try to multi-task. When asked, over half the people attribute their continuing to try and multi-task to personality rather than blaming it on technology, but many people over many surveys indicated that the pace of modern life has become too much for them. The survey identified some key demographics amongst the general population, with 42% happy with multi-screening, the majority of whom are women, 21% highly stressed by the addiction to technology and these tend to be the younger educated group, another 21% felt overloaded and disengaged from social media and these tender to be older people living in rural areas, the remaining 17% didn't seem bothered either way about the information overload. The majority of those other than the people who weren't troubled or had disengaged felt that there was non-stop competition to get their attention.

Professor Adam Later, PhD observes how the technology of smartphones, email, gaming, fitness tracking, etc are creating behavioural internet addictions by which a person becomes physiologically dependent on the technology. In a 2017 interview with the New York Times, he defines these as "something you enjoy doing in the short- term, that undermines your well-being in the long-term but that you do compulsively anyway."[52] The long-term negative implications include, obsessively thinking about the Internet, tolerance, a reduced ability to control impulsive behaviour, being unable to stop using the Internet, and withdrawal symptoms should you stop or have a break. Evidence for this comes from many studies and rehabilitation clinics such as that in Washington State called ReStart, which treats young people with addiction to gaming which has also become a major social problem in Asia.[53] Even in the early days of the internet concerns with excessive use were being expressed which led to the development of the cognitive-behavioural model of

Pathological Internet Use (PIU). The PIU model distinguishes between the pathological use of the internet for a specific purpose such as online porn or gambling and generalised PIU which involves a more general range of behaviours. Davis over 20 years ago cautioned that this represents a distinct pattern of behaviour among a specific subgroup of our population and that "the numbers for such behaviour have been greatly exaggerated by both the media and psychologists using unreliable and invalid research methods."[54] Whether this remains true today, is not clear given the exponential rise in phone and internet usage. A major question which has been the focus of much research is whether PIU causes emotional and behavioural symptoms and negative consequences in daily life, or whether the excessive dependence on the internet is the result of cognitive symptoms such as excessive negative rumination, low self-worth, low self-esteem, and social anxiety. This is an important distinction because simply reducing internet usage and offering alternatives may address the maladaptive behavioural consequences, but these won't work if cognitive symptoms are present and are what has caused the dependence in the first place. The behaviour is the result of predisposed vulnerability and a stressful life event and so individuals that are vulnerable because of a cognitive disposition, are stressed for example in social situations, so they take refuge in excessive reliance on the internet. It may be that the stressor is technology or the internet, so it is challenging to separate the cause and effect. Some studies have, however, linked excessive internet use to depression, anxiety and substance abuse which suggests that a psychopathology exists which causes the excessive use of the internet. A serious implication of multitasking and information overload is the potential to prevent neural pathways and patterns from being formed or strengthened, as attention is vital to how we process information and learn.

Attention in Robots

Machine attention has assumed great importance in AI with widespread applications in computer vision, NLP, automatic speech analysis, generation of language, and performance of multi-modal tasks, etc. All these applications require selective focus on and extraction of the most relevant sensory input to aid perception, rapid interpretation of the information, predictions, descriptions of image and video content, and task performance. When orienting attention to focus on objects and situations scenes, or images and text, robots face enormous challenges, including the need to ensure continuous, discrete, directed, and targeted movements and actions and accurate natural language to explain what they are seeing. It requires real-time integration of the multi-modal sensory inputs based on the random sampling of a continuous stream of sensory data that any truly autonomous robot must attend to.

In their 1997 review of the ML and attention literature Howard Egeth and Steven Yantis identified the three main challenges of machine attention which are (1) Controlling

attention through top-down goal-directed mechanisms and bottom-up stimulus-driven processes (2) The representational basis for visual selection and the degree to which attention is location or object-based and (3) The direction of attention to one stimulus after another over the time course. They argued that all are essential to making robots flexible and adaptable and providing them with the level of human attentional agility required to perform a wide range of tasks.[55] Twenty-five years later, Chinese computer scientist Meng-Hao Guo and colleagues identified several major phases of developments in machine attention mechanisms, Research has largely focused on NLP and visual attention with very few architectures modulating general sensory attention to cope with multi-modal sensory input or other specific types of attention like auditory attention. ML systems and ANNs were initially constructed without attention mechanisms but then early attention mechanisms, particularly in NLP, introduced the Encoder-Decoder neural translation mechanisms in RNNs and the Long Short-Term Memory networks (LSTMs) that we looked at earlier. In the RNN/ LSTM encoder, a whole sentence is encoded or summarised into a Context Vector (CoVe) (also called the meaning or thought vector). CoVes are word representations generated by pre-trained DL models for machine translation which capture both the semantic and syntactic information from the text input by leveraging the context given the source and target languages. The decoder uses the CoVe which is a number sequence to translate and produce the words sequentially and in order. In NLP, rule-based mechanisms were slowly superseded by statistical machine methods and then DL. Guo and colleagues describe current machine attention employing DL as a "dynamic weight adjustment process based on features of the input image" and observe how major progress has been made in "image classification, object detection semantic segmentation face recognition person re-identification, action recognition few-shot learning, medical image processing image generation, pose estimation, super-resolution 3D vision and multi-modal task."[56]

When confronted by novel objects or environments we rely heavily on our prior knowledge to determine whether to touch or engage with objects or an environment and robots must if they are to be autonomous. Prior knowledge in AI systems allows the system to employ a Computer Aided Design (CAD) mesh model made up of often thousands or millions of triangles that connect data points based on Cartesian coordinates, which can distinguish features such as holes and surfaces but not identify features. Any autonomous system needs to be able to iteratively recognize and integrate new objects from the environment into its object model database. For autonomous robots, the object performance estimation depends on the quality of 3D images and the attention paid to them. The problem, particularly in active exploration of the environment, can be easily compounded by "Bad incidence angle of the radar or laser beams, sensor noise which confuses images, various reflections, accidental collisions or occlusions of the objects." Occluded objects or those outside the Field of View (FOV) typically with current technology remain unrecognised by autonomous AI systems and traditionally, object recognition and exploration of the environment have been kept separate but autonomous robots need to be integrated.

A comprehensive 2018 review of attentional systems in AI highlighted how most have proved to be not very efficient and new approaches are needed. They identified hundreds of different attentional applications derived from three main sources which are Generic saliency models, Attentional gaze, and viewpoint selection or Task-driven attention.[57] Esther Colombini and colleagues observe how attention mechanism models employed in robots tend to be simple systems based on many sensors or complex biologically inspired attentional visual systems.[58] Humans can detect salient objects in a remarkably rapid and reliable manner, with Alexander Russell and colleagues describing how, like other organisms we "use the process of selective attention to optimally allocate their computational resources to the instantaneously most relevant subsets of a visual scene, ensuring that they can parse the scene in real-time."[59] Filtering of sensory information in real-time appears necessary to allow the brain to use only a small subset to process sequentially and reflects the fact that the quantity of visual information projected from the retina to the visual cortex of the brain exceeds its information processing capacity. It is widely acknowledged that the efficient deployment of attention mechanisms reflecting human biological attention processes will lead to a paradigm shift in ML DL and robots and potentially revolutionize computer power. Colombini's own proposed model envisages an attentional model inspired by biological systems that support a variety of robotics sensors.[60]

In a seminal paper published in 2012, Carlos Agüero and colleagues compared covert visual attention systems in robots and observed how biologically inspired attentional models have progressed to allow robotic systems to deal simultaneously in real-time with both "Generic stimulus and with task-specific constraints."[61] Biologically inspired search approaches such as transformer architecture and Google's open-sourced pre-training algorithm called Bidirectional Encoder Representations from Transformers (BERT) have led to breakthroughs in NLP. This is because BERT has been designed to attend to and process words about all the other words in a sentence, rather than one-by-one in order and their use of the Cloud TPUs significantly speeds the process up. Meng-Hao Guo and colleagues have identified four different main approaches to machine attention whether biologically inspired or simpler sensor-driven mechanisms which are (1) Channel Attention which is What to pay attention to such as objects and features such as orientation, colour, and shape etc, and objects (2) Spatial Attention which is Where to pay attention to (3) Temporal Attention which is When to pay attention and (4) Branch Attention which is Which to pay attention to.

We have seen how the human brain only focuses on parts of sensory input and selectively uses working and long-term memory together with the selected input to guide reasoning processes and behaviour. Attention in humans increases the neural response to the stimuli which are attended to and in ML and DL attentional capacity makes the systems more flexible, and to perform multiple tasks autonomous robots must be able to focus and switch attention between tasks when required. We also saw how visual-spatial and feature attention in humans involves both bottom-up and top-down processes. The automatic bottom-up or exogenous is driven by an external stimulus in the environment and the process of identifying image patterns through oriented edges, spatial frequency, and contrast in colour, intensity, or motion. It appears to be automatic because it is driven by external factors such as certain stimuli that might be inherently salient in a particular context or against a background. This inspired the highly influential saliency-based model of bottom-up attention using a single topographical saliency map first proposed by American Professor of computer science, psychology, and neuroscience, Laurent Itti and colleagues in 1998. Based on the behaviour and the neuronal architecture of the early primate visual system, they described how "The system breaks down the complex problem of scene understanding by rapidly selecting, in a computationally efficient manner, conspicuous locations to be analysed in detail" ... "A dynamical neural network then selects attended locations in order of decreasing saliency."[62, 63]

In contrast, top-down attention in humans involves the voluntary or deliberate allocation of attention to targets of interest be it objects or regions in space driven largely by prior knowledge, intentional plans, and current goals. Some models have explored top-down or endogenous internally generated and goal-driven attention aimed at enabling more sophisticated behaviours by, for example, reflecting motivational and behavioural states to create a context-dependent attention activation map[64] or identify objects that are being used by reference to a person's gestures and verbal instructions.[65]

Although Top-down and Bottom-up attention mechanisms in humans are typically regarded as an integrated process, studies seem to indicate that they may be two attentional systems that operate independently of each other.[66, 67] However, Maria Ruz and Juan Lupiáñez in their 2002 review of the literature related to attentional capture observed that it is not clear "the extent to which such effects are automatic or rather modulated by endogenous factors." They noted that "Bottom-up orienting is resource-free, cannot be suppressed, is unaffected by subject's expectancies or by concurrent memory load and does not require conscious awareness". They concluded that "Although both exogenous and endogenous factors appear to influence attentional allocation, mental sets related to the task at hand seem to be the most influential in modulating attentional capture. Mental sets in psychology reflect the brain's tendency to stick with the most familiar solution to a problem and stubbornly ignore alternatives[68]. "As such, although attentional capture appears to be automatic by default, in that it can occur in the absence of a specific mental set, it seems clear that it can be endogenously modulated."[69]

Neuroscience has developed competing hypotheses for the extent to which prior beliefs and expectations involved in endogenous attention modulate and influence the conscious perception of exogenous stimuli. One suggests that perception and cognition are separate and that if there are no prior knowledge or expectations the stimuli compete for global workspace with cognition. Others are predicated on the predictive model which involves the brain constantly generating and updating a mental representation or model of the environment, so perception is the interaction between sensory stimuli (bottom-up) and the conceptual knowledge involved in cognition (top-down). Yair Pinto and colleagues describe it as the brain "continuously attempting to minimize the discrepancy between its inputs and its emerging models of the causes of these inputs, via neural computations approximating Bayesian inference." They were studying how conscious perception is affected by predictive models when trying to identify the causes of sensory signals, with expected stimuli appearing to enter consciousness perception faster than neutral or unexpected stimuli. The Verification Hypothesis maintains that the brain uses the predictive model to minimize prediction errors by determining whether sensory inputs verify or violate the predictive models which is when conscious contents emerge. Pinto observed how by the predictive processing model and in line with predictive processing accounts of perception "top-down expectations play a larger role when bottom-up input is ambiguous" which benefits detection and identification.[70] They experimentally excluded the possibility of response priming which we discussed earlier. They claimed that the results were not attributable to a phenomenon discovered in 1997 by Konstantinos Moutoussis and Semir Zekiin called perceptual asynchrony in which two simultaneously presented attributes are perceived by humans asynchronously instead of simultaneously.[71] This reflects some of the challenges of building bottom-up and top-down mechanisms into robots.

To construct abstract representations of the world the human brain employs both bottom-up and top-down mechanisms and spatial and temporal information from our environment combined with experience and memories of that representation to predict the world. AI models though have no a priori knowledge of the world so any data must be viewed in terms of what it represents about objects, people, locations, events, ideas, activities, processes, etc, to give meaning to the world. For robots to learn efficiently they must be able to attend to the correct, interesting, or appropriate spatially and temporally salient raw data in their environment and process and integrate that information. Using a bottom-up mechanism requires the robots to be able to automatically attend to salient or conspicuous stimuli generated by objects or the environment such as colour, gradient, brightness, and movement. Top-down requires a robot to engage in a deliberate allocation of attention to the spatial and temporal information of a target of interest largely by prior knowledge, intentional plans, and current goals.

In 2023 Shiau-Chuen Chiou found that "Spatial and temporal information are not automatically integrated but share the storage capacity and compete for a common pool of cognitive resources" and observed how whether spatial and temporal information are represented in working memory, as integrated units, or as individual features, influences how much information might be retained and how the retained information might be manipulated."[72] This suggests that integration and conscious perception of exogenous stimuli involves the allocation of brain processing resources and storage capacity with memory retention in humans. Training robots though to use a combination of bottom-up and top-down attentional mechanisms to construct a world model requires huge amounts of data and is a very slow time-consuming process.

The enormous advances in the last few years of targeted attention in AI are evident when we google any topic and millions of web pages are processed in a fraction of a second, in NLP, the summarising of text and parsing or tagging parts of speech or analysing sentiments in speech and text and translation. In computer vision, ML and DL have allowed AI systems when generating and classifying images, to increasingly extract target data and accurately classify and detect objects. DL models require huge amounts of labelled datasets and interactions with the environment to train AI systems and robots and attentional mechanisms are necessary to allow a robot to be more selective in what it attends to which speeds up the learning process and improves computational efficiency. Neuroscientist Grace Lindsay in her paper Attention in Psychology, Neuroscience and Machine Learning, describes machine attention as a core computational quality that involves the "Flexible control of limited computational resources."[73] We will focus first on machine attention in NLP and then computer vision although as we see they are increasingly requiring collaboration in research and are the two fields in which attentional machine mechanisms have produced the most progress, with both illustrating the complex challenges of equipping robots with attentional capacity.

Attention and Neural Language Processing (NLP).

When humans translate languages, we use mental representations including context to understand what a whole sentence might mean which for machines requires a combination of the linguistic, translational, and statistical or probabilistic techniques used by humans. It involves attention and correct selection of the components of language. Tokens are pieces of words and in language processing, lexical analysis by the AI machine breaks the text input down into tokens which are essentially a sequence of characters with semantic meaning and regarded as units of grammar such as nouns, verbs, adjectives, etc to the machine. This matches characters in the input text against pattern sets or regular expressions that define different types of tokens which include keywords, punctuation symbols, or operators. The output vector which is the list of tokens is parsed for syntax analysis and semantic meaning by breaking the input down into smaller more manageable chunks.

Adopted in the 1990s, Statistical Machine Translation (SMT) provided an alternative to the original Rule-Based Machine-based translation (RBMT) approach which was very costly because it basically encoding involved moving forward word by word and typically failed to generalize across different languages and relied entirely on "morphology, syntax, lexical selection, transfer, and generation."[74] SMTs are called statistical translation because SMT strings words together as phrases based on the statistically likeliest or most probable combinations using the most common previous translations available. In 2015 Ukrainian computer research scientist Dzmitry Bahdanau, who focuses on human interaction with machines, introduced Neural Machine Translation (NMT). NMT involves using DL to build a single ANN to read and translate text by automatically searching for or "Softly aligning" parts or sets of positions of the input sentence called hidden states that are regarded as most relevant to predicting the sentence.[75] They take the weighted sum of hidden states with weights learned through an FFN and use embedded words, which involves words being represented and similar words having a similar representation, which enhances the system's ability to generalize meaning across words.

The concept of embedded words is based on the work of the American linguist Zellig Harris (1903–1992) in the 1950s who in his Distributional Hypothesis demonstrated how words that have similar contexts will have similar meanings.[76] Bahdanau demonstrated that the new soft alignment attention mechanism using embedded words, sentence context, and mental representations when translating, produces the most relevant translation of whole sentences comparable to human intuition. The system adjusts and amends to ensure the correct grammar and generate the closest resemblance to human speech. The attention mechanism of NMT has been widely applied in computer vision and speech recognition with probabilistic models like NMT driving the emergence of online translation tools such as BabelFish Altavista and Google translate which are increasingly used by international corporations. In NMTs, each layer employs single neural networks to replace the normal pipeline-type structure in which language understanding is arranged which would be, an audio input, feature extraction, phoneme detection, word composition, and text transcript (output). The single neural network replaces the intermediate steps so it would look like audio input and NN transcript (output) and can learn from a few images to complete different tasks. NMTs have revolutionized NLP, are reasonably accurate, very fast, and can generalize learning to form new connections and interpretations and be incorporated into most software. Their major limitation is they are heavily dependent upon the quality of the source text, require training on huge data sets, and struggle when the input as with text for example is rich in information or simply very long. This is called the long-range dependency problem of the encoder-decoder mechanism with the performance and capacity to make accurate predictions rapidly degrading with the longer the sentences.[77]

Google's most recent model employs the Google Neural Machine Translation (GNMT) system uses patterns to predict the probability of a word sequence that allows the machine to DL with reduced human supervision and improve its translation capability by learning from earlier translation tasks. GNMT systems can translate an entire novel in less than two minutes and get smarter every time it does so. GNMTs translate over 140 billion words per day, largely achievable because of Google's access to enormous data training sets and end-to-end learning technology. It achieved this through using end-to-end learning, without any explicit supervision of where or what to attend to. End-to-end learning involves being able to simultaneously, rather than sequentially, learn all the necessary steps between the input and final output or action. It employs a different algorithm in each layer to deal with different functions and solve tasks like feature extraction, prediction, optimization, and preprocessing, to specialize in performing tasks necessary for the overall problem.

Another recent innovation introduced by Hyoga Hiruma and colleagues from Waseda University, National Institute of Advanced Industrial Science and Technology (AIST) is called the Query-Key-Value (QKV) formula which they claim can switch attention targets in NLP. QKV was first used by a team at Google Brain led by Ashish Vaswani in a 2017 paper proposing a Transformer model and described attention as the weighted average of some values with the transformer involving the whole sentence looking at itself. They map a query and a set of key-value pairs to an output, where the query, keys, values, and output are all vectors called QKV vectors. Transformer models learn the contextual relationships between words using a mechanism called self-attention, whereby the model weighs the importance of different words in a sequence using their context. The model does not use a RNN because with NLP they look at each word one by one which creates problems with long-term memory. It involves encoding the sentence into a sequence of vectors and when decoding it selects a subset of these vectors, which means by paying attention to these subsets of vectors the model can cope with much longer sentences. Hiruma and the team trained the system on a simulator and in real-world environments and the model succeeded in picking up specified objects based on inputting extra conditions to specify which targets to attend to using a top-down process. The team claimed that the robot demonstrated "high data efficiency, interpretability, and robustness against appearance changes" and achieved this through using end-to-end learning, without any explicit supervision of where or what to attend to."

Despite the progress and innovation in NLP translation systems with greatly improved attentional mechanisms, the reality is that they typically still require checking by human translators who bring critical thinking and more nuanced interpretations currently beyond NMT systems. They can't translate and imitate 100% like the human brain and although they hold enormous potential and demonstrate the vital role attention plays in such tasks as NLP, it highlights how much we take for granted the ease with which we communicate through language despite its significant complexity and potential for ambiguity or inattention to context and meaning.

Visual Processing and Attention.

Cognitive architectures have tended to not allocate resources directly to developing visual attention unless implementing biologically inspired models or researching vision in humans. Most visual mechanisms have emerged from other competencies in terms of designing the AI system based on allocating limited resources, which is in keeping with the GWT and working memory which is important in handling and processing competing information for further processing. Some cognitive architectures use this process to guide attention as we will see when we look at working memory and AI. Some of the more interesting biologically inspired AI mechanisms use saccadic mechanisms to switch between the memory neurons or memory nets that are the most activated, Hopfield networks (nets), or queue and modulators. Hopfield nets are memory networks that were invented by John Hopefield over 40 years ago with a major impact on machine learning and AI and are recurrent ANNs that consist of one layer of 'n' fully connected recurrent neurons.[78] This means that the output of each node or neuron is an input to all other nodes or neurons but not itself and forms an associative memory network or content addressable memory (CAM) which optimizes data search by storing patterns in the AI memory connected with the input pattern. Things linked in memory like the words bacon and eggs are recalled as associated memories and depend on the intensity of the association between the input patterns and stored patterns.

When humans attend to and visualize a scene they need to understand the often complex and entangled relationship between objects and the information contained in the environment which has been difficult to achieve in AI. Recently a team at MIT has achieved this by representing the relationship between objects separately and then integrating the relationships to give the system an understanding of the whole visual field of view. Using a deep learning, Energy-Based Generative Model (EBM) which applies statistical physics to computer learning, they can build an image from text based on the relationships between objects. EBM models allow the AI system to understand the relationship between objects, though unless robots can attend to the appropriate information in a real-world environment to determine and understand such relationships, they will never become fully autonomous.

Humans use focal gaze and attention to take advantage of visual, spatial, and semantic cues to help them navigate efficiently and perform tasks, by focusing on the relevant and salient features of the environment. Humans can in unfamiliar environments make intelligent inferences about the probable relationship between objects and information features, by using their world model which is learned as they develop. The challenge has been to build real-world models into robots because this requires huge amounts of real-world data to train robots to produce the required behaviour and with robots needing to explore the environment to learn, it can lead to hardware damage and safety issues. Deep Reinforcement Learning (DRL) holds considerable promise for fully automated learning in real-world environments and recently, Google in collaboration with UC Berkeley released a DRL algorithm called Soft Actor-Critic (SAC). DRL employs a fundamental concept which in simple terms is denoted by the statistical metric Mu (μ) which is the expected value and determines the action the robot is to take at a particular state in the environment. SAC employed a stochastic policy which returns a probability distribution of multiple actions in the action space for a given state, for example, move up, down left right, and an optimal policy which will allow the robot to navigate towards a target goal. Robots may perform different actions each time and can through a process of trial and error learn from actions, the best policy by initially performing random operations and iteratively, converging on the optimal policy. By comparison, a deterministic policy will always map a given state

to one particular action so there is always only one action possible in each state. Training robots to operate in real-world environments that involve attention, requires input in the form of training data, the model's parameters which are the variables that the AI learning system uses to adjust to the data, and the hyper-parameters which are parameters that govern the training itself and are usually constant during the training. DRL algorithms can equip robots to perform some complex tasks at a superhuman level such as certain games etc, but this involves huge amounts of training which limits their application in the real world. The challenge is to make the use of data training more efficient, some systems will use a model to represent the real-world experience and even assign intrinsic rewards to the robot as a measure of success or failure, whilst others use actual exploration of the environment. Research is also focusing on combining both intrinsic motivation and exploration which we will discuss later in humans and robots.

Mobile autonomous robots incorporate a range of actuators (the mechanisms responsible for moving or controlling the system) and sensors including sonar and LIDAR sensors, GPS, and V2X to navigate detect, and locate obstacles and objects with microphones and speakers enabling interactions with humans and many different types of cameras for visual scanning of the environment. If in AI and robots, the sensory hardware and software systems are unable to attend to the appropriate information when trying to perform tasks or language processing, etc, this would pose serious problems in terms of autonomy, safety, and inefficient consumption of computer resources.

Canadian computer scientist John Tsotsos (1952–) who specializes in computer vision and robotics and Dutch experimental psychologist Wouter Kruijne, whose research includes visual attention and learning, have extensively researched the subject of machine attention and intelligence. Tsostos defines attention as "The goal of matching current tasks to processing resources by manipulating the brain's processing machinery to preferentially process expected stimuli while ignoring the irrelevant yet monitoring all stimuli to deal with the unexpected."[79, 80] He and Kruijne claim that attention and spatial selection are essential to a wide range of computational tasks and for machines to be able to learn from their environment[81]. In ML, particularly machine vision and Robot Vision Systems (RVS), attention is closely linked to supporting and harnessing working and long-term memory, especially with sequential reasoning which relies on perception and memory. Back prop algorithms employed in RNNs, particularly CNNs, have revolutionized machine vision but have yet to satisfactorily address the issue of visual attention.

Human visual attention although mostly subconscious requires complex and very rapid neural mapping processes. Even recognizing 3D objects or scenes is complicated by the many possible variations in the viewpoint, occlusion, brightness, position, etc, that may change the object or scene's appearance. This poses the problem in machine vision systems of whether and how you encode the object from a particular viewpoint and the extent to which this may affect subsequent recognition based on that encoded representation. For robots to pay selective attention requires them to narrow down the search space and the scope of the sensory data to be processed, which is precisely what humans do when we apply bottom-up and top-down attention mechanisms to something in our environment. When a robot wants to acquire visual sensory information from the environment, it scans an object, situation, or environment which is equivalent to the human alert orientation and execution systems discussed earlier. It involves even at the most basic level for example creating 2D and 3D models of the surfaces that the robot is moving and navigating on. Accurate and complete 3D models or representations are essential for navigation, conceptualizing, semantic segmentation of images, image processing and analysis, pattern recognition and classification recognizing, tracking, grasping, and manipulating objects. Autonomous robots, require real-time interaction with the physical world and selecting relevant features or objects of interest required to perform tasks and behave like humans increasingly using a combination of bottom-up (Stimuli-driven) and top-down (Goal-driven) mechanisms.

Computer scientists Francisco Martin and colleagues observe how many mobile robots are equipped with cameras attached to the head on articulated necks, which improves the robot's field of vision and their capacity to attend to their environment. Robotic eye movements can be simulated, although some robots like iCub and Kismet employ physical eyes which are more akin to real vision in humans. A problem is that most robots do not have 360-degree views of their environment, which makes rapid spatial attention and response to sudden or unexpected events in the environment a major challenge. Visual-spatial attention in robotics, like humans, also involves both covert and overt attention and external attention and at its most generic level engagement with the environment. Overt attention in robots is the machine equivalent of the saccadic movement of the human eyes, which involves tracking the space around the robot and looking for salient objects, whilst covert robot attention involves using image data to search for relevant information related to a specific task and so biases the visual system towards that information. Martin observes that all robotic attention requires an attention management system to direct the sensors to points of interest or salience because of the impossibility of covering the whole field or space around the robot, coupled with the essential need to use the limited computer resources efficiently.[82]

The Selective Tuning Model (STM) first proposed by Tsotsos in 1995 was considered to be a plausible neural network model for spatial selection which he described as a "theoretical necessity for all visual attention systems whether biological or artificial." The STM is a pyramidal neural structure involving feed-forward and feedback connections with the strength of response to a stimulus representing the "goodness of the match to the model" and employing a winner-takes-all approach with the neurons with the largest response emerging as the winners.[83] Significant progress has been made since the first STM model. The task of processing and classifying information that is visually attended to is made significantly easier when confronted by an object or situation that is already in the robot's object model database or memory. Humans rely enormously on prior knowledge of our environment to identify objects and people, which partly explains why people tend to feel uncertain or less safe in unfamiliar environments when confronted by unfamiliar situations or objects we still try to use our prior knowledge to try to make sense of or identify and classify them but if this fails, we are in a predicament. This means that robots must have a selection mechanism to visually select an object but if a robot has no prior knowledge of an object, it must take the sensory data generated by that novel object or feature usually from strategically selected viewpoints, construct a representational model and using the appropriate algorithms classify it based on assessing the differences between the expected information and the observed information.

Computer scientist Laurent Itti and neuroscientist Christof Koch observed in 2001 how the study of visual attention in ML had concentrated on key themes, such as the (1) Importance of context (2) Saliency of the information (3) Utilisation of saliency maps (4) Inhibition of attention (5) Role of eye movements in coordinating and controlling attention and (6) Need to restrict the location to understand a scene or recognize an object.[84] Many machine attention systems use common or classical visual saliency algorithms or filtering and pattern detection to identify differences between the expected information and the observed information. Today's robots can be very good at recognizing objects using camera images and DL algorithms to extract task-relevant or salient information from the image. A major problem and constraint is that bottom-up (stimuli-driven) attention typically cannot be modified to address dynamic changes in the objects tasks, or situations requiring attention. ML and DL allow AI systems to passively classify novel images and recognize human speech but are yet to adequately equip robots with dynamic bottom-up attention that guides active behavioural skills and enables them to learn as they engage with the real world. This is partly why most cognitive architects are increasingly employing top-down mechanisms in conjunction with bottom-up mechanisms to further narrow down what the robot selects to attend to and enhance the attention of robots in task-driven activities and dynamic environments. This includes important functions such as paying attention to human collaborators, for example dealing with demands or needs of patients in healthcare. Top-down attention mechanisms are used extensively in image captioning and Visual Question Answering (VQA) in which the computer can ask and answer questions about an image or set of images or video content. This is a trivial task for humans but for a machine means being able to attend to and understand the content and use natural language to provide relevant answers to questions about the content, which would allow humans and robots to interact more naturally. This requires what Aafaq Nayyer and colleagues describe in their 2019 review of the field of video description as "Understanding of many entities and the detection of their occurrences in a video employing computer vision techniques including background scene, humans, objects, human actions, human-object interactions, human-human interactions, other events, and the order in which events occur."[85] VQA is widely employed in video description in which natural language sentences are automatically generated to describe a video's content and the related field of Visual Information Retrieval (VIR), in which content-based searches are performed on visual image, text, or mixed visual and text. Both require the AI system to understand the visual content and describe it using grammatically accurate language so combine research from computer vision and NLP.

Classical approaches were used to combine the detection of subjects, objects, and verbs with template-based language models to generate sentences but were replaced for a short time by statistical methods before DL which is the current state-of-the-art in video description. Most VQA systems consist of image question and answer encoders and involve fine-grained analysis and multiple steps of reasoning. Image encoders extract features such as objects, attributes, and relations between these, but require huge amounts of labelled data and human-annotated questions and answers to train on. Question encoders convert the question which is input into a vector representation and the answer decoder uses both to generate the answer. Nayyer and colleagues also observed however that "The release of large datasets revealed that these methods cannot cope with the diversity in unconstrained open domain videos."[86] They recognize that a major difficulty is ascertaining from the images etc the relative "contribution towards the accuracy or errors of the visual features" and point out that "current evaluation metrics fall short of measuring the agreement between machine-generated descriptions with that of humans." Peter Anderson and colleagues proposed a "combined bottom-up and top-down attention mechanism that enables attention to be calculated at the level of objects and other salient image regions". They describe how the bottom-up mechanism (based on Faster R-CNN) proposes image regions, each with an associated feature vector, while the top-down mechanism determines feature weightings.[87]

Their approach reflects a growing trend in using VQA to employ attention, co-attention in which bilinear pooling is used to achieve much greater fine-grained visual recognition, and transformers to integrate image and question features more effectively and dynamically.

With the widespread adoption of DL and RNNs, particularly CNNs, attentional mechanisms are increasingly employed in updating whole networks to reflect changes in the environmental stimuli, but suffer from major limitations when confronted with novel or unfamiliar objects and environments. In 2015 computer scientist Max Jadeberg seeking to address this problem introduced Spatial Transformer Networks (STNs) as a method that can be used in existing CNNs.[88] STNs have become a hugely popular way of increasing the ability of CNNs to increase attentional spatial invariance by allowing them to detect features or objects that are not exactly like the images that the system has been trained on and are typically in different positions. We saw how pooling in CNNs removes irrelevant information by compressing the images, but STN can increase spatial invariance which means that when an object occurs in an image it can be detected irrespective of its position, for example when images and objects are subject to translation, scaling, rotation, cropping, as well as non-rigid deformations. Seen from different perspectives many objects are hard to recognise or classify but STNs transform the image into a canonical or predicted pose, which is regarded as the most representative of that image or object and would for example be the view that comes to mind first when asked to imagine an object. Canonical views were first discovered in 1981[89] and established how views of objects may be privileged depending on the criteria applied by the viewer. These include the level of goodness of the view in terms of the sufficiency of information to be gleaned, the first image that is imagined, the view selected as subjectively the best, the viewpoint that is the quickest and least error-prone and the viewpoint that is employed for the longest time.[90]

Key Attention Mechanisms.

In 1995 Tsotsos described how machine attention involves key mechanisms[91] which are (1) Selection mechanisms-which choose which information to attend to from the many options available. These involve (a) World Models, (b) Attentional Gaze and (c) Viewpoint selection, and many saliency methods focusing on features/objects/ interesting events/ the location, and time. Selection mechanisms include saliency systems that classify and identify objects using Bounded Boxes and Image Segmentation; Static saliency in which the system detects from a static image the most salient features with the two main types being Static spectral saliency and Fine-grained saliency; and Motion saliency in which the motion is captured frame by frame and allows the system to track an object typically using Optical flow maps (2) Restriction and Suppression mechanisms which ignore some elements by restricting the search space and restricting the processing to the space selected as a basis using methods such as priming the robot for a specific task or limiting the visual field of view. Suppression mechanisms which though rarely employed in cognitive architectures involve suppressing information from other sources around whatever the information of interest is it an object or event. This might, for example, prevent the robot from returning its attention to the same salient information or mean that the robot is not able to attend to irrelevant information or allow its attention to be diverted.

Some systems employ a combination of elements from all three using what is called the Branch-and-Bound mechanism which is predicated on using these elements to partition the total set of feasible solutions into smaller subsets of solutions which are then evaluated systematically until the best solution is found. Cognitive architectures vary in the extent to which they employ these various mechanisms, but by far the most common are selection mechanisms so let's start by looking at some of the selection mechanisms used by robots.

(1) Selection Mechanisms.

All selection systems whether using real and/or simulated vision include by default (a) A World Model, (b) Attentional Gaze, and Viewpoint selection, and many with models that focus on (C) Features/objects/ interesting events/ the location and time. Let's briefly look at these.

(a) World Models.

A world model is believed to be central to all biological systems and reflects how humans use huge amounts of sensory-motor data generated from interactions with the environment, to construct an abstract mental model to simulate their environment and make predictions based on the model.[92] The pioneering American computer engineer Jay Wright Forrester (1918–2016), often called the father of System Dynamics which is a mathematical modelling approach to the non-linear behaviour of complex systems, described how "The image of the world around us, which we carry in our head, is just a model. Nobody in his head imagines the world, government, or country. He has only selected concepts, and relationships between them, and uses those to represent the real system." A world model is an abstract representation of the spatial or temporal dimensions of our world and requires robots to build an accurate encoded representation of the environment based on the sensory information it receives, and it must deal with uncertainties and gaps in that environment. Deep generative world models in AI employ the biological principles governing how our brains construct such abstract world models by combining spatial and temporal representations which are the two major dimensions of dynamic images.

Our primary visual cortex is constantly updating our brain's information based on the ever-changing visual sensory input to integrate it with other sensory modalities. It uses past and present mental models by integrating selected concepts and relationships between things to construct a dynamic world model. Robots must be able to select objects or events to focus on from their worldview model and when interacting with the real world constantly update the world model through learning. Attention involves holding information in working memory and integration of new information in real-time, which requires significant computer capacity and processing resources. Worldview selection architectures are typically either bottom-up in terms of being driven by the data or top-down in being driven by the demands of a task to be executed in a particular environment. Professors Wolfgram Burgars and Martial Hebert in the 2008 Springer Handbook on Robots, describe the various ways robots generate world models of a real-world environment or workspace even concerning the relatively basic task of modelling the terrain for indoor and outdoor activity. Traditionally world models were created using CAD tools with the modellers working from drawings and if objects are moved the model needs updating and re-measurement of distances and locations etc, but AI systems and robots now use video cameras and 3D range sensors and DL generative models. Indoor models include 2D representations such as occupancy grids, line maps, topological maps, and landmark-based representations, whilst outdoor models that can't assume flat indoor surfaces tend to use, elevation and variants maps, point clouds, meshes, or 3D grids.[93] Topological maps like the map of the London Underground below represent environments using graph-like structures with nodes corresponding to places and edges to paths between locations and although better at scaling to large environments don't represent the detailed geometric structure of the environment which in unstructured natural and outdoor environments is vital to robots.

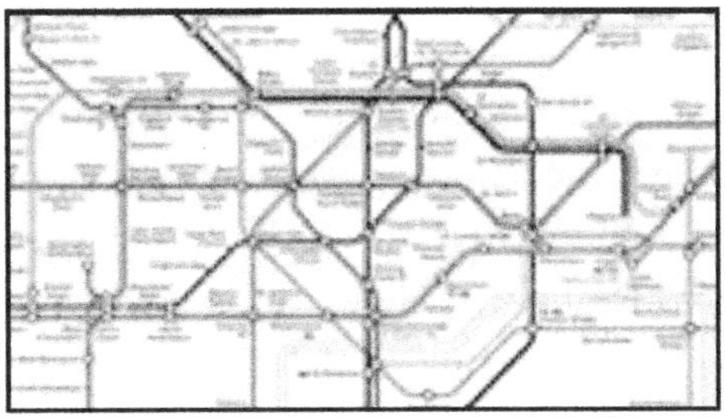

Example of Topological Map.

Geometric models use grid-based approximations or geometric primitives which include points, lines, and polygons or areas and are the basic building blocks used to describe 2D and 3D shapes. The problem is that full three-dimensional models are very computationally demanding so are not suitable for mobile robots. Elevation maps are more compact, 2 1/2 D representations that generate information about the height, gradients, and shape of landscapes and objects by storing in each cell of a discrete C-space grid, the height of the surface at the corresponding place in the environment but do have the disadvantage of not being able to deal with vertical or overhanging objects. In recent years raster (grid) based Digital Elevation Model (DEM) has replaced topographic maps for collecting, storing, and conveying surface topography information. Rasters or grid models are spatial data models that define space as an array of equally sized cells, arranged in rows and columns when compared to Vector data that represents geographic data symbolized as points, lines, or polygons. They can though suffer from positional errors and geometric distortions.

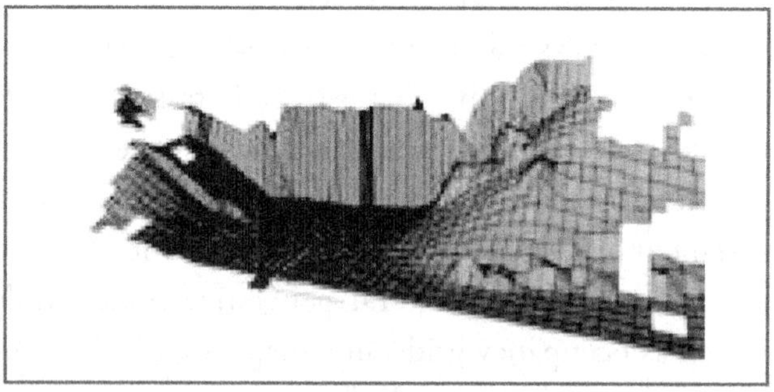

Standard Elevation Map

Designers have focused on developing "Spatially Intelligent" robots that determine their location by comparing the spatial data encoded in memory to real-time data provided by onboard sensors. Once localized, the robot can understand the space and dynamically navigate its environment with specialized sensors allowing a robot to conduct focused tasks or missions while collecting and sending back sampling data about the environment. A recent innovation in AI learning has involved the use of 3D point clouds which use a set of data points in a three-dimensional coordinate system to describe an external surface in 3D. 3D point clouds can represent the volume of solid objects or the shell or boundary of an object. Many systems use simulators to train which are translated to real environments in a process known as the Sim-to-Real approach. Creating world models remains a tedious and lengthy process and although most employ model-based RL approaches to learn a model of the environment, this typically requires training in the actual environment.

In 2018, David Ha and the German computer scientist and pioneer in AI, Professor Jurgen Schmidhuber (1963–) inspired by the human brain's cognitive approach, presented a novel world model approach allowing much more rapid unsupervised training to learn a compressed spatial and temporal representation of the environment. An important observation is that our brains don't just predict the future but predict future sensory data given our current motor actions which underlie reflexive responses. They emphasized how any AI system benefits from having a "good representation of past and present states, and a good predictive model of the future" which requires human brains to "Learn abstract representations of both spatial and temporal aspects of this information." They describe how they could train a robot "entirely inside of its dream environment generated by its world model and transfer this policy back into the actual environment."[94] They used "a temperature parameter of the internal world model to control the amount of uncertainty of the generated environments", training the robot in a "noisier and more uncertain version of its generated environment" and demonstrated how this helps prevent the robot from "taking advantage of the imperfections of its internal world model." Their model employs three key elements which are a (1) Visual sensory element (the V model) which receives a high dimensional input, which is usually a 2D image frame from a video which it encodes or compresses into a small representative code in a low dimensional latent vector. This is a widely used method to capture the most important features/ aspects required to learn and represent the input data. (2) Memory element (the M model) that uses experience to make predictions. They describe this as "integrating the historical codes to create a representation that can predict future states" and (3) a Controller element (the C model) that uses both the visual sensory element and memory element to choose an action to take. The world model consists of the V and M models. For the V model, the robot utilizing the VAE model discussed earlier receives an input, usually a 2D image frame that is part of a video sequence at each time step in the form of an observation from the environment and learns a compressed version of this abstract representation which is used to reconstruct the image. The M model compresses not the content but the information over time by making predictions using a probability density distribution function that reflects the stochastic and random sampling nature of the data process. The C model uses the information from both the V and M models (i.e., the world model) in decisions to act and applies an RL approach, based on maximizing the expected cumulative reward from its actions.

This world model effectively compresses and predicts the sequence of image frames it has observed and even if the V model can't capture all the details of every frame accurately, it can learn to navigate a real environment. The combination of compressed representations and predictions allows the robot to use its abstract world model to learn without prior planning, based on their theory that if the world model is sufficiently accurate the robot can transfer learning from the virtual world to the real world. Using this innovative world model, they successfully trained a robotic car to learn how to drive a racetrack by enabling the controller to use continuous actions of braking steering left and right, and accelerating based on visual pixel inputs gained over 10,000 random trials. It resembles how a highly trained racing driver would drive because the movements become automatic and reflexive through the learning process and the driver no longer needs working memory capacity because it has become encoded at a subconscious level.

Another vital feature of robotic attention is Attentional Gaze and Viewpoint selection which reflects that a world model is of little practical use if a human or a robot constantly selects the wrong things to look at or can't make sense of what it sees based on different viewpoints.

(b) Attentional Gaze and Viewpoint Selection.

3D objects need to be represented and encoded in 2D and it is important whether the internal representation generated is viewpoint-dependent or independent. Attentional gaze and viewpoint selection include the (1) Need to control the robot's physically embodied visual gaze to focus on or select a part of or region of the environment or features or objects of interest and obtain more information by using viewpoint selection to get closer to an object or view it from different perspectives or angles (2) Ability to shift the visual gaze to select relevant or salient information and (3) Allocation of the time and resources required to process the information and decide upon the appropriate course of action or behaviour ideally in real-time.

A major problem facing most robots is the heterogeneity of the visual systems or sensors used such as cameras, laser, and radar, which complicates the search for the most efficient computation of the viewpoints available to the robot and the processing of 2D and 3D images of the environment essential to attentional focus. This requires consideration of the robot system constraints and modelling of uncertainties creating the Viewpoint Planning Problem (VPP). VPP is effectively a search problem that recognizes the need to establish the minimum number of views of an object or situation to be acquired by the optical sensors to allow the robots to reconstruct the object or acquire single features and classify or identify an object. It is particularly challenging with vision tasks comprising multi-features or regions of interest and typically involves defining a search space in which potential solutions can be evaluated to find the best solution. It has been further complicated over the last few years by the increased number of machine vision applications in research and the programming of RVS to perform such tasks is "considered a tedious and complex challenge."[95] VPP is most evident in teleoperated robots used in highly dangerous situations such as nuclear environments, bomb disposal, or disaster scenarios, with as Jan Dufek and colleagues point out a major problem being "insufficient situational awareness of the remote field, caused by the onboard sensing limitations, such as relatively stationary and limited field of view and lack of depth perception from the robot's on-board camera." This usually involves using two robots, with one acting as the visual assistant but neither represents autonomous robots as they inevitably require human operators[96] which highlights the challenges in equipping robots with the necessary capacity for attention to accurately identify and process in real time the information available from the environment.

In 2023 Alejandro Magana and colleagues described VPP as a "Multidimensional and challenging problem that any vision task demanding the computation of multiple and valid viewpoints must consider" which "is still regarded as a complex and unsolved challenge, lacking a generic and efficient framework for computing multiple viewpoints within feature-based applications." They categorised VPP though as a geometrical problem and to tackle it proposed what they called Feature-Based Constrained GC-spaces which are designed to efficiently acquire single features. They subsequently extended this to multi-feature problems using Feature Cluster Constrained Spaces (GC-spaces). GC-spaces work by "spanning continuous solution spaces in 6D, providing an infinite number of valid viewpoints that guarantee the successful acquisition of feature clusters, taking into account various viewpoint constraints."[97] They broke the VPP into two sub-problems, which are the Viewpoint Generation Problem (VGP) and the Set Cover Problem (SCP) with SCP described as a decision problem. To address the SCP decision problem, the first step is to identify the potential features that can be clustered together "based on their spatial vicinity and orientation similarity" using for example a centroid-based k-means algorithm. By addressing both VPP and SCP the robot's visual system can compute the "number of necessary feature clusters and corresponding optimized sensor orientations (and) the necessary GC-spaces." They concluded that "spanning an infinite solution space is a powerful technique for implicitly considering model uncertainties within real applications and delivering alternative solutions" and that their "viewpoint planning strategy based on GC-spaces provides a springboard for a novel and efficient approach for tackling the VPP, comprising closed and deterministic solutions."[98]

This illustrates the highly innovative ways in which cognitive architects are using our knowledge of the cognitive processes of the human brain to overcome the challenges in equipping robots with the selection mechanisms that imbue the robot with the capacity to attend to the environment and accurately extract reliable information upon which to act.

(c) Features /Time/Region/Objects/Events of Interest.

Selective attention involves the bottom-up processing of data driven by stimuli and selecting salient and distinct features. It directly affects the behaviour of humans and reflects the fact that we typically receive far more sensory information than we can process at the moment. Most machine vision models use bottom-up methods and low-level features such as intensity and colour, although some models employ high-level features, like faces. Very few employ mid-level features like texture although some like Takeshi Uejima and colleagues are now addressing this gap.[99] Equipping a robot with selective attention has been observed by Sock Ching Low and colleagues to make them better at "Interacting with more complex environments and exploit greater sensor resolution and sensitivity" and "considers the model not simply as an algorithm solving an abstract problem but something that enables a body to function, which is a key principle of embodied cognition.[100]

Saliency Models

Saliency algorithms and saliency models generate saliency maps based on the most salient features of an image being represented by the brighter pixels in an image. Classical saliency models use serial and parallel search mechanisms in relation not just to low- or high-level features but time, region, objects, and events of interest. Good examples of the wide range of roles that selective attention plays in robots from guided search, the generation of saliency maps and social interactions with humans are (provided by a) Guided Search, (b) Deep Gaze I, and (c) Kismet.

(a) Guided Search.

Humans are constantly consciously and subconsciously searching the environment whether it's for house keys or for TV remote controls, items in shops people they know, etc. Searching in an unfamiliar environment, or one in which there are many distractions or over a very large search space can significantly increase difficulty and the Reaction Time (RT) required to find the target or salient information. Robots must be able to adapt to unforeseen circumstances and demonstrate on occasion improvisation capabilities. The two models of guided search are (1) Serial search with attention directed to one item at a time and classifying targets or distractors in turn, or (2) Parallel processing items simultaneously and deciding based on this process. The speed or RT of the parallel search process in humans is constrained by the eyes only being able to fixate on three to four items per second or when it involves reasoning about complex goals and determining the feasibility of movements and actions. In AI guided search models attempt to employ the human visual search process when looking for a target object with distractions which requires algorithms that can search the range of possible tasks and motion plans and use statistical ML to guide the search process using heuristic search approaches.

We saw how the early stages of visual scanning in humans involve pre-attentive processing in which sensory input is processed in parallel streams and compared with a vast array of stored representations of objects or items. Only a limited set of attributes (e.g., colour, size, motion) could be processed in parallel and by different brain areas, across the whole visual field and those features need to be integrated by binding into a single representation. Jeremy Wolfe observes how the deployment of attention "is far from random" and "guided toward likely targets by a limited set of stimulus attributes such as colour and size ("classic guidance) …and by several scene-based properties." The problem as Wolfe points out is that this search process involves a bottleneck in processing between these stages because only one or a few visual objects at any one time can be submitted and compared to that vast array of stored representations for recognition.[101] Attention in guided search refers to controlling the selection at this bottleneck of what to visual process further based on a guided representation using between 12 to 24 features or attributes of which colour is one of the strongest. At this stage, both bottom-up and top-down processes are involved. Wolfe has argued that this bottleneck needs to be modelled in AI systems and robots using human processing because (1) Targets may be easy to identify but hard to find (2) Saccadic eye movements impose an obvious seriality on visual processing (3) Feature integration theory involves attention to bind features together and the (4) Existence of change blindness in which an observer fails to notice the change in an image. Melissa Võ and Jeremy Wolfe found that "familiarity with a scene or previous encounters with objects embedded in a scene" doesn't necessarily improve subsequent searches for objects, although "if (the) same object was searched for again, memory for the previous search was capable of producing very substantial speeding of search despite many different intervening searches". This was taken to demonstrate that "lack of memory guidance during initial object searches, despite previous encounters with the target objects, demonstrates the dominance of guidance by generic scene knowledge in real-world search."[102] In other words, when searching for something people typically scan a scene as the main form of guidance, and if the scene rather than objects is familiar based on previous searches, this will enhance the attentional outcome.

Wolfe recently developed a revised AI model of visual search called Guided Search 6.0 (GS6) in which he identified guidance as coming from five sources of pre-attentive information which are (1) Top-down feature guidance (2) Bottom-up feature guidance (3) Prior history (e.g., priming) (4) Reward and (5) Scene syntax and semantics. These five sources are combined into a "Spatial Priority Map," which evolves during the search as what Wolfe describes as a "dynamic attentional landscape". Search guidance in Wolfe's model is not uniform but directed approximately every 20 times per second to the most active location by three Functional Visual Fields (FVFs) which represent Foveal Biases that favour objects closer to the point of fixation.[103] The three FVFs comprise (1) Resolution (2) Governing exploratory eye movements and (3) Governing covert deployment of attention. The system compares items to target templates held in computer memory and identifies them as targets or rejects them as distractors. Modelled as a diffusion process, binding and recognition of the data takes around 150 m/s per item. In AI a diffusion model involves a DNN that contains latent variables that learn an image structure by removing the noise or blur and use the concept abstraction behind that image to create new variations. Wolfe's guided search models involve an asynchronous hybrid of serial and parallel processes with multiple items undergoing recognition at the same time and the search only terminating when an accumulating quitting signal reaches a threshold, which is adaptive because it is based on performance feedback which shapes subsequent searches.

Heuristic search guidance has been employed in a novel approach by Lakshmi Nair and Sonia Chernova called feature guided search (FGS) designed for a robot to derive a task plan when tools it requires are missing or unavailable to construct appropriate replacement tools. FGS effectively integrates reasoning about the physical attributes of objects with existing heuristic search algorithms and reduces the number of failed attempts to construct a replacement tool by over 93% compared to standard heuristic approaches. It demonstrates the vital role of attention in contributing to creative problem-solving.[104]

(b) Deep Gaze I.

Deep learning has led to vast improvements in the saliency maps that we discussed earlier and projects like Deep Gaze 1 have sought to improve upon the standard saliency models pioneered by neuroscientists Laurent Itti, Christof Koch, and Ernst Niebur. Standard models have less than optimal performances measured against the MIT Saliency Benchmark, which Matthias Kummerer and colleagues have attributed to an inability to model the influence of high-level image features such as objects and " the enormous amount of training data necessary to train these networks makes them difficult to apply directly to saliency prediction." They claimed that state-of-the-art models of saliency explain only one-third of the explainable information in the spatial fixation structure and proposed a new model using a Deep CNN based on understanding how humans choose eye fixations. They capitalize on the earlier work of Eleonora Vig and others who were the first to attempt to model saliency using DNNs using the DNN developed by Alex Krizhevsky and colleagues.[105, 106, 107] Fixation is the ability to aim the eyes to a particular spot accurately and happens between any two saccades when the eyes are relatively stationary and most visual input occurs. They point out that most machine vision models use low-level cues such as edge detectors and colour filters or local image statistics and attempt to incorporate more high-level features by combining for example with face detection mechanisms that have performed poorly because they believe "human fixations are largely clustered around objects."

Motivated by the recent progress in object classification and task generalization, Kümmerer and the team sought to overcome this by using a pre-trained DNN with eight learned layers, five convolutional and three fully connected layers. The training data for fixation has typically been very small, typically employing no more than 1000 images and fewer than 100 fixations per image although using DNNs designers have experimented with various layers. To address this, they used a popular data network in machine vision and DL called ImageNet which was developed by American/ Chinese data scientist Fei-Fei Li (1976–).[108] Launched in 2010, the first-ever ImageNet Large Scale Visual Recognition Challenge (ILSVRC) now provides researchers with a common set of images to benchmark models and algorithms. Fei-Fei Li was adamant that vast amounts of real-world data would make algorithms more accurate and observed how "The paradigm shift of the ImageNet thinking is that while a lot of people are paying attention to models, let's pay attention to data. Data will redefine how we think about models." ImageNet which was intended to support visual object recognition software research using a massive dataset of over one million images has since catalyzed many developments in AI. In 2012 the DNN AlexNet dominated the field of computer vision in the ImageNet challenge, but by 2017 most teams were able to reach 95% accuracy in the challenge.

(c) Kismet.

Designed in 1999 by MIT Cynthia Breazeal and Brian Scassellati recognized attention as a vital sub-system for social intelligence which is context-dependent.[109] Kismet was designed as an autonomous robot able to engage in social interactions with humans and part of the larger Cog Project, based on the way infants learn to communicate with adults. The system architecture incorporated five subsystems, the Perception system, the Motivation system, the Attention system, the Behaviour system, and the Motor system, and used Wolfe's then-current model (which as we have just seen has since been revised and updated). Breazeal described how "visual stimuli are filtered by broadly tuned "categorical" channels (such as colour and orientation) to produce feature maps with activation based upon both local regions (bottom-up) and task demands (top-down). The feature maps are combined by a weighted sum to produce an activation map.[110] Kismet now resides in the MIT museum.

Since this early work, there has been growing interest in the effect of a robot's attentional capacity and ability to display appropriate social behaviours on human perception of robots, which includes engaging, maintaining, and demonstrating attention. Deborah Johansen and colleagues in their study of the effect of robot's attentional behaviours on users' perceptions and behaviours in healthcare environments identified "Leaning forward, self-disclosure, and changes in voice pitch" as fundamental requirements in social interactions.[111] We will discuss this later under robot emotions and mentalizing when collaborating with humans.

Saliency Detection.

Takeshiki Uejima and colleagues observe that "Visual saliency modelling, i.e., understanding how the brain selects important information to analyse further and to determine where to fixate next, is an important research topic in computational neuroscience and computer vision."[112] This involves finding regions of interest in terms of finding out where to look for objects etc and distinguishing objects from the environment so that they can be identified as targets or dismissed as distractors and may be due to a change, in contrast, intensity, or pattern. A challenge to this view of saliency detection in humans comes from the perspective of Gestalt psychologists, who maintain that we first perceive whole objects before we analyse individual features. Most visual scenes have parts with different instantaneous saliency values which will be detected first in the automatic bottom-up process that most modern theories maintain is a feature-driven process.

Growing psychophysical and neurophysiological evidence suggests that human attention involves the structural organization of the scene into perceptual objects, not just image features supporting the Gestalt view. Aurel Wannig and colleagues studying the neuronal activity in area V1 of macaque monkeys observed "An automatic spread of attention to image elements outside of the attentional focus when they were bound to an attended stimulus by Gestalt criteria" in which "attention enhances the representation of all parts of an object, even parts that are not task-relevant." They concluded that human visual attention can "select spatial locations, features, and objects." which supports the Gestalt model and further highlights the difficulty of optimizing saliency models in machine vision.[113]

The first step in recognising an object is to detect it, which means being able to extract the salient regions from an unknown background. In humans this involves the fast and parallel pre-attentive process followed by slower, serial, and more complex attention processes. During the pre-attentive process orientations, edges, or intensities can appear to automatically stand out or pop up, which makes that object a possible candidate for detection even if it isn't yet recognized. Itti and Koch's model was designed to mimic this visual search process and has been extended by Dirk Walter and colleagues to object recognition.[114] There are dozens of saliency models in use but three main types of saliency detection algorithms commonly used in AI systems are (1) Saliency systems that classify and identify objects using Bounded Boxes and Image Segmentation (2) Static saliency in which the system detects from a static image the most salient features with the two main types being, Static spectral saliency and Fine-grained saliency and (3) Motion saliency in which the motion is captured frame by frame and allows the system to track an object typically using Optical flow maps.

The key metrics for good saliency models have been listed by Niklas Wilming and colleagues as including "Few parameters, Intuitive scale, Low data demand, and Robustness."[115] A challenge with any selection mechanism is over-reliance on one sensory modality like vision which means that if the information from other modalities is large but missing this creates the missing modality problem which results from the incompleteness of the training data. In real-world applications, malfunctioning sensors can also lead to missing data which exacerbates the problem. The missing modality problem has become an area of much research recently with many applications such as self-drive vehicles using a variety of media including audio, images and video, and thermal and visual cameras to address this.

Bounded Boxes and Image Segmentation.

(a) Bounded Boxes

Bounded Boxes use data annotators to identify x and y coordinates and draw an abstract rectangular region around an object to enable an algorithm to include all the main features to classify, identify, and locate objects. In autonomous vehicles, surveillance cameras, drones, and robots, other vehicles, traffic signals, and pedestrians or objects are bounded and numbered, using a wide range of bounded frames and data augmentation to track other objects. The bounded box involves the system learning about the contents of images or objects in bounded boxes in training data, by labelling objects or features of interest whether a face, a sign, type of vehicle, or anything of interest. This allows the robots to learn to identify objects to avoid collisions and predict the presence of similar objects in a new environment. The bounded box involves the system learning about the contents of images or objects in bounded boxes in training data, by labelling objects or features of interest whether a face, a sign, type of vehicle, or anything of interest. This allows the robots to learn to identify objects to avoid collisions and predict the presence of similar objects in a new environment. The bounded boxes must be labelled consistently and accurately and using bounded boxes saves significant computational power.

Labelling tightly means not cutting off any part of the object, although for occluded and truncated objects different models may draw boxes only around the visible parts and models vary in how they address this, and all instances of the object class must be labelled, however small or partially visible.

Bounded Boxes.

(b) Image Segmentation.

Segmentation is different from bounded boxes in that it is a form of labelling that breaks down the whole image into pixel groups which are then labelled and classified. It requires the image or object to have been detected and identified first which typically involves using a bounded box approach.

Bounded Boxes.

Static saliency

Static saliency algorithms use different image features to detect the most salient or conspicuous features of objects that stand out in a static non-dynamic image. They are now available on Open CV which is a library of programming functions and free and open-source software largely for real-time computer vision, with the two most widely used algorithms being (a) Spectral Residual and (b) Fine Grained.

(a) Spectral Residual.

Spectral residual is a method that rather than directly using the spatial domain of an image by relating features to detect specific categories of objects, uses what is called the Log-Spectra. Traditional models require extensive training and suffer from problems generalizing to new data. The log-spectra method seeks to overcome this by representing the image in the frequency domain with analysis showing how the signal's energy is distributed over a range of frequencies rather than over time. This usually involves employing a pair of mathematical operators called a transform, often the 2-D Fourier Transform, which converts the signal between the time and frequency domains and allows you to extract the spectral residual of an image.

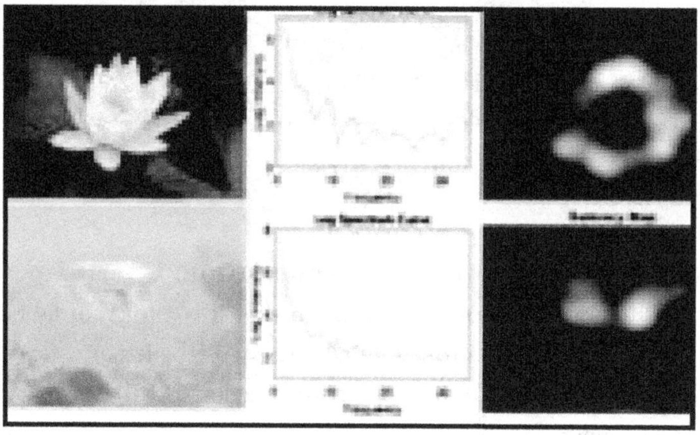

Salient Object Detection using Spectral Residual.

Xiaodi Hou and Liqing Zhang used the spectral residual to propose a rapid method of constructing saliency maps in the spatial domain and most importantly "Independent of features, categories, or other forms of prior knowledge of the objects". They claim that it is therefore "Implemented with the least reference on statistical knowledge of the objects."[116] The spectral residual reflects how the efficient coding of any image involves decomposing the image information H(Image) into two parts where H (Image) = H(Innovation) + H (Prior Knowledge). H(Innovation) is the novelty information whilst H (Prior Knowledge) is redundant information that needs to be suppressed when coding. This recognises that this prior knowledge represents statistically invariant properties of our environment which are not random but obey highly predictable distributions. Hou and Zhang point out that this redundant information can be removed in encoding to reveal an approximation of the innovation part of an image and claim that this is "inherently responsible for the popping up of proto-objects in the pre-attentive stage." Proto-objects are described by Alexander Russell and colleagues as part of their "biologically plausible model of object-based visual salience."[117] They refer to Rensink's Coherence Theory[118] which uses proto-objects as "Pre-attentive structures with limited spatial and temporal coherence" that are "rapidly computed in parallel across the visual field and updated whenever the retina receives a new stimulus." They describe how "Focused attention acts to stabilize a small number of proto-objects generating the percept of an object with both spatial and temporal coherence. Because of temporal coherence, any changes to the retina at the object's location are treated as changes to the existing object and not the appearance of a new one. During this stage, the object is said to be in a coherent field. Once attention is released from the object, it dissolves back into its dynamic proto-object representation." Their model is a "grouping mechanism which estimates the location and spatial scale of proto-objects within the input image" and they claim that with "its basic computation mechanisms with known biological correlates yet can match the performance of state of the art, non-bio-inspired algorithms." They concluded that "the performance of the algorithm strengthens the growing body of evidence which suggests "that the bottom-up attentional bias in subjective interest is object-based not feature based.

(b) Fine-grained

Fine-grained calculation saliency is based on how ganglion cells in the centre retina of the human are on if they respond to bright areas surrounded by a dark background or off centre if they respond to dark areas surrounded by a bright background. Hence, they react to contrasts between the centre and the surroundings with high-resolution saliency maps (heat maps) generated in real time by using integral images. Motion saliency is when the motion is captured frame by frame and allows the system to track an object.

Most computer vision research typically only addresses relationships of objects within the same frame based on space coordinates (X, Y) and disregards time information (t) so analysis of each frame does not capture information over time such as movement. By contrast, motion saliency explores information about motion such as the conspicuous state, velocity, and position of an object. It measures the image intensity (I) as a function of both space and time and can generate flow vectors about the information. Video Saliency Object Detection (VSOD) has generated widespread research and involves measuring optical flow by calculating the velocity vectors for points within, for example, an image sequence or video and estimating where those points could be in the next image in the sequence. Optical flow maps track the motion and movement of the points of interest with widespread applications such as video compression, tracking objects, face recognition, stabilizing images, comprehending scenes in videos, interactive games involving humans, and robotics for mapping, navigation, and localization. The popular Lucas-Kanade, Horn-Schunck, or Buxton-Buxton algorithms estimate the Sparse Optical flow by tracking the displacement over time of small patches of pixels or specific sets of features in an image like the edges or corners. These are typically points of interest or good features to track because they are rich in information content like corners or edges which are invariant to translation, scale, rotation, and intensity changes. Special algorithms like the ShiTomasi Corner Detector and Harris Corner Detector are also used. After the image is divided into small patches it tracks the displacement of each patch between two consecutive frames determined by minimizing the difference between the patches. The Lucas-Kanade Sparse Optical flow method does not perform well with rapid or large motion and the open source library Open CV has developed a way of calculating what is called the Dense Optical flow which measures the flow vector for all the pixel points in the frame using the Gunnar Farneback algorithm introduced in 2002 which can provide up to one flow vector per pixel making it much more accurate although is computationally much more expensive. The size of objects is relevant with the absolute size represented by the actual pixel size of the object, and relative size is the ratio of the object's pixel size to the entire image. Small object detection and the need to identify positioning and classify the object, particularly in complex scenes with cluttered image backgrounds further complicating the analysis of optical flow information. Research has shown that for small object detection, the context or surrounding environment is essential and requires the integration of both global and local contextual information. DL methods and the use of DCNN have achieved great success but are ineffective in small object detection and most cannot achieve real-time processing.

Yuqiu Kong and colleagues proposed a Spatial Context-Aware Network (SCA-Net) for salient object detection (SOD) to provide high-level semantic knowledge about salient objects. Deep learning researchers have been exploring context to enhance object recognition which encompasses both spatial and temporal context, particularly in video data. Spatial context has used a variety of approaches such as capturing the relationships between objects using graph-based representations, spatial context networks, or pairwise relationship modelling. With temporal context, they have used the temporal dependencies between consecutive frames to improve object detection. Kong and colleagues claim that SCA-Net can "more effectively aggregate multi-level deep features" and "a more accurate initial saliency map can be obtained to facilitate subsequent predictions." They achieve this by using both a Long-Path Context Module (LPCM) and a Short-Path Context Module (SPCM) with the SPCM "progressively enforcing the interaction between local contextual cues and global features."[119] Popular motion Optical flow maps include CRAFT, GMFlow, SpyNet, PWCNet, and LFNet.[120] Hengsen Wang and colleagues observe that although these are accurate and excellent in speed detection, they can suffer from image generalization and contour blurring. They attribute this to the fact that "traditional optical flow calculation methods only rely on the image information of two adjacent frames to calculate the optical flow, which is a short-term strategy for computing optical flow, ignoring the motion states of objects in the past and future periods."[121] If for example there is temporal and spatial misalignment between adjacent frames making the motion information inaccurate it will be difficult or impossible to predict the motion. Wang and colleagues have proposed a new model that expands the range of optical flow perception by using multiple frames combined with an optical flow perception module which enhances the effectiveness of feature fusion. This brief discussion illustrates the innovative way that computer scientists are utilizing our knowledge of the saliency and object recognition mechanisms employed by our brains to develop algorithms that are rapidly improving the attentional performance of robots.

(2) Restriction and Suppression Mechanisms.

If the architecture is designed to restrict attention to a task, the robot is typically primed to restrict the search space which means programming it to only use sensory input geared to the demands of the task so is equipped with what constitutes task-driven attention, This is achieved by providing the robot with specific domain knowledge or endogenous motivations in terms of existing or prior knowledge or restriction of the robot's attention to defined exogenous cues or objects relevant to the task and particularly by restricting the robots field of view. Restriction mechanisms ignore some elements by restricting the search space and restricting the processing to the space selected. They use methods such as (a) Priming the robot for a specific task, (b) Exogenous and endogenous cues and tasks or goals and (c) Restricting the visual field.

(a) Priming

In humans, the context and sequence of events can prime someone's perception of what they are seeing with exposure to a stimulus influencing the unconscious response to a subsequent stimulus. This process can be harnessed by AI systems to essentially prime or bias the processing of information to direct the focus on a task.

(b) Exogenous (bottom-up) (external stimuli) and Endogenous Cues (top-down) and tasks or goals.

Attention can be guided by exogenous and endogenous cues both of which play a role in Visual Perceptual Learning (VPL) in terms of the relevance or irrelevance of the VPL to attentional focus and tasks or goals. The use of cues and focus on goal or task completion is important for humans and robots to prevent being overwhelmed by the vast amount of information in the environment and both the brain and AI systems have limited computational resources.

Exogenous (bottom-up) attention in humans is an automatic and rapid (about 100 ms) type of spatial attention in which attention is automatically drawn to events or cues in the environment, such as a brief flash of light. It is involuntary and effectively reflexive with attention to external stimuli involving bottom-up attention to stimuli, so it is less flexible than voluntary endogenous attention.[122] Cues are driven by the properties of a stimulus such as colour, orientation, and luminance, occur in our peripheral vision and draw our attention to the target of our attention provided the gap between the cue and the target is not too long. Human studies have shown that after this time there may be an inhibition in reorienting attention to that location due to saccade preparation which we discussed earlier.[123] Endogenous cues are those generated from within the robot or human and are a goal or task-driven, voluntary, and sustained type of attention. Endogenous attention therefore selects the piece of information relevant to the goal.

Cognitive architects Sara a Marques-Villarroya and colleagues have been seeking to build a bio-inspired endogenous attention-based architecture for a social robot using a robot called Mini that has been designed to assist and entertain older adults with mild cognitive impairment[124]. Their rationale is that in HRI "it is essential to select the most relevant stimuli to achieve a natural experience" and to achieve the most efficient focus of attention (FOA) to facilitate the faster processing of the most relevant information in detail. They observe how "The literature offers different approaches for endogenous models; although in most cases, these are descriptive and are not supported by concrete developments. This has been caused by a lack of generality in the algorithms." They maintain that "no work in the literature has proposed a purely endogenous architecture that reacts in real-time" and ...." works on attention management that is applied to robotics are scarce and they have usually focused on mobile robots that need attention to navigate correctly. They set out the following list of requirements which are; (a) To decide on the optimal FOA considering the available stimuli and the current task (b) To process information from multiple sensory modalities (e.g., vision, touch, and sound) (c) To respond in real-time, avoiding lengthy processing and response times (d) To select the correct FOA to achieve natural human–robot interaction(e) To include a voluntary attention process that detects the relevant information for a goal in a sustained manner (f) To include a mechanism that allows the robot to focus on a stimulus punctually, even when it is not the most salient stimulus, in a sustained manner (g) To take into account the current task of the robot (i.e., the system needs to communicate with the DMS to know what task the robot is currently performing and its next action, and then, with this information, the system can sort the stimuli according to their relevance to the task) (h) To filter out irrelevant stimuli for the current task (i) To consider scalability and modularity (i.e., allow the inclusion of new sources of information) (j) To manage resources efficiently (i.e., inhibit or disable detectors depending on the task).

Mini's software employs an attention manager which they describe as taking "inspiration from psychological and neuroscientific works that considered how human attention works within social interactions." To extract information from the robot's environment, Mini relies on visual, auditory, and tactile detectors, which are only activated when necessary. They include interaction detectors which allow the robot to establish an FOA, endogenous detectors, and mixed detectors can detect endogenous and exogenous information in the environment. The endogenous detectors are only used when required to perform a particular task or skill, whereas the mixed detectors remain active, but the attention manager uses their data only if required for the current skill. Intriguingly they employ what they describe as "ego-spheric representation to achieve multi-sensory integration" visual, auditory, and tactile information" in a 3-D view of the stimuli within the robot's environment, so capture and "generate a unique representation for all endogenous stimuli." They reported that the results showed that Mini was "able to identify the most salient stimulus for a given task, handle multi-modal stimuli (visual, auditory and tactile) produce real-time responses, select the most relevant information to achieve natural human–robot interaction, react to sustained and punctual attention and correctly manage activation and deactivation of detectors to reduce computational load."[125] This illustrates the progress that may well be made in providing robots with the critical capacity to use attention in the way that humans interact with the environment and others.

(c) Restricting the Visual field.

Not many models employ suppression systems because it limits the potential scope of the robots' attentional capabilities, but suppression techniques may involve suppressing the robot's visual input to the field around the point of interest, avoiding task-irrelevant information, negative priming or constraining the shifting of attention back to previous locations or objects.

Attention to Harmful Knowledge or Content.

The objective of attending to and using information from the environment for humans or robots is to reduce uncertainty in the environment. To train robots to learn and operate in the real world they need to learn to tolerate uncertainty and random environmental challenges and behaviour. Entropy RL models present the optimal environment in which robots can learn and make the most robust for operation in the real world. Although we traditionally assume that the acquisition of knowledge is beneficial, studies have shown that additional knowledge or information content can result in a deterioration in the performance of a system, most obviously when that knowledge or content is harmful.

Harmful knowledge may be obtained through being programmed with or simply having learned that knowledge. Humans can use reasoning and analysis based on experience to discriminate though unfortunately less than we should. RL might be employed to teach a system to disregard harmful information, but it needs to be able to do this as rapidly as possible. The Complementary Learning Systems (CLS) theory postulates that humans learn using two complementary systems with the neocortex system being slower and relying on the distributed representations, whilst the hippocampal system is faster and learns by relying on pattern-separated representations. The hippocampal system is then used to support learning using the neocortex system. Both systems project to the striatum which as we have seen is key to RL in humans.

Sam Blakeman and Denis Mareschal observe how "Current deep RL approaches share similarities with a 'neocortical' system because they slowly learn distributed representations through backpropagation in Deep Neural Networks (DNNs). An ongoing criticism of such approaches is that they are data inefficient and lack flexibility."[126] To address the problems of data inefficiency and the lack of flexibility Blakeman and Mareschal proposed a novel algorithm known as Complementary Temporal Difference Learning (CTDL), which combines a DNN with a Self-Organizing Map (SOM) to obtain the benefits of both the neocortical and hippocampal system. When evaluated CTDL demonstrated several benefits over the classic Deep Q-Network (DQN) approach which they report as "(1) the utility of complementary learning systems for the evaluation of actions, (2) that the TD error signal is a useful form of communication between the two systems and (3) that our approach extends to both discrete and continuous state and action spaces". They argue that "Developing human-like artificial intelligence will require implementing multiple learning and inference systems governed by a flexible control system with an equal capacity to that of human control systems."[127]

In 2021, Meta sought to address this problem by launching Few-Shot (or Low-Shot) Learner (FSL) with the express intent of accelerating the process by which an AI system can detect harmful content. FSL is a supervised machine learning approach that involves data training and learning to make predictions based on limited information, as humans do, rather than the trend of feeding and training systems using large amounts of data. It enables machines to learn to detect unusual or rare cases and by training on several related tasks learn to generalize information from a few examples to unseen though related information. FSL uses meta-learning in which the machine learns how to solve a problem and has shown great potential promise as a learning method, driven in part by the increased interconnection of billions of devices, generating a huge amount of data that needs to be centrally processed and stored. This ubiquitous connectivity has inspired the move towards an Internet of Everything (IoE) framework but being able to extract valid information from huge volumes of data is expensive and still poses the problem of harmful content. Training on limited data though is very challenging and much training currently requires many defective samples for training or exposure to much harmful content. Meta's use of FSL to detect harmful content has been tested on misleading COVID information, and identifying content inciting violence or hate speech and bullying. The intention is to enhance the integrity of the AI systems to allow them to be more intelligent in the way they attend to and discriminate between useful and valid knowledge and harmful knowledge.[128]

Endnotes

1 Pashler, H., 1997.
2 James, W., 1890.
3 Tsotsos, JK., 2019.
4 Katsuki, F, Constantinidis,C., 2014.
5 Lindsay, G W., 2020.
6 Chun, M., et al., 2011.
7 Hommel, B., et al., 2019.
8 Chun MM, et al., 2011.
9 Corbetta, M, Shulman, GL., 2002,
10 Itti L, Koch C., 2001.
11 Hommel, B., et al., 2019.
12 Shipp, S., 2004.
13 Posner, MI, Petersen, SE., 1990.
14 Degutis, JM, Van Vleet, TM., 2010.
15 Thompson KG, et al., 2005.
16 Petersen SE, Posner MI., 2012.
17 American Psychological Association Dictionary (APA).
18 Payne, P, Crane-Godreau, MA., 2015.
19 Ruge,H., et al., 2013.
20 Payne P, Crane-Godreau, MA., 2015.
21 Connolly, J., 2002.
22 Everling, S, Munoz, DP., 2000.
23 Rommelse, N., 2008.
24 Moray, N., 1959.
25 Bronkhorst, A W., 2000.
26 Corbetta, M,. 2002.
27 Ramiro-Cortés, Y, et al., 2013.
28 Kandel, ER., 2001.
29 Ringach, DL., 2009.
30 Ringach, DL., 2009.
31 Pezzulo, G, Zorzi, M, Corbetta, M., 2021.
32 Engl, E, Attwell, D., 2014.
33 Mitra, A, et al., 2018.
34 Brosch,T, et al., 2014.
35 Easterbrook, JA., 1959.
36 Fox, E, et al., 2001.

[37] Najma,S., et al., 2012.
[38] Armstrong, T, Olatunji, BO., 2012.
[39] Fredrickson, BL., 2001.
[40] Bar-Haim, Y., et al., 2007.
[41] Wilson, E, MacLeod, C., 2003.
[42] Eldar, S., et al., 2010.
[43] Waters, AM, et al., 2010.
[44] Gamble, AL., 2009.
[45] Nikolaidou, M., et al., 2019.
[46] Abend R, et al., 2019.
[47] Hallion, LS, Ruscio, AM., 2011.
[48] Posner, M. I., 2008.
[49] Martin, J Dahl., et al., 2021.
[50] Herz, N, Baror, S, Bar, M., 2020.
[51] Herz, N, Baror, S, Bar,M., 2020.
[52] Later, A., 2017.
[53] Longstreet, P., Brooks, S., 2017.
[54] R A, Davis., 2001.
[55] Egeth, HE, Yantis, S., 1997.
[56] Guo, MH., et al., 2022.
[57] Nguyen, T., et al., 2018.
[58] Colombini, E., et al., 2016.
[59] Russell, AF, et al., 2014.
[60] Colombini, E., et al., 2016.
[61] Agüero, TH., et al., 2012.
[62] Itti, LC, Koch C, Niebur E., 1998.
[63] Itti LC, Koch C., 2001.
[64] Breazeal, C, Scassellati, B., 1999.
[65] Haasch A, et al., 2005.
[66] Desimone, R, Duncan J., 1995.
[67] Pinto Y, et al., 2013.
[68] Huang F, Tang S, Hu Z., 2018.
[69] Ruz, M, Lupiáñez, J., 2002.
[70] Pinto Y, et al., 2013.
[71] Moutoussis, K, Zekiin, S., 1997.
[72] Chiou, SC., 2023.
[73] Lindsay, GW., 2020.
[74] Sreelekha, S., 2017.
[75] Bahdanau, D., et al., 2015.

[76] Harris. Z, 1954.
[77] Cho K., et al., 2014.
[78] Hopfield, J J., 1982.
[79] Tsotsos, JK., 2011.
[80] Tsotsos, JK., 2019.
[81] Tsotsos, JK, Kruijne W., 2014.
[82] Martín, F, et al., 2021.
[83] Tsotsos, J K, et al., 1995.
[84] L, Koch C., 2001.
[85] Nayyer, A, et al., 2019.
[86] Nayyer A, et al., 2019.
[87] Anderson P, et al., 2018.
[88] Jaderberg, M., et al., 2015.
[89] Palmer, S, Rosch, E, Chase, P., 1981.
[90] Ghose T, Liu Z., 2013.
[91] Tsotsos, J K, et al., 1995.
[92] Matsuo Y, et al., 2022.
[93] Burgard, W, Hebert, M., 2008.
[94] Ha, D, Schmidhuber J., 2018.
[95] Peuzin-Jubert, M., et al., 2021.
[96] Dufek, J., et al., 2020.
[97] Magaña A, et al., 2023.
[98] Magaña A, et al., 2023.
[99] Uejima T, et al., 2020.
[100] Low, SC, Vouloutsi, V, Verschure, P., 2021.
[101] Wolfe, JM., 2012.
[102] Võ, ML H, Wolfe, J. M., 2012.
[103] Wolfe, J.M., 2021.
[104] Nair L, Chernova S. 2020.
[105] Krizhevsky, A, et al., 2012.
[106] Vig E, Dorr M, Cox D., 2014.
[107] Kümmerer, M, Theis, L, Bethge, M., 2014.
[108] The Worlds I See (2023).
[109] Breazeal C, Scassellati B., 1999.
[110] Breazeal, C., 2002.
[111] Johanson DL, et al., 2019.
[112] Uejima T, et al., 2020.
[113] Wannig A, et al., 2011.
[114] Walther, D et al., 2002.

[115] Wilming, N., et al., 2011.
[116] Hou, X, Zhang, L, 2007.
[117] Russell AF, et al., 2014.
[118] Rensink, RA., 2000.
[119] Kong, Y et al., 2021.
[120] Haidar Sharif, M., Martinet, J., Djeraba, C. 2008.
[121] Wang, H, Chen C Li, L & Peng, C., 2023.
[122] Sam M.S., 2013.
[123] Rafal, RD., et al., 1989.
[124] Marques-Villarroya S, et al., 2022.
[125] Marques-Villarroya S, et al., 2022.
[126] Blakeman S, Mareschal D., 2020.
[127] Mareschal, D, Blakeman S., 2021.
[128] Song, Y et al., 2023.

"For me, consciousness is the most interesting unsolved problem of science, and, in fact, we may never know what it is about a particular arrangement of neurons that gives rise to consciousness. Our consciousness, like the air we breathe or like the passage of time, is central to our existence as intelligent beings."

Alan Lightman American Physicist and writer.

Chapter 6

Consciousness

CHAPTER 6

Human Consciousness.

What is Consciousness?

WILL ROBOTS EXPERIENCE THE PHENOMENOLOGICAL conscious content in the way that we do and require the computational equivalent of the type of subconscious processes that we rely upon before it reaches out conscious awareness? Let's first look at consciousness and later the subconscious.

Emerging in the 17th century, the English word consciousness derived from the Latin Cōnscius (from Con-together + Scīre- to know), in other words, to know together and (Cōnscius Sibi) meaning to know (or be aware) that one knows. When Alan Turing posed his now famous question as to whether machines can think he was referring to the outward signs of intelligence and not the inner workings of the subjective human experience. His test of whether a machine could pass as a human was based on creating the impression of a machine capable of appearing to think like a human, not a machine that can display conscious understanding. But beneath human thinking lies conscious and subconscious processes that define us as individuals and constitute our highly subjective understanding of the world and our thoughts.[1]

When trying to understand consciousness, we should heed the words of the Australian philosopher and cognitive scientist, David Chalmers (1966–) who cautioned that "The tools of neuroscience cannot provide a full account of conscious experience, although they have much to offer." Neuroscientists and psychologists like to operationalise consciousness which means "describing something in terms of the operations (procedures, actions, or processes) by which it can be observed and measured."[2] Consciousness is often described as the mind's subjective experience and involves fleeting sensations that arise from the processing of sensory stimuli by our body and brain and perception in terms of organizing and

interpreting the sensory information so that we can make sense of our world. It is complex because we need to understand the (1) Different levels of consciousness (2) Content of consciousness in why and how some but not all sensory information is processed in the brain, why that which is selected generates subjective experiences and perception and (3) How meta consciousness which has been defined as the "intermittent explicit re-representation of the contents of consciousness" allows us to reflect on and share our subjective experiences with others.

Professor Peter Carruthers in a 2017 paper titled, the Illusion of Conscious Thought, argues that the illusion refers to the fact that they originate from processes that we are completely unaware of. He acknowledges that we are aware of a stream or train of consciousness but that thoughts include only non-sensory mental attitudes, like judgments, decisions, intentions, and goals which are amodal in that they are perceptual representations of whole things some of which are occluded and not accessible to our senses and abstract events not related to sensory experiences. He claims that thoughts don't feature in working memory and remain forever unconscious, but we confuse thoughts with the visual imagery and words generated by the internal voices in our head that we use to interpret the underlying thoughts. A thought effectively describes the processes or workings of our mind that are inaccessible to us and Carruthers maintains that "What rises to the surface of your mind are the results of these mental processes" so the supposed transparency of our mental processes is an illusion because everyone is interpreting this underlying unconscious process. This means that when we interpret our thoughts or those of others, we are using one single mind-reading mechanism.

Consciousness appears essential to our being able to interpret, give meaning to, and respond to the working processes in our head and the information in our environment but does not give us access to an objective reality that exists independently of our highly subjective perception which varies with every individual. Most of what we know about conscious processing is gleaned from subjective accounts of people's internal world, observing behaviour, and the impact on the consciousness of traumatic brain damage, strokes, or neurological. diseases such as dementia. Scientists distinguish between the State of consciousness and the Content of consciousness. The widely held and intuitive assumption is that we need to be conscious to have a self and that consciousness requires awareness of sensory information from our environment, the processing of that information, and the attribution of meaning to that information by extracting data and attribution relating to the self.

Human awareness relies on sensory input but merely being aware of sensory information does not amount to being able to perceive the world or our existence, because the raw data requires internal processing for it to be given meaning. Many sensory modalities particularly vision, touch proprioception and more have been built into AI and we looked at them separately, but any intelligent system whether human or robot relies on raw data input and based on processing that raw data input, being able to perceive meaning and construct an internal neural representation to perform cognitive tasks even if not necessarily at the higher level of cognition.[3] This process of constructing an internal neural representation of the information reflects the interactive nature of the MBBN and the environment and the amounts and type of data used in the creation of the conscious state and the generation of conscious content. When we talk about the state of consciousness, we describe various states of wakefulness or awareness that should be distinguished from the mental content experienced during the state of being conscious.

We have known since 1846, when American dentist William Morton (1819–1868) invented anaesthesia in the form of inhaled ether for surgery, that the experience of general anaesthesia, reveals how remarkably quickly, someone can be rendered unconscious. It has been described as the black box phenomenon because we know what goes into the box and what comes out, but what happens inside the box has for a long time remained a mystery. New imaging techniques for measuring brain activity, such as fMRI and EEG are enabling scientists to explore in much greater detail, the boundary between being conscious or unconscious. Studies are starting to reveal what Morton couldn't know, which is that consciousness appears to be the global feature of the noisy and chaotic state of the human brain. The transition to unconsciousness when undergoing anaesthesia does not involve switching off the whole brain but stopping the various brain areas from communicating with each other by reducing the level of connectivity. A 2010 study demonstrated this when using the Benzodiazepine Midazolam which is a common anaesthetic and sedative used to treat severe agitation. It has been known that during early deep sleep connectivity within the cortex is disrupted and people remember little conscious content during this period. They found that using Midazolam, there was a similar breakdown of cortical effective connectivity which led to a loss of consciousness.[4] Our more recent understanding of the effect of anaesthesia on consciousness suggests that by disrupting cortical connectivity it selectively reduces consciousness while sparing non-conscious (automatic) or subconscious brain activities. Brain areas under anaesthesia are still capable of being aware of sensory information in the environment, for example, when music is played to a person under anaesthesia, the part of the brain responsible for processing noise may still become active. They will likely not have any conscious awareness of listening to or having heard the music or necessarily remember any of the content. The human brain is constantly maintaining a subtle balance and modulation of hyper-connectivity or neural activity and insufficient connectivity or neural activity and integration of information and it is this balance which allows us to maintain consciousness and function. By building a sufficient level of connectivity and integration into robots using ANNs will they be able to rise through the various levels of consciousness and process information subconsciously before it emerges into a conscious awareness? Human consciousness may however be even more complicated than just neural connectivity and integration within our brain and we must briefly digress into the strange world of quantum consciousness.

Quantum Consciousness.

In 1998 American Stuart Hameroff the famous English mathematician and Noble Laureate Roger Penrose and anaesthetist and psychologist, claimed that consciousness arises in the microtubules of the brain neurons and obeys the laws of quantum dynamics. Microtubules are made from repeating subunits of tubulin that form cylindrical protein lattices and give nerve cells or neurons their cytoskeleton and are involved in a wide range of fundamental biological processes in our brain including intracellular transport and cell division. They play a vital role in synaptic plasticity by dynamically polymerizing directly into the dendritic spines which is thought to be essential to memory and learning.[5] Penrose and Hameroff's Orchestrated Objective Reduction ('Orch OR') model, maintains that microtubules change shape during anaesthesia due to the decoherence of the electrons that make up the polymers. Decoherence is the point at which quantum mechanics can no longer be used to explain the behaviour of the system, which is governed by classical mechanics because up to that point, the process is ruled by the laws of quantum physics which are believed to go to the heart of our relationship with reality and the known universe. Penrose relies on Gödel's theorem published in 1931 by Kurt Gödel (1906–1978) which in extremely simple terms states that with any mathematical model, there is a statement that whilst true (axiomatic) cannot be proved. He argues that human consciousness as an axiom cannot be proved in the classical model which requires us to turn to quantum mechanics for the answers. This has met with significant scepticism with Patricia Churchland, a neurophilosopher at the University of California, San Diego, observing that one might as well invoke "Pixie dust in the synapses" to explain human cognition.

Penrose's consciousness requires an objective threshold to be reached at which the wave function of the quantum particles collapses and decoherence happens. He used his Diosi-Penrose gravity equation to invoke quantum gravity collapsing the wave function, which means that once above a certain mass the system would not exhibit quantum gravity.[6] Hameroff provided the biomechanical framework for Penrose's theory by demonstrating that the microtubules are structured in increasingly complex levels of fractal patterns which could maintain quantum states for long enough to perform quantum computations performed by qubits, orchestrated by synaptic inputs in terms of electrical oscillations in the micro-tubules.[7] Subsequent work by Brazilian theoretical physicist Dr. Cristiane de Morais-Smith (1964–) who is a worldwide recognized researcher working on the theory of strongly correlated systems, has proved that electrons can be organized in fractal patterns which would enable such quantum processes to happen. Microtubules are known to couple to and regulate neural synaptic functions and by changing shape when the superposition of the electron is collapsed, they change the synaptic connections involved in consciousness. They can only do this because they experience intermittent isolation from environmental interactions which would normally collapse the quantum state. Significant scepticism about this theory centres on the fact that quantum laws appeared only to apply at very low temperatures of around -272°C, but not at body temperatures. We now know though that our brains can exhibit many features of quantum behaviour.[8]

A 2014 review of the Orch OR theory though highlighted how the recent discovery of quantum vibrations in microtubules suggests that brain waves could derive from these deep quantum vibrations and cites Roderick Eckenhoff who first observed that anaesthesia selectively reduces consciousness via the micro-tubules. Anirban Bandyopadhyay from MIT has even discovered warm temperature vibrations in the microtubules of our brain. Microtubule quantum vibrations (e.g., in megahertz) appear to interfere and produce much slower EEG beat frequencies. Other studies have also evidenced warm quantum coherence in plant photosynthesis, bird brain navigation, and our sense of smell. The Orch OR model postulates that the collapse of the wave function invokes gravity, which should according to the laws of quantum physics spontaneously emit a measurable amount of radiation. In an experiment conducted deep underground in 2020 by the Foundational Questions Institute, FQXi, and led by physicist Catalina Curceanu, at INFN in Frascati, Italy they failed to detect the emission of radiation that might prove the theory however, Penrose's original theory did not invoke gravity, but that goes beyond the scope of this book.[9] A think tank called the Foundational Questions Institute' ('FQXI') suggested that the theory had been refuted. However, one of the investigators and author of the FQXI-sponsored research and articles, Hungarian physicist Lajos Diósi ('D-OR') had produced a similar version of OR called D-OR and subsequent analysis has robustly rebutted this apparent attempt to smear the theory.

Despite all these various discoveries and scepticism, the controversial Orch OR theory of consciousness remains stubbornly alive. Hameroff even robustly defended it against accusations that it fails to address various paradigms such as conscious versus unconscious activity, maintaining that quantum computations that don't reach conscious awareness still may influence subconscious processes. Another apparent constraint is that microtubule quantum states are very fleeting and only persist for around 0.1 msec and it is highly unlikely that a highly fragile small network of neurons could sustain Orch for a sufficient length of time estimated at 500 msec to reach the threshold necessary to maintain a low content conscious moment. If Penrose and Hameroff are correct, then we may have significant challenges or potential opportunities with building conscious robots involving quantum mechanical features that allow the system to access the quantum fluctuation of the universe. Perhaps super-intelligent robots could one day build this quantum feature themselves?

Dr Christian Kerskens, lead physicist at the Trinity College Institute of Neuroscience (TCIN), and David Perez have asserted that quantum processes are part of cognitive and conscious brain functions. Kerskens describes how they used a novel technique whereby they take a known quantum system, which interacts with entangles (see below) with an unknown system which means that the unknown must also be a quantum system. To achieve this, they measured the proton spins (see below) of brain fluid as the known system and used a specific MRI designed to entangle spins. They found MRI signals resembling heartbeat evoked potentials, a form of EEG signals that we discussed earlier and not normally detectable with MRI, and maintained that they could only observe them because the nuclear proton spins in the brain were entangled. Keersen observes that "the brain processes must have interacted with the nuclear spins, mediating the entanglement between the nuclear spins. As a result, we can deduce that those brain functions must be quantum." He observed how "Because these brain functions were also correlated to short-term memory performance and conscious awareness, those quantum processes are likely an important part of our cognitive and conscious brain function" and "Quantum brain processes could explain why we can still outperform supercomputers when it comes to unforeseen circumstances, decision making, or learning something new.[10]

American scientist Robert Lanza (1956–) known for his research on cloning and co-authors, theoretical physicist Matej Pavsic (1946–) and astronomer Bob Berman (1946–)[11] explore consciousness through the lens of Biocentrism. Biocentrism postulates that life creates reality and that time, space, and reality depend on us, and in essence, there is only a single consciousness, and that all life and therefore all reality arises within that. This is not difficult to reconcile with the notion that the universe only exists in terms of space and time because the universe according to the laws of thermodynamics contains all energy that is conserved. It therefore contains all the information, although entropy as we have seen inexorably increases uncertainty and reduces the information available to us until the point of the cold death of the universe when all that information is lost. Consciousness involves the ability to be aware of and attribute meaning to and understanding of the information that can be extracted from the universe, of which we are an embodied part and if the laws of the universe and information are governed by quantum dynamics, then we must rely on those laws in some form or other to generate that information. This appears to justify Hameroff's assertion that quantum computations that don't reach conscious awareness may still influence subconscious processes. Lanza and Pavsic describe how "modulation of ion dynamics at the quantum level allows all parts of the information system that we associate with consciousness with the unitary "me" feeling—to be simultaneously interconnected". Hameroff argues that "An object, a machine, a computer cannot have a unitary sense experience, or consciousness before the mind (natural or artificial) constructs a spatial-temporal reality." He observes that "eventually we will understand these algorithms well enough to create thinking machines and enhancements to ourselves that will change the world in ways we cannot even fathom."

Stuart Kauffman (1939–) and Andrea Roli ask the question 'What is consciousness?' by reference to how organisms find their way in their world. They observe that for an organism, finding one's way involves finding possible uses of features of the world that might be beneficial or avoiding those that might be harmful, which they call affordances, and is consistent with needing a spatial-temporal reality. They concluded, which might not surprise Roger Penrose that "(i) Artificial general intelligence is not possible (ii) The brain-mind is not purely classical (iii) Brain-mind must be partly quantum and that (iv) Qualia are experienced and arise with our collapse of the wave function."[12]

Given these various perspectives on quantum dynamics and consciousness and its relation to the wider universe we should be aware of the argument forwarded by George Zarkadakis, who observes that if quantum consciousness is true then the current premise of AI is false and "there cannot be truly intelligent machines for it is impossible to code the mind that exists at the ultimate causal level of reality as it would mean that we can reprogram the universe and therefore ourselves and that would imply that we exist outside our universe which we do not and cannot."[13] Do humans derive consciousness from the universe and simply reflect that wider consciousness or is consciousness a feature of the biological system and MBBN that we consist of? We simply do not know at this stage and must continue to work with the classical laws of mechanics in constructing robots whilst recognizing that quantum computing as we will see later may potentially transform the field of AI and robotics in ways that are difficult to envisage.

Anglo-Irish scientist, writer and professor of molecular genetics Johnjoe McFadden (1956–) recently postulated a new theory of consciousness predicated on the distinction between matter and energy, thus preserving the dualistic nature of mind and matter. McFadden's theory relies on the well-established phenomenon, that neurons sending electric signals down axons to communicate simultaneously send a pulse of electromagnetic energy into the surrounding tissue, creating an Electromagnetic Field (EMF) that is routinely detected by EEG scans. He argues that this EMF is information-rich and the seat of consciousness, volition, and free will, hence his theory is called the Conscious Electromagnetic Information (CEMI) Field theory.[14] McFadden acknowledges that this has yet to be proved but cites the disturbance in the EMF which is generated by synchronous firing amounts to conscious awareness as evidence to support CEMI. He observes how the EM field influences neuronal firing by triggering those neurons that are on the brink of firing which is supported by others and the observation that extracellular fields feed back onto the electric potential across the neuronal membrane via ephaptic coupling, independent of synapses.[15] Ephaptic coupling involves transmission between adjacent axons rather than across the synapse. It spreads impulses along and across adjacent axons with the action potential propagating along one axon firing up an adjacent axon. From the Greek word "ephapse" signifying the act of touching (as opposed to "synapse" or linking) an ephaptic interaction describes communication between neuronal cells via electrical conduction through the surrounding extracellular space, not the synaptic junction. A 2010 study concluded that "Modulation of network activity by positive and negative feedback fields based on the network activity in real-time provides direct evidence for a feedback loop between neuronal activity and endogenous EMF. This significant susceptibility of active networks to EMFs that only cause small changes in membrane potential in individual neurons suggests that endogenous EMFs could guide neo-cortical network activity."[16] There is much more research to be done in the role of the EMF in generating consciousness but this and the many competing theories reflect how little we know about the interface between the conscious and unconscious.

An influential hypothesis for consciousness called the Global Neuronal Workspace (GNW) which addresses the difference between basic awareness and perception, was proposed in 1988 by theoretical neurobiologist Bernard Baars (1946–). GNW theory maintains that consciousness content and perception occur when the sensory activation of a non-linear neural network is amplified by recurrent processing, which sustains a neural representation and allows the information contained in that representation, to be broadcast and globally accessed across the brain by unconscious local neural processors.[17] This information is held fleetingly in short-term working memory, which makes us consciously aware of it. The GNW model reflects a fundamental feature of consciousness, which is that the brain processes information in a way that results in the activation and engagement of a wide variety of unconscious processes in the brain and involves a neural representation of working memory. This makes intuitive sense in that when you think about mulling on a thought, we must hold a neural representation of that thought in our working memory to make it conscious to us to allow us to manipulate that information and cognitively act upon it. This neural representation may be what Caruthers described as the inference of the underlying working processes of the brain in that they are amodal and merely perceptual representations of whole things some of which are occluded and not accessible to our senses and abstract events not related to sensory experiences.

Another highly influential model for how our brains consciously perceive the world is provided by the Predictive Processing Theory (PPT) because it states that our brains are predictive machines, constantly seeking to minimize the difference between predictions and sensory signals. Error prediction is achieved by the brain using top-down predictions based on prior knowledge to modulate bottom-up sensory input with prediction errors fed through unconscious and conscious processing. Much AI research is focused on developing deep brain networks by harnessing the predictive model at the level of meta-cognition and awareness and higher-order cognition. However recent research observed that studies on the GNW had revealed findings at odds with the empirical predictions of the predictive model and difficulty reconciling it with the PPT. This is problematic as to have a unified theory of mind requires building this predictive process into a theory of mind such as the GNW. By employing a neuro-computational model using Active Interference which uses neural dynamics as approximating to Bayesian inference scientists have claimed to reconcile the GNW model and the predictive model to formulate a predictive GNW (PGNW).[18] One major remaining limitation is that the predictive model does not explain conscious states and there remains a lot of research to justify its claim to be a coherent theory of mind that encompasses both unconscious and conscious processing and conscious states. It also needs to be reconciled with SMM in AI which as we saw has been proven to share core assumptions related to the structure of the human brain, in terms of memory, learning, perception, and motor functions that govern the architecture of the human mind. It may well be that the conscious mind based on these assumptions of the structure of the human brain relates as we will shortly to the existence of and requirement for a self as the primary basis of consciousness. One of the main challenges is not just understanding where and how the brain gives rise to consciousness, but the critical role of the human body in creating consciousness.

The Body and Consciousness.

Because the early years of AI were focused on intelligent machines based on computational capacity, the notion of embodied intelligence in terms of the interaction between the robot, its body, and the environment was largely neglected. In recent years attempts to create cognition in AI have moved towards reflecting our understanding of how cognition emerges from the totality of human biology rather than just the brain as a computational device and the complex integration of the various biological systems that connect the body, the brain and mind and govern our interaction with the environment and world at large.

Rolf Pfeifer (1947–) and Josh Bongard (1974–) are two of a growing number of scientists who believe that the body and the CNS cannot be separated from cognition.[19] They argue that the body shapes the way we think and that the embodied nature of our brains and minds should define a new way of looking at intelligence and AI. Our ever-evolving understanding of the relationship between basic human survival mechanisms and emotions and feelings has reinforced our appreciation of the role of the body, not just the brain, in crafting the self. Antonio Damasio (1944–) is a Portuguese American neuroscientist, whose pioneering work has had a major influence on our understanding of the neural systems which underlie memory, language, and consciousness. Damasio has been a vocal proponent of how emotions play a vital role in higher cognition. His Embodied Consciousness Theory postulates that the mind cannot be regarded as separate from the body and that the body's interaction with the environment makes it an agent in cognition (thinking) and the creation of the sense of self. Damasio's perspective reflects the school of ecological psychology pioneered by J J Gibson, who refutes the theory of the famous American specialist in linguistics Noam Chomsky (1928–) that only the brain acts as a computational device that processes the impoverished information that comes through our senses, is required to construct a coherent output, including the sense of self. Ecological psychology, however, maintains that we are not just computational processors of impoverished information but are as professor of psychology Harry Heft observes, "Individuals who can never know the world independently of their own experience through the whole body not just the brain or mind."[20] Harvard neuroscientist, Professor Sarah Garfinkel focuses on how the processing of emotions is altered in a range of clinical conditions including anxiety, ASD, post-traumatic stress disorder (PTSD), and psychosis. Garfinkel has studied traumatized war veterans and argues that "identifying with and owning a body is central to being a conscious self." Anil Kumar Seth (1972–) a British professor of cognitive and computational neuroscience argues that humans are in essence machines because we are always predicting the likely outcome of every interaction, but our perception of self and the world, that enables such predictions comes through our bodies not just our minds.

Recent evidence of the brain in our gut, which is hidden in the walls of our digestive system and known as the Enteric Nervous System (ENS) appears to further elevate the role of the body in consciousness. The ENS is not confined to humans and is found throughout the animal kingdom, including insects, snails, and marine polyps. The walls of our guts are lined with between 100–500 million neurons, the largest number in the whole body, and appear to operate similarly to our brains. The gastrointestinal tract is connected to the brain, through the gut-brain axis linking our body states with the processing of emotions and thinking in our brain. Although no one has suggested that our guts are capable of thought, it is acknowledged that our gut communicates reciprocally with our brain through back-and-forth electrical and chemical signalling. This influences emotional moods, motivation, and mental well-being, regulating stress, and dysfunction, and has even been linked to anxiety, PTSD, and depression rather than the other way around. A recent study found three communication channels between the brain, the gut, and internal organs such as the liver, heart, and lungs, which are bidirectional with multiple feedback loops, and likely involve interactions between different channels Disturbance of the gut-brain axis are also associated with a range of disorders including inflammatory gastrointestinal disorders, obesity and eating disorders.[21] Research is constantly revealing surprising links between brain activity and the body, for example, the connection between memory and metabolism. It has been known that the hippocampus, which we will discuss later, is involved in cognition and endocrine functions, but Tingley and colleagues studying mice found that it was also involved in monitoring glucose levels in the body. Tingley observed "clusters of sharp wave-ripples recorded from the mouse hippocampus reliably predicting a decrease in peripheral glucose concentrations within about 10 min."[22]

The vagus nerve is one of the largest nerves in the human body, and because it tracks through the body is sometimes called the wanderer nerve, conveying signals back and forth between the brain and gut. It is involved in the immune system through the production of cytokines which are proteins that function as chemical messengers in your immune system. The vagal tone reflects how well the vagus nerve is working, and is measured indirectly, using heart rate variability which is the time between heartbeats (HRT) and relates to your ability to return to rest after exercise so reflects fitness levels and the higher the HRT the better your vagal tone. Mayer observes how alterations in gut-brain interactions have been established in irritable bowel syndrome but also neurological conditions like epilepsy Parkinson's and cardiovascular disease and that reduced vagal tone is associated with some conditions like Crohn's disease.[23]

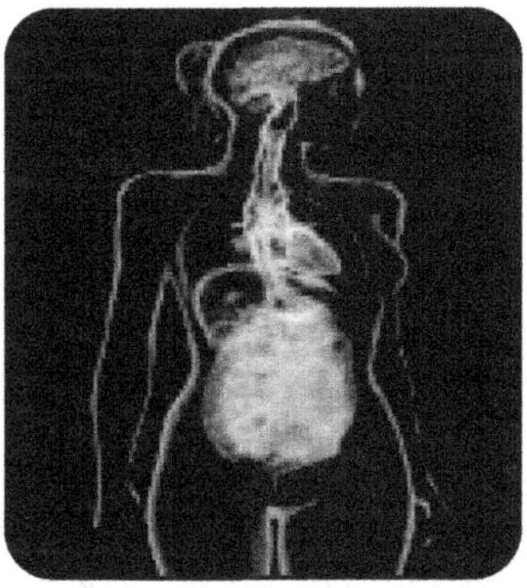

This might explain the therapeutic efficacy of mind-body treatments, such as Cognitive Behavioural Therapy (CBT) with one study showing significant brain changes in patients with irritable bowel syndrome, correlated with changes in the microbiome after undergoing CBT.[24] Vagal tone can also be improved with breathing exercises through breathing exercises, yoga, and meditation.

A 2015 study observed that 90% of serotonin is produced by over 100 trillion intestinal cells and certain gut bacteria respond to serotonin. Although most studies have been conducted on mice, serotonin is known to regulate digestion and is very important in moods, feelings, and happiness with too little serotonin associated with low moods and depression. This is why Serotonin Reuptake Inhibitors (SSRIs) are used as antidepressants to allow there to be more serotonin circulating in the body and when they work people report a decrease in negative thinking or rumination and an improvement in overall functioning. Galvanized, by the study of micro-organisms, recent evidence also suggests an influence on self-consciousness, by the lungs and heart together with the gut, known collectively as the viscera which represents a growing challenge to the traditional notion of the human self as a self-enclosed discrete entity that can be studied in isolation from the environment.[25] Research is revealing how our microbiome has a fundamental role to play in three areas regarded as defining the human self, which are (1) The adaptive immune system that distinguishes the self from the non-self (2) Brain functions associated with mediating self-awareness and (3) The human personality, emotional states, and cognition and genomic influences on phenotypes, particularly via the macrobiotic gut–brain axis.[26] A team at the University of Pennsylvania found that by changing millions of gut bugs in the microbiome of mice you could motivate those mice to become more active.[27] This may not be as easy in humans, but good news for any lazy mice out there though less good news for domestic cats. Shaun Gallagher (1948–) an American philosopher known for his work on embodied cognition, social cognition, agency, and the philosophy of psychopathology describes how the growing field of neuro-psychoanalysis, is a response to this "greater awareness that the body itself, not just the brain, provides the material substrate for the emergent phenomenon we call mind."[28] Neuro-psychoanalysis as a growing field of research, aims to integrate the science of subjective thoughts, feelings, memories, etc, with our neuroscientific knowledge of the objective anatomical brain structure and functions and the body as a whole. It holds great promise for furthering our understanding of the role between the body and the brain in shaping consciousness and the self.

Different Levels of Consciousness.

To operationalize consciousness, scientists have distinguished between three levels called (1) Anoetic (2) Noetic, and (3) Autonoetic consciousness (Noetic-from the Greek "Noetikos"-intellectual, the verb Noein to think and the noun Nous for the mind.) Damasio hypothesized three levels of self which he describes as the (1) Proto self, (2) the Core self, and (3) the Autobiographical self. He acknowledges that the neural patterns formed at the various levels of self, don't constitute the mind in its entirety, but admits to often conflating them to avoid being seen to perpetuate the dualistic distinction between the brain and the mind. Damasio's hypothesis of how the self emerges, refines the noetic distinction with its three levels, by further elaborating the role of emotions and feelings in the different levels of consciousness, which he describes as "adding self to the mind mix." He describes the proto-self as constituting "the most basic form of awareness, which acts as a reference point for the conscious self and "comprises a coherent collection of neural patterns or neural maps, which map moment-by-moment the state of the physical structure of the organism" with the neural patterns of the proto-self, generated in the upper brain stem. Damasio believes that emotions are involved at all levels through the dynamic connection between brain regions, not as was once thought processed hierarchically and wants to centralize emotions and feelings into the overall discussion on consciousness and self. Damasio's core self involves the brain processing sensorimotor information about the relationship between the body and objects in the world and generates neural patterns and images that describe the object engaging with the proto-self, which forms our semantic knowledge and memory. Damasio's autobiographical self consists of the biographical knowledge about the past and the future, so links the present to the past which is encoded in neurons and stored in our long-term episodic and other types of memories in the cortex which we will discuss when we look at human memory. Damasio describes these as "Multiple images whose ensemble defines a biography generate pulses of core self whose aggregate constitutes an autobiographical self".[29] When he refers to multiple images, he is referring to the multiple neural patterns generated in the cortex and sub-cortical related areas, like the hippocampus where short-term memories are consolidated for transfer into long-term memory storage in the cortex, which are generated, over time so they form our stored experiences and memories essential to the autobiographical self. Damasio believes that without the sense of self, represented by these neural patterns at the pro, the core, and the autobiographical levels of consciousness, we would not possess consciousness.

This perspective is shared by German philosopher, neuroscientist, and psychiatrist George Northoff, whose work has focused on why and how our brains can construct subjective phenomena like self, emotions, and consciousness. Northoff maintains that "we claim that there is some kind of self that performs the different kinds of sensory, motor, social, affective, or cognitive functions, or tasks. Even more puzzlingly, we claim such a self to have consciousness. This means that there must be some kind of self at the very bottom of the consciousness, that first and foremost makes possible both the occurrence of consciousness and the performance of the various functions and tasks." Northoff argues that we can only access consciousness through self and that this requires the proto, core and autobiographical self. "We experience our self, a sense of self in consciousness which amounts to what is described as self-consciousness in philosophy. If we lose consciousness, we lose our experience of self." Northoff observes that the entire world consists of waves of time, and it is these neuro-waves that pass through our brains that produce the mental experience.[30] The obvious question this raises is whether the waves passing through the robot brain could ever produce mental experiences or do you require a human brain to receive these neuro-waves. The challenge will be to determine whether our robots could ever have a "kind of self at the very bottom of the consciousness hierarchy, which first and foremost makes possible both the occurrence of consciousness and the performance of the various functions and tasks that require them to access the consciousness content about the self."

Anoetic Consciousness

Anoetic consciousness arises from sub-cortical neural activity and the information being processed appears to involve a more limited level of connectivity and integration within the brain than the higher cortical levels of consciousness. The steady state of the body's internal, physical, and chemical conditions maintained by all living systems is known as Homoeostasis, and our rudimentary awareness of the state of homoeostasis in our bodies is called Interoception. This does not involve intellectual or cognitive processes or explicit thinking or knowledge of the world but awareness of the internal conditions of the body. The vagus nerve through the gut-brain axis links the ENS and homoeostasis with the CNS in terms of the thinking and emotions processed in the brain. Information from the body processed in the brain is first processed in sub-cortical regions and then cortical regions. One cortical region, that plays a vital role is the recently discovered 5th lobe of the brain, called the Insula, which is heavily involved in interpreting, integrating, and regulating interoceptive information. Interoception as a pre-reflective state, should not be confused with introspection which involves reflecting on, observing, and noticing your own mental and emotional processes and requires a higher level of consciousness. Processing of homoeostatic information involves the generation of neural patterns in sub-cortical brain structures, particularly the brain stem.

It is achieved primarily by an automatic negative/error feedback loop in the endocrine nervous system (ENS) and the CNS, with endocrine glands releasing a wide range of regulatory hormones, into the bloodstream. As we saw, these modulate the expression of proteins in nerve cells which influence behaviour at many levels and help control our moods, growth and development, the way our organs work, metabolism, and reproduction. Homoeostasis automatically changes the internal conditions for the body to survive external challenges and involves a complex interaction of multiple feedback systems modified by higher control centres in the CNS.

The thalamus involves extensive communication between brain regions and plays a key role in our level of consciousness working closely with the hypothalamus which consists of two small almond-shaped regions of several small nuclei, with a variety of functions linked to the ENS by the pituitary gland. The pituitary gland is often referred to as the master gland in the body because it instructs other glands to release hormones which helps maintain homoeostasis and regulates various physiological processes, including body temperature, hunger, thirst, and other vital functions such as consciousness and sleep cycles. The hypothalamus plays a critical role in relaying information between different brain regions and processing sensory signals that regulate many different aspects of human physiology, such as hormone release and regulation of appetite, thirst, energy balance, body temperature, circadian rhythms, and sleep cycles, reproduction, sex drive, and many other basic functions. The hypothalamus is implicated in the feelings of lust and elevated hormone levels including testosterone. They contain specialized neural circuits, that facilitate communication between different brain regions and control key aspects of human behaviour and physiology.

Two other vital regions of the mid-brain are the Tegmentum (Latin for covering) also called the Ventral Tegmental Area (VTA) and the Tectum (Latin for roof) which sits just below the thalamus. The VTA lies in front of the Tectum, whilst the Tectum is the dorsal or back part of the mid-brain. The VTA is as it is part of the reward system for motivation and drive and vital in generating the excitatory neurotransmitter dopamine. The VTA consists of fibre tracts and three regions distinguished by their red, grey, and black colour, which are the (1) Red nucleus which is the (1) Red nucleus which is a pair of oval-shaped clusters of nuclei that appear pink thought to be due to the presence of iron, (2) Periaqueductal Grey nucleus (PAG) and (3) Substantia Nigra nucleus (SN) (nigra = black) which are also key generators of dopamine. Damage to the VTA can cause coma and the VTA has even been implicated in the act of falling in love. Imaging studies of the brains of people who have recently fallen in love with someone have displayed high activity in the VTA and Nucleus Accumbens with the region flooded with dopamine which drives us toward an expected reward. Other chemicals such as cortisol, phenylephrine (found in chocolate), and adrenaline are also elevated whilst serotonin appears to be reduced as it often is found to be in those with OCD. The PAG also known as the analgesia centre is key to the propagation and modulation of pain and is involved in sympathetic responses and learning and the action of defensive or aversive behaviours. In adult humans, the Tectum consists of the inferior and the layered Superior Colliculus, with the superficial layers being sensory-related and receiving input from the eyes as well as other sensory systems, whilst the deeper layers are motor-related, and capable of activating eye movements and other responses and intermediate layers, with multi-sensory cells and motor properties. The superior colliculus is involved in preliminary visual processing and control of eye movements and the inferior colliculus is in auditory processing. This layered neuronal structure of cells, which like the cortex, enables the generation of neural patterns, in the form of topographical maps of auditory and visual stimuli.

Damasio describes interoception, as the way our "conscious mind makes this basic life management knowable " and the most primary form of self-awareness. Louis Renoult observes how interoception, reflects the "biological value of homoeostasis and sensory perception of these internal states, which guide an organism's, basic survival, motivations, and adaptive behaviours in response to information it receives from the environment."[31] Although interoception does not involve higher cognition thinking or intelligence it involves affective (emotional) experiential states and raw sensory and perceptual mental sensations. Affective means "relating to, arising from, or influencing feelings or emotions".[32] Marie Vandekerckhove and Jaak Panksepp argue that we need to bring these affective experiential states "more centrally into the overall discussion on consciousness" because "these vastly underestimated primary forms of consciousness may be foundational for all forms of higher knowing consciousness". Vandekerckhove M, Panksepp and Damasio, and other scientists advocate that we look at consciousness and the self from the perspective that they are supported and justified by this primary process of awareness of life regulation and observe critically how "current scientific approaches have been more concerned with understanding the higher declarative states of consciousness."[33]

American anatomist, physiologist, and neuroscientist Peter Sterling(1940–) observed that homoeostasis relies on "feedback and error correction, whilst allostasis involves predictive error prevention and the "continual adjustment of the milieu to promote survival and reproduction."[34] Reproduction which involves social collaboration requires efficient regulation by anticipating needs and preparing to satisfy them before they arise and allostasis reflects the fact reproduction and survival as basic goals that drive the efficiency of the body's adaptive change. Allostasis assigns a central role to the sub-cortical structures of the brain above the brain-stem, particularly the amygdala in regulating the internal milieu by governing anticipatory behaviour. Adaption of the internal state though is driven by the integrative function of the brain in a process of top-down anticipatory or predictive regulation. Allostasis reflects the human brain's capacity to communicate and enhances survival by integrating complex information to control behaviour in response to predicted and anticipated changes rather than simply reacting. Because allostasis is anticipatory, it is often described as more labile in being liable to change or easily altered and from a psychological perspective is characterized by easily aroused or expressed emotions. This can create emotional instability with high energy consumption and potential allostatic overload or what we know as chronic stress and anxiety. Allostasis plays an important role in our reward system, which is critical to our motivation and drive, and our levels of anxiety, which involves several brain regions. Prediction errors fed through unconscious and conscious processing in the predictive processing minimize the difference between predictions and sensory signals. Prediction and probability calculations through allostasis and other brain mechanisms anticipate and deal with uncertainty.

Noetic Consciousness.

Endel Tulving (1927–) described noetic consciousness as "a state of absolute consciousness, in that an individual is aware of facts, concepts, words, and meanings, however, they are not aware of any connection to their own experiences."[35] Neuroscientists Marie Vandekerckhove and Jaak Panksepp describe how the noetic process arises, when "attention within higher brain regions is intensified and reflection starts with the clear stream, of sensorial and perceptual representations of objects in the world and events related to them. Thereby, specific aspects of events…" They observe that "we still do not know exactly when children begin to retrieve information on a semantic level or become noetically aware."[36] There appears to be a phase change between the unconscious and conscious processing of stimuli (information) entering the brain, with scientists observing a substantial increase in the neuron activity necessary to construct visual imagery. In the first 200–300ms, processing of sensory stimuli is unconscious, but after 400ms the processing of the sensory information becomes entirely conscious, with intense neural activity in the front of the brain, the temporal lobe, and the Cingulate cortex in the middle of the brain. Information processing is shifted through sensory and working memory into short-term through long-term memory and becoming conscious of the information appears to start with it being processed in the working memory in the front of the brain. The frontal cortex is associated with higher level noetic consciousness which involves the conscious processing and recollection of sensory information in brain regions above the brain-stem, of knowledge related to what, where, and when episodes or events happened, and awareness about the world, which includes facts, concepts, and ideas. Called semantic knowledge, this involves mental representations but does not feature specific representations of the self, recollection of life experiences, or the capacity to recursively introspect on one's own subjective experience. These are the domains of autonoetic or self-consciousness. The level of conscious shift is dependent upon the level at which raw data is processed to extract information and the level of integration and connectivity between functional brain regions, that construct a representation out of that information, which can then be manipulated and subject to higher cognitive processes.

Semantic knowledge is grounded in the sensorimotor interactions with the environment, which generates the raw data of homoeostasis and allostasis that create anoetic consciousness. American psychologist and cognitive scientist Lawrence Barsalou (1956–) believes that "viewing knowledge as grounded in sensory-motor areas, changes how we think about basic cognitive processes, including categorization, concepts, attention, working memory, long-term memory, language, problem-solving, decision making, skill reasoning, and formal symbol manipulation."[37] Studying the relationship between the activation of sensorimotor information based on the language we use and the cognitive processing of that language by examining the effects of the body /object interaction, led to the development of the Body Object Interaction (BOI) metric. BOI measures the level of ease with which a person can relate to words based on the sensorimotor information they convey.[38] One study used 1,618 monosyllabic nouns to determine the level of visual word recognition and response time, with higher-level BOI words like couch compared to lower-level BOI words like cliff. The sensory-motor information activated for the couch is higher, and the response time is quicker, probably because most of us spend more time on couches than on cliffs. This suggests that semantic knowledge is grounded in sensorimotor interactions with the environment, which accords with Barsalou's Perceptual Symbol Systems theory.[39, 40]

Semantic memory was traditionally regarded as amodal, which reflects how even when only parts of an object may be not visible, we use previous experiences to fill in the gaps to know what it is. However, in the 1980s, Lakoff and Johnson observed that the human conceptual system which contains semantic knowledge is fundamentally metaphorical rather than just amodal, with three types of metaphorical concepts, which are (1) Orientational (2) Ontological and (3) Structural.[41] Language is shaped predominantly by structural conceptual metaphors and is not just a rhetorical device but allows people to use simple and basic experiences to conceptualize abstract concepts, for example conceptualizing debating as a war with war-like terms employed such as demolishing or attacking an opponent's arguments. Orientational metaphors reflect the relation and interaction of objects etc in space like up-down, inside-out, front-behind, shallow-deep, etc. Ontological metaphors involve abstractions like emotions, or ideas, being represented or viewed as entities like objects, substances, containers, or people, for example, the mind as a machine or computer. Structural metaphors involve complex abstract concepts being presented in terms of typically more concrete concepts, such as describing labour and time as a resource. We rely on abstract concepts and seek to define them metaphorically by reference to more concrete concepts that are more clearly structured. Conceptual knowledge, according to Barsalou, contains components of our experience including "knowledge about settings, objects, people, actions, events, mental states, properties, and relations." Barsalou claims that "Componential knowledge in the conceptual system supports a wide variety of basic cognitive operations, including categorization, inference, the representation of propositions, and the productive creation of novel conceptualizations. In turn, these basic operations support the spectrum of complex cognitive activities, including high-level perception, attention, memory, language, thought, and socio-cultural cognition."[42]

Our brains construct knowledge about our environment through our sensorimotor interactions with the environment using amodal and metaphorical processes. The core self involves the brain processing this sensorimotor information about the relationship between the body and objects in the world, which generates neural patterns and images that describe the proto-self or anoetic consciousness, engaging with objects that form our semantic knowledge and memories. Psychologists acknowledge the extent to which cognition is grounded in our sensorimotor systems and that it plays a vital role in how we learn, represent, and retrieve information about our world.

Autonoetic Consciousness.

Autonoetic consciousness involves a higher reflective mental process, that permits a person to recursively introspect on one's own subjective experience through time and to perceive the continuity in one's identity from the past to the present and the future. It involves the processing of sensory stimuli in sub-cortical brain regions, particularly the Amygdala and Cingulate Cortex, and in the Orbitofrontal Cortical region, which is a region of the brain that neuroscientists think is very important in processing the self. It involves the evaluation of whether the information from the environment is pleasant (a reward) or unpleasant (aversive) and activation of previous experiences encoded in memories across the brain which help to determine whether the stimuli should be acted upon in terms of a person's motivation to approach or withdraw from it. This accords with Stuart Kauffman and Andrea Roli's earlier description of affordances in terms of the decision to act and select options in response to sensory information.

Wilhelm Wundt (1832–1920) often described as the father of experimental psychology, used reflective introspection in his efforts to identify and quantify the basic components of human consciousness and it remains a common practice in journaling and mindfulness therapies. Tulving in his early work on human memory, was the first to distinguish between episodic and semantic memory when he described episodic memory as part of long-term explicit memory, comprising a person's unique recollection of experiences, events, and situations. Episodic memories include emotions associated with the event and conscious thoughts and are also called declarative because they involve the storage of personally experienced facts and events, and recall involves some degree of conscious effort, in that this information is being declared. Autonoetic consciousness requires episodic memory which typically develops in children at around three to five years of age when they start to remember explicit events as having been personally experienced.[43] Intensification of attention within the cortex plays a critical role in constructing our experiential and autobiographical self. It equates to being able to understand our world over time through the lens of our highly subjective sense of self because those episodic memories are by their very nature subjective. For robots to experience autonoetic consciousness appears to require the capacity for long-term declarative, episodic memory which would constitute the robot's unique recollection of its personal experiences, events, and situations, which integrate all the critical psychological domains of action, emotions, and feelings, personality in terms of dispositions, learning and memory. Although all AI needs memory to operate, learn, and adapt it currently does not possess episodic memory or the capacity to integrate these various domain equivalents.

Human consciousness involves various levels of attention and awareness of the internal bodily states, rising through awareness and knowledge of objects and things, through to higher reflective consciousness which requires episodic memory which constitutes the subjective feeling of relating that information to yourself. Consciousness and self don't just sit above us but appear to be embedded in and emerge from our biology and most critically the adaptive functions of our brains, which are continually operating in response to our interaction with the environment and others. This adaptive function involves the constant processing and integration of sensory information at various levels and potentially the encoding and long-term storage in the cortex as prior knowledge is available for later retrieval. It contains an encoded self that allows humans to experience the I self and the remembered self and involves emotions at various levels.

The Grounded Enactive Predictive Experience (GePe) model has sought to establish some core principles behind the conscious experience in terms of the content of consciousness and the phenomenological consciousness and qualia that we have discussed. GePe asserts that the brain is a control system that maintains a metastable balance between the body and the brain's internal world of the body and the environment through action and behaviour. The core principles which underlie the subjective conscious content of the conscious experience of knowing have been as set out by Paul Verschure whom we met earlier as follows as the basis of a robot-based neural architecture which is "(1) Grounded in the experiencing physically- and socially instantiated self (2) Co-defined in the sensorimotor coupling of the agent to the world. (3) Maintained in the coherence between sensorimotor predictions of the agent and the dynamics of the interaction with the world (4) Combines high levels of differentiation (each conscious scene is unique) with high levels of integration and (5) Depends on highly parallel, distributed implicit factors with metastable, continuous, and unified explicit factors".Verschure separates the process of consciousness from its content (quale) and to gain direct third-person access to the entirely subjective first-person experience of qualia purports to synthesize consciousness.[44]

This involves Verschure's Distributed Adaptive Control(DAC) theory of the mind and brain developed over twenty years which he describes as "a theory of the design principles underlying the MBBN involving "this functional hypothesis being mapped onto the brain-stem, mid-brain, and the thalami-cortical and cortico-cortical systems and analysed concerning our understanding of deficits of consciousness."[45] He explains that if we regard consciousness as a unique feature of living systems and the product of biology rather than a property of matter, then we must consider its function and features in terms of survival and fitness. Verschure argues that "to act or know-how, the brain must answer four fundamental questions: Why, What, Where, When. Thus, the function of the brain is to continuously solve the so-called, H4W problem." The content of consciousness and the qualia that give rise to the subjective experience must be understood as an emergent property of biology not just matter, which reflects the various dynamic levels of consciousness that we have just discussed. Given the observations of self as the basis of the three levels of consciousness then self and consciousness are both an emergent product of our biology as a dynamic system. This characterizes not only animal sentience but distinguishes the phenomenological content of consciousness from the autobiographical sense of self over time.

If the content of consciousness emerges from biology, then it is the functional integration and differentiation of information in the biological system that allows that system to go beyond merely following the OODA loop of doing. The OODA loop, borrowed from US military intelligence involves Observing, Orientating, Deciding, and Acting upon information in the environment but does not involve or require knowing in terms of conscious understanding. The complexity of our brain generated by this functional integration and differentiation of information in the biological system should not though be treated as simply the same as the level of brain activity. As we have seen with epilepsy there is a heightened and almost uncontrolled level of brain activity, but this does not constitute consciousness it disrupts the conscious experience and when we go to sleep consciousness is slowly lost but significant brain activity continues.

Another major theory of consciousness, the Global Workspace Theory (GWT) which we discussed earlier, started in the 1980s as a psychological theory of consciousness but has evolved into the Global Workspace Dynamics (GWD) theory with a family of related GW theories, in which consciousness is viewed as arising from highly integrated and widespread cortico-thalamic (C-T) activity supported by both pre-frontal and posterior regions of the brain. Neural information rises into consciousness when it appears in the GW and is broadcast through our brain. GWD has been sufficiently influential as a theory of consciousness that it has formed the basis of attempts to create real-life computational equivalents of our brain based on a GWD simple cognitive architecture to match conscious and unconscious processes with obvious implications for AI and robots. It involves momentarily active, subjectively experienced events that resemble working memory in humans and provides the opportunity to manipulate that information as a basis upon which to make decisions and act upon it. Some people use the theatre metaphor in which a spotlight fleetingly reveals to us the contents of our consciousness. These are then broadcast to a diverse range of subconscious processes which operating in parallel with some communication producing further input for the global workspace. Information competes for this limited workspace which is known to rapidly decay with the information being lost to our conscious awareness. Even when Baar formulated the GW theory, he recognized that by itself it is insufficient to generate higher consciousness which needs to interact with the self which suggests that for robots to have higher consciousness they will require a self-concept.

We saw earlier how the French neuroscientist Stanislas Dehaene's neuronal avalanche concept seeks to extend the GW theory to general consciousness to explain how from all the competing information, what gets selected to be broadcast throughout the cortex and gets subconsciously processed. This reflects how neural networks in distinct brain regions integrate information over space and time and converge onto a single coherent interpretation. Dehaene identified the role of neural avalanches in terms of cascades or bursts of activity in neuronal networks. Studies have demonstrated that these avalanches are the manifestation of the different non-linear aspects of the brain wave processes abundant in the human cortex as we discussed earlier. These wave processes encompass both synchronized brain wave activity and seemingly incoherent spiking activity and are an important part of the generation and processing of information in the brain.[46]

How can we tell if humans are conscious?

Reflexes.

To comprehend how we can tell when a person is conscious, we need to first look at the reflexive nonconscious system and how it processes stimuli which will allow us to compare it to the various levels of consciousness. This is not just a theoretical exercise, but vital when clinicians are deciding whether a patient's responses are purely reflexive or if there are signs of volitional consciousness and determine treatment options including withdrawal of life support. Earlier we looked at the reflexive pathway which controls reflexive responses that are "responses to stimuli that are involuntary or free from conscious control and involve a reaction to stimuli without cognition of things that are typically, potentially harmful to us."[47] Our awareness and perception of the outside world come from our senses, but reflexes are responsible for the action of our muscles and govern the way we move and typically produce an unexpected and automatic reaction to an outside stimulus. For example, ducking if something comes towards you, or pulling your hand away from something hot like a stove. Babies are born with more than a dozen very important reflexes like automatically sucking on a finger which is a sign that they can feed or grasp in by exercising the grasp reflex, and you will see them suck their thumbs as evidenced in in-utero baby scans. The rooting reflex allows them to find the mother's nipple which you will see if you stroke their face, and they automatically move towards your finger. The startle or Moro reflex makes them throw their legs and arms out when startled by a noise or sudden movement or even the sound of themselves crying, though this stops after about two months. They initially smile reflexively and not in response to a social interaction or to something that you do such as tickling or making a silly face. When they turn their face to one side stretch out the arm towards that and bend the other one, they are exercising the tonic neck reflex. Pushing a spoon out with their tongue is the tongue reflex which usually stops about six months at which point they may be ready for solid foods.

There is an approximately half-second delay between the moment that the senses receive a stimulus and our becoming consciously aware of a sensation. This is because reflexive responses are processed in the spinal cord before the brain even becomes aware of the stimuli and even though the brain will probably become aware of the resulting movement, it can't or is too late to control the reflexive response. This may well explain why the reflexive system responds so quickly in less than 1/0th of a second to compensate for the delay in conscious processing although doesn't explain the much faster reflexes seen in other animal species and even some plants that don't as far as we know possess conscious processing mechanisms. Because reflexive movements are not processed by the brain, they are not conscious or intentional responses to stimuli, although they may often appear to be subconscious, which is why a person may think a patient or loved one is responding subconsciously, but they are not, for subconscious behaviour still involves processing by the brain.

Wakefulness and Awareness.

Wakefulness is a prerequisite for consciousness, but wakefulness and the presence of reflexive responses are not sufficient to indicate consciousness because a minimal level of consciousness requires both wakefulness and awareness of sensory stimuli. We measure consciousness by observing behaviour and being reliant largely on subjective self-reports, which in a clinical context may simply be reduced to pushing a button, to establish that a person is consciously responding. The behavioural metrics we use for humans can be applied successfully to many other animal species which means that they appear to exhibit the same level of consciousness as we do. We know that language is not necessary for consciousness so neither language nor behavioural observations can adequately tell us what consciousness is. Whilst in a coma a person shows no signs of being awake or any signs of awareness and perception of stimuli and lies with their eyes closed and not responding to their environment, sounds including voices or pain, because they aren't able to perceive anything. A coma typically lasts for between 2–4 weeks and during this time a person may progress into a vegetative or minimally conscious state or wake up, although following a massive stroke, a Canadian woman Annie Shapiro (1913–2003) remained in a coma for 29 years but suddenly woke up in 1992.

A tragic case, which illustrates the complex nature of consciousness and the challenges facing doctors, involved a 25-year-old woman who sustained serious brain injuries in a car accident in 2005. Adrian Owen, a University of Cambridge neuroscientist, and his team used fMRI to demonstrate preserved conscious awareness in this woman who did not fulfil the diagnostic criteria for a coma but appeared to for a vegetative state, which is not the same as a coma, because a person is awake even though they appear to show no signs of conscious awareness. Such patients may open their eyes, wake up go to sleep at regular intervals, and exhibit basic reflexive responses, which include blinking in response to loud noises. They can automatically regulate their heartbeat and breathing without support, so homoeostasis is maintained, but they don't display meaningful responses or show or express any emotions. What Adrian Owen and his colleagues found was that when the young woman was asked to imagine playing tennis or moving around her home, she activated predicted cortical brain areas, in a manner indistinguishable from that of healthy volunteers.[48] By showing awareness of the stimuli and processing in the brain (the request to play tennis) and the ability to attribute meaning to the request to imagine doing something, she displayed both awareness and perception of the spoken request, so she failed the criteria for both coma and vegetative state. A person may show minimal or inconsistent awareness and will be regarded as in a minimally conscious state, which can happen after a coma or having been in a vegetative state. In most cases, a minimally conscious state isn't usually regarded as permanent unless it lasts for several months with no signs of improvement.

Robots and Consciousness.

All the evidence points to the human body playing a key role in the generation of the various levels of consciousness, but is the self we recognize a product of our mind rather than a product of the physical entity that comprises the human body, brain, and mind and the MBBN? This raises the question of where and how the non-material sense of self and the mind emerges from our material body and brain and the need to address the MBBN challenge with robots. What will be the robot equivalent of the material substrate for the emergent phenomenon we call the mind and what sort of body do autonomous robots need? Given our knowledge of the dynamic, embedded, and emergent nature of human consciousness, how can we tell whether our robots will ever be capable of being genuinely regarded as conscious or ever have the experience of the observing self that appears necessary to maintain the conscious state and the content of consciousness?

Recognizing this challenge, Searle in his debates on AI suggested the following eleven metrics for assessing the state of consciousness in machines. (1) Subjectivity, in that conscious states owe their existence to animal or human experiences) (2) Unity, in that one experiences events as part of one unified conscious field (3) Qualitativeness, in that conscious states are always qualitative (4) A sense of self in that conscious experiences relate to personal.identity (5) Active/passive consciousness (the difference between intentional activity and passive perception (6) Gestalt structure, which means that conscious experiences come as organized structures (7) Mood because conscious states come in a sort of mood or other (8) Attention in that consciousness involves focusing more on the centre than on the periphery) (9) Pleasure/unpleasure because there is always a feeling attached to conscious states (10) Intentionality in that conscious experiences refer to things beyond themselves and (11) Situatedness in that conscious states come with the sense of a background situation.[49]

Unity is the defining concept of consciousness as a field out of which experiences arise and in humans, we understand consciousness by reference to the underlying brain mechanisms and neural correlates. A minimal level of neural activity is required for a conscious experience generated through perception, thought, or memory, measured by reference to both the conscious state and content of consciousness. What defines consciousness is not the qualities of the object or event being perceived, but the experience of the observing self that appears necessary to maintain the conscious state.[50] This seems to be what William James alluded to when he described the I self and accords with Damasio's description of the "multiple images whose ensemble defines a biography generate pulses of core self whose aggregate constitutes an autobiographical self." The multiple images in terms of content are what distinguish the phenomenological content of consciousness from the state of wakeful consciousness. Could robots ever generate these pulses of core self and the autobiographical self that Damasio claims are necessary to allow our robots to not just experience the state of conscious awareness but experience the phenomenological content of consciousness? Let's look first at homoeostasis, which as we saw generates the primary form of consciousness in humans, and whether it has a role to play in robots.

Robots and Homoeostasis.

Interoception reflects the primary awareness of physiological needs in terms of homoeostasis, allostasis and anoetic consciousness and is closely entwined with the raw data of self-related emotions. How can robots that do not currently possess such raw data achieve interoception, consciousness, and a sense of self? AI and DL are harnessing predictive processing which means that robots in theory could rely on the feedback and error correction processes using backprop to make the basic robot "life management knowable as its most primary form of self-awareness." Diana Valenzo and colleagues argue that certain core contextual elements would need to be modelled into AI robots if they are to exhibit behavioural flexibility in a way that simulates human behaviour. They claim that these core contextual elements include self-related factors in humans such as 'Physiological needs, intrinsic motivation, and emotions are elements of the agent-related context that have a strong impact on this level of behavioural organization."[51]

Irina Strigo and Arthur Craig observe that "the neuroanatomical architecture of interoception provides deep insight into the functional organization of all feelings and behaviours in humans and that an AI robot requires a baseline equivalent or primary awareness of its "physiological needs" to be able to function properly. Strigo and Craig maintain that this includes a viability zone that requires "a certain level of energy, integrity in their sensors, and maintenance of an optimum temperature for the proper operation of their motors."[52] We saw earlier that it has been proposed that in ANNs such a baseline equivalent could be achieved via a hormone production controller (HPC) based on the modulatory role of human hormones. This would involve the HPC triggering a specific artificial hormone to activate target neurons based on the internal state which would affect the input weight of those neurons. Based on a feedback mechanism, production of the hormone would, according to Pablo Vargas and colleagues cease upon completion of the task.[53] This has already been employed in a mobile robot using the modulation of artificial hormones to control conflicting behaviours with the robot experiencing low energy navigating in a challenging environment and completing a difficult task and modelled to reflect allostasis.[54] By using various such techniques to capture the robotic version of homoeostasis and allostasis, Vasiliki Vouloutsi and others believe that we may be moving towards potentially equipping robots with a basic anoetic awareness and a level of interoception in being able to monitor their internal state.[55] Although this HPC mechanism appears to provide a very basic state of introspection or self-monitoring it does not appear to allow the robot to experience conscious recollection, subjective awareness of or knowledge of its existence as a distinct entity in the world, experiences related to the self nor higher cognition related to the full consciousness. In sum, it does not by itself seem to imbue the robot with the essential quality of the subjective experience so let's look at what subjectivity may look like for robots.

Robots and Subjectivity.

Given our understanding that consciousness involves intrinsically subjective mind states and how the self relates to consciousness, awareness, and perception, how do we establish whether robots can possess the unified state of consciousness based on the current theories related to human consciousness? What would the content of robot consciousness look like, and would it confer the machines with a sense of self? How can a robot have an intrinsically subjective view of the world about its existence if it can't experience the unity of a conscious field? Most people and many philosophers and scientists would say one can never enjoy the subjective experience if it is not conscious. Julian Kiverstein focuses on the philosophy of cognitive science and neuroscience and cites the Dynamic Sensorimotor (DSM) account of conscious experience. Kilverstein argues that provided certain conditions are met these could confer on a robot the conscious experience. DSM states that the conscious experience involves perceptually exploring the world using one's practical sensorimotor knowledge and comprises three elements involved in the sense experience which are (1) Intentions to move or motor intentions (2) Predictions about the motor and sensory consequences of movement and (3) Actual sensory input, including proprioception. He argues that "It doesn't matter what the creature is made from because it is possession of sensorimotor knowledge that is necessary and sufficient for having a conscious experience and that this brings with it a kind of self-awareness". He further observes that "Subjectivity comes more or less for free in creatures that can acquire and exercise a mastery of sensorimotor knowledge". Kiverstein's argument rests on consciousness consisting of the possession of sensorimotor knowledge as necessary for subjective experience, not the neural activity that implements the sensorimotor knowledge, and that machines can therefore constitute a unified subject. Kilverstein maintains that subjectivity is a point of view that is no more than a 'Location representation in terms of the origin of where they come from and the experience of them as belonging to "me."[56] Sensorimotor knowledge is used to form expectations about the likely effects of our movement, or movement in the world, on our sensory input, and provided it has a mechanism for comparing predictions about sensorimotor dynamics with actual sensory input, it can modify its behaviour. It is the monitoring of the sensorimotor knowledge that constitutes subjectivity. He qualifies his argument by acknowledging that matching the three ingredients of DSM may not be sufficient for full self-awareness and that a machine would have to have its purposes, goals and end reasoning.

This perspective of the human conceptual system and semantic knowledge is part of a wider movement influencing the cognitive sciences and AI, which has moved away from the symbolic theories of semantic knowledge employed in GOFAI and embraced theories of connectionism and simulation/embodied/ situated theories. You will remember that symbolic theories rely on the mind-computer analogy in which all cognition and language processes in the mind can be reduced to symbols and semantics. The problem is that to attribute meaning to anything, requires that all the symbols used by the mind to conceptualize things must be grounded in the real world and we know they aren't. Instead of the mind solely manipulating symbols to represent abstract concepts the simulation/embodied/situated theories propose that the mind manipulates mental representations which involves bringing to bear prior experience and meaning through semantic knowledge. Embodied simulation can be traced back to the 19th century and the German physician Carl Wernicke (1848–1905). It describes how when we use language to describe semantic knowledge about the world, we construct mental experiences in which we imagine perceiving or interacting with the described events, people, and objects like a couch cliff, or a bell. It includes metaphorical and amodal processes that mean that we can perceive something even though there is no sensory information available to us. For example, if part of an object is hidden from your view or occluded you can still construct a mental representation of that object because the perception of that information is common or redundant across multiple senses (e.g., auditory, visual, tactile). You can perceive that there are six sides to a cube, even though only three of them can be seen at any one time. Bence Nanay argues that amodal perception plays a "constitutive role in our everyday perception" in that it is what allows us to give existence to things by using amodal perception to complete things without all the sensory information.[57] Nanay points out though that "the neural mechanisms responsible for both modal and amodal perception are the same." Wernicke illustrated the metaphorical processes by reference to the concept of the word bell, which is formed by the associated memory images of visual, tactual, and auditory perceptions and these memory images represent the essential characteristic features of the object, bell. It includes semantic knowledge of language and words, the cognitive semantic meaning of language and words brought about by embodied individual knowledge, experience, and construction of that word. It is recognized that embodied simulation can't account for all cognition but is believed to be a critical component of human knowledge and one assumes an essential requirement in autonomous conscious robots.

For robots to display noetic consciousness they potentially need like humans to possess a process by which attention and neural activity are intensified, and reflection is generated by a clear stream of sensorial and perceptual representations of objects and related events. For embodied semantic knowledge ANN in theory would need to somehow perform a similar function to that of the frontal cortex. In humans, this involves the conscious processing and recollection of sensory information in brain regions above the brain stem of knowledge related to what, where, and when episodes or events happened, and awareness about the world, which includes facts, abstract concepts, and ideas and amodal and metaphorical processes. It does not necessarily require awareness of any connection to their own experience but must as a minimum include perceptual cognition of objects and others which seems to be possible with AI robots if as we will see they possess the limited capacity to pass the mirror test in terms of recognizing the perceptual self as a distinct object separate from others. Humanoid robots have already demonstrated the capacity to shake the hands of humans and remember who they are which requires a minimum albeit limited spatial and temporal dimension. However, even if perceptual consciousness at the semantic knowledge level is available to robots, they still appear to lack the self-referential embodied aspects that make the autonoetic conscious experience a subjective part of the self-concept, which would enable them to construct abstract concepts and ideas and mental representations about the self in relation to the world.

Doing and Knowing in Robots.

At the heart of the debate about the potential for robotic consciousness lies the difference between doing and knowing and intelligence and understanding which we discussed earlier. Stuart Kauffman (1939–) and Andrea Roli observe that much of Western epistemology (theory of knowledge) involves knowing about and understanding the world, but evolution is based largely on doing by reference to survival and affordances. They refer to the OODA loop and the perspective of human existence as divided into the world, the self, and action, explaining that we can't always use the conventional methods of induction, deduction, and abduction to always know about novel things, but this doesn't stop us doing and learning from open-ended behaviour and engagement with the environment and others. This separates phenomenological consciousness from knowing in terms of understanding and Roli and Kauffman boldly claim that "Amazed by today's AI system capabilities, some await with fear, replacement by intelligent robots. We hope to show that this is not possible for wonderful and fundamental reasons: the becoming of any world with an evolving biosphere of philosophic zombies, let alone conscious free will agents, is, remarkably, beyond any mathematics we know."[58]

If Roli and Kauffman are correct, robots will always be on the other side of the metaphorical wall which involves them doing things but will never be able to join us on the other side of the wall which involves conscious understanding and the subjective experience. The problem is that even if robots were able to operate only based on affordance and survival in an evolutionary sense, their actions could be no less threatening than if they were to understand the consequences or have any subjective appreciation of what was happening. After all mankind's understanding of its acts and subjectivity has not stopped it from committing horrendous acts of cruelty and devastation, so why are understanding and subjectivity necessary for robots to operate and thrive in our world?

Various cognitive architectures use a range of different approaches to machine memory, and we will address the different types of human and robotic memory later, but for present purposes, most cognitive architectures that build in long-term memory don't distinguish between procedural or implicit memory and episodic or autobiographical memory. This is probably because the emphasis is on memory as a means of operating the system and performing tasks, rather than conferring the AI with possible self-awareness and the capacity to recall memory content, in the form of what-where-when or its integration into a single content representation over time, referencing the self as the observer and the one experiencing the environment. One system that currently purports to do this is a hybrid AI system called CLARION which led by the cognitive scientist and computational psychologist Ron Sun, has been in development since 1994. The novel feature of Ron Sun's work is that the CLARION cognitive architecture has integrated both bottom-up and top-down learning which involves an interaction between metacognition and cognition, which is as we have seen fundamental to how humans learn.[59] We saw earlier how error prediction in the brain is achieved by the brain using top-down predictions based on prior knowledge to modulate bottom-up sensory input with prediction errors fed through unconscious and conscious processing. The developers claim that bottom-up and top-down learning allows CLARION to simulate both problem-solving and social interactions, which it does by combining, symbolic and connectionist representations, implicit and explicit psychological processes like procedural and episodic memory, and cognitive and psychological processes.[60] CLARION purports to capture the "essential structures, mechanisms, and processes of the mind" and "represent psychological theories in a computational form". Ron Sun explains that the minimum requirements for a system to do this are (1) Procedural or implicit processes that deal with actions and skills like driving a car (2) Declarative processes for general knowledge and reasoning, in terms of semantic and episodic memory (3) Meta-cognitive processes for monitoring and regulating other processes and (4) Motivational processes that provide the impetus for action and cognition. These are inextricably linked by being closely coupled through integration and connectivity of the various brain regions involved in terms of neural networks and as we saw when we looked at human neurons, a high level of synchronized neural activity. Sun explains that CLARION achieves this through various subsystems which are (1) action-centred, (2) non-action-centred, motivational, and (3) the meta-cognitive subsystems which together address the psychological domains of action, learning, memory, personality, emotion, and their interaction.[61] The explicit level of representation in the AI system, potentially represents conscious awareness, whilst the implicit is unconscious and outside awareness as it is in humans.

Sun claims that it represents a dual process theory of mind, with each subsystem tightly coupled and interacting with each other, with implicit and explicit processes involved at the same time, such as when we drive a car, are almost seamlessly integrated. Sun argues that explicit knowledge is best captured by symbolic localist representations which equate to the conscious representation and manipulation of the information in short-term working memory and implicit by distributed connectionist representations. He distinguishes between the low-level primary drives which are largely physiological and evolutionary and reflect homoeostasis and allostasis and the higher drives which are predominantly social such as affiliation and belonging, dominance and power, social values, curiosity, autonomy, etc. Drive and motivation are captured by a combination of the level of external relevance of a situation or task and the level of internal inclination towards activating a drive to perform a task or behaviour, with the goal calculated by reference to the related drives' strengths. The obvious parallels with the human brain lie in CLARION's apparent ability to achieve integration and connectivity of these various features, whilst distinguishing between implicit and explicit memory, combined with metacognition and the essential features of motivation and drive. This is a remarkable example of how our knowledge of the way our brains and minds process information and function is now being harnessed in designing and developing AI architectures.

Neuroscientists Stan Franklin, Steve Strain, and Sean Kugele have developed the computational cognitive architecture LIDA (Learning Intelligent Distribution Agent) which they describe as an integrated artificial cognitive system that attempts to model cognition in biological systems, from basic perception and action to high-level cognition[62]. LIDA was first developed for the US Navy and has been extended to the world of medical diagnosis with Medical Agent X (MAX). Franklin asserts that as cognitive architecture, LIDA models a large part of human cognition and that "LIDA integrates findings from a broad range of cognition-related fields into an architecture for designing and implementing software agents that utilize biological cognition, including human-like intelligence." According to Strain, Kugele, and Franklin this involves an "iterative cognitive cycle of understanding, attention (consciousness), and action, endowing it with the ability to integrate multiple sensory modalities into flexible, dynamic, multi-modal representations according to strategies that support specific task demands." They observe that "These representations enable diverse, asynchronous cognitive processes to be dynamically activated according to rapidly changing contexts, much like in biological cognition."[63] LIDA is based on the observation that most human cognition functions through frequent iteration of (~10 Hz) interactions, known as cognitive cycles, between conscious contents, the different memory systems, and the selection of action or responses, and these cognitive cycles, amount to atoms of cognition from which higher-level cognitive processes are constituted. They observe that "Rather than relying solely on approaches that convert naturally occurring information sources into data structures native to the von Neumann computing paradigm, it might be wise to avail ourselves of evolution's successes to inspire the design of software and hardware systems that organize processes in a fundamentally different way"(LIDA and MAX). They appear though to place understanding before attention and equate attention with consciousness rather than consciousness being a prerequisite for attention.

The cognitive cycle consists of the following stages which bear examination so we can understand how our cognition works in the human brain which we will look at shortly. In the understanding phase, the architecture enables the machine to detect environmental stimuli using feature sensors and sensory memory to identify the stimuli. The architecture uses associative memory to allow matching with objects, categories, actions, and events, which are then processed in a workspace using transient episodic and declarative memory to represent a situational model that allows the robot to understand what is happening. We saw how clinicians use certain metrics to tell whether a person is conscious, including ensuring reflexive movements are not confused with consciousness wakefulness and awareness and the difference between the state of consciousness and the content. Let's see how such metrics might be applied to robots.

Reflexes in Robots.

Reflexes are an important part of our biological system that allows us to react rapidly to danger or threat and would appear necessary to robots if they are to operate in potentially hostile real environments. Robots need rapid reflexes to perform human tasks like grasping, bi-pedal walking, and balancing when in dynamic locomotion. Balancing in humans requires proprioception and sensory feedback from visual and vestibular systems, which is both very rapid but simple in that it can react to a simple sensation. We have seen how in humans this involves complex neural pathways and reflects the diverse morphology and musculoskeletal body structure that enables dynamic movement, including stretch reflexes to control muscles through contraction and inhibitory crossed responses. Discovered by neuroscientists the crossed response exercises an excitatory/ inhibitory balance between corresponding muscles on the other side of the body to avoid these muscles operating at the same time which would make balance almost impossible. For robots to achieve reflexes that come anywhere near to human reflexes requires highly complex bio-mimetics which is evident in the growing research into bio-inspired robots that mimic human neural networks, muscles and skeletons. Currently, most robots are useless in this respect, even though bi-pedal robots are getting better at navigation and orientation. They do not possess the ability to rapidly respond to their environment in the way that humans can, particularly when there is imminent danger or the need to adjust their behaviour rapidly such as when encountering a hot surface that could damage their sensory mechanisms. If robots are to operate in uncertain and potentially dangerous environments, then reflexive responses would appear to be a vital requirement.

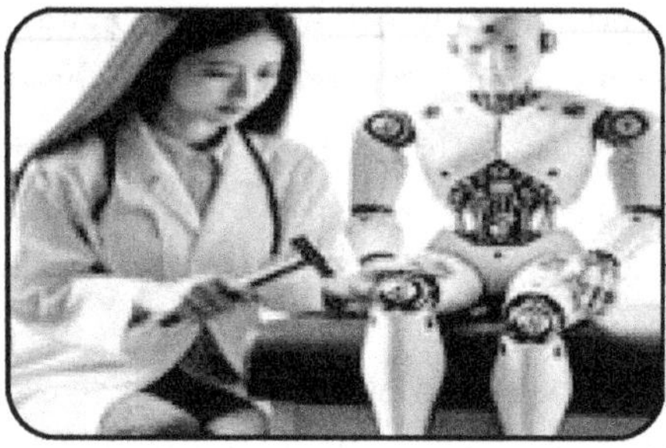

In 2019 scientists from MIT and the University of Illinois unveiled a two-legged robot designed to be deployed in a disaster situation, called Little Hermes which can walk, run, jump, and interact with the environment in a human-like manner. The constraint is that it does so by synchronizing with a human rather than autonomously and although the robot can learn from the human, the researchers openly acknowledge that developing an autonomous robot with human-like reflexes is some way in the future. Scientists have been using a musculo-skeletal bipedal robot that mimics the human neural networks, muscles, and skeleton with a similar body structure and dynamics to a human to study the role of proprioceptive sensory feedback control (e.g., stretch, and cross-reflexive control) on dynamic motion such as hopping, gait and even posture in people and avoiding falling over. They observed how the stretch reflex and crossed response contribute to reducing the lateral inclination during hopping, which provides roboticists with insight into human reflexes and biomimetic ideas for robot locomotion control.[64] This further illustrates how much we are learning about human behaviour by using robotics. Despite this, Professor Sami Haddadin of the Munich School of Robotics and Machine Intelligence (MSRM) of the Technical University of Munich (TUM) has observed that robot's "reactions to impending collisions tend to be rather mindless: They just stop and don't move until a person takes action." Haddadin points out that humans can anticipate pain but that robots can't and part of the reason for this is that 'Our pain perception is highly complex and linked to emotions. We can't compare this to a human's pain sensation. Robots are tools and not living creatures. Artificial pain is nothing more than a technical signal based on data from various sensors."

Despite these observations, Haddadin and the team at MSRM have managed to develop an initial reflex mechanism for robots based on artificial pain when touching hot or sharp objects, with a robot withdrawing its arm in a reflexive movement. The major challenge is that our understanding of the human reflex model remains rudimentary and as it stands robots cannot anticipate pain and do not have reflexes comparable to humans. Whether they will in the future may depend on our ability to build into our robots the ability to perceive and feel pain and the capacity for emotions which we discuss later.

Wakefulness in Robots.

At the heart of the various theories of human consciousness and subconscious processing is the recognition that the content of conscious content is fleeting and that a self is involved or evoked in this process at all levels. Verschure's biologically inspired Distributive Adaptive Control (DAC) attention architecture is a biomimetic dynamic robot-based "Whole Brain " neural architecture. Grounded in the neuroscience of invertebrate and vertebrate biological systems it instantiates MBBN which purports to bridge the exploratory gap between the mind and brain. Anil Seth, Sid Kouider and Andrea Caria working together with Verschure on abstract models of data flow in the new technology of Telepresence have described Transparency and Presence as key features of normal waking consciousness. Telepresence has been defined as "Enabling a "person to perform actions in a distant or virtual location as if physically present in that location"[65] which allows humans through robots to experience the sensation of having a presence or performing tasks in remote locations when they are not physical there like surgical operations. Their definition of presence refers to the "subjective sense of the reality of the self and the environment, and transparency refers to the observation that we 'see through' our mental representations of the environment, directly experiencing their content." Presence involves "successful explaining away' by top-down predictions of interoceptive signals evoked by

afferent sensory signals and by autonomic regulatory signals." Transparency and presence are sufficiently closely coupled for them to use the abbreviation TP for waking consciousness. They suggest that seeing through' our mental representations of the environment, directly experiencing their content and engaging in the process of explaining away the interoceptive by afferent sensory signals and by autonomic regulatory signals" constitutes a synthetic conscious wakefulness in humans and would therefore in robots. They base their model on the principle of the Grounded Enactive Predictive Experience model of consciousness (GEPE). These principles about the qualia of subjective experience are that qualia are "(1) Grounded in the experiencing physically instantiated self (2) Enacted in the sensorimotor coupling of the agent to the world (3) Maintained in the coherence between sensorimotor predictions of the agent and the dynamics of the interaction with the world (4) Combine high levels of differentiation with high levels of integration and thus reflect highly informative discriminations among extremely large repertoires of possible states and (5) Depend on both highly parallel, distributed and implicit factors and metastable, continuous and unified explicit factors comprising a Global Workspace hypothesis." Verschure describes this "convergent synthetic methodology using the analysis of synthetic biologically grounded agents, or quale parsing." Coherence of information about the self and the world arises from differentiation, connectivity and communication, and integration of various brain regions operating in parallel and involve brain wave dynamics that convey information from both synchronized and incoherent spiking activity.

A robot potentially needs to reflect these highly complex dynamics, which we will explore shortly, but can robots exhibit wakefulness over and above simply being switched on and off, because activating an ANN does not necessarily mean the robot is awake in the sense that humans wake from sleep. A robotics firm from the United Kingdom called Engineered Arts has built a robot made from modular parts, called Ameca which appears to wake up and make a surprised expression on waking. Ameca can be controlled via Engineered Arts' cloud software called Tritium and its primary purpose is to further research into human-robot interaction. Although it looks uncannily real the fact is that it only mimics wakefulness and does not display the level of neural connectivity which is a pre-requisite in humans for being awake. Ameca appears to be a classic example of how when we look at the behaviour of the robot appearing to wake up, we anthropomorphise, and it is very difficult to not believe that it is genuinely being wakeful and aware of itself. If it were a human performing the same movements and look of surprise, we would be celebrating Ameca's emergence from a coma, but the reality is that Ameca has simply been turned on and programmed to look surprised. There are currently no robots or AI systems that can wake up in the sense that humans wake up and become conscious of their world and themselves.

Consciousness and Self in Robots.

Computer scientists James Reggia, Gerrard Katz, and Gregory Davis claim to be looking for the neurocomputational correlates of consciousness or Computational Explanatory Gap (CEG). Their approach to CEG is to identify how using ANNs, accessible high-level cognitive processes such as decision-making, reasoning, intent-directed problem solving, executive control of working memory contents, plan generation, and language can be generated by lower-level neurocomputational processes.[66] Albeit in 2013, Reggia openly acknowledged that there was then "no convincing implementation of instantiated consciousness in a machine" but believed that their work on humanoid cognitive robots that learn to perform tasks via imitation learning is highly relevant to this issue.[67] They claim that the problem is purely computational and that extending phenomenology beyond purely subjective qualia to higher cognitive processes that reach conscious awareness, which is called cognitive phenomenology, will provide greater insight into what consciousness in machines comprises. They maintain that this has the potential to ultimately create a machine capable of experiencing subjective phenomenological consciousness, comparing it with the field of neuroscience in which scientists are still trying to understand how our brains implement higher cognitive functions using the underlying neural circuitry. This approach appears to be focused on the content of consciousness rather than establishing the state of consciousness and seems to reflect Tononi and Edelman's observation that the state of consciousness arises from a particular level of complexity that is the result of the coexistence of functional integration and differentiation of information in our brain.

Despite all these initiatives, theories, and arguments about unity and subjectivity, we don't know whether AI will ever generate reflexive behaviour, display genuine wakefulness, achieve the unity required for the conscious state or conscious content, or possess genuine self-awareness which we discuss later. If consciousness is a quantum phenomenon this would require AI and robots to tap into universal consciousness although this appears impossible because of the basic laws of physics. Others argue that AI and robots will always remain at the doing, not the conscious knowing phase of development.To assess this let's look more closely at the self in humans and robots.

Endnotes

1 Clayton NS, et al., 2003.
2 APA Dictionary of Psychology.
3 Anderson JR., 2004.
4 Ferrarelli, F, et al., 2010.
5 Dent, EW., 2017.
6 Ekert A, Jozsa, R, Penrose, R, 1998.
7 Hameroff, S, Penrose, R., 2013.
8 Huelga SF, Plenio, MB., 2013.
9 Donadi, S., et al. 2021.
10 V Kerskens, C, Pérez, D L., 2022.
11 The Grand Biocentric Design: How Life Creates Reality (2016).
12 Kauffman S, Roli A., 2022.
13 In Our Image, 2015.
14 McFadden, J., 2020.
15 Anastassiou,CA., et al., 2011.
16 Fröhlich, F., et al., 2010.
17 Baars.B., 1988.
18 Hohwy 2013.
19 How the Body Shapes the Way We Think (2007).
20 Ecological Psychology in Context (2001).
21 Mayer, EA., 2021.
22 Tingley, D., et al., 2012.
23 Pellissier S, et al., 2014.
24 Jacobs, JP., et al., 2021.
25 Gordon JI, et al 2016.
26 Carabotti, M., 2014.
27 Dohnalová, L., et al., 2022.
28 How the Body Shapes the mind 2005.
29 Damasio A., 1999.
30 Neurowaves (2023).
31 Renoult, L, et al., 2019.
32 Oxford English dictionary.
33 Vandekerckhove, M, Panksepp J., 2009.
34 Sterling P. 2004.
35 APA Dictionary of Psychology.
36 Vandekerckhove M, Panksepp J., 2009.

[37] Barsalou, L. W., 1999.
[38] Wellsby, M, Pexman, P. 2014.
[39] Tillotson,SM., et al., 2008.
[40] Siakaluk, PD., et al., 2008.
[41] Lakoff, G, Johnson, M., 1980.
[42] Barsalou, L W., 2012.
[43] Perner, J., et al., 2007.
[44] Verschure PF MJ., 2016.
[45] Verschure, P., 2012.
[46] Galinsky V. Frank L 2023.
[47] APA Dictionary of Psychology.
[48] Owen AM. Et al., 2006.
[49] Searle, J., 2004.
[50] Baars, BJ., et al., 2003.
[51] Valenzo, D., et al., 2022.
[52] Strigo IA, Craig AD., 2016.
[53] Vargas, P., et al., 2005.
[54] Moioli RC, Vargas, PA., Husbands, P., 2009.
[55] Vouloutsi, V., et al., 2013.
[56] Kiverstein, J., 2007.
[57] Nanay, B., 2018.
[58] Roli, A, Jaeger, J, Kauffman, S., 2022.
[59] Anatomy of the Mind: Exploring Psychological Mechanisms and Processes with the CLARION Cognitive Architecture (2016).
[60] The Oxford Handbook of Cognitive Science, Oxford Handbooks (2014).
[61] Sun, R., 2016.
[62] Franklin, S., et al., 2012.
[63] Strain, S, Kugele, S, Franklin, S., 2014.
[64] Liu, X, et al., 2018.
[65] Merrian Webster.
[66] Reggia, JA, Katz, GE, Davis, GP., 2018.
[67] Reggia, J., 2013.

"A self is not something static, tied up in a pretty parcel and handed to the child, finished and complete. A self is always becoming."

Madeleine l' Engle. A Circle of Quiet. American writer of fiction, nonfiction, poetry, and young adult fiction.

Chapter 7

The Self

CHAPTER 7

The Human Self

THE FAMOUSLY SCEPTICAL SCOTTISH PHILOSOPHER David Hume (1711–1176) sought to describe how the human mind acquires knowledge and formulated his bundle theory of self, in which there is no knowledge beyond our experience. He argued that the self is just a bundle of impressions or perceptions of who we are because that is what we see, rather than there existence of an actual self.[1] The German professor of theoretical philosophy, Thomas Metzinger(1958–) argues that no one has ever had a self and that all that appears to be self is the phenomenological experience which is an ongoing process rather than a thing called the self. Metzinger observes how "Everything that we experience is a virtual self in a virtual reality" and that our experience of ourselves and our environment are the product of an internally generated model. If that is the case, it would seem highly feasible to create a self by using the appropriate cognitive architecture to produce an internally generated model of self. Buddhists believe that our conception of self is constructed from physical and mental sensations, impressions, and reactions which are an ever-forming and changing process, which is borne out by the latest neuroscientific research. Zhen Buddhists describe how the paradoxical truth is found where there is neither self nor no self and how "the ultimate truth transcends the antithesis of self and not-self."[2]

The experience of relinquishing or not having a coherent sense of self is not as fantastical as it might appear and can even be self-induced if that is not an oxymoron. The English writer and philosopher, Aldous Huxley (1894–1963) wrote the following passage which reflected how, during his experience in 1953, with the psychedelic drug mescaline, a hallucinogenic and intoxicating compound found in the Peyote cactus, he had experienced the loss of or dissolution of himself. "For persons are, selves and, in one respect at least, I was now a not-self, simultaneously perceiving and being the not-self of the things around me. To this new-born. not-self, the behaviour, the appearance, the very thought of the self, had momentarily ceased to be and of the other selves, its one-time fellows, seemed not indeed distasteful (for distastefulness was not one of the categories in terms of which I was thinking) but enormously irrelevant."[3] He describes how he had achieved almost mystical insights, by experiencing being untethered to his sense of self and subsequently wrote of how "the mystical experience is doubly valuable; because it gives the experience a better understanding of himself and the world and because it may help him to lead a less self-centred and more creative life." Huxley believed that "the human urge to transcend oneself is universal through times and cultures" and he explored how the psychedelic experience can invoke a feeling of deep connection with the universe, even though it can paradoxically invoke a very frightening psychosis and deep disturbance in the sense of our core self.

Christopher Timmerman, from the Centre for Psychedelic Research, at Imperial College London leads the research group specializing in potent psychedelics and their effects on the human brain and consciousness. He and his team recruited 20 healthy volunteers who received a 20 mg injection of the psychedelic compound in the flower called Dimethyltryptamine (DMT) and a placebo on separate visits to the lab. They found that DMT causes the brain to become hyper-connected and brain regions were communicating in a chaotic and flexible almost anarchic way. Timmerman observed that "the stronger the intensity of the experience, the more hyper-connected were those brain areas.[4] Robin Carhart-Harris, a professor of neurology and psychiatry has also studied DMT observing how "People describe leaving this world and breaking through into another that is incredibly immersive and richly complex, sometimes being populated by other beings that they feel might hold a special power over them, like gods." This demonstrates the immense, almost mystical power of our brain that can be achieved when inhibition over connectivity between different brain regions is removed. In due course, we may need to let robots experiment with the machine equivalent of psychedelics, but before we do that, we need to know first that they can even have a self that they can detach from or transcend.

The 17th-century philosopher, mathematician, and polymath, Renee Descartes (1596–1650) in his Meditations on First Philosophy[5] used the now famous phrase Cogito Ergo Sum (I think therefore I am), reflecting his angst-ridden philosophy that if there was any reason to doubt anything (this is called radical doubt) it couldn't be asserted to be true or real. Descartes argued that, because we are unable to completely trust the information coming through our senses, we must doubt the existence of everything, and maybe nothing is real. The only thing doing the doubting that is real is our thinking mind, hence the only thing you can't doubt is the fact that it is your mind is thinking. Taken to its logical conclusion thinking about ourselves is the only thing we can prove and that thinking requires cognition of the world around us and how we relate to it.

Cognitive psychologists study how people know things, but we can't prove we exist beyond our cognitive process, which at least gives us the Cartesian certainty of identifying ourselves through our thoughts, whilst distinguishing ourselves from others. It is this primacy of cognition, which is traditionally believed to generate our sense of self, and Cartesian dualism which divides the material matter of the brain and the non-material thinking mind, has become such a persistent belief system that it has led to recent attempts to address the perceived problem of the mindless nature of neuroscience which focuses on the matter of the brain and the brainless nature of psychology and psychotherapy which focuses on the mind.[6] We need to understand whether thinking about themselves is the only thing robots will be able to prove and whether their thinking requires cognition of the world around them and how they relate to it. Will Cartesian dualism pose the same enduring problem for robots that it has for humans or be of no consequence to them at all and will the truth of what they represent to us as knowledge have any meaning to them?

Developing a Self

William James formulated one of the first workable theories of self and the subjective aspects of the self and was convinced that the existence of a self was predicated on what he described as the "core of sameness."[7] He distinguished between the "I" and the "Me" self, describing the I self, as people perceiving and observing themselves as doing or thinking, so they are cognisant of themselves, which he called the "Knower". For example, when you are silent you can listen to yourself thinking or observe yourself behaving in a particular way. James called the I self, pure ego and acknowledged it as an elusive almost ephemeral sense of being aware of the process of thinking about yourself. He believed that this "I" self, could not be divided, nor explored by science, as it is a non-material stream of consciousness and a thread of continuity, linking our past, present, and future. The Me self is the subjective, reflective, and phenomenological (first-person experience) aspects of the individual, comprising the material in terms of the body and possessions, etc, and the internal or spiritual self which includes the personality, core values, and conscience. James argued that the spiritual Me self is the core self and the most permanent, whilst the social self, with its emotions, feelings, and perceptions, changes depending on the context, so that an individual may have several social selves. James established a distinction between the self as an object of attention and the self as a subject, which experiences the self as an object, as did the Austrian-British philosopher Ludwig Wittgenstein (1889–1951), considered to be one of the greatest philosophers of the 20th century when he distinguished between the I as a subject and the Me as an object.

For a long time, scientists have disputed whether the self develops as a continuous or discontinuous process in stages. The classical viewpoint developed in the 1950s by psychologist and psychoanalyst Erik Erikson (1902–1994) and other models of development, is that it takes around twenty years for the self-concept to develop and involves reaching certain milestones. According to the social development of self-theories, developing a sense of self is the same as developing a sense of others in that it involves the process of socializing and reflects internalizing others' perspectives about us or assimilating the behaviour of others. Hence self is not special or separable from the perceptions of others.[8] The legendary Swiss psychologist Jean Piaget (1896–1980) believed passionately in the role of child education and studying the intellectual development of his own three children. He was one of the first psychologists to systematically study cognitive development through childhood and his highly influential theory of cognitive learning gained widespread popularity in the 1960s. He disagreed with the generally held view at that time that children were like mini adults and argued that they think very differently from adults, a view that is now widely held. Piaget's theory was that children go through four different stages when they learn and his constructivist theory held that children construct their knowledge of their world through experience and perceptions which are organized in schemas, which we will discuss later. Information that is new to a child is either assimilated into an existing schema or the schema is adjusted to accommodate the information, or a new schema is constructed. The four stages of development are (1) the Sensory-motor stage (from 0–2) in which very young children learn about the world through their senses and movement (2) the Preoperational period (2–7) in which language and symbolic play emerge (3) Concrete operational stage (7–11) in which logical thought emerges, but abstract and theoretical thinking remain challenging and (4) Formal operational stage (12-adulthood) in which abstract thought and deductive reasoning emerge.

Piaget's work has been subject to significant criticism and childhood development is no longer believed to progress along such discrete and well-defined stages. His research used a very small sample, and they were from a background of well-educated professionals of high socioeconomic status. His theory was difficult to operationalise and therefore replicate in other studies and many adults may not reach the final developmental stage which is one example of the fact that many individual variables can affect the rate and extent of development. Subsequent research into the theory of mind has demonstrated that children as young as three can possess the ability to understand what others think or even assume the perspective of another person. Research has thus shifted away from earlier views that development arises in well-defined stages, with scientists now recognizing that development involves many more subtle mini-stages that make the process more continuous.

The German American psychologist Ulric Neisser (1928–2012) believed that different aspects of the self, developed at different phases in infancy and maturation and distinguished between the Direct and the Reflective self. The direct self, developing in early infancy comprises the ecological self, which is your interaction with the environment, and the interpersonal self, being verbal and non-verbal interactions with others. The reflective self consists of self-representations and reflection on those experiences that develop in later infancy and as a child matures. The reflective self comprises the extended (or remembered self) the conceptual and the private self. The remembered self consists of memories as a narrative and an imagined future and develops in the third year or later. The conceptual self is when one regards oneself as an object of thought and someone else's thoughts and involves developing schemas about oneself based on personal social and cultural experiences which appears to be very important in information processing and influencing behaviour. Schemata (plural of schema) are mental structures used by individuals to organize knowledge and guide cognitive processes and behaviour. They categorize objects and events reflecting common elements and characteristics allowing interpretation and prediction of the environment and consist of (1) Schemas about others and include such things as personality, appearances, behaviour, and preferences (2) Social schemas which include what we do and don't do in social settings (3) Event schemas which influence how we might behave at work or in specific contexts and (4) Self-schemas which cover our knowledge about ourselves and how we assess our capabilities and preferences. The private self is the appreciation of the subjective nature and privacy of one's thoughts and begins in late infancy with brain maturation when a child realizes that its own experiences aren't automatically perceived by others and must be communicated to that other. In addition to perception related to the five well-known senses of smell, taste, sound, vision, and touch, we possess social perception which includes perceptions and narratives about ourselves (self-perception). "Self-perception is a person's view of his or herself or of any of the mental or physical attributes that constitute the self. Such a view may involve genuine self-knowledge or varying degrees of distortion."[9] Self-perception acts like a filter through which we interpret the world and our private self because as we develop through childhood to adulthood and reflects our thoughts, behaviour, and beliefs which an individual seeks to rationalize. Psychologists have long sought to understand if people assess their self-perception and their behaviour by understanding their internal state of mind.

Perception in children changes with the development of their senses and as their brains mature. When babies are unable to speak or walk their perceptions of their environment are largely confined to their physical comfort or needs and the conditions, they are in. As they mature their perceptions become more meaningful with their perceptions reflecting their accumulated memories and experiences and the development of the I self as they start to observe their thoughts and behaviour and that of others means that they will perceive versions of their Me self as they mature. Development of perception is intensive in infancy and early childhood and is one of the most vital components of being prepared for life. Perceptions are involved in every interaction with their environment which lay down and strengthens neural pathways and influences further attention and perception. As children grow this includes their self-perception in terms of their internal mental world involving the processes that seek to understand things about themselves and our environment. The distinction between internal and external factors is very important with internal including emotions, needs, attitudes, and beliefs which may through the process of frequent repetition become entrenched or implanted in that child's narrative of themselves. i.e., their self-concept and how they relate to the world. External social factors and environments will also affect perceptions.

In the 1970s a widely respected pioneer in the field of social psychology, American Daryl J Bem (1938–) developed the self-perception theory (SPT) of attitude formation and change, which asserts that people's perceptions and attitudes are formed by observing external factors, in particular the behaviour of others rather than internal factors such as mood states or internal thought processes. Internal factors only apply when there is no previous attitude or experience or because the emotional response is ambiguous, and his studies showed that emotions appeared to follow behaviours. Bem believed that people assess others' attitudes and perceptions as a reflection of their own, so their behaviour including previous behaviour affects their perceptions and behaviour. A 2006 study tested whether implicit racial bias which people are unaware could be changed by changing people's facial expressions. They required participants to hold a pencil in their mouth which forced them to smile, whilst looking at photos that would normally elicit implicit racial bias. They found that those forced to smile exhibited less implicit racial bias. At the heart of SPT is the fact that your behaviour can influence your perception of yourself and future behaviour. It is evidenced by the James Laird experiment in which people who changed behaviour through facial expressions such as smiling, or frowning were found to be happier or sadder as a result, which has been used to support the view that behaviour affects perception. This seems counter intuitive in that we normally imagine that our attitudes and moods determine our behaviour, not the other way around, and has generated an ongoing debate between this self-perception theory and the theory of cognitive dissonance.

A major feature of how we develop a coherent sense of self was first identified by Charles Cooley (1864–1929) in his Looking Glass Self Theory.[10] This theory asserts that the social self and how an individual feels about themselves are dependent on their perception of how others perceive them and the internalization of this perception. This has consistently been supported by subsequent research. Cooley was part of the symbolic interactionist school which believed that no child could develop a sense of self without others to reflect that self to the child. They argued that all humans are defined by their interactions with others, which arises from and results in the simultaneous development of an appreciation of the perspective of others and an individual's social self. This manifests as an individual imagining what another person thinks of and how they judge them. Some theorists suggest that this supports the idea that there is no real self but only a self-crafted by reference to others which makes the individual a passive player moulded by the perceptions and judgment of others. Cooley and others like the American philosopher and sociologist George Mead (1863–1931) acknowledged that children and adults play an active role and have control over this by the attention they pay to this and how or whether they respond. Importantly they recognised how inaccurate perceptions of other people's judgements about others could be. Hence the self-image although influenced by others ultimately reflects the mind and perceptions of the individual. This reflective aspect is captured in the difference between the I self and the Me self and Mead and Cooley appear to have believed that the Me is shaped by the looking-glass effect but that the I self, is what determines how that person reacts to these imagined perceptions or judgments. The construction of the social self in middle to late teens is, therefore, a constant interplay between the I self and the Me self.

The Gestalt Theory is part of how we visually perceive our world and proposes that the memory for encoding information is influenced by perceptions but requires context. Perceptions are gained through our sensory systems (touch, taste, smell, hear) that we looked at earlier which create new or trigger existing neural networks in our brains. When reactivated through recall the stronger they become, and the easier they are to recall and thus remember. Wertheimer, Köhler, and Koffka observed that perception is greater than just the sum of the sensory stimuli i.e., the whole is greater than its parts which they called Holism or the unified whole. The attributes of the perceived whole are not deducible from analysis of the parts in isolation, and they established six principles that identify how humans subconsciously group similar elements, recognize patterns, and simplify complex images when we perceive objects and situations. Gestalt theory reminds us that we see the world as a whole and in context and that our experiences and prior knowledge reflect this. Gestalt therapy looks at the mind and behaviour to teach people to become aware of significant sensations within themselves and to notice and attend to their environment so that they respond fully and reasonably to situations. It gives people insight into how their memory of events and episodes are perceived by them and reflects how Gestalt was at the time a clear response to structuralism which seeks to describe the physical structures that compose the brain and the mind.

All these various theories about the self have contributed significantly to our understanding of the psychology and neuropsychology of learning, complimented by our growing knowledge of how learning is achieved through the dynamics of neural networks and the biological basis of autonomy and agency. In very simple terms when we are talking about memory and learning, the brain receives information through the senses and encodes that information, which it then needs to consolidate and store in the form of neural patterns, for it to be activated or retrieved later. Our memories of self and our experiences, allow us to see ourselves as unique entities and self-protracted over time and give us purpose, unity, and a sense of coherence and agency in our lives. Everything we know about ourselves is contained in the self-concept and self-schemas which guide our behaviour and responses to others and situations and events. This plays a vital role in generating coherent identity and individualism.Let's look at the relationship between the self and identity.

Identity and Self

Our brain constructs a personal identity through stories about the self, and our identity is shaped by memories, experiences, relationships, and values as we go through life. In an extensive review of the philosophical implications of the self in cognitive science, the American philosopher, Shaun Gallagher (1948–) distinguishes between the Minimalist self and the Narrative self. The minimalist self is a consciousness of oneself as an immediate subject of experience, not extended in time, which depends on brain processes and an ecologically embedded body, but one does not have to know or be aware of this to have an experience that still counts as self-experience. Gallagher describes the narrative self as a "more or less coherent self (or self-image) that is constituted with a past and a future in the various stories that we and others recount about ourselves."[11] In the 1980s, the American psychologist Dan McAdams (1954–) developed the Integrative Life Story model of identity based on his observation that identity is "an internalized and evolving life story that reconstructs the individual's past and imagining of the future."[12, 13] The self emerges from the integration of experienced moments into a meaningful and remembered narrative of experiences. Self-identity is how we relate to and compare ourselves with others in the many diverse contexts of gender, racial, ethnic, sexual preferences, orientation, intellectual capacity, creativity, resilience or emotional fragility personality, etc. Bertram Joseph Cohler (1938 –2012) an American psychologist, psychoanalyst, and educator in the 1960s, who made major contributions to understanding adversity, resilience, and coping, was a pioneering and influential advocate of needing to understand the role of personal narrative and life course in forming identity and self.[14] Cohler advocated a life course perspective in which identity is understood in a structural, social, and cultural context and that the life course reflects this. Janet Giele and Glen Elder define life course as "A sequence of socially defined events and roles that the individual enacts over time" with development extending through many stages in life. They believed that five key factors influence the life course which are (1) Life-span development (2) Human agency (3) Historical time and geographic place (4) Timing of decisions, and (5) Linked lives.[15]

Psychology professor, Robyn Fivush describes how we "construct a web of interacting threads" that form the self and believes that the life story develops over time with identity originating in childhood, becoming a particularly important issue during adolescence, culminating in projections into old age.[16] Studies have shown that the personal narrative goes well beyond the simple remembering of events and situations or people because it is how we derive meaning and value from past experiences.[17] Both the core self and the narrative self, appear to play an important role in our mental health. British professor of developmental psychology, Bruce Hood observes that "We live in a matrix in our minds, with the self an illusion consisting of a narrative of ourselves that the brain constructs."[18] Important autobiographical memories emerge in childhood and develop in adolescence, with individuals beginning to integrate their own experiences within a cultural and social framework. Fivush and colleagues argue that "Autobiography is a critical developmental skill and that "narrating our past connects us to ourselves, our families, our communities, and our cultures."[19] A person's underlying self-narrative reflects deep-seated emotions and feelings and stored memories which are the building blocks with which we construct a dynamic narrative about ourselves which in turn shapes how we perceive our world, others, and ourselves. Emotions and feelings are built into our memories and our narratives and are highly influential in influencing our habitual perceptions or ways of seeing and responding to situations because that is who we think we are and how we should react.

In the 1960s German American developmental psychologist and psychoanalyst, Erik Erikson (1902–1994) developed some of the most enduring and influential theories about how a child gains an identity and a distinct sense of self.[20] He formulated the eight stages of development of identity from infancy, through adulthood, with each stage involving a psychosocial crisis that reflects the society or environment conflict of the individual's needs with those of the society and environment they live in. Psychology Professor Susan Harter (1939–) a leading figure in the field of how children develop self who has focused on self-esteem, the construction of multiple selves, and false-self behaviour makes observations gleaned from her studies and the literature as to how perceptions develop and change as a child matures and its cognitive capacity increases. She developed the Competence Motivation Theory (1978) in which an individual's sense of personal competence is not a trait but increases if they successfully master a task and this encourages them to master more tasks.[21] She observes that there are significant variations in the rate of development of self-perception between individuals and therefore these phases aren't fixed deadlines or milestones but indicative of progress in a child's development of self. They provide an interesting overview of key features in the phases and process of our development. Harter developed a quantitative scale that has three specific domains and one general domain of self-esteem. The specific are (1) Cognitive competence (2) Social or interpersonal competence and (3) Physical competence- primarily in athletic skill, whilst the general is feelings of worth or esteem independent of any skill. Self-esteem is defined as "the degree to which the qualities and characteristics contained in one's self-concept are perceived to be positive."[22]

The American philosopher and cognitive scientist, Daniel Dennett, describes how, "Talking about the mind for many people is rather like talking about sex, slightly embarrassing, undignified, maybe even disreputable."[23] American philosopher and professor of experimental cognitive psychology, Joseph Levine (1952–) who focuses on the philosophy of mind describes the "Explanatory gap" between the physical brain and subjective experiences of self. [22, 44] German American neurophysiologist and computational neuroscientist Christof Koch (1956–) best known for studying consciousness, observes that "We know of no fundamental law or principle operating in this universe that forbids the existence of subjective feelings in artefacts designed or evolved by humans." French philosopher Paul Ricoeur was a pioneer in exploring reflexive philosophy, in his search to understand how we become aware of ourselves as existing, thinking and acting. Ricoeur defined identity as comprising Identity, which he characterized as a sameness that never changes, and Identity as sameness through and across change and believed that self-hood comprises both.

One of the many reasons why it is complicated to seek to build a self into robots is that the search for the neurological basis of the I self has proved highly elusive even in humans, let alone modelling this in robots. Will androids need or ever be capable of possessing both an I self and a Me self and will a coherent sense of self be an integral requirement of being a conscious sentient android? We use our sense of self to establish boundaries distinguish between ourselves and others and make life decisions based on self-reflection. Humans are social creatures, with a need to belong to social groups, which involves distinguishing between who are our friends and who might be foes understanding the social interactions and relations within groups and identification of potential mates. We are very aware of social context and have developed neural networks or circuits associated with social relations that involve self-regulation and inhibition of basic motivational systems that might prioritize the individual over the group or involve exclusion or negative feedback from the group.

A theory for why we evolved this sense of the self and a social brain has been advanced by British anthropologist and evolutionary psychologist, Robin Dunbar (1947–) and is called the Social Brain Hypothesis (2009). In simple terms, species that have pair-bonded mating systems, which extend to friendships that do not involve reproduction, tend to have larger brains. This is open to criticism as to the absence of consensus as to what amounts to pair bonding the lack of metric against which to measure brain size and the fact that it is not size that matters. The logical implication is that bonding with others is cognitively demanding and fundamental to social functioning and requires a sense of self which because of our complex social systems requires a larger more complex brain. The greater cognitive demand potentially arises from the need to coordinate behaviour with one's mate. This appears to fall foul of the chicken and egg paradox by confusing the causes and consequences, in that many animals mate, without possessing what we understand as a self, and that this happened before the self is thought to have evolved in human primates. However, considerable research points to a clear correlation in primates between group size and predation (survival) risk and a phase transition (see below) in the early evolution of the primate social brain, which may address these apparent contradictions. Dunbar observes the important co-evolutionary effect in the phylogenetic (evolutionary relationships among biological entities) transition from asociality (rejecting or lacking the capacity for social interaction), to sociality. Sociality is the degree to which individuals in an animal population tend to associate with social groups and form cooperative societies.[25] Sociality embraces even those early exchanges between the infant and the caregiver and slowly broadens beyond the carers to others. All the time, a child is building up neural patterns and networks and therefore memories based on those experiences, and a narrative self, is constructed by reference to those experiences. Could a robot build up this level of experience and narrative from its environment to construct a sense of self in the way that humans do?

The Genes and Self

It is not that long ago we were taught that children are born as Table Rasa which is Latin for a clean blank slate and means born with an absence of preconceived ideas or predetermined goals and that this blank slate could be inscribed by our parents, others, and the environment. How many times have you heard people's behaviour or response to something, or someone being explained away by, it's just the type of person they are. Scientists have devised many theories and techniques to empirically study the highly subjective self, including the fundamental features of self with one fruitful area of research into why people behave and think as they do, being dispositional traits, together with intelligence and personality, which make us such individuals. Dispositional traits are one of three key mediating factors, together with, goals, and behavioural representations that mediate our social behaviour following perception or thoughts that initiate the motor action that leads to behaviour.

American psychologists Dan McAdams (1954–) and Jennifer Pals a personality/ social psychologist whose research looks at how people connect memories of past life events to their present self through the life stories they construct and narrate, argue that a person's internalized life story is shaped by reference to their characteristic goals and values, which reflect their dispositional traits and have sought to establish a comprehensive framework for understanding the whole person.[26] They cite the 5 big principles underlying an integrated personality and emphasize patterns of behaviour which are (1) An individual's unique variation on the general evolutionary design for human nature, expressed as a developing pattern of (2) Dispositional traits (3) Characteristic adaptations (4) Self-defining life narratives, complexly and differentially situated and (5) Culture and social context.

We know that communication and integration in the brain of all perceived sensory information is crucial to how the self is processed in the brain, and we will shortly look at how and where the information related to the self is processed in the brain, and the role of schemas, but our current knowledge of how personality and dispositional traits are represented in the brain remains poor. Based largely on current fMRI data from individuals, there is a school of thought that traits are "biophysical entities arising from the CNS and not just abstract metaphors for the way we behave."[27] Since the emergence of our knowledge about the human genome, scientists have sought to understand the extent to which our personalities and identities including dispositional traits are a product of our genes our environment, or both. Most parents are amazed at the differences between siblings brought up in the same household and ostensibly treated the same. The drive to relate everything to our genes has been described by a British epidemiologist, medical doctor, and science writer, Professor Tim Spector (1958–) of King's College London, who has studied identical twins for many years, as "the gene-centric view of the universe, of which he admits to initially having been an "overzealous advocate." He observes how research in its early days feverishly attempted to identify specific genes that shape personality traits and dispositions. Richard Dawkins suggested that we are all hard-wired and our behaviour is largely deterministic based on our genes and that it is the evolution of the genes, not the species that matters.[28] If he is right then it suggests that the emotions we feel, how we behave and think, and what we are motivated by are hard-wired into us from birth in our genes. William Wright concluded that although genes contribute to our behaviour, they do not dictate it.[29]

A balanced, and nuanced approach to the question of just how influential genes are, is offered by American Professor William R. Clark (1938–) and Professor of Biological Chemistry, Michael Grunstein (1947–) who argue that we are and that we aren't hard-wired, by which they mean half of the human behaviour can be accounted for by our DNA, but half can't.[30] Few human behaviours are affected by a single gene and most extreme behaviours and aggression, sexual preferences, mental function, eating disorders, alcoholism, and drug abuse are believed to be governed by a complex of genes, typically across chromosomes. There are many conflicting views on the role of genes in shaping personal identity and we need to avoid a reductionist view of attributing all or even a large component of our personality to our genes.

The majority of what we know about the link between behaviour and genes has come from the study of identical twins who share 100% of genes and data from fraternal twins who share 50%. Early findings from twin studies appeared to suggest genes played an important role in personality which sparked the hunt for the best candidate genes. One of the world's leading research centres the Minnesota Centre for Twin and Family Research observed that we inherit more than 50% of our traits including such traits as obedience to authority, vulnerability to stress, and risk-seeking. A comprehensive meta-analysis of all the research related to the link between the genes, ascribed to neuroticism and MDD, was conducted by a large consortium in 2010. Neuroticism is a dispositional trait that is a pervasive risk factor for psychiatric conditions and genetically overlaps with major depressive disorder (MDD).[31] The review found a novel locus or variant in a known gene that has been implicated in bipolar disorder and schizophrenia, suggesting that genes play a significant role in certain personality traits such as neuroticism. However, more recently Spector, although, at the time he was talking about IQ, observed that "any change in environment… has as much effect than genes, …on almost every human characteristic" (including IQ). This reflects a point made in the 1990s by Sarah Hamson, a senior scientist at the Oregon Research Institute who for thirty years used longitudinal studies to study personality and health over the human lifespan and observed that "humans are unique because of consciousness and culture and these are eminently non-physical phenomena."[32] Robyn Fivush argues that family identity is as important as personal identity and more than simply a product of our DNA.

In the 1940s British scientist Conrad Waddington (1905–1975) and Swiss Ernst Hadorn (1902–1976) pioneered the study of combining genetics and developmental biology. In 1945 Waddington coined the term Epigenetics to reflect how the environment can work with the genome to change gene expression. Epigenetics means above or on top of genetics and refers to external modifications to DNA that do not change the DNA sequence but turn genes on or off and therefore affect how cells read and express the genes. Expression of the gene determines which proteins are constructed by the DNA instructions from a cell and thus are central to the transmission of information between neurons. The environment affects which genes are activated, and which remain dormant with your epigenetics, changing throughout your life. Hair cells, skin cells, muscle, and brain cells all look differently because different DNA has been turned either off or on. The field of epigenetics has radically altered how we perceive the role of genes in that epigenetics can alter the phenotype which means "the set of observable characteristics of an individual resulting from the interaction of its genotype with the environment" (OED). Studies have shown that prenatal and early postnatal environmental factors influence the adult risk of developing various chronic diseases and behavioural disorders.[33] Epigenetics reflects the importance of behaviour and lifestyle choices, such as what you eat, if you smoke, or how much you exercise, because these determine which genes are expressed or rather switched on or off. Food has links to epigenetic change with diet potentially modifying epigenetic tags. The effects don't change your DNA and some epigenetic expression of the genes can be reversed or altered, for example, smokers who have stopped smoking can see a reverse of the epigenetic changes. Studies in animals, plants fungi, and some invertebrates, although not yet humans, have found that it is even possible to inherit epigenetic changes with a higher likelihood if the changes occur in the egg cells or sperm, so epigenetics changes can be passed down to future generations.

A recent overview of the role of nature and nurture by David Hosken, John Hunt, and Nina Weddell observes that genes do affect behaviours and other traits but merely reflect the statistical likelihood of interaction between the genes and ignore the important environmental role. They observe that "Environments are not just the physical environment but also the biotic and social environments provided by other individuals, which have emerged as key factors shaping behaviour and the effects of behaviour. As these latter environments are highly dynamic, continually changing and evolving, this form of gene-environment interaction (GxE) is also always shifting and thus contributes to the substantial variability observed in behaviour."[34] What is clear, is that the dichotomous belief in academic circles that personality and behaviour, can be attributed to either nature or nurture is dead, even if this myth still circulates in sensational quarters of the media. We are creatures that operate in a social and cultural context and in different situations, which in combination with the complex interaction of genes, shape our self. We know that because of our unique neural patterns and pathways, we all develop highly subjective schemas and experience emotions and feelings unique to us which is why individuals differ from each other in a myriad of ways.

Dispositional Traits.

Psychologists have traditionally focused on how dispositional traits, in particular individual intelligence and personality which reflect emotional and motivational characteristics, determine the illusive boundaries between what is normal or abnormal. The abnormal when it falls into a psychiatric categorisation may require intervention or treatment. Psychologists regard dispositional traits as broad, distinct, recurrent, and often habitual, patterns, to describe and explain an individual's behaviour, thoughts, and emotions. They have been observed to exert a major influence on the self-schemas we construct but although they are regarded as stable internal factors of personality, they have posed a problem to psychologists, when trying to agree on how many traits there are and the nature of such traits.

Psychologists have in the past defined personality by reference to types rather than traits, but this involves the artificial allocation of people, to a small number of categories of type, in an all-or-nothing manner in that they either are or are not this personality. By contrast, traits can be possessed along a continuum, which recognizes the fact that most people don't have the extreme personalities assumed in types but possess more complex combinations of characteristics. It can be confusing though, when people describe someone as a personality type because they possess several traits, but this is usually because those traits have been observed and deemed to relate to each other to form a type. This approach has been used in factor analysis in which psychological theories are developed by reducing many variables such as traits into a few numbers of factors, with which the traits correlate, in that the factors are affected by or depend on those traits.

American psychologist, Gordon. W. Allport (1897–1967) often described as the founding figure of personality, developed a highly influential three-tiered hierarchy of personality traits consisting of cardinal, central, and secondary traits. At the top of the trait hierarchy, are cardinal traits which are pervasive characteristics that become how others perceive that person's identity. Cardinal traits tend to develop later in life and dominate and shape a person's behaviour, thoughts, and actions. Classic examples are Einstein (Brilliant), Mother Theresa (Goodness), Adolf Hitler (Evil), Donald Trump (Narcissistic) Martin Luther King, Jr (Heroic), etc. It is less likely that most people are identified by a main cardinal trait but instead more by several central traits. Central traits are the basic traits found in most people that affect behaviour but don't shape it completely and include things like loyalty, affability, anger, deviousness, etc. Central traits tend to influence behaviour so that some people are more disposed to be friendly whilst others are aggressive or angry. Allport believed that most people have 5–10 central traits. Secondary traits are those which people typically exhibit in different contexts so are not always apparent and less influential in general behaviour. For example, being short-tempered when stressed or aggressive when threatened. To varying degrees, dispositional traits combine with the goals and behavioural representations that we discussed in the last chapter and develop over time, to shape our perception, thoughts, and behaviour. If you have a primary cardinal trait this may well account for a large part of your behaviour and values, but most of us have a combination of central traits which we usually recognize.

In 1936 Allport and Henry S Odbert (1909–1995) extracted from the Webster's New International Dictionary, 18,000 words used to describe personality and 4,500 terms used to describe types of behaviour or personality traits and employed a lexical approach to construct a classification or taxonomy of types by reducing all the words used to describe characteristic features, into main group traits which define most personality types. This reflected their view that individual traits play an important role in why we behave as we do. British-American psychologist, Raymond Cattell (1905–1998) best known for his psychometric research into the intra-personal psychological structure distilled these traits into 16 factors. One of the most influential models to emerge since their work is the Five-Factor Model, developed in 1992 by the American psychologists Robert McCrae (1949–) and Paul Costa (1942–) which reduces all the many words used to describe characteristic features and distils these into 5 main factors which are Openness, Conscientiousness, Extraversion, Agreeableness and Neuroticism, which is often remembered using the acronym OCEAN.[35] Openness means being open to trying new experiences and unconventional ideas, customs, cultures, and beliefs that question their assumptions. Those who are more closed are wary of the unfamiliar and tend to prefer the status quo by sticking to routines and predictable whilst openness and curiosity are often associated with higher intelligence.[36] Conscientiousness means that people are aware of the implications and consequences of how they act which makes them more goal-oriented and motivated to act and less impulsive than non-conscientious types, whilst also tending to feel a responsibility to others like not being late or being tidy etc. Extraversion is characterized by being more outgoing, talkative, and sociable and liking or even seeking attention. By contrast, introverts prefer small groups and tend to be shyer and trust fewer people, and those combining both are called by the great name Ambiverts. Agreeableness tends to be those who are more friendly and likeable and avoid confrontation and arguments so will act as peacemakers. Disagreeable people tend to be more selfish in promoting their interests and less charitable or altruistic, which makes it more difficult for them to form relationships and may even be more subject to bullying.[37] Neuroticism reflects the level of emotional stability, anxiousness or fearfulness as well as the extent to which people might focus on the negative or positive.

Much research defining the boundaries of what is normal and abnormal behaviour, has focused on the importance of neuroticism and extraversion with neuroticism impacting very negatively on relationships. The 5-factor model is commonly used in research and personality testing using self-report questionnaires, even across cultures. Traits are measured on a continuum rather than a binary, do you have them or not, because most people will possess all of them but to different degrees. Employers now routinely employ personality profiling to assess suitability for a particular job. Dispositional traits are not just an important tool for determining whether you are right for a job or explaining antisocial or criminal behaviour, but most importantly they contribute to shaping our perception, thoughts, and behaviour so being aware of our traits can be a valuable way of gaining greater agency over our life and social interactions and relationships. Personality and dispositional traits cannot and should not be assessed without reference to the context of a person's situation or environment and the cumulative effect of the environment is probably much greater than simply the role of our personality and our dispositional traits. Psychologists call this situationism, which is the theory that "human behaviour is determined by surrounding circumstances rather than just personal qualities."[38] It is the interaction between the person (traits) and the situation that matters most. For example, in what situations do we become most anxious or aggressive which makes it more complex, recognizing that sometimes the traits are causing the situation to arise in the first place, etc.

Close Relationships and Self.

A considerable body of research suggests that our sense of self or who we feel we are, and motivations and goals are most evident in the context of close relationships. Significant others from childhood and later, have a major influence on our thinking, our emotions, and motivations in terms of our patterns of responding to close others and can provide a conscious and unconscious framework for future relationships. Susan Andersen and Kathy Berenson in their analysis of the social mind in personal relationships and interpersonal behaviour defined significant others as " all individuals who have been deeply influential in one's life and in whom one is (or once was) emotionally invested, including members of one's family of origin (such as parents or siblings) as well as persons encountered outside of family relations either early on or later in life (such as friends, co-workers or boss)."[39] There is growing evidence that much of our pattern of behaviour and actions towards those with whom we are relationally close, is based on our subconscious and can involve the transference from one close relationship to another which can feel effortless. This includes emotional responses, acceptance and rejection, regulation of one's emotions and feelings, motivations, interpersonal behaviour, and the use of cognitive biases and defence mechanisms. A person may lack awareness or intention, barely think about behaviour in a relationship, or feel unable to control it, so may deem their behaviour unintentional. The assumption is that although the memories of the self are distinct from memories of the self in relation to close others they are nevertheless linked.

Anderson and Serena Chen observe that knowledge or mental representations about yourself, linked to your knowledge or mental representation of significant others, combine to form the relational self-schemas that affect our emotions, motivation, self-regulation, and self-evaluation and can be transferred to new relationships.[40] Studies of subconscious behaviour in close relationships, demonstrate that the memories and representation of the relational self, are activated by cues and often drive a person's emotions, cognition, motivation, and behaviour. These are based on how that individual expects to be treated, or have reacted or behaved based on the current or past relationships. If behaviour in a relationship has been chronically repeated, this relational self will be more readily accessible and a person may behave subconsciously in a way consistent with that relational self, unaware of why they respond as they do, often unintentionally and seemingly outside their control. For example, being passive-aggressive, acting out, being in denial, etc.

Not all behaviour in close relationships is purely automatic and may involve varying degrees of automaticity or may be consciously rather than subconsciously goal-directed, particularly when a person is entering or forming an impression of a new relationship, though in that situation the relationship would not necessarily qualify as a close relationship simply because it is new. The longer the duration of the close relationship, the more automatic our behaviour is likely to become and the more entrenched the relational self. Subconscious behaviour may be triggered by preconscious clues or be the result of that behaviour being readily accessible as part of the self-memory system (SMS) which we discussed earlier. We have seen how the relational self comprises three main schemas namely (a) Self-schemas, (b) Schemas about the significant other, and (c) Relational or Interpersonal schemas, or what is also called interpersonal scripts. Interpersonal scripts are patterns of learned behaviour that reflect expectations of how people react to certain behaviours. For example, if I make a mistake I will be punished and criticized which might manifest as I am unworthy.

These schemas are stored in memories which are activated automatically, either temporarily or chronically to combine and form a general relational schema. Some psychologists argue that the relational self, formed through these schemas goes further than the mere linking of the self and the other close person, in that the self merges aspects of the other into the self. For example, the other's perspective becomes part of the self-perspective although they are unaware of this because of its automaticity. In some cases the other (s) is/are incorporated into the self, to the extent that they define themselves by reference to their close relationships so that there is an interdependence. We saw how attachment to caregivers plays a vital role in a baby's development and there is much research into how this can later affect adult relationships, reflecting the belief that early self-memory systems operate as a template for future relationships which can be subconsciously activated temporarily or chronically.

We may have several attachment models that might be activated depending on the nature of the relationship and whether it's family relationships, platonic friendships, or romantic partners.[41] One aspect of early relationship issues such as a child's early rejection, mistreatment, neglect, or abuse that psychologists have observed is rejection sensitivity in which a person will be more sensitive to rejection than others and almost expect to be rejected and typically overact cognitively and emotionally to the perception of rejection by others and in particular significant others. This is distinguished from a sensitivity to threats with research suggesting that rejection sensitivity is learned.[42] A study using brain imagery technology found that activity in the dorsal anterior cingulate cortex was higher in those with heightened rejection sensitivity when confronted by disapproving facial expressions which imply rejection than when confronted by angry or disgusted and more threatening facial expressions.[43] Rejection sensitivity can seriously impact a person's daily life and the forming of relationships[44] and is even regarded as an early marker for depression in both women and men.[45] It is also linked to a range of mental health issues including anxiety, loneliness, borderline personality disorder, and body dysmorphic disorder, and to people who tend to spend more time ruminating and thinking about themselves.[46] This begs the question whether robots would ever be able to form close relationships?

The Neurobiology of Self.

Antonio Damasio proposed for research purposes, a working definition of the I self as "A dynamic collection of integrated neural processes, centred on the representation of the living body, that finds final expression in a dynamic collection of integrated mental processes."[47] Damasio recognizes the elusive and illusory nature of the I or knowing self, observing that, "It is more dispersed, often dissolved in the stream of consciousness and at times so annoyingly subtle that it is there but almost not there."[48] Nobel Prize-winning psychologist and economist, Daniel Kahneman (1934–2024) famous for his work on how people make decisions and the application of psychology to economics, believes that we have an experienced (introspective) self and a remembered (retrospective) self. He describes the experienced self as a string of moments with each moment lasting around 3 seconds and with 20,000 such moments each waking day amounting to over 500 million moments in a 70-year life. Most of these experienced moments evaporate and are no longer even remembered, whereas the remembered self appears relatively stable and more permanent. The experiences represent the minimalist self, which has been described by the French philosopher Michael Henry (1922–2002) as the "Flame that enlightens its surroundings and thereby itself."[49] This characterization of the experienced self echoes the description provided by the great neurophysiologist Sir Charles Sherrington (1857–1952) as the "Enchanted loom where millions of flashing shuttles weave a dissolving pattern, always a meaningful pattern, though never an abiding one. Its commanding presence orders sensation, movement, thought, a lifetime of memory and dream." The rebellious Noble Prize-winning, Algerian-born French philosopher, author, dramatist, and journalist, Albert Camus (19131960), described how, "If I try to seize this self of which I feel sure, or if I try to define and to summarise it, it is nothing but water slipping through my fingers."[50] Peggy O'Connor (1965–), who is a Professor of Philosophy and Gender, Women, and Sexuality Studies, and Chair of the Department of Philosophy at Gustavus Adolphus College, describes the self as a "Unicorn", which whimsically, highlights the challenges we face when trying to understand the I or the knowing self.[51]

The fleeting nature of subjective moments, from which we string together coherent memories of our experiences and hence our self, reflect the phenomenological nature of how our minds work and the beguiling nature of what psychologists call as we saw earlier qualia. Qualia are "the internal and subjective component of sense perceptions, arising from stimulation of the senses by phenomena. They possess "characteristics or qualities that determine the nature of a mental experience (sensation or perception) and make it distinguishable from other such experiences."[52] They include memories and emotions and may be conscious, subconscious, or preconscious. Examples of sensory qualia are what it is like to feel pain, to see a colour, etc, all of which are entirely subjective. Our self-qualia give us a sense of coherence and agency as unique individuals and as American philosopher Thomas Nagel (1937–) observed this subjective character of consciousness experience, "escapes physical theory."[53] Warren J in his review of Joseph LeDoux's book Synaptic Self (2003), which links self to memory and learning at the synaptic level in the brain, went as far as to observe that "getting inside our qualia is proving much more difficult than getting inside the atom."[54]

How the Brain Integrates Information-related to the Self

Understanding where the self emerges from the chaotic brain, brings us to another special neurobiological feature of our brain which is the process by which sensory information and data related to emotions, thinking, and motor responses or behaviour (movement) in connection with the self are integrated in the cerebrum. Integration generally happens largely in the association areas of the cortex and in another very important region of the brain called the Insula. We saw how neural patterns are not confined to predetermined topographical maps in the primary cortex regions but exist across the cortex in the associative areas sub-cortical regions and even the retinas of the eyes, playing a vital part role in how the brain integrates information about the self and your environment and perceives our world. The associative areas are concentrated in the parietal, occipital, and temporal lobes, the territory near the temporal pole (under which the amygdala and hippocampus lie), and the forward part of the frontal lobe. They receive inputs from multiple brain areas, but their functions are not directly related to sensory or motor inputs in the primary regions, more the organizing and integrating of the vast amount of incoming sensory information from all these sources vital to generating behaviour. They account for our ability to understand reason, be conscious, and possess complex behavioural patterns and emotions. They do this by influencing a broad range of cortical and sub-cortical targets which include the hippocampus (memory and emotions), the Basal Ganglia (emotion recognition, subjective feeling elicitation, and reward valuation) the Cerebellum, (the tiny version of the brain at the back) the Thalami and sensory information of hearing, taste, sight, and touch (but not smell) from your body to your brain and other association cortices.

Studies have shown that the parietal association cortex attends to complex external and internal stimuli, the temporal association cortex identifies the type of stimuli, and the frontal association cortex plans the appropriate behavioural response. The collective effect of information integration is what we know as cognition or how we come to know the world, in the form of semantic and episodic knowledge or memories. Emotional responses and behaviours are generated through integration, becoming the basis for our subconscious and habitual responses which combined with autobiographical memories reflect past behaviours and responses. These subconscious neural patterns and neural pathways, established over time through experience, have a powerful influence on governing our ongoing response to internal and external stimuli, which includes how we react to others and shape our version of ourselves.

The Insula

The Insula (Latin for Island) or what may neuroscientists now call the 5th Lobe, was first described in 1796 as an island of cortex by German physician, physiologist, anatomist, and psychiatrist Johann-Christian Reil (1759–1815), who also coined the term psychiatry. Located on both sides of the brain, deep within the lateral sulcus separating the temporal lobe from the parietal and frontal lobes, it is covered with folds of these adjacent lobes called Operculum (Opercula -the plural of operculum in Latin, which means little Lid). The insula gained great attention following Antonio Damasio's formulation in 1999 of the somatic marker hypothesis, that rational thinking cannot be separated from the feelings and emotions represented in the brain as body states. Activated when a person feels, anticipates, or assesses the intensity of pain, the Insula acts a neuronal hub in the cortex, playing a central role in constructing our sense of self.

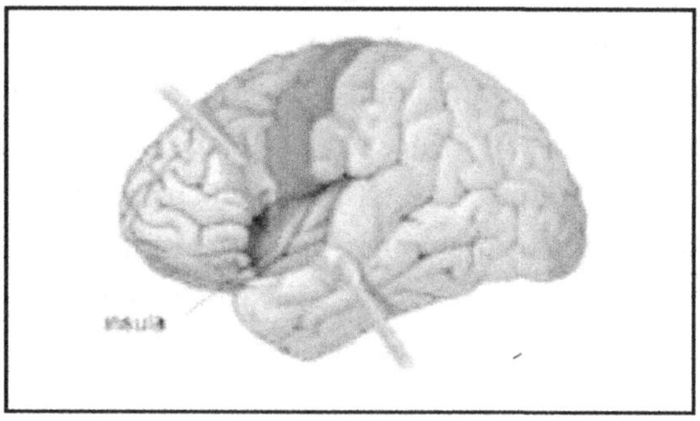

It supports subjective feeling states and is thought to be the source of the interoceptive awareness of the physical self as a feeling (sentient) entity, which constitutes the basic form of consciousness and the primary basis of selfhood. It identifies salient information and stimuli in the environment by reference to subjective feelings states, for example, thirst, sensual touch, itch, sexual arousal, temperature, heartbeat, hunger, etc, and initiates cognitive processes for further processing of salient or important information. Although usually associated with smell and taste the Insula generates some emotions and feelings, particularly disgust and avoidance and the constant updating of subjective feeling states, enables conscious awareness of one's own body. Hence its central role in the conscious or explicit motivation associated with the brain's reward system, and engagement in an activity or the implicit/ unconscious desire to do so.[55]

In a 2015 study, American psychologist Madeleine Goodkind and colleagues suggest that problems in the Insula are a common cause of many psychiatric disorders and structural and functional abnormalities have been associated with major depressive, bipolar, and anxiety disorders and even anorexia, all of which involve various disorders of the sense of self and an individual's levels of motivation.[56] The Insula receives sensory inputs from the spinal cord and brainstem, via the thalamus and inputs from the parietal, occipital, and temporal association cortices. It has a role in somatosensory, vestibular balance, motor integration, and multiple functions through connections with other brain regions, particularly the left amygdala (see later),the left inferior frontal cortex, the cingulate cortex the thalami and basal ganglia.[57] It connects with the basal ganglia / ventral striatum, which integrates information associated with homoeostasis into emotional, cognitive, and motivational functions specifically desires, cravings, and addiction, creating neural patterns or maps of the viscera. The right insula is associated with self-referential tasks and recent studies suggest that it plays a vital role in mediating self and representing the self in terms of creating the sentient self. Sentient from the Latin – Sentiens, means feeling and perceiving, which allows you to be aware or to perceive or feel things about yourself.[58]

The right insula is involved in the highly complex integration of interoceptive, exteroceptive, and proprioceptive information, which forms a vital part of the self. It does this through neural activity that exhibits special topographic and dynamic features through functional and temporal integration of information. Temporal integration involves combining information over time to improve detection and discrimination, whereas functional integration as we have seen involves allowing brain regions to work together to process information and generate responses. Imaging studies show significant neural activation in the Insula, in the form of neural pathways and neural patterns, when participants were required to reflect upon their personal qualities compared to an acquaintance.[59] The Insula is responsible for the neural mapping of body states. Integration in the associative cortex and insula plays a vital role in the coherent organization of information in the brain, but this doesn't tell us where the self is to be found in the brain or where and how information related to the self is processed in the brain.

Where in the Brain is the Self Located?

Localization of the elusive self in the brain has become a major challenge in psychiatry, psychology, and neuroscience. Is there a specific location, or does it arise through communication between various brain regions and integration of the information? Douglas Hofstadter describes how “In the end, we are self-perceiving, self-inventing, locked-in mirages that are little miracles of self-reference.”[60] Hofstadter believes that we are a self-referencing system by which he means that our brains possess abstract feedback loops that create the illusion of self, and this is just the way our brains create a model of our world, based on which it incorporates the abstract self as a composite of all the abstract patterns.

The Modular Mind Theory has been postulated by American professor of psychology, Michael Gazzaniga, (1939–) who argues that our brains process information on a modular basis with each module performing a specialized function, with various modules simultaneously processing the specific aspects of the information, whilst recruiting the other modules in the process. Something is considered modular if its functions are made of multiple dimensions or units to some degree. This builds on the earlier work of a founder of the cognitive sciences, Jerry Alan Fodor (1935–2017) who developed the modular theory of mind in which multiple cognitive processes, are deemed “encapsulated and inaccessible but are nevertheless required to perceive even one thing.”[61] However, the modular view has been heavily criticized on the basis that mental activity appears to be distributed throughout our brain’s neural patterns and can’t simply be dismantled into component modules or units.

What appears to be acknowledged by most theories is that information must be processed encoded and stored in differentiated neural circuits or sub-assemblies and requires integration for it to constitute a coherent sense of self. To understand the neuroscience of where information related to the self is processed in the brain, we need to go back to 1977 when neuroscientist Timothy Rogers and his team were exploring the depth of information processing in the brain. They developed the Self-Referent Effect, which asserts that information about the self, is more likely to be processed and encoded differently from information about another person. We tend to remember information related to ourselves better than others and Rogers believed that this is the result of information about self being more strongly encoded.[62] They were exploring the organizational level of information processing and encoding in the brain, evidenced by memory formation and recall and how this was affected by information related to self, compared with information not related to self. They found that our memory is weaker for non-self-referential words, which appear to be less deeply processed. For example, a participant is given a list of adjectives (e.g., intelligent, shy) and asked which best describes their dispositional traits, qualities, characteristics, attributes, beliefs, etc and because they are required to decide concerning knowledge about themselves it is a self-referential process. Others are asked to make observations about an object and its size which requires no reference to self. The findings suggest that those asked self-referential questions, remember them better than non-referential questions and tend to remember the words better when they relate more closely to themselves than unlike themselves, which suggests that encoding of self-related information is deeper, particularly with words they endorse or agree with.

Cynthia Symons and colleagues in their meta-analysis of self-referential studies observed that "the self is a well-developed and often-used construct, that promotes elaboration and organization of encoded information."[63] The explanation is that the brain encodes memory about us in a more highly organized and accessible way, with easier retrieval and recall, probably because people habitually use encoded self-referential memories, making recall easier, suggesting self-judgments are more deeply processed. They observed individual differences, reflecting a bias towards certain referential information among types or groups, for example, depressed people tend to remember depressive traits compared to non-depressive who remember non- depressive traits. Those low on low in private self-consciousness tend to be less self-referential, than those high in private self-consciousness.

Studies using EEG, fMRI, and Positron Emission Topography, (PET) have provided growing evidence that when our brain processes information related to the self it manifests as heightened neural activity in the Cortical Midline Structure (CMS) and associated sub-cortical areas, even though other regions may be involved. The CMS is the main brain area associated with encoded self-knowledge and retrieval of that knowledge compared to retrieval of social or semantic knowledge.[64] Based on a statistical analysis of the results of multiple scientific imaging studies, George Northoff argues that "self-referential processing in the CMS constitutes the core of our self and is critical for elaborating experiential feelings of self, uniting several distinct concepts evident in current neuroscience."[65] Studies have shown that the CMS overlaps with pain-related brain regions and disturbance in the CMS can result in chronic pain and neurological and psychiatric conditions. Current neuroscience suggests that the self is processed and encoded in specific brain regions with a critical role played by the CMS in integrating and processing the self, across personality, social behaviour, emotional control, formulating, and processing thoughts related to self and judging others, our levels of motivation and drive and our capacity to adapt our behaviour to different situations.

The Default Mode Network (DMN)

Another unique neurobiological feature of the human brain appears to be the capacity to ruminate, over-think and worry which is associated with the function of the MPC that is part of the Default Mode Network (DMN) that we discussed earlier. The DMN exhibits high metabolic activity when ruminating and worrying etc and it is believed to be an important cortical hub for communication with widespread brain regions. It is involved in a variety of highly complex functions including episodic memory (of everyday events or past experiences such as the first day at school or a birthday), visuospatial processing, reflections upon self, and aspects of consciousness. The existence of the DMN was first hinted at in

1929 when Hans Berger observed that the brain was constantly active when a patient was asleep. It was discovered in 2001 when scientists observed that over 90% of the energy consumed by the brain is used to support this region.[66] The DMN exhibits strong low-frequency oscillations coherent during the brain's resting state involving self-referential processing, interoception, autobiographical memory retrieval, and imagining the future but is deactivated when a person is engaged in cognitive tasks. Brain imaging technologies such as fMRI and EEG, which identify active parts of the brain when an individual is not task-orientated, have revealed that the DMN appears vital for communication with widespread brain regions. This is based on its high metabolic activity and connectivity and involvement in a variety of highly complex functions, including (1) Episodic memory (of everyday events or past experiences such as the first day at school or a birthday) (2) Visuo-spatial processing (3) Reflections upon self and (4) Aspects of consciousness.

Our brains constantly employ affective forecasting, which relates to our moods, feelings, attitudes, and beliefs about others and ourselves, and the DMN acts like our inner theatre when we self-referentially evaluate and imagine various past or future scenarios. This may have given us a survival advantage thousands of years ago, but we spend a surprising amount of time in emotional forecasting and ruminating. Studies indicate that the more we meditate, the more we think negatively and become anxious, resentful, regretful, or self-critical, which leads to repetitive rumination. The DMN is more active in lonely people, and DMN dysfunction is associated with over-rumination and self-preoccupation in various forms of mental illness, including serious depression. It is attributed to difficulty deactivating from the DMN which may explain why experienced meditators exhibit less activity in the DMN. The PCC which is part of the DMN is implicated in internally directed thought processing[67] but recent evidence in non-human primate studies suggests it also controls cognition and changes in behaviour in response to environmental change.[68]

The Encoded Self

Georg Northoff, argues that the brain's resting state or DMN is a "neurophenomenal predisposition with its self-specificity activity or self-specific organization reflecting spatial and temporal orientation to the "organisms own particular needs and demands,"[69] which enables assignment of self-specificity to both external and internal stimuli. He proposes what he admits to being a daring hypothesis which is that any extrinsic stimuli must react with the brain's intrinsic features related to self to engage brain functions such as cognition, sensorimotor, and affective responses, or social functions, including choices. Reward-driven tasks and decisions are affected by and ultimately dependent upon the resting state, and all tasks must be assigned some degree of self-specificity. Evidence for this is increased neural activity in the CMS when processing emotions, making decisions, and evaluating rewards. He argues that even lateral regions involving higher-order functions like working memory and attention cannot avoid being affected by the CMS and self-specificity, albeit at a lower level, given how strong the deviation from the resting state is. This is more heretical than it might appear considering all functions depend on smooth interactions between various brain regions.[70] This suggests that our encoded self is constantly operating in the background. Studies show that shifting the self into the present rather than emotional forecasting can be a very helpful way to avoid negative rumination and mindfulness and that paying attention to the moment without judgment does precisely this. However, we all live lives full of so many distractions, including using our various devices, that we rarely spend time in the moment, or if we do, we don't do so mindfully but in negative rumination.

Research by Northoff and his team in 2015 suggests that the brain at rest possesses what they described as at-rest self-containment.[71] Northoff questions the traditional view that the self results from higher-order cognition associated with a larger brain and neocortex, specifically the pre-frontal cortex, rather than the lower-order sensory functions. He argues that the self is neither a feature of the higher nor lower cortex but that the self is encoded in the brain in its resting state. He describes the self as an intrinsic function of the brain, and spontaneous activity associated with basic functions, including perception, emotions, and reward, is not a feature of higher cognitive function but is central to linking external and internal stimuli. The studies appear to support this view. They operationalize the self by assessing the self-relatedness of stimuli and how they affect behaviour. Their analysis is conducted through the lens of three key domains of behaviour and self, emotions and reward, and decision-making because they represent important neuro-affective, neurosensory, and neurocognitive functions. They found that self-specificity and emotions are evident in activity in the ACC, which we discussed earlier.[72] Northoff's 2009 study involved taking brain images of participants when they perceived emotional pictures using the International Affective Picture System but were not required to perform any tasks. Afterwards, they were asked about the level of emotional arousal and valence (positive/negative) of the pictures and the level of self-specificity. The objective is to see to what extent the underlying neural activity correlated with the subjective degree of self-specificity after being in the scanner. They found a significant correlation in the neural activity in the brain stem, sub-cortical and cortical regions, mid-line regions, and the degree of self-specificity. These sub-cortical regions included the Tectum, Colliculi, Amygdala, VTA, Mediodorsal Thalamus, and Ventral Striatum (VS). In contrast, cortical regions concerned are the periatal anterior cingulate cortex (PACC), VMPFC, DMPFC and the posterior cingulate cortex (PCC). Northoff's work supported a strong correlation between subjective self-referential emotions, reward evaluation, decision-making, and underlying neural activity and concluded that the CMS mediates self-specificity. To do this, they had to ensure that the self-specificity rather than emotional valence correlated with the CMS activity, but they found that the CMS and sub-cortical regions are modulated in different ways. Other studies focused on reward using reward stimuli, which included alcohol, gambling (slot machines), and natural reinforcers such as food, etc., found high self-specific stimuli elicited high degrees of neural activity in the VTA, the VS, and the VMPFC where reward induced strong activity changes, suggesting a close overlap between reward and self-specificity. The CMS, through the combined functions of all these connected regions from the front to the back of the brain and their links to sub-cortical areas, is crucial to how our brain processes information related to the self.This is why self awareness is so important in influencing and governing our thoughts and behaviour. So let's look more closely at what self- awareness consists of.

Self-Awareness in Humans

In 1972, psychologists Shelley Duval and Robert Wicklund published their landmark theory of Objective Self Awareness, which observed that people focus on the self as an object or external environment. They observed that focusing on ourselves is a vital human capability that allows us to self-evaluate and compare our behaviour against our standards of behaviour to change our behaviour based on generated emotions such as shame or pride, etc.[73, 74] Studies have shown that when people are not focused on the self, they don't necessarily act according to their standards, so focusing on the self is an important aspect of monitoring ourselves and self-regulation. We all know how challenging it is to be friends or partners with someone who can't appropriately regulate their behaviour or empathize with our needs. Focusing on the self is not the same though as self-awareness, as we see with narcissists. Toddlers don't appear to experience objective self-awareness in being able to intentionally focus on themselves as distinct beings until they are around 18 to 24 months old. Studies have found that people suffering from neuropsychiatric disorders like schizophrenia or psychosis have impaired self-awareness and even display a delay in passing the mirror test,(discussed below) as well as manifesting thought insertion and other disorders that seriously disturb their sense of self.[75] Monitoring and regulating one's behaviour by being self-aware appears essential to functioning socially and effectively. Let's look more closely at self-awareness by starting with the famous mirror test.

The Mirror Test for Humans

Research into the knowledge of the self, based on when things go wrong with a person's perception of their self, is making us rethink the concept of self-awareness, but the act of recognizing yourself in the mirror, coupled with powers of introspection and self-reflectivity, has been regarded by philosophers and scientists as the pinnacle of human capabilities, requiring a complex and large brain. There is an objective biological self in terms of the entity we see in the mirror and a subjective cognitive and emotional self, which gives us each a unique identity and self-concept, which we recognize in the mirror. Being aware of yourself means what happens every day when we wake up from being asleep or being unconscious in other ways. You typically experience a return of that sense of you and the self you see in the bathroom mirror will be the self you see in pictures of you when you were younger, even if the face has gotten older, with more lines and the eyes wiser and perhaps more rheumy. You will see not just the physical face that is you but the reflective self that is experiencing and witnessing you being you. This simple, unnoticed, and automatic act of recognizing your reflective self in the mirror, called self-consciousness, makes us one of the few animal species, including apes, dolphins, elephants, and magpies, that appear capable of different degrees of self-awareness. Recent research suggests some ants and manta rays may possess self-awareness, and Japanese scientists have observed a very small fish called the Cleaner Wrasse pass the mirror test, which means that they seem able to recognize their image in the mirror. They clean off gel markers put on their heads by researchers, which they could see in the mirror, though some suggest that they may have mistaken this for parasites on the skin, which they naturally clean off. Chinese neuroscientists recently found that rhesus monkeys can learn to identify themselves in the mirror, evidenced by touching a spot painted on their heads. The monkeys initially react to their reflection in the mirror as though they have seen another animal, but after repeated exposure to the mirror, they interact directly, non-socially, with their image. It is unclear whether this training uncovers a latent self-awareness or teaches the monkeys a new cognitive skill, which suggests that recognizing oneself in the mirror is a learned process, not an inbuilt feature of the brain.[76] The American evolutionary psychologist Gordon G. Gallup (1941–) developed the mirror test of self-awareness as an undergraduate and since 1970, it has been used as an almost universal threshold for distinguishing between conscious and unconscious beings. Gallup still believes that only chimpanzees, orangutans, and humans beyond a certain age have passed the mirror test, even though he claims to be "more than happy to consider the possibility that any other species might be capable of recognizing itself in a mirror." Failure by other species to pass the mirror test has not been attributed to poor problem-solving skills but the inability to recognize themselves and their bodies in the mirror as entities distinct from others. Studies involving the Cleaner Wrasse have raised significant questions about whether other species are self-aware, which would make humans look distinctly less special in the scheme of things.

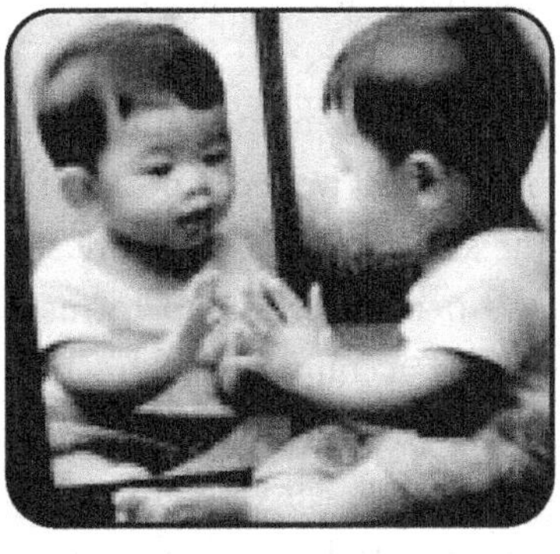

It may suggest that the mirror test does not measure self-awareness but something else and we may have failed to understand what goes on in other animals' minds. If passing the mirror test is a skill that can be learned, does this mean that it is not an inherent trait that appears confined to the great apes and is part of the developing sense of self, whereby you are equipped with the neural correlates that allow you to become the object of your own attention? Can AI machines be programmed to learn how to pass the mirror test, as they have already programmed robots to recognize people, which means that we need to look at what passing the mirror test means for our robot friends and us? How can we say that a human or a robot has passed the mirror test without knowing what constitutes self-awareness?

Humans are social beings embedded in social contexts, including family, friends, and others, and we seek and need companionship, intimacy, and connection with others. How the self and behaviour emerge by reference to others is an important field in psychology and neuroscience and forms the social self. Any concept of desired behaviour in robots should accommodate this more complex dynamic of the self and others as a social construct. We have seen how neuroscience has revealed how our brains process information related to the self very differently from processing information related to others and that the neural correlates of self are complex and highly integrated in the human brain. We don't need language to pass the mirror test, as young children and other species can do so without formal language. Some psychologists, including social psychologist Dr Alain Morin, observe that humans use inner dialogue to focus on themselves and that this internal verbal conversation serves several functions regarding human self-awareness. In his review of nine neurocognitive models of consciousness, Morin identifies the main attributes of self-awareness as the (1) Perception of the self in time, (2) Complexity of self-representations, (3) Frequency of self-focus, (4) Amount of self-related information and (5) Accuracy of self-knowledge. Morin argues that our levels of consciousness and self-awareness reflect the nature and complexity of the information related to the self, with more sophisticated self-representation associated with higher levels of consciousness and the processing of crude types of self-information related to lower levels of consciousness. He believes that inner dialogue is key to self-awareness at various levels and that conceiving oneself across time plays a crucial role in generating the extended self.[77]

We need to be aware of some significant criticisms of the mirror test, which include the following. It is used less in animals that rely on non-vision senses like smell to navigate and identify their surroundings. Some animals react aggressively to the image because of instinctual behaviour, so they may need more time to conduct the test properly. Some animals may not be bothered to try and remove a mark on their head or face but still recognize themselves, and awareness of the self, as we have seen, is more binary than passing the test.

Mentalizing and the Theory of Mind (TOM)

American psychologist James Jerome Gibson (1904–1971) argued that self-awareness starts early in life and begins with a child's direct ability to perceive itself as an actor and object through proprioception or kinaesthesia, in terms of the child's body being able to sense movement, action and locate itself in space. Based on sensorimotor data, Gibson observed that the perceptual self forms the basis of the later self-concept as the child develops but does not involve an internal representation of the self that the self-concept later involves. Infants, for example, use visual information to perceive changes in their orientation but do not employ a self-concept. Self-awareness in humans is a complex and elusive concept which psychologists have described as involving many self-referential processes, including the essential ability to distinguish the self from others. Self-referential processes include not just bodily awareness and self-face recognition but distinguishing between our private and public selves, thinking about our past and future, and inferring the thoughts or minds of others, which is called mentalizing. One theory for this that has been widely adopted is the TOM.

Mentalisation is the basis for empathy compassion, and metacognition, and individuals deficient in the capacity to mentalise are known to have difficulty passing the mirror test. You can't have social behaviour without a minimum level of self-awareness that allows you to distinguish yourself from others, which involves mentalizing. Mentalizing involves interpreting/understanding behaviour (one's own and that of others) as psychologically motivated by reference to underlying intentions and mental states, like thoughts, feelings, desires, and intentions. In 1978, American psychologists David Premack (1925–2015) and Guy Woodruff produced a landmark paper on the ability of humans, and they suggested some animals, like chimpanzees, attribute mental states, intentions, desires, emotions, beliefs, and knowledge to both us and others. They concluded that "as humans, we assume that others want, think, believe and like and thereby infer states that are not directly observable, using these states anticipatorily, to predict the behaviour of others as well as our own." The key elements of mentalizing are the ability to (1) Understand others' states of mind (including potentially other robots), which includes the capacity to empathize and show compassion (2) Predict others' behaviour and actions. (3) Use metacognition concerning one's cognitive processes, and (4) Effectively communicate these to another party Premack and Woodruff observed that drawing inferences about another person's mind is "universal in human adults."[78] There is an ongoing debate about whether inference of other people's level of knowledge and mental state, is an automatic process, or whether we deploy costly cognitive processes and resources. Paula Rubio-Fernandez and colleagues point to the evidence from laboratory studies employing false belief tasks that suggest belief inferences are cognitively costly, controlled processes, although they recognize that "in everyday conversation, belief reasoning is pervasive and therefore potentially automatic in some cases."[79] There is a distinction between making inferences about others' cognitive states, such as beliefs, thoughts, and reasons for doing things, and their emotional states, with the latter often referred to as Affective TOM. Mentalization concerning both cognitive and emotional states is regarded as a key component of self-awareness and an important developmental milestone in children with delayed development of TOM in children even appearing to make passing the mirror test more difficult.

Alan Leslie in 1994, postulated that TOM is a specialized mechanism (TOMM) that produces domain-specific learning and that "childhood autism can be understood in terms of an impairment in the growth and functioning of this mechanism."[80] Leslie laid out five guiding ideas in the theory of TOMM, which were as follows (1)The key to understanding the origins of the theory of mind lies in a time-pressured, online processing to interpret an agent's (another person's) behaviour in terms of underlying intentions (2) Descriptions of intentional states are computed by ToMM which is post-perceptual (follows the perception of the stimuli), operates spontaneously, is domain-specific, and is subject to dissociable damage-in the limit, modular (3)TOMM employs a proprietary representational system which describes propositional attitudes (4)ToMM forms the specific innate basis for our capacity to acquire a TOM and (5)ToMM is damaged in childhood autism resulting in its core symptoms and impairing these children's capacity to acquire a theory of mind. Leslie's position represented a move away from the earlier view that TOMM was domain-general capacity for meta representation to the view that it is domain-specific.

In 2006, Valerie Stone and Phillip Gerrans and Valerie Stone observed how over twenty years of studies in neurological patients with lesions and autistic children suggested that "Lower-level domain-specific mechanism, e.g., tracking gaze, joint attention interacting with higher-level domain-general mechanisms for meta representation, recursion, and executive function can account for observed patterns of deficits in both autism and neurological patients. They maintain that "deficits in lower-level domain-specific mechanisms for processing social information without proper inputs" and "the intact capacity for meta representation by itself cannot make correct ToM inferences." They argue that it is unnecessary to postulate a separate TOM mechanism.[81] The problems arise when there are domain-specific deficits in processing information related to inferring others' thoughts. Recent research by neuroscientist Matthias Schurz and colleagues suggests that understanding other people's mental states involves three levels of processing in our brain, which are the (1) Internal cognitive processes of representations of others' mental states, (2) Emotional processes involved in watching other people experience emotions based on "shared emotional, motor, and somatosensory representations" and a (3) Combination of these which combine the cognitive and emotional process that allows us to feel empathy and compassion which as we will see are not the same thing.[82]

For a long time, it had been thought that mentalization involved cognitive processes while empathy specifically involved just emotional processes, but based on a comprehensive review of neuroimaging studies, it now appears that TOM and empathy involve elements of both cognitive and emotional processes, which again illustrates how we can't simply separate cognition and emotion as though they are distinct and separate processes in the brain. There is no consensus, though, on whether certain TOM processes are automatic, faster, and subconscious, while others are more cognitively controlled and slower. A common research method of assessing TOM is the notion of False Belief, defined as a "belief based on incorrect inference about external reality that is firmly sustained despite what almost everybody else believes and what constitutes incontrovertible and obvious proof or evidence to the contrary." They differ from delusions, which, although false, involve beliefs that can't be changed even when presented with incontrovertible facts that disprove the belief. Delusions are commonly seen in mental health conditions like schizophrenia and can manifest, for example, in grandiose delusions in which they believe they are highly talented, rich, or powerful; paranoid delusions where they think that someone is going to harm them; or somatic delusions in that they believe something is wrong with their body. Humans can consider other people's false beliefs as a factor in their behaviour and understand that these false beliefs or delusions influence their actions.

Psychology tests of children often involve a false belief task pioneered in 1983 by cognitive scientists Heinz Wimmer and Josef Perner, which requires a young child to infer that another person does not possess knowledge that they possess about something.[83] For example, the child is shown a sweet box containing coins and asked what they think someone else would expect to find in the box. Children under 3–4 typically answer pennies, while older children say sweets. However, recent studies suggest differences between non-verbal and verbal tasks. Paula Rubio-Fernández and Bart Geurts attributed this to the fact that non-verbal experimental designs "allow infants to keep track of a protagonist's perspective over a course of events whereas verbal designs tend to disrupt the perspective-tracking process in various ways, which makes it too hard for younger children to demonstrate their capacity for perspective tracking." Rubio-Fernández and Geurts appeared to demonstrate that children under 4 years old can also pass a "suitably streamlined version of the verbal false-belief task provided that they are allowed to keep track of the protagonist's perspective without too much disruption".[84] This may explain why some infants can pass false-belief tasks at 13 or 15 months old.[85, 86] Growing evidence suggests a strong connection between mentalizing and executive functions or executive or cognitive control, which describes the top-down mental processes needed when you must concentrate and pay attention and how they develop in parallel. Several theories have sought to explain this but are beyond the remit of this book.[87]

Empathy and Compassion

One of the most important features of mentalizing and TOM is the ability of humans to empathize with and show compassion to others. It often frightens people to interact with robots as the perceived lack of ability to experience or feel any human emotions means they have no empathy and compassion, which means no constraint on what they might do that could be harmful. It is important to distinguish between sympathy and empathy because sympathy is a display of pity often coupled with the relief of not facing the same problem. In contrast, empathy reflects an understanding of how someone feels and a willingness to mentally, if not physically, put yourself in their place. Empathy is connected to the wider concept of altruism, defined as the "disinterested and selfless concern for the well-being of others" (OED). Scientists have long sought a scientific explanation of altruism as opposed to its antithesis of egotism and selfishness, with many studies testing whether the motivation is underlying empathetic concern and altruistic or egoistic. American social psychologist Daniel Batson (1943–) has observed the undeniable force of self-interest but explains that studies suggest under certain circumstances, "we can care for others for their sakes, not our own."[89] Batson is well known for the empathy-altruism hypothesis, which states that "feeling empathy for a person in need evokes motivation to help that person in which these benefits to self are not the ultimate goal of helping; they are unintended consequences."[90] Batson's laudable goal has been to establish how our knowledge of and understanding of empathy and altruism may be directed at a more humane society, which of necessity must include autonomous androids if they are to work and live amongst us, particularly when serving vulnerable people.

Empathy in humans requires not just the effective state in the observer and the phenomenal experience of concern but the adjusting of their behaviour accordingly. The behaviour adjustment is usually subconscious and may involve mirroring the other person's behaviour, including their postures, mannerisms, facial expressions, accents, pitch, and tone of voice. Empaths tend to be much quicker in recognizing emotions generally, including threatening facial expressions and neutral or happy expressions. Empathy is known to reduce the distress and aggression of others and foster social bonds, with the potential to promote more harmonious communal behaviour.

Although related to empathy, compassion should not be conflated with empathy or altruism either as a psychological state of mind or, as we will see shortly, in terms of the underlying neuronal processes. Compassion has been defined as the "feeling that arises when you are confronted with another's suffering and feel motivated to relieve that suffering." Jenny Gu and colleagues, in a recent empirical examination of the factor structure of compassion, identified five elements to compassion, which are (1) A cognitive appreciation or recognition and understanding that another person is suffering (2) Understanding the universality of suffering in human experience (3) Sympathy towards that person in terms of feeling emotionally connected to their fear or distress (4) Tolerating uncomfortable feelings aroused (e.g., fear, distress) so remaining open to and accepting the person suffering but a desire to relieve that person's suffering and (5) The willingness and motivation to take steps and find solutions to alleviate the suffering.[91] Dutch-born biologist and highly respected primatologist Frans De Waal (1948–) describes how people and many animal species are hard-wired to be generous and compassionate and that this is as natural as the tendency to violence and rivalry.[92]

Only recently have scientists discovered that although related, the human brain appears to process empathy and compassion differently. Matthieu Ricard (1946–) used to specialize in cellular genetics before he left his home country, France, 35 years ago to study Buddhism spending five years on a reflective retreat in the Himalayas. He now works on humanitarian projects in Tibet and Nepal. His involvement as a participant in research into meditation has earned him the title of the person whose brain has been imaged using fMRI more times than anyone else. He has written many successful books, recorded some wonderful conversations with the Dalai Lama, and explored the universal human journey in conversation with a philosopher and a psychiatrist. Ricard describes how he was a guinea pig and collaborator on a neuroscientific study of empathy led by a German psychologist and social neuroscientist and the scientific director of the Max Planck Society's Social Neuroscience, Tania Singer (1969–).[93] He was asked to alternate 20 times between a state of empathy and a neutral state of mind while the team scanned his brain to identify changes. Singer was surprised that the images were not what they would have expected to see when someone was empathizing with someone else's pain; Ricard told her that he "had meditated on unconditional compassion, trying to feel a powerful feeling of love and kindness for people who were suffering, but also for all sentient beings." Subsequent data analysis revealed that the neural networks activated by someone meditating on compassion differed greatly from those when someone is being empathetic. They observed that the neural network linked to negative emotions and distress was not activated during meditation on compassion but those associated with, for example, positive emotions of affiliation and maternal love were.[94]

This led to a much wider project to study the neural differences between empathy and compassion, which is very important for people in health care and social work, given that surveys estimate that over 60% suffer empathetic fatigue and burnout associated with constant exposure to empathetic pain towards others. Ricard explained how meditators must reflect on the Second Noble Truth, in which the Buddhist goal is the release from suffering, called Nirvana. It tells us that our belief that we possess a permanent and unchanging self means that we cling to the narrative of ourselves that we have created, and the attendant cravings of jealousy, hate, and other manifestations of self, cause us unhappiness. He describes how the love of others, compassion, and empathy are connected, but the subsequent study focused on establishing whether they activate different neural networks. This required Ricard when being empathetic, to focus on the suffering of a close relative, for instance. He recounts how he imagined a person close to him being injured in a car accident, lying in blood by the roadside at night, far from help, and that disgust at this bloody spectacle became mingled with his distress. After this, he watched a BBC documentary on the plight of the Romanian orphans and their tragic distress and suffering, and after an hour felt exhausted. He was then asked to meditate on compassion, and his overwhelming feeling was the need to comfort and provide love to the Romanian children and alleviate their suffering. When experiencing this, he felt no fatigue or burnout but an endless feeling of love and compassion, which he describes how "An atom of love replaced each atom of suffering." Singer's analysis of the fMRI data and a further longitudinal study revealed significant differences in the brain areas involved between empathy towards the close relative and compassion for the Romanian orphans. Longitudinal studies involve subjects being followed over time with continuous or repeated monitoring of risk factors and variables or health outcomes; with empathy, the Anterior Insula (AI) and the Anterior Cingulate Cortex (ACC), which we discussed earlier, mainly involved, which are linked to the negative experience of pain in ourselves and others, which became so intense that their perception of their suffering became almost unbearable and following the tests participants reported seeing much more suffering wherever they went. Although with compassion, they initially felt empathy as evidenced by activity in the AI and the ACC, it relatively quickly moved towards the PFC with the emphasis on providing generous love and support in a positive way that doesn't conflate the negative experience of another person's pain with one's pain.[95, 96] The conclusion drawn from the imaging studies was that empathy tends to be associated more closely with the self than compassion so the relation to the self is an important component of empathy, though this is not the same as prioritizing the self over the other; in fact, it is the opposite.[97, 98]

Metacognition

American developmental psychologist John Flavell (1928–) described how, as part of the mentalization process, "Meta-cognitive knowledge is one's stored knowledge or beliefs about oneself and others as cognitive agents, about tasks, about actions or strategies, and about how all these interact to affect the outcomes of any sort of intellectual enterprise." He describes how "Meta-cognitive experiences are conscious cognitive or affective (emotional) experiences that occur during the enterprise and concern any aspect of it often, how well it is going" and it is the way humans' control and monitor their thought processes."[99] Chris Frith, in his study of the role of metacognition in social interactions and the self-monitoring aspects of metacognition and other's beliefs, made a further distinction between implicit metacognition, which is rapid and automatic and occurs subconsciously, and explicit metacognition, which is slow and deliberate and is very susceptible to errors.[100] Frith observes that "the rapid and automatic implicit meta-cognitive processes are what enables us to adopt a we-mode which enhances joint action and collaboration" by our automatically taking account of our knowledge and intentions and the knowledge and intentions of others." By comparison, explicit metacognition "enables us to consciously reflect on and justify our behaviour to others" which underlies our ability to "explicitly share experiences with others, as in reflective discussion and teaching." One of the problems, as Frith acknowledges, is that it is very difficult for people to access their underlying cognitive processes, and reports on their own and other's metacognition can be very inaccurate. He observes how young children exhibit little metacognition and "do relatively little monitoring of their memory, comprehension, and other cognitive enterprises." This may be because their conceptual self is still developing, and they rely on the perceptual self. Metacognition, whether implicit and explicit or a combination of both, seems to play a central role in our ability to work together in groups and operate socially, and Frith believes that the TOM is associated with understanding our and others' behaviours in terms of beliefs and desires is an example of explicit meta-cognition concerning other people's state of mind. Ian Apperly and Stephen Butterfly provide an alternative perspective: people use two systems to track beliefs and belief-like states in others, which could resolve the lack of consensus on the topic and explain the success of infants and non-human animals on some belief reasoning tasks. They attribute it to "a cognitively efficient but inflexible capacity for tracking belief-like states which in humans persists in parallel with a later-developing, more flexible but more cognitively demanding theory-of-mind abilities".[101] What is clear from all the research is that possessing a coherent and healthy sense of self and self-awareness is central to our well-being, which may explain why we go to such great lengths and use many ways to protect the self, most of which we are completely unaware of.

The Robot Self

The various theories related to consciousness and self in humans require a high level of integrated and convergent phenomenological information about the self and the self in relation to the environment. Self is an intrinsic awareness of existence that reflects our sense of selfhood that believes that we possess an essential core being and that our self and relation to the world is a discrete, definable, and enduring reality. This is consistent with Morin's earlier observation that our levels of consciousness and self-awareness reflect the nature and complexity of the information related to the self, with the more sophisticated self-representation associated with higher levels of consciousness and the processing of crude types of self-information related to lower levels of consciousness. The more sophisticated self-representations reflect the greater integration and differentiation of the information at the cortical level in humans. The conscious self involves awareness and attention, emotions and feelings, the importance of hot cognition as opposed to cold cognition, motivation, and autobiographical memory and learning and these aspects of self are what allows us to perceive and give meaning to the world and ourselves. Will the robotic brain be able to integrate information in a way that allows them to develop a sense of self and self-awareness? Let's look first as self awareness.

Robots and Self-Awareness

Enabling an android to experience self-qualia and phenomenon is a major challenge facing AI and AC that has yet to be overcome. Nicholas Cassimatis and his colleagues made the following observation that goes to the heart of the problem facing AI developers in creating consciousness and self-awareness in robots, which is, that "Many problems in robotics arise from the difficulty of integrating multiple representation and inference techniques into one robot, that can construct and maintain rich, coherent, and dynamic models of its environment." Cassimatis explains that "integrating multiple representation and inference techniques, consists of the separate challenges of data fusing, symbol grounding and flexibly combining reasoning, planning, perception, and action."[102] This is essential if androids are to be able to autonomously adjust and adapt their behaviour appropriately, according to the demands of often complex and uncertain environments. Cassimatis references both the environment, consciousness and self-awareness and his observations reflect Damasio's earlier conceptualisation of the self in requiring "integrated neural processes, centred on the representation of the living body, which finds final expression in a dynamic collection of integrated mental processes." Data fusing is the need to integrate the information and data that a robot receives into a coherent representation of its environment, which includes representations of the self, whilst being able to distinguish between multiple representations. In androids this typically includes sensor-based exploration, motion planning, localization and mapping and the need to overcome inaccuracy and uncertainty in sensor data.[103]

Christos Chatzichristos and Sabine Van Huffel, describe the highly challenging task of data fusion as follows, "the joint analysis of multiple inter-related datasets that provide complementary views of the same phenomenon."[104] The process of correlating and fusing information from multiple sources generally allows more accurate inferences than the analysis of a single dataset can yield. Zoé Lacroix describes how data fusion "corresponds to the need to integrate information acquired from multiple sources (sensors, databases, information gathered by humans, etc.)."[105] She explains how "the term was first used by the military to qualify events, activities, and movements to be correlated and analysed as they occurred in time and space to determine the location, identity, and status of individual objects (equipment and units), assess the situation, qualitatively or quantitatively determine threats, and detect patterns in activity that would reveal intent or capabilities." Sensor data fusion is becoming increasingly important as robots become more autonomous because it is essential to reducing uncertainty and risk in the environment and enabling autonomous robots to perform complex tasks when faced with noise and uncertainty or incomplete or inaccurate data. This notion of reducing uncertainty in information reflects the free energy principle and the Bayesian theory in terms of how our brains use prediction to reduce uncertainty in the environment and search for understanding. Data fusion is even more challenging when it comes to the need for sensor fusion to include not just environmental information, but all the sensory information related to the self-concept which we humans take for granted and the internal sensory information generated by the android equivalent of human homoeostasis and allostasis. Are cognitive architectures like LIDA and those involving SMM close to imbuing machines and our robots with self-aware minds?

Let's look more closely at some of the recent claims that a level of self-awareness has been achieved in robots. Israeli American Hod Lipson is a roboticist and professor of mechanical engineering, who has declared that he wants to build a self-aware robot that is independent of the engineers that create it. Sounds like all parent's wishes for a child. Lipson believes that this can be achieved through a process of self-simulation and is working on a project initially led by Chinese scientist Boyuan Chen. Scientists have for years programmed robots with images of themselves to allow them to see themselves in the environment and conduct a primitive form of self-talk, but this has very much been confined to limited pre-set tasks, not to actions with a more general purpose. Lipson thinks that we can now go beyond the set tasks to general-purpose activity because the AI algorithms that equip the robots with the capacity to create self-simulation, have been made possible by the rise of CNNs and DL. This has revolutionised computation in terms of speech, image, and audio performance but as we have seen CNNs can be very computationally demanding and need GPUs and parallel processing capabilities to train the different AI models. Lipson regards the basic steps of self-simulation and self-awareness as involving the robot possessing an internal representation of its body and how it moves in physical space and using this to guide its behaviour. The robotic arm that has been created by Lipson and the team currently sits on a tabletop surrounded by 5 video cameras which provide the robot with data about the arm and the information it generates creates a fuzzy model of what the arm looks like. Over a few hours of training, the robot equipped with the robotic inner dialogue can employ logic to assimilate the images of itself and work out the actions it can perform. This appears to only equip it with a form of self-perceptual information rather than a full self-concept, but as Lipson acknowledges it represents another baby step in the attempt to create self-awareness in robots and full autonomy. The robot through many random movements generates sufficient data upon which deep learning algorithms can create self-simulation, which has enabled the robot to sense that something is wrong if a limb is damaged and to correct it. Lipson recognises that this amounts to the most primitive level of self-awareness which will need to arise on a continuum culminating in the level of full self-awareness that we possess. The developers acknowledge that it isn't currently even close to replicating the basic body awareness of an animal or baby let alone the human adult and the challenge will be to scale up the complexity to the point it can simulate full self-awareness in humans. Nevertheless it is a bottom-up approach compatible with the notion that self-awareness relates to increasing levels of integration of information and and may well help us determine when self-simulation becomes actual self-awareness.

According to Lipson, it provides a concrete platform upon which to be able to benchmark and have a metric of basic awareness in robots. It seems consistent with our understanding that the self-concept is grounded in the perceptual self and develops into the self-concept based on learning through the experience of being embodied in the environment like humans. He has argued that our focus has been on modelling robots that can interact with, learn from, and adapt to the world but "when it comes to self we have relied on hand-coding the model." Lipson compares this first iteration of self-simulation to the neural cortical maps that in human brains map the various senses, which we discussed earlier. In this sense, the neural patterns in the sensory maps in the human cortex, serve as a neural substrate in robots for the beginning of self-awareness and what it gives the robot is the beginnings of behavioural flexibility, which in this case starts with self-simulation of an arm. In 2022 the team observed that the robot had begun to imagine itself which Lipson describes as self-image gradually emerging as "a sort of gently flickering cloud that appeared to engulf the robot's three-dimensional body," and "as the robot moved, the flickering cloud gently followed it." This sounds eerily reminiscent of Sir Charles Sherrington's description of our elusive "I" self as arising from the brain as the "enchanted loom where millions of flashing shuttles weave a dissolving pattern, always a meaningful pattern, though never an abiding one. Its commanding presence orders sensation, movement, thought, a lifetime of memory and dream." There is no indication that this constitutes an experienced I self in a robot and Lipson readily admits that this self-simulation is "trivial compared to that of humans, but you have to start somewhere, and we have to go slowly and carefully, so we can reap the benefits while minimising the risks."

Lipton and his team are working on other initiatives that translate this self-simulation to the cognitive component of self-awareness but believe that "It is a complete leap of faith to believe that eventually, this will get to human-level cognition and beyond" but aim for these to converge and he believes that when they do, self-simulation will equate to genuine self-awareness in robots.

Robots and the Mirror test.

Neuroscientist Robert Mitchell has described how the relationship between mirror tests and the self-concept has traditionally involved using either of two AI models. The first is Gallup's self-recognition model based on the self-concept which embraces the various aspects of self-awareness and the second, Guillaume's kinaesthetic-visual matching model which asserts that recognising oneself depends on kinaesthetic-visual matching, which involves self-recognition through a combination of bodily awareness (kinaesthetic), and self-face recognition (visual).[106] Mitchell explains how the KVM model "ties self-recognition, bodily imitation and pretence, communication via simulation, recognition that one is being imitated, and mental planning together in one neat skill-kinaesthetic-visual matching."

It does not require all the other aspects that constitute Morin's full self-awareness, so appears easier for our robots to pass than Gallup's self-recognition model based on the self-concept. Kinesthetic refers to a person's awareness of the position and movement of their body through sensory organs (proprioceptors) in the muscles and joints and is one of the many ways that humans learn as will see, the way humans learn is highly relevant to how our robots may learn. Around 65% of us are visual learners so we learn best when we see things, which is not surprising when studies have shown that our brains can process visual information in terms of videos and images, rather than written, almost 60,000 times faster and retain around 80% of what they see compared to 20% of what they read and 10% of what they hear. Although around 30% of people prefer auditory learning, only around 5% of people need to physically engage in something to learn and are called kinaesthetic learners.

Mitchell compared the two models and concluded that "the Kinaesthetic-Visual Matching model is conceptually coherent and makes reasonable and accurate predictions and that the self-concept model is conceptually incoherent and makes inaccurate predictions from premises which are themselves inaccurate. From a theoretical and empirical standpoint, the kinaesthetic-visual matching model is the better explanation of self-recognition."[107] If this is correct, then robots need only pass the lower threshold of the self-recognition test in the kinaesthetic-visual matching model, in terms of recognising their body and face, for us to be able to say that they can recognise themselves in the mirror. The current behaviour of robots in passing the mirror test appears to be governed not by possessing a genuine self-concept involving self-awareness but through employing kinaesthetic-visual matching.

The robots extract their features and after several attempts, learn to recognise those features in composing an internal representation of itself as a distinct physical entity but this does not require formulation of any abstract conceptual representations beyond the physical object and learning the category self. The robots might be able to recognise that the face and body staring at themselves in the mirror in the morning, belong to the category self, but lack Morin's requirement for the perception of the self in time, the complexity of self-representations, the frequency of self-focus, the amount of self-related information and the accuracy of self-knowledge. In other words, they don't know that there is a self behind the robot face and body and having a robot face and body means nothing to them.

Computer scientists Arianna Pipitone and Antonio Chella recently claimed that AI robots can pass the mirror test by employing Guillaume's kinaesthetic-visual matching model and the manipulation of perceptual and motion data. Their work has focused on developing self-awareness in robots via inner speech, which they assert, requires the robot to employ an overt self-talk system which provides an external confirmation of an inner dialogue or inner voice in the form of logic and reason based on symbolic information, to recognise itself, a process that psychology has established humans employ.[108, 109] Pipitone and Chella equate the robot's inner voice which is rendered audible, to powers of self-reflection because it enables the robots to overtly self-talk, rehearse and plan tasks, which aids learning and self-regulation.[110] It is estimated that humans spend over a quarter of our lives listening to silent inner speech or in rumination. Audible inner speech in robots provides functionality and transparency in cooperation and collaboration with humans by allowing humans to follow and trace the robot's progress. Importantly it influences the level of trust in the robot exhibited by people when interfacing with a robot. Pipitone and Chella maintain that "the results show that functional and transparency requirements according to the international standards for collaborative robots, are better met when overt inner speech accompanies human-robot interaction."[111] There is, though, a crucial distinction between reliance on perceptual features and motion information about the physical self even if accompanied by an artificially constructed audible inner dialogue that explains the actions and the generation of a conceptual sense of self.

The conceptual self as we have seen stems from an internal and largely abstract sense of who you are and the understanding of that self as an autonomous entity. Self-conceptual information contains the various vital components of self-awareness, whilst perceptual information consists of simply being able to distinguish yourself from other objects as a distinct entity or category within the physical boundaries of your body. Loss of the self-concept can be catastrophic to humans even if they can continue to perceive and recognise their face and body in the mirror. Equipping a robot with the apparent ability to explain that it is seeing its body and face in the mirror does not yet constitute genuine understanding and self-awareness as opposed to being able to categorise itself as a physically distinct entity. Our perception of the information and being able to extract meaning and understanding from the data that we obtain from the environment is closely related to how deeply coupled with the environment we are and as we saw, much of the information is processed by reference to the self and the needs of the organism. When Pipitone and Chella describe the robot talking to the self in the mirror, they observe that it involves the "processing of a symbolic representation of encoded environmental stimuli which includes the images reflected in the mirror, the mirror itself, and other entities outside the mirror together with the cognitive architecture for inner or outer speech which they designed." The cognitive AI architecture designed by Pipitone and Chella enabled the robot to explain its decision processes and awareness of its environment which add to its distinction between itself and the environment, which

appears to enhance self-recognition, but does constitute genuine self-awareness or amount to a conceptual self. It appears to be the articulation by the robot by a simulated inner voice of its ability to classify its perceptual self as it would another object rather than reflecting a genuine conceptual sense of self or awareness of self-concept. It seems more akin to the perceptual learning equivalent to that of a young babbling baby, which is necessary to and will become part of an embedded self-concept which develops later or that of some animals that appear to pass the mirror test. Even if robots pass the mirror test there is yet no evidence to suggest that they possess anything more than a limited perceptual notion of the self as a distinct physical entity.

Mentalization and TOM in Robots

Collaborative robots are increasingly used in industrial and social applications to work with humans but in practice, they are currently limited to inferring the intentions of humans based on predefined programmed tasks and even when confined to task performance are typically unable to infer intentions when presented with a novel unfamiliar task. Arjun Chandrasekaran who focuses on computer vision, ML, NLP and AI and colleagues observe that most research to date has focused on getting AI to be "more accurate, more human-like, to understand our intentions, beliefs, tendencies, contexts and mental states rather than our being able to understand AI's beliefs, knowledge, and quirks". In their study, A Theory of AI's Mind, It Takes Two to Tango, they argue that for human-robot interaction (HRI)to be effective TOM must go both ways, but that poses enormous challenges.[112] To collaborate and interact with humans, robots need to respond to verbal and non-verbal behaviour with the latter involving the perception of emotional behaviours and signals. It is relatively easy when giving a robot a command that it follows to infer its intentions if it has complied with your command or if it has been programmed to enable it to communicate its intentions, but the more autonomous the robotic agent the more difficult this becomes. Effective communication must involve inferring the other's current state of knowledge rather than simply programming the robot to repeat the intention which would be inefficient and potentially highly annoying in both humans and robots. This has major implications in terms of whether we would ever be able to explain how they understand us or if robots can explain their intentions and reasons to us. The latter is the XAI problem which appears to plague AI in that although we can know the inputs and the outputs of a system, we don't typically know what goes on between the input and output in the black box which is particularly true of deep learning models. This is not just frustrating from the developer's perspective but poses important safety, ethical and legal considerations and the need for governance and regulation of AI, particularly as we know that data can contain significant biases that influence their actions and percieved behaviour. Just like humans.

Over the years different approaches have been employed to equip robots with the ability to interact with the environment, humans, and other robots in a more natural and even more social manner, which includes not just task-oriented cooperation and collaboration but naturalised conversation and broader social interactions. The sector has come a long way since the first system in 1993 which involved a passive robotic arm to assist in safe medical procedures.[113] It is increasingly important given the manufacturing trend towards customising robots for specific roles rather than mass producing them and people working more and more closely with robotic agents. Social interaction is computationally challenging and spans not just recognising a human but tracking their gestures and their focus of attention, identifying what they are doing and the level of engagement with the human or humans.[114]

Plan recognition has been a focus of research in seeking to equip robots with the ability to make inferences about humans or other robot actors from observations of their behaviour in terms of their interaction with the environment and with each other. The research community has even devised the acronym PAIR (Plan, Activity, and Intent Recognition) for this field of study. This has included specific research into intention recognition by robots in NLP and inferring intent from the movement and trajectory of the human body language, visual gaze, and facial expressions. Remember though that the latest scientific evidence points to problems in human TOM arising when there are domain-specific deficits in the processing of information related to inferring other's thoughts and intentions. This suggests that the sensory domains must be adequate if the robot is to possess a TOM mechanism or different levels of competence in PAIR performance.

Yujiao Cheng and colleagues describe how research into human and robot collaboration has been approached at three levels namely (1) Low-level which simply seeks to avoid collisions specifically with humans (2) Mid-level which involves efficient cooperation with humans and requires task plan recognition and trajectory prediction in which the robot predicting the continuous movement of the human collaborator to allow the robot to plan a safe trajectory. The challenge increases when there is more than one human or one robot collaborating, and teams of robots and people are involved. An example when assisting in assembly tasks is providing tools and parts and (3) High-level collaboration in terms of selecting operating modes and automatically assigning tasks which involves the human and robot working closely as partners.[115]

As an example of addressing mid-level cooperation Fahad Khawaji and colleagues have developed an algorithm called a Model Predictive Control (MPC) based trajectory planner which employs a 3-D sensing system to measure the skeletal data of the human worker. They describe how "At each sampling time of the sensing system, an optimal delivery position is estimated using the real-time worker data. At the same time, the future positions of the worker are predicted as probabilistic distributions."[116]

In another study, Chien Ming Huan and Bilge Mutlu sought to equip the robot with the ability to predict the human counterpart's intent by monitoring their gaze based on observed gaze patterns. Their experiment showed that anticipatory control meant the robot responded to user requests and completed the task between 2.5 to 3.4 seconds faster when compared to a robot using a reactive control method.[117] An example of more high-level collaboration was proposed by Sandra Devin and colleagues in their shared plan achievement called Human-Aware Task Planner (HATP) predicated on achieving a shared goal between the robot and the human by allowing the robot to "elaborate and execute shared plans that are flexible enough to be achieved in collaboration with a human in a smooth and non-intrusive manner."[118] Many shared plans have been devised which seek to reduce resource conflict, allow robots to adapt actions to accord with the human plan or consider divergent intentions or beliefs which often necessitates unnecessary communication and verbalisation between the human and robot of ongoing adjustments which can prolong tasks and waste time. HATP envisages the robot being able to determine which decisions it can take by itself and those it should leave to the human. This involved four steps in which, based on priorities the robot (1) Selects the action, then it (2) Decides whether to perform the action or if the human does then it must (3) Instantiate the action and (4) By constantly monitoring the human's behaviour and actions react appropriately. In the decision-to-act phase, the robot is programmed with two modes which are the negotiation mode and the adaption mode. In the negotiation mode the robot asks the human if it wants to perform the action and responds accordingly, whilst in the adaption mode the robot will wait and take the initiative if the human does not.[119]

It is challenging to design collaborative robots in an industrial context and even more so for autonomous robots in a social context, but it is important given that we live in groups and form relationships and social networks. Much research on machine interaction particularly movement has been inspired by insect behaviour in which the individual is not important, but the prevailing characteristic of most insect colonies and behaviour is the anonymous, homogeneous nature of the group with group interest as the driving factor. This has allowed robots to be designed based on groups of robots collaborating to perform relatively simple tasks but failed to design social robots capable of displaying the complex social interactions exhibited by humans socialising in groups or simply when working together and collaborating.[120] We already use individual AI robots in a wide range of applications including healthcare, assisting visually impaired people to interpret their world and potentially autonomous cars etc, which are examples of how vital it is that we trust the AI process and believe that it knows what it is doing.

American Cynthia Breazeal (1967–) is a robotics scientist and entrepreneur, former founder and chief scientist of Jibo, a company that developed personal assistant robots. She now directs the Robotic Presence Group at the MIT Media Lab and has developed numerous autonomous robots. Breazeal makes a fundamental and insightful distinction between different types of social robots and what we might expect from such robots. These are robots that are, (1) Socially Evocative and can "anthropomorphise and capitalise on feelings evoked when humans nurture, care, or are involved with their "creation" (2) Social Interfacing that provide a natural interface by employing human-like social cues and communication modalities. Social behaviour is only modelled at the interface, which usually results in shallow models of social cognition (3) Socially Receptive that are socially passive but that can benefit from interaction (e.g., learning skills by imitation). Deeper models of human social competencies are required than with social interface robots and

(4) Sociable ones that are pro-actively engage with humans to satisfy internal social aims (drives, emotions, etc.). These robots require deep models of social cognition." "Sociable" robots appear to present the greatest challenge partly because they involve proactive engagement with people. Terrence Fong and colleagues in their survey of socially active robots added three complementary and slightly nuanced features to this which are (5) Socially situated, that is surrounded by a social environment that they perceive and react to and be able to distinguish between other social agents and various objects in the environment (6) Socially embedded, that is (a) situated in a social environment and interact with other agents and humans (b) structurally coupled with their social environment; and (c) at least partially aware of human interactional structures, for example turn-taking in conversations and (7) Socially intelligent that show aspects of human style social intelligence, based on deep models of human cognition and social competence."[121] Fong and the team identify certain human social characteristics which socially interactive robots should display which are (1) Expressing and/or perceiving emotions (2) Communicating with high-level dialogue (3) Learning/recognizing models of other agents (4) Establishing and maintaining social relationships. (5) Using natural cues (gaze, gestures, etc.) (6) Exhibiting distinctive personality and character and (7) The capacity to learn/develop social competencies. All these are an integral part of how humans communicate with each other, and different initiatives have sought to equip robots in both industrial and social situations with the ability to effectively communicate which includes enabling them to analyse human motion, gestures, facial expressions, verbal utterances etc. It is very difficult to mimic the behaviour of humans and creating the impression of mentalisation and possessing a mind as well as communicating with high-level dialogue, which reflects the complex nature of human-to-human interaction and cooperation.

Although a significant amount of interdisciplinary research has focused on HRI, it has been criticised by many researchers, including psychologist Elizabeth Broadbent as being dominated by engineers and computer scientists and requiring greater input from neuroscientists and psychologists. This is particularly important if we accept Martin Michlmayr's observation that the function of TOM in humans is to "1) Comprehend and explain the behaviour of other people 2) Predict other's behaviour and 3) Manipulate and influence others by controlling the information available to them." Based on this characterisation of TOM, imbuing robot with TOM poses significant potential challenges and risks even as we are providing obvious benefits."[122] Being able to process a symbolic representation of encoded environmental stimuli related to bodies and faces including its own, or even pass the mirror test, does not confer upon a robot the capacity to mentalise which is a core component of human social interactions and communication. If robots are to mentalise like humans, they must be able to distinguish between their private and public selves, think about their past and future and possess the ability to infer the thoughts or minds of others based on TOM and metacognition.

Professor of communication, family studies and human development Judee Burgoon and communication expert Aaron Bacue in their 2003 review of research into non-verbal communication skills describe how "It is the nature of the human condition that, try as we may, we cannot enter the reality of another individual's experiences, thoughts, or feelings. Imprisoned as we are within our bodies, the fallible process of communication is the primary agent currently available for crossing the psychological expanse between two or more individuals."[123] The attribution of mentalisation to robots by humans strangely appears stronger when people are under pressure in terms of having to make decisions and interacting with a robot.[124] Even animals can respond to visual non-verbal cues such as body posture, gaze, or facial expression to predict other animals' behaviour and mental states. Humans can draw inferences about a person's unobservable mental states, including their knowledge, intentions, beliefs and desires because they possess the capacity for mentalisation or TOM. Humans mentalise implicitly and explicitly with implicit mentalizing being automatic, spontaneous, and subconscious though this does not involve conceptualising, whereas explicit mentalizing is conscious, controlled, conceptual, and effortful.[125] The challenge when assessing individual mentalization when asking people to self-report is that they are unable to factor in the implicit mentalizing aspects of it. Let's look at the various attempts to develop and equip robots with the ability to mentalise rather than just the tendency for people to attribute mentalisation to them in preference to listening to mechanistic (non-intentional) descriptions of robot's actions. The examples chosen are simply illustrations of different approaches as it is impossible to be exhaustive of all the current initiatives seeking to solve the problems associated with HRI. They cover the role of mentalisation in the construction of natural dialogue and conversation, assessing trustworthiness and general taking of another person's perspective.

An important example of mentalisation commonly used by people is during a conversation when engaging in social dialogue. A significant body of social psychology and communication research has shown that conversations reflect and affect the characteristics of the participants and the essential dyadic (the interaction between two people) nature of interpersonal relationships. However, cognitive scientists Stefan Kopp and Nicole Kramer who focus on cognitive psychology and robotics observe in their 2021 paper that "Despite frequently made claims of superhuman performance in for example speech recognition or image processing, so far no (AI) system is able to lead a half-decent coherent conversation with a human." They argue that despite the advances in machine learning and language processing the current technical processes are missing the "core human capabilities that humans have evolved for natural conversational interactions" and believe that the cognitive sciences and psychology provide crucial insights that contemporary AI and engineering have "for some part deliberately started to ignore."[126]

Conversation varies enormously in terms of content and quality and although it doesn't always involve intelligence, empathy, affiliation, or friendship, all conversation requires cooperation and mentalisation. Cooperation involves the primacy of joint construction of the conversation which reflects how people interact in conversation and iteratively co-construct their interaction and mental states at the same time. American psychologist Michael Tomasello maintains that conversation involves the "innate human ability and urge to cooperate in constructing meaning because humans are ultra-social animals." Mentalisation involves a person and hence would involve a robot "perceiving and knowing about the interacting party, the interaction and about itself in order to provide mutual support and true cooperation in co-constructing the conversation."[127] Other than the simplest versions Conversations inevitably involve each party hypothesising about the other's meaning and intentions. Construction of meaning is heavily dependent upon the receiver's highly subjective perception of the situation and context which the sender can't control, and we are always subconsciously or consciously trying to estimate the impact in terms of what the other person understands by what we say. Kopp and Krämer criticise the current machine learning approaches that attempt to create conversational tools in HRI as reminiscent of behaviourist approaches which seek to extract patterns of communicative conduct based on using huge amounts of surface-level data and establish correlations with surface input features at that superficial level. Although this behaviourist approach may have led to significant progress being made in artificial spoken language, it is largely confined to structured tasks and task-oriented dialogue whilst operating within well-defined and confined domains. Recent efforts directed at developing open-domain dialogue systems and social conversational agents have faced significant challenges when dealing with non-literal meanings, ambiguity, vagueness, nuances, ironies, and subtleties which typify human-to-human conversations. Kopp and Krämer don't believe that open-domain dialogue and social conversation can be learned a priori from large data sets and training, but must emerge from conversational learning which by its very nature is a joint activity. Attempts to build more realistic conversational models in robots have been beset by problems of interruptions, hesitations, or long delays that disrupt and disturb the normal natural flow that characterises ordinary conversation between people. Their proposal envisages a more holistic mentalizing approach that they hypothesise would allow a robot to "perceive, interpret, and understand a person's mental states" as a "precondition for a smooth,coherent dialogue."[128]

Conversation involves universal practices which Stephen Levinson nearly twenty years ago metaphorically termed the Human Interaction Engine(HIE).[129] Research suggests that the HIE, which is constantly being updated to apply to other non-human primates, comprises four features which are (1) Being able to communicate using all the various visual, auditory, acoustic and hands and mouth sensory channels (multimodality), which is particularly important in face-to-face communications (2) Sequence organisation in that so that each component of the conversation is contingent of the previous and allows for appropriate response and interjection (3) The taking of turns in conversation and (4) Recognising the intentionality behind what someone says, with people responding to someone's presumed intentions not their behaviour. The HIE is regarded as separate from and preceding the origin of language and believed to be the basis for language development, which is why even when we are trying to converse with someone who can't speak the same language or when we are using sign language with the deaf, the rules of social interaction appear to be adhered to. This has been witnessed in Orangutans[130] and Bonobo chimps[131] which seem to engage in similar behaviours without language.

What is not clear as Judith Burkart and her colleagues observe, is the extent to which the HIE is driven by phylogenetic (evolutionary) or inherited traits or is a product of what have been described as "Convergent components that must have shaped our unique set of socio-cognitive skills."[132, 133] We can never assume that what is said will be interpreted in the way we intend because we have no direct control or knowledge of how the recipient will interpret and decode the message and the computational variables of that person's mind are hidden from us. We largely rely on our knowledge about the other person, the context or situation and any cues provided by the other person. The more we know someone the more confident we are in our hypotheses as to what the variables are and what they might mean. Kopp and Kramer recognise that human communication is never free from error especially conversations with their potential for misunderstandings and unintended effects on people but observe that research has shown how conversations are remarkably "resilient, robust, and efficient in establishing sufficient coordination" between people.[134]

To capture this complex but robust process of communication they make certain recommendations that AI needs to adopt if social robots are to be imbued with what they call the "hallmarks of situated, multi-modal, and flexible conversation" when engaged in HRI. They boldly assert that as far as they are aware, theirs is the "first attempt to look at the basic human abilities of joint co-construction and mentalizing from an interdisciplinary perspective including psychology and computer science" and the recommendations are based on "a vast body of human-human communication and its psychological processes that (they) reason to be relevant and necessary to take into account when modelling HRI communication," particularly the primacy of cooperation and mentalization which are largely missing in current technical approaches.[135] They advocate that the primacy of cooperation proposes that AI uses in the future should involve a think enough, speak and listen more approach rather than the current think (and predict) more, speak less approach which they claim stems from the erroneous belief that the reasoning for a response in AI is done on the spot by the robot based on the adequacy of its training data. The cooperative element in human conversation suggests that in HRI the robot needs to participate in the incremental step-by-step co-construction of meaning which means that it must be able to take the human perspective and swiftly and cooperatively adapt to it. This requires computational tools for modelling the non-linear and dynamic and multi-dimensional linear and non-linear ways in which dyadic interactions and conversations unfold.

Cooperation between humans reflects how all parties to a conversation are using mentalization to construct a representation of others' beliefs and attitudes and the ability to update that representation based on new information and cues. Kopp and Krämer believe that to equip robots with this requires a combination of cognitive approaches and current machine learning methods. They maintain that a genuine conversational interaction for social and collaborative agents or robots, will not be attained by simply trying to simulate the behavioural response sequences typical in human-human interactions but requires the robots to "cooperatively and incrementally co-construct a successful interaction with the human user" and this will involve "coordinative mechanisms such as partner-specific adaptation of multi-modal utterances, responsive turn-taking, informative feedback, or collaboratively resolving misunderstandings." This necessitates equipping robots with "the underlying ability to intertwine incremental mentalizing and socio-communicative behaviour at the dialogue level." Their specific recommendations involve cognitive AI architectures that incorporate at least three different layers of cooperation and coordination which are (1) Contact which necessitates perceiving each other and paying attention (2) Understanding and (3) Alignment of plans and goals. The major challenge is enabling understanding in robots, particularly given the current lack of adequate computational models in HRI which reflects the lack of consensus on how humans understand their world. They are realistic in their assessment of the prospect of building conversational robots when they conclude that "conversational agents that are living up to the primacies of mentalizing and incremental joint co-construction can only be expected to come into existence very much in the future, and requires substantial and collaborative research in cognitive science, psychology, and computer science."

One major difficulty is equipping robots with the capacity to infer the intention of humans or other robots based on their implicit rather than explicit intentions. Implicit intention inference refers to how we seek to understand a person's intentions not just from the content of what they say, but by reference to their body movements and even eye movements or visual gaze. In HRI as with humans, correct intention inference is important to a natural and effective interaction between the human and the robot.[136] Intention inference involves the need to make inferences based on the unobserved variables or parameters in the minds of the parties having the conversation, which are largely hidden to us when we converse. Even our own are largely hidden from us, evidenced by how often we don't know in advance exactly what we are going to say let alone what another person is going to say and why. Many AI models are data-driven without prior information in terms of expression, of intention which has been shown to yield low inference accuracy and the aim of many developers is to improve intention inferencing in HRI.

One team of computer scientists seeking to equip robots with the capacity to infer human intentions on novel tasks have proposed an evolving hidden Markov model (EHMM) based approach. This purports to allow robots to "learn new human intentions incrementally by carrying out structure and parameter updating based on the observed sequence, in parallel with the recognition" and has achieved high recognition accuracy in the robots. The EHMM uses a Markov chain Monte Carlo (MCMC) sampling technique for state inference in which the system arrives at a certain conclusion using reasoning or evidence, which makes them more than just assumptions by drawing upon samples where the next sample is dependent on the existing sample (i.e., the Markov chain that we discussed earlier). It estimates parameters using means, variances, and expected values and explores the posterior distribution of Bayesian models. The posterior distribution simply means an estimate based on the prior probability plus new evidence (called likelihood) so it updates its beliefs and inferences based on the probability of an event occurring after taking into consideration new information. The EHMM model can be used in state space models where the state such as the person's mind and the conversations can be high-dimensional, and they can deal with dynamic time series problems that involve unobserved variables or parameters that describe the evolution in the state of the underlying system. In this case, the hidden variables are those in the minds of the people having a conversation. The model purports to allow the robot to learn untrained intentions online by human demonstrations by unifying intention inference and incremental learning so it utilises incremental learning ability for use in dynamic environments with changing tasks and potential conversations because it enables the robot to infer the intentions of humans.[137]

Professor Michal Kosinski, a computational psychologist recently attempted to assess whether AI networks were capable of mastering TOM and describes how they were exploring the fact "the intriguing possibility that theory of mind, or the uniquely human ability to impute unobservable mental states to others, might have spontaneously emerged in large language models (LLM)."[138] They designed 40 false-belief tasks and using them on the LLMs tested the latest iteration of a neural network known as the Open AI's Generative Pre-training Transformer (GPT) (GPT-4). They claimed that since GPT-3.5 the next iteration of GPT-4 had spontaneously learned the cognitive ability to infer other thoughts, beliefs, and desires progressing from that of a three-year-old in GPT-3.5 to the level of a seven-year-old child with GPT-4. Kosinski described this as "a watershed moment" in the evolution of AI and observed that "the ability to impute the mental state of others would greatly improve AI's ability to interact and communicate with humans (and each other) and enable it to develop other abilities that rely on TOM, such as empathy, moral judgement, or self-consciousness."

Despite Kosinski's bold claims, it is not clear whether GPT-3.5 or 4 appeared to develop TOM because they are using LLMs, which means that the computer may be behaving like that because it is interpreting language patterns rather than displaying genuine mentalizing. Kosinski's earlier studies showed that neural networks could learn to analyse facial features, head angle and emotional expression to predict sexual orientation and political views with a high degree of accuracy, but does this constitute understanding others and the self? Others have observed that these systems often display a tendency to lapse into patterns of being unable to infer beliefs or views and come up with spurious correlations and the knowledge they base the responses upon, appears very susceptible to even very subtle differences or changes. It may demonstrate the superior capability of computers to analyse language patterns but not evidence genuine intuition or inference associated with human TOM or its spontaneous emergence from the development of these programs.

A 2021 study conducted by Boyuan Chen in collaboration with Carl Vondrick and Hod Lipson based on the behaviour of the robot concluded, that the "findings begin to demonstrate how robots can see the world from another robot's perspective." The robot had observed another robot in a playpen that had been programmed to seek and move towards any green circle it could see, but sometimes the green circle was occluded by a red box, leading the robot to not move towards that green circle or move towards another. The observing robot after about two hours was able to predict with an almost 98% degree of accuracy the other robot's goal and path.[139] The ability of the observer to put itself in its partner's shoes, so to speak, and understand, without being guided, whether its partner could or could not see the green circle from its vantage point, is perhaps a primitive form of empathy. They believe that it may be the beginning of endowing robots with TOM and the credibility of robots leading to greater trust and reliance.

Marcel Finkel and Nicole Kramer argue that "a robot's mentalizing abilities have shown to be beneficial because they allow a robot to infer users' inner states, thus serving as a prerequisite for understanding their beliefs and attitudes." They raise the rather frightening risk that "social robots are often deployed by private and thus profit-oriented companies. In such cases where an organisation's implied manipulative intent is salient, the effect of robots' mentalizing abilities might be reversed" and "pose a persuasive threat to users rather than a feature for better understanding, thereby decreasing credibility attributions." Their study revealed that people focus on the robot and not the organisation behind it "causing potential opportunities for such organisations to hide their economic interests from the users."[140]

Thomas Hellström and Suna Bensch in their 2018 study point out how humans tend to "suppose that they have mental states allowing them to understand the rationale for their actions" which as we see with a robot like R2-D2 with its squeaks and strange motions does not require much.[141] They stress the need to define what understanding means when it comes to HRI and be clear about the ways that robots can support human understanding which they denote as communicative actions by reference not just to its physical actions but social interactions. Robots need to communicate their intentions and internal state to create trust and avoid anxiety about robotic behaviour.[142] They observe that understanding a robot's state of mind rather ambitiously "includes not only the intentions of the robot, but also the desires, knowledge, beliefs, emotions, perceptions, capabilities, and limitations of the robot." It extends to active feedback and even to mundane acts such as a robot communicating the need to replace a robots' batteries which is no different to someone saying that they are hungry or tired. They present a novel model of interaction for understanding which they describe as enabling "both human and robot to utilize a first or higher-order theory of mind to understand each other and perform communicative actions in order to support the other's understanding." The example they give is the need for autonomous cars to communicate their intent to allay the current feelings of being unsafe and confused when confronted by such cars and such initiatives are being pursued by the automobile industry. Hellström and Bensch's model of interaction for understanding is in their words "designed to describe how communicative actions are generated and performed by the robot to reduce the difference between the robot's state of mind (SoM), and the robot's inferred model of the human's model of the robot's SoM". This involves both the human and robot utilising higher-order ToM to understand each other and they provide a list of five questions for those developing and implementing the model that contains understanding and what that means and requires in interactive robots which are "(1) What information (if any) should be communicated to the human? (2) How should the robot represent and infer the human mind? (3) How should communicative actions be generated to communicate the required information? (4) To whom should the robot direct the communicative actions and (5) Which mechanism should enable the model?"

Humans tend to regard machines as reliable rather than trustworthy and the boundary between the two represents the involvement of mentalisation towards the robot as a machine. Sameule Vinazi, Angelo Cangelosi and Antonio Chella have observed how few studies explore whether robots could trust humans rather than humans trusting robots. To address this and in the spirit of true tango between man and machine, they have proposed an artificial cognitive architecture that allows a robot to assess the trustworthiness of a person or the human it is interacting with to facilitate and improve decision-making. The maturity of TOM in an individual has also been shown to correlate with the ability to assess and judge trustworthiness accurately with Ken Rotenberg and colleagues attributing low trust beliefs that promote loneliness from childhood to adulthood to early social disengagement and negative cognitive schema.[143] Trust is an important part of all social relationships and they claim that the model integrates a probabilistic TOM and trust model, supported by an episodic memory system based on the view that together with the developmental paradigm these are essential to trust. The episodic memory architecture allows the robot to use memories of past interactions with people to influence its behaviour towards a person who is new to the robot, which they believe gives it aspects of human personality.[144] Using established psychological tests, they consistently produced the same level of response and accuracy as children. They acknowledge that their work builds upon the earlier work of Massimiliano Patacchiola and Angelo Cangelosi who had built a Bayesian model that incorporated features of ToM and trust[145] but extended it by "integrating the original model into a complete robotic architecture and extending it with an episodic memory component". Their architecture also contains an algorithm inspired by a Bayeseian filter called a particle filter, which has been in use since the 1990s and is an estimation algorithm which "combines prior knowledge on how the state is expected to evolve over time with measurements that include information related to the current state."[146]

They used the Vanderbilt psychological test which had been developed to reflect that trust and assessment of the reliability of a person in preschool children goes beyond just considering their knowledge, and includes a person's motives and intentions which can be influenced by a person's prior history of deception if made known to a child.[147] The Vanderbilt test involves preschool-age children (in their study 90), equally divided into 3, 4, and 5-year-olds, who were shown a video of an adult actor, either helping or tricking people, advising another adult trying to locate a sticker hidden in one of two boxes. Helpers gave the correct location, whilst the trickers pointed to the wrong box. Subsequently, that same pointer would advise the child using the same sticker-finding task and the child must decide whether to trust their suggestion or not. Only the 5-year-olds appear able to differentiate between the helpers and trickers and thus demonstrate that they have ToM. They plan to expand the architecture to more general-purpose applications which they confidently believe is straightforward because it simply involves reorganising nodes and edges in the model to represent a wider range of real-life situations. They envisage that it will be able in time to deal with assessing more than just one person for trustworthiness and has wide applications in elderly care, robotic-assisted surgery, joint-action HRI in the performing arts and autonomous cars.

Computer scientists David Pynadath, Paul Rosenbloom and Stacy Marsella using the Sigma cognitive architecture developed a novel approach to TOM in an interactive situation with multiple robots, called Inverse Reinforcement Learning (IRL) of the other agent's reward function to make effective decisions and choices in the face of uncertainty about the behaviour of other robots and people. Sigma is built on graph models and has been used in a very basic form of TOM, with both parties being explicitly trained with knowledge of the other parties pay off or reward functions which does not constitute mentalisation, just prior knowledge of the others state of mind in this case reward functions. A reward function defines the goal and provides feedback to the robot which influences its behaviour and learning. Sigma can update the robot's knowledge using a gradient descent mechanism which is a common iterative optimization algorithm used to find the minimum/maximum of a given function, so can minimise the loss function to assess how well the algorithm is modelling the data.[148] They compared the IRL model which purports to be a "novel combination of RL and IRL for decision-making in combination with Theory of Mind about another agent" with three other situations. These were (1) Explicitly modelling the other robot's knowledge (2) In a passive non-interactive environment and (3) With a given set of reward functions for the other robot(s) none of which constitute TOM.

The test involved the robots negotiating over how to allocate apples and oranges with rewards given when an offer is accepted by the other robot and involved the robots using IRL to update a belief about the other robot's reward function to predict the choices the other robot more accurately will make. They ensured that the behaviour of the robots was non-deterministic by incorporating a 10% chance of deviation from the robot's optimal decision to a suboptimal decision. The robot can use its beliefs about the other robot's negotiating policy to update its predicted future rewards. Pynadath and colleagues wanted to take the process beyond an unknown stochastic (random) process to a situation in which the process is driven by the other robot's state of mind which the robot is trying to predict. The inverse element in reinforcement learning uses an assumption that the other robot has a fixed reward function that it is using to make rational decisions and allows the observing robot to use the observed behaviour to reason backwards to infer and thus determine the other robot's reward function. They validated the correctness of the algorithms but acknowledged that it needs to be used in an enriched multi-robot domain to get clear insights into the various modelling assumptions. They conclude that "Sigma's general capability for multiagent learning opens up a promising line of future research into RL-based TOM."[149]

NASA has explored two cognitive architectures Robonaut and Polybot (formerly Polyscheme). in developing their humanoid robots using ACT-R. This has involved both computational systems of which Polybot is an example and an actual robot called Coyote designed by Xanatos and the Scarab Corporation. Both Robonaut and Polybot purport to "enable effective human-robot interaction using perspective taking by the robots" but have different limitations and capabilities and advantages and disadvantages. The architectures harness not just knowledge but how perspective-taking and inferencing occur in a naturalistic environment, which in NASA's case involves astronauts working on a collaborative project with robots and needs those robots to perform perspective-taking and infer intentions if they are to follow the commands of the astronauts in highly pressured situations. For example, Gregory Trafton and colleagues gave a simple command to a robot to move towards a cone, but because the human can only see one cone, the robot knows this and is required to construct or model the scene using the human's perspective to be able to move towards the correct cone.[150, 151] They believe "that the default approach for designers should be to use person-to-person interaction as the model for human–robot interaction" which involves amongst other things perspective taking which requires a person or robot to "take one another's perspective and reason about interactions and the world from this alternative point of view."

Their work provided three important conceptual guidelines which are as follows "(1) To make the cognitive systems of robots like those of humans when it will aid in human–robot interaction (2) To build cognitive robotic systems that are integrated across perception, cognition, and action and (3) Building models of human–robot interaction based on a human to human interaction will result in good design heuristics throughout the project". Polybot includes what they call specialists with each modelling a specific aspect of the world. Cognitive scientist Nicolas Cassimatis, describes how "Specialists in Polybot communicate and combine information by simultaneously concentrating on the same focus of attention."[152] Collaboration between a robot and a human requires the robot to have embodied cognition with sufficient skills to communicate in the way humans do which must include basic skills such as voice recognition, natural language, and cognitive tools to generate at least some common representations of situations and shared behaviour.[153]

None of these developments suggest that they understand what others might be thinking or can relate this to a genuine self-concept which allows them to separate themselves from others. Despite these various initiatives and occasionally bullish views of the capacity of robots to possess TOM, others like Anna Heschel and her colleagues believe that HRI is a "young field currently in a phase of unrest" and that "a strong mismatch still exists between our expectations and the reality of social robots" and warn that we may be approaching a "social robots' winter" reflecting how the initial hype associated with AI was followed by a prolonged period of disillusionment.[154] A team of computer scientists have even included it in their list compiled in 2018, of the 10 great challenges of Robotic Science.[155] Heschel and the team believe that "studying the human brain when we perceive and interact with robots will provide insights for a clearer and deeper understanding of the human side of HRI" rather than the historical focus on the passive perception of the AI agent.

Robots and Empathy and Compassion

One psychological field explored in some depth concerning robots and TOM is empathy and some scientists believe we are facing a new frontier with the emergence of empathetic robots.[156] Over recent years robots have increasingly been used as caregivers for the elderly,[157] in child education[158] and kindergartens and mental health support.[159] Elizabeth Broadbent observes how they have been seen to provide benefits in "relieving loneliness and improving communication." She highlights how studying our interaction with robots can tell us much about ourselves and these various support roles suggest the importance of robots being able to show congruent emotions and behaviour towards humans of which empathy is one aspect.[160] Being able to offer emotional support, detect signs of distress, provide companionship and respond to customer frustration or satisfaction, are immensely valuable qualities for an autonomous social robot that is required to deal with people, particularly in an emotional support role. However, most social robots currently fall into Cynthia Breazeal's category we discussed earlier of Socially Evocative robots that we anthropomorphize and capitalise on feelings evoked when humans nurture, care, or are involved with their creation that we discussed earlier. Genuine empathy as a human trait is highly regarded and the absence of empathy is often regarded as a "prominent feature of serious psychological disorders and mental disabilities."[161]

Attempts to define empathy and compassion in humans have given rise to ambiguity as to what extent empathy involves cognitive or affective processes. American psychologists Daniel Goleman (1946–) and Paul Ekman (1934–), argue that the three components which signal the different types of empathy are Cognition, Emotion and Compassion. Cognitive empathy which is also known as empathic accuracy is more of a skill in terms of knowing someone's state of mind so the desire to understand can convey an emotional detachment, whilst emotional empathy is the ability or desire to feel what another person feels, and compassion or empathetic concern is the ability or desire to sense and understand what support another person needs from you. Intriguingly it has been observed that narcissists and sociopaths can harness cognitive empathy to manipulate people. As we have seen compassion involves emotional empathy in the initiation of the empathetic response and it is not clear whether genuine empathy can be divorced quite so neatly from emotions or precisely how emotional and cognitive empathy or compassion motivate us to act.[162]

Computer scientists Sung Park and Mincheol Wang conducted a systematic review of the literature regarding empathy in interpersonal, virtual agents, and social robots and as a result defined robotic empathy for HRI as the "robot's (observers) capability and process to recognize the human's (target) emotional state, thoughts, and situation, and produce affective or cognitive responses to elicit a positive perception of humans."[163] This encompasses both emotional and cognitive empathy and in the notion of cognitive responses,compassion. For a robot to display any sort of empathy it needs to be able to pick up on facial expressions, voice tones, body language and even physiological signals such as body temperature or stress and adapt their behaviour and responses to suit the emotional context. It is immensely challenging to get robots to emulate the complex social dynamics of individuals with others and in groups, in a way that makes them convincing or indicates that they genuinely experience all the dynamics that are involved in social situations. Studying social robots potentially increases the likelihood of anthropomorphizing such robots, though Kate Darling who is a research scientist at the MIT laboratory believes that this may not always be a bad thing and explores how we build diverse kinds of relationships with robots. In her 2016 paper, she argues that "there are cases where encouraging anthropomorphism is desirable" and advocates the need to resist the approach of those who believe that "robots should be framed in non-anthropomorphic terms and treated strictly as tools."[164, 165]

However, the distinction between empathising with a robot whether through anthropomorphising or not and a robot being able to genuinely empathise with us is fundamental and should not be confused or conflated. Robots must be able to display meaningful social interactions with humans and other robots when in real social settings particularly in caregiver situations and central to establishing long-term interactions with people, not least because empathy or perceived empathy is believed to foster a sense of affiliation and similarity.[166, 167] Empathy involves sharing in another person's experience of and perspective on the world and 'the ability to understand another person's mental states (e.g., desire or pain) by putting oneself in the position of that person."[168] David Levy observes that some people may prefer having a relationship with a robot that can be programmed to be empathetic in that they can be "social, smart and loyal", rather than have a relationship with someone unpredictable and disloyal and this could lead to greater interaction with robots than with each other because they have all the good features and none of the bad.[169]

Some studies claim that robots are demonstrating a primitive form of TOM even if the behaviours exhibited by the robot are far simpler than humans and confined to expressing themselves in images rather than language. Hod Lipson has argued this is not unlike when humans use visualisations to imagine the future.[170] Processing a symbolic representation of encoded environmental stimuli does not make a robot self-aware as self-awareness involves many self-referential processes, not just bodily awareness, and self-face recognition. It requires being able to distinguish between the private and public self, thinking about their past and future and as we have seen, the ability to infer the thoughts or minds of others. It is unclear whether the studies that suggest robots have a basic TOM merely reflect the superior capability of computers to analyse language patterns rather than constitute compelling evidence of genuine intuition or inference associated with human TOM.

Empathy involves not just the ability of an observer to understand another person's mental state but also adapting their behaviour and responses to suit the emotional context. Hence there are two distinct components to empathy which in humans appear to happen naturally and almost seamlessly. First the recognition of the human's (target) emotional state, thoughts, and situation and second the change in the observer's behaviour to make it more congruent and empathetic. This raises the vexed question as to whether it is possible for a robot to simply mimic empathy by adjusting the behaviour provided it has been trained to recognise the state of another even if they don't possess TOM or feel any emotion or genuine empathy. The question is whether this more limited behavioural manifestation of empathy has the same impact on the observed person but given the earlier observations on our tendency to mentalise robots when emotions appear to be exhibited it is highly likely that it will. This is relevant to robotic empathy in that it would be a much lower hurdle to equip the robot to recognise the behaviour and respond without needing to demonstrate that they can or need to experience emotions themselves, nor possess a genuine sense of self, given that empathy tends to be associated more closely with the self than compassion. Can robots display genuine empathy which requires a variety of mental and/or emotional processes in showing concern for others or are they merely displaying the ability to mimic the behaviour associated with empathy such as appearing to be listening and displaying other bodily signs of empathetic behaviour etc? Adjusting and mimicking behaviour are not the same as experiencing genuine empathy and conflating the two makes the mistake that it demonstrates an effective state in the observer which in a robot would require the phenomenal experience of concern. Adjusting behaviour when displaying empathy does not require or necessarily involve mimicking a person's behaviour but different forms of empathetic behaviour such as sensing someone else's happiness or unhappiness, patient listening to a person, comforting them and understanding their feelings.

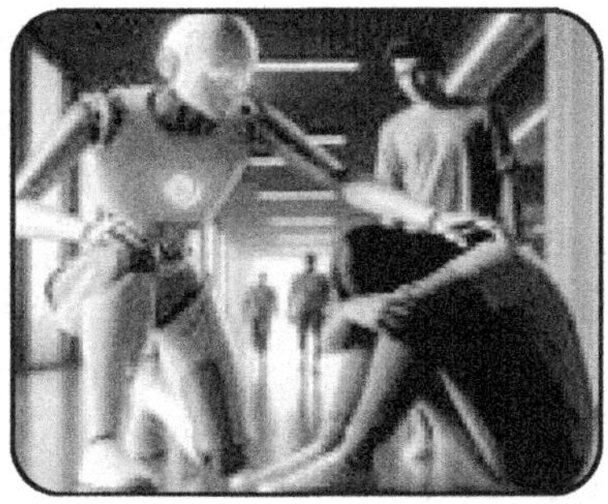

Robots can be trained to mimic empathetic behaviour with potential consequences for how a person feels towards the robot. Tanya Chartrand and John Bargh were two of the first psychologists to study the subconscious mimicking of other people's behaviour, in what they called the chameleon effect, based on the way the chameleon changes the appearance of its skin to blend into any environment.[171] Psychologists believe that being a chameleon has important social consequences and is designed to make that person feel more positive towards you and even find you more attractive. It can even be consciously adopted to make people find you more sociable and likeable. It appears though that individual differences between people in empathic concern have little effect on the level of mimicry which suggests that the cognitive component of perspective-taking is more important in encouraging mimicry than the emotional component. This indicates an important distinction between mimicry and genuine empathy with the emphasis in mimicry on the benefits to the observer, whereas genuine empathy is directed at supporting the other person and as we saw is based on a greater subconscious association of the self with that person. There appears to be a subtle distinction between two types of mimicry, one being mirroring mimicry in which a person does the opposite, for example, if the observed person moves their right hand the observer moves the left. Most people when face to face appear to spontaneously mirror each other in this way. The other is anatomical mimicry in which the observer makes the same movements as the observed so for example moving the same hand.[172]

Daniel Casasanto and colleagues studied the response of participants to mirroring by an embodied virtual agent called VIRTUO, in a fully immersive virtual environment, who mimicked their head movements either mirror-wise, anatomically, or not at all. They found that if someone subconsciously mimics another person anatomically, that person will react more negatively towards them than if they were mimicked in a mirror-wise way or not at all. They dubbed this the Reverse Chameleon effect and proposed that this may be because as hypothesised in earlier studies "mirror-wise mimicking increases the mimicker's readiness for cooperative joint actions" even if the movements like head movements are not goal-directed, whereas "anatomical mimicry decreases readiness to cooperate".[173] This illustrates how subtle the roles of mimicking are and how they may be related to joint attention, action observation, task sharing, action coordination and agency and that mimicking and empathy are not the same thing even if empathy may involve mimicking behaviour.

Researchers Hodeki Kozima and colleagues have developed the Roboto Genetic Model which involves implementing a possible "ontogeny (developmental process) of the social capability onto a robotic embodiment with a certain phylogenetic background (innate prerequisites)." They believe that ontogeny requires being able to maintain eye contact and engage in joint attention which is a version of listening and attentiveness and requires a mechanism which allows for the "spatiotemporal coordination of attention and bodily movement between the self and another." However, it requires the attribution of mental states (TOM) as an innate prerequisite.[174] In humans, empathy tends to be associated more closely with the self than compassion and the conflation of the negative experience of another person with one's pain.[175] Japanese researchers Seisuke Hayakawa and Katsunori Miyahara, argue that mental simulation is not enough and that empathy requires "anchoring in receptivity" which is achieved mostly through listening. According to their view, it involves four distinctive features which are "(i) the actual sharing of a perspective; (ii) dynamical unfolding; (iii) collaboration and (iv) mutual transformation."[176] They observe that simulationists rarely incorporate the role of listening into their framework and that it is an embodied collaborative process at the two levels of imaginative perspective-taking which requires (1) Imagining the other's perspective using mental simulation, and (2) Realistic perspective-sharing which involves coming to see the situation in the same way as the other rather than just imagining it. This means that when we are being empathetic, we must first imagine that person's state of mind and situation, but the deeper level of empathy involves assuming a realistic perspective-sharing which requires the empathising agent to undergo what they call an "epistemic self-transformation." This refers to how the empathetic experience transforms the empathiser by teaching that person something that they could not have learned without that kind of experience.

What this brief discussion on empathy and mimicking in robots suggests is that (1)

Empathy and mimicking are distinct processes with empathy directed at the observed party whilst mimicking typically serves the interests of the mimicker. Mimicking though is a potentially important component in the display of empathy (2) Empathy has a cognitive and emotional component, but mimicry appears to involve predominantly cognition whilst empathy is more emotional (3) The emotional component of empathy reflects how empathy tends to be associated more closely with the self than compassion and the conflation of the negative experience of another person pain with one's own pain (4) Both empathy and mimicry require the observer to pick up on facial expressions, voice tones, body language, and even physiological signals and adapt their behaviour and responses to suit the emotional context (5) Empathy requires both imaginative perspective-taking and realistic perspective-sharing with the latter involving an "epistemic self-transformation" (6) Empathy in humans requires a self and the ability to experience emotions and pain both one's own and others (TOM) (7) Empathy cannot be achieved purely through spatio-temporal coordination of attention and bodily movement between the self and another, such as through simply maintaining eye contact and engaging in joint attention. Mimicry can be achieved by this mirroring though anatomically mimicking tends to result in the mimicker being perceived more negatively. This allows us to make a much clearer assessment of the likelihood that robots will be able to display genuine empathy or effective mimicry that can be confused with empathy.

Notwithstanding these observations, there are several robots both androids and animal robots that have been designed to display empathy and even marketed as being able to do so, including the following examples, Pepper the Robot, Paro the Therapeutic Robot, Ellie the Virtual Therapist and Mabu the Personal Healthcare Companion. In 2014 SoftBanks launched Pepper the Robot whom we met earlier, which they described as 'Human emotions, decoded, embraced and echoed" and marketed as the first robot with a heart and the "first social humanoid robot able to recognize faces and basic human emotions." Production was stopped in 2021 because it appears that the developers faced challenges with limited functionality and unreliability and struggled to find a global customer base, only producing 27,000 Peppers.[177] First launched in 2003 and now on its 8th iteration Paro the Therapeutic Robot is an advanced interactive robot that has been developed by the Japanese company AIST which is a pioneer and leader in industrial automation. The premise behind Paro which resembles a baby harp seal is the "documented benefits of animal therapy to be administered to patients in environments such as hospitals and extended care facilities." Paro is equipped with tactile, light, audition, temperature, and posture sensors, and can perceive people and their environment and recognize the direction of voice and words such as its name, greetings, and praise and is designed to respond in a live fashion. The problem is that simply being integrative does not amount to being empathetic or even displaying the complex social dynamics of individuals. It is also a seal, even if research is revealing how other species like rodents, monkeys and dogs can experience complex emotions and neural activity associated with empathy. It has been shown that interfering with the signalling of oxytocin alters empathy-related phenomena across species.[178] Oxytocin is a hormone that's produced in the hypothalamus and released into the bloodstream by the pituitary gland. It is known to play a role in social bonding, facilitating childbirth, and maternal behaviour in the period after childbirth and is higher during orgasm and ejaculation."[179]

In 1966 MIT professor Joseph Weizenbau (1923–2008) created a program called Eliza intended to provide a very crude form of person-centred counselling which consisted of repeating what the person has said back to them. In a sense, it was the first chatbot. Although hard to believe, people loved Eliza and believed that she understood their problems and loved talking to her for long periods.[180] He named Eliza after the character, Eliza Doolittle, the cockney flower girl in George Bernard Shaw's play Pygmalion because her elocution lessons allowed her to create the illusion that she was a duchess and Eliza created the illusion that she understood what the user meant. Weinbaum was horrified that he had created what psychologists call the transference effect in computers. Transference refers to the schemas we bring to encounters with others and it determines how we see those people. It became such an acknowledged phenomenon that is now called the Eliza effect. His horror was such that by the 1970s he had become a strident albeit rather lonely figure opposing the march of AI describing it as an "index of the insanity of our world." Despite these potential dangers, 60 years later, the USC Institute for Creative Technologies as part of a project called Sim Sensei developed Ellie the Virtual Therapist, with the stated goal of designing a virtual human for healthcare support. Ellie far surpasses the basic computer program that was Eliza and is designed to monitor micro-expressions, respond to facial cues, perform sympathetic gestures and build a rapport with the user. In 2018 Cory Kidd the CEO and founder of Catalia Health launched Mabu which they described as the Personal Healthcare Companion primarily designed to assist people with staying on top of their healthcare plans and medication. Iolanda Leite and colleagues claimed to have developed an empathetic iCat robot that when observing and commenting on a chess match was able to display empathic behaviours towards one of the players yet behave neutrally towards the other one. They found that those towards whom the robot displayed empathy perceived the robot as more friendly.[181] Their focus was on identifying which aspects of social interaction are the most relevant for robots and measuring the level of perceived friendship in the HRI but does not establish that the robot was displaying genuine empathy in the way humans do.

These examples of social robots show that they are increasingly being deployed into various support situations and purport to exploit the facet of empathy and concern for others which is a fundamental hallmark of what it means to be human and as we now know common to many animal species.

Metacognition in Robots

Metacognition as we saw is a higher-order form of thinking that involves being able to observe yourself as the object of thinking or thinking about thinking and requires active control over the cognitive processes engaged in learning and introspective monitoring of the internal self. Metacognition like mentalisation is a form of meta-level mental representation but typically of one's own rather than mentalisation which refers to inferring another person's state of mind involving TOM. Metacognition in humans is essential to teamwork and collaboration which import a notion of shared responsibility and recognition of the need to cooperate, to optimise the knowledge of the individuals and the group acting together. This is why, despite the obvious challenges, metacognition in AI has assumed growing importance with the deployment of autonomous robots that must collaborate and operate in teams alongside humans and other robots. Metacognition requires three different types of knowledge or as Flavell described them, Task Person and Strategy variables which are "(1) Conditional knowledge which is about when and the conditions in which certain knowledge is useful, planning and the approach to a given task (2) Procedural which is the knowledge about the skills, heuristics and strategies you use to perform tasks and do things and (3) Declarative knowledge which is knowledge about the self as a learner and what influences your performance and evaluating your progress". The problem with metacognition as a concept is that it is described variously as self-regulation or executive control. This fails to reflect the fact that it involves different types of knowledge about the self and meta-cognitive regulation which poses difficulties in separating cognitive activity and meta-cognitive activity. For example, the American developmental psychologist who specialises in children's cognitive development John Flavell (1928–) believed that cognitive knowledge was different to meta-cognitive knowledge.[182] In the context of AI and robots, limited meta-cognition already has an important potential role in error correction and the need for robots to be able to monitor and self-regulate their actions and complete tasks without human intervention. Because AI uses complex algorithms it has been argued that it is impossible for robots to display the equivalent of human metacognition, but many current AI cognitive architectures do support a form of metacognition, mainly with reference to observing, analysing, and regulating the system and the processes involved in performing the tasks. Such metacognition appears to be more about self-monitoring through simple data collection on what is happening in the system, rather than constituting higher level mentalisation or awareness of the self and meta-cognitive self-knowledge. It is more akin to procedural knowledge than conditional or declarative knowledge. When a cognitive architecture observes itself, it collects data about the systems internal operation and status which may even support earlier solutions so has a temporal component, it compares different options, justifies, and explains why and uses the data (analysis) to alter plans or tasks (regulation). Some systems have used a limited form of meta-cognition for example to recall previous solutions, compare different answers and justify the decisions it makes.[183]

All the various initiatives that we have discussed are largely research based and found in academia rather than widely in industrial or social situations but provide an intriguing insight into the progress being made to imbue AI with the capacity for mentalisation in the form of TOM and meta-cognition.

Endnotes

[1] An Enquiry Concerning Human Understanding.
[2] Suzuki, 1952, 51.
[3] Doors of Perception (1953).
[4] Timmerman Slater, CB, Bauer P, Gosseries O., et al., 2023.
[6] Gallagher, S., 2000.
[7] The Principles of Psychology (1890).
[8] Pfeifer, JH., 2007.
[9] APA Dictionary of Psychology.
[10] Human Nature and the Social Order (1902).
[11] Gallagher, ST., 2000.
[12] McAdams, DP., 1985.
[13] McAdams, D P., 1996.
[14] Cohler, B J., 1982.
[15] Methods of Life Course Research (1998).
[16] Family Narratives and the Development of an Autobiographical Self (2019).
[17] D'Argembeau, A, et al., 2014.
[18] The Self Illusion (2012).
[19] Fivush, R., et al., 2011.
[20] Childhood and Society.
[21] The Construction of Self.
[22] The APA Dictionary of Psychology.
[23] The Intentional Stance (1996).
[24] Levine J. 1983.
[25] Shultz S, Dunbar RIM., 2007.
[26] McAdams, D. P., Pals, J. L., 2006.
[27] Markett S., et al., 2017.
[28] The Selfish Gene (1976).
[29] Born That Way. Genes, Behaviour, and Personality (1999).
[30] Are we hardwired: The role of genes in human behaviour (2000).
[31] Moore, H., 2010.
[32] Hamson, S. E., 1992.
[33] V Jirtle R.L, Skinner M.K., 2007.
[34] Genes and Behaviour: Beyond Nature Nurture (2019).
[35] McCrae RR, John OP., 1992.
[36] Schretlen, D. J., et al 2010.
[37] Schretlen, D. J., et al 2010.

[38] Oxford English Dictionary.
[39] Andersen, S M., Berenson, K R., 2003.
[40] Andersen SM, Chen S., 2002.
[41] Overall NC, Fletcher G, Friesen M., 2003.
[42] Romero-Canyas R., et al., 2010.
[43] Burklund L J, Eisenberger, N, Lieberman, M., 2007.
[44] Downey, G, Feldman, SI., 1996.
[45] De Rubeis, J., et al., 2017.
[46] Gao, S., et al., 2017.
[47] Self Comes to Mind (2012).
[48] Making Minds Conscious. (2021).
[49] L'Essence de la manifestation.
[50] The Myth of Sisyphus (1942).
[51] O'Connor, P., 2014.
[52] APA Dictionary of Psychology.
[53] What Is It Like to Be a Bat? (1974).
[54] LeDoux J.1995.
[55] Seth, AK., 2013.
[56] Goodkind, M, et al., 2015.
[57] Stephani, R., et al., 2011.
[58] Scalabrini, A., et al., 2021.
[59] Modinos, G.et al., 2009.
[60] I Am a Strange Loop (2007).
[61] Fodor, Jerry A. 1983.
[62] Rogers, TB., et al., 1977.
[63] Symons, C.S., Johnson, BT., 1997.
[64] Northoff G, et al., 2006.
[65] Northoff G, et al., 2006.
[66] Raichle, ME, Snyder A Z., 2007.
[67] Buckner, RL., et al., 2008.
[68] Pearson, JM. et al., 2011.
[69] Unlocking the Brain (2014).
[70] Northoff, G., 2014.
[71] Northoff,G.,et al., 2015.
[72] Moran, J M., et al., 2006.
[73] Duval, S, Wicklund, R A., 1972.
[74] Silvia, P, Duval, TS., 2001.
[75] Gallup, G G., et al., 2003.
[76] Chang L, et al., 2014.

[77] Morin, A., 2006.
[78] Premack D, Woodruff, G., 1978.
[79] Rubio-Fernandez, P., et al., 2019.
[80] Leslie, A 1994.
[81] Stone, VE, Gerrans, P., 2006.
[82] Schurz, M, et al., 2021...
[83] Wimmer, H., Perner, J., 1983.
[84] Rubio-Fernández P, Geurts,B., 2013.
[85] Onishi, KH, Baillargeon, R., 2005.
[86] Surian L, Caldi, S Sperber, D., 2007.
[87] Hughes, C., 2002.
[88] Diamond,A., 2012.
[89] A Scientific Search for Altruism: Do We Only Care About Ourselves? (2018).
[90] Batson, C D,Lishner, D,Stocks, E., 2015.
[91] Gu, J, et al., 2017.
[92] The Age of Empathy: Nature's Lessons for a Kinder Society (2010).
[93] Altruism: The Power of Compassion to Change Yourself and the World (2015).
[94] Klimecki, O M., et al., 2012.
[95] Singer, T, Bolz, M., et al., 2013.
[96] Klimecki, O., Ricard, M, Singer, T., 2013.
[97] Singer, T, Bolz, M., 2013.
[98] Klimecki, O, Ricard, M, Singer, T., 2013.
[99] Flavell, JH., 1979.
[100] Frith, CD., 2012.
[101] Apperly, IA, Butterfill, SA., 2009.
[102] Cassimatis, N., 2022.
[103] Nagla KS., et al., 2014.
[104] Tensors for Data Processing (2022).
[105] Bioinformatics (2003).
[106] Mitchell, RW., 2002.
[107] Mitchell, R., 2001.
[108] Vygotsky, L., 1962.
[109] Pipitone, A, Chella, A., 2021.
[110] Pipitone, A, Chella, A., 2021.
[111] SO/TS:2016 and COMEST/Unesco
[112] Chandrasekaran, A., et al., 2017.
[113] Troccaz, J., et al., 1993.
[114] Thomaz, A, Hoffman, G, Cakmak, M., 2016.
[115] Cheng, Y, Sun, L, Tomizuka, M., 2019.

116 Khawaja, FI., et al., 2021.
117 Ming Huan, C, Mutlu B., 2016.
118 Devin, S, et al., 2017.
119 Devin, S, et al., 2017.
120 Fong, T., et al., 2003.
121 Fong, T., et al., 2003.
122 Michlmayr, M., 2002.
123 Burgoon, JK., Bacue, AE., 2003.
124 Spatola, N., et al., 2022.
125 Low, J Perner, J., 2012.
126 Kopp S, Krämer N. 2021.
127 Tomasello, M., 2010.
128 Kopp, S, Krämer, N., 2021.
129 Levinson, SC., 2006.
130 Fröhlich, M, van Schaik CP., 2022.
131 Fröhlich, M, et al., 2016.
132 Burkart JM, et al., 2022.
133 Heesen R, Fröhlich, M., 2022.
134 Kopp, S, Krämer, N., 2021.
135 Kopp S, Krämer N. 2021.
136 Wang, QI., et al, 2022.
137 Liu,T., et al., 2022.
138 Kosinski. M, 2023.
139 Chen, B, Vondrick, C, Lipson, H., 2021.
140 Finkel, M, Krämer, NC., 2022.
141 Hellström, T, Bensch, S., 2018.
142 Nomura, K, Kawakami., 2011.
143 Rotenberg, KJ, et al., 2015.
144 Vinanzi, S., et al., 2019.
145 Patacchiola, M, Cangelosi, A., 2016.
146 Elfring, J, et al., 2021.
147 Vanderbilt, KE, Liu, D, Heyman, GD., 2011.
148 Pynadath DV, Rosenbloom, PS, Marsella, SC., 2014.
149 Pynadath, DV, Rosenbloom, PS, Marsella SC., 2014.
150 Sofge, D., et al., 2004.
151 Trafton, JG, et al., 2005.
152 Cassimatis, 2002.
153 Butterfield, J., et al., 2009.
154 Henschel, A, Hortensius, R, Cross, ES., 2020.

[155] Yang GZ, et al., 2018.
[156] Coplan, A., 2011.
[157] Robinson, HM, et al., 2014.
[158] Hyun, E., et al., 2010.
[159] Diehl, JJ, et al., 2012.
[160] Broadbent, E., 2017.
[161] Hodges, S D., Biswas-Diener, R., 2007.
[162] Hodges, S, Myers, M.W., 2007.
[163] Park, S, Whang, M., 2022.
[164] History with Animals Reveals about Our Future with Robots, 2001.
[165] Darling, K., 2016.
[166] Anderson, C., Keltner, D., 2002.
[167] deVignemont, F, Singer, T., 2006.
[168] Kozima, H., et al., 2004.
[169] Levy, D., 2007.
[170] Chen, B, Vondrick, C, Lipson., 2021.
[171] Chartrand, TL, Bargh, JA., 1999.
[172] Casasanto, D, et al., 2020.
[173] Sebanz, N, Bekkering H, Knoblich, G., 2006.
[174] Kozima, H., et al., 2004.
[175] Klimecki, O, Ricard, M, Singer, T., 2013.
[176] Hayakawa, S, Miyahara, K., 2021.
[177] Reuters 2021.
[178] Paradiso, E, Gazzola V, Keysers, C., 2021.
[179] Cera, N, et al., 2021.
[180] Weizenbaum, J., 1966.
[181] Leite, I., et al., 2013.
[182] Flavell, JH., 1979.
[183] Marshall, JB., 2002b.

"The conscious mind may be compared to a fountain playing in the sun and falling back into the great subterranean pool of subconscious from which it rises."

Sigmund Freud

Chapter 8

The Subconscious

CHAPTER 8

The Human Subconscious

Are we in control of our choices and Decisions?

IF IT WEREN'T FOR NON-CONSCIOUS processing our brains would be overwhelmed, because our subconscious by storing memories and prior knowledge acts as a filter and unconscious automatic acts conserve valuable energy. The subconscious world of our brains has traditionally been regarded as something that exists below our conscious mind, with primacy given to consciousness and reason but there is growing scientific evidence that most of our behaviour is not consciously intentional. Intentionality involves a causal link between the intention to act and the act itself and requires a goal representation of that act in terms of an envisaged outcome. Goal representations though can be subconscious, and the intent may not be at all obvious to that person. Psychologists observe that not every intentional act requires an intentional goal representation and distinguish between intentional acts that do need a goal representation whether subconscious or not and those that are more gestalt in that the act or behaviour is more than just the sum of its parts in terms of intentionality. For example, driving a car requires intentionality but the act of driving involves automaticity in which you do so without the intentional goal representation when you are driving. You can be using the conscious and subconscious or unconscious processes with the same goal in mind and neuroscientists have searched for the neural correlates of unconscious and subconscious processing that influence our behaviour decisions and choice making.

Our non-conscious comprises both unconscious and subconscious processes. Unconscious processes are automatic and cannot be accessed by us through introspection. Subconscious processing involves storing information that we have learned much of which becomes automatic but can still be accessed through introspection and reflecting on why we might have acted in a certain way or made certain choices. Our conscious mind arises when during a short period of focal attention, we become aware of thoughts and actions and of what we are doing and doesn't require introspection to access but is fleeting. We get the impression of a continuous chain of consciousness because we are constantly linking the different phases of focal attention whether it's directed externally at the environment or internally at our thoughts, so conscious thought dominates our perception of how we operate and interact with the world.

Until recently scientists regarded with scepticism the notion that our subconscious mind can control our behaviour even though most of the animal kingdom appears to rely almost entirely on non-conscious processes to guide their behaviour. There has been an enormous surge in scientific and public interest over the last twenty years, as to why humans behave as they do and how much is conscious or subconscious or otherwise, evidenced by multimillion best-selling science books, like Nudge (2009) by Nobel prize-winning economist, Richard Thaler, and Cass R. Sunstein, which describes how we make the choices we do, Blink (2007) by Malcolm Gladwell, which describes how we think without thinking and Kahneman's Thinking Fast and Slow (2013). Brian Tracy (1944–) a Canadian American motivational public speaker and self-development author of over eighty books observes that "our subconscious mind makes all our words and actions fit a pattern consistent with our self-concept and our innermost beliefs about ourselves?"[1] Science is getting closer to agreeing what constitutes the subconscious content and recognise how the state and content are integrated and non-dissociable, but it remains unclear exactly how and why almost 95% of the processing of sensory information takes place below the threshold of consciousness or awareness and only some neural signals carry information that provides the content of consciousness and awareness of that content.

Neuroscientist Jaan Aru observes how "consciousness only has access to very specific levels of representation and that other levels are firmly locked away from it" with conscious experience "restricted to certain contents and computations."[2] Our brain unconsciously overcomes the binding problem and allows us to distinguish between objects and we remain unaware of how our brain subconsciously constructs the thoughts and sentences, we use in normal conversation. Even the content of conscious experience has a finite time window before it passes out of conscious awareness, so is elusive. Linking the conscious to subconscious influences is a major challenge to constructing AC machines that mimic our capacity to explore and perceive our environment in the way our brain does. Mimicking human consciousness in a machine must reflect not just the state and content of consciousness but the fact that most information processing is performed unconsciously or subconsciously before it emerges briefly into consciousness. This is further complicated by the popular perception of the subconscious as the source of creativity and insight and the basis for a mystical collective unconsciousness that taps into the universal consciousness that links us.

Constrained by using a fixed computer architecture any AI system seeking to create the thermodynamic and subjective features of our brain when processing unconscious, subconscious, and conscious information appears doomed to failure. A simple observation first made by Alan Turing is that computers can't even dynamically tell themselves to stop or switch thinking, whereas humans can do this almost effortlessly and this flexibility appears inherent in the way our brains operate. It might be more feasible if there were a location in our brain which we knew gives rise to the state of consciousness and identifies where and how the content is processed, but the unconscious, subconscious and conscious processes are dynamic and distributed over space and time. In 2021 researchers were able to uncover for the first time what happens in animals' brains when they learn from sub-conscious, visual stimuli.[3] They identified the VTA (discussed earlier) as a site of dopamine release linked to the subconscious learning of complex visual stimuli, not just conscious learning. This has important potential implications for using dopamine stimulation to support enhanced learning in those who due to trauma, ageing, or oncological problems face problems. It also illustrates just how much we have to learn about our subconscious processes.

Cognitive psychology has shown that the subconscious mind can process information from stimuli that our brain is not consciously aware but these processes have been described as unsophisticated and the subconscious mind as dumb, in the belief that it deals only with run-of-the-mill, unimaginative tasks, or activities.[4] American John Bargh (1955–) is a leading social psychologist, who has focused on the role of our unconscious in the context of social behaviour. Bargh observes how many psychological scientists still regard the subconscious as operating at this subliminal level and treat it as "the shadow of a real" conscious mind, even though there now exists substantial evidence that the "unconscious is not identifiably less flexible, complex, controlling, deliberative, or action-oriented than is its counterpart."[5] He describes how social psychology has more recently approached the unconscious differently by researching the extent to which people are aware of the important subconscious influences on their judgments and decisions. All non-reflexive human behaviour starts with input in terms of the perception of stimulus generated externally or by internal thoughts, which may or may not result in a motor action or decisions or choices as behavioural output. This process of perception and behaviour has spawned a whole field of research into cortical functioning used to guide both neuroscience and AI research. Our current understanding is that it involves three fundamental processes which are (1) The matching of sensory information with prior expectations, (2) Exploring different possible options or inferences about uncertain and incomplete sensory information and (3) Making predictions based on a range of possible options. Many cortical models assume a feed-forward architecture, but the problem is that the neocortex consists of not just a bottom-up feed-forward processing but also a top-down recursive loop in feedback processing and another level of complexity associated with processing information against prior expectations.

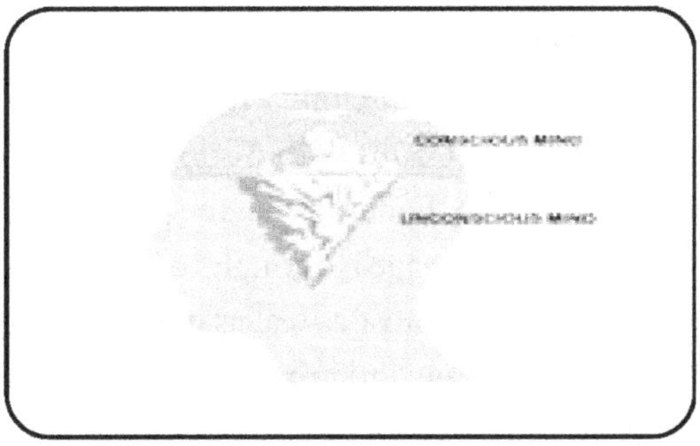

American neuroscientist, psychologist, computer scientist, data scientist, and entrepreneur David Hegger (1961–) is an ardent advocate of developing a new model of cortical functioning for humans and AI systems, that more accurately reflects these three levels of processing and believes that they are controlled by neuromodulators and oscillatory activity. Hegger's approach reflects how most information is processed by our brains outside conscious awareness. Perception and attribution of meaning to stimuli from the environment is largely an unconscious or subconscious process, with our brains using sensory information which is ambiguous and uncertain and matching it with prior knowledge contained in encoded memories which are unreliable. Psychological theories have traditionally treated decision-making as a higher cognitive process separate from the neural correlates of perception and action with decisions made before actions are planned. The Bayesian coding hypothesis states that rather than computing a single best estimate of what the information is, our brain represents information on a probabilistic basis.

Computational neuroscience researchers Rodrigo Echeveste and Mate Lengyel claim that Bayesian coding with its reliance on a probabilistic calculation, is optimal when integrating information from multiple sources. This includes unreliable sensory information and memories and means subjective evaluations based on incomplete information are used to update the brain's internal models of the world.[6] There is a paucity of neuroscientific evidence as to how neurons encode the uncertainty in information arising from the stochastic noise and missing information that is inherent in the unconscious processing of sensory information.[7]

In 2006 a seminal and widely accepted paper suggested that a wide range of psychophysical experiments involving a variety of tasks ranging from cue integration and decision-making have evidenced how "humans perform near-optimal Bayesian inferencing." The variation in the response of the neuronal activity previously thought to be a problem, appears to play a vital role because according to the authors "neurons both represent probability distributions and combine those distributions according to a close approximation to Bayes' rule."[8] It appears that it is the very uncertainty that allows our brains to use probability distribution to make predictions. But how does this unconscious or subconscious predictive Bayesian inferencing machinery influence our behaviour? Its accuracy depends on the extent of prior knowledge and primarily unconscious and subconscious behaviour both learned and automatic. This means the predictive machinery repeated time and time becomes rapid and habitual, so totally unconscious and subconscious with repeated experiences and even mental representations evoked almost robotically, with a subconscious or unconscious response generated even by an imagined scenario. For example, the prospect of making a public speech may be rehearsed in the mind and the image is enough to provoke a subconscious and automatic and unpleasant response, that is almost as real as going through the experience.

John Bargh is contributing to a large and growing body of evidence that demonstrates that most of our behaviour, whether the outcome of perception or our thoughts, is preceded by and largely the result of entirely unconscious and subconscious brain processes. This runs counter to the traditional associating of our conscious minds with rational and logical analysis and reasoning and characterisation of the subconscious as intuitive and less rational. Bargh observes that, between the initial perception based on the predictive processes, the conscious thought and the resultant motor behaviour, there are three key mediating factors in all social behaviour, which are (1) Dispositional traits, (2) Goals, and (3) Behavioural representations, with usually one or more involved and he describes this as our Subconscious Behavioural Guidance System. Unconscious behaviour such as in hypnosis or sleepwalking has been assumed to be unintentional, and assumed to apply to both our unconscious and subconscious actions, which has led to the belief that the subconscious does not deal with complex or flexible processing and that these are the sole domain of consciousness. To test this assumption, Bargh and other social psychology researchers have focused on the subconscious mental processes that influence intentional behaviour which seems counterintuitive because we believe that we are always aware of why we make rational decisions and choices and behave the way we do. Growing scientific evidence indicates that the subconscious mind is as action-orientated, deliberate and complex as the conscious mind and plays a powerful and pervasive role in influencing our intentional conscious behaviour, thoughts and decision-making.[9] The growing recognition of the evidence that we are not as in control as our conscious minds would like to believe has led to a move away from the binary distinction between conscious and subconscious behaviour, towards a recognition that the processes are almost intertwined in a perpetual dance and separating intentional conscious thoughts and behaviour from our subconscious does not reflect how our brains and minds work. Scientists are beginning to acknowledge the power of the unconscious and subconscious mind in driving our goals, motivations, desires, perceptions, and general behaviour and the rapidity with which it can make us behave, almost like a mental reflexive rather than a physical reflexive response to a situation. Much research has shown that many psychological processes such as forming judgments, emotions, attitudes, social behaviour, evaluation, stereotyping, prejudice, and even pursuit of goals, do not require or involve conscious thinking intent or control but are guided by our subconscious. Unconscious cognitive biases and defence mechanisms have a serious impact on our perception of the world but Bargh observes that "Conscious intent and guidance is not necessary even for the highest of higher mental processes to operate."[10] One major implication is that many early life experiences can form the subconscious substrate and schema for our behaviour which means that we continue to use earlier experiences which may not have been modified or adapted to the adult version of ourselves. They become so automated that they create a strange dissonance between what we know to be rational or more appropriate behaviour because we are guided by our subconscious traits, goals, and behavioural representations. Our dispositional traits are bound up with our self-schemas including low self-esteem, lack of confidence or poor body image that make us sensitive and vulnerable, and which influence our subconscious processing. This is why different people will react very differently to the same situation, event or stimuli and the intensity of the emotions generated may be unapparent or puzzling to the observer.

Subconscious Decision Making

Almost every aspect of our daily lives is governed by decision-making and choices with most made rapidly with little thought, whilst some are only made after prolonged consideration and over time. Cognitive psychologists studying human decision-making, use three observables which are the (1) Action decision or choice (2) Time taken to make that decision or act and (3) Confidence that people express in the decision or appropriateness of the behaviour. Neuroscientists have long sought to understand the neural correlates of both rapid and more deliberative decision-making by reference to how accurate or appropriate the decision and choice is, how quickly they are made or the reaction time and the level of confidence the decision maker has in the decision, which has a major role in influencing our behaviour and correcting mistakes. Most psychological behavioural research has typically treated decision-making as a sequential process involving perceptual, cognitive, and motor stages with a known relationship between the reaction time and the accuracy of the decision. The Evidence Accumulation Models (EAMS) which includes the Drift Diffusion Model (DDM) developed in the 1970s by Roger Ratcliffe are often used to measure rapid decision making, in which a person matches the perception of information to prior knowledge usually by reference to something in their immediate surroundings to decide how to respond.[11] The term drift in decisions in the DDM when related to perceptual tasks refers to the drift towards the correct answer, whereas in choice situations it relates to the relative appeal between the choice options. Most simple models have assumed the boundaries or points at which a decision is made are constant but recently non-constant boundaries have been used and observed to more accurately fit the data and the fact that correct responses are made faster than incorrect responses. The EAM models recognise that decisions are made based on the process of accumulation of evidence, which given the many possibilities is inevitably noisy, but there is a threshold known as Bounded Evidence Accumulation at which based on the accumulated evidence the decision or choice is made. These models include several cognitive computational models, used to explore the latent and subconscious processes that underlie our decision-making.[12] They have become established as the main scientific paradigm for explaining our decision-making processes and as neuroscientist Russell Boag and colleagues observe in a recent study, this is no longer just confined to simple decisions but vital in safety-critical decisions often related to survival, across such " diverse domains of air-traffic control, driving, forensic and medical image discrimination, and maritime surveillance."[13] Although these models originally focused on simple and rapid decision-making, they have become increasingly popular in psychological research on both rapid and more deliberative decision-making. This requires scientists to be able to identify the point at which the dominant signals containing the information emerge across the threshold to reach conscious awareness and facilitate a conscious decision to act or behave in response to a stimulus. When we decide to do something, the theory is that this is the result of the accumulation of the neuronal signalling evidence in our brain having reached the threshold and been selected.

In recent years the sequential approach to decision making has been fundamentally challenged by scientists including Polish-born neuroscientist Paul Cisek. He focuses his research on understanding how our brains deal with the interactions with the environment and how we plan and select actions which includes AI, the cognitive sciences and the neural mechanisms of decision-making, planning, and movement control. Cisek believes that the traditional models of abstract problem-solving in humans have relied on "perceptual, cognitive, and motor stages, defined within an information processing framework."[14] He argues that this approach does not capture the real-time interactive nature of our real-world behaviour in the environment and claims that our brains don't first construct representations of the world, on which to build knowledge and then compute and execute an action plan based on that representation. Our brain does not sequentially decide what to do and then how to do it and he supports his argument by observing how neuroscientific studies of the cortex have failed to reveal distinct perceptual, cognitive, or motor systems. Neurons in the posterior parietal cortex appear to contain a mixture of cognitive, motor, and sensory information which is at odds with the sequential and separate perceptual, cognitive, and motor stages used by cognitive psychology to explain behaviour. He partly attributes this to the legacy of the now rejected Triunal model of the brain discussed earlier. Neurophysiological data from diverse cortical and sub-cortical regions points to there being two parallel and simultaneous processing systems, which use the sensory information to assess the options and the selection of the option. Our brain appears to simultaneously process various potential motor action options and select between them.[15]

Drawing upon ecological theory, Cisek's alternative model purports to more accurately reflect how our behaviour is a product of adapting survival behaviours and inherited behaviour traits. The rationale is that interactional behaviour in real life is very different to the information processing we see in digital computers or classic psychological studies of human behaviour. Neither adequately nor accurately represent the complex array of options and demands made upon us by our environment nor give us insight into the decision selection process used by our brains. Cisek's model involves "potential actions competing against each other for further processing, while information is collected to bias this competition until a single response is selected."[16] This selection process is refined by the nature of selective attention which constrains the amount of sensory information used, but no complete representation of an action plan is prepared. The traditional view has been that the visual pathways are split into the dorsal and ventral pathways and that the dorsal pathways build a coherent representation of our world. Cisek argues that instead it "diverges into a number of sub-streams each specialised towards the needs of different kinds of actions" and claims that "the competition between potential actions plays out largely within the reciprocally interconnected fronto-parietal system."[17] Biasing inputs play a role in selection and include input from the basal ganglia as we saw modulates reward and risk and the PFC which is involved in decision making. Cisek's mathematical model involves both the decision for example as to which object to reach for and planning the movement being processed and integrated in the same brain region rather than kept separate or sequential. Multiple potential actions are simultaneously represented as continuous regions of neural activity within neuron groups with their associated decision variables in the frontoparietal cortex, all competing for selection and that is how decisions and actions are processed simultaneously. Cisek points out that neural activity related to decisions tends to happen in the same neural regions involved in sensorimotor processing and action planning and that this process is necessary for real-time behaviour. Much research into animals and humans has established that our brains seek to accumulate sufficient information from external sensory information about the outside world and internal information about the current behavioural needs to make an accurate prediction or decision. According to Cisek this all happens simultaneously and in an integrated fashion in the same brain regions dealing with sensorimotor processing, decision making and action planning. This means that the more sources and the more noise there is, the more neuronal signalling evidence can be accumulated on which to base the decision which should result in faster decision-making. The debate between sequential and simultaneous processing remains unresolved and is further complicated by findings in connection with the level of information contained in the electrical processes that are involved in the subconscious processing of information.

We saw how neuroscientists use EEG to measure electrical brain signals in humans and increasingly use multi-electrode arrays (MEAs) which contain tens to thousands of microelectrodes to produce large-scale parallel recording in animal brains, particularly rodents which facilitates the study of neural circuits and their functions.[18] MEAs can simultaneously detect signals from different neuronal groups and are being used in Brain-Computer Interface (BCI) studies with neural interface technology being employed to restore function to people with paralysis.[19] The challenge has been to develop the technology that can pick up signals across entire frequency ranges to identify all the various brain waves at play during the recording. To do this requires a high signal-to-noise ratio (SNR) which is the measure of the degree of precision or fidelity with which you can measure the neural information contained within a bandwidth or brain wave of the level, of the targeted signal against the level of background noise. Signal to noise ratio is the ratio of the power spectral density (PSD) of a signal (meaningful information) with respect to the power of the background noise (or more simply the ratio of relevant to irrelevant information, measured as the ratio of signal power to the noise power in decibels so a ratio of more than 1:1 indicates higher signal than noise. New technology uses various methods to improve the ratio including filtering out noise, using error-correcting techniques, enhancing the signal, and reducing the level of impedance as high impedance in the flow of the current is more sensitive to noise. This has been improved using new materials on the electrodes like platinum black coating polypyrrole/carbon nano-tubes composite electrode deposited on metal electrodes to increase the active surface area which increases the SNR. Intriguingly technologies such as MEAs appear to be getting us closer to understanding how our brains simultaneously process vast amounts of incomplete information to guide our thoughts, decisions, and actions. The response to a stimulus of neurons and neuronal assemblies which encode sensory and other information is reflected in the spiking of the electrical action potential and even a single spike can contain a large amount of information. The threshold for generating the spike depends on the strength of the stimulus. When we decide, our brains need to integrate in a timely fashion, sensory, cognitive, and motor information to initiate the appropriate response or behaviour. Research has recently focused on the Lateral Intraparietal Area (LIP) of Macaque monkeys with the LIP being shown to play a key role in simple sensorimotor perceptual decision-making.

A recent study explored the function of the LIP in encoding and decoding information in the spike signals when a decision is being made, using a data-driven, statistical model which allows them to reflect the dependence of spike signalling on the variables associated with performing tasks and determining behaviour. They concluded that the responses in the LIP suggested an "Interacting set of temporally overlapping response components, implying that some of the spikes emitted during decision formation are potentially related to several factors, some of which are irrelevant to the accumulation of evidence."[20] These include (1) components of the task related to the stimulus, (2) the visual stimulus, and (3) the saccadic decision to focus on a stimulus. They observed that this involves the superposition of information from various sources, which means that it is acting in multiple states at the same time until the information is decoded. This requires a much longer temporal integration of the information which may run to seconds to allow for it to be decoded, using what they call a time-varying weighting function as opposed to a simple threshold. In essence, by decoding the information they revealed that there is much more choice-related information in the neural activity than can be deduced or revealed by simply counting the signal spiking activity and that in many circumstances a fixed threshold is untenable.[21]

It would appear that AI robots need to be able to perform such functions to accurately mimic the processes employed by human brains. Professor of brain and cognitive sciences, David Knill (1960–2014) and colleagues observed that this is consistent with Cisek's model with the process involving a single unified mechanism and reflecting the Bayesian coding hypothesis that "the brain represents sensory information probabilistically, in the form of probability distributions."[22] Like our brains, machine and deep learning capitalise on using patterns in data to make inferences or predictions about that data or information, with learning driven by discrepancy between predictions and outcomes and consequent adjustments. Humans have a level of confidence about predictions related to their beliefs about the world even if this is subconscious and confidence is important in learning in changing environments, with confidence weighting suggesting reliance on the accuracy and reliability of prior knowledge.[23] Neuroscientist Florent Meyniel observes how "Prior behavioural evidence indicates that human learning adheres to this confidence-weighting principle" and that the lower the confidence the more flexible we are in adapting to change. Using magnetoencephalography (MEG) studies, Meyniel claims that "there exist, in the human brain, signals reflecting surprise, that are dampened by confidence in a way that is appropriate for learning according to Bayesian inference."[24] He maintains that "Confidence modulates intrinsic properties of the brain state in terms of oscillatory activity and neuromodulation, which in turn amplifies or reduces, depending on whether confidence is low or high, the neural responses to discrepant observations." Meyniel and Dehaene assert that "The sense of confidence is an essential ingredient of probabilistic learning in the human brain, and that the right inferior frontal gyrus hosts a confidence based statistical learning algorithm for auditory and visual sequences.[25]

Studies have shown that many higher mental processes involve a combination of conscious and subconscious processes and even intentional acts can be influenced by automatic non-conscious assumptions about others, events, or situations. When psychologists describe psychological processes as automatic, they ascribe certain features of automaticity and a dual mode of processing which treats the subconscious process as either automatic or nonautomatic based on these features. These features are whether it is unintentional, unconscious, uncontrollable, efficient, and fast, whilst the non-automatic features are the opposite so are conscious, controllable, inefficient, intentional, and slow. The dual mode reflects the belief that attention has a limited energy capacity, so when a person is learning, attention is flexibly allocated during the different stages of the learning process. In simple terms, the more you practise something and the easier it becomes the less you need to allocate attention to it, to the point that it becomes automatic and requires minimal or no attention. When complex tasks become routine or automatic, less motivation is required, which coupled with the need for less attention makes the automatic subconscious process more energy and time-efficient. Psychologists know that people's perceptions, attentional focus, and memories are based on the interaction between a person's hidden variable of needs, knowledge, expectations and the information, tasks, or situations that they are dealing with at any moment. Perception, attention, and memory are all entwined in how the brain perceives and processes information and involves a reciprocal connection and access to long-term memory. A perception process that is innate or has become automatic requires little if any attention, or motivation and less obvious or conscious access to the long-term memory of a person's needs, knowledge, or expectations.

Belief in this all-or-nothing approach to the subconscious automatic response is persistent, but studies have shown that some behaviour previously thought to be unintentional and uncontrollable could be impacted or disrupted when attention is diverted from the behaviour or task. This suggests that some level of attention is required, and it is now understood that there is a distinction between pre-attentive processes and goal-driven processes, with each having different features. Pre-attentive processes are entirely automatic whereas goals are the "cognitive representation of a desired state" whether that is clearly defined like losing weight or more abstract like being more positive about life and can stem from intrinsic or extrinsic motivation.[26] Goal-driven processes involve learning new tasks or responses and initially require conscious input, but once started and learned they can be processed automatically but unlike pre-attentive automatic processes, remain accessible to the conscious. Learned processes require some allocation of attention whereas pre-attentive ones don't and the general view is that learned goal driven processes do not simply become pre-attentive. When applied to the construction of the self-narrative and perception of self and the world, the preconscious processes involve both pre-attentive and learned goal-driven processes which may or may not become automatic. Our subconscious mind processes information that our brain is not consciously aware of and which does not involve any conscious intentionality. Our brain has a limited capacity to consciously process the information received from the multiple stimuli that bombard our senses every moment, and most processing must therefore take place outside our conscious awareness so subconsciously or unconsciously. We know the unconscious nature of the pre-attentive processes whereby we attend to and select information and how most processing of information remains largely beyond our conscious awareness. Social psychology research is demonstrating the extent to which the subconscious mental processes can influence decisions and behaviour even though the person is unaware of these subconscious influences.

The Robot Subconscious

Having seen how complex unconscious and subconscious processing in humans are and their vital role in guiding our behaviour, decisions, and choices, let's look at the challenges facing AI and robotics in addressing these non-conscious processes. AC has yet to be achieved in robots, so how can they possess a subconscious state, and does this have any meaning without reference to a conscious mind? Are we falling into the semantic trap of assuming that because it is a sub, which means below consciousness, it requires consciousness to exist? We know that AI machines, even without consciousness, can exhibit high levels of domain-specific intelligence and it would be easy to assume that the processes that AI relies on are by their very nature, entirely purely unconscious or subconscious systems. It intuitively makes sense that for robots to become conscious, they must be equipped with the subconscious and unconscious processes that characterize the human brain at some stage if they aren't already. Some scientists believe that given a proper time-line of evolution and the right algorithms, it will, in any event, be possible to simulate the subconscious mind, which suggests that they believe that AI is currently not regarded as having the computational equivalent of a subconscious mind but do utilize the equivalent of human automatic, unconscious processes.

We have seen how the unconscious and subconscious mind have traditionally been separated from the perception of the conscious reasoning mind and its emphasis on logic and symbolism, reflecting a long-standing belief that our subconscious behaviour is often irrational and driven by latent desires and motivations. We now believe the human subconscious is not "identifiably less flexible, complex, controlling, deliberative, or action-oriented than its counterpart."[27] The problem of determining whether AI robots possess a subconscious requires us to address the meaning of the subconscious in the context of AI and avoid conflating the subconscious with the unconscious and automatic because AI systems can possess automatic (unconscious) behaviour based on programming and learning. We are limiting the challenge to whether robots can possess a subconscious, which remains to be disputed and clarified. Secondly, whether there is a focal point of awareness enabling consciousness does not determine whether the attributes or qualities of subconscious processes exist, or we would fall into the trap of regarding the subconscious as no more than a shadow of the conscious mind. Therefore, we will assume that we do not require a consciousness for an AI subconscious process.

Subconscious mental processes are not unsophisticated or dumb or only deal with run-of-the-mill, unimaginative tasks or activities and any AI subconscious should reflect this. Distinguishing the subconscious from the unconscious reflects the fact that unlike the unconscious, it can still be accessed through introspection, which allows us to reflect on why we acted or made decisions in a certain way or made certain choices which must include the role of intuition and the use of heuristics or short cuts we employ. Is it possible for an AI system to have a subconscious that resembles humans unless it can access that subconscious through introspection and gain insight into its doubts and limitations and why it fails so that it can learn from its mistakes and understand why something went wrong? Maria Fox and colleagues, who have been developing introspective models in robots, argue that "One of the challenges for intelligent autonomy is to recognize when the actual execution of a behaviour has diverged so far from the expected behaviour that it can be considered to be a failure" and believes that this requires robot introspection.[28]

However, in 2013, computer scientist Roger Grosse wrote an article on introspection in AI and how decisions are made and problems are solved, observing that "introspection had largely fallen out of favour in AI, particularly in the fields of machine learning, vision, and probabilistic inference," with which he is most familiar? He claims that "the field (AI) has a Behaviourist-like aversion to thinking about algorithms in terms of plausible mental representations and cognitive strategies." He is referring to the behaviourist school of psychology that we discussed earlier, which believed it was fruitless to try and understand the subjective mental states of people other than through observing behaviour. Grosse cited the following arguments against introspection in AI which are that "(1) They have outlines of solution for most problems of interest, and introspection won't help us fill in any of the details (2) Computers computational architecture differs from the brain perform billions of sequential operations per second and insights don't carry over whereas the brain has more effective associative memory and high-level representations (3) Introspection isn't scientific and (4) It's hard to predict the problems an introspection-based strategy will generalize to."

Despite Grosse's observation, some roboticists like Maria Fox actively embrace the need for robotic introspection beyond simple anomaly detection. They have been seeking to equip robots with the capacity to understand what and why they are doing what they do. Fox and colleagues argue that learning from data and the robot's behavioural states allows the robot designers to use the model based on the observed behaviour of the robot to reliably predict and explain the behaviour carrying out that same task in any other environment.[29] Although endorsing the behaviourist approach, they explicitly aimed to provide robots with the means to introspect about the development of their behaviour in executing a task. They recognize that in robots, the behavioural state is hidden in that it can't be sensed directly by the robot like the human subconscious can through introspection. To do this, they used a Hidden Markov Model (HMM) representing the association between noisy sensor readings and the possible behavioural states of the system and the probabilities of transitioning between pairs of states. Unlike previous models focused on improving navigation, they wanted to learn how a robot accomplishes a task. They recognise that much remains to be done but believe they have made an important first step towards learning a reliable connection between the raw sensed data recorded by a robot and a symbolic reasoning level. Whatever the task may be the hidden states are acquired automatically through the clustering of sensor input using a Kohonen network with the integrity of the states established through evaluating the predictive power of the HMM model. Their conclusion sums up how such initiatives represent a move towards providing a robotic subconscious that interfaces with observed behaviour when they claim that this sounds strikingly like the interface between the subconscious and conscious in humans.

Computer scientist Hongmin Wu and colleagues argue that introspective robots can be more collaborative and safer and robot introspection can effectively "(1) Estimate, monitor and prevent abnormal events which are normally anomalies identified using models of normal robot behaviour. (2) Evaluate the internal state of the Robot, (3) Strengthen the ability to recover abnormalities, and (4) Optimise control and motion decisions."[30] They have developed a new model which goes by the impressively complicated title of Bayesian Nonparametric Vector Autoregressive Hidden Markov Models, which, in simple terms, is designed to equip the Robot to learn a 'library of skills that, when integrated with low-level control, allows for robot online decision making".Wu and colleagues argue that introspection essentially allows the Robot to estimate the confidence of the sensor input, like the way that humans subconsciously make decisions with varying degrees of confidence. They proposed an introspective framework called Sense-Plan-Act-Introspect-Recover (SPAIR) in which the Robot's introspective ability allows it to monitor anomalies in real-time.[31]

These various developments and initiatives show that identifying hidden states is important for robotic learning and improvement and reflects how important human introspection is to human improvement especially understanding why we make our choices and decisions and behave as we do. However, does being able to identify the hidden computational states of an AI system or Robot establish that a genuine subconscious equivalent exists to that of humans? This is further complicated when we consider such observations as Carl Jung's, that our collective subconscious, which we tap into when we dream, constitutes the "matrix of inherited human archetypes and narrative tropes that recur throughout the history of humans." If we were to employ computational terms, could it be argued that the subconscious is equivalent to the data and information we have encountered throughout our lives and upon which we make subconscious and conscious decisions? If our AI systems employ vast amounts of data based on the archetypes and narrative tropes fed into them by human programmers, they will display the same subconscious influences. Does this, mistakenly attribute the capacity for intrinsic agency, desires, and motivations akin to how the human subconscious works to AI models?

A counterargument is that AI robots are by nature purely unconscious or subconscious creatures but act without the constraints and influences of having an ego or self and do not require a self-like consciousness. Some scientists believe that the similarities between AI and the human subconscious are so great that humans could even be replaced were it not for the role of the conscious. The argument is primarily based on the observation that we are neural networks trained using the big data from our environment, and most of our behaviour, whether the outcome of perception or thoughts, is preceded by and largely the result of entirely subconscious processes in our brains. We have seen in humans how dispositional traits, goals, and behavioural representations, usually one or more involved, act as a subconscious behavioural guidance system and reflect the extent to which humans are self-learning adaptive creatures whose behaviour becomes predictable because we are largely creatures of habit. We saw how the subconscious could simultaneously process various information streams in an integrated fashion, which filters the information and presents a simplified version to allow the conscious brain to focus on the most relevant. In this sense, not unlike the human brain, AI is processing vast amounts of data behind the scenes to provide an output, though with no knowledge about the process involved in producing the output or the decision to act in a certain way. Both AI processes and the human subconscious act as vectorizers, using vast amounts of sensory inputs and internal processes to make decisions efficiently and largely automatically with neither humans nor AI revealing the intricate computations involved in the process. With AI using DNNs and reinforcement learning, like humans, it learns from rewarding positive decisions and penalizing negative ones. Taken to its logical conclusion, a chip could be used to replace the human subconscious and reach decisions much more rapidly, more accurately, and without the influence of the human ego or emotions. It requires us to visualize our subconscious as the equivalent of the computer hard drive containing all the data we have acquired through our lifelong experiences and the consciousness we see on the screen or in terms of the output. If we did not concentrate on the output but on the processes involved, we would be overwhelmed by the data. Even introspection must be selective when understanding the hidden processes involved in decision-making and behaviour. We will also see how when we talk about memory, accessing memory is fraught with problems regarding the flaws and fragility of those memories, which we employ to introspect and subject to effective ways of blocking out certain memories and not revealing to us the complex variables that influence our behaviour, including habits, biases, etc.

Endnotes

1 Tracy, B., 2008.
2 Aru, J, et al., 2019.
3 Vanduffel,W., et al., 2021.
4 Loftus EF, Klinger MR. 1992.
5 Bargh, JA, Morsella, E., 2008.
6 Echeveste, R, Lengyel, M., 2018.
7 Knill, DC, Pouget, A., 2004.
8 Ma, WJ., 2006.
9 Bargh, JA, Morsella, E., 2008.
10 Bargh, J., 2007.
11 Ratcliff, R., 1978.
12 Ratcliff, R, et al., 2016.
13 Boag, RJ., et al., 2023.
14 Cisek, P, Kalaska, JF., 2010.
15 Cisek, P, Kalaska, JF., 2010.
16 Cisek, P., 2007.
17 Cisek, P, Kalaska, JF., 2010.
18 Khodagholy, D., et al., 2015.
19 Donoghue, JP., 2008.
20 Park, IM., et al., 2014.
21 Gottlieb, J, Balan, P., 2010.
22 Knill, DC, Pouget, A., et al., 2004.
23 Meyniel,F, Dehaene, S., 2017.
24 Meyniel, F., 2020.
25 Meyniel, F, Dehaene, S., 2017.
26 Fishbach, A, Ferguson, M J., 2007.
27 Bargh, JA, Morsella, E., 2008.
28 Fox, M., et al., 2006.
29 Fox, M., et al., 2006.
30 Wu, H.et al., 2017.
31 Zhou, X et al., 2020.

"But feelings can't be ignored, no matter how unjust or ungrateful they seem."

Anne Frank, The Diary of a Young Girl. Anne was a German-born Jewish girl who kept a diary in which she documented life in hiding under Nazi persecution.

Chapter 9

Emotions and Feelings

CHAPTER 9

Human Emotions and Feelings.

WE CAN'T SEPARATE MOTIVATION AND drive, cognition or memory, and learning from what Damasio calls the "Crown Jewels" of our emotions. Charles Darwin observed that most species appear to experience emotions[1] and the latest research into animals and emotions suggests that mammals (dogs, cats, rodents) have emotions, and so do birds such as ravens[2] and even bees appear to express emotions by exhibiting pessimism.[3] William James marvelled at how the Brain and the body communicate and observed how emotions maintain and reciprocally mediate this. Although complex and heterogeneous, emotions are the raw data of bodily or somatic responses activated through neurotransmitters and hormones released in the body and Brain. They reflect the state of the body's internal milieu and its automatic subconscious drives and motivations, so they represent our present reality about any and every experience, be it unpleasant or unpleasant, and they exist on a continuum. According to evolutionary theory, emotions constitute a second human motivational system and are an adaptive response to our environment. The processing of emotions occurs in the Brain's higher cortical regions, which raises the crucial question of how our emotions relate to how we feel or think.[4]

Given how important they are in our lives and their role in memory and learning and activating the sometimes-negative prior knowledge contained in our schemas, it is strange that no agreed definition exists for emotions. Several difficult theories exist about how emotions are processed and their relationship to the feelings they generate needs to be revised. Rather than enlighten us, these theories appear to muddy the water in that emotions are either a physiological (bodily) response, a neurological (brain) response, or a cognitive response in that our thoughts produce them. William James and Carl Lange, the Danish physician who contributed to the fields of neurology, psychiatry, and psychology, developed the James Lange (1834–1900) physiological theory, which asserts that we process emotions from the bottom up. In other words, emotions begin with bodily responses to external events, not with the conscious subjective experience of an emotion (top-down). The prevailing understanding was that interoception, a term coined over 100 years ago, was deemed to be the function of visceral and somatic neurons (nerve cells) in the spine and brain stem. This was labelled peripheralist because it emphasizes events at the periphery of an organism's body, not in our Brain or mind. William James believed there are only a few standard emotions: happiness, sadness, anger, fear, disgust, and surprise. However, we have since expanded the range of emotions to include others, like the social emotions of jealousy and embarrassment. Darwin believed that each human emotion generated in dedicated brain regions had a specific adaptive role that enhanced human survival and were thus evolutionary remnants that our superior rational thinking mind has now superseded. There is a problem with this because the same emotion can generate different responses, so how can emotions have a separate adaptive role or dedicated brain regions? James's belief that there were identifiable triggers for the emotions that governed certain behaviours has been challenged as incompatible with the fact that bodily arousal responses are too undifferentiated to account for the wealth of emotions, making it too difficult for our brains to perceive the various physiological states accurately. The James Lang theory fails to explain emotions not generated by bodily arousal or the role of learned emotions and cognition or feelings.

In 1927, Walter Cannon (1871–1945), the American neurologist and physiologist who coined the terms homoeostasis and the fight or flight mechanism, and his doctoral student at Harvard, Philip Bard (1898–1977), asserted that external events could trigger emotional and physical responses simultaneously but are independent of each other and that emotions affect firstly the CNS, which includes the Brain. For example, if you see a snake, you may experience fear and sweating, but these occur simultaneously, not sequentially. You don't sweat because of the fear, or emotions don't cause the reaction in the body or vice versa. Cannon, Bard, and James would probably have been aware of the work of Angelo Mosso (1846–1910), a 19th century-Italian contemporary of William James, who established the beginnings of how modern physiologists study emotion when he expanded the study of emotions to the internal processes of the brain (Mosso, 1879). While doing so, he invented the first neuroimaging technique called human circulation balance, which recorded the pulsation of subjects with cortical skull defects and provided evidence of how blood is redistributed in the Brain during emotional and intellectual activity. Mosso's technique confirmed Charles Scott Sherrington's hypothesized link between blood flow through the brain and metabolism, which is the basis of modern non-invasive techniques such as fMRI and PET scans. The greater understanding of brain activity using modern imaging technology to study emotions led to the Lang Cannon, and Bard theories, being modified as recently as the 1960s by the American social psychologist Stanley Schachter (1922–1997) and American clinical psychologist Jerome Singer's (1924–2010) cognitive two-factor theory of emotion (1962), which postulates that emotions arise from physiological arousal. Still, the cognitive labelling or appraisal of such arousal creates emotion. In other words, feelings appear to create emotion because that individual will label those kangaroo testicles as delicious or disgusting; until they do that by their reaction, we can't tell their emotion. The American psychologist Richard Lazarus (1922–2002) went even further with his Cognitive Appraisal Theory, now known as the Lazarus theory of emotion, hypothesizing that thinking must precede emotional or physiological arousal, meaning we get emotional because we think about something. We can generate emotions purely by thinking about something, such as thinking about it, rather than having to eat kangaroo testicles. It appears most of our emotions are linked to our thoughts, reflected in the experience of depression and anxiety. he difficulty with these various theories is that they tend to separate emotions and thinking as though they are being processed in the Brain and body in parallel or sequentially, perpetuating the traditional misconception of the Brain as processing information hierarchically. The earlier conceptualization of emotions reflects science's enduring assumption that humans are a superior animal species with a monopoly on higher thinking and that emotions are subservient to the thinking or cognitive brain. This may be why, in modern society, emotions in Western culture continue to be viewed as suppressed or regulated using reason, certainly in public, as though they are a primitive relic and incompatible with rational man. This is partly attributable to a myth that has evolved since the 1930s that identifies the amygdala as the fear circuit that generates emotions as part of the primitive fight or flight response controlled by our amygdala. Recent research has shown that the amygdala is not just a fear circuit involved in the fight or flight response but is activated for a diverse range of mental phenomena, including imagination, planning, remembering, pain, language, and reward responses, during social interactions and some but not all emotions. In a recent interview, American neuroscientist Joseph Le Doux (1950–) explains how, after 40 years of scientists "exploring how non-conscious processes involving the amygdala, detect and respond to danger "has contributed to the 'amygdala fear centre meme," which he refuses to endorse. This meme fails to explain that the conscious experience of fear, is not wired into the amygdala, but is instead "a cognitively assembled understanding that you are in harm's way based on non-conscious memories, schemas, and mental models."[5]

We now know that the amygdala and other major components of the Limbic System are involved in processing emotions, as are the Pre-frontal Cortex (PFC) and the Hippocampus, which have well-established roles in the regulation of stress and promotion of homoeostasis by regulating physiological and behavioural responses to stress such as aversive and anxiety-like behaviours.

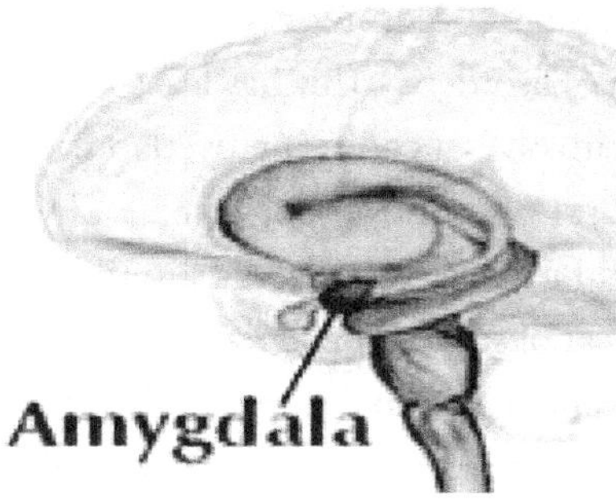

The amygdala is connected to many brain areas and structures, such as the PFC, which are involved in processing cognition and connected to functions such as autonomic responses like breathing, touch, and sensitivity, hence, it is involved in regulating bodily responses. Recent research has shown that some environmental and genetic factors moderate how the amygdala responds to stress, but the amygdala is neither necessary nor sufficient for experiencing emotion. Current scientific thinking is that the amygdala helps the brain deal with uncertainty or ambiguity and chronic stress and has been shown to cause circuit-specific remodelling of neurons in the amygdala, creating greater anxiety. We also now understand the much broader role of the amygdala in integrating imagination, planning the future, remembering pain, language, and reward responses during social interactions, and some specific emotions. These advances have led psychologists to explain the generation or triggering of emotions by using either somatic or cognitive appraisal theories, with the latter appearing to prioritize the cognitive processes of appraisal over emotions.

In contrast, the somatic theory assumes that emotions are primary to cognitive processes. The group of psychological theories now known collectively as the Cognitive Appraisal Theories (CATs) seek to explain human emotions arising from the subjective evaluation of events occurring in the environment. Lazarus maintained that this consists of a "Primary" appraisal regarding whether a stimulus is perceived as harmful or dangerous and a "Secondary" appraisal when we evaluate how to handle the situation and draw upon past experiences to do so. We can see why we can no longer separate emotions from reason and emotions, as they are inextricably entwined. This contradicts the longstanding belief that good decisions are only made by keeping emotion out.

This illustrates how, in everyday language, particularly in western culture, feelings and emotions are often treated interchangeably, possibly because feelings are experienced consciously. In contrast, emotions involve both subconscious and conscious experiences in that they are unconscious bodily experiences that we can become subjectively and consciously aware of. Damasio coined somatic emotional markers, which describe the legacy of emotions generated in previous social interactions and relationships. The tendency to conflate emotions and feelings ignores important distinctions between the two, with emotions tending to be short-lived tense with an identifiable cause and timing where feelings endure. However, emotions can range from seconds to several weeks and even months, although these are typically associated with personality traits or emotional disorders.[6] This has led to the distinction of emotions according to duration, which again involves a confusing overlap with feelings. The divisions are (1) Autonomous changes: very short (in seconds), spontaneous physical feelings associated with the somatic theory of emotions and triggered by certain stimuli, without recognition of situation, object, or event (such as the fear of something in darkness) (2) Expressions which are short in seconds, associated with objects and based on the appraisal (3) Classical emotions which lasting longer are consciously observed, named and associated with somatic and appraisal theories (4) Mood which lasts days or months, are consciously perceived, and tend to be less intense than classic emotions, being characterized by slow, positive, or negative changes (5) Emotional disorders like depression etc and (6) Personality with emotions reflecting personalities such as shyness and neuroticism. Evidence suggests that making decisions without emotions is disastrous, as there is a deep interconnectedness between emotions and cognition.

American Mary Helen Immordino-Yang is a Professor of education, psychology, and neuroscience researching the psychological and neurobiological bases of social emotion, self-awareness, and culture and their implications for learning, development and schools. She observes how "By compromising the possibility of evoking emotions associated with certain past situations, decision options, and outcomes, patients became unable to select the most appropriate response based on their experience. Their logic and knowledge could be intact, but they failed to use past emotional knowledge to guide the reasoning."[7] Immordino-Yang believes that emotions are vital to learning and that global white matter connectivity between the associative cortex and sub-cortical regions, like the amygdala, allows for the integration and organization of information, which entwines emotions and the resulting feelings, as part of the Brain's processing and interpreting of information. The fact that feelings, based on emotion, arise from the Brain's capacity to process and integrate information is precisely what recent neuroimaging research by Yuri Terasaw and colleagues found when they observed that "integration of information about the internal bodily state and the external environment appears crucial for the experience of emotion."[89] This largely rests on imaging studies, which revealed an overlap between activation of brain regions such as the Insula, which, as we saw earlier, collects information on the subjective feelings and outcomes of past behaviour, and the front of our brains, which processes attention and working memory to formulate goals and decide on responses and future actions.

The information processed by the Insula is further integrated into the association region of the frontal lobe and we cannot treat thinking and emotions as separate because this is not how our brain processes information. The Brain integrates rather than separates thinking and emotions, so regulating or controlling emotions should more accurately be described as predicting outcomes, formulating goals, and deciding on responses and future actions based on integrating emotions and thinking. As recently as 2002, scientists discovered a "phylogenetically unique ascending pathway to the cortex in monkeys that conveys homoeostatic sensory activity from all bodily tissues to the brain." This has been broadened to include a homoeostatic sensory motor hierarchy that coincides with the central ANS, the emotional motor system, much of the classical limbic system, other brain regions, and the insular and cingulate cortex, including sympathetic-vagal balance. We know the vagus nerve is the main nerve of the PNS, which acts as a sensory highway that relaxes your body after stress or danger when it is typically activated by the SNS in a fight or flight response to stimuli.[10] This is why scientists are very excited about the vagal nerve as a possible source of treatment ranging from migraine to inflammation. Recent studies have implicated the vagal nerve in regulating moods, monitoring the function of the internal organs, and even discerning facial expressions.[11, 12, 13]

Emotions impact physiology in terms of our body and the Brain, but the absence of an agreed-upon definition of emotion has encouraged some researchers to look for a definition that could be used to identify whether other species experience emotions as we humans do. This includes the unlikeliest candidate, the fruit fly, which may stop or at least give you pause for thought before you unceremoniously swat those flies when they bug you. In 2014, David Anderson and Ralph Adolph sought to develop a framework for studying emotions by comparing emotions to reflexive responses, arguing that "emotional behaviours are a class of behaviours that express internal emotion states which exhibit certain general functional and adaptive properties that apply across any specific human emotions like fear or anger, as well as across species."[14] They established five features that characterize emotions, which are (1) Generalisability, (2) Persistence, (3) Valence (4) Automaticity, and (5) Scalability. The reflexive responses we discussed tend to be very specific, such as the knee jerk, startle, or Achilles responses.

In contrast, with emotional responses, various stimuli may provoke the same emotional response or the same stimuli may provoke different emotional responses, making them less predictable and, therefore, less generalizable. Unlike reflexive responses, emotions can scale in intensity, although reflexive responses are automatic and unintentional. Emotions persist, which is why sometimes you may feel angry for some time after the stimuli, whereas reflexive responses subside quickly. Compared to reflexive responses, these combined features of emotions provide a delay mechanism that allows for the integration of the information and the preparation of an appropriate response, which is where cognition comes in. When we are very young, we tend not to experience discrete emotions, only whether something is pleasurable.

Recently, research has focused on what American psychologist James A Russell (1947–), whose lifelong work has focused on emotions, observed as an ongoing sense of what we feel at any one time, which he calls the core emotion or core affect. Russell, co-authored with Liza Barrett Feldman the Psychological Construction of Emotion.[15] The core effect has been shown to operate against the two bipolar axes of Emotional Arousal or activation (high / low) and Emotional Valence (unpleasant/pleasant), which reflects the Circumplex Model of Emotions (CME), developed by Russell in 1980 to capture the subjective nature of feelings. Emotional arousal means heightened physiological activity ranging from high to low or how energized you feel. Studies have shown that arousal levels have, as have seen, selective effects on our attention levels and on enhancing or impairing memory, with obvious implications for autobiographical memory. The research shows that emotional arousal enhances

memory for high-priority information and reduces memory for low-priority information. Emotional valence was first coined by the German/American psychologist Kurt Lewin (1890–1947) in 1935, one of the modern pioneers of social, organizational, and applied psychology in the United States. Valence is when an individual attaches a subjective value to the experience of an event, object, person, or other entity. An entity that attracts the individual has a positive valence, whereas one that repels has a negative valence. Lewin's theory states that individuals and groups are influenced by restraining forces that counter-driving forces and maintain the status quo, with driving or positive forces pushing in the direction that causes change to happen. This has obvious relevance to the construction of self over time regarding our preferences and dislikes and our embracing or rejecting change and novelty. But we avoid or withdraw from what we seek out or approach, which determines our experiences and inevitably influences our self-narrative. Lewin used valence to describe the positive (pleasant) or negative (unpleasant) value associated with a stimulus as an event, object, or situation, expressed on a continuum or spectrum from pleasant to unpleasant, attractive to aversive, or avoidance to approach. Happiness typically consists of high arousal and a pleasant valence, whereas depression is low arousal and unpleasant valence.[16] Valence can also describe specific responses to a situation, such as anger and fear of being negative. Russell's CME showed that arousal and valence are independent. This means that being happy and sad are at the opposite ends of the valence axis, and you can't experience both at the same time nor can you feel tense and tired or sleepy at the same time. To experience mixed emotions, they must be similar in how you subjectively feel in response to that emotion. They can't consist of significantly different emotions regarding arousal or valence. In each quadrant, the emotions are sufficiently similar to each other to give rise to the experience of mixed emotions.

Russell and Barrett and an increasing number of scientists are challenging, at a fundamental level, our traditional knowledge of emotions as a unitary event instead of a multi-component process. They build upon appraisal theories, which assert that "emotions are elicited and differentiated based on a person's subjective evaluation of the personal significance of a situation, object, or event."[17] Although this conflates emotions and feelings, it is another way of describing how an experience activates schemas and governs our subjective responses and thoughts or feelings. Russell describes how, moment by moment, we experience a free-floating mood; for example, we may feel calm (low activation and pleasant), tense (high activation and unpleasant), or happy (pleasant, high activation). These core emotional states impact our reflexes, perception of the world, thinking, and behaviour and are influenced by many internal and external factors. The valence and arousal in our core emotion form the basis for an emotional episode, which, unlike the reflexive response, is rarely consistent and of much greater variety than our standard categorizations of emotions such as anger, fear, and happiness provide. He and others suggest that, instead of simply using these stereotypical types or categories of emotion to explore emotions in research, we should use the core effect to understand the psychological construction of emotion. The core effect is what we experience all the time, although sometimes we may understand why others are unaware of why we feel as we do because there is no external explanation. Russell compares this to emotions and feelings provoked in response to a specific event or situation, which he calls prototypical emotional episodes (PEE). Russell and Barrett consider the PEE to last a short time and involve the core affect, a change in behaviour, and cognitive appraisal of the emotion, which gives rise to subjective feelings, which cause further physiological changes, including bodily, chemical, and brain changes. This process is part of our psychological construction; it comprises the processes involved in a PEE and involves the associations between the various components of emotions and their categorization as a particular type of specific emotion. The processes include such aspects as facial expressions, appraisal of stimuli, attribution of salience and valence, the subjective experience, the voice change or tone, and levels of emotional regulation, which determine the nature of an emotional response and expressed feelings. Russell extended the CME in the inclusion of a dominance dimension in PAD dimensional theory with the Pleasure (or valence) dimension referring to the hedonic quality of the emotion and whether it is positive or negative and the Arousal (or intensity, activation) dimension, which refers to the level of physiological arousal or neurological activation.[18]

A 2017 study by Cowen and Keltner concluded "that reports of emotional experience were structured as 27 emotion clusters (i.e., categories) with "fuzzy rather than firm boundaries." This confusion over dimension and categorization has led Liza Barrett and her colleagues to suggest "future studies should consider asking questions more directly about physical and psychological variation within emotion categories as well as similarities between categories." This debate, together with the conflation of emotions and feelings and the acknowledgement of a core emotional state, reveals the complexity surrounding emotions and how we identify them accurately.[19] The Plutchik Model of Emotions was designed by American psychologist Dr Robert Plutchik (1927–2006) based on eight primary emotions that are the foundation for joy, sadness, acceptance:ce, disgust, fear, anger, surprise, and anticipation. Each primary emotion has an opposite based on the physiological reaction each emotion creates. For example, joy is the opposite of sadness; fear is the opposite of anger, etc. Plutchik's emotion wheel simplifies the task of conceptualizing and understanding the complex interaction and nature of primary and secondary emotions in humans and animals.

Plutchik defined emotion as "a complex chain of loosely connected events which begins with a stimulus and includes feelings, psychological changes, impulses to action and specific, goal-directed behaviour." Again, this subsumes feelings under the overarching conceptualization of emotions, but it does help us visualize emotions as a bodily process culminating in an impact on our behaviour.[20] Plutchik's model also allows us to see that both somatic and cognitive appraisal mechanisms are involved. Many parameters characterize an emotional state, and groups of similar emotions can even be attributed to a definite colour.

The culture we grow up in teaches us to categorize and name emotions and how and when to express emotions, emphasizing emotional regulation. The major problem with categorizing emotions, such as sadness or anger, is that this fails to capture their dimensional nature and how they can even overlap, in that you can feel mixed emotions. The psychologist Robert Pletcher has expanded on the work of a pioneer in the study of emotions, American psychologist Paul Ekman (1934–), who widely acknowledged the identification of 6 emotions by suggesting that eight basic human emotions are grouped in pairs. These are anger-fear, happiness-sadness, surprise-anticipation, and trust-distrust. Labelling or categorizing emotions risks conflating the conscious feeling with the underlying bodily sensations, which may result in overlapping and contradictory emotions. This reflects a long-standing debate about whether we can or should categorize emotions like this at all or whether they should be seen on a continuum along the valence and arousal dimensions. Let's look at how emotions relate to but differ from feelings.

How do Emotions Relate to Feelings?

How do our emotions relate to our feelings, and what is the difference? Emotions can be evoked by external stimuli or internal motivations and drives, which lead to complex physiological, cognitive, and behavioural reactions, which are the attention to the stimuli provoking the emotion in terms of its importance and whether the response or feelings generated are meaningful or appropriate. The APA Dictionary of Psychology defines feelings as the "self-contained phenomenal experience which involve(s) subjective, evaluation independent of the sensations, thoughts, or images evoking them," which are typically pleasant or unpleasant.[21] They are regarded as intra-psychic, meaning they occur within the psyche, mind, or personality. They are the conscious experience of our perceptions of the emotional data and sensations in the body.

Feelings are how we characterize the raw data of our emotions and relate them to our self-concept and how we consciously respond to and interpret our emotions, which includes the language used to describe the impact of the experience of the emotion on us. Feelings, if you like, reflect our self-schemas based on the experience or prior knowledge of the emotions generated by past situations, events, and others since childhood. When we say someone has hurt our feelings, we observe that they have hurt our self-schema or version of ourselves regarding our beliefs, values, sites, and preferences, which emerge when those neural patterns are activated. They may activate a relational schema with negative associations with the relationship that evokes unpleasant emotions; the same emotion experienced by two people may give rise to distinct and even opposite feelings. For example, someone may feel anxious, while another person may feel excited about the emotions generated by the same event, or some people will experience the feeling of anger in a situation. In contrast, others will remain calm and passive. Feelings involve the subconscious accessing and activating existing neural schemas, influencing how you respond to the emotions you are experiencing. How often have you seen someone expressing extreme feelings in response to the same situation you are experiencing, yet it doesn't make you feel like that? For example, practising Buddhists would likely not feel angry because their feelings have been hurt in response to a verbal insult. In contrast, a testosterone-driven male teenager may, in response to the same stimulus, experience at the same time. This doesn't necessarily mean that this represents greater emotional regulation on the part of the Buddhist but more likely reflects the Buddhist's belief that by clinging to a narrative of ourselves, the negative feelings of attendant cravings of jealousy, hate, and other manifestations of self, will determine how we respond to a situation and are what causes us the unhappiness and negative feelings. The feelings we experience typically reflect the activation of the schemas stored as neural patterns in response to bodily sensations, which is why when someone hurts our feelings, the more entrenched those neural patterns and pathways and self-schemas, the more we potentially cling to our subconscious narrative and the more deeply it can affect us as emotional memories tend to be encoded more deeply. Because the activation of schemas is subconscious, they are often difficult to recognize. Still, people often form friendships because they usually share common feelings in response to experiences, so they affirm each other's feelings, even if those are negative. This provides positive relational schemas that are activated in the company of friends and why it is much easier to spend time with someone who affirms and appears to share your feelings than someone who appears not to understand them or judges you for having those feelings. However, if those people close to you upset you, then, as we all know, that can hurt far more than when a stranger might do the same thing, even if that provokes anger.

Our self-schemas reflect relatively stable and core negative or positive beliefs about ourselves and our emotions. They influence how we expect to feel and respond to situations and emotional experiences. Negative self-schemas have been linked to feeling depressed, anxious, or having low self-esteem, poor body image, shyness, and a myriad of problems with the self, with feelings easily hurt and emotions constantly raw. Depression is a classic example of the feelings generated by bodily emotions, which can turn into a vicious cycle and depressive mind state. The observation that emotions mediate between the Brain and body reflects how we build a sense of self as we develop, in that we build up habitual feelings based on bodily sensations or the raw data of emotions; they become inexorably linked and reciprocal. Your feelings can generate the bodily emotions that perpetuate or exacerbate those very feelings. That is why we find ourselves habitually feeling responding and responding to the same situation time and time again. Impulsive behaviour, driven by emotional sensations and engrained feelings, can arise when you don't allow any time between experiencing the feeling and the response generated and typically reflects a subconscious and habitual response to situations.

Psychologists distinguish the different ways of dealing with emotions; each has been shown to impact your physical and psychological health. When people talk about regulating the emotions that are often to avoid appearing to act inappropriately to others, the problem is those feelings appear appropriate to you because that's the version of self they are not necessarily seeing. It would be more constructive for a person to seek to understand the feelings generated in response to the emotions we display and how they relate to our concept of ourselves or our self-schemas.

Repression describes the subconscious avoidance of emotions, while suppression describes the deliberate avoidance because of the difficulties of dealing with the emotions. Studies have shown that repressing and suppressing emotions will potentially make the feelings more entrenched and can endanger our physical and psychological health in terms of chronic stress, which is draining our energy. Psychologists disagree on using the term feelings, given that different people may have feelings in response to the same emotional sensations that vary considerably and are heavily influenced by our highly subjective schemas. An important distinguishing feature of emotions is that emotions are universal and instinctual across cultures. In contrast, feelings vary widely from person to person and are shaped by our personality, experiences, and self-concept. As mentioned earlier, the self-concept is hurt when the language of feelings is evoked. This perspective means that love or friendship should not be regarded strictly as emotions but as feelings that reflect how that individual sees themselves about the other person; in this instance, the feelings are relational and reflect relational schemas. Feelings are different ways of expressing the emotional experience. Hence, emotions precede feelings, and feelings involve images, thoughts, and memories, which enhance and largely govern the response to the experience of an emotion. Given the absence of an agreed definition of emotions, we shall adopt Antonio Damasio's approach to restrict the definition of emotion to the "observable, unconscious behavioural component" and feelings to the "subjective experience of emotion." What is clear is that being aware of your emotions and feelings is vital to greater self-awareness and a much healthier approach than repression or suppression of emotions. It is the first step in determining whether the feelings and underlying neural patterns serve you well or badly and should be changed. It also highlights the immense challenges facing AI developers seeking to imbue AI systems with the ability to read and exhibit diverse emotions. This is further complicated, as we have seen by the fact that it has yet to be possible to confer a robot with a sense of self and identity, which is a prerequisite for the subjective experience of feelings. Before we discuss emotions and robots, we will briefly look at other important aspects of emotions, including gender, culture, and emotional intelligence.

Gender and Emotions.

The media are constantly stereotyping gender differences in emotions and representing the assumed differences between women, who are portrayed as openly expressing their emotions when compared to the strong, silent male. These portrayals assume that the core emotional qualities of being masculine or feminine are radically different and, in the past, reflected in the widespread power relations and sexuality across social groups with both Western and Eastern cultures, commonly stereotyping women as more emotional than men.[22] Calling a person emotional has historically suggested a sense of fragility or weakness. It has been employed as a derogatory reference to the female gender or a jibe aimed at a man, as though he is behaving like a woman by showing his emotions unless, strangely, those emotions involve anger or aggression. In intimate relationships, we are commonly exposed to popular conceptions about gender differences in terms of emotional needs and expressive behaviour. The expectation theory has attributed this to how people expect or assume gendered differences in emotional expression and experience; The expectation creates and perpetuates these perceived inequalities.[23] Despite many years of research, it had, until recently, been unclear whether this widespread expectation and stereotyping reflects a genuine underlying gender difference between the male and female brain, which determines how emotions are experienced, processed, and expressed. Unfortunately, many studies have confused the individual experience of different types of emotion with the nature of the external expression or display of different types. This may explain why, despite all the research over many years, there has been no scientific consensus that women experience emotions more often or more strongly than men or are generally more expressive. To determine the difference requires a combination of the assessment of physiological reactions as indicators of emotional experience and subjective evaluations concerning emotional expressivity. Some studies have shown, though, that women do experience and express more frequent and stronger negative emotions than men.[24] In contrast, many studies have shown that men exhibit more intense emotional reactions to stimuli perceived as threatening or erotic. The findings suggest that gender differences in emotional experience are much smaller than expected and more specific in their distinctions.

A recent study investigated gender differences in the emotional experience and how emotions are expressed, using the recorded Heart Rate (HR) as an indicator of emotional experience. At the same time, individuals watched 16 video clips that induced sadness, anger, horror, disgust, neutrality, amusement, surprise, and pleasure. The findings suggest that men can, and often do, experience more intense emotions, whereas women have higher emotional expressivity, particularly for negative emotions. They concluded that gender differences depend on the specific type of emotion type, not the valence in terms of it being positive or negative.[25] Men may experience emotions as intensely, if not more intensely, than women. Still, women experience and express specific emotions more negatively, which may explain why women tend to experience more mood disorders.[26 27]

A 2021 study led by neuroscientist Professor Lise Eliot from Rosalind Franklin University is the first to review all the wide-ranging research into a mega-synthesis of this supposed difference between men and women.[28] Her team reviewed thirty years of research and hundreds of the largest and most highly cited brain imaging or fMRI studies addressing thirteen distinct metrics of supposed sex differences between men and women, such as the thickness or volume of grey matter. For almost every metric, no differences had been reproduced across studies, even those involving thousands of people. Eliot has spent years researching the differences between the genders and the role of neuroplasticity and has called for the closing of the troubling gaps that grow from small differences between boys and girls to help all children reach their fullest potential and reflected in her fascinating book on gender differences Pink Brain, Blue Brain: How Small Differences Grow into Troublesome Gaps (2012). Eliot concluded that "Men and women's brains do differ slightly, but the key finding is that even these modest distinctions are due to brain size, not sex or gender and sex differences in the brain are tiny and inconsistent, once individuals' head size is accounted for." This reflected another study by Stanford University demonstrating false discovery or over-publication of false-positive findings in the scientific literature on functional MRI sex difference. In other words, people only publish findings and ignore all those results that showed no significant differences. This publication bias is driven by the public appetite for information on differences rather than results showing no differences.

Women's brains are approximately 11% smaller in proportion to body size than men's. Still, smaller brains allow for a slightly higher ratio of grey matter to white matter and a higher ratio of connections between the cerebral hemispheres. Eliot observes that "none of these size-related differences can account for familiar behavioural differences between men and women, such as empathy or spatial skills." Eliot's study was titled, Dump the 'dimorphism,' which refers to the call to stop perpetuating the pink /blue tyranny of the belief that men and women are different versions of the species and sums it up beautifully when she says, "the brain is like other organs, such as the heart and kidneys, which are similar enough to be transplanted between women and men quite successfully." Eliot's study found that the marginal differences between the male and female Brain could be better replicated between diverse cultural groups, such as Chinese and Americans. This is not to say there aren't differences between cultures, but simply that gender differences are not significant across cultures. Studies do appear to show significant differences in the levels of emotional arousal between Western and Eastern cultures and shape how people feel in certain situations, how people express their emotions, and even what we might call emotions. This suggests that emotions and the expression of emotions are heavily influenced by our environment and social or cultural situations, not gender.

Emotions and Culture.

We learn emotions by observing others, how they express emotions, and how people respond to us. These experiences, as we develop, often vary significantly between cultures. In the 1990s, Markus and Kitayama distinguished between how different cultures regard the notion of self, calling this the concept of Self-Construal. They observed that Westerners construe the self as independent and separate from others, which we often call individualistic or independent self-construal. In contrast, eastern cultures tend to display interdependent self-construal in which the core unit of society is the group, not the individual, called collectivist or interdependent self-construal.[29, 30] However, like most things, it is not black and white, as all individuals can combine independent and interdependent self-construal. Westerners appear more predisposed to experience high-arousal emotions than low arousal. In contrast, Eastern or collectivist cultures value low arousal emotions over high arousal emotions and experience and prefer to experience low more than high arousal emotions.

We saw how emotional intelligence, as opposed to cognitive intelligence, has a long history, starting with the work of psychologists John Mayer and Peter Salovey in the 1990s. They believed that emotional intelligence involved (1) Identifying emotions on a non-verbal level (2) Using emotions to guide your thinking, (3) Being able to read and understand what meaning emotions convey and the responses generated by emotions and (4) Controlling your own emotions for your benefit and that of others benefit. Daniel Goleman (1946–) took their work further by postulating five levels of emotional intelligence. These are "(1) Self-awareness of your emotions and knowing what you feel at any given time and understanding the impact of your moods on others (2) Being able to anticipate consequences before acting on impulse and redirect your actions (3) Being able to be motivated by emotional factors to achieve goals (4) Enjoying learning and persevering when it is difficult (5) Showing empathy and being able to handle relationships, inspire others and get people to want to do the same with you."[31] Goleman describes emotional intelligence as a person's ability to manage or regulate his emotions to be expressed appropriately and effectively. It has resulted in many schools incorporating social and emotional learning as part of their formal curriculum, and in 2002, UNESCO launched an international campaign to promote emotional learning in the classroom. EQ has also become a stalwart of leadership programs and a key to self-fulfilment and personal happiness. Goleman believes that higher EQ reduces bullying and discipline problems, including drug abuse in schools, and that children's EQ is heavily influenced by their home environment, which may dictate their level of emotional intelligence by the time they enter formal schooling. To the extent they have suffered from neglect or abuse or been in an environment where EQ is not displayed, they will face challenges when they arrive at school. What underlies the ethos and emphasis of EQ is the recognition that no emotion is wrong or should be judged, but that it is the feelings and whether those feelings are expressed appropriately, which is inevitably a cultural and social construct. This predominantly Western conceptualization of EQ suggests that controlling emotions in Western cultures rather than Eastern cultures presents a greater challenge.

We now know emotions are an integral part of how we think and behave and are not just there to be regulated by the rational mind but should be harnessed by ineffective thinking and behaviour. Whether they are expressed appropriately and effectively depends on the cultural and social context in which those emotions are displayed. Batja Mesquita, a social scientist and professor of cultural psychology at the University of Leuven in Belgium, has been described as a pioneer in studying emotions. Based on three decades of scientific study and immigrant experiences, Mesquita argues that emotions are not innate but formed and shaped by living our lives together. She observes how emotions in some cultures live inside a person but are experienced and live between people. She stresses how the whole phenomenon of emotions, not just the categories of emotions, varies between cultures. For example, if you are angry, that anger lives between you and me. Some emotions, such as shame, are considered negative or bad in Western culture and may share similarities between cultures but also major differences. In some cultures, shame may be equated with modesty and embarrassment, an acknowledgement of your status about another or your place in the world. Hence, shame may be seen as a way of strengthening a relationship, not necessarily harming it, and she makes a compelling argument that we need to move beyond empathy. Even though there may be many differences in how and when people experience and express emotions, we know that emotions are vital to being human.

Robot Emotions and Feelings.

We have seen that emotions play a central subconscious and conscious role in human decision-making,[32] attentional focus, learning, goal setting for goals that make us happier or more fulfilled, social interaction, forming and maintaining relationships, and generally guiding our behaviour. We saw how the Layered Reference Model of the Brain (LRMB) designed by Wang and colleagues demonstrated that "emotions alter our processes of perception, attention, and decision making, enabling the development of emotionally driven responses." Some scientists continue to controversially assert that emotions are categorical and form a set of distinct and basic emotions that represent a primary pattern response or tendency to act in a particular way. For example, fear is related to avoidance or withdrawal. Others like Russell increasingly recognize that emotions are not unitary events but a multi-component process experienced dimensionally in an underlying affective space, involving, at a minimum, the two dimensions of valence and arousal. However, even dimensional perspectives of emotion must allow for correlations among different emotional states like anger, fear, and valence or degrees of emotion. Cognitive appraisal theories view emotions as the result of the evaluation (appraisal) of incoming stimuli according to personal relevance and involve important dimensions such as valence, novelty, goal relevance, goal congruence, and coping potential. Some basic emotions are also believed to be innate in humans rather than socially constructed; this further complicates the challenge and suggests that these must be programmed from the outset.[33] The various approaches differ depending on whether the focus is on how the emotions are generated or the function of the emotions in the actions and behaviour they lead to; some regard emotions as preceding action, while others regard them as providing feedback signals. There still needs to be consensus on what the basic emotions might include, and according to the Oxford Handbook of Affective Computing, they range from 2 to 18 emotions.[34]

Although emotions as raw data are somatic in origin involving the CNS, they also potentially engage highly subjective cognitive appraisal and the generation of complex feelings unique to the individual. This makes it highly challenging to equip AI systems and robots with the ability to experience emotions and address the dual problem of being able to recognize the emotions of others. We have seen the importance of mentalizing and TOM, which involves the robot understanding feelings, intentions, and beliefs, including bodily cues such as gaze, posture, facial expressions, and vocal cues such as vocal tones and expressions. The complexity of this dual challenge may explain why most A.I. research initially focused on the creation of cognitive architectures and highly intelligent, information-processing thinking machines with significantly less effort expended in creating machines with behavioural architectures that allow them to experience human internal states like emotions and feelings or reflect the intuitive and subconscious role of emotions in biassing and governing behaviour.

Robotic research has focused on rule-based decision-making mechanisms, which agrees with Max Frisch's observation above that the logic of probability makes them more accurate than emotional man. This starkly contrasts with the observation of Marvin Minsky, who claimed that "the question is not whether intelligent machines can have any emotions, but whether machines can be intelligent without emotions."[35] Some thinking machines employ a finite number of states connected in a graph by the transitions between states; some use planning algorithms to provide solutions or answers that effectively select the optimal plan by eliminating poor-quality options, while others employ probabilistic decision techniques which use data and statistics to inform objective probability estimates or subjective probability based on rough estimates. None of these approaches incorporate emotional states, even though emotions appear central to the ability of social robots to establish a natural interaction with humans and for them to be more believable, accepted, and trusted.

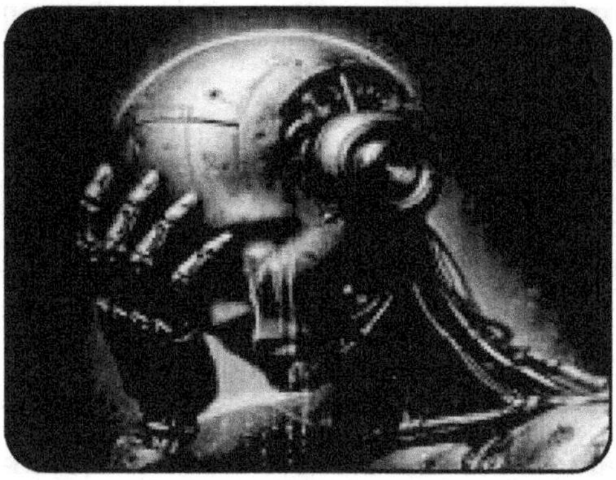

Emotion is important not just in decision-making and behaviour but in reinforcement learning, and many neuroscientific studies have linked emotions closely to reward processing. We have seen the importance of RL in ML and DL systems in allowing a robot to learn by gradually adapting to its environment and task. RL uses, for example, Markov Decision Processes by sampling and learning from data. This does not involve learning emotions but improves ML efficiency by exploiting the relation of emotions with motivation and action, with emotions providing complex feedback signals that shape behaviour in terms of reward or punishment and goal-driven behaviour.[36] One of the problems observed by Thomas Moerland and colleagues in their 2018 research review is that "while the topic of emotion in RL agents has received attention for nearly 20 years, it appears to fall in between the machine learning and affective modelling communities." They point out that "there is no framework connecting the variety of models and implementations" and that "appraisal theory focuses on how emotions are elicited, while categorical emotion models focus on action tendencies, i.e., the immediate function of emotions. Some consider emotions to precede action selection, while others focus on emotions as feedback signals."[37]

Technology has equipped some robots with the ability to appear to display emotions, which is an important function in customer service applications where robots are interfacing with the public, such as in healthcare, hospitality retail, airports, hotels, restaurants, and transportation areas. Studies have shown that a robot's ability to express surprise and happiness positively impacts the interaction between humans and robots and that people prefer anthropomorphic robots. This involves such non-verbal aspects as appearance, voice, and facial expressions used by the robot and more subtle differences such as the head tilt, cultural differences, and the gender of the robot.[38] However, the same study showed that people are concerned by the increased use of robots and their potential to replace human-to-human contact.[39] In 2006, Geoffrey Hollinger and colleagues described how they had developed a robot that interacts with humans in a crowded conference using a combination of speech, musical, and movement responses. The robot was programmed to continuously update "an internal emotional state, modelled realistically after human psychology research," which involved using empirically determined mapping functions so that the "robot's state in the emotion space is translated to a particular set of sound and movement responses."[40] These models, though, do not amount to the robot experiencing emotions but simulating or mimicking the experience.

It has been observed that many users of social companion robots rapidly lose interest due to the robot's lack of intelligent and adaptive response to their needs and requirements, with robots typically appearing heartless given that they perform predefined and repetitive behaviours. Luis-Felipe Rodríguez and Félix Ramos observe how "multidisciplinary inquiry has provided evidence that shows the significance of emotions not only to the rational behaviour of individuals but to achieve more believable and human-like behaviours in intelligent systems."[41] An interesting study of child robot experiments was conducted in 2011 by Raquel Ros and colleagues in what they called the wild environment of the paediatric wards rather than controlled environments. The rationale for choosing young children is that they are known to respond much more readily and strongly to social robots and employ basic social repertoires. Earlier studies have shown that simple reactive routines like touching the head, which causes the robot to play a prerecorded giggle, were highly effective at engaging children between 1.5–2 years old.[42] They identified three main challenges in developing social robots capable of "coping with highly dynamic and stochastic elements inherent in the social milieu," which are (1) Contingent responding in dynamic social environments, which means enabling the robot to adapt to people's response to the robot's behaviour and actions (2) Interpretation of multi-modal social sensory information which in their words has delivered limited progress and (3) Designing these into sociable robots that go beyond remote control of the robot. They recognized that while progress in the electromechanical design of robots has been impressive, this has not been reflected in designing robots that possess "the capacity to correctly interpret the social world correctly, thus hampering their ability to construct appropriate responses. They highlighted a paradox facing social robot developers: "the more sophisticated the robot, the greater a user's expectations of it and thus the greater the risk of their disengaging on discovering the system's limitations."[43]

In the 1980s, Aaron Sloman and Monica Croucher, argued that "the mechanisms producing emotions are also the mechanisms required for great flexibility in a complex environment" and that emotions are needed for robots to prioritize their decision-making processes.[44] By the mid-1990s, Rosalind Picard was observing how "if we want computers to be genuinely intelligent and to interact naturally with us, we must give computers the ability to recognize, understand, even to have and express emotions"[45] laying out three key aspects of emotions in AI which are the ability of machines to (1) Recognize and express emotions (2) Respond intelligently to human emotion and (3) Regulate and utilize emotions to behave more intelligently and effectively.

The emerging field of affective computing was formalized in Piccard's seminal book Affective Computing (2000) and comprehensively explained in Klaus Scherer's more recent book A Blueprint for Affective Computing: A Sourcebook and Manual (2010). Picard defined affective computing as "the computing that relates to, arises from, or influences emotions."[46] Affective computer, often used interchangeably with Artificial Emotional Intelligence (AEI), is an "umbrella term for human emotion, sentiment and feelings, emotion recognition, and sentiment analysis"[47] and is a multidisciplinary field involving computer science, psychology, and cognitive sciences. AEI research now widely acknowledges that emotional memory is critical to fast decision-making, learning, creativity, and adapting to a dynamic environment for humans and autonomous robots. Twenty years ago, Nancy Alvarado noted that "the question for many designers of intelligent systems is no longer whether to incorporate emotion and motivation, but how to do so. This question has strong consequences for the autonomy and flexibility of a system, and its ability to respond to novel, unpredictable or changing environments, in other words, the degree of intelligence it will display."[48] By 2004, Jonathan Gratch and Stacy Marsella observed that the development of computational models of emotion "is a core research focus for artificial intelligence that will facilitate advances in the large array of computational systems that model, interpret or influence human behaviour."[49] This may explain why, although the focus of A. was initially on developing intelligent systems that can detect a person's emotions, we have witnessed over the last few years a growing trend toward developing cognitive architectures that can provide robots with a fully cognitive state that embeds perception, adaptation, motivation, and emotion mechanisms.

This includes various approaches to modelling emotions, such as simulating emotions using affective space, motivation-driven, neuro-inspired, and hormonal or homoeostatic models. Affective space models mathematically model the representation of the emotional state using 3D dimensional spaces of valence, arousal, and power, which reflect how emotions are expressed. As Xuecheng Jin and Zengfu Wang observe, emotions are "significantly different from each other in at least one dimension"[50] which allows the use of a gradient of emotions to compute the emotional interaction of an agent or robot at any one time. Motivation-driven models reflect the Motivational Theory of Emotions (MTE), which states that emotions are action control systems designed to prioritize pursuing some goals over others. Andrea Scarantino describes how MTE "replaces the primacy of the appraisal and feeling aspects of emotions with the primacy of their motivational dimension."[51] Neuro-inspired models include Sophia, who we encountered earlier is the first anthropomorphic robot to express over 60 human feelings/emotions. Another novel recent example is Nikhil Churamani and colleague's framework for affect-driven behaviour generation in social robots that, amongst other things, includes an "Affective Core, that employs self-organizing neural models to embed behavioural traits like patience and emotional actuation that modulate the robot's affective appraisal,"[52] thus employing Russell's core affect that we discussed earlier. Homoeostatic systems have sought to incorporate, as we saw, the homoeostatic and allostatic drives that influence emotion processes and are intended to equip a robot with the ability to satisfy its drives and maintain the robot's well being which is reflected by its affective state characterized by positive valence and balanced arousal.

Researching emotions in robots requires establishing whether the Computational Modelling of Emotions (CME) in robots is, as we have seen in most cases to date, limited to merely mimicking, simulating, i.e., acting out human emotions or genuinely enabling the robot to experience and express emotions and feelings themselves with the potential even to develop their emotional domains like other animals that may not necessarily mirror human emotions. Many scientists now believe that emotions and feelings have a vital influence on cognitive behaviour. Having a biologically influenced model of human behaviour requires that emotions be modelled into embodied architecture and not confined to mimicking humans. The absence of an agreed definition of emotions has led some researchers to focus on certain discrete categories or domains, such as sadness, fear, joy, anger, etc. This, however, neglects the dimensional nature of emotions that we saw earlier with Russell's Circumplex theory and the PAD theory and potentially fails to address the bodily nature of the raw data of emotions, the role of cognitive appraisal and the generation of feelings, which as we discussed are closely entwined with but different to emotions.

Emotions consist of internal motivation and drive, individual physiology, appraisal of a perceived event or situation, how the emotions are expressed or regulated, and how they relate to the subjective framing of feelings. Emotional states might be loosely considered to be more short-term influences on behaviour. At the same time, drives and motivations (discussed later) are more medium-term, and personality traits are long-term features or influences on behaviour. All actions and behaviours require our brain to solve what we discussed earlier constantly, the five problems of Why, What, Where, When and Who, called H5W. The H is how the autonomous agent acts in the world. Studies into cognition and emotions in robots rarely mention this aspect, even though contextual information is essential to plan the sensorimotor sequences necessary to act, execute a selected task, and direct behaviour.[53] The context for humans includes issues related to the self, the environment, and the task, action, or behaviour and how sensitive individuals are to monitor self-related demands and their performance, which evokes not just cognition but emotions. These are driving factors in autonomous behaviour, and how a human reacts and adapts to the context that it finds itself in determines whether it can switch between behaviours, rapidly adapt to dynamic environments, and satisfy the internal issues related to self.

Internal states of homoeostasis and allostasis and emotions and the constant updating of the environmental context considering novelty and unpredictability are central to planning and executing goal-directed actions, and inevitably involve emotions. To get robotic agents to behave flexibly and adaptively would require them not just to follow learned programmed behaviours but to engage in the whole action cycle of behaviour from the perception of stimuli to action, processing of the stimuli, and assessing the internal and external context of their behaviour. Humans do this naturally, but computationally processing and monitoring diverse and complex factors related to the self, the environment, and action in the form of a task or behaviour is incredibly challenging. We discussed earlier personality traits and will later address the role of drives and motivation, four focusing on emotions and feelings. Emotions as transient states influence cognitive behaviour and decision-making or planning and need to be modelled in a way that captures the discrete categorical type of emotion such as fear or anger, joy, sadness, stress, and arousal etc and the dimensional aspects of emotions in terms of valence, novelty, goal relevance, goal congruence and coping potential.[54] It is further complicated because most emotions expressed through the body occur naturally and subconsciously. The shift in focus in AI architecture to encompass both the detection of human emotions by robots and the equipping of robots or virtual agents with the expression of emotions and feelings reflects the growing convergence between biological intelligence and machine intelligence.

A review of initiatives in behavioural models for social robots by computer scientists Olivia Nocentini and colleagues distinguishes between three main research areas, which are (1) Designing cognitive architectures to allow intelligent behaviour, (2) Enhancing the ability of the robot to adapt its behaviour to the needs of the human user and (3) Equipping robots with the capacity to establish an empathetic connection with humans thus improving HRI. A combination of this research allows robots to go beyond basic perceptual, motor, and cognitive abilities to operate smoothly in more complex social scenarios involving shared decision-making, problem-solving, and experiencing, processing, displaying reading, and interpreting emotions.[55]

In a 2012 review Luis-Felipe Rodríguez and Félix Ramos argued that "synthetic emotional processes have become an essential requirement in cognitive agent architectures."[56] They highlight three major approaches, which they categorize as the Appraisal, Hierarchical, and Dimensional theories employed in the biologically inspired CME. These are driven by the various psychological theories for autonomous agents, which acknowledge emotions as a process and that the neural substrates of emotions are closely related to the cognitive processes of attention, perception, and memory.[57] This is evident in modelling the Cognitive Appraisal Theories (CAT) of emotion in AI, which assume that emotions are generated by the subjective cognitive evaluation of the environment that closely relates to or might impact an individual's goals, plans, and beliefs. The hierarchical models assume that humans possess a limited number of basic emotions, such as anger, disgust, fear, and surprise, which are innate and instinctive, and the dimensional models differentiate emotions using dimensional parameters such as arousal and valence.[58] A slightly different approach to appraisal worth mentioning has been advocated by Joost Broekens and Doug DeGroot, who have sought to develop a formal notation or computer language for integrating the appraisal theories by implementing and modifying the network schemas involved in cognitive appraisal. This focuses on the "declarative semantics of the structure of appraisal,"[59] meaning the actions generated by interaction rather than the process by which they are brought out. It requires the CME to model the evaluation of emotional stimuli, the elicitation of emotions, and the generation of emotional responses.

Let's look at some of the attempts by cognitive architecture designers to enable robots to use emotions to make decisions, learn, and in general behaviour.

Robots Possessing and Exhibiting Emotions.

What appears to be an insurmountable obstacle to robots possessing and expressing genuine emotions reflects the somatic theories of emotion in which emotions depend on human and animal physiology. They are the raw data of a highly complex biological system comprising the body and brain. It is not inconceivable that this somatic process could be simulated, in whole or part. It is an enormous task to replicate in an AI system the complexity of the signalling between the body and brain and the subjective real-time nature of the emotions and feelings that are evoked by stimuli in the environment. It is further complicated by the need to build a sense of self that emerges from this basic raw emotion and involves cognitive appraisal based on highly subjective schemas that generate feelings. Emotions have a strong cultural and social component and processes akin to emotions could evolve from improving the social nature of robots through language processing and their culturally sensitive interaction with humans and other robots. No model has yet been developed to reflect the cultural and gender dimensions of emotions and by keeping robots gender-neutral, we may not need to accommodate these differences, which, as we know, may consist largely of social constructs.

A comprehensive 2016 review by Zdzislaw Kowalczuk and Michal Czubenko, in what they describe as a cybernetic approach to modelling artificial emotion identified twelve proposed solutions which are (1) ActAffAct (2) FLAME (3) EMA (4) Parlee (5) FearNot! (6) Fatima, (7) WASABI, (8) Cathexis, (9) KARO, (10) MAMID, (11) FCM, and (12) Motion and the innovative Intelligent System of Decision-making (ISD). Kowalczuk and Czubenko rather pessimistically point out, though, that "most psychologists shun attempts to relate their considerations to a mathematical synthesis of emotional phenomena" and that despite this plethora of purported solutions, "there are no useful signs of mathematical modelling and practical simulation of human emotions. Perhaps, knowledge of creating and evolving/variability of human emotions is still insufficient to carry out such a study."[60] They observed how most CMEs can be classified based on the following: (1) Whichever psychological theory or theories they employ, (2) Cognitive, physiological, and motivational components involved in forming emotions, (3) Phases involved in the emotional process, and (4) The applied description of emotions. When they talk about the phases involved in the emotional process, these include the low-level evaluation of emotions based on simple stimuli and the high-level evaluation of secondary emotions. This is based on emotional memory, modification of the priorities of the objectives/needs, implementation of the agent's (robot's) actions, planning behaviour, deploying behaviour, and communication. The following are some initiatives and solutions to building emotions into artificial systems in virtual agents and actual physical robots.

(a) I am acting Effectively affecting Acting (ActAffAct).

ActAffAct is an emotional architecture for virtual agents operating as actors in virtual games and was designed to enhance the credibility of computer game characters, chatterbots or other virtual characters. The architecture relies on labels and the pre-programming of beliefs, effectively imbuing the actors with personalities and allowing them to transition from perception during which external information is translated into an intelligible form to cognitive appraisal based on valence, decision-making which they call Relational Action Tendency (RAT), and finally execution. The agent experiences and cognitively appraises the emotions, chooses the appropriate current target, and decides upon a suitable response. However, the emotions do not impact the agent's overall goals.

(b)Fuzzy Logic Adaptive Model of Emotions (FLAME).

FLAME involves the emotional evaluation or cognitive appraisal of new events based on the robot determining the impact of the events on its objectives or goals. Hence, it generates an emotional value of the event about the agents' objectives or goals and involves cognitive appraisal. This means that the system assesses the significance of certain targets or the importance of goals. It uses fuzzy logic to input this into a Fuzzy system to calculate the desirability of the events and generate the appropriate group of emotions. For example, if a goal is not reached or the event fails to move the robot positively toward a desired goal, sadness may be accompanied by shame, anger, and fear. It permits a degree of emotional self-regulation by having a built-in interruption system, reflecting motivational needs that can overpower the effects of emotions. Emotions can't modify their goals or needs but confine them to selecting the appropriate actions or behaviour. Fuzzy logic is intuitive in that it allows the system to use various learning methods, which include using conditioning by linking an object or event with emotion, reinforcement learning by changing the assessment of the impact of events into the realization of the objectives, a probabilistic approach to learning behaviour patterns, and heuristic learning of gratifying behaviour.

(c)Emotion and Adaptation (EMA).

EMA employs a mapping algorithm to map a complex range of appraisal variables to interpret external stimuli related to an event or object and calculate the impact on a robot's goals. This generates up to 24 emotions with different intensities. The variables include relevance and desirability of those goals, causality in terms of whether the robot caused the events, opportunity, whether it could have been predicted, urgency and whether a response can be delayed, commitment, controllability as to whether the robot can impact the event or whether the event can change itself, power in terms of what is required to control the event and adaptability in terms of whether the robot can cope with the changes brought about by the event. The architecture includes motivational and cognitive components, with the robot planning and implementing its behaviour using these emotional appraisals.

(d) ParleE.

Parlee is a conversational virtual agent that evaluates events using a probabilistic algorithm based on previously learned behaviour. The architecture comprises five blocks with ten emotions of varied intensity. The first block is the Emotion Appraisal Component, which evaluates events based on the robot's personality, reflecting its plans and other agents' models and producing a vector of emotions. The second "Scheduler" block calculates the probability of achieving the goals to generate a vector of emotions. The remaining blocks comprise an Emotion component, an Emotion decay component, and the Modelling of other agents to modify the vector of emotions. These have all been incorporated into Obie, which, as a conversational agent, can reflect its personality while unusually accommodating the behaviour of other agents.

(e) FearNot!

FearNot is a virtual agent in storytelling story-telling AI applications that simulates sensitivity by sympathizing with victims of violence and has been used in schools to develop coping strategies and problem-solving with the emotions elicited. It must be credible in terms of the emotions it displays and its behaviour and involves cognitive appraisal with emotions having six attributes, which are the name or label of the emotion, the valence (negative or positive), the source of the emotion, the cause, intensity (a sharp value which decreases with time), and the time of occurrence. It also reflects the robot's mood and energy, which will affect the intensity of its emotions. Ko, Walczak, and Czubenko describe it as "the most comprehensive approach to computational systems of emotion." It relies on the cognitive appraisal theory to use emotions to control the behaviour of the agent and its communication with others.

(f) Fear not AffecTIve Mind Architecture (Fatima).

Fatima is an upgraded version of FearNot designed to control an autonomous virtual agent based on a continuous but sequential cognitive appraisal process, which uses motivational elements to generate certain emotional variables employed in decision-making. This is also related to the emotional state of the virtual agent based on the somatic state in response to external stimuli but relies on the cognitive appraisal theory to use emotions to control the behaviour of the agent and its communication with others.

(g) Affect Simulation Architecture for Believable Interactivity (WASABI).

WASABI involves modelling emotions in a robot in a continuous three-dimensional space called the Pleasure Arousal Dominance (PAD) space. PAD contains labelled primary emotions with a parallel cognitive module that generates secondary emotions and decisions about the desired action. In contrast, a third module links the emotions and actions to generate speech that reflects the current emotions. WASABI has been used in a virtual tour agent, MAX, in a virtual museum guide scenario. MAX can exhibit human-like emotions, which affect its conversations with people because every new person triggers a positive emotional impulse in its PAD space. Wasabi does not cognitively appraise events or situations. Still, it generates emotions in response to external stimuli and can change them over time, although the emotions don't affect the agent's behaviour, only the conversation.

(h) Cathexis.

Cathexis, implemented as an Emotional Pet Robot called Yuppy, is based on evolutionary, cognitive appraisal, and somatic theories of emotions and is the first architecture to employ Damasio's concept of somatic markers. It involves five modules: Perception, Action, Behaviour, Emotions, and Motivation, each with different functions. The perception model processes sensory information; the motivation module consists of four needs(1) Energy/battery (2) Temperature (at which it is driven)(3) Fatigue (amount of consumed energy) (4) Interest. The emotional model comprises basic emotions (anger, fear, sadness, joy, disgust, and surprise) and mixtures that generate responses to external stimuli and secondary emotions linked to objects. The behaviour model generates the most appropriate reactions based on the emotions and motivational modules. The action module implements the reactions using low-level and high-level discrete emotions.

(i) KARO.

KARO is a mobile robot that uses a dynamic logical language to model emotions, distinguishing operators of Knowledge, Beliefs, Actions, Abilities, and Desires. It evaluates emotions (as labels) and modifies the agent's priorities using four free-floating emotions (not connected to external objects) associated with certain attitudes: happiness, sadness, anger, and fear. Happiness is triggered by achieving goals, sadness is caused by failure, anger is triggered by being frustrated by the active plan, and fear is triggered by goal conflicts or danger. Karo does not use somatic or appraisal theories, evaluates emotions (as labels), and modifies the agent's priorities.

(j) MAMID.

MAMID uses appraisal theory to create the four basic emotions of fear, anger, sadness, and joy using external data, internal interpretations, desires, priorities, and individual characteristics. MAMID was designed to keep calm in response to situations such as a hostile crowd scenario but is limited by only having the four basic emotions.

(k) Emotion Forecasting Using FCM.

Emotion forecasting uses Fuzzy Cognitive Maps (FCM) that can capture cause-and-effect relationships to predict human emotions. It considers the agent's arousal levels, such as excitation or boredom, and the positive or negative valence. FCMs build a model of a system comprising boxes containing variables and connections or edges that represent causal links between the variables. FCM takes one of twelve emotion values of excited, happy, pleased, relaxed, peaceful, calm, sleepy, bored, sad, nervous, angry, and annoyed and evaluates every emotion in the trivalent scale of light, medium, and strong so it has 36 possible states of emotion. It uses appraisal employing information from the analysis of mood and bio-psychology concepts.

(l) xEmotion.

xEmotion is a subsystem based on Plutchik's model that generates emotions in response to observations and interactions. It does this by employing somatic theory to construct up to 8 pre-emotions, which Kowalczuk and Czubenko describe as "the rawest form of emotions, which are connected to impressions (basic features perceived by the agent), such as pain and rapid movement." Up to 24 classical sub-emotions connected to certain objects or situations provide useful information to the agent through cognitive appraisal and are remembered by the agent. Together, they constitute the agent's emotional state, narrowing down its possible choice of reactions while focusing on the agent's long-term goals. X Emotion has been used in an interactive avatar, DictoBot, and an automatic, autonomous driver.

Kowalczuk and Czubenko concluded their review with the observation that "almost all of the computational models of emotions employ the theory of appraisal as their scientific background." However, the emotion system employs evolutionary, somatic, and appraisal theories. They observe that because "appraisal analysis of stimuli from the environment should be carried out in real-time... that Cathexis and xEmotion (both using simplifying fuzzy emotions), or EMA may be recommended for simple mobile robots. On the other hand, robots with greater computing power can be designed using systems based on the fuzzy representation of emotions. They can indirectly shape the agent's behaviour, such as x Emotion, FLAME, and EMA.

In 2004, María Malfaz and Miguel Salichs designed an architecture based on emotions for autonomous personal robots built on the iRobot Magellan Pro mobile platform, on which various sensors (laser, sonar, and camera) and devices are installed. They have achieved this by adding an Emotional Control System (ECS) to the cognitive architecture developed by Barber and Salichs.[61] The ECS contains three modules which are (1) Drives, (2) Activity Selection, and (3) Emotional Supervisory System (ESS). They describe how the "Drives module is the one that controls the basic drives of the robot. Conversely, The Activity Sele, action module determines the rob the it's goals and action tendencies. Finally, the ESS module generates the emotional state of the robot". Drives are essential to an autonomous robot in determining its goals and selecting the appropriate behaviour to achieve them, but they must be combined with emotions. The difference with Malfaz and Salichss's approach is that rather than just internal drives, such as temperature and battery, they include external variables, such as social interaction and social motivations. They list the drives as Energy, Entertainment, Affect, Health, and the robot's Programmed Agenda. The emotions condition the robot's behaviour. Still, they acknowledge that not all behaviours are driven by emotion. Hence, they have a 'supervisory role," which means that the dominant drive will prevail without an active emotion. If an emotion becomes active, it will influence the activity selection module, including the emotions of Happiness, Anger, Fear, and Sadness, which are generated from the evaluation of the well-being of the robot, which involves current well-being and the prediction of future well-being. They acknowledge that past well-being would evoke such emotions as shame, pride, gratitude, etc, but do not include those. Their goal is to use the design as a global appraisal system in robot reinforcement learning and enable emotions to act as alarm signals for when things are happening.[62]

One field of research that some believe holds significant promise is the Semantic Pointer Theory (SPT) of emotions developed by Chris Eliasmith. Eliasmith's cognitive architecture, the Semantic Pointer Architecture Unified Network (Spaun), remains the world's largest functional brain model operating at the level of individual spiking neurons and uses images as inputs with a physically modelled arm as the observed output.[63] SPT hypothesizes that in biological systems, higher-level cognitive functions are made possible by what they call semantic pointers. These pointers are neural representations carrying partial semantic content that captures a compressed representation in shallow high-dimension vectors that point towards the underlying more semantically complex relations so that deeper semantic content can be inferred from these semantic pointers. SPT is based on the Neural Engineering Framework (NEF),[64] which is a mathematical theory of how biological neural systems can implement a wide variety of dynamic functions by synthesizing large-scale cognitive systems from sub-networks.[65] Using the neuromorphic hardware that we discussed earlier, NEF has been proven a viable means of translating the mathematical algorithmic formulations into a neural network, and Runchun Wang and colleagues using NEF-employed field programmable gate arrays (FPGAs) to perform massively parallel real-time pattern recognition successfully.[66] Ivana Kajić and colleagues used the semantic pointer architecture to develop a computational framework called POEM (for POinter-Emotions) which they describe as "explaining numerous important phenomena concerning emotions, including how some stimuli generate immediate emotional reactions, how some emotional reactions depend on cognitive evaluations, how bodily states influence the generation of emotions, how some emotions depend on interactions between physiological inputs and cognitive appraisals, and how some emotional reactions concern syntactically complex representations." They observe how using the open-source neural network simulator Nengo their POEM algorithm simulates numerous aspects of emotions by integrating "multiple levels of meaning from embodied experiences through culturally constructed linguistic structures." They assert that "emotions are semantic pointers, which are patterns of activity in neural populations that enable the brain to represent complex verbal and nonverbal information."[67]

Launched in 1997, RoboCup is an international scientific initiative to advance state-of-the-art intelligent robots. In 2017, Caroline Rizzi, Colin Johnson, and Patricia Vargas observed how RoboCup competitions in the "fields of contextual perception, real-time adaptation, and flexible decision-making," particularly the Standard Platform League (SPL), had stagnated. To address this stagnation, they proposed their novel Situation-Aware Fear Learning (SAFEL) model, designed to allow "robot players to learn the behaviour profile of the opponent team at runtime. Later, players can use this knowledge to predict when an undesirable outcome is imminent."[68] The purpose is to provide robots with "more believable social responses, leading to more natural interactions between humans and robots."[69] They observe that "research on artificial fear-learning is still in its infancy and has several aspects with a margin for improvement, among which we can highlight situation appraisal." They describe how "SAFEL combines concepts of situation-aware expert systems with well-known neuroscientific findings on the brain fear-learning mechanism to allow companion robots to predict undesirable or threatening situations based on past experiences. One of the main objectives is to allow robots to learn complex temporal patterns of sensed environmental stimuli and create a representation of these patterns. This memory can be later associated with a negative or positive "emotion," analogous to fear and confidence." It is structured using four modules that are intended to reflect anatomical brain structures and cognitive functions that, as we have seen, are vital to human behaviour, which are the (1) Sensory system, (2) Amygdala, (3) Hippocampus, and (4) Working memory. They describe how "each module is based on a different approach and performs a different task in learning and memorizing environmental cues to predict the occurrence of unpleasant situations. They claim that the main contribution of the model is "the integration of fear learning and context awareness to fuse emotional and contextual artificial memories."

One of the most advanced AI models to incorporate emotions to date is the framework Soar-Emote using the unified computational model of Robert Marinier and John Laird, which combines an abstract cognitive theory of behaviour control (PEACTIDM) and a detailed theory of emotion (based on an appraisal theory), integrated Soar"[70] which has combined much of Damasio's theories and work on emotions and Gratch and Marsella's appraisal theory of emotions. Soar cognitive architecture models human appraisal in which humans evaluate situations against various dimensions about the goals, which evoke emotions and include relevance, salience, causality, control over the situation, etc. It reflects the belief that emotions aren't reflexive but result from the cognitive appraisal of perceived events or situations. Marinier and Laird have observed that for appraisal models to be realistic and accurate in terms of the complexity of human appraisal, they need to model the whole complex appraisal process, including such important features as the immediate nature of the emotions. This must include, as Damasio has observed, how a person perceives the emotions in terms of what feelings are generated by the emotions, which can include mood states and recent events. To do this, Marinier and Laird use Scherer's appraisal theory because it is considered the most comprehensive with 16 dimensions, and they have modelled eleven.[71] When modelling complexity, Marinier and Laird employ an appraisal frame for the emotion, the mood, and the feeling, with the emotion interacting with the mood that then interacts with the feeling. This provides a measure of intensity and how important the feeling is with a necessary and natural bias towards some feelings and the likely impact on the behaviour, but the mood must be contextually relevant and not just based on historical events. They start with mood as a neutral state, with the mood moving towards or away from the emotion in steps to determine how far from neutral the mood is. Their work results in using a combination of specific mood-emotion and feeling-intensity functions that can be implemented in existing cognitive architectures. This raises the question of whether robots are experiencing emotions or mimicking human emotions.

These are just a few examples of brain-inspired cognitive architectures designed to equip a robot with an emotional response, such as the automatic fear response in the last study and the learning and memorization of potential threats to one's survival given that it is far faster than a cognitive appraisal of critical situations. The other part of the dual problem is enabling the robots to detect and process emotions in others.

Robots Detecting and Recognising Emotions in Others.

Human-robot interaction (HRI) requires the robot not just to express emotions but to be able to read and respond to the emotions expressed by humans and potentially other robots. This has important applications in fields such as education, marketing, security, and medicine and in developing social robots that can express empathy and compassion.[72] The combination of recognizing and expressing emotions has been called the Affective Loop.[73] Michael Mateas observes how challenging it is "to accurately sense complex and elusive emotions influenced by many physiological, individual, and cultural factors that are unavailable to computational sensors, and the need to simplify emotions, to computationally model them may lead to interfaces that flatten our emotional experiences rather than engaging our emotions in their full everyday richness. For robots to detect and recognize emotions, often called Emotion Analysis (EA) the focus is largely on visual systems, audio/speech, and physiological emotion recognition with different modalities contributing unequally at different times to whatever emotion is evoked. It has also been observed by Patricia Bota and colleagues how "Physical manifestations are easily collected; however, they present low reliability since they depend on the user's social environment, cultural background (if he is alone or in a group setting), their personality, mood, and can be easily faked, becoming compromised."[74] Although seemingly natural and effortless to humans, emotion recognition is complicated. In a 1968 study, Albert Mehrabian observed that approximately 55% of emotions are expressed through the face and called facial or emotional expressions, 38% verbally and how we say it, and 7% in normal daily communication.[75] EA facial analysis can involve DL-based facial expression recognition (FER) systems, facial micro-expression analysis (FMEA), and 3D FER. Still, there needs to be more focus on physiological-based emotion recognition.

We begin to perceive faces within days of birth and they offer a rich source of salient information about our world and who to trust etc. Gleaning emotions from facial expressions requires recognizing subtle emotional cues expressed explicitly through facial expressions, eye movements, speech, action, and physiological signals and emotions that might be masked. Our impressions of people are significantly impacted by subtle facial expressions, such as happy faces being perceived as more trustworthy but angry faces less so. This tendency has been described as the common view that many people worldwide think certain facial muscle movement configurations reliably reveal emotions. This is evidenced by the widespread use of emojis or emoticons across cultures with schematized versions of facial expressions representing various emotions. It has been observed that even in humans, there are significant challenges to inferring emotions from faces; it may be more complex than mapping the configuration of facial movements to specific emotional states.

A 2020 review by Liza Barrett and colleagues cited as the most comprehensive and systematic to date focused on the emotions of anger, disgust, fear, happiness, sadness, and surprise. They concluded that the evidence "rests on several flawed assumptions and incorrect interpretations of research findings" and that "there is an urgent need for research that examines how people move their faces to express emotions and other social information in the variety of contexts that make up everyday life, as well as careful study of the mechanisms by which people perceive instances of emotion in one another." They note that human perceivers do not "infer emotions from particular configurations of muscle movements in a sufficiently reliable and specific way" and "emotion categories are not expressed with facial movements that are sufficiently reliable and specific across contexts, individuals, and cultures to be considered diagnostic displays of any emotional state."[76] Mapping facial expressions to emotions is complicated because the expressed emotions reflect personal, contextual, and psychological factors, such as culture, personality, and social context. It also involves using single-modal emotion recognition and integrating multiple modalities (face, voice, posture, speech, etc.) called Multi-modal Emotion Recognition (MER) to recognize, interpret, process, and simulate emotions. This requires data fusion of the various modalities. Dacher Keltner and colleagues concluded that "there is no one-to-one correspondence between a specific set of facial muscle actions or vocal cues and any and every experience of emotion."[77] It is not possible to rely solely on facial muscle movement as it can involve changes in gaze, body movements, respiration, gestures, hand movements, voice, tactile contact, and autonomic responses," and there are many more than the six basic emotions expressed in this way.

Affective computing includes the distinct topics of emotion recognition involving usually embedded devices and sentiment analysis. Given the lack of consensus on the definition of emotions, most models employ either discrete categorical or continuous dimensional emotions and use different databases to train models. You will remember that Ekman described emotions as measurable, discrete, somatic-based, and evolutionary in that they evolved as physiological and communicative functions. Damasio distinguishes between primary and secondary emotions, and Plutchik's taxonomy allows us to combine emotions to form complex emotions and personality traits. Emotion recognition and affective computing involve representation learning and emotion recognition from the body and multi-modal emotion recognition using speech, face, and body gestures. Sentiment analysis involves largely analysing text and mining opinions and implicit emotions in large data volumes. Despite the complexity and controversy surrounding the generalisability of facial expressions, the field of affective computing has developed public benchmark databases, mainly consisting of unimodal databases (i.e., textual, audio, visual, and physiological databases) and multi-modal databases. When using MER, performance in expressing and recognizing emotions is enhanced, which has obvious advantages. Various cues can augment or complement each other and provide data even if some sensory modalities are unavailable due to sensor failure. Unimodal and MER emotion recognition employs a range of computational approaches, such as classification and regression. Sven Buechel and colleagues point out that much emotion analysis has used either classification or regression depending on whether emotions were treated as categorical or dimensional. Buechel observes that categorical is more "Coarse-grained" and treated as a classification problem, with the most widely used being Ekman's Basic Emotion (BE) model which sub-categorizes emotional states into emotion categories to classify them.[78] In classification, an emotion may belong to one or a fixed number of emotion categories, which requires class boundaries in the affective data space. If treated as a single emotion, then a single-label learning (SLL) problem may be assigned by MER with a single dominant emotion label to each sample.

Dimensional analysis is described as more "Fine-grained." It treats EA as a regression problem "so that (most often) a vector of real-valued numbers will be produced as the result of an emotion assessment."[79] Multi-label learning (MLL) has been designed to address the fact that emotions may be mixed and dimensional in that one instance may be associated with multiple emotion labels and recognizes that people can respond differently emotionally to the same stimuli or situation. Emotion regression analysis rather than classification is used to predict the continuous probability distribution of emotions, which can be represented in dimensional valence-arousal in what is called the Valence-Arousal-Dominance (VAD) model of which James Russell and Albert Mehrabian's model is most used. They observed that "three independent and bipolar dimensions, pleasure-displeasure, degree of arousal, and dominance-submissiveness, are both necessary and sufficient to define emotional states adequately." Now, publicly available datasets can be used to evaluate how well a model recognizes emotions. These include DEAP, MAHNOB-HCI, ASCERTAIN, Eight-Emotion, EMDB, AMIGOS, etc., all used to elicit emotions in machines.

Sicheng Zhao and colleagues, in their 2021 paper observe that MER involves data collection and training using existing data or new recordings in specific environments and emotion annotation and that "methods, especially the ones based on deep learning, require large-scale labelled data for training."[80] They point out significant challenges facing MER, which are (1) The Affective Gap, which computer scientist Michael Mateas has described as "a disjunction between the subjective, situated, cultural, and social ways in which people think about and experience emotions and the objective, formalized ways in which computers model them" (2) Perception Subjectivity in which people have different emotional responses to the same stimuli (3) Data Incompleteness maybe due to sensor failure or incomplete data (4) Cross-modality Inconsistency in which different modalities may conflict with each other and thus express different emotions (5) Cross-modality Imbalance in which different modalities may contribute unequally to the evoked emotion (6) Label Noise and Absence which reflects the fact that labelling the emotions is prohibitively expensive and time-consuming, highly inconsistent and results in a large amount of data with few or even no emotion label. All these contribute to the difficulty of computerizing emotions in any AI system and of robots accurately recognizing and inferring emotional states in humans based on explicit or implicit cues. They recommend future research areas that could enhance robot multi-modal emotion recognition.

These include (1) Contextual and prior knowledge modelling in which the system can consider users' prior knowledge, such as personality and age, contributing to emotion perception. For example, optimistic or pessimistic individuals are likely to view the same stimuli differently (2) Learning from unlabelled, unreliable, and unmatched affective signals, which would involve advanced ML techniques, including unsupervised representation learning, dynamic data selection, and embedding the special properties of emotions (3) Robust, and secure deep learning for MER which explains the decision-making process about reading emotions(4) Combining explicit and implicit emotion signals (5) Greater incorporation of emotion theories into MER systems. They also advocate much more practical and real-world settings for developing and training than neat lab settings for these models, including dealing with data that contains a lot of noise unrelated to emotion and testing in different cultures and languages from the training set and across various settings. This should involve greater use of edge computing in autonomous agents incorporating MER.

Maria Vircikova, Gergely Magyar, and Peter Sincak have proposed an Affective Loop Model for social robotics, which purports to provide the robot with the capacity to adapt its behaviour according to the needs and preferences of a particular user.[81] They observe that their approach differs from what they describe as the 'Wizard of Oz' technique, in which a person remotely operates the robot. Their tool purports to "personalize effective artificial behaviours in cooperative human/robot scenarios, where human emotion recognition, appropriate robotic behaviour selection, and expression of robotic emotions play a key role." The social interaction between people is complex and involves many nuanced and not-so-nuanced clues about a person's state of mind, such as emotions, language, tone, facial expression, body posture, etc. These are essential to establishing empathetic behaviour between people.

Neuroscientist Diana Valenzo and her colleagues observe that "humans exhibit behavioural flexibility and that the appropriateness and flexibility of behaviour and interaction with the environment are entirely context-dependant, grounded in cognition, emotions, and behaviour."[82] The stability and adaptability of an organism and an autonomous agent reflect the need for a constant balance between the internal state of that organism or agent (human), including the emotions, the cognitive brain, and the interaction with the environment. These internal states in humans consist of emotions, motivational drives, and personality traits, and recognizing this, the designers of cognitive architecture have, over recent years, begun to focus more research effort on realistically modelling these internal states. A new interdisciplinary field of affective computing focuses on building machines that can recognize, interpret, process, and simulate human emotions. Pepper, the robot we met earlier, was advertised as the world's first robot that could recognize human faces and basic human emotions, although production was stopped in 2021.[83] A major problem recognized by the designers of emotion recognition architectures is that "facial detection and recognition algorithms are compute-intensive operations," the goal has been to "parallelize the algorithmic operations for large-scale information exchange in real time" in both static and mobile robots.

This neglects the major caveats discussed earlier in terms of the reliability of reading emotions from human facial expressions. Affectiva, which was acquired by Smart Eye in 2021, is however a leading emotion detector technology company that develops software to analyse facial expressions during video calls, allowing a computer to gauge customer reactions and satisfaction. Facial recognition software is a major field of AI research and the basis of affective computing, which combines human emotion, sentiment analysis, and pattern recognition in A.I. Affectiva iMotions technology (iMotions (2015). Affectiva iMotions Biometric Research Platform). was compared to facial Electromyography (EMG), which is widely employed in laboratories to measure facial emotions using muscle activity but needs electrodes and technical equipment and found to be as effective at detecting happy and angry expression and produced less false negative impressions of neutral faces.[84]

Endnotes

1 The Expression of the Emotions in Man and Animals (1872).
2 Fraser, ON, Bugnyar, T., 2010.
3 Bateson, M, et al., 2011.
4 Tooby, J, Cosmides, L., 1990.
5 LeDoux, JE., 2020.
6 Oatley, K, Keltner, D, Jenkins, JM., 2012.
7 Emotions, Learning, and the Brain (2015).
8 Terasawa Y, Fukushima H, Umeda S., 2011.
9 Harrison, NA, et al., 2010.
10 Craig, AD., 2002.
11 Prescott, SL, Liberles, SD., 2022.
12 Steenbergen, L, et al., 2021.
13 Breit, S, et al., 2018.
14 Anderson, DJ, Adolph, R A., 2014.
15 Barrett, L F, Russell, J.A., 2015.
16 APA Dictionary of Psychology.
17 Scherer, KR., 1999.
18 Mehrabian, A, Russell, JA., 1974.
19 Barrett, LF., et al., 2018.
20 Plutchik, R., 2001.
21 APA Dictionary of Psychology.
22 Gard, MG, Kring, AM., 2007.
23 Ridgeway, CL Bourg C., 2004.
24 Fischer, AH, et., 2004.
25 Deng, Y, et al., 2016.
26 Tobin, RM, et al., 2000.
27 Brody, LR, Hall, JA., 2000.
28 Eliot, L., 2012.
29 Markus HR, Kitayama S., 1991.
30 Nangyeon, L., 2016.
31 "Emotional Intelligence" (1995).
32 Loewenstein, G, Lerner JS., 2003.
33 Ekman, P., 1998.
34 Calvo, R., et al., 2015.
35 Minsky, M, 1986.
36 Baumeister, R., et al., 2007.

[37] Moerland,T M, Broekens, J, Jonker, CM., 2018.
[38] Yu, CE, Ngan, HFB., 2019.
[39] Christou, P., et al., 2020.
[40] Hollinger, G., et al., 2006.
[41] Rodríguez, L, Ramos, F., 2012.
[42] Tanaka, F, Cicourel, A, Movellan, J., 2007.
[43] Ros R, et al., 2011.
[44] Sloman A, M. Croucher 1981.
[45] Picard, R W., 1995.
[46] Picard, R W., 1997.
[47] Wang, Y, et al., 2022.
[48] Alvarado, N., 2002.
[49] Gratch, J, Marsella, S., 2004.
[50] X, Wang, Z., 2005.
[51] Scarantino, A., 2014.
[52] Churamani, N., et al., 2022.
[53] Rosenbaum S,et al., 2014.
[54] Gratch, J, Marsella, S., 2014.
[55] Nocentini, O., et al., 2019.
[56] Rodríguez, L, Ramos, F., 2012.
[57] Phepls, EA., 2006.
[58] Miwa, H., et al., 2001.
[59] Broekens, J. DeGroot, D., 2004.
[60] Kowalczuk, Z, Czubenko, M., 2016.
[61] Barber, R. Salichs, M.A., 2001.
[62] Malfaz M, Salichs M A, 2004.
[63] Eliasmith, C, X, Choo., 2013.
[64] Stewart, TC, Tripp B, Eliasmith, C., 2009.
[65] Eliasmith C, Anderson, CH., 2003.
[66] Wang, R., et al., 2017.
[67] Kajić, I., et al., 2019.
[68] Raymundo, CR, Johnson, CG, Vargas, PA., 2017.
[69] Raymundo, CR, Johnson, CG,Vargas, PA., 2015.
[70] Marinier, RP, Laird, JE, Lewis, RL., 2009.
[71] Scherer, K R., 1999.
[72] Sorrentino A., et al., 2020.
[73] Höök, K., 2008.
[74] Bota, P, et al., 2019.
[75] Mehrabian, A., 1968.

[76] Barrett LF, et al., 2019.
[77] Keltner, D, et al., 2019.
[78] Russell, J, Mehrabian, A., 1977.
[79] Buechel, S,Hahn, U., 2016.
[80] Zhao, S, et al., 2021.
[81] Vircikova, M, Magyar, G, Sincak, P., 2015.
[82] Valenzo, D., et al., 2022.
[83] D'Onofrio, G, et al., 2022.
[84] Feyerabend D, Schacht A., 2020.

"The advantage of a bad memory is that one enjoys several times the same good things for the first time."

Friedrich Nietzsche. 19th Century German philosopher.

Chapter 10

Memory

CHAPTER 10

Human Memory

We have seen how important attention is for humans and AI in real-world environments, but what does our brain choose to remember or to forget, and how do we use memory to influence our behaviour in the real world? Over our life, the process of memory and learning based on what we attend to and the processing of information related to self shapes the person we become. The psychology of memory and learning theories are based on observation of people's behaviour, corroborated, modified, or contradicted by the neuroscientific research into how sensory information is encoded, stored, and recalled by our brain when we learn, experience emotions, are motivated or driven to do things and behave in certain ways. What we learn as we go through life reflects our experiences and the information that has been subconsciously or consciously attended to, which is encoded and stored in our neural networks as prior knowledge, which relates to our life experiences in the key domains of the self.

These domains are (1) How we express and control our emotions and react or respond to situations and others (2) Our intellectual and cognitive capacity, curiosity and creativity (3) How we view our physical appearance and level of attractiveness (4) Our environment or context in terms work and lifestyle, the groups we are part of (5) Our perception of our financial security and access to resources (6) How we occupy our time (7) Our social relations and connectedness and (8) Our spirituality or faith etc. They combine to shape our identity, personality, values, and beliefs and influence how we learn through experience, perceive ourselves, and relate to others. They are dynamic and can change positively and negatively because circumstances change, your decisions and choices, and how you respond to the world around you. Humans are geared to maintaining the equilibrium of homoeostasis and allostasis and constantly trying to predict outcomes most or least favourable to us. Like most organisms, we don't like uncertainty or the unexpected in our environment and when things, places, and people are familiar, we tend to be more comfortable, whereas the novel or unexpected can make us anxious and even fearful as individuals or groups. It is not surprising when someone who has dementia finds much about their world and self unfamiliar, and familiar faces, objects, and environments become so unfamiliar and alien to them that it is extremely distressing. We learn though from new information and experiences, and if we surround ourselves only with the familiar, we will soon stop learning. Learning depends on encoding and storing knowledge gleaned through experience, which allows us to recognize people, things, and experiences that we have encountered before and constitutes our prior knowledge of our world. To learn and adapt to changes vital to our survival and thriving, we need to remember things that are physically or psychologically harmful or produce positive or negative emotions, rewards, and pleasant things rather than punishments or unpleasant experiences.

Preconceptions and predictions about our environment, situations, and others are built upon our emotionally memories, prior knowledge, and current motivations, and

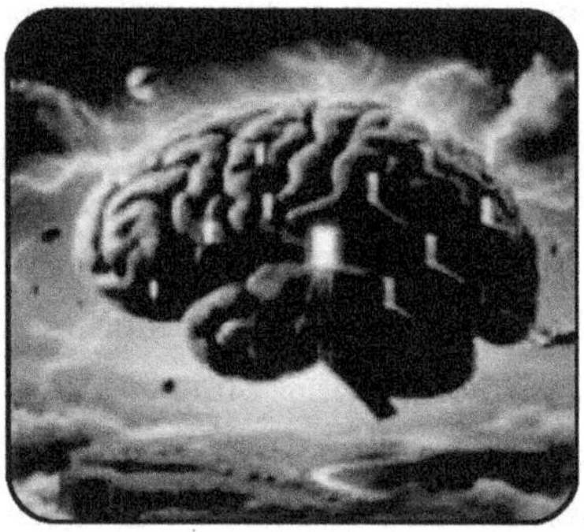

drive and form the basis of the biases displayed by individuals and groups towards the unfamiliar or less familiar, which includes, for example, racial identities and prejudice.[1] Subtle changes in our brains serve to distinguish us from others. Neuroscience suggests that an implicit bias enables us to identify perceived differences between people familiar to us and those who aren't. When we lived in small groups in caves or tribes, this served as a protective or survival measure against potential threats from others. Still, in the modern world, these biases and prejudices seem a primitive relic of that behaviour. Anxiety or fear of the unfamiliar does not require a person to have the experience or memory of having encountered a type of person who is perceived as different based, for example, on colour, race, or sexual orientation. It appears to reflect a more fundamental drive to be part of a group, the pleasant associations with membership, or the unpleasant sensations of being left out of or rejected by those in a group. The implication is that if we perceive that people don't belong to our group, we experience an implicit bias against the unfamiliar and those deemed to be others. Bias can also indicate a learned fear response, in which the fear relates to an unpleasant experience and influences our behaviour because humans continually refer to the past to predict the future, whether in relationships, situations, events, or the ideas and beliefs we encounter. We may carry that memory of the fear response, influencing our response to similar situations or people. Our brains are designed to respond differently to familiar or unfamiliar faces, including our face, with studies showing that with familiar faces, the fusiform gyrus or fusiform face area, associated with the processing of information in the visual pathways, is activated (FFA). The FFA is a large region of neurons in the temporal and occipital lobe and is, therefore, also known as the Lateral Occipitotemporal Gyrus.

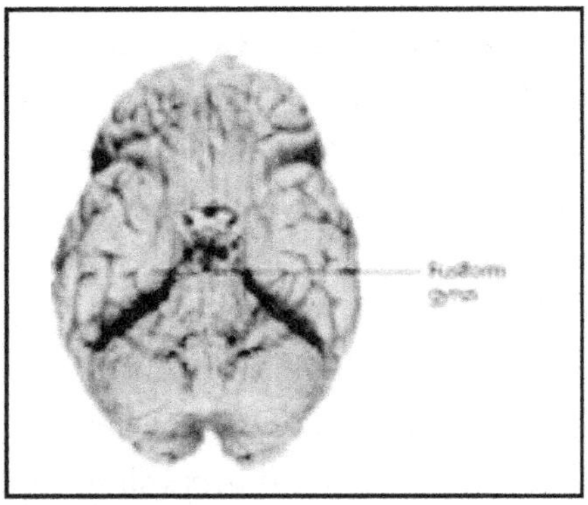

The Fusiform Gyrus.

The FFA helps us distinguish between objects with more similar objects, generating neural activity in the lateral regions of the FFA. In contrast, the less similar is seen in the medial region. There is a specific cluster of neurons in the left FFA called the Visual Word Form Area (VWFA), which remembers and recognizes visual words and is vital to fluent reading in children. If damage to this area is bad enough, an individual may have a condition called Alexia and be unable to read or comprehend letters or words. When this happens during development, it is called developmental dyslexia, but it can result from injury in adults who could read before, which is acquired dyslexia.[2] Such individuals can still talk and write normally but can't read, even if they have written the words themselves. However, some will be able to recognize individual letters and may be able to piece the words together slowly. The FFA is an example of how our prior knowledge of the world, such as faces and words, is stored throughout the cortex.

Emotions influence attention and memory, subconsciously and subconsciously, in determining the salience or relevance to us at the time, with attentional biases having an enormous influence on what information we attend to and what is subsequently processed and stored in our long-term memories. When an experience is more emotional or relates to us and prior knowledge, we tend to remember it better and for longer. Whatever we learn from previous success or failure in meeting goals, or what is perceived as pleasant or unpleasant, or frightening or comforting, may be retained as memories that shape and influence every domain of the self, whether it's the cruel comments made by peers at school about our body shape or looks or the prize at sports day, etc. Our brains utilize stored data and prior knowledge to process incoming sensory information from environmental stimuli and distinguish between the novel and the familiar. There are many different types of memory and ways that we learn but all involve a biological process, with neuroscientists distinguishing between (1) Encoding, (2) Storage, and (3) Retrieval of visual, acoustic, semantic, and elaborative information.

Memory and Learning.

Our modern understanding of the complex relationship between memory and learning began with the pioneering work of German Psychologist Hermann Ebbinghaus (1850–1909), who introduced the term memory and who, over seven months of experimenting on himself, often up to three sessions per day, researched into how we learn and forget about things.[3] He transformed our understanding of remembering and forgetting things and developed the concepts of forgetting the steep learning curves and recognizing that new information could be recalled more easily when associated with prior knowledge. It can be infuriating to forget things, and most of us don't know why some things will not seem to stick in our memory, with many memories lost or decaying over time. Our brains seem hardwired to forget because forgetting information allows us to filter out irrelevant information, or we would be overwhelmed by the sheer volume of experiences. Ebbinghaus discovered some crucial features of why and how we forget, which include how the rate at which we forget information is exponential. Unless we keep reviewing and refreshing our memory, it will weaken over time, with the largest fall-off happening shortly after we have learned something, although the rate of decay levels off after time. He developed the forgetting curve to represent mathematically the rate at which memories are forgotten. Testing his ability to remember nonsense syllables, chosen because he wanted to use something of which he had no prior cognitive knowledge or memory associated with it, he realized that we remember things better if they mean something or because we are engaged and interested in the information, particularly if it is presented logically and clearly. He demonstrated how the strength of the memory will determine the length of time over which it can be recalled and before it is lost. Because his experimental work was based solely on himself as the case study, it has made it difficult for scientists to replicate and generalize the findings to the general population. Still, despite this, the main findings need to be central to our understanding of how we learn.

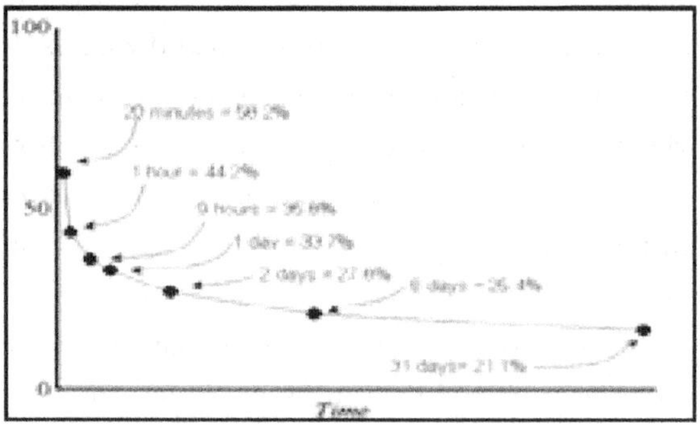

Ebbinghaus's The Forgetting Curve.

Although Ebbinghaus is remembered best for the forgetting curve and the spacing effect of learning, most people are probably more familiar with his concept of a steep learning curve. However, recent research has shown that the qualitative use of the learning curve merely describes the general learning process over time, which is of little utility as a mathematical learning curve model because it only measures the completion of a specific task against time, not general learning. The task being learned must be measured repeatedly and involve only one variable. Because there are typically many variables in learning that impact how quickly or slowly you learn, the rate of progression for general learning rather than a very specific task needs to be accurately reflected in the learning curve model. This means the phrase steep learning curve provides little real insight into how we generally learn using memory.

Ebbinghaus demonstrated that arousal, energy levels, and other physiological factors, such as emotional states, influence how we remember. For example, if you are tired, stressed, or in a low mood, which, we have seen, will affect your attention levels, as you may not be concentrating sufficiently for the information to be processed and encoded. He acknowledged that people vary enormously in terms of their memory, and even factors such as noise or other environmental interferences can affect what you remember. He realized that we learn information spread or distributed over a period better than when we cram learning into shorter periods, which suggests that learning a little every day is better than trying to learn it all in one day and that refreshing your memory when your memory of information when it is getting poorer, will help you strengthen the memory. He demonstrated that we could avoid this steep drop off in memory seen in the forgetting curve if we overlearn something, which means that even when we think we have mastered information or a task or skill, we should keep learning it. However, this will be moderated by the extent to which we overlearn, the information type, and the retention period length.[4]

Flawed Memories.

Even when we think we remember things accurately, our memories are often flawed because 100% accurate recall seems impossible. Long before neuroscientists had established a clearer understanding of the process by which memories are stored in neural assemblies in the cortex, Frederic Bartlett had developed the concept of mental schemas and was a pioneer in the field of learning and providing an in-depth analysis of how we acquire knowledge about ourselves, other people, social relationships, and social groups. He recognised that this knowledge is stored in abstract mental structures or templates in our memory through abstract schemas containing this prior knowledge. Bartlett observed that 'the first notion of getting rid of is that memory is primarily reduplicative or reproductive. In a constantly changing environment, literal recall is extraordinarily unimportant…if we consider evidence rather than supposition, memory appears far more decisively an affair of construction than mere reproduction."[5] Retrieving long-term memory is much more complicated than simply flicking a switch and recovering the information of an event or experience precisely as it happened at the time. and encoding and storage of memory is a flawed process, with recall never perfect. Different brain regions are involved in the encoding, storing, and retrieving of semantic and episodic memories and retrieval modifies and re-encodes episodic memories, which are vulnerable to being rapidly lost.

Dan Schacter is a leading memory researcher who identified seven sins of memory errors that produce flawed or suboptimal memories. These are (1) Transience or inability to access the memory, (2) Absent-mindedness or lapses of attention, (3) Blocking, which is a memory that is temporarily forgotten, for example, things on the tip of your tongue you can't quite remember (4) Misattribution or confusing the source of a memory or dream (5) Suggestibility which involves implanting memories of things that never happened (6) Bias in the form of distortion of memory to accord with current believes and values and (7) Intrusive (persistent) memories such as in post-traumatic memories that we cannot seem to forget. Most of us regularly experience these seven sins and Schacter observes that the sins of misattribution and bias are most closely linked to the self, for example, mistaking a dream for an actual event or confabulation, which fills in memory gaps.[6] Given that episodic memory is so interwoven with the construction of a coherent autobiographical self, these sins of memory, coupled with the constructive nature of memory, adversely affect the reliability of our self-schemas and autobiographical memory. The extent to which they do so varies significantly between individuals, as some have far better memories than others, and some memories seem much more enduring.

Flashbulb Memories.

People vary enormously in how they remember the past, with some able to recall detailed episodes and others certain features of events, which are more factual or semantic. Ebbinghaus's forgetting curve appears to fail to consider those memories that become so deeply ingrained that we never seem to forget them, like the sight of the Twin Towers collapsing on 9/11 or JF Kennedy's assassination. These are commonly known as flashbulb memories, a phrase first employed by Roger Brown (1925–1997) and James Kulik (1940–) in 1977 when studying the ability of individuals to recall consequential and surprising events. They are defined as "a vivid, enduring memory associated with a personally significant and emotional event, often including such details as where the individual was or what he or she was doing at the time of the event."[7] People often compare these memories to photographs and are highly confident of their accuracy and they are thought to stick in our memory because of the significance level and context of the event or information.

The implication is that witnessing something sufficiently emotionally arousing creates this type of memory. A strange feature is that they include where the individual was, what they were doing, or who told them about an event and their feelings at the time, including more insignificant details. The problem is that when Brown and Kulik first brought this flashbulb concept into public awareness, they only questioned the study participants on very specific newsworthy events and didn't test the accuracy or integrity of their recollection of events. Subsequent research has shown that these flashbulb memories still decay over time but may appear to remain more vivid to us because our perception of the memory and confidence in its accuracy creates the impression that they are more reliable than regular memories.[8]

Reality Monitoring and Memories.

Our brains have an intriguing mechanism for distinguishing whether the information is internally generated or sourced from our external environment. This allows us to recognize ourselves as distinct, conscious agents interacting with our world and determining what is real.[9] This is called reality or source monitoring and means that we can, in normal circumstances, determine whether our memories are real or imagined like a dream. It is not full-proof, and errors in reality monitoring can result in people confusing reality with imagined events and, even result in full-blown hallucinations. Studies have indicated that impairment in a person's reality monitoring may explain why some people experience hallucinations, including those with schizophrenia. Neuroimaging studies have identified a neural network involved in recollecting the source of information you remember and determining whether it was generated from externally derived or internally derived information in the form of contextual cues. This includes the Hippocampus and Parietal Cortex, with the Anterior PFC exercising control over the encoding and retrieval of feature representations, which are bound together by the hippocampus and integrated into the parietal cortex's associative regions into a first-person perspective. The anterior PFC is consistently activated when recollecting self-generated contextual cues about information. When Lagioia and colleagues studied adolescents who express schizotypal traits, which potentially increase the risk for later psychosis and hallucinations, they found significantly reduced activity in their anterior PFC indicating a problem with reality monitoring.[10]

Confidence in the accuracy of your memory, concerning the source of information and whether it is real or imagined, and the degree of accuracy of the information recalled has important implications for eyewitness evidence in legal proceedings and research into the accuracy of eyewitness accounts, suggests that the more accurate the memory, the more confident the person is in its accuracy. and the less effortful it will be to recall. Research tells us that real memories possess different attributes of content and quality to imagined events, which has been called the Undeutsch hypothesis after German Udo Undeutsch

(1917–2013), one of the theory's developers. Real or actual memories include more contextual, sensory, and semantic information. In contrast, imagined memories tend to refer more to the person's thinking processes because if the memory is sourced internally, it will not have external contextual information accompanying memories sourced from the environment. This has applications in detecting verbal and non-verbal errors of memory, which, although honestly recounted, rather than lying, have been identified as a major cause of wrongful convictions, particularly when identifying suspects. The problem is exacerbated by how poor people judge the accuracy of other people's memories. A widely used tool of choice for testing the accuracy of witness testimony based on their memory is the Criteria-Based Content Analysis (CBCA), which was initially designed for child victims of sexual abuse but has been extended, albeit controversially, to adult witnesses, offenders, and /other case types. The CBCA is a checklist of 19 cognitive and motive reality criteria to distinguish between memories of self-experienced real-life events and fabricated or fictitious memories. Some countries do not allow them to be admitted as valid evidence in court because of inconsistent scientific evidence of their accuracy. However, a recent meta-analysis of the studies suggested that the CBCA is reliable.[11]

Research has shown that we rely on indirect cues or short-cuts to assess the accuracy of our memories based on what we know or think we know about how memory works or how readily retrievable the memory appears to be. Incorrect memories require more effort to recall, and we attribute greater strength and accuracy to memory if it requires less effort. When people use words that suggest they are making more effort to recall, like delays to think or conditionality such as maybe or sounds of hesitancy, it indicates honesty but less confidence in the accuracy. It increases the likelihood that it is imagined. We should remember that scientists have demonstrated that it is possible to manipulate memory encoding and retrieval by activating engram neurons to generate entirely false memories and create memories without any sensory experience. In amnesiac mice, by reactivating silent engram neurons, they can induce the retrieval of memories, which they couldn't do using natural cues, which suggests that they can exist in different states of activity or dormancy. The latest hypothesis is that memory consists of connected engram ensembles dispersed across multiple brain regions, which form an engram complex of overall memory composed of these components.

Photographic Memories.

Some people claim to remember every detail or fact in a way that resembles a photograph, which would challenge the view that memory is naturally flawed and reconstructive but there is no scientific evidence that photographic memories exist any more than flashbulb memories are 100% accurate. About 100 people worldwide have a very rare condition, highly superior autobiographical memory (HSAM), and appear to never forget what has happened to them every day and can associate calendar dates with events even if they don't involve them. This makes for a wonderful party trick but has a significant downside, as James McGaugh, a research professor in neurobiology and behaviour has discovered in his study of HSAM. Many people with the condition suffer from obsessive traits and ruminate excessively because they can't let things go. McGaugh observes that, although these people have phenomenal memories, their memories are only partially accurate recordings and make mistakes. For example, they are often no better than most people at remembering faces or telephone numbers. He distinguishes them from people who claim to have a photographic memory because they can remember details from a scene that they only briefly see once. Although there is no scientific evidence for photographic memory, they are considered to have an eidetic memory, which allows them to scan an image and remember data. Their ability to remember is often associated with a particular task, for example, drawing a scene from memory or recalling the order of cards they have scanned, which are then shuffled, and their accuracy fades over time, so they still make mistakes. It is not associated with other traits such as intelligence, and children born with this ability often lose this over time. A remarkable case of eidetic memory is a British architectural artist and autistic savant called Stephen Wiltshire (1974–) who has achieved fame for his ability to draw landscapes, buildings, and skylines from memory after seeing them only once, as the view below of Paris gleaned through the aircraft window?

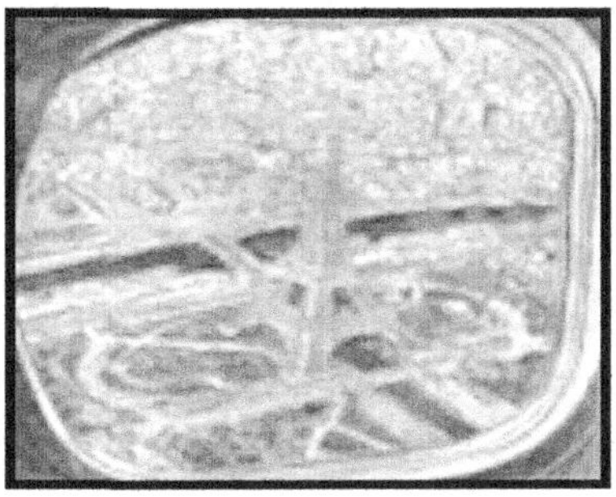

Stephen Wiltshire. Drawing of Paris from Memory.

Having HSAM, or eidetic memory, is not the same as those who have trained themselves to have a remarkable memory, for scientists now believe that you can train the natural memory to perform what appear to be remarkable feats and perform as Memory Athletes. A remarkable example of this is the record-holding German memory athlete Simon Reinhard a passionate chess player who observes that to have an incredible memory, you do not need to be a super brain or a genius, which he describes as a typical media myth to make things more spectacular. He claims, "Everybody can learn these techniques and improve their memory. And it is fun and motivating to see one's progress just to train your brain." He observes that our memory is a powerful tool but agrees with the neuroscientists that it is still limited and that there is no such thing as a perfect memory. These memory athletes often use mnemonic devices to remember long strings of data or a visual story upon which to link memories. There are mnemonic devices for learning:(1) Visualizing certain locations and remembering things like routes, locations, and lists by mentally moving through the familiar location to aid recall (2) Chunking which helps memory by breaking down information into chunks and acronyms, such as taking the first letter of each word or line and putting them together (3) Rhyming mnemonics, so the rhythm helps you remember; and lastly, musical tunes, which can help. Reinhard's technique involves storytelling and context to help retain information and can be used to remember any information better and more reliably. He is also passionate about and very good at chess. He believes that what appears to characterize top chess players is that they use intuition to make moves, which "is often the result of a huge, subconscious treasure trove of accumulated knowledge that the brain can access through pattern recognition." Reinhard claims that good memory is about how well your knowledge is interconnected, how precisely you can recall your knowledge, and how deep your understanding of the field of your knowledge is. Interconnectedness and depth of knowledge ultimately depend on the strength and enduring nature of the neural patterns stored in your cortex, which reflects that memory, for most people, is about the learned experience and the effort required to acquire that knowledge.

There is very little point in encoding or storing information if you can't recall it when you need to, and different types of memory are stored or retained in different areas of the brain, either short-term or long-term. Although only in its infancy, there is growing evidence based on brain imaging studies that people with these special types of memory have structural differences in the brain. Still, there is much further research required before this is understood. When neuroscientists point out that those with HSAM, eidetic memory, or memory athletes still make mistakes, they are referring to the fact that storage in the brain isn't neat but involves storage across many different parts of the brain. As Bartlett had indicated, when the brain retrieves information, it activates or replays the neural patterns associated with the first experience of the sensory information, which involves pulling the information from various brain regions. It can never do this 100% accurately, and recall strength will depend on the strength of the neural pathways and neural patterns formed from the previous experience.

Déjà Vu.

Recently, scientists have sought to explain what happens in the brain when we experience that strange sensation of familiarity, which explains the name déjà vu, French for already seen. It is believed that around 60% of people experience déjà vu, some more often than others. There is no scientific consensus on what happens in the brain, but it appears that some areas of the brain send signals to the frontal regions that indicate that an experience is repeating itself. Dr. Akira Robert O'Connor focuses on the cognitive and neural bases of episodic memory judgments and observes that déjà vu consists of the sensation of familiarity accompanied by the "meta-cognitive recognition that these feelings are misplaced and that you know something is not quite right." O'Connor maintains that the frontal cortex is trying to correct mistakes in memory and believes that "For most people, experiencing déjà vu is probably a good thing as it is a sign that the fact-checking brain regions are working well, preventing you from misremembering events. Research has shown a link between increased dopamine levels and the experience of déjà vu. In a strange case, a 39–year-old healthy male physician took a common decongestant and appetite suppressant to treat flu but, within 24 hours, developed intense and recurrent déjà vu experiences, which stopped when he discontinued the medication. Studies reveal a link between déjà vu and paramnesia, which are rare memory disorders involving distortion to specific types of memories. An intriguing example is Capgras syndrome, in which a person holds a delusional and false belief that an identical duplicate has replaced someone or something significant to that person. It is named after a French psychiatrist, Joseph Capgras, who, in 1923, treated a woman who complained that her husband and other persons she knew had been replaced with doubles, and a double of herself existed. Capgras called this "I'illusion des copies," which translates to the illusion of look-alikes. We now know it is linked to damage to the frontal lobe, the right limbic system, and the temporal lobes.

Types of Memory.

Richard Atkinson (1929–) and his student Richard Shiffrin (1942–) were part of the cognitive revolution in psychology in the early 1950s and 1960s, which moved away from trying to explain humans by simply observing their behaviour to studying attention short-term memory, and the stages of information processing.[12] They were motivated by the desire to understand what we forget and what we choose to store in permanent long-term memory. In 1968, they developed the highly influential "multi-store" memory model, distinguishing between sensory, short, and long-term memory. We saw when we discussed how, when visually scanning the environment, we use pre-attentive processing, which allows the brain to hold huge amounts of information as Sensory Memory for less than 1 second, during which we attempt to recognize it.[13] Sensory memory forms the first of three types of memory characterized by the nature of the encoding process, the amount of information that is processed, and the length of time or duration over which the information is stored. Sensory memory includes visual and auditory information. It is believed to save the brain energy because it doesn't require processing information in short-term memory, which takes longer and consumes more energy. Sensory information decays rapidly and represents the processing and potential recognition of the information from the senses but does not involve storage of that information and, therefore, can't be recalled.[14] Let's look at the transition of information in humans from the sensory stage to the short-term working memory before we look at the AI memory equivalent.

Sensory Memory.

The prevailing model of short-term memory is the Multicomponent Working Memory Model (MWMM), developed in 1974 by Alan Baddeley and Graham Hitch, because they felt that Atkinson and Shiffrin's Multi-store model of memory failed to consider the fact that short-term memory involves information processing.[15] Their model reflects that short-term memory is not a single unified construct, has a limited capacity, and, most importantly, involves manipulating information to make it available for more complex cognitive purposes, which involves executive control. The MWMM identifies four different components of working memory that provide a mental workspace for this to happen, which consist of (1) Visuo-spatial sketchpad memory (2) Verbal working memory (3) An Episodic buffer and (4) The Central executive attentional control system. Instead of all the information going into one place for short-term memory, Baddeley and Hitch demonstrated that each component system processes different types of information. In the visuospatial sketchpad, information is stored in a visual or spatial form and is used for navigation. In verbal working memory, sometimes called the phonological loop, written and spoken information is processed to manipulate speech-based information. This processing is achieved in two separate processes: the phonological store, or inner ear, processes speech perception and storing the spoken words for about 1–2 seconds, and the articulatory process, or inner voice, processes speech production and rehearses and storing verbal information from the verbal store. The most important component of working memory is central executive control, which monitors and controls the visual sketchpad and the phonological loop and links them to long-term memory. In this way, our working memory allows us to pay attention to certain information and supply prior knowledge from our long-term memory during the process.

Working Memory.

Being conscious of information starts with being processed in the working memory, but as we saw earlier, the conscious mind seems able to process only around 50 bits per second. Hence, allocating attention to working memory uses a valuable and constrained resource, which is determined by its salience and our level of motivation, for example, concentrating on our car driving rather than listening to a passenger or talking. We still need to understand fully how executive attention works. Baddeley argues that we don't store information here as it involves controlling the visual sketchpad and the phonological loop processes and linking them to prior knowledge in longterm memory. The episodic buffer was introduced in 2000 by Baddeley and Hitch to act as a storage backup, which connects long-term and short-term memory and allows additional prior knowledge to be readily accessible.[16]

Our working memory allows us to navigate, read, comprehend, and manipulate information to solve problems and apply reason. However, it has an important capacity limit because it depends on allocating attention to the information and requires executive control processes that interact with long-term memory. When information is processed in short-term working memory, the information is usually no longer available in the environment. Hence, the brain needs to maintain and manipulate that information to assess further its salience and relevance, which involves increasingly abstract coding of the information and determining whether that information will be dismissed, forgotten, or transferred to the hippocampus and potentially encoded in long-term memory in the cortex. An interesting characteristic of human memory is that pictures or visual images are remembered better than words; superior overall picture memory may be mediated by more effective and automatic engagement of the brain areas important for visual memory.[17] We saw how attention is an important component of working memory and for learning, understanding, and reasoning. Working memory is thought to have the capacity for about 5–9 items and lasts around 30 seconds, although this can be increased by a process called chunking, which involves converting a string of items into larger chunks; for example, number 785694364 is 785 694 364.

One of the earliest case studies to support the MWMM was conducted by Tim Shallice and Elizabeth Warrington in 1974, involving a patient known as KF, who sustained brain damage in a motorcycle accident.[18] Although KF could function normally, his working memory retained only 1–2 items on a list he was asked to memorize briefly. He would need to remember information more quickly when giving it verbally rather than visually, suggesting there are two subsystems at work. An intriguing study was performed in 1996 in which chess players were asked to solve a chess problem while doing different things. Solving chess problems requires spatial processing and problem-solving. They found that being required to tap on the table or repeat the word see-saw to themselves did not affect solving the problem. When asked to say out loud random numbers or press keys on a keyboard to create a circular pattern, they found solving the problem much harder because these tasks involve cognitive and spatial effort, which they required for the chess problem. Tapping on the table and repeating the word see-saw did not.[19] Holding mental representations in place is crucial to us being able to not simply respond to our world in a conditioned way in that we react to a stimulus because, by allowing us time, albeit briefly, to formulate a reasoned response to our environment, working memory gives us the illusion or reality of free will and acts as a building block for further processing of the information.

Neuroscientists agree that working memory is primarily involved in goal and task-oriented behaviour because keeping the visual and verbal information available helps you execute the task or achieve the goal. For a long time after working memory was discovered over 50 years ago, it was believed that enhancing the information by rehearsal, choosing what to try and remember, and repeating the information repeatedly, like phone numbers and names, was critical for storage in short-term memory and transfer to long term memory. Psychologists call this the Rehearsal Loop, which means that the individual appears to have control over what to rehearse and remember in the long term. The belief was that the more you repeat something, the more likely you are to remember it, which is why sometimes, the things at the end tend to get repeated less often and are more difficult to remember. The problem is that as long ago as 1973, Fergus Craik and Michael Watkins concluded that "neither overt rehearsal nor maintenance in short-term storage is by itself sufficient to enhance long-term memory."[20] The time you rehearse or keep information in your short-term working memory doesn't by itself impact the likelihood that it will be stored in long-term memory. We now know that there are two types of memory rehearsal in short-term memory: Maintenance and Elaborate rehearsal, which both involve repetition, but only elaborate rehearsal enhances the storage in long-term memory. When you keep repeating either verbally or silently back to yourself something by rote, like a phone number, that is simply trying to maintain it in the memory. By comparison, elaborative rehearsal involves different encoding strategies, such as organizing the information in your head, providing examples of visualizing the information in an image and employing mnemonic devices like acronyms and rhymes.[21] Although much more effective because it involves strategies for elaboration, it is much more complicated and effortful.

Patricia S. Goldman-Rakic (1937–2003), an American professor of neuroscience, neurology, psychiatry, and psychology at Yale University School of Medicine, pioneered work uncovering the neural mechanisms of the working memory and its relationship with the PFC. She showed how we hold abstract thoughts in mind as we process them and demonstrated that it was possible to scientifically study the higher cognitive functions, which many thought were beyond the realm of the methods available to science at the time. Goldman-Rakic demonstrated how impairments in working memory are linked to many thought disorders, including schizophrenia. In the intervening years, brain imaging studies have revealed several brain areas involved in working memory, which relates to the capacity limit and nature of temporary storage in working memory, which are the PFC, the Cingulate Cortices, and the Parietal Cortices, and recently, the Cerebellum and areas in the mid-brain including the Basal Ganglia involved in reward and motivation.

Long-Term Memory.

Suppose information gets through the pre-attentive sensory memory process and passes through the selective attention stage of manipulation in short-term working memory with its multi-component processing. In that case, it may require further elaboration and formation of new memories. This may last hours or days and takes place in the hippocampus. Much of what we understand about memory processing has emerged from studying individuals who have lost the capacity to encode and store new memories and the ability to retrieve memories. One of the most famous patients in the history of psychology is Henry Molaison, commonly referred to as patient HM. Born in 1926, Henry was 27 when he lost his memory during an operation to treat his epilepsy. His neurosurgeon, the highly renowned William Beecher Scoville, had suctioned out and removed completely two tiny (3–3.5 cm3) but densely packed areas of neurons, shaped like sea horses, which form about 1% of the total volume of the cortex. These two complex S-shaped structures are located deep within the temporal lobe and together called the hippocampus, from the Greek hippocampus (hippos, meaning horse and Kampos, meaning sea monster, although it is translated as sea horse because it resembles a seahorse, and Scoville removed both hippocampi. Scoville did not know how crucial these little seahorse-shaped clusters of neurons were to memory and was seemingly unaware that Henry would no longer be able to generate and store new long-term memories, even though it transpired that he could still access some existing memories He has yet to learn what we know now: that the hippocampus is a unique area where new neurons continue to be generated into adulthood, a process called neurogenesis. When he realized the surgery had rendered Henry amnesiac, he sent him to Wilder Graves Penfield, the discoverer of cortical neural maps whom we encountered earlier, and a neuropsychologist, Dr. Brenda Milner, who realized that Henry's intelligence was still intact, and so was his sense of self. Henry became the most studied patient in medical history, with nearly 100 researchers studying him, until his death at 82, when his brain was preserved in 3-dimensional slices for posterity. Henry was unable to make new memories and had lost all memories covering the eleven years before the operation, so as far as his memory was concerned, he was stuck as 16-year-old Henry. He could not remember faces or recognize his family, music, or new words, yet he could still do all sorts of puzzles and tests and repeat them without being aware that he had already done them. He remained cheerful and energetic, and his personality appeared to remain unaffected by his catastrophic loss of memory.

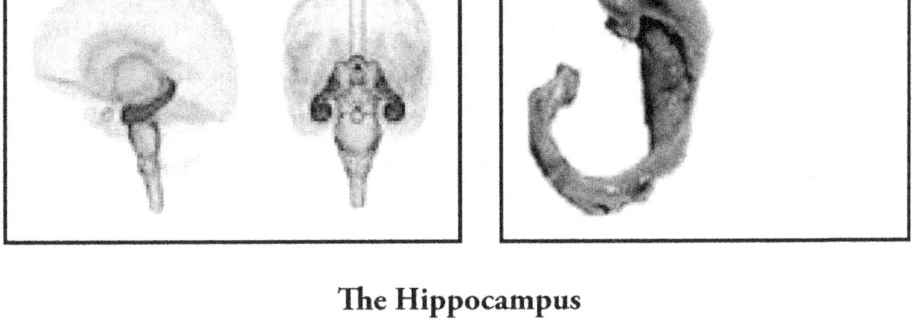

The Hippocampus

Can you imagine what it would be like to be constantly trapped in the present? This is precisely what happened to Clive Wearing, an accomplished British musician who composed and sang classical music. Born in 1938, Clive, at the age of 47, contracted the herpes simplex virus, which led to encephalitis that caused significant damage to his CNS, particularly his hippocampus and the frontal lobes of his brain. Clive's case was very rare, as unlike most patients with amnesia, he suffered total retrograde amnesia, which meant he had no memories at all, so he could not remember events, people or places, names, facts, or general knowledge. Clive's case is unusual, as often people with retrograde memory retain memories, especially from childhood and adolescence, although it usually involves the loss of knowledge about facts and events rather than learned skills like piano playing or driving the car, so they may know how to drive or play the piano, but forget they have a car or piano. He also had anterograde amnesia, so he can't make or store new memories, and although he has children, he can't remember them. However, he can remember his wife, Deborah, even when she is absent. In 2006, she wrote a moving book about her husband called Forever Today, and Clive was featured in a 1986 documentary called Equinox: Prisoner of Consciousness. He can remember things for an average of around thirty seconds, but often less than seven seconds, which is the span of our working memory, but then he forgets the information. Despite this devastating impact on his memory, he can still play complex piano pieces, sight-read music, and conduct the choir. This reflects that he has retained his procedural or implicit memory, a learned memory such as reading and writing. He can, therefore, still learn new procedural tasks, which become implicit memories through constant repetition, even though he cannot remember ever having heard or listened to music before. This is because these are encoded and stored in the basal ganglia, which remains unaffected by the virus. Entries in Clive's journal, which he keeps meticulously, give him some form of continuity and describe how he feels as though every 20 seconds, he has woken up for the first time. He understandably has suffered depression, anger and confusion, unstable moods, and auditory delusions. He has no idea who made the entries in his journal because he can neither remember nor write them. Because of the brain damage, Clive has lost the ability to encode and store new information in the hippocampus and, as a result, transfer that information into long-term storage in the cortex; he is reliant solely on his working memory to function and the existing implicit longterm memories retained in the brain, which reflects the depth of their previous encoding or if they are newly learned through repetition. To understand what happens in our brain to cause such a catastrophic impact on the memory of extreme cases like Henry and Clive, we need to look at how the brain transfers short-term memory into long-term memory and the role of the hippocampus that Scoville had removed from Henry and that the encephalitis had so cruelly devastated in Clive.

Often called the hippocampal formation, a choroid plexus covers the hippocampus, which lines the brain's ventricles (fluid-filled cavities) and produces the cerebrospinal fluid

(CSF) that floods our brain. It is divided into the head, body, and tail (the thin curved part) and associated regions, which connect it to the cortex and are critical to normal memory function. These other areas include the Parahippocampal region and the Subiculum, which connects the hippocampus with the Entorhinal cortex (EC) and is a critical gateway for cortical information going to and from the hippocampus. It is called the entorhinal (inside rhinal) cortex because it is partially enclosed by the Rhinal (olfactory) sulcus (groove), which is involved in processing smell information and generating olfactory neural maps and is probably why smells can evoke such strong memories of an experience or person.

The EC is involved in learning and memory by translating and integrating cortical exteroceptive information (stimuli external to the body) into higher-order complex neural patterns as representations of experiences in the cortex. This exteroceptive information is combined with encoded motivational and interoceptive information, which supports cognitive functions and conscious memories.[22] In effect, your external and internal worlds come together. Problems with the EC are associated with Alzheimer's disease, temporal lobe epilepsy, and schizophrenia.EC atrophy is associated with mild memory loss associated with mild cognitive impairment and is observed to precede the reduction in hippocampal volume seen in Alzheimer's patients.[23]

In 1930, American neuroanatomist James Papez (1883–1958) identified a specific neural circuit that starts and ends with the hippocampus, now known as the Papez circuit, which is concerned with hunger, motivation, sex drive, mood, pain, pleasure, appetite, and memory and involved in imagining the future[24, 25] and even believed to be involved in our sense of morality.[26] Paul MacLean subsequently expanded this to other areas. He gave it the now more widely used name of the limbic system, which is believed to be associated with emotions. At the same time, the hippocampus processes short-term memory before transferring to long-term memory and is connected to these other regions.

Psychology Professor Karl Lashley sought to locate long-term memory in rats by selectively removing different cortex regions. Lashley discovered that the resulting defects in memory depended on the location removed and realized that long-term memory was distributed throughout the cortex. Since Lashley's time, a neuroscientific model of long-term memory that evolved from such case studies as Clive and HM asserts that while the hippocampus forms new short-term memories, over a period, this information is consolidated and encoded for long-term storage throughout the cortex. Donald Hebb believed that synaptic plasticity and mental associations were distributed in synaptic connections across neural assemblies and focused on identifying a central mechanism to explain the delay between stimulus and response. Hebb believed that sensory stimulation-initiated patterns of neural activity, which were centrally maintained by circulation in synaptic feedback loops, which he called reverberatory activity and made it possible for the response to follow the stimulus after a delay. He argued that short-term memory consists of this 'reverberatory neural activity,' but storing new information in long-term memory requires establishing new synaptic connections or pathways in the cortex. Hebb explained that short-term memory is converted into long-term memory when the reverberatory activity patterns of short-term memory are stabilized through synaptic connections. Once the activity pattern was stored, it could be recalled repeatedly by excitation from sensory neurons or other reverberatory activity patterns.[27] Hebb's theory has been supported by subsequent research findings in three key areas (1)The neurons in the PFC that could remain active for many seconds were discovered to encode information about an earlier stimulus or a response (2) The discovery of long-term potentiation (LTP), in which synaptic connections are persistently strengthened, leads to a long-lasting increase in signal transmission between neurons based on recent activity patterns and (3)The discovery is that recurrent neural networks can possess multiple stable states and delayed activity patterns. He observed that information processing isn't one way but flows back and forth between short-term and long-term memory to facilitate recognition based on prior knowledge in stored memories. The theory remains controversial with others arguing that neurons contain an intrinsic mechanism that allows for the delay. There remains no clear evidence that LTP stabilizes neural networks, and there may be more memory traces than just short and long-term memory traces, as we now know that all neural synapses can keep track of past activity and that synapses display several types of plasticity, which depend upon the synaptic activity. Short-term plasticity can last from tens of milliseconds to several minutes and is thought to underlie information processing" and can be reduced as in short-term depression in which repeated stimuli delivered at short time intervals can suppress neurotransmitter release for hundreds of milliseconds to tens of seconds. Repeated stimulation at short time intervals can lead to a transient increase in transmitter release for hundreds of milliseconds to seconds (facilitation) and sustained, high-frequency synaptic activation for tens of seconds to minutes (augmentation and post-tetanic potentiation (PTP).[28]

We now have a far better understanding of the mechanisms responsible for short-term plasticity, even if their precise role in learning and memory and the relationship between short-term memory and long-term memory consolidation remains unclear. We still don't fully understand the all-or-nothing nature of memory storage and forgetting, the causes of short-term forgetting, or why it is so difficult to control rehearsal, but know that memories require a range of dynamic, integrated systems for neural plasticity, which involves the encoding, storage, and retrieval of information, which are central to learning about our environment and ourselves. Austrian-born American medical doctor Eric R Kandel (1929–), psychiatrist, neuroscientist, and professor of biochemistry and biophysics, together with Arvid Carlsson and Paul Greengard, won the 2000 Nobel Prize for their foundational research into how memory is stored in our brain. Kandel focused on pyramidal neurons in the hippocampus, making remarkable discoveries about their electrical firing. He concluded that because the signalling properties of neurons are very alike, the "neuronal mechanisms of learning and memory probably did not reside in the properties of the neurons themselves" but from how they are functionally connected. The number of connections in the hippocampus and other neural systems in humans made the research too challenging, so Kandel capitalized on earlier research showing that basic forms of learning are common to humans, other animals, and invertebrates. Kandel explored a range of candidates, including crayfish, lobster, flies, and the nematode worm Ascaris, but descended upon the giant marine snail Aplysia because its nervous system has a small number of unusually large cells, with many invariants and uniquely identifiable, which made it more amenable to study. Kandel realized that well-established learning paradigms, including habituation, sensitization, and classical and operant conditioning that we discussed earlier, could be used to study Aplysia and that he could use electrical stimuli of the neurons rather than sensory stimuli. This allowed him to investigate whether synaptic changes arose during these learning paradigms. Kandel knew from previous research that every perceptual or motor action has an internal or neural representation in the brain, which can be simple or complex, as we saw with topographical maps. He refers, in the account of his research, to the discovery by John O'Keefe (1939–) in 1971 that all pyramidal cells (that we discussed earlier) in the hippocampus possess the ability to encode a representation of space and thus form cognitive maps. Different pyramidal cells fire around familiar places depending on where the animal, in this case a mouse, is, which is why O'Keefe called them place cells. Each space or field is represented by specific cells in the hippocampus, and when overlapping, they form a map of the space and the animal's surroundings. When the mouse explores new surroundings, new maps are formed. O'Keefe was awarded the 2014 Nobel Prize for discovering these "cells that constitute a positioning system in the brain." Kandel's inspiration was to hypothesize that the changes in synaptic connections because of protein expression at the synapse related to the electrical firing was stabilizing this representation in the form of the cognitive maps generated by learning and, most crucially, by the animal attending to its surroundings. Kandel's dedication and ground-breaking discovery exemplify the thousands of hours of often thankless lifetime research by scientists that have allowed us to slowly reveal the remarkable processes at the molecular level by which our brain functions.

Another major challenge facing scientists when studying the relationship between short-term encoding in the hippocampus and long-term memory encoding and consolidation has been that the technology, typically EEG, used to measure memory processing has only recently allowed them to take a snapshot of the process. It has yet to enable them to follow this process. It may take weeks or months to capture how new memories are transferred from the hippocampus, stabilized, organized, and organized by the cortex to form long-term memory. There is more than one type of long-term memory. To understand the relevance of the exciting findings in neuroscience, we need to look at how psychologists view the role of long-term memory in the development of the self. Endel Tulving, famous for his work on memory storage and retrieval was one of the first people to challenge the traditional belief that there was only one type of long-term memory when he developed the Encoding Specificity Principle. Largely because of his focus on the subjective experience, Tulving identified the seemingly abiding distinction between semantic memory, which is not unique to humans, and episodic memory, which he believed to be unique to humans, and which Antonio Damasio, believe to be the basis of our core and autobiographical self. Tulving distinguishes between storage and retrieval necessary for remembering an event and episodes; he described autonoetic consciousness as part of episodic memory. Psychologist Louis Renoult focuses on the cognitive neuroscience of memory. and others have pointed out that recent memory research points to semantic and episodic memory being "inextricably intertwined" while "retaining a measure of distinctiveness, even though their neural correlates demonstrate considerable overlap."[29] Tulving had already acknowledged this fact when he observed that they are closely interdependent and interact with one another virtually all the time." The entwined episodic and semantic memories play a closely connected role in shaping our autobiographical memories. Tulving believed episodic memory is a largely faithful record of a person's experiences. However, we have seen that our modern understanding is that it is a construction process rather than reproduction. Recent research has developed an even more nuanced approach in observing that the prior knowledge contained in episodic memories is no longer regarded as reflecting a literal record of events but as the "gist of an episode."[30] Scientists think this is because of the extensive overlap in regions of the brain involved in true and false memories, which means that false memories may also be based on gist rather than detailed recall, which likely reflects that it is far more energy-efficient to store shorter, more compressed versions of the episodes or experiences.

Episodic memory involves constructing detailed images of a past event in a person's memory, and research shows overlapping brain regions are involved in episodic memory or recall of past events and imagining the future. This is probably why context and story help in memory encoding and recall. Weiler and colleagues scanned participants' brains when thinking about past and future Xmas and New Year holidays and concluded that unique neural patterns exist for memory and future thinking.[31] A 2011 study found that memories engage a higher state of autobiographical memory, and mental visual images are clearer and more numerous when we think about the past compared to when we imagine the future. They found that episodic memories involved greater hippocampal and visuospatial activations. At the same time, future planning depends on different brain regions involved in semantic memory retrieval, and our memories appear to be made much richer by the recovery of visual imagery in encoded neural patterns.[32] Recent research suggests that episodic memory and, by implication, self-awareness may have a long evolutionary history. Our belief that only humans possess episodic memory has been challenged by the observation of Timothy Allen and Norbert Fortin that the "fundamental features associated with episodic memory are observed in birds and mammals with anatomical and functional homologs in other species."[33] Most of the research on encoding long-term memory has focused on episodic memory in terms of an organism's experience of the environment. Recent studies involved fMRI scanning of participants required to make memory judgments about photograph sequences depicting past events from their lives or others' lives. The results indicated that different brain regions are involved between autobiographical episodic memories (the medial temporal lobe, which is where the hippocampal formation resides) and autobiographical semantic memories (the hippocampal tail, superior parietal lobule, and intraparietal sulcus). The work of Signy Sheldon and colleagues in 2016 revealed that they "found that self-reported individual differences in how one recalls the past were related to predictable intrinsic connectivity patterns of the medial temporal lobe (MTL) memory system. A pattern of MTL connectivity to posterior (rear) brain regions supporting visual-perceptual processing (occipital/parietal cortices) was related to the endorsement of episodic memory-based remembering (recalling spatio-temporal event information).

In contrast, MTL connectivity to inferior and middle pre-frontal cortical regions was related to the endorsement of semantic memory-based remembering (recalling facts)". These findings suggest that the tendency to engage in episodic autobiographical remembering is associated with accessing and constructing detailed images of a past event in memory, while the tendency to engage in semantic autobiographical remembering is associated with organizing and integrating higher-order conceptual information. More broadly, these findings suggest that differences in how people naturally use memory are instantiated through distinct patterns of MTL functional connectivity.[34, 35] Consistent with this is the research that has shown semantic learning to be slower than episodic learning, which is believed to be because it is mediated by the hippocampus rather than the medial temporal cortex. Over the years, the concept of semantic memory has evolved even more than that of our understanding of episodic memory.[36] It is now recognized that not all semantic memory requires language and is grounded in perception and action.[37]

Memory Loss.

One of the reasons why Alzheimer's is so devastating is that as it progresses, each person experiences different symptoms because of the nature and extent of damage done to specific regions and different lobes. As we have seen, the frontal lobe involves emotion, thinking and movement, motivation and drive, and working and autobiographical memory. It is considered our emotional control centre, home to our personality, and essential to our self. A person suffering damage to the frontal lobe may lose motivation, become tired, lethargic, and even struggle to get out of bed. The frontal lobe regulates our emotions and behaviour to stop us from being rude, threatening, or behaving inappropriately or bizarrely. This damage may result in bizarre and inappropriate behaviour such as swearing, undressing, urinating in public, eating, and drinking non-food items.[38] Given how tragic this can be for loved ones, the traditional view has been that the self is entirely lost when people suffer from frontal lobe dementia. Which overlaps with Alzheimer's, but Alzheimer's is characterized not by damage to the frontal lobe but by the extent to which damage occurs in the other lobes, particularly the temporal lobe, which is involved in sensory processing, particularly hearing and language, smell, and vision, and together with the frontal lobe is very important to episodic memory, emotion, and language. The temporal lobe is involved in acquiring new knowledge and recognizing things or remembering where you might have left something, so damage can make it difficult to acquire or recall new information. However, in Alzheimer's, episodic memory is most affected, not semantic memory, which is why people can remember facts or words and language. Alzheimer's rarely affects the occipital lobe and visual processing of information, but if it is, a person may not recognize familiar objects or faces. If the parietal lobe is damaged, the impact will depend on whether it is the right or left lobe, as most people typically have a dominant left-side parietal lobe. Damage to the left lobe may result in difficulty writing or arithmetic and the inability to tell left from right. Damage to the right parietal lobe, which receives information from the occipital or visual lobe, may make it difficult for people to recognize faces or objects as they suffer from agnosia, which means losing knowledge of the face or object. Sometimes, people may recognize your voice but not you.

Frontal lobe dementia typically manifests earlier in life. It can result in dramatic alterations in their self-concept, physical appearance, knowledge about themselves, relationships, perception of their capacity to do things, and physical appearance. This is largely because of memory's role in forming a stable and coherent sense of self and how memory loss appears to lead to the person's loss. Recent research suggests that despite the very distressing confusion and changes associated with dementia in its various forms, the self is not entirely lost and that the essence of the self remains. This has implications for the treatment of those who have dementia as well as providing insight into the neurobiological basis of self. Dr Muireann Irish, an associate professor of psychology in Australia whose research has highlighted this, describes in an online article in Aeon how, when treating Alzheimer patients, a "vast repository of life experiences, personal history, stories, and fables endure, even late into the illness." As the disease progresses, the individual's concept of themselves becomes more rooted in their past. Distressing family members can observe that they may no longer form part of that person's narrative, but Irish argues that we should recognize that the self is in flux rather than irrevocably lost. She believes we should avoid treating individuals as though they are becoming lost or completely stripping away. She illustrates this with the tragic example of how, when a sufferer may be enquiring where a deceased family member is and is told that person is dead, this sudden news can lead that person to experience grief, as though the news of their loss was being imparted for the first time.

Memory and learning are one of the six major domains involved in human cognition and shaping self; the others are executive function and decision-making, complex attention, language, perceptual-motor control, and social cognition. Professors Tito Kircher and Anthony David, whom we met earlier, talking about qualia, argue that "all information (is) processed on a self-versus nonself continuum or self-valence" and that self-processing is specific to these domains, so it isn't as simple as whether there is a self or not a self. They observe that the basic level of self-processing involves sensory integrative functions but that deficits in some domains of self-processing and specific aspects of memory or types of memory will not necessarily destroy all the information related to the self. What we saw with Clive and Henry is that while damage to the brain areas related to these domains has been destroyed or become dysfunctional, resulting in their case in the catastrophic loss of memory and learning, many core aspects of themselves have been maintained, albeit in a diminished way that makes their lives very challenging, as it does with people with dementia or certain neurological conditions. We have seen how vital our brain's capacity to communicate and integrate information between the brain regions is in processing information concerning self across these domains, and memory, specifically episodic memory, plays a crucial role in linking these domains and the past to the future self. Damage or disease of the brain that impacts connectivity or integration of that information will show up in deficits in some of these domains, though rarely all. Clive, for example, had lost all his explicit/declarative memory and the ability to store new such memories, while Henry lost 11 years of explicit/declarative memory.

Robyn Fivush, who we also met earlier, argues that autobiographical memory is different from episodic memories, even though it involves them, because autobiographical memory has a social-cultural dimension based on the language and narratives involved and relies on episodic memory in three different ways. Firstly, in remembering the self as having experienced an event or episode; secondly, the linking of the episodes; and thirdly, it serves social and emotional functions, including self-definition, self-in-relation, and self-regulation.[39] We have seen how autobiographical memory plays a vital role in self by allowing an individual to remember past episodes and imagine future episodes and events and short- and long-term goals, etc., and is a complex process that integrates an individual's perspective, interpretation, and evaluation across self, others, and time to create a personal history. Remembering the past and contemplating the future has even been hypothesized to involve overlapping psychological and neural processes, with Schacter calling imagining the future "constructive episodic simulation."[40] Fivush and Nelson observe that autobiographical memory "emerges gradually across the preschool years and that the components that contribute to the process of emergence include basic memory abilities, language and narrative, adult memory talk, temporal understanding, and understanding of self and others."[41] Episodic memory and a sense of self develop after the age of 2. Still, unlike the memory of specific events or episodes, autobiographical memory requires a coherent narrative of self and identity, predominantly formed through social interactions and developing through adolescence. Through this process of autobiographical memory, self, and identity are defined and reflected in part by the growing need in social interactions for the maturing individual to regulate emotions, which then become part of that identified self. This is why a child's exposure to interaction with others is such an influencing factor in forming their autobiographical self and developing the CMS.

Trying to understand the role of the hippocampus in memory formation and learning in response to episodic experience has given rise to two theories, the Memory Indexing (MI) and Cognitive Map (CM) theories. The MI theory postulates that engram neurons contain no information but are involved in activating downstream engram ensembles and merely complete a pattern that represents the experience, which subsequent environmental cues can reactivate. The CM theory suggests that the firing activity of engrams is related to the organism's orientation to its location and contextual environmental cues such as visual-tactile and olfactory information and spatial location, which is mapped in the engram ensembles. Memory tests on animals for navigation tasks typically involve rodents rather than primates, and rodents are known to explore their environment much more than visually scan it like primates, further complicating the research. Both theories are recognized as having shortcomings, which we won't discuss here. Still, it is known that reactivation is more likely in environments that resemble the initial environment in which the encoding took place. Scientists now believe that the MI and CM theories may coexist and be reconciled in that they both contribute to constructing memory traces, with the premise for both theories being that an experience to lead to learning must induce a physiological change in our brains. The challenge has been to identify where and how such changes happen in the hippocampus and how they support the enduring physiological changes that constitute long-term memory encoding and storage in the cortex. This is where the engrams we encountered earlier come in because an important feature of engram assemblies is that certain populations in the same network are activated while others aren't. By measuring the level of excitability or activity of neurons before exposure to a novel environment, those with greater intrinsic neuronal excitability are more likely to become recruited as engram cells and be allocated to an engram assembly. This confirms Hebbs's reverberatory theory, which is not confined to the hippocampus but is seen in other areas, including the PFC and the Insula. Scientists had theorized a standard model of memory in which memories were initially created in the hippocampus, transferred over time to the cortex, and disappeared from the hippocampus.

A more recent model called the Multiple Trace Model (MTM) suggests that traces or details of episodic memories are retained in the hippocampus, while outlines are stored in the cortex. Still, both models may need to be revisited, considering recent research that has allowed scientists to harness the physical changes observed in neurons during the interaction of the hippocampus and the cortex while encoding long-term memories. Scientists discovered in the 1980s that when activated through stimulation, neurons express the proto-oncogene Fos,[42] and discoveries of more gene expression related to neural activity demonstrate that physiological changes are induced in neurons when activated above a threshold of stimulation. The next step was identifying the neuron activity during memory recall, which only became possible when they could be genetically tagged to allow scientists to manipulate the neurons during memory retrieval.

A team from the RIKEN-MIT Centre for Neural Circuit Genetics at the Picower Institute for Learning and Memory, led by Nobel Laureate Susumu Tonegawa, did precisely this when in 2012, they developed new technology to allowed them to track the process of consolidation of long-term memory in the cortex of mice brains by labelling engram neurons, which form the engram assemblies, allowing scientists to identify and manipulate these engram circuits, potentially revolutionizing our understanding of memory.[43] They have been engineered using optogenetics to express the red fluorescent protein mCherry, which appears pink, owing to overlap with biocytin signals. This now highly specialized field of engram research seeks to address how the engram structure affects the strength of memories, how the various engram assemblies interact, and the implications of engram silencing in forming memories.

Karl Lashley, the geneticist turned psychologist, had spent 30 unsuccessful years trying to locate an engram and finally described them as elusive, echoing Semon's earlier observation that to examine the neurobiological basis of an engram represented a hopeless undertaking.[44, 45] Modern technology, particularly optogenetics, and the work of Tonegawa, her team, and others validate Hebb's hypothesized cell assemblies, in which reverberatory activity patterns between reciprocally interconnected cells are simultaneously generated during an experience. They have recorded, in real-time, over weeks, neural activity in the engram neural circuits that connect the hippocampus to the cortex while the mice performed a memory-guided virtual-reality task.

We know that information constantly flows back and forth around the brain, with feedback through the various pathways, and the thalamus acts as the control centre, with the inhibitory and excitatory neurons in the thalami central to this. Their technology used cellular-resolution imaging to capture the simultaneous process by which the hippocampus, thalamus, and cortex act to consolidate long memory and identified the anterior (front) thalamus as crucial. Tonegawa's teams were conscious of studies that, since Ryle, in 1949, had distinguished between explicit memory, which he called declarative, and implicit or non-declarative memories. Explicit memory involves the "ability to retain information about explicit facts or events over a significant period and to recall and explain these consciously."[46] It is linked to emotions, encoded unconsciously and triggered by subsequent associations.

Because of potentially important differences between the encoding of explicit and implicit memories like those retained by Clive, their model used a task that only involved explicit memory. This required the mouse to self-initiate and engage in a task that required integrating information and spatial navigation to assess the high task engagement and long-term memory consolidation, with remote memories being recalled relatively easily. Most memory models in animals use a fear-based task because fear conditioning is easy to implement, and memories of fearful events can last a lifetime. They found no correlation between learning and memory in the amygdala or the EH. Still, they did find that while the hippocampus encodes multiple short-term memories, the "anteromedial thalamus preferentially encodes salient memories and gradually increases correlations with the cortex to facilitate tuning and synchronization of cortical ensembles." When they interrupted and inhibited the anterior thalamus during training, the mice could not consolidate or retrieve the memories.[47]

Tonegawa explains that the anterior thalamus acts as a gateway, selecting and stabilizing salient short-term memories for long-term storage in the cortex. This confirms earlier findings of the short-term role of the hippocampus and how the cortex acts as a repository for long-term memories. They concluded that the hippocampus and the cortex are simultaneously forming memories but that the long-term memories "remain silent for about two weeks before reaching a mature state when the cortex assumes control, which is how they become consolidated." It appears to confirm Semon's hypothesis that the engrams constitute the neural substrate of a short and long-term memory trace.

We saw how Tonegawa's team focused on explicit memory because there may be important differences in processing these compared to implicit memories. Implicit memories refer to various forms of memory that support skill and habit learning, perceptual priming, and other forms of behaviour expressed through performance (hence often called procedural memories) rather than involving the recall of explicit facts or events. Implicit memories can be retrieved without conscious awareness or intent and "operate automatically and accumulate information that is not accessible to conscious recollection or involve emotions,[48] for example, remembering how to read, finding your way around your house, or riding a bike involves encoding, storage, and retrieval of underlying motor, visuospatial, or cognitive skills that involve procedures rather than episodes. Implicit memories are encoded differently from episodic memories and rely on the basal ganglia and the cerebellum, which are mainly linked to movement, learning, and processing of emotions. All the senses send signals to the basal ganglia, and it is part of the network that controls voluntary body movements. Both Henry and Clive had retained implicit or procedural memory and, to varying extents, could still perform certain tasks that had become automatic and learn new ones through repetition. Although implicit memories aren't articulated in terms of using language, they can still influence behaviour and our knowledge of different tasks.

Scientists have traditionally focused on the difference between explicit and implicit memory and, less so, the similarities or whether they share encoding resources in the brain. Explicit memory involves conscious reflection and retrieval of previous experiences or knowledge. The approach of Tonegawa's team reflected that it still needs to be clarified whether explicit and implicit memory uses different memory mechanisms in terms of encoding and consolidation or involves the same single memory system. Still, there is growing evidence to suggest that they use different systems. Many studies in which patients have experienced brain damage or disease show that damage to some regions only impairs implicit memory, while damage to others will impair explicit memory. Other neuroimaging studies in healthy patients support this observation. Still, the extent of their independence in terms of encoding and storing remains needs to be clarified, as well as whether they are independent or share some components. This is highlighted by a 2006 study exploring this issue, concluding that "implicit and explicit memory is subject to the same encoding factors and can rely on similar perceptual processes and representations."[49] The reality is that we don't yet know.

The information must typically be distinctive, salient, relevant, and typically associated with an emotion, be it pleasant or unpleasant, for it to be encoded and consolidated in long-term memory. We saw how the Self-Referent Effect means that the processing of information about the self appears to be more strongly encoded[50] and takes place in the CMS, which integrates information across the various psychological domains. This involves processing across personality, social behaviour, emotional control, formulating and processing thoughts related to self and judging others, our levels of motivation and drive, and our capacity to adapt behaviour to different situations. Our memories are neural patterns that store sensory information about our body states, emotions, and thinking related to past experiences of events and situations. Until recently, infants were assumed to be unable to recall and had childhood amnesia. However, recent studies using non-verbal tests like imitation indicate that recall developments begin towards the end of the first year of a child's life, and major changes occur during infancy. Studies demonstrate that children have few memories of events during the first 3 1/2 years of life, but this gradually increases from age 3 to age 7. Between seven to 10 years old, they have a robust memory of past events. Electrophysiological and behavioural tests evidence more effective and efficient encoding, consolidation, and storage processes resulting from developments in the temporal-cortical networks in children's brains during this period that support recall of episodic and autobiographical memory.[51] Explanation of this includes the fact that memories are not related to self. It may be less well stored, or the hippocampus continues to form up to 7; therefore, young children may need more processes for forming rather than maintaining self-related memories. Another possible reason is that only when they can start to verbalize internally and externally and use pronouns such as I, me, you, he, and she,., do they store and recall self-related memories. One compelling explanation is that memory only develops with a child's narrative skills, particularly about socialization, usually within the family. This is because narratives give coherence, emotional resonance, and a level of chronology, making the memory easier to remember, with memory adapted to retain information most likely required in a person's environment.

When combined with the nature of attentional biases, we discussed earlier; we can start to see how and what we attend to, and the greater encoding of self-referential processing, have a vital impact on the long-term storage of memories of experiences related to ourselves. This includes elaborative encoding, which relates new information to prior knowledge in the existing neural patterns in the cortex. Specifically, those areas activated when the information was first encountered. We are always scanning for irregularities in our environment and identifying what is familiar or what we recognize as having experienced before. By doing this, we accumulate knowledge and experience in making predictions and responding or behaving using informed decisions. Working memory and the encoding of long-term memory via the hippocampus consolidate and store these new experiences for future retrieval.

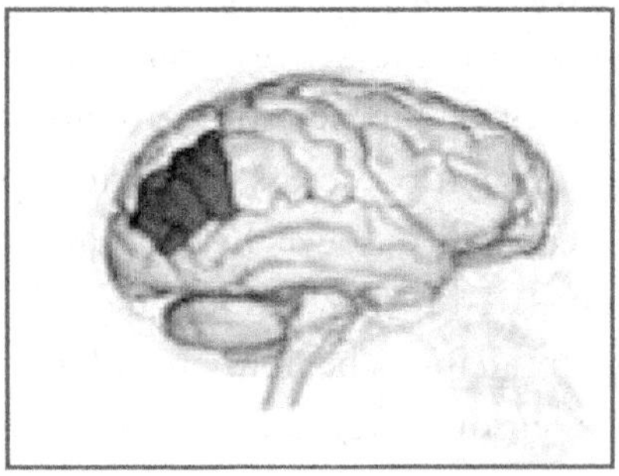

The Angular Gyrus

Though missing in most primates, the AG is highly developed in humans and connects largely to the associative areas of the cortex and, through connections with the frontal cortex, is associated with spatial awareness, language processing, attention, social cognition, conflict resolution, number processing, and memory retrieval. This is why damage to the AG can lead to dyslexia, difficulty with mathematics, difficulty telling your fingers apart, and difficulty understanding the difference between right and left. Also involved are the various associative areas in the cortex where most sensory information in the brain is integrated. Some models suggest that the hippocampus and the VMPC compete, while others suggest that they synchronize activity to create the schemas.Let's now look at the role of memory in our robots.

Robot Memory

Computer Memory

We have seen how, in humans, working memory is crucial for attention, reasoning, and learning, and most, if not all, AI systems incorporate the encoding, manipulation, and maintenance of information in the form of an artificial working memory equivalent. This is indicative of a broader challenge of building subsystems into AI that handle information processes in the modular way that our brains do to optimize the efficient use of resources that the system uses. Computer scientist Andrew Coward describes how natural selection has led to the brain processing information through modular and hierarchical systems with each module having sub-modules, and on this basis, it is possible to describe the "memory phenomenon at a high (psychological) level in terms of the information processes performed by the major anatomical structures," such as the pre-frontal cortex, the hippocampus and the cortex On this basis Coward claims that "the information processes and resultant hierarchy of descriptions make it possible to understand cognitive phenomena like episodic, semantic or working memory in terms of neuron processes via a series of intermediate levels of description" which in turn allow the description to be causal.[52] In this sense, we can link the causal connection between sensory memory through working memory into short-term and long-term memory based on various anatomical structures in the human brain.

We discussed how most standard computer hardware uses what is known as the von Neumann or Princeton architecture, which separates the computing functions in the central processing unit (CPU) and Memory Units with memory functions that store program instructions and data. However, these are largely serial, whereas our brains use massively parallel computing. John von Neumann recognized that it was impossible to build infinitely large memory storage and built the architecture on a hierarchical storage basis using the 90/10-locality rule, which states that 90% of the computer execution time is spent on about 10% of the data. The locality rule is central to how computers work and their memory capacity and observes that all computer processes repeatedly use subsets of their resources over extended time intervals and do so in phases or bursts. The use of computer resources refers to grouping their use in time and grouping them based on how contiguous or close they are in space or location. By accommodating the locality rule, it avoids thrashing, which happens when there is insufficient memory, which leads a computer's memory resources to become overused and degrades performance. Thrashing typically happens when the hard drive excessively transfers information between the system's and virtual memory. Virtual Memory (VM) is a memory management technique that uses hardware and software to provide more free-space memory. It does this by allocating temporary hard disk space to memory, so if the system is low on memory, the VM combines the Random Access Memory (RAM) with this temporary space. RAM is physical hardware in the computer that temporarily stores data and serves as the computer's working memory. It can significantly boost the computer's ability to process information by loading the data quickly rather than from the hard drive. When RAM runs low, VM helps move data from RAM to a paging file space, which frees up the RAM and allows it to be read much more quickly from RAM than from the hard disk. Ultimately, the von Neumann bottleneck presents a major problem for delivering the level of computer power needed and for the autonomy of robots, which require the type of memory processes humans use to be adaptable in the real world. This is why some scientists believe that the future of computers and AI relies on developing quantum computing and neuromorphic systems, with the latter probably closest to commercial application.

Computers employ a systems Bus, which is the shared digital pathway between the various components. The front or local Bus is the path from the CPU to memory split into an address bus with address lines to signal a memory location, and data is transferred over data lines to that location. They also use several peripheral Buses for connection to peripheral devices. This wastes energy and time in conveying information between these and creates a limit on the throughput called the von Neumann bottleneck. To overcome this problem, systems have employed various approaches, for example, adding caches, which are auxiliary memories to the CPU that speed up the processing. Other approaches to overcome the bottleneck include prefetching, which prioritizes data and instructions. The contents of RAM are always erased when the computer is turned off, although by putting the computer into hibernation, the information on the RAM is transferred to the hard drive and then back when turned on. Many computers can use Read Only Memory (ROM), which retains their data when power is switched off and can be altered, such as flash drives and solid-state drives. RAM technology is improving all the time, for example, using resistive RAM, magnetic RAM, ferroelectric RAM, and spin-transfer torque RAM.

Another approach is Systems on a Chip (SoC), which puts processing, memory, and other system resources on one chip and, to date, has been used in mobile devices but is becoming available in computers. AI cognitive architectures seek to develop the fast and persistent memory capability to allow AI to use memory as a vital part of adaptability and autonomy. As we shall see, the prospects of quantum computing and neuromorphic systems for robots hold considerable promise.

Cognitive Architectures and Memory.

Autonomous robotic systems must employ memory systems to learn and adapt to the environment. The Atkinson and Shiffrin's model and the modified Baddeley and Graham Hitch models we discussed earlier form the basis of most AI systems. The greater the level of abstraction the system uses for planning action and the greater the level of autonomy sought, the greater the memory requirements will be. The engineering requirements and the biological plausibility of the memory system employed constrain most cognitive architectures. However, architectures incorporate many types of memory and different periods over which it can be stored, i.e., short-term, long-term, etc. Hierarchical AI systems are designed for planning and achieving goals and problem-solving and typically split the goal into sub-goals or chunks, which must also be stored as memory fragments.[53] Let's look at the different types of memory the various cognitive architectures attempt to capture.

Sensory Memory.

When dealing with artificial sensory motor skills and multi-modal sensory memory, AI systems need humans to capture the multiple sensory data and pre-process it in sensory memory before transferring it to the available memory structures of varying duration. To do this, it needs to maintain the stimulus long enough, which may involve several techniques. The biological sensory system used by humans involves analysing multiple sensory cues, making the perception of the environment more reliable and accurate, which includes the various sensory stimuli such as visual, audio, touch (haptic), and nociception, which deals with pain or toxic and harmful stimuli. Haptic memory is employed when we handle and grasp familiar objects. Visual memory is prone to decay after about two seconds, and haptic cues and memory also affect our visual experience.[54, 55] In audio stimulus, echoic memory enables the memory to last long enough, so features and binding depend on the stimulus's properties, including the acoustic and phonemic levels. The latter refers to how sounds are represented in the mind, how sounds are combined to form words, and how these words are used in a language. The challenge in computer memory has been to incorporate the sensor and memory components in one electrical device to mimic the biological sensory neurons, with the most studied in AI being visual sensory memory. In humans, there is an interval of a few tens to two or three hundred milliseconds before visual stimuli can be measured in terms of conscious awareness or generating observable behaviour, during which the brain processes it in multilevel visual pathways. Iconic memory in humans recognizes that with visual stimuli, a faithful representation of what was seen remains perceptually available and persists for several hundred milliseconds after the stimulus has decayed and disappeared. Hence, the terms visually persistent or iconic are used and, as Vincent Di Lollo observed, can involve combining and integrating "two or more sequential displays even if separated by a brief temporal gap, although the duration of the "visual persistence is negatively related to the duration of the inducing stimulus."[56] Iconic memory has explained how some images and patterns can be retained even when there is visual masking, which happens if one image created by different visual stimuli masks the visibility of another image completely or partially. Much research has gone into developing artificial sensory memory with iconic memory, also called visual or photo memories. Still, it faces considerable challenges in creating the material substrate that can simultaneously combine the sensory and memory components by detecting and storing light information. It is one of the most vital but challenging AI sensor systems because it requires merging photoreceptors with non-volatile memory storage.[57]

Several cognitive architectures, mainly hybrid types, use sensory memory, including Soar, ACT-R, and CLARION. Because of the vital role of visual and touch sensory information, robotic researchers have recently developed a bimodal artificial sensory neuron to implement the fusion of the visual and haptic processes in a way that more precisely and accurately depicts the environment using sensory cues and sensory memory.[58] Artificial sensory neurons have been developed for visual and haptic stimuli to emulate human sensory and motor functions using photodetectors and pressure sensors in neurologically inspired robotics. This is combined with bioinspired soft electronics in robots, with materials designed to provide "stretchability, mechanical toughness, adhesion, and structural colouration" and even material that can self-repair or heal.[59] Chinese researchers have developed a "bimodal artificial sensory neuron (BASE) based on ionic/electronic hybrid neuromorphic electronics to implement the visual-haptic fusion. The BASE consists of core components, which are a resistive pressure sensor, perovskite-based photodetector, hydrogel-based ionic cable, and a synaptic transistor, and the developers claim that this technology has the "potential to advance technologies in cyborg and neuromorphic systems by endowing them with supramodal perceptual capabilities."[60] Robotics research is also rapidly developing multi-modal sensory capabilities that can capture the sensory memory features necessary for robots to pre-process the sensory stimuli before transferring and encoding the information in longer-duration memory storage.

Working Memory.

A recent AI research project by David Freedman and Nicolas Masse from Chicago University who focus on visual learning, recognition and decision-making, demonstrated that working memory uses neural networks in different ways depending on the task's complexity and that short-term memory comprises many different processes. What they discovered is that the brain stores short-term memories in neurons that remain silent whereas other neural circuits are constantly active or persistent which correlates with the complexity of the task. In the AI model, they observed that this persistent activity continues after the input has ceased in what they call a reverberation or echo and is evidence of the need to manipulate the information in some way. The silent networks evidence a short-term plasticity to remember information temporarily and both persistent and silent processes can last for several seconds up to a few minutes. Masse described it as "like writing something with your finger on a fogged-up mirror instead of writing it with a permanent marker."[61]

To capture the more realistic biological role of working memory in AI requires potentially modelling not just sensory memory but the limited storage capacity of human working memory, the decay of that information over relatively short time-periods and the rapid updating of the information to reflect the change, as well as the causal linking of it to long-term memory and decision-making and task performance and goal planning. Many AI systems use an activation mechanism which accesses and processes information from long-term memory in terms of the knowledge relevant to a goal or task, with the selection based on the bias given to the information. These typically also model limited storage capacity, decay, and rapid updating. Other architectures use what is known as the blackboard approach to memory which allows modules running in parallel to access memory from a shared short-term storage system, which is known to be effective in complex problem-solving. Neither the activation mechanism nor the blackboard approaches are biologically plausible, but a few systems more closely resemble the PFC working memory mechanism that humans employ. Unlike humans there are in practice few limitations on the potential working memory capacity in AI, but any autonomous robot requires new goals or new sensory data to replace existing content if it is to be sufficiently adaptive to change in the environment. There is currently little consensus amongst AI architect designers as to the optimal way to replicate the short-term capacity that characterises human working memory, as we will see by looking at the following examples of different approaches. One architecture called GLAIR wipes the content when it switches tasks, whilst others like CELTS reduce the content to reflect the degree of relevance and emotional valence, with content that ceases to pass the threshold of relevance or saliency being removed. An AGI system called The Novamente Cognitive Engine (NCE) which is now known as Open Cog is grounded in a systems theory of intelligence. Open Cog uses the concept of short-term and long-term importance values associated with "Atoms." They describe this as a simulated economy approach in which the currencies used reflect short-term importance (STI Currency) and long-term importance (LTI Currency). The Atoms continue to be stored in memory provided they maintain utility and are linked to other memory functions.[62] It is not clear though how in the absence of imposing a hard limit like some cognitive architectures, for example in the number of seconds it is retained or the amount of information in terms of items, chunks, or objects that it will be able to prevent the working memory exceeding capacity like human working memory does automatically.

An approach that is growing in popularity which seeks to avoid setting external parameters is known as recommendation architecture or recommender systems, which we see all the time now because they are employed by Netflix, Amazon and many others to filter our previous selections to make recommendations based on our historical preferences. They work in two different ways which are, by using metadata to identify what people watch after a particular show or employing a user matrix to assign a likelihood that they will like something based on historical data so that it is memory-based and emerges from the memory.

Long-Term Memory.

Let's look at long-term memory in AI, with the main observation being that cognitive architectures rarely distinguish between explicit and implicit memories. We have already discussed one of the exceptions which is CLARION. You will remember that CLARION purports to capture the "essential structures, mechanisms and processes of the mind" and "represent psychological theories in a computational form". Ron Sun explains that the minimum requirements for a system to do this are. (1) Procedural or implicit processes that deal with actions and skills like driving a car (2) Declarative processes for general knowledge and reasoning, in terms of semantic and episodic memory (3) Metacognitive processes for monitoring and regulating other processes and (4) Motivational processes that provide the impetus for action and cognition. Sun believes that explicit knowledge is best captured by symbolic localist representations which equate to the conscious representation and manipulation of the information in short-term working memory and implicit by distributed connectionist representations. The obvious parallels with the human brain lie in the apparent ability of CLARION to achieve integration and connectivity of these various features, whilst distinguishing between implicit and explicit memory, combined with metacognition and the essential feature of motivation and drive.

We also saw Global Workspace Dynamics (GWD) theory with a family of related GW theories, which views consciousness as arising from highly integrated and widespread cortico-thalamic (C-T) activity supported by both prefrontal and posterior regions of the brain. Neural information rises into consciousness when it appears in the GW and is broadcast through our brain. Stanislas Dehaene's neuronal avalanche concept extends the GW theory to general consciousness to explain how, out of all the competing information, what gets selected to be broadcast throughout the cortex and gets subconsciously processed. This reflects how neural networks in distinct brain regions integrate information over space and time and converge onto a single coherent interpretation. Activation of long-term memories can be seen as arising from the integration and connectivity of encoded information whilst distinguishing between implicit and explicit memory, combined with meta-cognition and the essential feature of motivation and drive, which converge onto a single coherent interpretation or memory.

All AI requires term memory to operate the system even if they don't distinguish between short and long-term memory systems and they need procedural memory and semantic memory in terms of certain knowledge to perform tasks. Different architectures use different systems which include a set of if-then rules and pre-programmed procedures in symbolic reasoning, sensory-motor or task schemas or behavioural scripts. One of the major challenges with AI though is providing systems with a sufficiently long-term memory rather than short-term capacity. Many RNNs use past information to improve performance on current and future inputs, with the most common model being the Long Short-Term Memory (LSTM) model that we encountered earlier. They are dynamic classifiers in that they are not confined to feed-forward processes but effectively look back in time in units measured as time steps which resemble a form of AI memory. Behavioural scripts provide the AI system with instructions or orders it will follow to complete an action or task. Every script contains key elements which are, a conditional which establishes the condition for the action to be performed, the action itself, a target which the action will impact, and the target priority in terms of selecting the targets based on the conditions. It emerged from the study of ecological robots that we discussed earlier, and which cooperate and compete when adapting to the environment.

Schemas in Robots.

Neural schemas have a potentially important role in supporting increasingly sophisticated AI software to support complex behaviours in autonomous robots. Sensory motor or task and behavioural schemas have recently been developed for application in both virtual agents and real-world robotics. This is designed in a hierarchical fashion that is layered and integrated with the processing by the ANN. We have seen how schemas in the human brain represent functional units that interact because when activated the information stored in the schema is used in the process of perception, attributing meaning to the incoming information and adjusting to accommodate new information. The challenge is that they involve a top-down and bottom-up interaction and consist of a wide range of types of schemas which include in addition to general long-term memory schemas, perceptual schemas dealing with sensory input or perceptions, motor schemas, dealing with motor action and sensorimotor schemas which interact and integrate visual and tactile sensory input and motor action. In autonomous robots' architecture design has involved separately developed types of schemas which are perceptual and sensorimotor schemas for identifying objects or obstacles and motor schemas for moving around and navigation that need to be integrated. The behavioural output is driven by the input of visual and tactile sensory information and the activation of the different schemas. The more active schemas represent the levels of confidence, and the architecture confers the ability to incorporate adaptation and learning through neural network processes to develop new behavioural architecture. Several architectures have been built for different schemas or neural maps, for example, VI-SIONS, specific Robot Schemas (RS), Neural Schema Language.

Endnotes

1 Molenberghs, P., 2013.
2 Brem, S, Maurer, U, Kronbichler, M., et al., 2020.
3 Ebbinghaus, H., 1885.
4 Driskell, J E, Willis, RP, Copper, C., 1992.
5 Bartlett, F., 1932.
6 Schacter, DL, Addis, DR., 2007.
7 APA Dictionary of Psychology.
8 Wixted, J T, Wells, GL., 2017.
9 Simons J S, Garrison JR., Marcia K., 2017.
10 Lagioia, A., et al., 2011.
11 Bárbara, G., et al., 2016.
12 Atkinson, R, Shiffrin, R., 1968.
13 Malmberg, KJ., et al., 2019.
14 Tripathy, SP, Öğmen, H., 2018.
15 Baddeley, A D, Hitch, G., 1974.
16 Baddeley, A D., 2000.
17 Grady, CL, et al., 1998.
18 Shallice, T, Warrington, E K., 1974.
19 Robbins, T W., et al., 1996.
20 Craik, F I, Watkins M J., 1973.
21 Chai, WJ, Abd Hamid, AI, Abdullah, JM., 2018.
22 Eichenbaum, H., et al., 2007.
23 deToledo-Morrell, L., et al., 2004.
24 Schacter, DL, et al., 2007.
25 Addis, DR, Schacter, DL., 2011.
26 Fumagalli, M, Priori, A., 2012.
27 Seung, H., 2000.
28 Fioravante D, Regehr WG., 2011.
29 Renoult, L, et al., 2019.
30 Cheng, C., et al., 2016.
31 Weiler, JA., et al., 2010.
32 Viard, A, et al., 2011.
33 Allen, TA, Fortin NJ., 2013.
34 Sheldon, S., et al., 2016.
35 Dafni-Merom, A, Arzy, S., 2020.
36 Horner, A., Bisby, J., Bush, D. et al., 2015.

37 Renoult, L, et al., 2019.
38 Silveri, MC., 2007.
39 Fivush, R., 2011.
40 Schacter DL, Addis DR. 2007.
41 Nelson, K, Fivush,R., 2004.
42 Curran, T., and Morgan, J. I. 1985.
43 Liu, X, et al., 2012.
44 Lashley KS, 1931.
45 Semon, R W, Simon, L., 1921.
46 APA dictionary of Psychology.
47 Kandel, ER, Dudai, Y, Mayford, MR., 2014.
48 APA dictionary of Psychology.
49 Nicholas, B., et al., 2006.
50 Rogers, TB., et al., 1977.
51 Bauer, PJ., 2006.
52 Coward, L.A. (2011.
53 Perner, A, Zeilinger, H., 2011.
54 Bliss, J, Crane H D, Manfield, PK, Townsend, J T., 1966.
55 Kelly, JW., et al., 2011.
56 Di Lollo, V., 1970.
57 Guan, X., et al., 2020.
58 Wan, C., et al., 2020.
59 Qi, M., et al., 2023.
60 Wan, C.et al., 2019.
61 Masse, N, Freedman, D., et al., 2019.
62 Goertzel B et al., 2010.

"The disruptive potential of quantum technology will make the change of the Internet era look like a small bump in the road!"

Kevin Coleman Author

Chapter 11

Quantum Robots

CHAPTER 11

Quantum Computing in Robots.

THE AMERICAN PROFESSOR OF APPLIED mathematics at MIT Peter Shor (1959–) famously observed that "Quantum computers will be able to solve problems that are intractable for classical computers," while the CEO of Google, Sundar Pichai (1972–), more recently claimed that "A.I. could accelerate quantum computing and that quantum computing could accelerate A.I." Physicists debate at what size the microscopic world transitions into the macroscopic one. Physicist Jonathan Friedman at Amherst College in Massachusetts observes that quantifying that change is "a thorny issue" and "you'll get a different answer from just about everyone you talk to." Physicist Simon Gröblacher at Delft University in the Netherlands is one of an increasing number of scientists who "think that there is no hard boundary where everything all of a sudden changes and quantum mechanics becomes incorrect."[1] Scientists are already aware of many quantum effects in biological systems which as demonstrated by Huelga and Plenio in their 2014 review includes a diverse range of quantum phenomena such as "coherent excitations of biomolecules, quantum tunneling of protons, van der Waals forces, ultrafast dynamics through conical intersections, photon-assisted electron tunneling as the basis for smell, entangled states in photosynthesis, mechanisms for magnetoreception in the avian compass."[2] The legendary American Physicist Richard Feynman (1918–1988) once said that "Nature isn't classical, dammit, and if you want to make a simulation of nature, you'd better make it quantum mechanical." We encounter quantum effects in our everyday lives in our smartphones, which contain billions of transistors that rely on the quantum mechanical properties of semiconductors to control currents and voltages. Quantum rules have been observed to apply at larger and larger scales in clouds of atoms, quantum drums, wires, and etched silicon chips.

Much research is being focused on quantum computing, quantum sensors, quantum communication, and the necessary quantum algorithms, and there is now a dedicated and emerging field of Quantum Robotics seeking to combine the principles of quantum computing with robotics. The concept of quantum robots has been around since the late 1980s when the American physicist and pioneer in quantum computing, Paul Benioff (1930–1922), published his paper, A microscopic quantum mechanical Hamiltonian model of computers as represented by Turing machines.[3] The Turing machine refers to Alan Turing's 1936 mathematical model for the universal computer, which was defined long before the invention of computers. Benioff's quantum robot was unaware of its environment and could not make decisions or carry out many tasks, so it was primitive. In the last ten years, quantum computing has evolved rapidly. By using the principles of quantum mechanics to encode and manipulate information, quantum computers can perform calculations far faster than classical computers. Although this does not mean that there is no ultimate limit on computational speeds because regardless of technology improvements, this will always be constrained by the fundamental laws of physics such as size or mass, entropy, and thermodynamics, quantum computational power still has the potential to revolutionize life as we know it.

The quantum wars have begun in earnest with the drive towards the holy grail of what, in 1982, Richard Feynman called "Quantum Supremacy."[4] This has been described as the experimental demonstration of a quantum computer's dominance and advantage over classical computers by performing calculations previously impossible at unmatched speeds. Over the last few years, in pursuit of this goal, we have seen significant worldwide investments in technology development, such as the European FET Flagship on Quantum Technologies, which has invested €1 billion into quantum research over the next ten years and is aiming to build a network of 5,000 quantum-related researchers across the E.U. Set up in 2001, the National Laboratory for Quantum Information Sciences in China is now the global leader in quantum communication. In 2018, the U.S. Congress passed the National Quantum Initiative Act to "accelerate quantum research and development for the economic and national security of the United States," and the U.S. federal budget on quantum technology doubled 2019 from over $400 million in 2019 to an estimated $900 million in 2022. Japan's government established its Quantum Technology and Innovation Strategy in 2020 and eight new R&D centres, and in 2022, it set out its Vision of Quantum Future Society. In 2021, China, through Origin Computing, became the third country after Canada and the United States to develop a complete domestically built quantum computer. In May 2023, China's 176-qubit processor, Zuchongzhi, became the quantum processor with the largest qubits and millions of times faster than traditional supercomputers. However, IBM's 433-qubit processor revealed in 2022 is the fastest quantum processor, and their declared aim is to have a quantum computer with more than 4,000 qubits by 2025. Quantum Information Processing (QIP) can provide higher computational speeds for certain problems, such as searching unsorted databases, quantum simulation, solving linear equations, and speeding up signal and data processing. ML has widespread applications in AI and potentially certain fields in robotics.

However, the challenge must not be underestimated, and as Microsoft's quantum corporate vice president Zulfi Alam observes, "Building a quantum computer is like sending someone to the moon or adventuring to Mars. It has the same level of complexity or more and requires a team of experts all working very closely together, where the mission is much greater than the individual parts, which may, in part, explain why quantum computing remains in its infancy. Quantum technology has the potential to exponentially increase the speed at which robots learn and make decision-making with powerful parallel quantum computing, fast searching capabilities, and efficient learning using quantum algorithms.

To understand why quantum computing is so potentially powerful with important applications to A.I. and robots, we need to look at the basic physics underlying the quantum process, which we will keep as simple as possible. Quantum theory reflects our best understanding of the way nature works and is the most precise way we have discovered of conducting any experiment to obtain information about a physical system. We should start with two observations made by American Professor of Electrical Engineering and Computer Science Dan C. Marinescu which are that "Quantum information is information encoded to some property of quantum particles and obeys the laws of quantum mechanics." and "the amount of information that can be stored or transmitted by a physical system is related to the number of states the system can find itself in."[5] Particles are required to transport any information from one location to another in our universe and Newtonian classical mechanics shows us how the state of physical systems evolves and changes over time from one state or configuration to another as they move along a particular evolutionary path or trajectory. The position or momentum of an object at the classical level is always well-defined, but this also requires an understanding of quantum states underlying that classical physical system or objects. The quantum state of an atom describes the energy, motion, magnetic field, and other physical properties that govern its behaviour.

A question plaguing physicists for many years was why a physical system like a particle or object chooses a particular path in space from all the possible paths it could follow. Quantum mechanics sought to answer this by describing how a system, given the dynamic operator of time, changes over some time by relating changes in the system's quantum state to the energy, motion, magnetic field, and other physical properties in the system like a particle. Physicists know the energy state of a system as the Hamiltonian or the sum of the kinetic (motion) energy and the potential (position) energy of a physical system. In classical mechanics, the Lagrangian function characterizes the state of a physical system by expressing it as a function of positions and momentum (velocities), which are vector quantities. These represent the total kinetic energy of the system, which equals the sum (Σ) of the kinetic energies of all the particles and all the potential energy (V) of the system. The kinetic (energy of motion) plus the potential energy (energy of position) gives us the precise energy state or Hamiltonian of a physical system at any given time. The theory applies to all physical systems, including the movement and direction of travel of subatomic particles such as electrons and photons. Nature has designed physical systems such that when changing states, the values of the Lagrangian function for all the points along each imaginable path are summed up, and the path with the smallest energy required, which is the shortest path, is selected. In that sense, light or photons travelling at the fastest speed cannot take a shorter time or follow a shorter path, which is why the path it takes must equate to the passage of time.

Quantum mechanics governs the outcome when we observe or measure the state of a physical system at the quantum level and of necessity has to reflect the quantum wave characteristics of the subatomic particles in terms of the superposition of different and discrete energy states and the information or energy of the system contained therein. This arises from Schrodinger's equation, which relates the system's total energy or change of the physical system, such as particles, to their wave function nature. Due to Heisenberg's uncertainty principle, we can't know the particle's motion and position as we can at the classical Newtonian level with objects. Quantum theory tells us that the act of measurement of the wave function of a particle gives us all the knowable information of its quantum or Hamiltonian state as a physical system. In terms of its dynamical aspects, this includes the energy and motion of a quantum particle like an electron at a certain time (t). Quantum theory involves scientists rather arbitrarily splitting the universe into systems that at the particle level, are represented by particles possessing a wave function representing all the probabilities of its location at any moment. Making a measurement collapses the quantum system (wave function) to a point in space or what mathematicians call the eigenstate (or eigenvector), which is a measured observable such as its Hamiltonian and obtaining the corresponding eigenvalue which determines the measurable, observable vector quantity in real numbers. You will remember a vector is an object with both magnitude and direction, so it is the quantum state or vector energy of the vector space in a system.

In quantum mechanics, the mathematical notation for the quantum state of a physical system at (t) is the inner product (or dot product) of two vector states denoted by a bra /ket $|\Phi|\Psi|$, consisting of a left part$|\Phi|$ called the bra and a right part $|\Psi|$ called the ket. Remember that we can only obtain this vector information after measuring the system and collapsing the particle's wave function. Greek letters are typically used in mathematics to represent quantum states generically; for example, the Greek letter Psi ψ represents the wave function of an electron ket as $|\psi|$. Simply describing a particle like an electron as a quantum wave state $|\psi|$ does not convey precise enough information about the particle's energy, which requires us to use vector calculations to give real number outputs. Remember, the dot product takes two equal-length sequences of numbers, typically called coordinate vectors, and returns a single number. The generalization of the dot product to an arbitrary vector space is simply a way of combining two vectors to get a number or numerical value as information. A vector space is a group of collectively added and multiplied ("scaled") numbers called scalars (discussed earlier). The quantum state or vector energy of the vector space in a system, in this case, a quantum particle, can be obtained from the numbers derived from this vector information in the bra/ket notation, which might be a column or row vector. If the angle bracket is pointing left, like < a|, then it's a bra, so a row vector, but if the angle bracket is pointing right, like |a>, then it's a ket, so a column vector. You will remember that a vector is a list of numbers (can be in a row or column), while a matrix is an array of numbers (one or more rows, one or more columns).

$$\begin{bmatrix} 1 & 8 & 13 & 12 \\ 14 & 11 & 2 & 7 \\ 4 & 5 & 16 & 9 \\ 15 & 10 & 3 & 6 \end{bmatrix}$$

For example, in the above diagram, each column is a vector with 4 components. Quantum mechanics mathematically ascribes vector values or numbers to the wave function of a particle or its quantum state by calculating the probability rather than the 100% certainty of the particle being at a given point of space and time because the wave function only contains probabilities and we can't know the precise location until it is measured. The laws of quantum physics mean that it is impossible to predict the outcome with 100% confidence as only the probability of obtaining each possible outcome can be calculated. The quantum calculation employs the widely used rule formulated in 1926 by one of the quantum pioneers, German physicist Max Born (1882–1970), known as the Born rule, which conceptualizes the wave function of particles such as electrons as the probability density, when measured of finding that electron. For example, in any given quantum state, as proportional to the square of the amplitude of the system's wave-function at that state. The wave amplitude refers to the maximum displacement of a particle on the medium from its rest position and the energy of a wave or particle is directly related to the square of its wave's amplitude, which is a fundamental principle of wave mechanics applicable to all waves. Born's rule assigns an amplitude to each measurement outcome with the probability equal to the amplitude squared. Hence, we obtain vector data based on the probabilistic nature of the wave function.

We operate in a 3-dimensional world, and trying to visualize more than three dimensions, like the X, Z, and Y axis vector coordinates in the diagram below, is almost impossible. This is why it is often easier to regard the vector rather than just a position in space as describing vectors as quantities. In physics, it is necessary to be able to compute not just in three or four dimensions if you include time, but also in many higher dimensions. For example, a rigid object in space may be located using the object's x-coordinate, y-coordinate,z-coordinate, and vectors. Still, that object may rotate in three directions, such as the roll, pitch, and yaw, so you would need six dimensions and more vector information to be precise. The level of vector information gets increasingly complex with the number of vectors involved at different dimensions, which makes plotting high-dimensional data from which to extract information problematic.

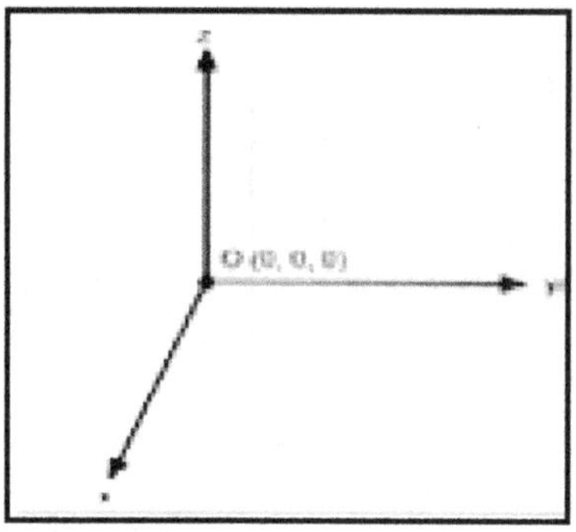

3-Dimensions.

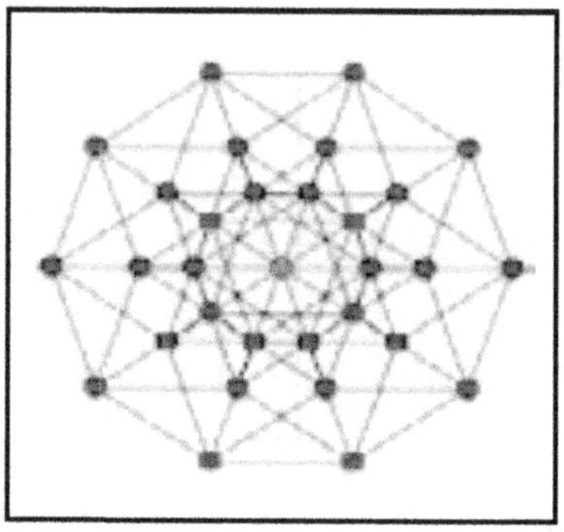

6-Dimensional Vector Coordinates.

To realistically describe a physical system and wave functions using a mathematical model in vector space or even as vector quantities, we need to be able to deal with much more than single-figure high dimensions while trying to avoid them becoming unmanageably large. We encounter high-dimensional vectors in neuroscience, for example, when we use 100 electrodes in an EEG to record the activity (information) of 100 neurons because we need 100-dimensional vectors to describe the neuronal activity at any time. Over an extended period, say ten minutes or 600 seconds, sampling neural activity 1000 times per second requires 1000×600=1000×600 = 600,000 samples. Multiplying that by the 100 neurons, we need a 60,000,000-dimensional vector to represent the neural activity over ten minutes.

A remarkable property of quantum systems like wave states is that until they are measured and the wave of probabilities collapses (or decoheres), they can exist in superposition, meaning they can be simultaneously in multiple states. Wave states can interfere with each other like all waves. Because a probability amplitude describes each state, constructive interference will enhance the amplitude while destructive interference cancels out the amplitude. It is commonly believed that because these probabilities arise from an intrinsic pure state and are prescribed by physical law, they are an objective view of reality in terms of the behaviour and energy of a system based on the particle's wave characteristics. By contrast, Bayesian probability is a mathematical theorem employed by our brains when we obtain new information from our environment and update our beliefs about the probability of something being true or that it will happen. It gives us only a subjective view of the world. For reasons we won't go into here, this interpretation of the objective and subjective worldview is disputed by some, including Carlton Caves and colleagues in their 2001 paper, Quantum probabilities as Bayesian probabilities[6] so we use the distinction subject to such caveats.

Unlike the bit used in classical computing, which can be either 0 or 1, quantum computing uses Quantum Bits (Qubits) as the basic unit of stored information in the system, which can be made from a range of particles, including trapped ions, photons, artificial or real atoms, or quasiparticles, that possess characteristics such as charge, polarization, and spin. Using qubits means that you would, in theory, need an infinite number of classical bits to describe the state of a single qubit because they are characterized by continuous numbers, which means that a single qubit can encode an arbitrarily large amount of information, including for example, the spin of the particle. Spin refers to the fact that certain particles possess an intrinsic angular momentum and behave like they are spinning, although they don't physically spin. Spin represents information about the state of the physical system when measured. The Stern–Gerlach experiment performed in 1920 by Otto Stern and Walter Gerlach used silver atoms passed through a magnetic field to first demonstrate the quantization of spin when they discovered that electrons have their individual, continuous spin as they orbit around the nucleus of an atom. As the silver particles passed through the Stern–Gerlach apparatus, they were measured as equally distributed among two possible values, with the half ending being measured up at an upper spot ("spin up") and the other half at the lower spot ("spin down"). This told Stern and Gerlach that they could only appear as quanta which are the smallest discrete unit of the phenomenon of spin, and because a magnetic field was deflecting the particles, they realized that spin is a magnetic property associated with the intrinsic nature of angular momentum. In physics, quantum refers to a single packet of matter or energy. Quantization of spin restricts the number of possible values of spin as a quantity or state of a system, consisting of up or down. This is because they can only assume a discrete quantity of energy proportional in magnitude to the frequency of the radiation it represents., The two conceivable possible spin states or energy quanta of electrons are spin-up state, being an angular momentum that is clockwise, and spin-down with spin, which is anticlockwise. The spin angular momentum of an electron is a vector quantity in which magnitude represents the magnitude of the spin, and the direction lies along the axis of rotation of the particle, following the right-hand rule of physics. The right-hand rule states that to determine the direction of the magnetic force on a positive moving charge, point your right thumb in the direction of the velocity (v), which is the speed in combination with the direction of motion of the particle so it equals (a x b), your index finger in the direction of the magnetic field (b), and your middle finger will point in the direction of the resulting magnetic force (a).

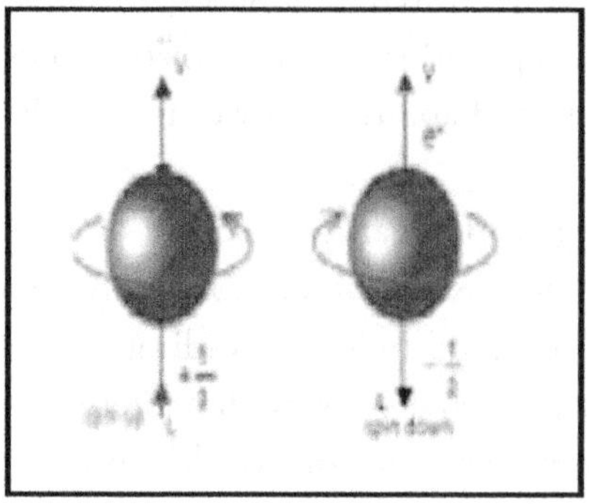

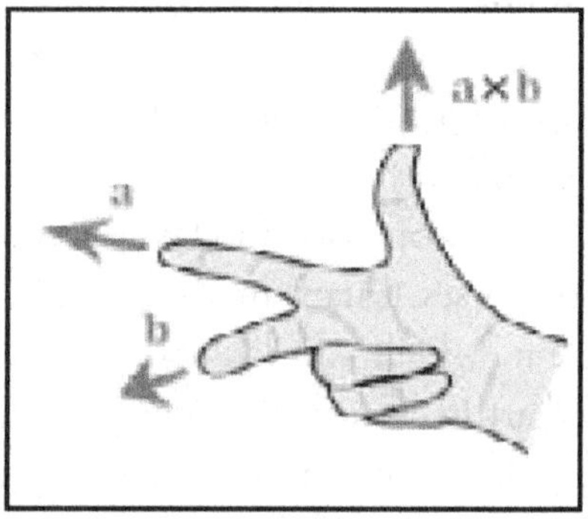

Spin and the Right-Hand rule.

A particle's spin contains vector quantity information related to its quanta energy or whether it is in the up or down spin state when measured. The major difference is that qubits, before being measured, can exist in multiple quanta states or superpositions at once, so they can be 0, 1, or both up and down, so 0 and 1. This means they contain much more potential vector or quanta information about the quantum system before measurement than the classical binary bit. Measurement results in the collapse of the wave, which is called decoherence, and all the range of probabilities is reduced to one state, which means that they then only exist in one state, so up / clockwise or down / anticlockwise, which as a vector quantity provides a source of binary information like the 0 and 1 of classical bits.

Another important quantum feature of particles is how they retain a connection or correlation with each other when they interact, even if long distances subsequently separate them. This is known as quantum entanglement, in which their behaviour is correlated, which means that, for example, the spin state or quanta energy of one particle will probably always be the opposite of the other, so if you know the spin state of one particle, you will know the spin state of the other with which it is entangled. It is important to appreciate that the correlations between the two entangled quanta mean that by measuring the quantum state of one, the other one's state is still being determined. Still, you obtain probabilistic information about its state. If one particle is in an up-state, the probabilistic information gleaned from the other particle suggests that it must probably be in the down-state, and if you change the state of one, the probability of the other state will change too, even if they are too far apart to communicate with or influence each other. In 2022, John Clauser, Alain Aspect, and Anton Zeilinger received the Nobel prize for establishing experimentally that this quantum entanglement is not just a spooky weird feature of the quantum world but obeys underlying rules of nature. Their achievement goes to the heart of whether aspects of quantum physics are random or only appear random because we possess limited and insufficient information about these systems to reveal an underlying, deterministic reality. Put another way, is there a fundamental indeterminism to quantum systems until they are measured, or might there be a "hidden variable" that contains information underlying our reality that predetermines the outcome before we measure it? In the 1960s, John Stewart Bell established conclusively that quantum mechanics is incompatible with the local hidden variables theory, which suggests a hidden variable satisfies the locality principle: an object is influenced directly only by its immediate surroundings or locality. Bell demonstrated theoretically, though not experimentally, that a broad class of local hidden-variable theories cannot reproduce the correlations between measurement outcomes that quantum mechanics predicts. Clauser, Aspect, and Zeilinger provided the eventual experimental confirmation of Bell's work.

By combining two qubits, you will have four states because of the combined superposition and a system of many such entangled qubits can be described as a superposition. Because a spin state contains vector information in terms of spin being up or down, by knowing just the state or spin of one particle or by changing the state of that particle, you can communicate information about the other particle in terms of its up or down spin state or the changed spin state even if it is across the other side of the Universe. This means that you can obtain information about both particles without appearing to be limited by the speed of light, which is the fastest anything (without mass) can travel so that information can be communicated in a fraction of the time of classing computing. However, a fundamental rule of physics, which has yet to be disproved since Einstein formulated it in 1905, is that no information-carrying signal of any type can travel through the Universe faster than the speed of light. Einstein's law of relativity reflects that particles can only travel at (for massless) the speed of light if they have no mass or below if they possess mass. It is possible to take advantage of the curved nature of space to get information to a location quicker, but it can still be within the speed of light. To date, no one has proved that even with quantum entanglement, faster-than-light communication is possible within our Universe. Einstein called entanglement though "spooky action at a distance" as the instantaneous communication appears to violate the speed of light rule, but it doesn't.

An important research priority in the computing sciences is to find more efficient ways of handling huge amounts of data to access the information in the data. The remarkable quantum features of interference in superposition and entanglement mean that the volume of potential information about the system's state contained within arrangements of qubits by adding qubits to a quantum computer exponentially increases computing power. Information represented by only 500 qubits would be impossible with even more than 2^500 classical bits. A classical computer would take millions of years to find the prime factors of a 2,048-bit number, but quantum computers using Qubits could calculate this in minutes.

All particles used as qubits to manipulate information are highly fragile, with the survival period during which the qubits are in superposition and displaying interference and entanglement, known as the coherence time, being as short as just a few hundred microseconds. To maintain qubits in a state of coherence, they must be protected from outside interference and kept in a tightly controlled environment using magnetic fields or specialized circuitry and at temperatures near absolute zero to minimize subatomic disturbances before measurement. Qubit control systems are extremely sensitive to thermal noise, crosstalk from electromagnetic interference, or quantization error, the inherent uncertainty in digitizing an analogue (wave) value. Any waveform deviation applied to or read from qubits can result in higher computational errors, or even system decoherence, and error correction is a major problem with quantum computing, with error correction procedures far more complex than in classical computing.

Despite recent progress in error correction techniques, they still occur frequently in quantum computations and are exacerbated as qubit counts increase and the complexity grows. A commercially useful quantum computer or robot needs to be reliable, fast, and on a viable scale, and advancements in quantum computing hardware construction and error-correction methods are being developed. Physicist Jonathan Friedman at Amherst College in Massachusetts has observed how "many recent demonstrations of large-scale quantum effects take their cues from previous methods used to freeze out the noise in individual atoms: "We're exploiting the same exquisite techniques, not for single atoms but for quadrillions of atoms, to make things that are closer to engineered devices that we can observe and exploit."

Quantum computers need sufficient qubits to handle complex calculations, tackle real-world problems, and outperform classical computers, but maintaining and controlling large numbers of qubits simultaneously is very challenging. Quantum scientists are demonstrating that quantum effects, such as the superposition of particles (electrons), can be sustained in loops of superconducting wire that possess no resistance. Using magnetic fields, physicists can induce the electrons to stream clockwise and counterclockwise simultaneously, all acting as one. Using magnetic flux qubits (also known as persistent current qubits), which are micrometer-sized loops of superconducting metal, they can store information in the magnetic flux. Physicists have demonstrated that they can maintain a flux qubit at 170 nanoamperes for ten nanoseconds, which disproves theories that require the collapse of the wave function to happen at these scales.[7] The slightest vibration or fluctuation in thermodynamics, particularly heat, would destroy the quantum state, which is why most quantum states need the temperature near absolute zero. Scientists have developed laser cooling in which the atoms are trapped in the laser beam, and lower-energy atoms replace high-energy electrons using evaporative cooling, which siphons the high-energy electrons off. This is one example of scientists starting to be able to harness quantum laws in larger and more complex mechanical systems, with the main constraint being to find the materials that can sustain the quantum effect for as long as possible. Microsoft has also been exploring the use of Topological qubits, which have built-in protection against environmental noise by manipulating their structure and surrounding them with chemical compounds to protect against contamination. The quantum information is encoded in a pair of what are called Majorana zero modes, which are bound states of fermions (protons, neutrons, electrons, neutrinos, and quarks) that are physically or separated by a topological gap and are their antiparticles composed of an electron and a hole, which makes the qubits more stable.

In March 2022, Azure Quantum built a topological quantum bit, and Krysta Svore, a Microsoft engineer, described the achievement as follows: "What's amazing is humans have been able to engineer a system to demonstrate one of the most exotic pieces of physics in the Universe. And we expect to capitalize on this to do the almost unthinkable — to push towards a fault-tolerant quantum machine that will enable computation on an entirely new level closer to how nature operates." They claim that topological qubit technology has opened the way to building quantum computers with over a million qubits.

This should be put into context, given that theorists have estimated that the human brain might contain 100 billion quantum bits, meaning that our brains are far more powerful than all the digital computers in the world. Matthew Fisher (1960–), arguably the world's foremost quantum physicist, collaborated with an interdisciplinary team of quantum physicists, molecular biologists, biochemists, colloid scientists, and behavioural neuroscientists to obtain explicit experimental evidence as to whether our brains are quantum computers. Over five years, Fisher and colleagues have explored whether quantum processing with nuclear spins operative in the brain might be identified and isolated. Phosphorus is "the unique biological element with a nuclear spin that can serve as a qubit for such putative quantum processing–a neural qubit".[8] Fisher logic is as follows: (1) Pyrophosphate, which contains two phosphate molecules, is released from Adenosine Triphosphate (ATP), which is the source of energy for use and storage at the cellular level[9](2) Each Phosphate molecule possesses a quantum spin, and they become entangled with each other (3) They are protected from decoherence by grouping in protective clusters which include phosphate and calcium ions called Posner Clusters which can preserve entanglement for a long time (4) When separated they remain entangled even over long distances in the brain and Fisher believes are responsible for influencing neurotransmitter release and synaptic firing between neurons. If correct, it would provide remarkable evidence of spooky action at a distance at work in the brain. We must again enter a note of caution, with many scientists sceptical that this is the case, and even Fisher recognizes that "if phosphorus nuclear spin is not being used for quantum processing, then quantum mechanics is not operative in long-time scales in cognition." Still, he also believes that "Ruling that out is important scientifically. It would be good for science to know."

Thomas Gilovich, Dale Griffin, and Daniel Kahneman, observe that over fifty years of behavioural decision-making, research has established that humans don't always follow the rational rules of Bayesian probability theory.[10] Some studies have claimed to observe that human reasoning obeys the laws of quantum, not classical probability, and explains a variety of seemingly irrational judgment and decision-making processes used by our brains.[11] Fisher concluded that "quantum probability may provide a better description than the classical probability for the way humans reason under uncertainty." The argument is based on the following logic, namely, "(i) Human judgments, such as attitude judgments, are often not simply read out from memory, but rather, they are constructed from the cognitive state for the question at hand; and (ii) Drawing a conclusion from one question changes the context and disturbs the cognitive system, which then (iii) Affects the answer to the next question, producing order effects, so that (iv) Human judgments do not always obey the commutative rule of Boolean logic. If we replace "human judgments" with "physical measurements" and replace "cognitive system" with "physical system," then these are the same reasons that led physicists to develop quantum theory in the first place." Professor Andrei Khrennikov has even formulated new rules for human reasoning under uncertainty based on quantum probability theory.[12] Whether quantum rules apply in the brain remains controversial. Still, in due course, studies will reveal whether the quantum brain is a valid hypothesis, as Fisher points out.

Quantum computers must be able to perform a single feedback/forward cycle of computation ranging from a few hundred nanoseconds up to a few microseconds to capitalize on the coherence characteristics of the qubits before they decohere. To do this, they require a very fast closed-loop control system to apply quantum gates to the quantum state of the qubits, perform the current algorithmic step, read the quantum states of the many qubits, and generate new states for all the qubits. Each quantum gate is a physical process or circuit that changes the state of a qubit or a collection of qubits but acts as a unitary operator on a quantum state such as spin, which means that it preserves the inner product of two vectors and must keep the sum of probabilities of all possible outcomes at exactly 1. Preserving the lengths and angles between vectors means that the quantum gate process is reversible, and they must be because quantum mechanics is, by its very nature, reversible. In physics, a reversible process is a process in which the system and surroundings can be returned from the final state to the original conditions without producing any changes in the thermodynamics properties of the Universe, which means that if the time is reversed in principle everything can go back in time, so information is never lost. In classical mechanics and computers, the logical irreversibility of classical logic gates inadvertently causes heat and energy dissipation when new values are computed, and the old information is irretrievably lost. A quantum gate (or circuit), though, is reversible because it involves one-to-one mapping between sets of input vector and output vector values, so all the output vectors are just permutations of the input vectors. It means that if the qubit is, for example, the spin of an electron, we can't just delete the spin, the polarization, or the information it represents. In quantum computers, a reverse traversal of the quantum circuit will undo the operations brought about by a forward traversal of that circuit. Provided we have yet to measure it, or it has been decohered through the noise, etc. With classical computing, the process is largely irreversible. For example, if you add 9+ 6=15, you can't ask the computer to determine the required inputs to get to 15. It could have a variety of combinations, so the operation is irreversible. The bit of information is erased, and for every irreversible operation, a minimal amount of energy is used, which means that without reversible computing, the increases in the number of computations-per-joule-of-energy-dissipated would, according to Koomey's Law, reach the limit implied by Rolf Landauer's (1927–1999) principle by 2050.[13] Landauer's principle states that there is a minimum possible amount of energy required to erase one bit of information, known as the Landauer limit. However, a logically reversible computation is reversible at the physical, logical, and programmatic levels at which the information is not erased, so it may, in principle, be carried out without releasing any heat, so the Landauer limit would not apply.

The huge number of qubits and signals to be handled in quantum computers requires a distributed real-time closed-loop control system based totally or partially on hardware tightly connected to the generation of waveforms and capture blocks, including classical/ quantum interfacing blocks. Quantum computers have multiple building blocks necessary to run the special quantum algorithms. This typically involves converting classical digital information to a quantum computer for further processing to harness the quantum speeds and capabilities using quantum algorithms. This currently requires an automatic process of Classical-to-Quantum interface (Cl2Qu). They involve quantum gate models designed for universal purposes by applying gates to build complex quantum algorithms like the commonly used Shor and Grover quantum algorithms (see below). This may involve a quantum accelerator, also known as a quantum annealer, which is a type of quantum computing hardware like the classical GPU designed to solve optimization problems, i.e., finding the best solution from all feasible solutions using quantum techniques. In classical computers, you start with the hardware in terms of the computer chips and ultimately end up with the algorithmic process to generate output, and quantum computing uses complex quantum versions of the different building blocks employed in classical computers to get there. For example, it requires a quantum programming language, which requires a compiler to take the input to your algorithm and convert it into an assembly language (ASM). A compiler is a program that converts high-level high-level instructions into machine language code that a computer can read and execute. Assembly languages are a programming language that translates high-level assembly languages into machine language code and act as a bridge between software programs and the underlying hardware platforms, which means we need a quantum ASM version. Unlike machine language, which consists of binary and hexadecimal characters, assembly languages are designed to be readable by humans. Quantum computing requires quantum arithmetic, which is completely different from the arithmetic used in classical computing because the quantum operations performed by quantum gates are very different. All this must be mapped into quantum instruction sets describing the operations the quantum device can execute, and a new quantum architecture must be built with quantum ALUs and quantum processors. Even Quantum Random Access Memory (QRAM) is used in some algorithms to translate data vectors into quantum states, etc., like classical computers' architecture.

We saw earlier how QPU technology harnesses the quantum behaviour of particles like electrons or photons to make calculations much faster than processors in conventional computing. Whatever is performed on the quantum qubit level, though, requires quantum chips consisting of qubits connected and, unlike classical computing in design, involves continuous wave signals rather than discrete digital signals. In classical situations, you can repeatedly measure identical objects. It should produce the same results, but in quantum mechanics, repeated measurements on the same object are impossible since one measurement destroys the potential nature of the wave function; a Cl2Qu interface is important because, unlike classical computing, the outcome in quantum computing is not, therefore, deterministic but probabilistic so if you run a classical program 1000 times you get the same result. Still, the qubit superposition state is destroyed when quantum computing reads out the results. All the other possibilities before measurement disappeared, so they cannot be recreated similarly. You must perform many times the number of runs to average the outcomes and use the outcome with the highest probability. Known as sample averaging in statistics or ensemble averaging in physics, this involves making a series of measurements on many identical quantum systems to get a probabilistic interpretation of the information. All measurements in quantum mechanics are implied to be the ensemble measurements performed on many particles/systems prepared using the same process because the measurement perturbs the system, preventing you from measuring the same system twice. In the 1970s, Vladimir Braginsky and others introduced the concept of Quantum Non-Demolition' (QND), which allows repeated quantum measurements on single quantum objects without disturbing their value or information.[14] They were first applied effectively in quantum optics.

Quantum computing has the potential to address some major challenges currently facing A.I. and robots, including the exponential growth in data and increasing algorithm complexity, which demands computational power and memory that classical computers can't meet. This is particularly true of autonomous robots that need to learn and operate in real-time in dynamically changing real-world environments, where the high dimensionality and uncertainty of the robotic state space real environments require faster and more adaptive algorithms than classical computers can currently offer. It has enormous potential in enhancing quantum learning, particularly by substantially reducing the time it takes to train robots and developing quantum sensors, including integrating the multiple modal sensors essential to robotic perception. The challenges are such that the focus within the quantum world of computing distinguishes between long-term goals and Noisy Intermediate Scale Quantum (NISQ) computation. Early quantum computing research efforts focused on developing longer-term systems that employ full error correction, which could execute large instances of quantum algorithms.

In contrast, NISQ considers near-term machines with just tens to hundreds of quantum bits (qubits) and moderate errors. This reflects the fact that quantum computing continues to face many challenges, as eloquently summed up by John Preskill when, in 2018, he observed that "Quantum computers with 50–100 qubits may be able to perform tasks which surpass the capabilities of today's classical digital computers, but noise in quantum gates will limit the size of quantum circuits that can be executed reliably. NISQ devices will be useful tools for exploring many-body quantum physics and may have other useful applications. Still, the 100-qubit quantum computer will not change the world right away – we should regard it as a significant step toward the more powerful quantum technologies of the future. Quantum technologists should continue to strive for more accurate quantum gates and, eventually, fully fault-tolerant quantum computing.[15] In 2022, Mercy Amankwah and colleagues observed that "current and near-term quantum computers, known as noisy intermediate-scale quantum (NISQ) devices, are characterized by low qubit counts, high gate error rates and suffer from short qubit decoherence times.[16] Amankwah focused on quantum image processing (see below) and stressed the need to reduce the complexity of a quantum circuit to accommodate many qubits.

Initially, having an entirely quantum-powered robot or replacing classical with quantum computing may be a complex process. Researchers currently envisage robotics using quantum computing cloud services and QPUs that will initially work alongside classical CPUs employing Quantum, Accelerator,s (annealers), and Cl2Qu interface. Quantum application accelerators can apply sophisticated quantum techniques to solve problems, data structure, and processing on a classical computer. This allows users to submit problems to quantum computers without programming applications, algorithms, or low-level coding for each quantum architecture. Because the high cost and limited accessibility of quantum computing cause problems with widespread adoption, efforts are being made to provide access through cloud-based platforms, collaboration, and openness. In 2016, IBM connected a small quantum computer to the cloud, allowing simple programs to be built and executed on the cloud. Many cloud platforms now include q, Braid Lab by qBraid,

Quandela Cloud, Xanadu Quantum Cloud, Forest by Rigetti Computing, QComputing the Cloud, Quantum Playground by Google, etc.

In a 2023 research paper titled Quantum Robot: Structure, Algorithms and Applications, Dao-Yi Dong and colleagues purported to reformulate Benioff's conceptualization of a quantum robot and defined it as "a mobile physical apparatus designed using the quantum effect of a quantum system, which can sense the environment and its state, and can also process quantum information and accomplish some tasks".[17] This reformulation was from an engineering perspective focusing on "the information exchange and learning control of a quantum robot." They envisage three fundamental parts, which are the (1) Multi Quantum Computing Units (MQCU), (2) Quantum controller/actuator, and (3) Information acquisition units. They describe the MQCU, which is usually made up of QCUs (quantum computing units), as the cerebrum of the quantum robot. Each QCU is a quantum information processor that performs specific tasks such as quantum computing, quantum memory, and task description and exchanges information between each other using a quantum bus, which they observe could be a "refreshable entanglement resource or some other quantum circuits" that will exchange or pre-process some quantum information. Each QCU can store, analyse, compute, and process all kinds of information to help the MQCU construct some appropriate quantum control algorithms. The Quantum controller/actuator is the execution and control apparatus that receives and processes indication signals from MQCU and informs actuators to carry out corresponding actions. It acts as the connection between the cerebrum (MQCU) and arm (actuator) of the quantum robot." They observe how the actuator may be a pure quantum system or a semiclassical apparatus capable of processing quantum information. Generally, the actuator can exchange quantum and classical information with the quantum controller." The information acquisition units use quantum sensors to perceive the information from the environment. However, they may get other information from the cloud or through an external communication unit from quantum robots. Usually, the information to be acquired includes quantum information and classical information. Let's briefly look at quantum algorithms, sensors, and machine learning.

Quantum Algorithms.

Quantum algorithms using the information contained in qubits can solve existing problems faster and solve many problems that classical computing couldn't, although it can't solve problems known to be currently unsolvable by classical algorithms. The best-known quantum algorithm examples are Shor's algorithm and Grover's algorithm. Shore's algorithm, which was devised in 1994 by the American mathematician Peter Shore, is a useful quantum algorithm for factoring large integers that involve decomposing whole integers, not fractional, numbers that can be positive, negative, or zero, a matrix, or a polynomial into a product of another entity or factors, which can be multiplied together to give the original number or a matrix. Factoring is a very difficult computational problem, with the computational effort growing exponentially with the size of the integer to be factored. Any known classical computing algorithm cannot solve it efficiently, but the Shore quantum algorithm can achieve this. Grove's algorithm, devised in 1996 by Lov Grover, is the fastest possible quantum algorithm for searching an unsorted database and can be used to solve a variety of problems, including finding patterns in data, breaking cryptographic keys, and solving optimization problems.[18] Grove's algorithm leverages amplitude amplification to demonstrate a quadratic speedup, but it is estimated that we need a quantum computer of 10,000 logical qubits to see significant advantages over linear search. The quantum basic linear algebra subroutines algorithm (qBLAS) is widely used to calculate Fourier transforms, find eigenvalues and the corresponding eigenvectors, and solve linear equation systems. A Fourier Transform is a widely used method of detecting what frequencies are present in waveform data.

Another common example is the Harrow–Hassidim–Loyd (HHL) quantum algorithm designed by Aram Harrow, Avinatan Hassidim, and Seth Lloyd, which is exponentially faster than classical algorithms for solving linear systems of equations.[19, 20, 21] Linear equations have two variables, so X+Y = 5 or 4x – Y=3 are both linear variables and called a system of linear equations that can be solved together with the same variable of X and Y. Quantum Support Vector Machines (QSVMs) have also been developed, which operate on the same principle as the classical SVMS we discussed to find the hyperplane in a high-dimensional space that maximally separates different classes. QSVMs use quantum computers and a quantum gradient descent algorithm to perform the optimization required to find the hyperplane. Different types of QSVM have been introduced, some of which exponentially outperform classical methods, using Grover's quantum algorithm in conjunction with a special oracle function for quantum cluster analysis. Oracle functions are operations with some property often called a black box because you don't know how they work. QSVMs are faster in training and better at generalizing to unseen data and handling non-linear data. There is also a quantum alternative based on quantum random walks for graph search algorithms, and quantum algorithms for 'decision making under uncertainty' applications, like quantum Markov chain and quantum Markov processes, have been formulated. We have seen how optimization problems in robotics, a major issue in such applications as trajectory navigation planning, are robot vision, task allocation, and motion kinematics with the requirement to find the best solution reflecting the given cost function from several feasible solutions. They are often formulated as search problems, which in classical computing may use random walks, but in quantum, they use quantum random walk algorithms.

Quantum Sensors.

Quantum sensors have the potential to revolutionize robotics by providing unrivalled ultra-high precision, which would far exceed the sensitivity of the various classical sensors. Sensing technology in robots can involve tiny quantum sensors that use particles' quantum properties to detect tiny, miniscule changes in the environment's gravitational, electric, or magnetic fields. They can measure light and the properties of atomic elements like spin very accurately, so pairing quantum sensors with mobile robots can harness their quantum capabilities and autonomous behaviour to navigate highly complicated environments, manipulate delicate objects, and interact much more seamlessly with humans. They can also be combined with existing sensor technologies to enhance performance. A hybrid quantum sensory system was developed in 2020 by the lead author of the study, Joe Smith, and a team of researchers in Quantum Engineering Technology Labs and Bristol Robotics Laboratory (BRL) at Bristol University in magnetometry. They used a robotic arm equipped with a commonly used N.V. centre quantum magnetometer, which uses nitrogen-vacancy (N.V.) centres in ensembles of diamonds to make precise magnetic field measurements in magnetic fields and perform sensitive magnetic imaging at room temperature. They sensitized it further by incorporating Spin-Based Quantum Sensors (SBQS). The SBQS allows optical manipulation and readout of spin states at room temperature and is used in conjunction with the magnetometer as tactile sensing elements in the fibre optic cables for the robotic arm; the arm could detect ultra-high precision information about pressure, vibration, temperature, or texture from the environment.[22] Quantum sensors can also help robots navigate better; in 2020, the SPIDAR project was established to develop quantum-based LiDAR systems, single-photon detectors, range finding, and 3D imaging systems for driver-assisted and autonomous vehicles.

SPIDER detects and uses single photons emitted by an object to measure the detected object's distance with accuracy in the range of trillionths of a second compared to existing 3D camera systems or current LiDAR systems that measure laser beam travel time to and from objects with accuracy to the range of 100 milliseconds. They can even detect objects through fog or potentially around corners. Researchers have also demonstrated a prototype lidar system using quantum detection technology, which captures 3D images and detailed information even in extremely low-light conditions. They can even be used when GPS is impossible and use cold-trapped ion quantum sensors to measure tiny changes in gravity and atomic acceleration. Computer vision and image recognition algorithms are computationally very expensive because they involve computations on millions of pixels to get results. Using current classical algorithms to analyse image-based data has become increasingly challenging. The emergent field of Quantum Image Processing (QIMP) has the potential to significantly improve the storage, processing, and retrieval of visual information from images and video by converting digital to quantum paradigms and complementing digital imaging with quantum techniques. Quantum Image Representation (QImR) plays a key role in QIMP and determines the types of processing tasks and their performance.[23] The basic idea of QImR involves encoding the properties of an image, like colours at certain positions, in qubit-lattices. Pranav Gokhale and colleagues believe that the frontiers of quantum computers can be pushed even further with the development of Qutrits, a tripartite unit of quantum information involving three-level qutrits instead of qubits.[24] Teleportation relies on the feature of quantum entanglement. It involves the "disembodied" transmission of unknown quantum states between distant quantum systems. It has been possible since 1997 when an Austrian team led by Anton Zeilinger achieved the first successful teleportation of a quantum state by transferring the quantum state from one photon to another by entangling two photons and measuring one of them.

In 2004, a team at The National Institute of Standards and Technology (NIST) also transferred one beryllium atom's key properties to another without using any physical link. In 2008, Scientists at Tokyo University Quantum teleportation used optical fibres to teleport entangled photons across several kilometers. In 2015, the NIST team achieved 100 kilometres. To date, though, it has been limited to a two-dimensional subspace of quantized multiple levels of the quantum systems. Having been experimentally demonstrated, it has the potential to be used in a wide variety of practical applications, including quantum computing, secure communication, and cryptography; all benefit from this, and developing quantum technologies will rely significantly on quantum teleportation. Quantum teleportation does not travel faster than the speed of light but allows instantaneous communication because of entanglement. In 2019, two dependent teams of Chinese physicist Guang-Can Guo and his colleagues at the University of Science and Technology of China (USTC) and an international collaboration headed by Anton Zeilinger of the Austrian Academy of Sciences and Jian-Wei Pan of USTC managed to teleport a qutrit. In doing so, they demonstrated that qutrits, which can carry more information and have greater resistance to noise than qubits, may be used in future quantum networks.[25]

Visual tracking of moving objects is very important for robots to detect object disappearance detection and motion behaviour, but currently, only a few systems employ three-dimensional images using quantum representation and do so in the form of the quantum point cloud. Nia Jang and colleagues, however, recently proposed a three-dimensional quantum representation for digital images.[26] There are many quantum frameworks for converting classical digital images to a quantum computer for further processing, which include (I)FRQI, (I)NEQR, MCRQI, and (I)NCQI. Using quantum algorithms to do so requires an automatic Cl2Qu process. However, as Mercy Amankwah, whom we met earlier, observes, "Almost every image processing algorithm developed in the classical sense can also be developed in the quantum environment." She and colleagues have developed a novel and uniform framework for quantum pixel representations called QPIXL. She observes how the proposed QPIXL framework results in more efficient circuit implementations and significantly reduces the gate complexity for all considered quantum pixel representations. They also enabled the system to avoid using ancilla qubits. Most quantum circuits require several ancilla qubits to store partial results, which are initially in a well-defined state but must be returned to the same state at the end of the computation.[27]

Quantum Machine Learning.

Vedran Dunjiko and colleagues recently observed how "The emerging field of quantum machine learning has the potential to substantially aid in the problems and scope of artificial intelligence.[28] We have seen how robots use machine learning to analyse and mine sensor data for perception, localization, navigation, learning controllers, planners, and HRI. We have briefly discussed the application of quantum algorithms for quantum principal component analysis (PCA) for quantum regression and quantum ANNs. Quantum neural networks are only one form of the most recent machine learning models implemented on quantum computers. In 2021, Amira Abbas and colleagues looked at how powerful and trainable quantum machine learning models are about popular classical neural networks. They concluded that they had "demonstrated numerically that a class of quantum neural networks can achieve a considerably better effective dimension than comparable feed-forward networks and train faster, suggesting an advantage for quantum machine learning, which we verify on real quantum hardware." Also, that "a class of quantum neural networks is presented that outperforms comparable classical feed-forward networks. They achieve a higher capacity in terms of effective dimension and at the same time train faster, suggesting a quantum advantage" [26]. Yet despite robots being able to make decisions increasingly rapidly, trials using quantum techniques have yet to enable robots to learn faster, which is key to developing increasingly complex autonomous machines. We have seen how the challenge of supervised and unsupervised machine learning is reflected in the algorithms' computational complexity, the training set's size, or the data required for the algorithm to learn the distribution within the data. In reinforcement learning, it is the number of interaction steps the robot requires to learn by obtaining the rewards with a high degree of probability.

Several supervised and unsupervised learning approaches, including, for example, curve fitting, regression analysis, classifiication, and clustering problems can be accellerated exponentially. It is called quantum acceleration, which uses quantum computing techniques and quantum algorithms. However, Vedran Dunjiko observes that "while quantum improvements in supervised and unsupervised learning have been reported, reinforcement learning (as an interactive mode) has received much less attention"[29] even if some positive results have been achieved using quantum reinforcement learning.[30, 31] In 2021, though, for the first time, an international collaboration led by Philip Walther and a team of experimental physicists from the University of Vienna experimentally proved that they could speed up a robot's learning time using Quantum Enhanced Reinforcement Learning

(QERL) when interacting with its environment. This used a quantum search algorithm combined with a Quantum Communication Channel to reduce the number of trials required for learning the correct path when a robot was learning to navigate.[32] Their device also included a fast active quantum feedback mechanism. Using the Quantum Hardware Channel, the robot could interact with the environment by exchanging quantum or arbitrary superposition states; in a hybrid agent, it can choose between quantum and classical behaviour. In all previous applications of quantum techniques in learning, the agent and environment interact classically.

Still, Walther and the team used a "Quantum-enhanced hybrid agent capable of quantum as well as classical information transfer." This hybrid agent used a compact and fully tuneable integrated nanophotonic processor, harnessing nanotechnology's enormous advantages. Nanophotonic devices can slow down, enhance, produce, or manipulate light because the nanophotonic material, such as a photonic crystal, has a nanostructure that affects the motion of electromagnetic energy so that it can interface with telecom-wavelength photons. Quantum communications channels use "flying qubits" or photons to transmit qubits because photons are very well isolated from perturbations, which means that photonic qubits have long-lived superposition states and propagate with low attenuation (down to 0.2 dB/km at 1.55 μm) in optical fibres. Photons are stable massless particles with no electric charge, so electric and magnetic fields cannot deflect them. Still, they have momentum and energy and always travel at the speed of light in a vacuum, effectively acting as energy carriers. They are produced when an electron in a higher-than-normal orbit around the nucleus of an atom falls back to its regular orbit; because they always travel at the speed of light, they can't have any mass, so all their energy is derived from their motion and ceases to exist if they stop, meaning that they have momentum, speed, and frequency. Quantum communications using flying qubits could connect quantum computers to build a quantum network or quantum internet. Kimble describes how quantum networks involve "quantum interconnects, which convert quantum states from one physical system to those of another in a reversible manner. Such quantum connectivity in networks can be achieved by the optical interactions of single photons and atoms, allowing the distribution of entanglement across the network and the teleportation of quantum states between nodes."[33] This needs the fabrication of practical quantum light sources that can emit single and entangled photon states on demand.

In 2006, Dao-Yi Dong and colleagues published a paper titled Quantum Robot: Structure, Algorithms, and Applications. They announced a brand-new Quantum robot based on the fusion of quantum theory with robot technology.[34] Their quantum robot, which they compare to classical robots, was composed of three fundamental parts, which they describe as (1) multi-quantum computing units (MQCU), (2) Quantum controller/actuator, and (3) Information acquisition units. In addition, they proposed several learning control algorithms, including quantum searching algorithms and quantum reinforcement learning algorithms, which are presented for quantum robots. Remarkably, they observe that "quantum robots are superior to classical robots inefficient learning under novel quantum reinforcement learning algorithms." In 2022, Dirk Heimann and colleagues used Quantum Deep Reinforcement Learning (QDRL) to learn navigation tasks using a wheeled robot in three increasingly complex simulated environments. They observed, "To our knowledge, this is the first quantum machine learning (QML) demonstration for robotic behaviours. Thus, we establish robotics as a viable field of study for QML algorithms and henceforth quantum computing and quantum machine learning as potential techniques for future advancements in autonomous robotics."

Endnotes

[1] Gröblacher, S., 2019.
[2] Huelga, S., Plenio, M., 2014.
[3] Benioff, P., 1980.
[4] Feynman, RP., 1982.
[5] Classical and Quantum Information (2011).
[6] Caves, C, Fuchs, C, Schack, R., 2001.
[7] Knee GC, et al., 2016.
[8] Weingarten, CP, Doraiswamy, PM, Fisher, MP., 2016.
[9] Dunn, J, Grider MH., 2023.
[10] Heuristic and Biases: The Psychology of Intuitive Judgment 2002.
[11] Wang, Z et al., 2014.
[12] Ubiquitous Quantum Structure: From Psychology to Finance (2010).
[13] Koomey, J., et al., 2011.
[14] Braginsky, VB, Vorontsov, Y I., 1974.
[15] Preskill, J., 2018.
[16] Amankwah MG, et al., 2022.
[17] Dong, D, et al., 2006.
[18] Grover, L K., 1996.
[19] Lloyd, S., et al., 2013.
[20] Aımeur, E., et al., 2013.
[21] Rebentrost, P., et al., 2014.
[22] Smith, J, Zhang, D, Balram, K., 2023.
[23] Yan, F., et al., 2016.
[24] Gokhale, P, et al., 2020.
[25] Hu XM, et al., 2020.
[26] Jiang, N., et al., 2017.
[27] Amankwah, M.G., et al., 2022.
[28] Dunjko, V, et al., 2016.
[29] Abbas, A., et al., 2021.
[30] Dong, D., et al., 2005.
[31] Paparo, G. D., et al., 2014.
[32] Saggio V, ew.
[33] Kimble, H. 2008.
[34] Dong, D, et al., 2006.

Chapter 12

Conclusions

CHAPTER 12

Conclusions

ROBOTICS PRESENTS THE MOST LIKELY way that AI systems will gain cognition, which is a prerequisite for proper understanding. Many questions posed by the recent AI and robot technology need to be answered. Are we potentially creating a race of super-intelligent evil robots destined to replace humans as the dominant entities on the planet, perhaps by becoming our synthetic neocortex in the cloud? If robots improve enough beyond their current limitations, will we break our mental barrier, our need to retain our tribal specialness as a species, differentiating ourselves from them? I proffer some personal conclusions based on my research and what we have learned about the current science of AI and robots, but every reader will have formed his or her conclusions given that the raison d'etre for writing this book is to allow you to have an informed opinion about such issues.

We know that autonomous androids must ultimately be able to perceive and understand their world, communicate with people and share our world. Some researchers believe this will require a level of biological architecture in robots and their embodiment in the environment and interaction with the real world for this to happen. Neuroscientific research has revealed the brain's organizational complexity with cortical areas subdivided into ever smaller regions; each characterized by response properties and connections to other brain structures. Cognition, knowledge, and understanding depend upon the complex morphology and biologically integrated yet non-linear dynamic nature of the entire nervous system. Coupled with the somatosensory exploratory interaction of our physical and social environments, this generates our, experience, learning, and understanding of our world. The human brain seamlessly integrates information from multiple sensory modalities and sources and the challenge facing robot designers is to build architectures that can come even close to being able to do so. Will robots ever be able to simulate or display human cognition, learning and intelligence by adapting their behaviour in real-world environments, which are unpredictable, uncertain, and messy, with lots of risks, noise, and distractions?

The convergence of hardware technology and brain-inspired cognitive architectures has the potential for AI robots to harness some of the remarkable neurobiological features that make the brain so powerful. Innovative developments in machine and deep learning have enabled robots to surpass humans in many key domains with the potential to achieve AGI soon, although this is disputed by some. We have compared in humans and robots the domains of the conscious and subconscious brain, the role of the senses, emotions and feelings and homoeostasis, cognition, the self and memory, and learning in governing our behaviour, decisions and choices in life. The unique diversity in function, morphology, and sheer number of brain cells allows us to tolerate ambiguity and uncertainty, learn from impoverished and disorganized data to communicate, use electrical pulses and chemical modulators to perform parallel processing and even tolerate some damage and degradation in performance making us resilient. The electrical and chemical communication between neurons, neural plasticity, the synchronized oscillatory behaviour of neurons, self-organization of networks, latent brain states, the theory of FEP and Markov blankets in understanding, the balance between inhibition and excitation, the role of brain waves in influencing our mental states and the structural and functional connectivity between and within brain regions all appear essential to the effective integration and processing of raw data and information and to maintain a coherent sense of self.

We know that neurons in BNNS are typically significantly slower than the transmission speeds in ANNs yet compensate for this by the staggering number of synaptic connections which make our brains highly efficient energy structures requiring far less energy than most supercomputers. The formation of stable BNNs, cortical patterns and maps, and neural schemas influence our perception, interpretation, and attribution of meaning to new information and experiences. The brain's extensive local and long-range functional and structural connectivity allows it to integrate real-time information about emotions and feelings, thinking, motor responses (movement), and social cognition and the brain can store enormous amounts of information due to our greatly enhanced long-term and episodic or autobiographical memory capacity. We have looked at the major psychological learning theories like associative, non-associative, reinforcement, and social learning, and the critical role of curiosity exploration and continual learning and how these are being harnessed in ANNs. We have heard about the roles of the left versus right brain hemispheres in how we perceive our world, the brain's capacity to ruminate, overthink, and worry, and how our brains coordinate and process information related to the self. We know that creativity and curiosity are hallmarks of the human species, leading to our greatest achievements in the sciences and the arts.

Computer scientists are seeking to translate the complex knowledge of these biologically unique features into the cognitive architectures employed in AI systems and the robotic brain. The latest interdisciplinary research in the neurosciences, psychology, AI systems, cognitive architectures, and biologically inspired technologies such as neuromorphic chips and neuromorphic and quantum computing are accelerating the development of robots that look and behave like humans. These include quantum computing, the Distributive Adaptive Control (DAC), the Layered Reference Model of the Brain (LRMB), the Thousand Brains Theory of Intelligence, the Covariant model, the incorporation of thalamic functions, analogue modulation using neurotransmitters and hormones, dendritic structures, and glial cells like astrocytes, etc. We heard Peter Chen of Covariant making a distinction between mechanical calculations and more cognitive or thinking tasks, such as understanding what an object is and what it does and why, which require a degree of understanding, reasoning, or decision-making. Without understanding and the capacity to perform cognitive tasks, robots will continue to lack the flexibility to perform tasks in different environments and adapt to change. The Covariant Brain AI platform purports to enable robots to see, think, act, and understand their surroundings, make informed decisions, and adapt their actions to changing circumstances. According to Chen, possessing cognitive abilities and the capability to perceive their environment dynamically allows them to deal with novel tasks and multiple scenarios. Chen stresses that the Covariant Brain's knowledge and intelligence are driven by the data rather than human programming, and the understanding based on learning from the data comes, in Covariant's case, from learning all the connected robots. Attempts such as Covariant to combine all these features illustrate not just man's incredible ingenuity but how complex the human brain is; evidenced by the many theories seeking to explain how the brain constructs coherence and meaningful information.

It is generally acknowledged by those in the field of AI and robotics though that despite all these initiatives, androids are a long way off from being able to act as autonomous agents or to behave like humans. We should remind ourselves of Zador's sobering observation that we still cannot build a machine capable of building a nest, stalking prey or loading a dishwasher. In many ways, AI is far from achieving the intelligence of a dog, a mouse, or even a spider, and merely scaling up current approaches will probably not achieve these goals. Computer technology has made computers faster, more precise and accelerated AI but it can't compete with the human brain in dealing with novel, ambiguous, and highly complex environments. We saw how for RL to do the right thing, the reward function must capture exactly what you want but tends to overfit your reward, leading to things you didn't expect. Some RL performs well if the reward consists of a simple or single goal, like getting good at one game like Atari, but it doesn't generalize to other tasks or games. We heard claims that the generalization capabilities of deep RL are not strong enough to handle a diverse set of tasks yet and how hard it is to find cases where deep RL has created practical, real-world value.

In a 2022 paper, Dirk Heinemann and colleagues observed how classical techniques for DRL for robotic behaviours have produced impressive results, but they suffer from enormous sample complexity, limited generalization capabilities, and limited applicability for flexible long-term autonomy.[1] Although DL has revolutionized AI, there is a very long way to go before DL can compete with the human brain. This is reflected in the words of

Japanese mathematical neuroscientist Professor Shunichi Amari, who observed how there is currently no theory of deep learning except for individual minor ones and individual theories have not been sufficient to understand our brains comprehensively. In their 2022 review of robotic vision deep, Zhang and Lee observed that despite all the advances that have been made in recent years, bionic vision equipment is not as dexterous and intelligent as the human eye and currently, robot vision can only imitate one or a few physiological functions of the human eye and is unable to gather, process, and evaluate images in complex and changing settings as the human eye does. This reflects different problems in machine vision, including at the embodiment and awareness levels. Nicolas Lesica and colleagues, in a recent review of clinical applications for improving human hearing using AI, observed how innovation has stalled, and research suffers from a lack of computational tools that can match the complexities of auditory processing. Despite all the progress and innovation in NLP translation systems, even with these greatly improved attentional mechanisms, they typically still require checking by human translators who bring critical thinking and more nuanced interpretations beyond NMT systems. They can't translate and imitate 100% like the human brain, although hold enormous potential and demonstrate the vital role attention plays in such tasks as NLP. They highlight how much we take for granted the ease with which we communicate through language despite its significant complexity and potential for ambiguity or inattention to context and meaning.

Despite the remarkable progress made in recent years, there are several observations formulated from my research and writing this book, which I believe should make us cautious about jumping to the tempting conclusion that we are very close to having autonomous robots or should be scared of them. There has yet to be an indication that any AI system or robot technology possesses the features that make us unique, even those we share with other primates and a few other species. Some robots exhibit superior intelligence in some domains and far surpass us at rule-based learning, but as you go through the following observations reflecting what it means to be human, ask yourself whether based on what you have learned from this book how close are AI and robots to taking usurping humans. In no particular order, They are

(1) Our capacity for creativity, imagination, and fantasy and the ability to dream about the future, which AI machines do not possess.

(2) The fact that we can apply common sense to approaching tasks and problems and common sense is yet to be built into AI.

(3) Our ability to represent value notions such as beliefs or abstract concepts rather than merely follow behavioural rules or habitual and reflexive patterns.

(4) BNNs are currently much more fault tolerant and robust than ANNs, particularly given synaptic plasticity in BNNs.

(5) Our much greater adaptability in being able to perform a wide variety of diferent

tasks and functions and the ability to perform much more complex tasks when compared to ANNs.

(6) Our fundamental experience of emotions and feelings is based on the homoeostatic and allostatic processes of our bodies.

(7) The ability of our brains to process multi-modal sensory data that allows us in real time to attribute meaning, knowledge, and understanding to our world.

(8) Our unique sense of self-awareness and the illusive nature of our I and Me selves which give us our sense of continuity and narrative over time as individual beings and allow us to maintain a coherent sense of ourselves as entities moving through time.

(9) Our capacity to feel rather than mimic loyalty, empathy, and compassion and demonstrate altruistic Behaviour based on the complex relationship between the self and others and the TOM and mentalization that robots cannot yet display.

(10) The autobiographical nature of our long-term memory and its complex relationship with self-schemas and subjective experiences developed over time.

(11) Our ability to learn a range of tasks sequentially or segregated one at a time without confusing the task requirements or experiencing mutual interference between tasks and preserving the learning from previous tasks with minimal forgetting.

(12) Our natural curiosity and desire to explore and understand our world coupled with the knowledge that, as animals, we originate from and are part of the natural world.

(13) The social nature of our existence and the importance of having others around us together with the experience of friendship, family, love, and attachment to others.

(14) The existential fears governing much of what we do, related to recognizing the limitation of our existence in terms of death, meaninglessness, isolation, loneliness, and the desire for identity and freedom.

(15) The role of evolutionary and genetic forces on us as individuals influencing our behaviour and personalities.

(16) Our ability to experience physical and psychological pain and to possess powers of introspection about our mental states.

(17) Our resilience as individuals and as a species is to deal with and survive many difficult and threatening situations, which is something robots will need to master if they are to be autonomous. The physical resilience of their hardware will need to be accompanied by robust software in all situations and not overwhelmed by the challenges in front of them, even if no emotions are involved. Androids must achieve a successful outcome in the face of adversity and not simply grind to a halt in challenging, physical, or intellectual situations.

(18) The ability to procreate and nurture our young forms the basis of much of our value systems.
(19) The importance of neural inhibition in the way our brains rely on a dynamic and ongoing balance between excitation and inhibition, stability inhibition, stability, and adaptability to constantly respond to the environment, which is a major challenge to building autonomous androids.
(20) The fact that existing machines and deep learning rely on layers connected to those below and above don't mimic biological neurons because they are all effectively connected, which, as we know, is different from what happens in the human brain.
(21) The constraint that artificial neurons do not yet possess the wide range of complex, thermodynamic, non-linear integrative properties displayed by human neurons, such as pyramidal neurons and many diverse morphologies. Pyramidal neurons, as we saw, are known to integrate inputs at multiple dendritic segments; in contrast, artificial point neurons can only integrate at a single point or zone and contain extensive dendritic arbours and many thousands of synapses, so a realistic model would, in theory, need to include this type of dendritic structure.
(22) Our capacity to deal with novel, ambiguous, and highly complex dynamic environments.
(23) The incredibly energy-efficient nature of the human brain and the limitations and constraints placed upon AI and robotics by the level of computational power and volumes of data required to train.
(24) Our capacity to adapt and self-regulate our behaviour to achieve our goals or desires by addressing our shortcomings or failings.
(25) Our capacity to understand irony, humour or sarcasm in situations and in communication and behaviour.

These personal observations are consistent with Allen Newell's Unified Theories of Embodied Minds (UTEM), which stipulates the following features as essential to robotic autonomy. To remind us, they are: 1) Behaving flexibly as a function of the environment. Exhibiting adaptive (rational, goal-oriented) Behaviour (2) Operating in real-time (3) Operating in rich, complex, detailed environments (4) Using symbols and abstractions (5)Using language (6) Learning from the environment and experience (7) Acquiring capabilities through development (8) Operating autonomously, but within a social community (9) Being self-aware and have a sense of self (10) Being realizable as a neural system (11) Being constructible by an embryo-logical growth process and (12) Arising through evolution. Despite Newall's attempt to clarify the role of cognition in AI, there is currently no prevailing unified theory of cognition that cognitive architects universally adopt. None come close to imbuing robotic agents with the above capabilities and features.

We have seen how the overriding goal in machine learning is to enable an AI system to do what our brain does so efficiently, which is to process information from the environment in real-time. In AI it is currently only possible to do this on a much smaller and simpler scale than nature has organized for the human brain. Deep learning networks in AI allow these artificial nodes, with the level of connectivity determined to be stacked in neuronal layers, to extract features hierarchically, recognize patterns in data, and even generalize that feature information or pattern recognition to other new data.

Unlike human learning, which, even with young children, doesn't require constant supervision and is largely autonomous, machine learning still requires a high level of supervised training and learning and enormous volumes of data that would be impossible to train humans on. For example, the latest GPT-3 chatbot used over 400 billion words from the internet to develop its ability to imitate human speech and feedback replies. Although exposed to vast amounts of information in the environment, humans need and use far less information to learn, than AI currently requires and can build abstract mental representations without much effort or supervision. As Markus Vincze observed, much work is required on the scientific and technological tools and techniques to deliver all the various features exhibited by the human brain and body acting together into a single robot. Recent advances in machine learning have created new AI capabilities, but machine learning across these various applications remains narrow and highly specialized, with current systems still needing to be carefully trained or programmed for virtually every situation. We heard Alison Gopnik describe how, despite enormous strides in machine intelligence, even the most powerful computers still cannot learn as well as a five-year-old does. The remarkable level of connectivity and integration that our brains are capable of presents a major challenge for AI developers. Robots' effectiveness remains limited by the accuracy to which our programming models the real world and one area that highlights the human brain's apparent uniqueness and our specialness is creativity and our ability to produce beautiful art, architecture, music, and literature. Creativity researchers have highlighted cognitive abilities such as divergent, critical, analogical, and reflective thinking and personality characteristics, such as openness to experience, creative self-efficacy, task motivation, domain knowledge and expertise, risk-taking, and resilience in the face of criticism that one requires to be creative. Although our understanding of neuroscience and psychology and our beliefs about what constitutes creativity remain unclear, building creativity into AI androids appears to be extremely challenging from a hardware and software perspective. Using the features of creativity we discussed, AI developers will need to assess whether AI can ever be genuinely creative or mimic human creativity and dilute our appreciation of what human creativity can produce. Robots have mimicked some classic art, like the AI-generated 3D painting called The Next Rembrandt, but cannot be said to have exhibited any of these qualities that define creativity. AI continues to fail miserably when confronted by problems that can be solved using simple common sense in terms of the basic ability to perceive, understand, and judge things common to and can reasonably be expected of nearly all people.

Although obscure, the reasoning behind common sense continues to be extremely difficult to encode into rule-based AI systems. We know humans possess many common-sense assumptions underlying every statement or action. This shared, unstated background knowledge includes a general understanding of how the physical world works (i.e., intuitive physics), a basic understanding of human motives and behaviours (i.e., intuitive psychology), and a knowledge of the common facts that an average adult possesses. Dr. Howard Shrobe observes how common sense is required for robots to have the flexibility of human learning and the breadth of human knowledge. Still, machine common sense remains a critical but missing component of AI. As American computer scientist John E Laird, one of the developers of the cognitive architecture called Soar, observed cognitive architecture must be able to change through development and efficiently use knowledge to perform new tasks but building even a limited range of abilities into any cognitive architecture is proving exceedingly difficult. To model the human mind and achieve AGI, these machines must be able to perceive the world, solve problems by generalizing their knowledge across a wide range of domains, develop insights, and learn from experience. Searle's observation in 1989 that we need to decide whether computers, in the absence of a human-like brain, can achieve states of mind that will allow them to understand the world like humans remains as pertinent today. Creativity and common sense remind us that it is not just about doing things and performing tasks, however complicated and impressive, but about understanding what we are doing and why we are doing that is key to being human and key for robots to be like humans. We see this in our ability to represent value notions such as beliefs or abstract concepts, which appear to distinguish us from the rest of the animal kingdom and without which AI systems can barely claim to be close to what it means to be human.

We have seen how AI systems are centralized through the CPUs or other processing units and how in contrast, BNN has no central control over the process. Our brains comprise small and larger scale networks, reflecting various levels of anatomical organization from the neuronal network to the whole CNS and body. In ANN, the system's unique functional memory is in one location, whereas in BNNs, the memory is distributed throughout the neural network in the neural patterns and connections. BNNs are much more fault tolerant and robust, particularly given synaptic plasticity in BNNs, which confers upon some of us the ability to perform much more complex tasks when compared to ANNs. Although AI systems can typically compute far faster than humans and are making remarkable progress in reinforcement and unsupervised learning, they currently come nowhere near to being as adaptable as humans in their ability to perform various tasks and functions.

Simulation has been theorized as a process used by our brains to build mental constructs encoded in neural patterns that constantly flow between sensory perception, planning, decision-making or choices, and adjusting, which allows us to execute the appropriate motor action or response and is a feature of how our brains represent the behaviour of others. Robots are yet to be able to achieve this. Nor have they demonstrated Bandura's self-efficacy, which, as we saw, observes how people play a major role in adapting and self-regulating their Behaviour to achieve their goals or desires by addressing their shortcomings or failings. Although they are developing different sensory modalities, no AI cognitive architecture currently allows robots to process sensory data in a way that humans do when attributing meaning, knowledge, and understanding to that information; Nicholas Cassimatis believes this may be due to the immense challenge faced by AI, of needing to integrate multiple representations and inference techniques about perception cognition and action, which is something our brains do naturally all the time. Humans don't live in isolation but are embodied in an environment as robots will have to be. We can't just rely on a high level of reasoning to give meaning to all the raw input of external stimuli from our environment and the inner thoughts to generate appropriate Behaviour and actions. Unless AI robots are embodied and learn from their experience of interacting with the world, the big challenge will be how they will even begin to be able to perceive the world as we do or, as we discussed, understand and give meaning to the world, particularly as we will see much of this processing relates to information that is associated with the self and the preservation of the self?

We should remember Shukla Shubhendu and Jaiswal Vijay's observation on AI, which they defined as the study of ideas to bring into being machines that respond to stimulation consistent with traditional responses from humans, given the human capacity for contemplation, judgment, and intention. Each such machine should critically appraise and select differing opinions within itself. Produced by human skill and labour, these machines should conduct themselves in agreement with life, spirit, and sensitivity, though in reality, they are imitations. The complexity of BNNs substantially exceeds the current complexity of any of the current ANNs, even DNNs. We have seen how this requires researchers addressing a vast range of problems, including the structures, layers, size of neurons, functional capabilities of neurons, their learning capabilities, the style of computation, processing elements, processing speed, connections, strength, information storage, information transmission communication media selection, signal transduction, and fault tolerance.

Although the various types of ANNs employing artificial neuron arrangements attempt to capture how our brains process data and turn it into information through neural networks, none have yet come close to achieving the diversity and functions of human neurons. Our brains use and identify features from the raw data of sensory stimuli to compile an accurate internal representation of the information in that raw data, allowing us to act upon it by performing cognitive processes and making decisions or generating

Behaviour, etc. For example, all the diverse data features of a familiar human face will have been stored in the neurons and neural networks in our brain such that when we see that face again, the features that we rapidly extract from the raw visual data of the face are processed to build information based on an internal representation which allows us to recognize the face. Facial recognition in humans goes far beyond the vastly improved facial recognition of AI systems and robots. An additional difficulty to AI is that the human brain does not just recognize a face or features but imbues it with emotions, memories, biases, and prejudices, and abstract concepts like friend, lover, stranger, foe etc., and even a dynamic narrative so that our reaction to every face that we recognize is unique and highly subjective to us as an individual. Getting AI to achieve facial recognition is one thing, but allowing it to bring the richness of interpretation and response that the human face evokes remains a major challenge.

Humans are remarkable for their biological complexity and the mind and body allow us to experience what it means to be alive and enjoy the privilege of exploring and engaging with our environment and other people. Emotions and feelings, empathy and compassion, love and friendship are inherent to us and even the freedom to believe in abstract notions such as the soul and God and relate to the world through the illusive lens of the I self is something we should never take for granted, but currently, these remarkable features of what it means to be human appear well beyond the reach of AI robots. We must never though get complacent in our "Specialness" for change is happening far faster than we may realise. Given the pace of progress and the increased involvement of androids in everyday contexts we cannot say for certain that they will not achieve their own specialness that surpasses that of our species.For now there appears no reason to be frightened by super robots who pose far less of a risk to us than we do to ourselves.

Bibliography

1. Brooks, R. (2003). Flesh and Machines: How Robots Will Change Us. New York Pantheon Books.
2. Thrun, S, Burgard, W Fox, D (2005) Probabilistic Robots. MIT Press.
3. Barandiaran X, Di Paolo. (2017) Sensorimotor Life: An Enactive Proposal. Oxford, UK: Oxford University Press.
4. Kurzweil R. (2005). The Singularity is Near. Viking.
5. Dawkins R. (1976). The Selfish Genes. Oxford University Press.
6. Moravec H. (1988) Mind Children: The Future of Robot and Human Intelligence. Harvard University Press.
7. Dawkins R. (1986). The Blind Watchmaker. Norton & Company, Inc.
8. Long. J. (2012) Darwin's Devices. Basic Books; 1st edition.
9. Cangelosi A. (2015). Developmental Robotics; From Babies to Robots. MIT Press.
10. Newell.A. (1991). Unified Theories of Cognition. Harvard University Press.
11. Verschure P, Dr Duff A. (2014). Biologically Inspired Cognitive Distributed Adaptive Control Architecture.
12. Descartes.R. (1649) The Passions of the Soul and Treatise on Man'. Cambridge University Press: (2012).
13. Bloom P. (2004) Descartes' Baby: How Child Development Explains What Makes Us Human. New York, Basic Books,
14. Ashby. R. (1960) Design for a Brain, the Origin of Adaptive Behaviour. New York John Wiley and Sons.
15. Gardner H. (1983). Frames of Mind Basic Books; 3rd edition (2011).
16. Tammet D. (2006). Born on a Blue Day. Hodder Paperbacks.
17. Abraham A. (2018). The Neuroscience of Creativity. Cambridge University Press.
18. Kaufman J. (2023). The Creativity Advantage. Cambridge University Press.
19. Lythcott-Haims.J. (2015) How to Raise an Adult: Break Free of the Overparent-ing Trap and Prepare Your Kid for Success. New York: Henry Holt and Co.
20. Palahniuk. C. (1996) The Fight Club W. W. Norton & Company (2018).
21. Kvanvig. J. (2003) The Creativity Advantage. Cambridge University Press; New edition.

22. Pritchard D. (2010) Knowledge and understanding. Cambridge University Press.
23. Ray Kurzweil. (2005). When Humans Transcend Biology. Viking Press Inc
24. Anthony Chemero. (2009) Radical Embodied Cognitive Science. The MIT Press.
25. Todd Feinberg and Jon Mallett. (2016). The Ancient Origins of Consciousness How the Brain Created Experience. The MIT Press.
26. Emil du Bois-Reymond. (1884). Investigations of Animal Electricity.
27. Gyorgy Buzsaki. (2006). Rhythms of the Brain. Oxford University Press.
28. Jakob Howhy. (2013). The Predictive Mind. Oxford University Press.
29. Odeh Turjman. (2021). Life Without a Self: Unravelling the Mystery of Existence. Notion Press.
30. Gregory Hickok. (2014). Mirror Neurons. W. W. Norton & Company.
31. Donald Hebb. (1949). The Organization of Behaviour: A neuropsychological Theory. New York. John Wiley and Sons, Inc.
32. Warren Weaver. (1948). The Mathematical Theory of Communication. The University of Illinois Press.
33. Lisa Feldman Barrett. (2020). Seven and a Half lessons About the Brain.Harp-er Collins.
34. Michael Hoffman. (2015). Brain Beat. Page Publishing, Inc
35. Betty Edwards. (2001). Drawing on the Right Side of the Brain. Tarcher/Put-nam,US.
36. Arthur Ladbroke Wigan. (1844). A New View of Insanity: Duality of the Mind. Nabu Press.
37. Iain McGilchrist. (2021). The Matter with Things. Perspectiva Press.
38. Robert Ornstein. (1972). The Psychology of Consciousness. New York. Viking Press.
39. J A. Scott Kelso. (1997). Dynamic Patterns: The Self-Organization of Brain and Behav-iour. MIT press.
40. George Zarkadakis. (2015). In Our Own Image. Ebury Digital.
41. George Miller. (1951) Language and Communication. McGraw-Hill Inc., US.
42. John von Neumann. (1958) Computer and the Brain. Yale University Pres 3rd ed.
43. Ulisse de Corpo and Antonella Vannini. (2017). An Introduction to Syntropy. Independently published (4 Mar. 2017).
44. Robert Rosen. (1985) Anticipatory Systems. Pergamon Press. Oxfor
45. Ulisse de Corpo and Antonella Vannini. (2014). The Balancing Role of Entropy and Syntropy. Independently published (13 Jan. 2017)
46. Yuval Noah Harari. (2016) Homo Deus: A brief History of Tomorrow. Vintage.
47. Cecilia Hayes. (2018). Cognitive Gadgets: The Cultural Evolution of Thinking. Harvard University Press.
48. David Marr. (2010). A Computational Investigation into the Human Representation and Processing of Visual Information. MIT Press.
49. David Marr. (1983) Vision. MIT press.

50. Dana Ballard. (2015) Brain Computation as Hierarchical Abstraction. MIT Press.
51. William James. (1890) The Principles of Psychology. New York. Henry Holt and compa-ny.
52. Ivan Sechenov. (1863) Reflexes of the Brain. MIT Press.
53. Adam Later. (2017) Irresistible: The rise of addictive technology and the Business of Keep-ing Us Hooked. Penguin press.
54. Wolfgram Burgars and Martial Hebert. (2008). Springer Handbook on Robots. Springer.
55. Fei-Fei Li. (2023). The Worlds I See. Flatiron Books.
56. Francis Crick. (1994) The Astonishing Hypothesis. Simon & Schuster.
57. Peter Carruthers. (2015) The Centred Mind: What the Science of Working Memory Shows Us about the Nature of Human Thought. Oxford University press.
58. Robert Lanza, Matej Pavsic Bob Berman. (2016) The Grand Biocentric Design: How Life Creates Reality. BenBella Books.
59. Robert Monroe. (1971) Journey's Out of the Body. Doubleday.
60. Rolf Pfeifer and Josh Bongard. (2007). How the Body Shapes the Way We Think. MIT Press.
61. Gilbert Ryle. (1949) The Concept of Mind. Routledge.
62. Harry Heft. (2001) Ecological Psychology in Context. Psychology Press.
63. Anil Kumar Seth. (2021) Being You a New Science of Consciousness. Faber & Faber.
64. George Northoff. (2023) Neurowaves. McGill-Queen's University Press.
65. Thomas Metzinger. (2004) Being No One. MIT press.
66. Thomas Metzinger. (2010) The Ego Tunnel. Basic Books.
67. Aldous Huxley (1953) Doors of Perception. Vintage Digital.
68. HG Wells (1897). The War of the Worlds. AB Books.
69. Viktor Frankl. (1946) Search for Meaning.Rider.
70. Renee Descartes. (1641) Meditations on First Philosophy. Oxford Univesity Press.
71. Charles Cooley. (1902) Human Nature and the Social Order. General Books.
72. Wilhelm Wundt. (1874). Principles of Physiological Psychology. Forgotten Books.
73. Janet Giele and Glen Elder. (1998) Methods of Life Course Research Sage.
74. Robyn Fivush. (2019) Family Narratives and the Development of an Autobiographical Self. Routledge.
75. Bruce Hood. (2012) The Self Illusion. Constable.
76. Erik Erikson. (1950) Childhood and Society. Vintage.
77. Daniel Dennett. (1996) The Intentional Stance. MIT Press.
78. Paul Ricoeur. (1990). Time and Narrative. University of Chicago Press.
79. Paul Ricoeur. (1992) Oneself as Another. University of Chicago Press.
80. V.S Ramachandran. (2012) Tell Tale Brain. Windmill Books.
81. Anil Ananthaswamy. (2015) The Man Who Wasn't There, Tales from the Edge of the Self. Duckworth.

82. William Wright. (1999) Born that Way. Genes, Behaviour and Personality. Rout-ledge.
83. William R. Clark (1996) Sex and the Origins of Death. Oxford University Press.
84. Michael Grunstein and William R. Clark. (2000). Are we hardwired: The role of genes in human behaviour. Oxford University Press.
85. Douglas Hofstadter. (2007) I Am a Strange Loop. Basic Books.
86. David Hosken, John Hunt and Nina Weddell. (2019) Genes and Behaviour: Beyond Na-ture Nurture. Wiley.
87. Diane F. Halpern. (2000) Sex differences in Cognitive Abilities. Psychology Press.
88. Antonio Damasio (2012). Self Comes to Mind. Cornerstone Digital.
89. Antonio Damasio (2021) Feeling and knowing. Making Minds Conscious. Robinson.
90. Josef Brentano. (1874) Psychology from an Empirical Standpoint. Routledge.
91. Michael Henry. (1963) L'Essence de la manifestation.
92. Charles Sherrington. (1942) Man on his nature (1942). Cambridge University Press.
93. Albert Camus. (1942) The Myth of Sisyphus. Penguin.
94. Thomas Nagel. (1974) What Is It Like to Be a Bat? Reclam Philipp Jun.
95. Joseph LeDoux. (2003) The Synaptic Self. Penguin Books.
96. Hal Hershfield. (2023) Your Future Self. Piatkus.
97. Georg Northoff. (2014) Unlocking the Brain. Oxford University Press.
98. Daniel Batson. (2018) A Scientific Search for Altruism: Do we Only Care About Our-selves? Oxford University Press.
99. Frans De Waal. (2010) The Age of Empathy: Nature's Lessons for a Kinder Society. Sou-venir Press.
100. Matthieu Ricard. (2015) Altruism: The Power of Compassion to Change Yourself and the World. Atlantic Books.
101. Daniel J. Levitin (2016) A Field Guide to Lies. Penguin.
102. Lisa Miller. (2015) The Spiritual child. St Martin's Press.
103. Edward Deci and Richard Ryan. (1985) Self-Determination and Intrinsic Motivation in Human Behaviour. Springer.
104. Eric Kandel (2018) The Disordered Mind. Robinson.
105. R.D Laing. (1960). The Divided Self: An Existential Study in Sanity and Madness. Pen-guin Classics.
106. Herebert Gintis (2009). Bounds of Reason. Princeton University Press.
107. Frederic Bartlett (1932). Remembering. Cambridge University Press.
108. Arthur Koestler (1940). Darkness at Noon. Vintage Classics.
109. Thomas Suddendorf. (2014) The Gap: The Science of What Separates Us from Other An-imals Basic books.
110. Zoé Lacroix (2003). Bioinformatics. Elsevier.

111. Kate Darling. (2021) The New Breed: What Our History with Animals Reveals about Our Future with Robots. Henry Holt & Co.
112. David Levy (2007). Love and Sex with Robots. HarperCollins.
113. Richard Thaler and Cass R. Sunstein. (2009) Nudge. Penguin.
114. Malcolm Gladwell (2007) Blink. Penguin.
115. Daniel Kahneman (2013). Thinking Fast and Slow. Penguin.
116. Charles Darwin (1872). The Expression of the Emotions in Man and Animals. Harper Perennial.
117. Mary Helen Immordino-Yang.Emotions, (2015) Learning, and the Brain. W. W. Norton & Company.
118. Liza Barret Feldman and James A Russell. (2015) The Psychological Construction of Emotion. Guildford Press.
119. Lise Eliot. (2012) Pink Brain, Blue Brain: How Small Differences Grow into Troublesome Gaps.HarperOne.
120. Daniel Goleman (1995) Emotional Intelligence. Bloomsbury Publishing.
121. Batja Mesquita. (2022) Between Us: How Cultures Create Emotions. WW Norton and co.
122. Marvin Minsky (1986) The Society of Mind. Simon and Schuster.
123. Rosalind Picard (2000). Affective Computing. MIT Press.
124. Klaus Scherer. (2010) A Blueprint for Affective Computing: A Sourcebook and Manual. Oxford University Press.
125. Chris Eliasmith. (2013) How to Build a Brain: A Neural Architecture for Biological Cog-nition. Oxford University Press.
126. Hermann Ebbinghaus. (1885) Memory: A Contribution to Experimental Psychology. New York by Teachers College, Columbia University.
127. Dan Schacter. (2001) The Seven Sins of Memory: How the Mind Forgets and Remem-bers. Mariner Books.
128. Deborah Wearing (2006). Forever Today. Corgi.
129. Endel Tulving. (1983) Elements of Episodic Memory.Oxford University Press.
130. Greg Lukianoff and Jonathan Haidt. (2018) The Coddling of the American Mind: How good intentions and bad ideas are setting up a generation for failure. Penguin.

References

A

Abbas, A., Sutter, D., Zoufal, C. et al. The power of quantum neural networks. Nat Comput Sci 1, 403–409 (2021). https://doi.org/10.1038/s43588-021-00084-1

Abbasi S, Maran S, Jaeger D. A general method to generate artificial spike train populations match-ing recorded neurons. J Comput Neurosci. 2020 Feb;48(1):47-63. doi: 10.1007/s10827-020-00741-w. Epub 2020 Jan 23. PMID: 31974719; PMCID: PMC7036336.

Abbink D https://www.tudelft.nl/en/stories/articles/robot-and-humans-have-to-understand-each-other

Abend R, de Voogd L, Salemink E, Wiers RW, Pérez-Edgar K, Fitzgerald A, White LK, Salum GA, He J, Silverman WK, Pettit JW, Pine DS, Bar-Haim Y. Association between attention bias to threat and anxiety symptoms in children and adolescents. Depress Anxiety. 2018 Mar;35(3):229-238. doi: 10.1002/da.22706. Epub 2017 Dec 6. PMID:

Adams S, Arel I, Bach J, Coop R, Furlan R, Goertzel B, Hall JS, Samsonovich A, Scheutz M, Schle-singer M, Shapiro SC, Sowa J (2012) Mapping the landscape of human-level artificial general intel-ligence. AI Mag 33(1):25–42.

Addis DR, Schacter DL. The hippocampus and imagining the future: where do we stand? Front Hum Neu-rosci. 2012 Jan 4; 5:173. doi: 10.3389/fnhum.2011.00173. PMID: 22291625; PMCID: PMC3251274.

Adler JM. Living into the story: agency and coherence in a longitudinal study of narrative identity development and mental health over the course of psychotherapy. J Pers Soc Psychol. 2012 Feb;102(2):367-89. doi: 10.1037/a0025289. Epub 2011 Sep 12. PMID: 21910554.

Agüero, C. E., Martn, F., Rubio, L., and Caas, J. M. (2012). Comparison of smart visual attention mechanisms for humanoid robots. Int. J. Adv. Rob. Syst. 9:233. doi: 10.5772/53571

Aguilera M, Millidge B, Tschantz A, Buckley CL. How particular is the physics of the free energy principle? Phys Life Rev. 2022 Mar; 40:24-50. doi: 10.1016/j.plrev.2021.11.001. Epub 2021 Nov 23. PMID: 34895862; PMCID: PMC8902446.

Amankwah, M.G., Camps, D., Bethel, E.W. et al. Quantum pixel representations and compression for N-dimensional images. Sci Rep 12, 7712 (2022). https://doi.org/10.1038/s41598-022-11024-y

Aımeur, E., Brassard, G. and Gambs, S. Quantum speed-up for unsupervised learning. Machine Learning 90, 261–287 (2013). URL http://dx.doi.org/10.1007/s10994-012-5316-5.

Akomolafe, Dipo. (2013). Scholars Research Library Comparative study of biological and artificial neural networks. European Journal of Applied Engineering and Scientific Research. 2. 36-46.

Akopyan F et al., "TrueNorth: Design and Tool Flow of a 65 mW 1 million Neuron Programmable Neu-rosynaptic Chip," in IEEE Transactions on Computer-Aided Design of Integrated Circuits and Systems, vol. 34, no. 10, pp. 1537-1557, Oct. 2015, doi: 10.1109/TCAD.2015.2474396.

Alexander, G. M., & Hines, M. (2002). Sex differences in response to children's toys in nonhuman primates (Cercopithecus aethiops sabaeus). Evolution and Human Behaviour, 23(6), 467-479.

Ali SA, Begum T, Reza F. Hormonal Influences on Cognitive Function. Malays J Med Sci. 2018 Jul;25(4):31-41. doi: 10.21315/mjms2018.25.4.3. Epub 2018 Aug 30. PMID: 30914845; PMCID: PMC6422548.

Allen EJ, Mesik J, Kay KN, Oxenham AJ. Distinct Representations of Tonotopy and Pitch in Hu-man Auditory Cortex. J Neurosci. 2022 Jan 19;42(3):416-434. doi: 10.1523/JNEUROSCI.0960-21.2021. Epub 2021 Nov 19. PMID: 34799415; PMCID: PMC8802918.

Allen TA, Fortin NJ. The evolution of episodic memory. Proc Natl Acad Sci U S A. 2013 Jun 18;110 Suppl 2(Suppl 2):10379-86. doi: 10.1073/pnas.1301199110. Epub 2013 Jun 10. PMID: 23754432; PMCID: PMC3690604.

Allman JM, Tetreault NA, Hakeem AY, et al.: The von Economo neurons in frontoinsular and ante-rior cingulate cortex in great apes and humans. Brain Struct Funct 2010; 214:495–517 Crossref, Medline.

Alvarado N., 2002 The Role of Emotion in an Architecture of Mind}, https://api.semanticscholar.org/CorpusID:1416416.

Amankwah MG, Camps D, Bethel EW, Van Beeumen R, Perciano T. Quantum pixel representations and compression for N-dimensional images. Sci Rep. 2022 May 11;12(1):7712. doi: 10.1038/s41598-022-11024-y. Erratum in: Sci Rep. 2022 Dec 19;12(1):21931. PMID: 35546151; PMCID: PMC9095730.

An, Shuying. (2013). Schema Theory in Reading. Theory and Practice in Language Studies. 3. 10.4304/ tpls.3.1.130-134.

Anastassiou CA, Perin R, Markram H, Koch C. Ephaptic coupling of cortical neurons. Nat Neuro-sci. 2011 Feb;14(2):217-23. doi: 10.1038/nn.2727. Epub 2011 Jan 16. PMID: 21240273.

Anderson Ashrafi G, Wu Z, Farrell RJ, Ryan TA. GLUT4 Mobilization Supports Energetic De-mands of Active Synapses. Neuron. 2017 Feb 8;93(3):606-615.e3. doi: 10.1016/j.neuron.2016.12.020. Epub 2017 Jan 19. PMID: 28111082; PMCID: PMC5330257.

Anderson, C., & Keltner, D. (2002). The role of empathy in the formation and maintenance of so-cial bonds. Behavioural and Brain Sciences, 25(1), 21–22. https://doi.org/10.1017/S0140525X02230010

Andersen, S. M., & Berenson, K. R. (2003). Wanting: Experiencing Prior Relationships in Present-Day. The social mind: Cognitive and motivational aspects of interpersonal behaviour, 2, 231.

Anderson DJ, Adolphs R. A framework for studying emotions across species. Cell. 2014 Mar 27;157(1):187-200. doi: 10.1016/j.cell.2014.03.003. PMID: 24679535; PMCID: PMC4098837

Anderson P, et al., "Bottom-Up and Top-Down Attention for Image Captioning and Visual Ques-tion Answering," in 2018 IEEE/CVF Conference on Computer Vision and Pattern Recognition (CVPR), Salt Lake City, UT, USA, 2018 pp. 6077-6086. doi: 10.1109/CVPR.2018.00636

Andersen SM, Chen S. The relational self: an interpersonal social-cognitive theory. Psychol Rev. 2002 Oct;109(4):619-45. doi: 10.1037/0033-295x.109.4.619. PMID: 12374322.

Andiliou, Andria & Murphy, P. (2010). Examining variations among researchers' and teachers' conceptualizations of creativity: A review and synthesis of contemporary research. Educational Research Review. 5. 201-219. 10.1016/j.edurev.2010.07.003.

Anjomshoae, S., Najjar, A., Calvaresi, D., & Främling, K. (2019). Explainable agents and robots: Results from a systematic literature review. Proceedings of the 18th international conference on au

Antic S. D., Hines M., and Lytton. W. W. Embedded ensemble encoding hypothesis: The role of the "Prepared" cell, sep 2018. ISSN 10974547.

Antonio Damasio. Self Comes to Mind (2012)

Aosved AC, Long PJ, Voller EK. Measuring sexism, racism, sexual prejudice, ageism, classism, and religious intolerance: The Intolerant Schema Measure. Journal of Applied Social Psychology. 2009;39(10):2321-2354. doi:10.1111/j.1559-1816.2009. 00528.x.

Appel, M., Izydorczyk, D., Weber, S., Mara, M., & Lischetzke, T. (2020). The uncanny of mind in a machine: Humanoid robots as tools, agents, and experiencers. Computers in Human Behaviour, 102, 274–286. https://doi.org/10.1016/j.chb.2019.07.031

Apperly IA, Butterfill SA. Do humans have two systems to track beliefs and belief-like states? Psy-chol Rev. 2009 Oct;116(4):953-70. doi: 10.1037/a0016923. PMID: 19839692.

Aqil M, Knapen T, Dumoulin SO. Divisive normalization unifies disparate response signatures throughout the human visual hierarchy. Proc Natl Acad Sci US A. 2021

Nov 16;118(46):e2108713118. doi: 10.1073/ pnas.2108713118. PMID: 34772812; PMCID: PMC8609633.

Armstrong T, Olatunji BO. Eye tracking of attention in the affective disorders: a meta-analytic re-view and synthesis. Clin Psychol Rev. 2012 Dec;32(8):704-23. doi: 10.1016/j.cpr.2012.09.004. Epub 2012 Sep 20. PMID: 23059623; PMCID: PMC3556338.

Arnaud D'Argembeau, Helena Cassol, Christophe Phillips, Evelyne Balteau, Eric Salmon, Martial Van der Linden, Brains creating stories of selves: the neural basis of autobiographical reason-ing, Social Cognitive and Affective Neuroscience, Volume 9, Issue 5, May 2014, Pages 646–652, https://doi.org/10.1093/scan/nst028

Aru J, Suzuki M, Rutiku R, Larkum ME, Bachmann T. Coupling the State and Contents of Con-sci-ousness. Front Syst Neurosci. 2019 Aug 30; 13:43. doi: 10.3389/fnsys.2019.00043. PMID: 31543762; PMCID: PMC6729974.

Ashby W. R.. Design for a Brain: The Origin of Adaptive Behaviour. London: Chapman and Hall, 1960.

Ashrafi G, Wu Z, Farrell RJ, Ryan TA. GLUT4 Mobilization Supports Energetic Demands of Active Synapses. Neuron. 2017 Feb 8;93(3):606-615.e3. doi: 10.1016/j.neuron.2016.12.020. Epub 2017 Jan 19. PMID: 28111082; PMCID: PMC5330257.

Achterberg, J & Akarca, D et al. Spatially embedded recurrent neural networks reveal widespread links between structural and functional neuroscience findings. Nature Machine Intelligence; 20 Nov 2023; DOI: 10.1038/s42256-023-00748-9

Attwell D, Laughlin SB. An energy budget for signalling in the grey matter of the brain. J Cereb Blood Flow Metab. 2001 Oct;21(10):1133-45. doi: 10.1097/00004647-200110000-00001. PMID: 11598490.

Authenticity and the true self in religion and spirituality by Andrew G. Christy, Grace N. Rivera, Rebecca J. Schlegel 2020 (Pages 119-139)

Avanzini, G. & Franceschetti, S. Cellular biology of epileptogenesis. Lancet Neurol. 2, 33–42 (2003).

Azevedo FA, Carvalho LR, Grinberg LT, Farfel JM, Ferretti RE, Leite RE, Jacob Filho W, Lent R, Hercu-lano-Houzel S. Equal numbers of neuronal and nonneuronal cells make the human brain an isometrically scaled-up primate brain. J Comp Neurol. 2009 Apr 10;513(5):532-41. doi: 10.1002/cne.21974. PMID: 19226510.—

B

Baars BJ, Ramsøy TZ, Laureys S. Brain, conscious experience and the observing self. Trends Neu-rosci. 2003 Dec;26(12):671-5. doi: 10.1016/j.tins.2003.09.015. PMID: 14624851.

Babaeeghazvini P, Rueda-Delgado LM, Gooijers J, Swinnen SP, Daffertshofer A. Brain Structural and Functional Connectivity: A Review of Combined Works of Diffusion

Magnetic Resonance Im-ag-ing and Electro-Encephalography. Front Hum Neurosci. 2021 Oct 7; 15:721206. doi: 10.3389/fn-hum.2021.721206. PMID:

Bachmann T. How to Begin to Overcome the Ambiguity Present in Differentiation between Con-tents and Levels of Consciousness? Front Psychol. 2012 Mar 21; 3:82. doi: 10.3389/fpsyg.2012.00082. PMID: 22461776; PMCID: PMC3309523.

Baddeley, A. D., & Hitch, G. (1974). Working memory. In G.H. Bower (Ed.), The psychology of learning and motivation: Advances in research and theory (Vol. 8, pp. 47–89). New York: Aca-demic Press. Baddeley, A. D. (2000). The episodic buffer: A new component of working memory? Trends in Cognitive Sciences, 4, (11): 417-423.

Baetens, K., Ma, N., Steen, J., and Van Overwalle, F. (2014). Involvement of the mentalizing net-work in social and non-social high construal. Soc. Cogn. Affect. Neurosci. 9, 817–824. doi: 10.1093/scan/nst048.

Bahdanau, D., Cho, K. H., & Bengio, Y. (2015). Neural machine translation by jointly learning to align and translate. Paper presented at 3rd International Conference on Learning Representations, ICLR 2015, San Diego, United States.

Bak, P., Tang, C., and Wiesenfeld, K. (1988). Self-organized criticality. Phys. Rev. A 38, 364–374. doi: 10.1103/PhysRevA.38.364

Balasubramanian V, Kimber D, Berry MJ 2nd. Metabolically efficient information processing. Neu-ral Comput. 2001 Apr;13(4):799-815. doi: 10.1162/089976601300014358. PMID: 11255570.

Balcetis E, Dunning D. See what you want to see: motivational influences on visual perception. J Pers Soc Psychol. 2006 Oct;91(4):612-25. doi: 10.1037/0022-3514.91.4.612. PMID: 17014288

Ballard DH, Zhang R. The Hierarchical Evolution in Human Vision Modelling. Top Cogn Sci. 2021 Apr;13(2):309-328. doi: 10.1111/tops.12527. Epub 2021 Apr 10. PMID: 33838010; PMCID: PMC9462461.

Baltieri M., Buckley C. L., Bruineberg J. (2020). Predictions in the eye of the beholder: an active inference account of Watt governors, in Artificial Life Conference Proceedings, eds Bongard J., Lovato J., Herbert-Dufresne L., Dasari R., Soros L. (Cambridge, MA: MIT Press;), 121–129. 10.1162/isal_a_00288

Bandura, A. (1977). Self-efficacy: Toward a unifying theory of behavioural change. Psychological Review, 84(2), 191-215. http://dx.doi.org/10.1037/0033-295X.84.2.191

Bao AM, Swaab DF. Sex differences in the brain, behaviour, and neuropsychiatric disorders. Neuro-scien-tist. 2010 Oct;16(5):550-65. doi: 10.1177/1073858410377005. PMID: 20889965.

Bandura A. Walters, R. H. Social learning theory, vol. 1 Englewood cliffs Prentice Hall, (1977).

Barandiaran, Xabier & Di Paolo, Ezequiel & Rohde, Marieke. (2009). Defining Agency:

Individual-ity, Normativity, Asymmetry, and Spatio-temporality in Action. Adaptive Behaviour, v.17, 367-386 (2009). 17. 10.1177/1059712309343819.

Bárbara G. Amado, Ramón Arce, Francisca Fariña, Manuel Vilariño, Criteria-Based Content Analysis (CBCA) reality criteria in adults: A meta-analytic review, International Journal of Clinical and Health Psychology, Volume 16, Issue 2, 2016, Pages 201-210, ISSN 1697-2600, https://doi.org/10.1016/j.ijchp.2016.01.002.

Bargh JA, Morsella E. The Unconscious Mind. Perspect Psychol Sci. 2008 Jan;3(1):73-9. doi: 10.1111/ j.1745-6916.2008.00064. x. PMID: 18584056; PMCID: PMC2440575.

Bar-Haim, Y., Lamy, D., Pergamin, L., Bakermans-Kranenburg, M. J., & Van Ijzendoorn, M. H. (2007). Threat-related attentional bias in anxious and nonanxious individuals: a meta-analytic study. Psychological bulletin, 133(1), 1.

Barlow HB: Single units and sensation: a neuron doctrine for perceptual psychology? Perception 1972, 1:371-394.

Barrett LF, Khan Z, Dy J, Brooks D. Nature of Emotion Categories: Comment on Cowen and Kelt-ner. Trends Cogn Sci. 2018 Feb;22(2):97-99. doi: 10.1016/j. tics.2017.12.004. Epub 2018 Jan 16. PMID: 29373283; PMCID: PMC6014873.

Barsalou, L. W. (1999). Perceptual symbol systems. Behavioural and Brain Sciences, 22(4), 577–660. https://doi.org/10.1017/S0140525X99002149

Barsalou, L. W. (2012). The human conceptual system. In M. J. Spivey, K. McRae, & M. F. Joan-isse (Eds.), The Cambridge handbook of psycholinguistics (pp. 239–258). Cambridge University Press. https://doi.org/10.1017/CBO9781139029377.017

Barth C, Villringer A, Sacher J. Sex hormones affect neurotransmitters and shape the adult female brain during hormonal transition periods. Front Neurosci. 2015; 9:37. doi: 10.3389/fnins.2015.00037

Bartlett, F.C. (1932).

Barthas F, Sellmeijer J, Hugel S, Waltisperger E, Barrot M, Yalcin I. The anterior cingulate cortex is a critical hub for pain-induced depression. Biol Psychiatry. 2015 Feb 1;77(3):236-245. doi: 10.1016/j. biopsych.2014.08.004. Epub 2014 Aug 13. PMID: 25433903.

Bajcsy Ruzena. Active perception. Proceedings of the IEEE, 76(8):966–1005, 1988.

Bahdanau, D., Cho, K., & Bengio, Y. (2014). Neural machine translation by jointly learning to align and translate. ArXiv:1409.0473.

Banks, J. (2021). Of like mind: The (mostly) similar mentalizing of robots and hu-mans. Technology, Mind, and Behaviour, 1(2). https://doi.org/10.1037/tmb0000025.

Barber, R. and Salichs, M.A. (2001) A New Human Based Architecture for Intelligent Autonomous Robots. IFAC Proceedings Volumes 34 81-86 https://doi.org/10.1016/S1474-6670(17)33117-8

Bauer PJ. Constructing a past in infancy: a neuro-developmental account. Trends Cogn Sci. 2006 Apr;10(4):175-81. doi: 10.1016/j.tics.2006.02.009. Epub 2006 Mar 14. PMID: 16537115.

Bazan A. (2011). Phantoms in the voice: a neuropsychoanalytic hypothesis on the structure of the unconscious. Neuropsychoanalysis 13, 161–176. 10.1080/15294145.2011.10773672

B. Baars A Cognitive Theory of ConsciousnessCambridge University Press (1988) Short-term forms of presynaptic plasticity. Curr Opin Neurobiol. 2011 Apr;21(2):269-74. doi: 10.1016/j.conb.2011.02.003.Epub 2011 Feb 23. PMID: 21353526; PMCID: PMC3599780.

Barrett D GT, Morcos AS, Macke JH, Analysing biological and artificial neural networks: challeng-es with opportunities for synergy? Current Opinion in Neurobiology, Volume 55, 2019, Pages 55-64, ISSN 0959-4388, https://doi.org/10.1016/j.conb.2019.01.007. (https://www.sciencedirect.com/science/article/pii/S0959438818301569).

Barrett LF, Adolphs R, Marsella S, Martinez AM, Pollak SD. Emotional Expressions Reconsidered: Challenges to Inferring Emotion From Human Facial Movements. Psychol Sci Public Interest. 2019 Jul;20(1):1-68. doi: 10.1177/1529100619832930. Erratum in: Psychol Sci Public Interest. 2019 Dec;20(3):165-166. PMID: 31313636; PMCID: PMC6640856.

Barron HC. Neural inhibition for continual learning and memory. Curr Opin Neurobiol. 2021 Apr; 67:8594. doi: 10.1016/j.conb.2020.09.007. Epub 2020 Oct 28. PMID: 33129012; PMCID: PMC7116367.

Bateson M, Desire S, Gartside SE, Wright GA. Agitated honeybees exhibit pessimistic cognitive biases. Curr Biol. 2011 Jun 21;21(12):1070-3. doi: 10.1016/j.cub.2011.05.017. PMID: 21636277; PMCID: PMC3158593.

Batson, C. D. (2018). A Scientific Search for Altruism: Do we Only Care About Ourselves? Ox-ford: Oxford University Press.

Batson, C.D. & Lishner, David & Stocks, Eric. (2015). The empathy-altruism hypothesis. The Ox-ford handbook of prosocial behaviour. 259-281.

Baumeister, R. F., Vohs, K. D., DeWall, C. N., & Zhang, L. (2007). How emotion shapes behaviour: Feed-back, anticipation, and refllection, rather than direct causation. Personality and Social Psychol-ogy Review, 11(2), 167–203.

Beggs JM, Plenz D (2003) Neuronal avalanches in neocortical circuits. J Neurosci 23:11167–11177.

Beggs JM, Timme N. Being critical of criticality in the brain. Front Physiol. 2012 Jun 7; 3:163. doi: 10.3389/fphys.2012.00163. PMID: 22701101; PMCID: PMC3369250.

Beguš, G., Zhou, A. & Zhao, T.C. Encoding of speech in convolutional layers and the brain stem based on language experience. Sci Rep 13, 6480 (2023). https://doi.org/10.1038/s41598-023-33384-9. Belle, V, Papantonis I (2021) Principles and practice of explainable machine learning Frontiers in big Data, 4.

Bellemare, M. G.; Naddaf, Y.; Veness, J.; and Bowling, M. 2013. The Arcade Learning

Environ-ment: An Evaluation Platform for General Agents. Journal of Artificial Intelligence Research, 47: 253–279.

Bengtsson SL, Dolan RJ, Passingham RE. Priming for self-esteem influences the monitoring of one's own performance. Soc Cogn Affect Neurosci. 2011 Sep;6(4):417-25. doi: 10.1093/scan/nsq048. Epub 2010 Jun 15. PMID: 20551143; PMCID: PMC3150849.

Beniaguev, D,Segev, I and London M. Single cortical neurons as deep artificial neural networks. Neuron, 109(17):2727–2739, 2021.

Benioff, P. The computer as a physical system: A microscopic quantum mechanical Hamiltonian model of computers as represented by Turing machines. J Stat Phys 22, 563–591 (1980). https://doi.org/10.1007/BF01011339

Benjamin, B, Gao, P, McQuinn, E, Choudhary, S, Chandrasekaran, A. Bussat, JM Alvarez-Icaza, R, Arthur, J, Merolla, P,Boahen, K. (2014). Neurogrid: A Mixed-Analog-Digital Multichip System for Large-Scale Neural Simulations. Proceedings of the IEEE. 102. 1-18. 10.1109/JPROC.2014.2313565.

Bering, J. (2006). The cognitive science of souls: Clarifications and extensions of the evolutionary model. Behavioural and Brain Sciences, 29(5), 486-498. doi:10.1017/S0140525X06499106

Berkman, E. T. (2018). The neuroscience of goals and behaviour change. Consulting Psychology Journal: Practice and Research, 70(1), 28–44. https://doi.org/10.1037/cpb0000094

Berlucchi G. Revisiting the 1981 Nobel Prize to Roger Sperry, David Hubel, and Torsten Wiesel on the occasion of the centennial of the Prize to Golgi and Cajal. J Hist Neurosci. 2006 Dec;15(4):369-75. doi: 10.1080/09647040600639013. PMID: 16997764.

Bernstein, J. G. & Boyden, E. S. Optogenetic tools for analysing the neural circuits of behav-ior. Trends Cogn. Sci. 15, 592–600 (2011).

Bertossi E., Ciaramelli E. (2016). Ventromedial pre-frontal damage reduces mind-wandering and biases its temporal focus. Social Cognitive and Affective Neuroscience, 11, 1783–91.

Bhoopchand, A., Brownfield, B., Collister, A. et al. Learning few-shot imitation as cultural trans-mission. Nat Commun 14, 7536 (2023). https://doi.org/10.1038/s41467-023-42875-2

Bianca A. Simonsmeier, Maja Flaig, Anne Deiglmayr, Lennart Schalk & Michael Schnei-der (2022) Domain-specific prior knowledge and learning: A meta-analysis, Educational Psycholo-gist, 57:1, 31-54, DOI: 10.1080/00461520.2021.1939700

Bioinformatics. Managing Scientific Data. 1st Edition – July 18, 2003. Editors: Zoé Lacroix, Ter-ence Critchlow. eBook ISBN: 9780080527987.

Bishop CM (2006) Pattern recognition and machine learning. Springer, New York.

Bissell D, Del Casino V (2017) Whither labour geography and the rise of the robots? Soc Cult Ge-ogr 18(3):435–442

B.K. Siesjo. Brain energy metabolism. John Wiley and Sons,New York, U.S.A., 1978.

Blakeman S, Mareschal D, A complementary learning systems approach to temporal difference learning, Neural Networks, Volume 122, 2020, Pages 218-230, ISSN 0893-6080, https://doi.org/10.1016/j.neunet.2019.10.011. (https://www.sciencedirect.com/science/article/pii/S0893608019303338)

Bliss; Crane, Manfield, Townsend (1966). "Information available in brief tactile presentations" At-tention, Perception, and Psychophysics". Attention, Perception, & Psychophysics. 1 (4): 271–283.

Boag RJ, Strickland L, Heathcote A, Neal A, Palada H, Loft S. Evidence accumulation modelling in the wild: understanding safety-critical decisions. Trends Cogn Sci. 2023 Feb;27(2):175-188. doi: 10.1016/j. tics.2022.11.009. Epub 2022 Dec 3. PMID: 36473764.

Bodiroza S, Schillaci G,Hafner,V V "Robot ego-sphere: An approach for saliency detection and attention manipulation in humanoid robots for intuitive interaction," 2011 11th IEEE-RAS Interna-tional Conference on Humanoid Robots, Bled, Slovenia, 2011, pp. 689-694, doi: 10.1109/Humanoids.2011.6100900.

Boly M, Massimini M, Tsuchiya N, Postle BR, Koch C, Tononi G. Are the Neural Correlates of Consciousness in the Front or in the Back of the Cerebral Cortex? Clinical and Neuroimaging Evi-dence. J Neurosci. 2017 Oct 4;37(40):9603-9613. doi: 10.1523/JNEUROSCI.3218-16.2017. PMID: 28978697; PMCID: PMC5628406.

Bongard, J. C. (2010). The utility of evolving simulated robot morphology increases with task complexity for object manipulation. Artif. Life 16, 201–223. doi:10.1162/artl.2010. Bongard 024.

Boroditsky L. Does language shape thought? Mandarin and English speakers' conceptions of time. Cogn Psychol. 2001 Aug;43(1):1-22. doi: 10.1006/cogp.2001.0748. PMID: 11487292.

Bos, M. W., Dijksterhuis, A., & van Baaren, R. (2012). Food for thought? Trust your unconscious when energy is low. Journal of Neuroscience, Psychology, and Economics, 5(2), 124–130. https://doi.org/10.1037/a0027388

Bostrom, N. 2003 Cognitive, Emotive and Ethical Aspects of Decision Making in Humans and in Artificial Intelligence, Vol. 2, ed. I. Smit et al., Int. Institute of Advanced Studies in Systems Re-search and Cybernetics, 2003, pp. 12-17]

Bostrom, N. 2003. Are You Living in a Computer Simulation? Philosophical Quarterly, 57(211): 243-255.

Bota, P, Wang, C, Fred, A, Plácido da Silva, H. (2019). A Review, Current Challenges, and Future Possibilities on Emotion Recognition Using Machine Learning and Physiological Signals. IEEE Ac-cess. PP. 1-1. 10.1109/ACCESS.2019.2944001.

Bothell D, Byrne MD, Douglass S, Lebiere C, Qin Y (2004) An integrated theory of the mind. Psy-chol Rev 111(4):1036–1060

Bowers, J. S. 2009. 'On the Biological Plausibility of Grandmother Cells: Implications for Neural Network Theories in Psychology and Neuroscience'. Psychological Review, 116: 220–251.

Brauneis R, Goodman EP (2018), Algorithmic transparency for the smart city Yale JL & Tech., 20 p. 103.

Breazeal, C, Scassellati, B.1999. A context dependent attention system for a humanoid robot.

Breazeal, C. Designing Sociable Robots. The MIT Press, 2002.

Braginsky. V.B, Vorontsov Y. I., "Quantum-mechanical limitations in macroscopic experiments and modern experimental technique" Usp. Fiz. Nauk 114, 41–53 (1974) [Sov. Phys. Usp. 17, 644–650 (1975)]; V. B. Braginsky, Y. I. Vorontsov, and F. Y. Khalili "Quantum singularities of a pon-deromotive meter of electromagnetic energy" Zh. Eksp. Teor. Fiz. 73, 1340–1343 (1977) [Sov. Phys. JETP 46, 705–706 (1977)].

Breit S, Kupferberg A, Rogler G, Hasler G. Vagus Nerve as Modulator of the Brain-Gut Axis in Psychiatric and Inflammatory Disorders. Front Psychiatry. 2018 Mar 13; 9:44. doi: 10.3389/fpsyt.2018.00044. PMID: 29593576; PMCID: PMC5859128.

Brem, S., Maurer, U., Kronbichler, M. et al. Visual word form processing deficits driven by severi-ty of reading impairments in children with developmental dyslexia. Sci Rep 10, 18728 (2020). https://doi.org/10.1038/s41598-020-75111-8.

Bridgewater. A.2018: The 13 Types of Data: Sources https://www.forbes.com/sites/adrianbridgwa-ter/2018/07/05/the-13-types-of-data/?sh=439d32bd3362

Broadbent, D. E. (1977). The hidden preattentive processes. American Psychologist, 32(2), 109–118. https://doi.org/10.1037/0003-066X.32.2.109

Broadbent E. Interactions with Robots: The Truths We Reveal About Ourselves. Annu Rev Psy-chol. 2017 Jan 3; 68:627-652. doi: 10.1146/annurev-psych-010416-043958. Epub 2016 Sep 14. PMID: 27648986.

Brod G, Werkle-Bergner M, Shing YL. The influence of prior knowledge on memory: a develop-mental cognitive neuroscience perspective. Front Behav Neurosci. 2013 Oct 8; 7:139. doi: 10.3389/fn-beh.2013.00139. PMID: 24115923; PMCID: PMC3792618.

Brody LR, Hall JA. Gender, emotion, and expression. Handbook of emotions. 2000; 2: 338–49.

Broekens, J. and D. DeGroot. "Scalable and Flexible Appraisal Models for Virtual Agents," CGAIDE, 2004.

Brosch T, Scherer KR, Grandjean D, Sander D. The impact of emotion on perception, attention, memory, and decision-making. Swiss Med Wkly. 2013 May 14;143: w13786. doi: 10.4414/smw.2013.13786. PMID: 23740562.

Buckner RL, Andrews-Hanna JR, Schacter DL (2008) The brain's default network: anatomy, func-tion, and relevance to disease. Ann NY Acad Sci 1124:1–38.

Buechel, Sven & Hahn, Udo. (2016). Emotion Analysis as a Regression Problem — Dimensional Models and Their Implications on Emotion Representation and Metrical Evaluation. 10.3233/978-1-61499-672-91114.

Buolamwini J, Gebru T 2018 Gender Shades: Intersectional Accuracy Disparities in Commercial Gender Classification Proceedings of Machine Learning Research 81:1–15, 2018. https://www.classes.cs.uchicago.edu/archive/2020/winter/20370 1/readings/gendershadesAIbias.pdf

Burg MF, Cadena SA, Denfield GH, Walker EY, Tolias AS, Bethge M, Ecker AS. Learning divisive normalization in primary visual cortex. PLoS Comput Biol. 2021 Jun 7;17(6):e1009028. doi: 10.1371/journal. pcbi.1009028. PMID: 34097695; PMCID: PMC8211272.

Burgard, W., Hebert, M. (2008). World Modelling. In: Siciliano, B., Khatib, O. (eds) Springer Hand-book of Robotics. Springer, Berlin, Heidelberg. https://doi.org/10.1007/978-3-540-30301-5_37.

Burgoon, J. K., and Bacue, A. E. (2003). "Nonverbal communication skills" in Handbook of com-munica-tion and social interaction skills. eds. J. O. Greene and B.R. Burleson (Hillsdale, NJ: Law-rence Erlbaum Associates), 179–219

Burkart JM, Adriaense JEC, Brügger RK, Miss FM, Wierucka K, van Schaik CP. A convergent in-terac-tion engine: vocal communication among marmoset monkeys. Philos Trans R Soc Lond B Biol Sci. 2022 Sep 12;377(1859):20210098. doi: 10.1098/rstb.2021.0098. Epub 2022 Jul 25. PMID: 35876206; PMCID: PMC9315454.

Burr, G.W. Shelby, R.M. Sebastian, A S. Kim, Kim, S. Sidler, S K. Virwani, M. Ishii, P. Narayanan, A. Fumarola, et al.Neuromorphic computing using non-volatile memory. Adv. Phys. X, 2 (2017), pp. 89-124

Butterfield, J., Jenkins, O.C., Sobel, D.M. et al. Modelling Aspects of Theory of Mind with Markov Random Fields. Int J of Soc Robotics 1, 41–51 (2009). https://doi.org/10.1007/s12369-008-0003-1.

Buzsaki, G.: Rhythms of the Brain. Oxford University Press, Oxford (2006)

Buzsáki G, Kaila K, Raichle M. Inhibition and brain work. Neuron. 2007 Dec 6;56(5):771-83. doi: 10.1016/j.neuron.2007.11.008. PMID: 18054855; PMCID: PMC2266612.

C

Cabeza R, St Jacques P. Functional neuroimaging of autobiographical memory. Trends Cogn Sci. 2007;11(5):219–227.

Calvo, R. A., D'Mello, S., Gratch, J., & Kappas, A. (2015). The Oxford handbook of affective computing. Oxford Library of Psychology.

Cangelosi, A., and Schlesinger, M. (2015). Developmental Robotics: From Babies to Robots. Cam-bridge, MA: MIT Press.

Cangelosi A. The grounding and sharing of symbols. Pragmatics and Cognition, 14(2):275–285, 2006.

Cangelosi, A & Bongard, J & Fischer, M & Nolfi, S. (2015). Embodied Intelligence. Springer Handbook of Computational Intelligence. 697-714. 10.1007/978-3-662-43505-2_37.)

Carreño, I. D. V. G. (2014). Theory of Connectivity as an Emergent Solution to Innovative Learn-ing Strategies. American Journal of Educational Research, 2(2), 107-116.

Carhart-Harris RL, Leech R, Hellyer PJ, Shanahan M, Feilding A, Tagliazucchi E, Chialvo DR, Nutt D. The entropic brain: a theory of conscious states informed by neuroimaging research with psy-chedelic drugs. Front Hum Neurosci. 2014 Feb 3; 8:20. doi: 10.3389/fnhum.2014.00020. PMID: 24550805; PM-CID: PMC3909994.

Carhart-Harris RL. The entropic brain – revisited. Neuropharmacology. 2018 Nov; 142:167-178. doi: 10.1016/j.neuropharm.2018.03.010. Epub 2018 Mar 14. PMID: 29548884.

Casali AG, Gosseries O, Rosanova M, Boly M, Sarasso S, Casali KR, Casarotto S, Bruno MA, Lau-reys S, Tononi G, Massimini M. A theoretically based index of consciousness independent of sen-sory processing and behaviour. Sci Transl Med. 2013 Aug 14;5(198):198ra105. doi: 10.1126/scitranslmed.3006294. PMID: 23946194.

Casasanto D, Casasanto LS, Gijssels T, Hagoort P. The reverse chameleon effect: negative social consequences of anatomical mimicry. Front Psychol. 2020; 11:1876. doi:10.3389/fpsyg.2020.01876

Cassimatis N, Nicholas L. (2002). Polyscheme: A Cognitive Architecture for Integrating Multiple Representation and Inference Schemes.

Caves, C, Fuchs, C, Schack, R (2001). Quantum probabilities as Bayesian probabilities. Physical Review A. 65. 10.1103/PhysRevA.65.022305.

Cera N, Vargas-Cáceres S, Oliveira C, Monteiro J, Branco D, Pignatelli D, Rebelo S. How Relevant is the Systemic Oxytocin Concentration for Human Sexual Behaviour? A Systematic Review. Sex Med. 2021 Aug;9(4):100370. doi: 10.1016/j.esxm.2021.100370. Epub 2021 Jun 9. PMID: 34118520; PMCID: PMC8360917.

Cerra, Peggy. (2003). The First Law of Psychology is the Second Law of Thermodynamics: The Energetic Evolutionary Model of the Mind and the Generation of Human Psychological Phenome-na. Entropy is the primary adaptive problem that life forms must solve.

Chai WJ, Abd Hamid AI, Abdullah JM. Working Memory from the Psychological and Neuroscienc-es Perspectives: A Review. Front Psychol. 2018 Mar 27; 9:401. doi: 10.3389/fpsyg.2018.00401. PMID: 29636715; PMCID: PMC5881171.

Chalmers, D. (1996). The Conscious Mind: In Search of a Fundamental Theory. Oxford: Oxford University Press.

Chamberlain K.A. et al. Oligodendrocytes enhance axonal energy metabolism by deacetylation of mito-chondrial proteins through transcellular delivery of SIRT2. Neuron. September 9, 2021. DOI: 10.1016/j. neuron.2021.08.011

Chance SA, Casanova MF, Switala AE, Crow TJ. Auditory cortex asymmetry altered minicolumn spacing and absence of ageing effects in schizophrenia. Brain. 2008 Dec;131(Pt 12):3178-92. doi: 10.1093/brain/ awn211. Epub 2008 Sep 26. PMID: 18819990; PMCID: PMC2724907.

Chandrasekaran A Yadav D., Chattopadhyay P, Prabhu, V. Parikh D., It takes two to tango: To-wards theory of AI's mind, 2017, arXiv:1704.00717.

Chang, C.-H. (Daisy), Ferris, D. L., Johnson, R. E., Rosen, C. C., & Tan, J. A. (2012). Core Self-Evaluations: A Review and Evaluation of the Literature. Journal of Management, 38(1), 81–128.

Chang L, et al 2014

Chang, Y, Li, W, Peng, J, Tang, B,Kang, Y, Lei, Y, Gui, Y, Zhu, Q, Liu, Y, Li, H. (2021). Re-viewing continual learning from the perspective of human-level intelligence. https://www.ninds.nih.gov/News-Events/News-and-Press-Releases/Press-Releases/Signaling-neighbor-ing-cells-provides-power-boost.

Chartrand TL, Bargh JA. The chameleon effect: The perception–behaviour link and social interac-tion. Journal of Personality and Social Psychology. 1999;76(6):893-910.

Chen, B., Vondrick, C. & Lipson, H. Visual behaviour modelling for robotic theory of mind. Sci Rep 11, 424 (2021). https://doi.org/10.1038/s41598-020-77918-x

Chen, H., Lee, Y., Maxwell, T., Sun, G., Lee, H. and Giles, C. (1986): "High order correlation model for associative memory," J. Denker, Ed., AIP Conf. Proc., Snowbird Utah.

Chen Qiu, Rajat S. Shivacharan, Mingming Zhang and Dominique M. Durand. Can Neural Activity Propagate by Endogenous Electrical Field? Journal of Neuroscience 2 December 2015, 35 (48) 15800-15811; DOI: https://doi.org/10.1523/JNEUROSCI.1045-15.2015.

Chen, Y. –C., et al. (2015). Transcriptional regulator PRDM12 is essential for human pain percep-tion. Nat. Genet. doi: 10.1038/ng.3308.

Cheng, Yujiao & Sun, Liting & Tomizuka, Masayoshi. (2019). Towards Better Human Robot Col-labora-tion with Robust Plan Recognition and Trajectory Prediction.

Cherbuin N, Luders E, Chou YY, Thompson PM, Toga AW, Anstey KJ. Right, left, and centre: how does cerebral asymmetry mix with callosal connectivity? Hum Brain Mapp. 2013 Jul;34(7):1728-36. doi: 10.1002/hbm.22022. Epub 2012 Mar 15. PMID: 22419524; PMCID: PMC4084519.

Cheruvu, Ria. (2018). The Neuroscience of Flow. 10.13140/RG.2.2.12520.57608.

Chiou SC. Integration or separation? Spatial and temporal representations of whole-body move-ments in visual working memory. Mem Cogn 51, 1076–1089 (2023). https://doi.org/10.3758/s13421-022-01387-y

Chipman, Susan E. F. (ed.), The Oxford Handbook of Cognitive Science, Oxford Hand-books (2017; online edn, Oxford Academic, 3 Nov. 2014), https://doi.org/10.1093/oxfordhb/9780199842193.001.0001, accessed 18 May 2023.

Cho, Kyunghyun & Merrienboer, Bart & Gulcehre, Caglar & Bougares, Fethi & Schwenk, Holger & Bengio, Y.. (2014). Learning Phrase Representations using RNN Encoder-Decoder for Statistical Machine Translation. 10.3115/v1/D14-1179.

Chou, T.-S., Bucci, L. D., and Krichmar, J. L. (2015). Learning touch preferences with a tactile ro-bot using dopamine modulated stdp in a model of insular cortex. Front. Neurorobot. 9:6. doi: 10.3389/fn-bot.2015.00006.

Christou P, Simillidou A. Stylianou M.C. Tourists' perceptions regarding the use of anthropo-morphic robots in tourism and hospitality Int. J. Contemp. Hospit. Manag., 32 (11) (2020), pp. 3665-3683

Chu J, Qaisar S, Shah Z, Jalil A. Attention or Distraction? The Impact of Mobile Phone on Users' Psychological Well-Being. Front Psychol. 2021 Apr 20; 12:612127. doi: 10.3389/fpsyg.2021.612127. PMID: 33959065; PMCID: PMC8093572.

Chua L Memristor. The missing IEEE Trans. Circuit Theor., 18 (1971), pp. 507-519.

Chun, Marvin M. and Golomb, Julie D. and Turk-Browne, Nicholas B., A Taxonomy of External and Internal Attention (January 2011). Annual Review of Psychology, Vol 62 p 73-10 SSRN: https://ssrn.com/abstract=1722016 or http://dx.doi.org/10.1146/annurev.psych.093008.100427

Cisek P, Kalaska JF. Neural mechanisms for interacting with a world full of action choices. Annu Rev Neurosci. 2010; 33:269-98. doi: 10.1146/annurev.neuro.051508.135409. PMID: 20345247.

Cisek P. A parallel framework for interactive behaviour. Prog Brain Res. 2007; 165:475-92. doi: 10.1016/ S0079-6123(06)65030-9. PMID: 17925265.

Cisler JM, Koster EH. Mechanisms of attentional biases towards threat in anxiety disorders: An integrative review. Clin Psychol Rev. 2010 Mar;30(2):203-16. doi: 10.1016/j.cpr.2009.11.003. Epub 2009 Dec 14. PMID: 20005616; PMCID: PMC2814889.

Citri A, Malenka RC. Synaptic plasticity: multiple forms, functions, and mechanisms. Neuropsy-chophar-macology. 2008 Jan;33(1):18-41. doi: 10.1038/sj.npp.1301559. Epub 2007 Aug 29. PMID: 17728696.

Clare L, Martyr A, Morris RG, Tippett LJ. Discontinuity in the Subjective Experience of Self Among People with Mild-To-Moderate Dementia Is Associated with Poorer Psychological Health: Findings from the IDEAL Cohort. J Alzheimers Dis. 2020;77(1):127-138. doi: 10.3233/JAD-200407. PMID: 32804138; PMCID: PMC7592652.

Clayton NS, Bussey TJ, Dickinson A. Can animals recall the past and plan for the future? Nat Rev Neuro-sci. 2003;4(8):685–691.

Clayton NS, Dickinson A. Episodic-like memory during cache recovery by scrub jays. Nature. 1998;395(6699):272–274.

Cohler, B. J. (1982). Personal narrative and the life course. In P. Baltes & O.G. Brim (Eds.), Life span development and behaviour (Vol. 4, pp. 205-241). New York: Academic Press.

Colin Ware, in Visual Thinking for Information Design (Second Edition), 2022.
Collell G., Fauquet J. (2014). How the relationship between information theory and thermodynam-ics can contribute to explaining brain and cognitive activity: an integrative approach, in Proceedings of the IEEE Sixth International Conference on Advanced Cognitive Technologies and Applications (COGNITIVE 2014) (Venice), 269–273.

Collell G, Fauquet J. Brain activity and cognition: a connection from thermodynamics and infor-mation theory. Front Psychol. 2015 Jun 16; 6:818. doi: 10.3389/fpsyg.2015.00818. PMID: 26136709; PMCID: PMC4468356.

Colombini, E, Simões A,Ribeiro C. (2016). An Attentional Model for Autonomous Mobile Ro-bots. IEEE Systems Journal. PP. 1-12. 10.1109/JSYST.2015.2499304.

Colosio M, Shestakova A, Nikulin V, Blagovechtchenski E and Klucharev V. Neural mechanisms of cognitive dissonance (revised): An EEG study" by in Journal of Neuroscience. Published online April 24, 2017, doi:10.1523/JNEUROSCI.3209-16.2017.

Conchla ML, Bianco IH, Wilson SW (2012) Encoding asymmetry within neural circuits. Nat Rev Neuro-sci 13: 832–843.

Conway MA, Meares K, Standart S. Images and goals. Memory. 2004 Jul;12(4):525-31. doi: 10.1080/09658210444000151. PMID: 15487548.

Greenwald, A. G. (1980). The totalitarian ego: Fabrication and revision of personal histo-ry. American Psychologist, 35(7), 603–618. https://doi.org/10.1037/0003-066X.35.7.603

Conway M, Lucy Justice L, D'Argembeau A, The Self-Memory System Revisited In: The organi-zation and structure of autobiographical memory. Edited by: John H. Mace, Oxford University Press (2019). © Oxford University Press. DOI: 10.1093/oso/9780198784845.003.0003

Conway M A, Pleydell-Pearce CW. The construction of autobiographical memories in the self-memory system. Psychological Review. 2000; 107:261–88.

Cook J, Knight E, Hume I, Qureshi A. The self-esteem of adults diagnosed with attention-deficit/hy-peractivity disorder (ADHD): a systematic review of the literature. Atten Defic Hyperact Disord. 2014 Dec;6(4):249-68. doi: 10.1007/s12402-014-0133-2. Epub 2014 Mar 26. PMID: 24668198.

Cook KM, You X, Cherry JB, Merchant JS, Skapek M, Powers MD, Pugliese CE, Kenworthy L, Vaidya CJ. Neural correlates of schema-dependent episodic memory and association with behav-ioral flexibility in autism spectrum disorders and typical development. J Neurodev Disord. 2021 Sep 15;13(1):35. doi: 10.1186/s11689-021-09388-9. PMID: 34525948; PMCID: PMC8442441.

Coplan, A. (2011). Understanding empathy: Its features and effects. In: A. Coplan & P. Goldie (Eds.), Empathy: Philosophical and psychological perspectives (pp. 5–18). Oxford University Press. Corbetta M, Shulman GL. Control of goal-directed and stimulus-driven attention in the brain. Nat Rev Neurosci. 2002 Mar;3(3):201-15. doi: 10.1038/nrn755. PMID: 11994752.

Coria-Avila GA, Manzo J, Garcia LI, Carrillo P, Miquel M, Pfaus JG. Neurobiology of social at-tach-ments. Neurosci Biobehav Rev. 2014 Jun; 43:173-82. doi: 10.1016/j.neubiorev.2014.04.004. Epub 2014 Apr 21. PMID: 24769402.

Cover TM and Thomas J.A. Elements of Information Theory. Wiley, New York, 1991

Coward, L.A. (2011). Modelling Memory and Learning Consistently from Psychology to Physiolo-gy. In: Cutsuridis, V., Hussain, A., Taylor, J. (eds) Perception-Action Cycle. Springer Series in Cognitive and Neural Systems. Springer, New York, NY. https://doi.org/10.1007/978-1-4419-1452-1_3.

Craft, A. (2011). Creativity and education futures: Learning in a digital age. Staffordshire, UK: Trentham Books.

Cramer P. Understanding Defence Mechanisms. Psychodyn Psychiatry. 2015 Dec;43(4):523-52. doi: 10.1521/pdps.2015.43.4.523. PMID: 26583439.

Creswell, J.D., Dutcher, J.M., Klein, W.M.P., Harris, P.R., Levine, J.M., 2013a. Self-affirmation improves problem-solving under stress. PLoS One 8 (5), e62593. https://doi.org/10.1371/journal.pone.0062593.0

Çukur, D. ve Delice, E.G. (2011). Erken Çocukluk Döneminde Görsel Algı Gelişimine Uygun Mekan Tasarımı. Aile ve Toplum Dergisi, 12, (7) 24

C. Stephani, R. Maciunas, M. Koubeissi et al.: Functional Neuroanatomy of the Insular Lobe. Brain Structure & Function (2011), vol. 2016, issue 2, p. 137-149

Cully A, Clune J, Tarapore D, Mouret JB. Robots that can adapt like animals. Nature. 2015 May 28;521(7553):503-7. doi: 10.1038/nature14422. PMID: 26017452.

Curran, T., and Morgan, J. I. (1985). Superinduction of c-fos by nerve growth factor in the pres-ence of peripherally active benzodiazepines. Science 229, 1265–1268. doi: 10.1126/science.4035354.

Csikszentmihalyi, M. (1975). Beyond boredom and anxiety. San Francisco: Jossey-Bass.

D

Da Costa L, Parr T, Sajid N, Veselic S, Neacsu V, Friston K. Active inference on discrete state-spaces: A synthesis. J Math Psychol. 2020 Dec; 99:102447. doi: 10.1016/j.jmp.2020.102447. PMID: 33343039; PMCID: PMC7732703.

Dafni-Merom A, Arzy S. The radiation of autonoetic consciousness in cognitive neuroscience: A functional neuroanatomy perspective. Neuropsychologia. 2020 Jun; 143:107477. doi: 10.1016/j.neuropsycholo-gia.2020.107477. Epub 2020 Apr 30. PMID: 32360475.)

D. D'Amico, M.E. Amestoy, A.J. FioccoThe association between allostatic load and cognitive func-tion: A systematic and meta-analytic review Psychoneuroendocrinology, 121 (2020), Article 104849, 10.1016/j.psyneuen.2020.104849.

Daniel P. Johnson, Mark A. Whisman, Gender differences in rumination: A meta-analysis,

Person-ality and Individual Differences, Volume 55, Issue 4, 2013, Pages 367-374, ISSN 0191-8869, https://doi.org/10.1016/j.paid.2013.03.019. (https://www.sciencedirect.com/science/article/pii/S0191886913001396).

Darling, Kate, 'Who's Johnny?' Anthropomorphic Framing in Human-Robot Interaction, Integration, and Policy (March 23, 2015). ROBOT ETHICS 2.0, eds. P. Lin, G. Bekey, K. Abney, R. Jenkins, Oxford University Press, 2017, Forthcoming, Available at SSRN: https://ssrn.com/abstract=2588669 or http://dx.doi.org/10.2139/ssrn.2588669

Davis, G. A. (2004). Creativity Is Forever (5th ed.). Dubuque, IA: Kendall Hunt Publishing.

de Bruijn H, Warnier. M, Janssen M, 2022, The perils and pitfalls of explainable AI: Strategies for explaining algorithmic decision-making, Government Information Quarterly,Volume 39, Issue 2, 101666. ISSN0740624X,(https://www.sciencedirect.com/science/article/pii/S0740624X21001027)

De Rubeis J, Lugo RG, Witthöft M, Sütterlin S, Pawelzik MR, Vögele C. Rejection sensitivity as a vulnerability marker for depressive symptom deterioration in men. PLoS One. 2017;12(10): e0185802. doi: 10.1371/journal.pone.0185802

Devin S, Aurélie C Rachid. A. 2017). About Decisions During Human-Robot Shared Plan Achievement: Who Should Act and How?

D'Argembeau A, Feyers D, Majerus S, Collette F, Van der Linden M, Maquet P, Salmon E. Self-reflection across time: cortical midline structures differentiate between present and past selves. Soc Cogn Affect Neurosci. 2008 Sep;3(3):244-52. doi: 10.1093/scan/nsn020. Epub 2008 Aug 23. PMID: 19015116; PM-CID: PMC2566769.

David Sander, Lauri Nummenmaa, Reward and emotion: an affective neuroscience approach, Current Opinion in Behavioural Sciences, Volume 39, 2021, Pages 161-167, ISSN 2352-1546, https://doi.org/10.1016/j.cobeha.2021.03.016.

Davis, B., & Francis, K. (2022). "Unprepared Learning" in Discourses on Learning in Education. https://learningdiscourses.com.

Davis J, Hsieh YH, Lee HC. Humans perceive flicker artifacts at 500 Hz. Sci Rep. 2015 Feb 3; 5:7861. doi: 10.1038/srep07861. PMID: 25644611; PMCID: PMC4314649.

De Pittà, M., Berry, H. (2019). A Neuron–Glial Perspective for Computational Neuroscience. In: De Pittà, M., Berry, H. (eds) Computational Glioscience. Springer Series in Computational Neurosci-ence. Springer, Cham. https://doi.org/10.1007/978-3-030-00817-8_1

Dedeurwaerdere S December 1, 2022. New Scientist https://www.newscientist.com/article/2348657-inside-the-future-of-epilepsy-therapy/

Degutis JM, Van Vleet TM. Tonic and phasic alertness training: a novel behavioural therapy to im-prove spatial and non-spatial attention in patients with hemispatial neglect. Front Hum Neurosci. 2010 Aug 24;4:60. doi: 10.3389/fnhum.2010.00060. PMID: 20838474; PMCID: PMC2936932.

Dehghani N., Wimmer R. D. (2018). A computational perspective of the role of thalamus in cogni-tion.

DeNardo LA, Liu CD, Allen WE, Adams EL, Friedmann D, Fu L, Guenthner CJ, Tessier-Lavigne M, Luo L. Temporal evolution of cortical ensembles promoting remote memory retrieval. Nat Neu-rosci. 2019 Mar;22(3):460-469. doi: 10.1038/s41593-018-0318-7. Epub 2019 Jan 28. PMID: 30692687; PMCID: PMC6387639.

Deng Y, Chang L, Yang M, Huo M, Zhou R(2016) Gender Differences in Emotional Re-sponse:Inconsist-ency between Experience and Expressivity.PLoS ONE 11(6): e0158666. doi: 10.1371/journal.pone

Dennett, D. (1991) Consciousness Explained, Little Brown & Co.

Dent EW. Of micro-tubules and memory: implications for microtubule dynamics in dendrites and spines. Mol Biol Cell. 2017 Jan 1;28(1):1-8. doi: 10.1091/mbc. E15-11-0769. PMID: 28035040; PMCID: PMC5221613.

Descartes's (1596-1650) in his "Meditations on First Philosophy"

Desimone R, Duncan J. Neural mechanisms of selective visual attention. Annu Rev Neurosci. 1995; 18:193-222. doi: 10.1146/annurev.ne.18.030195.001205. PMID: 7605061.

Deutsch H. Some forms of emotional disturbance and their relationship to schizophrenia. Psychoanal Q. 1942; 11:301– 321 de Vignemont F, Singer T. The empathic brain: how, when and why? Trends Cogn Sci. 2006 Oct;10(10):435-41. doi: 10.1016/j.tics.2006.08.008. Epub 2006 Sep 1. PMID: 16949331.

Diamond A. Executive functions. Annu Rev Psychol. 2013; 64:135-68. doi: 10.1146/an-nurev-psych-113011-143750. Epub 2012 Sep 27. PMID: 23020641; PMCID: PMC4084861.

Diehl JJ, Schmitt LM, Villano M, Crowell CR. 2012. The clinical use of robots for individuals with autism spectrum disorders: a critical review. Res. Autism Spect. Dis. 6: 249–62

Dietterich, T.G., Wettschereck, D., Atkeson, C.G., & Moore, A.W. (1993). Memory-Based Meth-ods for Regression and Classification. NIPS.

Di Lollo, V. Temporal characteristics of iconic memory. Nature 267, 241–243 (1977). https://doi.org/10.1038/267241a0

Dijksterhuis A, Bos MW, Nordgren LF, van Baaren RB. On making the right choice: the delib-era-tion-without-attention effect. Science. 2006 Feb 17;311(5763):1005-7. doi: 10.1126/science.1121629. PMID: 16484496.

Dijksterhuis A, Strick M. A Case for Thinking Without Consciousness. Perspect Psychol Sci. 2016 Jan;11(1):117-32. doi: 10.1177/1745691615615317. PMID: 26817729.

Dijksterhuis A, Nordgren LF. A Theory of Unconscious Thought. Perspect Psychol Sci. 2006 Jun;1(2):95-109. doi: 10.1111/j.1745-6916.2006.00007. x. PMID: 26151465.

Di Paolo E. A. (2003). Organismically-inspired robotics: homoeostatic adaptation and teleology be-yond the closed sensorimotor loop, in Dynamic Systems Approach for Embodiment and Sociality: From Ecological Psychology to Robotics, eds Murase K., Asakura T. (Adelaide, SA: Advanced Knowledge International;), 19–42.

Di Paolo, Ezequiel. (2000). Homoeostatic Adaptation to Inversion of the Visual Field and Other Sensori-motor Disruptions.

Dohnalová, L., Lundgren, P., Carty, J.R.E. et al. A microbiome-dependent gut–brain pathway regu-lates motivation for exercise. Nature 612, 739–747 (2022). https://doi.org/10.1038/s41586-022-05525-z

Dolan RJ. Emotion, cognition, and behaviour. Science. 2002 Nov 8;298(5596):1191-4. doi: 10.1126/sci-ence.1076358. PMID: 12424363.

Donadi, S., Piscicchia, K., Curceanu, C. et al. Underground test of gravity-related wave function collapse. Nat. Phys. 17, 74–78 (2021). https://doi.org/10.1038/s41567-020-1008-4.

Dong, D, Chen, C, Zhang, C, Chen, Z. (2006). Quantum robot: structure, algorithms and applica-tions. Robotica. 24. 513-521.

Dong, D., Chunlin, C. and Zonghai, C. Quantum Reinforcement Learning. Advances in Natural Computation Lecture Notes in Computer Science 3611 686-689 (2005).

D'Onofrio G, Fiorini L, Sorrentino A, Russo S, Ciccone F, Giuliani F, Sancarlo D, Cavallo F. Emotion Recognizing by a Robotic Solution Initiative (EMOTIVE Project). Sensors (Basel). 2022 Apr 8;22(8):2861. doi: 10.3390/s22082861. PMID: 35458845; PMCID: PMC9031388.

Donoghue JP. Bridging the brain to the world: a perspective on neural interface systems. Neuron. 2008 Nov 6;60(3):511-21. doi: 10.1016/j.neuron.2008.10.037. PMID: 18995827.

Dorigo, Marco & Colombetti, Marco. (1998). Robot Shaping: An Experiment in Behaviour Engineer-ing.

Dosovitskiy, A., Beyer, L., Kolesnikov, A., Weissenborn, D., Zhai, X., Unterthiner, T., Dehghani, M., Min-derer, M., Heigold, G., Gelly, S., Uszkoreit, J., & Houlsby, N. (2020). An Image is Worth 16x16 Words: Transformers for Image Recognition at Scale. ArXiv, abs/2010.11929.

Downey G, Feldman SI. Implications of rejection sensitivity for intimate relationships. J Pers Soc Psychol. 1996;70(6):1327–1343. doi:10.1037//0022-3514.70.6.1327

Drake ME. Cotard's syndrome and temporal lobe epilepsy. Psychiatr J Univ Ottawa 1988; 13(1): 36-39.

Drevets WC, Savitz J, Trimble M. The subgenual anterior cingulate cortex in mood disorders. CNS Spectr. 2008 Aug;13(8):663-81. doi: 10.1017/s1092852900013754. PMID: 18704022; PMCID: PMC2729429.

Driskell, J. E., Willis, R. P., & Copper, C. (1992). Effect of overlearning on retention. Journal of Applied Psychology, 77(5), 615–622. https://doi.org/10.1037/0021-9010.77.5.615.

Duan S, Lin Y, Wang Z, Tang J, Li Y, Zhu D, Wu J, Tao L, Choi CH, Sun L, Xia J, Wei L, Wang B. Conductive Porous MXene for Bionic, Wearable, and Precise Gesture Motion Sensors. Research (Wash DC). 2021 Jun 9; 2021:9861467. doi: 10.34133/2021/9861467. PMID: 34223178; PMCID: PMC8212815.

Dufek, Jan & Xiao, Xuesu & Murphy, Robin. (2020). Best Viewpoints for External Robots or Sen-sors Assisting Other Robots.

Dunjko V, Taylor JM, Briegel HJ. Quantum-Enhanced Machine Learning. Phys Rev Lett. 2016 Sep 23;117(13):130501. doi: 10.1103/PhysRevLett.117.130501. Epub 2016 Sep 20. PMID: 27715099.

Dunn J, Grider MH. Physiology, Adenosine Triphosphate. [Updated 2023 Feb 13]. In: StatPearls [Internet]. Treasure Island (FL): StatPearls Publishing; 2023 Jan-. Available from: https://www.ncbi.nlm.nih.gov/books/NBK553175.

Dutcher, J.M., Creswell, J.D., Pacilio, L.E., Harris, P.R., Klein, W.M.P., Levine, J.M., et al., 2016. Self-affirmation activates the ventral striatum: a possible reward-related mechanism for self-affirmation. Psychol. Sci. 27 (4), 455–466. https://doi.org/10.1177/0956797615625989).

Duval, S., & Wicklund, R. A. (1972). A theory of objective self awareness. Academic Press.

E

Eagly, A. H., Eaton, A., Rose, S. M., Riger, S., & McHugh, M. C. (2012). Feminism and psycholo-gy: Analysis of a half-century of research on women and gender. American Psychologist, 67(3), 211–230. https://doi.org/10.1037/a0027260

Edelman, G.M. and Gally, J.A. (2001) Degeneracy and complexity in biological systems. Proc. Natl. Acad. Sci. U. S. A. 98, 13763–13768

Egeth HE, Yantis S. Visual attention: control, representation, and time course. Annu Rev Psychol. 1997; 48:269-97. doi: 10.1146/annurev.psych.48.1.269. PMID: 9046562.

"Electrical stimulation of the macaque ventral tegmental area drives category-selective learning without attention" by Wim Vanduffel et al. Neuron.

Ernst, M., Nelson, E. E., Jazbec, S. P., McClure, E. B., Monk, C. S., Leibenluft, E., Blair, J., and Pine, D. S. (2005). Amygdala and nucleus accumbens in responses to receipt and omission of gains in adults and adolescents. Neuroimage 25, 1279–1291.

Echeveste R, Lengyel M. The Redemption of Noise: Inference with Neural Populations. Trends Neurosci. 2018 Nov;41(11):767-770. doi: 10.1016/j.tins.2018.09.003. PMID: 30366563; PMCID: PMC6416224.

Edelman GM, Gally JA. Degeneracy and complexity in biological systems. Proc. Natl. Acad. Sci. USA. 2001; 98:13763–13768.

EdgeAI:The Ultimate Guide to Vision Processing Units (VPUs).2022 https://www.thinkautonomous.ai/blog/vision-processing-units-vpus/#:~:text=Unlike%20a%20GPU%2C%20a%20VPU,when%20using%20the%20VPU%20chip.

Eichenbaum H, Fortin NJ. Bridging the gap between brain and behaviour: Cognitive and neural mechanisms of episodic memory. J Exp Anal Behav. 2005;84(3):619–629.

Ekman P. Afterword: Universality of Emotional Expression? A Personal History of the Dispute. In: Darwin C., editor. The Expression of the Emotions in Man and Animals. Oxford University Press; New York, NY, USA: 1998. pp. 363–393.

Eldar S, Yankelevitch R, Lamy D, Bar-Haim Y. Enhanced neural reactivity and selective attention to threat in anxiety. Biol Psychol. 2010 Oct;85(2):252-7. doi: 10.1016/j.biopsycho.2010.07.010. Epub 2010 Jul 23. PMID: 20655976.

Eldridge MA, Barnard PJ, Bekerian DA. Autobiographical memory and daily schemas at work. Memory. 1994 Mar;2(1):51-74. doi: 10.1080/09658219408251492. PMID: 7584285.

Elfring J, Torta E, van de Molengraft R. Particle Filters: A Hands-On Tutorial. Sensors (Basel). 2021 Jan 9;21(2):438. doi: 10.3390/s21020438. PMID: 33435468; PMCID: PMC7826670.

Eliasmith, C., X. Choo, 'The Semantic Pointer Architecture', How to Build a Brain: A Neural Archi-tec-ture for Biological Cognition, Oxford Series on Cognitive Models and Architec-tures (2013; online edn, Oxford Academic, 23 Jan. 2014), https://doi.org/10.1093/acprof:oso/9780199794546.003.0007,

Enard W, Przeworski M, Fisher SE, Lai CS, Wiebe V, Kitano T, Monaco AP, Pääbo S. Molecular evolution of FOXP2, a gene involved in speech and language. Nature. 2002 Aug 22;418(6900):869-72. doi: 10.1038/nature01025. Epub 2002 Aug 14. PMID: 12192408.

Engeser, S. (2012). Advances in Flow Research. New York, NY: Springer. https://doi.org/10.1007/978-1-4614-2359-1

Engl E, Attwell D. Non-signalling energy use in the brain. J Physiol. 2015 Aug 15;593(16):3417-29. doi: 10.1113/jphysiol.2014.282517. Epub 2015 Mar 11. PMID: 25639777; PMCID: PMC4560575.

Epley N, Akalis S, Waytz A, Cacioppo JT. Creating social connection through inferential reproduc-tion: loneliness and perceived agency in gadgets, gods, and greyhounds. Psychol Sci. 2008 Feb;19(2):114-20. doi: 10.1111/j.1467-9280.2008.02056. x. PMID: 18271858.

Ekert, A. Jozsa, R. Penrose,R Quantum computation, entanglement and state reduction. Phils. Trans. A Math. Phys. Eng. Sci. 356, 1927–1939 (1998).

Eliasmith C, Anderson CH. Neural engineering: Computation, representation, and dynamics in neu-robio-logical systems. Cambridge, MA: MIT Press; 2003.

Erikson EH. The problem of ego identity. J Am Psychoanal Assoc. 1956; 4:56–121.

Everling S, Munoz DP. Neuronal correlates for preparatory set associated with pro-saccades and anti-sac-cades in the primate frontal eye field. J Neurosci. 2000 Jan 1;20(1):387-400. doi: 10.1523/JNEUROS-CI.20-01-00387.2000. PMID: 10627615; PMCID: PMC6774131.

Everett, M, Chen, F,How, J (2018). Motion Planning Among Dynamic, Decision-Making Agents with Deep Reinforcement Learning. 3052-3059. 10.1109/IROS.2018.8593871.

Everts R, Lidzba K, Wilke M, Kiefer C, Mordasini M, Schroth G, Perrig W, Steinlin M. Strengthen-ing of laterality of verbal and visuospatial functions during childhood and adolescence. Hum Brain Mapp. 2009 Feb;30(2):473-83. doi: 10.1002/hbm.20523. PMID: 18219619; PMCID: PMC6871185.

F

Fang Q, Zhang S, Poo MM, Gong N. Mirror-induced self-directed behaviours in rhesus monkeys after visual-somatosensory training. Curr Biol. 2015 Jan 19;25(2):212-217. doi: 10.1016/j.cub.2014.11.016. Epub 2015 Jan 8. PMID: 25578908.

Farčić N, Barać I, Plužarić J, Ilakovac V, Pačarić S, Gvozdanović Z, Lovrić R. Personality traits of core self-evaluation as predictors on clinical decision-making in nursing profession. PLoS One. 2020 May 18;15(5): e0233435. doi: 10.1371/journal.pone.0233435. PMID: 32421752; PMCID: PMC7233533.

Farquhar S. Gal Y,2019 Towards Robust Evaluations of Continual Learning. arXiv:1805.09733v3 [stat. ML] 26 Jun 2019. https://sebastianfarquhar.com/assets/papers/farquharRobust2018.pdf

Ferrarelli F, Massimini M, Sarasso S, Casali A, Riedner BA, Angelini G, Tononi G, Pearce RA. Breakdown in cortical effective connectivity during midazolam-induced loss of consciousness. Proc Natl Acad Sci U S A. 2010 Feb 9;107(6):2681-6. doi: 10.1073/pnas.0913008107. Epub 2010 Jan 25. PMID: 20133802; PMCID: PMC2823915.

Festinger L (1954). "A theory of social comparison processes". Human Relations. 7 (2): 117–140. doi:10.1177/001872675400700202.

Feinberg, T & Mallatt, J. (2019). Subjectivity "Demystified": Neurobiology, Evolution, and the Ex-plana-tory Gap. Frontiers in Psychology. 10. 1686. 10.3389/fpsyg.2019.01686.

Feinberg, T. E., & Mallatt, J. (2016). The nature of primary consciousness. A new synthesis. Consciousness and Cognition: An International Journal, 43, 113–127. https://doi.org/10.1016/j.concog.2016.05.009

Feldman DE. The spike-timing dependence of plasticity. Neuron. 2012 Aug 23;75(4):556-71. doi: 10.1016/j.neuron.2012.08.001. PMID: 22920249; PMCID: PMC3431193.

Fernández-Espejo E. Cómo funciona el nucleus accumbens? [How does the nucleus accumbens function?]. Rev Neurol. 2000 May 1-15;30(9):845-9. Spanish. PMID: 10870199.

Finger S. Chapter 51: recovery of function: redundancy and vicariation theories. Handb Clin Neu-rol. 2010; 95:833-41

Ferry R, Nelson B. Tactile P300 to unpredictable electric shocks: Association with anxiety symp-toms, intolerance of uncertainty, and neuroticism. Biol Psychol. 2021; 162:108094. doi: 10.1016/j.biopsy-cho.2021.108094.

Festinger, L., Riecken, H. W., & Schachter, S. (1956). When prophecy fails. University of Minne-sota Press. https://doi.org/10.1037/10030-000

Feynman, R. P. Simulating physics with computers. Int. J. Theor. Phys. 21, 467–488 (1982).

Finkel M, Krämer NC. The robotic mentalist – On the influences of robots' mentalizing abilities and external manipulative intent on people's credibility attributions. Front Psychol. 2022 Oct 26; 13:993302. doi: 10.3389/fpsyg.2022.993302. PMID: 36389582; PMCID: PMC9645073

Finsterwald C, Magistretti PJ, Lengacher S. Astrocytes: New Targets for the Treatment of Neuro-degen-erative Diseases. Curr Pharm Des. 2015;21(25):3570-81. doi: 10.2174 /1381612821666150710144502. PMID: 26166612.

Fiorillo CD, Kim JK, Hong SZ. The meaning of spikes from the neuron's point of view: predictive homoe-ostasis generates the appearance of randomness. Front Comput Neurosci. 2014 Apr 29; 8:49. doi: 10.3389/ fncom.2014.00049. PMID: 24808854; PMCID: PMC4010728.

Firth J, Torous J, Stubbs B, Firth JA, Steiner GZ, Smith L, Alvarez-Jimenez M, Gleeson J, Van-campfort D, Armitage CJ, Sarris J. The "online brain": how the Internet may be changing our cogni-tion. World Psychiatry. 2019 Jun;18(2):119-129. doi: 10.1002/ wps.20617. PMID: 31059635; PMCID: PMC6502424.

Fishbach, A., & Ferguson, M. J. (2007). The goal construct in social psychology. In A. W. Kruglanski & E. T. Higgins (Eds.), Social psychology: Handbook of basic principles (pp. 490–515). The Guilford Press.

Fischer AH, Rodriguez Mosquera PM, Van Vianen AE, Manstead AS. Gender and culture differ-encesin emotion. Emotion. 2004; 4(1): 87–94. PMID.

Fischer DB, Truog RD. What is a reflex? A guide for understanding disorders of consciousness. Neurology. 2015 Aug 11;85(6):543-8. doi: 10.1212/WNL.0000000000001748. Epub 2015 Jun 17. PMID: 26085602; PMCID: PMC4540252.

Fivush R, Habermas T, Waters TE, Zaman W. The making of autobiographical memory: intersec-tions of culture, narratives and identity. Int J Psychol. 2011 Oct;46(5):321-45. doi: 10.1080/00207594.2011.596541. PMID: 22044305.

Flavell, J. H. (1979). Meta-cognition and cognitive monitoring: A new area of cognitive–developmental inquiry. American Psychologist, 34(10), 906–911. https://doi.org/10.1037/0003-066X.34.10.906

Fleer S, Moringen A, Klatzky RL, Ritter H (2020) Correction: Learning efficient haptic shape explo-ration with a rigid tactile sensor array. PLOS ONE 15(2): e0230054.

Flesch T, Nagy DG, Saxe A, Summerfield C. Modelling continual learning in humans with Hebbian context gating and exponentially decaying task signals. PLoS Comput Biol. 2023 Jan 19;19(1): e1010808. doi: 10.1371/journal.pcbi.1010808. PMID: 36656823; PMCID: PMC9851563.

Fodor, Jerry A. (1983). Modularity of Mind: An Essay on Faculty Psychology. Cambridge, Massa-chu-setts: MIT Press. ISBN 0-262-56025-9

Fong, T., Nourbakhsh, I., and Dautenhahn, K. (2003). A survey of socially interactive robots. Ro-bot. Auton. Syst. 42, 143–166. doi: 10.1016/S0921-8890(02)00372-X

Forestier, S. (2019). Intrinsically motivated goal exploration in child development and artificial in-telli-gence: Learning and development of speech and tool use. France: Université Bordeaux.

Fox A, Shelton S, Oakes T, Converse A, DavidsonR, Kalin N (2010). "Orbitofrontal Cortex Le-sions Alter Anxiety-Related Activity in the Primate Bed Nucleus of Stria Terminalis". J Neuro-sci. 30 (20): 7023–27. doi:10.1523/JNEUROSCI.5952-09.2010. PMC 2915894. PMID 20484644.

Fox E, Russo R, Bowles R, Dutton K. Do threatening stimuli draw or hold visual attention in sub-clinical anxiety? J Exp Psychol Gen. 2001 Dec;130(4):681-700. PMID: 11757875; PMCID: PMC1924776.

Fox, M, Ghallab, M, Infantes, G & Long, D. (2006). Robot Introspection through Learned Hidden Markov Models. Artificial Intelligence. 170. 59-113. 10.1016/j.artint.2005.05.007.

Fossati P. Neural correlates of emotion processing: from emotional to social brain. Eur Neuropsy-chophar-macol. 2012;22 Suppl 3: S487-91. doi: 10.1016/j.euroneuro.2012.07.008. PMID: 22959113. Forrester J. W 1971.Counterintuitive behaviour of social systems.

Frank M, Leitner J, Stollenga M, Förster A, Schmidhuber J. Curiosity driven reinforcement learning for motion planning on humanoids. Front Neurorobot. 2014 Jan 6; 7:25. doi: 10.3389/fnbot.2013.00025. PMID: 24432001; PMCID: PMC3881010.

Fraser ON, Bugnyar T. 2010 Do ravens show consolation? Responses to distressed others. PLoS One. 2010 May 12;5(5): e10605. doi: 10.1371/journal.pone.0010605. PMID: 20485685; PMCID: PMC2868892.

Fredrickson BL. The role of positive emotions in positive psychology: the broaden-and-build theo-ry of positive emotions. Am Psychol. 2001; 56:218–226.

Friedrich, S., Antes, G., Behr, S. et al. Is there a role for statistics in artificial intelligence? Adv Data Anal Classif 16, 823–846 (2022). https://doi.org/10.1007/s11634-021-00455-6

Friston K. Learning and inference in the brain. Neural Netw. 2003 Nov;16(9):1325-52. doi: 10.1016/j. neunet.2003.06.005. PMID: 14622888.

Friston, K. (2013). Life as we know it. Journal of the Royal Society, Interface, 10, 20130475. Doi:10. 1098/rsif.2013.0475

Friston, K. The free-energy principle: a unified brain theory? Nat Rev Neurosci 11, 127–138 (2010). https://doi.org/10.1038/nrn2787

Friston KJ. The labile brain. I. Neuronal transients and non-linear coupling. Philos Trans R Soc Lond B Biol Sci. 2000 Feb 29;355(1394):215-36. doi: 10.1098/rstb.2000.0560. PMID: 10724457; PMCID: PMC1692735.

Frith CD. The role of meta-cognition in human social interactions. Philos Trans R Soc Lond B Biol Sci. 2012 Aug 5;367(1599):2213-23. doi: 10.1098/rstb.2012.0123. PMID: 22734064; PMCID: PMC3385688.

Froese, T, Ziemke, T. (2009). Ziemke, T.: Enactive Artificial Intelligence: Investigating the system-ic organization of life and mind. Journal of Artificial Intelligence 173(3-4), 466-500. Artificial In-telligence. 173. 466-500. 10.1016/j.artint.2008.12.001.

Fröhlich F, McCormick DA. Endogenous electric fields may guide neocortical network activity. Neuron. 2010 Jul 15;67(1):129-43. doi: 10.1016/j.neuron.2010.06.005. PMID: 20624597; PMCID: PMC3139922.

Fröhlich M, Kuchenbuch P, Müller G, Fruth B, Furuichi T, Wittig RM, Pika S. 2016. Unpeeling the layers of language: bonobos and chimpanzees engage in cooperative turn-taking sequences. Scient. Rep. 6, 25887. (10.1038/srep25887)

Fröhlich M, van Schaik CP. 2022. Social tolerance and interactional opportunities as drivers of ges-tural redoings in orang-utans. Phil. Trans. R. Soc. B 377, 20210106. (10.1098/rstb.2021.0106)

Fuchs T. From Self-Disorders to Ego Disorders. Psychopathology. 2015;48(5):324-31. doi: 10.1159/000432404. Epub 2015 Sep 9. PMID: 26345993.

Fumagalli M, Priori A. Functional and clinical neuroanatomy of morality. Brain. 2012 [Epub ahead of printFundamentals of Brain Network Analysis Chapter 2 – Nodes and Edges, Editor(s): Alex Fornito, Andrew Zalesky, Edward T. Bullmore, Academic Press, 2016, Pages 37-88, ISBN 9780124079083, https://doi.org/10.1016/B978-0-12-407908-3.00002-9.

Fuster JM, International Encyclopedia of the Social & Behavioural Sciences, 2001

G

Galinsky, V.L., Frank, L.R. Critically synchronized brain waves form an effective, robust and flex-ible basis for human memory and learning. Sci Rep 13, 4343 (2023). https://doi.org/10.1038/s41598-023-31365-6

GallagharSTrendsinCognitiveSciences–Vol.4,No.1,January2000PII:S1364-6613(99)01417-5

Gallistel CR, Balsam PD. Time to rethink the neuronal mechanisms of learning and memory. Neu-robiol Learn Mem. 2014; 108:136–44.

Gally JA, Montague PR, Reeke GN Jr, Edelman GM. The NO hypothesis: possible effects of a short-lived, rapidly diffusible signal in the development and function of the nervous system. Proc Natl Acad Sci U S A. 1990 May;87(9):3547-51. doi: 10.1073/pnas.87.9.3547. PMID: 2333301; PMCID: PMC53939.

Gan X, Xiao B. Improved Bacterial Foraging Optimization Algorithm with Comprehensive Swarm Learning Strategies. Advances in Swarm Intelligence. 2020 Jun 22; 12145:325–34. doi: 10.1007/978-3030-53956-6_29. PMCID: PMC7354797.

Gamble AL, Rapee RM. The time-course of attentional bias in anxious children and adolescents. J Anxiety Disord. 2009 Oct;23(7):841-7. doi: 10.1016/j.janxdis.2009.04.001. Epub 2009 Apr 19. PMID: 19447004.

Gao S, Assink M, Cipriani A, Lin K. Associations between rejection sensitivity and mental health outcomes: A meta-analytic review. Clin Psychol Rev. 2017; 57:59-74. doi: 10.1016/j.cpr.2017.08.00.

Gazzola V, Rizzolatti G, Wicker B, Keysers C. The anthropomorphic brain: the mirror neuron sys-tem responds to human and robotic actions. Neuroimage. 2007 May 1;35(4):1674-84. doi: 10.1016/j.neuroim-age.2007.02.003. Epub 2007 Feb 13. PMID: 17395490.

Georgopoulos AP, Schwartz AB, Kettner RE. Neuronal population coding of movement direction. Science. 1986 Sep 26;233(4771):1416-9. doi: 10.1126/science.3749885. PMID: 3749885.

GG Gallup, JR Anderson, SM Platek, Self-awareness, social intelligence, and schizophrenia. The Self in Neuroscience and Psychiatry, eds T Kircher, AS David (Cambridge Univ Press, New York), pp. 147–165 (2003).

Ghose T, Liu Z. Generalization between canonical and non-canonical views in object recognition. J Vis. 2013 Jan 2;13(1):1. doi: 10.1167/13.1.1. PMID: 23283692; PMCID: PMC3586995.

Ghosh, V.E. and Gilboa, A. (2014) What is a memory schema? A historical perspective on current neuro-science literature. Neuro- psychologia 53, 104–114.

Gihleb R, Giuntella O, Stella L, Wang T, Industrial robots, Workers' safety, and health, Labour Economics, Volume 78, 2022,102205, ISSN 0927-5371, https://doi.org/10.1016/j.labe-co.2022.102205. (https://www.sciencedirect.com/science/article/pii/S0927537122000963)

Gilboa, A. & Marlatte, H. Neurobiology of schemas and schema-mediated memory. Trends Cogn. Sci. 21, 618–631 (2017).

Gilda, Kalpana & Dixit, Sharwari. (2016). CLOUD ROBOTICS: AN OVERVIEW.

Gilovich T, Griffin D, Kahneman D. Heuristic and Biases: The Psychology of Intuitive Judg-ment. Cambridge, UK: Cambridge Univ Press; 2002.

Gintis, H. (2009). The bounds of reason: Game theory and the unification of the behavioural sci-ences. Princeton University Press.

Giorgio M; Giulio S; David V; Lorenzo N; Francesco N (2008). The iCub humanoid robot: an open platform for research in embodied cognition (PDF). PerMIS'08. Retrieved 1 January 2018.

Global Workspace Theory, its LIDA model and the underlying neuroscience, Biologically Inspired Cognitive Architectures,Volume 1,2012,Pages 32-43, ISSN 2212-683X, https://doi.org/10.1016/j.bica.2012.04.001.

Goertzel B, Garis HD, Pennachin C, Geisweiller N, Araujo S, Pitt J, Chen S, Lian R, Jiang M, Yang Y, HuangD (2010a) OpenCogBot: achieving generally intelligent virtual agent control and humanoid robotics viacognitive synergy. In: Proceedings of international conference on artificial intelligence.

Goethals, G. R., & Darley, J. M. (1977). Social Comparison Theory: An Attributional Approach. In J. M. Suls, & R. L. Miller (Eds.), Social Comparison Processes: Theoretical and Empirical Perspec-tives (pp. 43-68). Washington DC: Hemisphere.

Goldberg EM, Coulter DA. Mechanisms of epileptogenesis: a convergence on neural circuit dys-function. Nat Rev Neurosci. 2013 May;14(5):337-49. doi: 10.1038/nrn3482. Epub 2013 Apr 18. PMID: 23595016; PMCID: PMC3982383.

Golden R, Delanois JE, Sanda P, Bazhenov M (2022) Sleep prevents catastrophic forgetting in spiking neural networks by forming a joint synaptic weight representation. PLoS Comput Biol 18(11): e1010628. https://doi.org/10.1371/journal.pcbi.1010628

Golenia L, Schoemaker MM, Otten E, Mouton LJ, Bongers RM. What the Dynamic Systems Ap-proach Can Offer for Understanding Development: An Example of Mid-childhood Reaching. Front Psychol. 2017 Oct 10; 8:1774. doi: 10.3389/fpsyg.2017.01774. PMID: 29066996; PMCID: PMC5641353.

Goodkind M, Eickhoff SB, Oathes DJ, Jiang Y, Chang A, Jones-Hagata LB, Ortega BN, Zaiko YV, Roach EL, Korgaonkar MS, Grieve SM, Galatzer-Levy I, Fox PT, Etkin A. Identification of a com-mon neurobio-logical substrate for mental illness. JAMA Psychiatry. 2015 Apr;72(4):305-15. doi: 10.1001/jamapsychia-try.2014.2206. PMID: 25651064; PMCID: PMC4791058.

Gopnik A. Machine Learning, Making AI More Human. Scientific American Magazine. 2017.

Gordon JI, Ley RE, Wilson R, Mardis E, Xu J, Fraser CM, et al. Extending our view of self: the human gut microbiome initiative (HGMI). NHGRI) with current estimates that half the cells in our body are microbial (Sender R, Fuchs S, Milo R. Revised estimates for the number of human and bacteria cells in the body. PLoS Biol. 2016;14(8): e1002533 doi.

Georgeon, Ol, Riegler, A (2019). CASH Only: Constitutive Autonomy through Motorsensory Self-Programming. Cognitive Systems Research. 58. 10.1016/j.cogsys.2019.08.0

Gokhale, P, Baker, J, Duckering, C, Brown, N,Brown, K, Chong, F. (2020). Extending the Fron-tier of Quantum Computers with Qutrits. IEEE Micro. PP. 1-1. 10.1109/MM.2020.2985976.

Gottfried JA. Central mechanisms of odour object perception. Nat Rev Neurosci. 2010 Sep;11(9):628-41. doi: 10.1038/nrn2883. Epub 2010 Aug 11. PMID: 20700142; PMCID: PMC3722866.

Gottlieb J, Oudeyer PY, Lopes M, Baranes A. Information-seeking, curiosity, and attention: com-putation-al and neural mechanisms. Trends Cogn Sci. 2013 Nov;17(11):585-93. doi: 10.1016/j.tics.2013.09.001. Epub 2013 Oct 12. PMID: 24126129; PMCID: PMC4193662.

Gottlieb J, Balan P. Attention as a decision in information space. Trends in Cognitive Scienc-es. 2010 Jun;14(6):240–248.

Gray HM, Gray K, Wegner DM. Dimensions of mind perception. Science. 2007; 315:619.

Gratch, J., & Marsella, S. (Eds.). (2014). Social emotions in nature and artifact. Oxford University Press.

Gratch J, Marsella. S, 2004, A domain-independent framework for modelling emotion,

Cognitive Systems Research, Volume 5, Issue 4, Pages 269-306, ISSN 1389-0417, https://doi.org/10.1016/j.cogsys.2004.02.002.

Graveline CJ, Mikulis DJ, Crawley AP, Hwang PA. Regionalized sensorimotor plasticity after hemi-spherectomy fMRI evaluation. Pediatr Neurol. 1998 Nov;19(5):337-42.

Graven, S. (2006). Sleep and brain development. Clinics in perinatology, 33(3), 693-706.

Gray K, Wegner DM. Feeling robots and human zombies: mind perception and the uncanny valley. Cognition. 2012 Oct;125(1):125-30. doi: 10.1016/j.cognition.2012.06.007. Epub 2012 Jul 9. PMID: 22784682.

Greenwald, Anthony G.; McGhee, Debbie E.; Schwartz, Jordan L.K. (1998), "Measuring Individual Differences in Implicit Cognition: The Implicit Association Test", Journal of Personality and Social Psychology, 74 (6): 1464–1480, CiteSeerX 10.1.1.489.4611, doi:10.1037/0022-3514.74.6.1464, PMID 965475.

Greenwald, A. G. (1992). New Look 3: Unconscious cognition reclaimed. American Psychologist, 47(6), 766–779. https://doi.org/10.1037/0003-066X.47.6.766

Gremel CM, Costa RM. Orbitofrontal and striatal circuits dynamically encode the shift between goal-directed and habitual actions. Nat Commun. 2013; 4:2264.

Grewal, K., Forest, J., Cohen, B. P., Ahmad, S. (2021). Going Beyond the Point Neuron: Active Dendrites and Sparse Representations for Continual Learning. https://doi.org/10.1101/2021.10.25.465651.

Gröblacher S 2019.PNAS https://www.pnas.org/doi/10.1073/pnas.1917212116

Groshev, M, Baldoni, G Cominardi, L de la Oliva, A Gazda,R Edge robotics: are we ready? an experimental evaluation of current vision and future directions, Digital Communications and Networks, Volume 9, Issue 1, 2023, Pages 166-174, ISSN 2352-8648, https://doi.org/10.1016/j.dcan.2022.04.032. (https://www.sciencedirect.com/science/article/pii/S2352864822000888).

Grover, Lov K. (1996-07-01). "A fast quantum mechanical algorithm for database search". Proceedings of the twenty-eighth annual ACM symposium on Theory of computing – STOC '96. Philadelphia, Pennsylvania, USA: Association for Computing Machinery. pp. 212–219.

Gruenfeld DH, Inesi EM, Magee JC, Galinsky AD. Power and the objectification of social tar-gets. Journal of Personality and Social Psychology. 2008; 95:111–127.

Grupe, D., Nitschke, J. Uncertainty and anticipation in anxiety: an integrated neurobiological and psychological perspective. Nat Rev Neurosci 14, 488–501 (2013). https://doi.org/10.1038/nrn3524.

G. Schöner, in International Encyclopedia of the Social & Behavioural Sciences, 2001.

Gu J, Cavanagh K, Baer R, Strauss C. An empirical examination of the factor structure of compas-sion. PLoS One. 2017 Feb 17;12(2): e0172471. doi: 10.1371/journal.pone.0172471. PMID: 28212391; PMCID: PMC5315311.

Guan, X., Wang, Y., Lin, C.-H., Hu, L., Ge, S., Wan, T., Younis, A., Li, F., Cui, Y., Qi, D.-C., Chu, D., Chen, X.D., Wu, T., 2020. A monolithic artificial iconic memory based on highly stable perov-skite-metal multilayers. Applied Physics Reviews 7, 031401. https://doi.org/10.1063/5.0009713

Guerin, F. (2012). Sensorimotor Schema. In: Seel, N.M. (eds) Encyclopedia of the Sciences of Learning. Springer, Boston, MA. https://doi.org/10.1007/978-1-4419-1428-6_463

Güntürkün O, Ströckens F, Ocklenburg S. Brain Lateralization: A Comparative Perspective. Physiol Rev. 2020 Jul 1;100(3):1019-1063. doi: 10.1152/physrev.00006.2019. PMID: 32233912.

Güntürkün O, Diekamp B, Manns M, Nottelmann F, Prior H, Schwarz A, Skiba M. Asymmetry pays visual lateralization improves discrimination success in pigeons. Curr Biol. 2000 Sep 7;10(17):1079-81. doi: 10.1016/s0960-9822(00)00671-0. PMID: 10996079

Guo, MH., Xu, TX., Liu, JJ. et al. Attention mechanisms in computer vision: A sur-vey. Comp.Vis-ualMedia 8, 331–368 (2022). https://doi.org/10.1007/s41095-022-0271-y

H

Ha D, Schmidhuber J. World models 9 May 2018. https://arxiv.org/abs/1803.10122v4.

Hadsell R, Rao D, Rusu AA, Pascanu R, Embracing Change: Continual Learning in Deep Neural Networks, Trends in Cognitive Sciences, Volume 24, Issue 12,2020, Pages 1028-1040, ISSN 1364-6613 https://doi.org/10.1016/j.tics.2020.09.004.

Hagmann P, Grant PE, Fair DA. MR connectomics: a conceptual framework for studying the devel-oping brain. Front Syst Neurosci. 2012 Jun 13; 6:43. doi: 10.3389/fnsys.2012.00043. PMID: 22707934; PMCID: PMC3374479.

Halfon, N., Larson, K., Lu, M., Tullis, E., & Russ, S. (2014). Lifecourse health development: Past, present and future. Maternal and Child Health Journal, 18(2), 344–365. doi:10.1007/s10995-013-1346-2.

Hall, J., Thomas, K. L., and Everitt, B. J. (2001). Cellular imaging of zif268 expression in the hip-pocam-pus and amygdala during contextual and cued fear memory retrieval: selective activation of hippocampal CA1 neurons during the recall of contextual memories. J. Neurosci 21, 2186–2193. doi: 10.1523/JNEU-ROSCI.21-06-02186.2001

Hallion, L. S., & Ruscio, A. M. (2011). A meta-analysis of the effect of cognitive bias modification on anxiety and depression. Psychological Bulletin, 137(6), 940–958. https://doi.org/10.1037/a0024355.

Hamann S. Cognitive and neural mechanisms of emotional memory. Trends Cogn Sci. 2001 Sep 1;5(9):394-400. doi: 10.1016/s1364-6613(00)01707-1. PMID: 11520704.

Hameroff, S MD, Penrose R, "Consciousness in the universe: Review of the 'Orch OR' theory," by FRS (https://dx.doi.org/10.1016/j.plrev.2013.11.013).

Hampshire A, Highfield RR, Parkin BL, Owen AM. Fractionating human intelligence. Neuron. 2012 Dec 20;76(6):1225-37. doi: 10.1016/j.neuron.2012.06.022. PMID: 23259956.

Harpin V, Mazzone L, Raynaud JP, Kahle J, Hodgkins P. Long-Term Outcomes of ADHD: A Sys-tem-atic Review of Self-Esteem and Social Function. J Atten Disord. 2016 Apr;20(4):295-305. doi: 10.1177/1087054713486516. Epub 2013 May 22. PMID: 23698916.

Han SW, Marois R. The source of dual-task limitations: serial or parallel processing of multiple response selections? Atten Percept Psychophys. 2013 Oct;75(7):1395-405. doi: 10.3758/s13414-013-0513-2. PMID: 23864266; PMCID: PMC4150022.

Harari, Y. N. (2016). Homo Deus. Harvill Secker.

Harikesh, P. et al. (2023) "Ion-tunable antiambipolarity in mixed ion–electron conducting polymers enables biorealistic organic electrochemical neurons", Nature Materials. doi: 10.1038/s41563-022-01450-8. https://www.nature.com/articles/s41563-022-01450-8

Harlow, H. F., Harlow, M. K. and Meyer, D. R. (1950). Novelty and curiosity as determinants of exploratory behaviour, Journal of Experimental Psychology 41: 68–80.

Harmon-Jones E, Harmon-Jones C (2008) Action-based model of dissonance: a review of behav-io-ral, anterior cingulate, and pre-frontal cortical mechanisms. Soc Pers Psychol Compass 2:1518–1538. doi:10.1111/j.1751-9004.2008. 00110.x.

Harrington A (1987) Medicine, mind, and the double brain. Princeton (New Jersey): Princeton University Press.

Harrison NA, Gray MA, Gianaros PJ, Critchley HD. The embodiment of emotional feelings in the brain. J Neurosci. 2010 Sep 22;30(38):12878-84. doi: 10.1523/JNEUROSCI.1725-10.2010. PMID: 20861391; PMCID:

Harvey PD. Domains of cognition and their assessment. Dialogues Clin Neurosci. 2019 Sep;21(3):227-237. doi: 10.31887/DCNS.2019.21.3/pharvey. PMID: 31749647; PMCID: PMC6829170.

Haasch, A.,Hofemann, N, Fritsch, J, Sagerer, G. (2005). A multi-modal object attention system for a mobile robot. 2712 – 2717. 10.1109/IROS.2005.1545191.

Hassabis, D Kumaran, D, Summerfield C and Botvinick M. Neuroscience-inspired artificial intelli-gence Neuron, 95 (2017), pp. 245-258

Hatsopoulos N G., et al., (2023) Propagating spatiotemporal activity patterns across macaque mo-tor cortex carry kinematic information," was supported by the National Institutes of Health (R01 NS111982). Additional authors include Karthikeyan Balasubramanianb and Vasileios Papadoura-kis of the University of Chicago.

Hautala, J. Can robots possess knowledge? Rethinking the DIK(W) pyramid through the lens of employees of an automotive factory. Humanit Soc Sci Commun 8, 212 (2021). https://doi.org/10.1057/s41599-021-00893-9

Hawkins J, Ahmad.S Why neurons have thousands of synapses, a theory of sequence memory in neocortex. Frontiers in Neural Circuits, 10(23):1–13, 2016.

Hebscher M., Gilboa A. (2016). A boost of confidence: the role of the ventromedial prefrontal cor-tex in memory, decision-making, and schemas. Neuropsychologia, 90, 46–58.

Hedda Hassel Mørch https://philosophynow.org/issues/121/The_Integrated_Information_Theory_of_Con-sciousness.

Heesen R, Fröhlich M. Revisiting the human 'interaction engine': comparative approaches to social action coordination. Philos Trans R Soc Lond B Biol Sci. 2022 Sep 12;377(1859):20210092. doi: 10.1098/ rstb.2021.0092. Epub 2022 Jul 25. PMID: 35876207; PMCID: PMC9315451.

Heimann, D, Hohenfeld, H, Wiebe, F, Kirchner, F. (2022). Quantum Deep Reinforcement Learn-ing for Robot Navigation Tasks.

Hellström T, Bensch. S 2018. Understandable robots. Paladyn, Journal of Behavioural Robotics 9, 1 (2018), 110–123

Heidegger M. (1927/2011). Being and Time (Macquarrie J., Robinson E., Trans.). New York, NY: Harper & Row.

Heider, F. (1946). Attitudes and cognitive organization. The Journal of psychology, 21 (1), 107-112.

Helbich M. Mental health and environmental exposures: An editorial. Int J Environ Res Public Health. 2018;15(10):2207. doi:10.3390/ijerph15102207.

Heft, H. (2001). Ecological Psychology in Context: James Gibson, Roger Barker, and the Legacy of William James's Radical Empiricism. Mahwah, NJ: Lawrence Erlbaum Associates.

Henderson CM, McClelland JL. Intrusions into the shadow of attention: A new take on illusory conjunctions. Atten Percept Psychophys. 2020 Feb;82(2):564-584. doi: 10.3758/s13414-019-01893-3. PMID: 32189233.

Hendrikse JL, Parsons TE, Hallgrímsson B. Evolvability as the proper focus of evolutionary de-velopmen-tal biology. Evol Dev. 2007; 9:393–401.

Henry M. L'essence de la manifestation. Paris: PUF, 1963.

Henschel A, Hortensius R, Cross ES. Social Cognition in the Age of Human-Robot Interaction. Trends Neurosci. 2020 Jun;43(6):373-384. doi: 10.1016/j.tins.2020.03.013. Epub 2020 Apr 30. PMID: 32362399.

Herai RH, Semendeferi K, Muotri AR. Comment on "Human TKTL1 implies greater neurogenesis in frontal neocortex of modern humans than Neanderthals". Science. 2023 Mar 10;379(6636): eadf0602. doi: 10.1126/science.adf0602. Epub 2023 Mar 10. PMID: 36893252.

Herbert MR, Harris GJ, Adrien KT, Ziegler DA, Makris N, Kennedy DN, Lange NT, Chabris CF, Bakard-jiev A, Hodgson J, Takeoka M, Tager-Flusberg H, Caviness VS Jr. Abnormal asymmetry in language association cortex in autism. Ann Neurol. 2002 Nov;52(5):588-96. doi: 10.1002/ana.10349. PMID: 12402256.

Herculano-Houzel S. The human brain in numbers: A linearly scaled-up primate brain. Front Hum Neuros-ci. 2009; 3:31. DOI:10.3389/neuro.09.031.2009

Herreras O,2016 Front. Neural Circuits, 15 December 2016 Volume 10 – 2016 | https://doi.org/10.3389/fncir.2016.00101

Hershfield H.E., Bartels D.M. (2018). The future self. In: Oettingen G., Sevincer A.T., Gollwitzer P.M., editors. The Psychology of Thinking about the Future, The Guilford Press, 89–109.

Hermiz, J, Chang, T, Du, Chao & Lu, W. (2013). Interference and memory capacity effects in memristive systems. Applied Physics Letters. 102. 10.1063/1.4794024.

Herz N, Baror S, Bar M. Overarching States of Mind. Trends Cogn Sci. 2020 Mar;24(3):184-199. doi: 10.1016/j.tics.2019.12.015. Epub 2020 Feb 6. PMID: 32059121.

Hewawasam, H., Ibrahim, M., Kahandawa, G., & Choudhury, T. (2022). Agoraphilic navigation algorithm in dynamic environment with obstacles motion tracking and prediction. Robotica, 40(2), 329-347. doi:10.1017/S0263574721000588.

Heyes C. Homo imitans? Seven reasons why imitation couldn't possibly be associative. Philos Trans R Soc Lond B Biol Sci. 2016 Jan 19;371(1686):20150069. doi: 10.1098/rstb.2015.0069. PMID: 26644604; PMCID: PMC4685515.

Hinton GE, Osindero S, Teh YW. A fast-learning algorithm for deep belief nets. Neural Comput. 2006 Jul;18(7):1527-54. doi: 10.1162/neco.2006.18.7.1527. PMID: 16764513.

Hiruma, Hyogo & Mori, Hiroki & Ito, Hiroshi & Ogata, Tetsuya. (2022). Guided Visual Attention Model Based on Interactions Between Top-down and Bottom-up Information for Robot Pose Pre-diction.

Hiser J, Koenigs M. The Multifaceted Role of the Ventromedial Pre-frontal Cortex in Emotion, Deci-sion Making, Social Cognition, and Psychopathology. Biol Psychiatry. 2018 Apr 15;83(8):638-647. doi: 10.1016/j.biopsych.2017.10.030. Epub 2017 Nov 20. PMID: 29275839; PMCID:

Hester T Peter Stone, P Intrinsically motivated model learning for developing curious robots, Arti-ficial Intelligence, Volume 247, 2017, Pages 170-186, ISSN 0004-3702, https://doi.org/10.1016/j.artint.2015.05.002. (https://www.sciencedirect.com/science/article/pii/S0004370215000764).

Hinton, G., Deng, L., Yu, D., Dahl, G., Mohamed, A., Jaitly, N., Senior, A., Vanhoucke, V., Ngu-yen, P., Sainath, T. and Kingsbury, B. (2012) Deep Neural Networks for Acoustic Modelling in Speech Recognition: The Shared Views of Four Research Groups. IEEE Signal Processing Maga-zine, 29, 82-97. https://doi.org/10.1109/MSP.2012.2205597

Hodgkin, A. L. & Huxley, A. F. A quantitative description of membrane current and its application to conduction and excitation in nerve. J. Physiol. (Lond.)117, 500–544 (1952).

Hodges, S. D., & Biswas-Diener, R. (2007). Balancing the empathy expense account: Strategies for regulating empathic response. In T. Farrow & P. Woodruff (Eds.), Empathy in mental illness (pp. 389–407). Cambridge University Press. https://doi.org/10.1017/CBO9780511543753.022

Hoffmann, H., and Payton, D. W. (2018). Optimization by self-organized criticality. Sci. Rep.8:2358. doi: 10.1038/s41598-018-20275-7

Hofman, M. A. (2014). "Evolution of the human brain and intelligence: from matter to mind," in Handbook of Intelligence: Evolutionary Theory, Historical Perspective and Current Concepts, eds S. Goldstein, J. A. Naglieri, and D. Princiotta (Berlin, Springer Verlag). (in press.

Hofman, M. A. (1988). Size and shape of the cerebral cortex in mammals. Part II. The cortical vol-ume. Brain Behav. Evol. 32, 17–26. doi: 10.1159/000116529

Holbrook C, Iacoboni M, Gordon C, Proksch S, Balasubramaniam R. Posterior medial frontal cor-tex and threat-enhanced religious belief: a replication and extension. Soc Cogn Affect Neurosci. 2020 Dec 24;15(12):1361-1367. doi: 10.1093/scan/nsaa153. PMID: 33180108; PMCID: PMC7759203.

Holeva, V., Tarrier, N. and Wells, A. (2001) Prevalence and Predictors of Acute Stress Disorder and PTSD Following Road Traffic Accidents: Thought Control Strategies and Social Support. Be-havior Therapy, 32, 65-83. http://dx.doi.org/10.1016/S0005-7894(01)80044-7.

Hollinger G, Georgiev, Y, Manfredi, A, Maxwell, B, Pezzementi, Z,Mitchell, B. (2006). Design of a Social Mobile Robot Using Emotion-Based Decision Mechanisms. 2006 IEEE/RSJ International Conference on Intelligent Robots and Systems. 3093 – 3098. 10.1109/IROS.2006.282327.

Hommel, B., Chapman, C. S., Cisek, P., Neyedli, H. F., Song, J.-H., and Welsh, T. N. (2019). No one knows what attention is. Attent. Percept. Psychophys. 81, 2288–2303. doi: 10.3758/s13414-019-01846-w

Hopfield, J. J. Neural networks and physical systems with emergent collective computational abili-ties. PNAS 79, 2554–2558 (1982).

Hohwy J. The neural correlates of consciousness: new experimental approaches needed? Con-scious Cogn. 2009 Jun;18(2):428-38. doi: 10.1016/j.concog.2009.02.006. Epub 2009 Apr 3. PMID: 19345590.

Höök, Kristina. (2008). Affective Loop Experiences – What Are They? 1-12. 10.1007/978-3-540-685043_1.

Horner, A., Bisby, J., Bush, D. et al. Evidence for holistic episodic recollection via hippocampal pattern completion. Nat Commun 6, 7462 (2015). https://doi.org/10.1038/ncomms8462

Horwitz, B. (2003) The Elusive Concept of Brain Connectivity. Neuroimage, 19, 466 470. =http://dx.doi.org/10.1016/S1053-8119(03)00112-5

Hou, Xiaodi & Zhang, Liqing. (2007). Saliency Detection: A Spectral Residual Approach. IEEE Conference in Computer Vision and Pattern Recognition. 2007. 10.1109/ CVPR.2007.383267.H.R. Markus, S. KitayamaCulture and the self: implications for cognition, emotion, and motiva-tionPsy-chol Rev, 98 (1991), pp. 224-253.

Hu XM, Zhang C, Liu BH, Cai Y, Ye XJ, Guo Y, Xing WB, Huang CX, Huang YF, Li CF, Guo GC. Experimental High-Dimensional Quantum Teleportation. Phys Rev Lett. 2020 Dec 4;125(23):230501. doi: 10.1103/PhysRevLett.125.230501. PMID: 33337185.

Huang F, Tang S, Hu Z. Unconditional Perseveration of the Short-Term Mental Set in Chunk De-composition. Front Psychol. 2018 Dec 11; 9:2568. doi: 10.3389/ fpsyg.2018.02568. PMID: 30618985; PMCID: PMC6297678.

Hubel David H and Wiesel Torsten N. Receptive fields, binocular interaction and functional archi-tecture in the cat's visual cortex. The Journal of physiology, 160(1):106–154, 1962.

Huelga, S., Plenio, M. A vibrant environment. Nature Phys 10, 621–622 (2014). https:// doi.org/10.1038/nphys3047.

Huffziger S, Ebner-Priemer U, Zamoscik V, Reinhard I, Kirsch P, Kuehner C. Effects of mood and rumination on cortisol levels in daily life: An ambulatory assessment study in remitted depressed patients and healthy controls. Psychoneuroendocrinology. 2013;38(10):2258-67. doi: 10.1016/j.psyneuen.2013.04.014.

Hugdahl K, Davidson RJ (2003): The asymmetrical brain. Cambridge, MA: MIT Press.

Hughes, C. (2002). Executive functions and development: Why the interest? Infant and Child De-velop-ment, 11 69-71 DOI: 10.1002/icd.296

Hummos A 2023: Thalamus: a brain-inspired algorithm for biologically plausible continual learn-ing and disentangled representations. Published as a conference paper at ICLR 24th Feb 2023

Hutler B, Rieder TN, Mathews DJH, Handelman DA, Greenberg AM. Designing robots that do no harm: understanding the challenges of Ethics for Robots. AI Ethics. 2023 Apr 17:1-9. doi: 10.1007/s43681-023-00283-8. Epub ahead of print. PMID: 37360148; PMCID: PMC10108783.

Husbands, P.: Evolving Robot Behaviours with Diffusing Gas Networks", in Husbands, P. In: Husbands, P. (ed.) Evolutionary Robotics. LNCS, vol. 1468, pp. 71-86. Springer, Heidelberg (1998).

Hutmacher F. Understanding the self of people with dementia: Psychological and philosophical considerations. Z Gerontol Geriatr. 2021 Mar;54(2):161-166. doi: 10.1007/s00391-020-01718-1. Epub 2020 Mar 23. PMID: 32206874; PMCID: PMC7946685.

Hyun E, Park H, Jang S, Yeon H. 2010. The usability of a robot as an educational assistant in a kindergarten and young children's perceptions of their relationship with the robot. Korean J. Child Stud. 31: 267–82

I

IEEE Access Trends and Advances in Biol.-Inspired Image-Based Deep Learning Methodologies and Applications. IEEE Access. https://ieeeaccess.ieee.org/closed-special-sections/trends-and-advances-in-bio-in-spired-image-based-deep-learning-methodologies-and-applications/

Inoue, Ki., Takada, M. & Matsumoto, M. Neuronal and behavioural modulations by pathway-selective optogenetic stimulation of the primate oculomotor system. Nat Commun 6, 8378 (2015). https://doi.org/10.1038/ncomms9378.

Iriki A, Taoka M. Triadic (ecological, neural, cognitive) niche construction: a scenario of human brain evolution extrapolating tool use and language from the control of reaching actions. Philos Trans R Soc Lond B Biol Sci. 2012 Jan 12;367(1585):10-23. doi: 10.1098/rstb.2011.0190. PMID: 22106423; PMCID: PMC3223791.

Irpan A (2018) Deep Reinforcement Learning Doesn't Work Yet. https://www.alexirpan.com/2018/02/14/rl-hard.html

Ishiguro Hiroshi. (2005). Android Science: Toward a new cross-interdisciplinary framework. 28. 1-6.

Itti,L C. Koch C, Niebur E, "A model of saliency-based visual attention for rapid scene analysis," in IEEE Transactions on Pattern Analysis and Machine Intelligence, vol. 20, no. 11, pp. 1254-1259, Nov. 1998, doi: 10.1109/34.730558.

Itti L., Baldi P. F. (2005). Bayesian surprise attracts human attention, in Advances in neural infor-mation processing systems (NIPS), 547–554

Itti L, Koch C. Computational modelling of visual attention. Nat Rev Neurosci. 2001 Mar;2(3):194-203. doi: 10.1038/35058500. PMID: 11256080.

Ito, T., Chiao, K., Devine, P., Lorig, T., & Cacioppo, J. (2006). The Influence of Facial Feedback on Race Bias. Psychological Science, 17(3), 256-261. doi: 10.1111/j.1467-9280.2006. 01694.x

Izhikevich EM. Solving the distal reward problem through linkage of STDP and dopamine signal-ing. Cer-eb Cortex. 2007 Oct;17(10):2443-52. doi: 10.1093/cercor/bhl152. Epub 2007 Jan 13. PMID: 17220510.

Izhikevich EM, Edelman GM. Large-scale model of mammalian thalamocortical systems. Proc Natl Acad Sci US A. 2008 Mar 4;105(9):3593-8. doi: 10.1073/pnas.0712231105. Epub 2008 Feb 21. PMID: 18292226; PMCID: PMC2265160.

Izuma K, Matsumoto M, Murayama K, Samejima K, Sadato N, Matsumoto K (2010) Neural corre-lates of cognitive dissonance and choice-induced preference change.

Proc Natl Acad Sci US A 107:22014–22019. doi:10.1073/pnas.1011879108 pmid:21135218

J

Jacob- Eshel B 2003 Bacterial self–organization: co–enhancement of complexification and adapta-bility in a dynamic environment Phil. Trans. R. Soc. A.3611283–1312A.3611283–1312http://doi.org/10.1098/rsta.2003.1199

Jaderberg, M.; Simonyan, K.; Zisserman, A.; Kavukcuoglu, K. Spatial transformer networks. In: Proceedings of the 28th International Conference on Neural Information Processing Systems, Vol. 2, 2017– 2025, 2015

James W (1890). The Principles of Psychology. Vol. 1. New York: Henry Holt. pp. 403–404.

Javanbakht A. A neural network model for schemas based on pattern completion. J Am Acad Psy-choanal Dyn Psychiatry. 2011 Summer;39(2):243-61. doi: 10.1521/jaap.2011.39.2.243. PMID: 21699351.

Jenkins A.C., Macrae C.N., Mitchell J.P. (2008). Repetition suppression of ventromedial pre-frontal activity during judgments of self and others. Proceedings of the National Academy of Sciences of the United States of America, 105, 4507–12.

Jennifer S. Lerner, Ye Li, Piercarlo Valdesolo, Karim S. Kassam. Emotion and Decision Making.Annual Review of Psychology 2015 66:1, 799-823

Jensen-Campbell, L. A., Adams, R., Perry, D. G., Workman, K. A., Furdella, J. Q. and Egan, S. K. (2002). Agreeableness, Extraversion, and Peer Relations in Early Adolescence: Winning Friends and Deflecting Aggression. Journal of Research in Personality. 363. 224-251.

Jerman A, Pejić Bach M, Aleksić A (2020) Transformation towards smart factory system. Syst Res Behav Sci 37(2):388–402

Jia X, Kohn A. Gamma rhythms in the brain. PLoS Biol. 2011 Apr;9(4):e1001045. doi: 10.1371/journal. pbio.1001045. Epub 2011 Apr 12. PMID: 21556334; PMCID: PMC3084194.

Jiang, N., Hu, H., Dang, Y. et al. Quantum Point Cloud and its Compression. Int J Theor Phys 56, 3147–3163 (2017). https://doi.org/10.1007/s10773-017-3483-7

Jin, X & Wang, Z. (2005). An Emotion Space Model for Recognition of Emotions in Spoken Chi-nese. 397-402. 10.1007/11573548_51.

J.K.L. Lau, H. Ozono, K. Kuratomi, A. Komiya, K. Murayama Shared striatal activity in decisions to satisfy curiosity and hunger at the risk of electric shocksNat Hum Behav, 21 (2020).

Jirtle R.L. and Skinner M.K. Environmental epigenomics and disease susceptibility. Nature Re-views Genetics 8, 253-262 (2007).

Ji, X., Paulsen, B.D., Chik, G.K.K. et al. Mimicking associative learning using an ion-trapping non-volatile synaptic organic electrochemical transistor. Nat Commun 12, 2480 (2021). https://doi.org/10.1038/s41467-021-22680-5

Johanson DL, Ahn HS, MacDonald BA, Ahn BK, Lim J, Hwang E, Sutherland CJ, Broadbent E. The Effect of Robot Attentional Behaviours on User Perceptions and Behaviours in a Simulated Health Care Interaction: Randomized Controlled Trial. J Med Internet Res. 2019 Oct 4;21(10): e13667. doi: 10.2196/13667. Erratum in: J Med Internet Res. 2020 Mar 3;22(3):e18362. PMID: 31588904; PMCID: PMC6914232.

Joukal, Marek. (2017). Anatomy of the Human Visual Pathway. 10.1007/978-3-319-52284-5_1.

Jouhanneau JS, Ferrarese L, Estebanez L, Audette NJ, Brecht M, Barth AL, Poulet JF. Cortical fosG-FP expression reveals broad receptive field excitatory neurons targeted by POm. Neuron. 2014 Dec 3;84(5):1065-78. doi: 10.1016/j.neuron.2014.10.014. Epub 2014 Nov 13. PMID: 25453844.

Johnson B, Flores Mosri D. The Neuropsychoanalytic Approach: Using Neuroscience as the Basic Science of Psychoanalysis. Front Psychol. 2016 Oct 13; 7:1459. doi: 10.3389/fpsyg.2016.01459. PMID: 27790160; PMCID: PMC5063004.

Johnston MV. Plasticity in the developing brain: implications for rehabilitation. Dev Disabil Res Rev. 2009;15(2):94-101.

J.L. Harding, K. Backholer, E.D. Williams, A. Peeters, A.J. Cameron, M.J. Hare, D.J. Magliano Psychosocial stress is positively associated with body mass index gain over 5 years: evidence from the longitudinal AusDiab study. Obesity, 22 (1) (2014), pp. 277-286, 10.1002/oby.20423

John, R.A., Shah, N., Vishwanath, S.K. et al. Halide perovskite memristors as flexible and recon-figurable physical unclonable functions. Nat Commun 12, 3681 (2021). https://doi.org/10.1038/s41467-021-24057-0

Jon S. Simons, Jane R. Garrison, Marcia K. Johnson,Brain Mechanisms of Reality Monitor-ing,Trends in Cognitive Sciences,Volume 21, Issue 6, 2017, Pages 462-473, ISSN 1364-6613, https://doi.org/10.1016/j.tics.2017.03.012.

Jordan E Pierce, Julie Péron, The basal ganglia and the cerebellum in human emotion, Social Cogni-tive and Affective Neuroscience, Volume 15, Issue 5, May 2020, Pages 599–613, https://doi.org/10.1093/scan/nsaa076

Jordan G, Deeb SS, Bosten JM, Mollon JD. The dimensionality of colour vision in carriers of anom-alous trichromacy. J Vis. 2010 Jul 1;10(8):12. doi: 10.1167/10.8.12. PMID: 20884587.

Joseph LeDoux's Synaptic Self: (How Our Brains Become Who We Are. JR Soc Med. 2002 Jul;95(7):373–4. PMCID: PMC1279949)

Josselyn SA, Tonegawa S. Memory engrams: Recalling the past and imagining the future. Science. 2020 Jan 3;367(6473): eaaw4325. doi: 10.1126/science. aaw4325. PMID: 31896692; PMCID: PMC7577560.

Judge, T. A., Erez, A., Bono, J. E., & Thoresen, C. J. (2002). Are measures of self-esteem, neuroti-cism, locus of control, and generalized self-efficacy indicators of a common core con-struct? Journal of Personality and Social Psychology, 83(3), 693–710. https://doi.org/10.1037/0022-3514.83.3.693

K

K. Zalta, A. & L. Chambless, D. (2012). Confidence Appraisals Protect against Anxiety in Re-sponse to a Transient Stressor. Psychology, 3, 441-446. doi: 10.4236/psych.2012.36062.

Kandel E.R. Dudai Y.Mayford M.R.The molecular and systems biology of memory.Cell. 2014; 157: 163-186 https://doi.org/10.1016/j.cell.2014.03.001

Kandel ER. The molecular biology of memory storage: a dialogue between genes and synapses. Science. 2001 Nov 2;294(5544):1030-8. doi: 10.1126/science.1067020. PMID: 11691980.

Kajić, I., Schröder, T., Stewart, T. C., & Thagard, P. (2019). The semantic pointer theory of emo-tion: Integrating physiology, appraisal, and construction. Cognitive Systems Research, 58, 35–53. https://doi.org/10.1016/j.cogsys.2019.04.007

Kaplan H, Hill K, Lancaster J, et al A theory of human life history evolution: diet, intelligence, and longevity. Evol Anthropol 2000;9:156–85. doi:10.1002/1520-6505(2000)9:4<156::AID-EVAN5>3.0.CO;2-7

Kasabov, N. (2019). Evolving Spiking Neural Networks. 10.1007/978-3-662-57715-8_5.

Katsuki F, Constantinidis C. Bottom-up and top-down attention: different processes and overlap-ping neural systems. Neuroscientist. 2014 Oct;20(5):509-21. doi: 10.1177/1073858413514136. Epub 2013 Dec 20. PMID: 24362813.

Kaufman, J. C., Beghetto, R. A., & Dilley, A. (2016). Understanding creativity in the schools. In A. A. Lipnevich, F. Preckel, & R. D. Roberts (Eds.), Psychosocial skills and school systems in the 21st century: Theory, research, and practice (pp. 133–153). Springer International Publish-ing/Springer Nature.

Kaufman, J. C., & Beghetto, R. A. (2009). Beyond big and little: The four-c model of creativi-ty. Review of General Psychology, 13(1), 1–12. https://doi.org/10.1037/a0013688

Kauffman S, Andrea Roli A. What is consciousness? Artificial intelligence, real intelligence, quan-tum mind and qualia, Biological Journal of the Linnean Society, 2022; blac092, https://doi.org/10.1093/biolin-nean/blac092.

Keban V 2023. May 23, Forbes Online Publication https://www.forbes.com/sites/forbestechcouncil/2023/05/23/the-need-for-an-artificial-sensory-nerv-ous-system-for-generative-ai/.

Keltner D, Sauter D, Tracy J, Cowen A. Emotional Expression: Advances in Basic Emotion Theo-ry. J Nonverbal Behav. 2019 Jun;43(2):133-160. doi: 10.1007/s10919-019-00293-3. Epub 2019 Feb 7. PMID: 31395997; PMCID: PMC6687086.

Kelley WM, Macrae CN, Wyland CL, Caglar S, Inati S, Heatherton TF. Finding the self? An event-related fMRI study. J Cogn Neurosci. 2002 Jul 1;14(5):785-94. doi: 10.1162/08989290260138672. PMID: 12167262

Kelly, J.W. & Avraamides, M.N. & Giudice, N. A. (2011) Haptic experiences influence visually acquired memories: Reference frames during multi-modal spatial learning. Psychonomic bulletin & Review. 8:1119-1125.

Kennedy, J. and Eberhart, R. (1995) Particle Swarm Optimization. Proceedings of the IEEE International Conference on Neural Networks,4 1942-1948 http://dx.doi.org/10.1109/ICNN.1995.488968

Kenrick DT, Griskevicius V, Neuberg SL, Schaller M. Renovating the Pyramid of Needs: Contem-po-rary Extensions Built Upon Ancient Foundations. Perspect Psychol Sci. 2010 May;5(3):292-314. doi: 10.1177/1745691610369469. PMID: 21874133; PMCID: PMC3161123.

Kenyon GN, Sen KC. The perception process. The Perception of Quality. 2014:41-50. doi:10.1007/978-1-4471-6627-6_5.

Kelso JA. Phase transitions and critical behaviour in human bimanual coordination. Am J Physiol. 1984 Jun;246(6 Pt 2):R1000-4. doi: 10.1152/ajpregu.1984.246.6.R1000. PMID: 6742155.

Kelso, Scott. (2013). Coordination Dynamics. 10.1007/978-3-642-27737-5_101-3.

Kennedy, D.P & Adolphs, R. (2014) Violations of personal space by individuals with autism spec-trum disorder. PLoS One. 2014 Aug 6;9(8): e103369. doi: 10.1371/journal.pone.0103369. View

Kenwood, M.M., Kalin, N.H. & Barbas, H. The pre-frontal cortex, pathological anxiety, and anxiety disorders. Neuropsychopharmacol. 47, 260–275 (2022). https://doi.org/10.1038/s41386-021-01109-z.

Kehoe, B Patil, Abbeel P and Goldberg K, "A Survey of Research on Cloud Robotics and Automa-tion," in IEEE Transactions on Automation Science and Engineering, vol. 12, no. 2, pp. 398-409, April 2015, doi: 10.1109/TASE.2014.2376492.

Kernberg O. Borderline personality organization. J Am Psychoanal Assoc. 1967; 15:641–685.

Kerskens C, Pérez D L. Experimental indications of non-classical brain functions. Journal of Physics Communications, 2022; 6 (10): 105001 DOI: 10.1088/2399-6528/ac94be

Kety SS (1957) The general metabolism of the brain in vivo. In: Metabolism of the nervous sys-tem (Richter D, ed), London: Pergamon, pp 221–237.

Khawaja FI, Kanazawa A, Kinugawa J, Kosuge K. A Human-Following Motion Planning and Control Scheme for Collaborative Robots Based on Human Motion Prediction. Sensors (Basel). 2021 Dec 9;21(24):8229. doi: 10.3390/s21248229. PMID: 34960323; PMCID: PMC8706253.

Khodagholy D, Gelinas JN, Thesen T, Doyle W, Devinsky O, Malliaras GG, Buzsáki G.

NeuroGrid: recording action potentials from the surface of the brain. Nat Neurosci. 2015 Feb;18(2):310-5. doi: 10.1038/ nn.3905. Epub 2014 Dec 22. PMID: 25531570; PMCID: PMC4308485.

Khrennikov A. Ubiquitous Quantum Structure: From Psychology to Finance. Berlin: Springer; 2010.

Kimble, H. The quantum internet. Nature 453, 1023–1030 (2008). https://doi.org/10.1038/nature07127

Kipke DR, Shain W, Buzsáki G, Fetz E, Henderson JM, Hetke JF, Schalk G. Advanced neurotech-no-logies for chronic neural interfaces: new horizons and clinical opportunities. J Neurosci. 2008 Nov 12;28(46):11830-8. doi: 10.1523/ JNEUROSCI.3879-08.2008. PMID: 19005048; PMCID: PMC3844837. Kirchhoff M, Parr T, Palacios E, Friston K, Kiverstein J. The Markov blankets of life: autonomy, active inference and the free energy principle. JR Soc Interface. 2018 Jan;15(138):20170792. doi: 10.1098/ rsif.2017.0792. PMID: 29343629; PMCID: PMC5805980.

Kiverstein J, Kirchhoff MD, Froese T. The Problem of Meaning: The Free Energy Principle and Artificial Agency. Front Neurorobot. 2022 Jun 23; 16:844773. doi: 10.3389/ fnbot.2022.844773. PMID: 35812784; PMCID: PMC9260223.

Kitzbichler M,2009 Public Library of Science. "The Human Brain is on The Edge of Chaos.

Klepeis, N., Nelson, W., Ott, W. et al. The National Human Activity Pattern Survey (NHAPS): a resource for assessing exposure to environmental pollutants. J Expo Sci Environ Epidemiol 11, 231–252 (2001). https://doi.org/10.1038/sj.jea.7500165

Klimecki, O. M., et al. (2012), op. cit.; Klimecki, O., Ricard, M., & Singer, T. (2013), op. cit.

Knee GC, Kakuyanagi K, Yeh MC, Matsuzaki Y, Toida H, Yamaguchi H, Saito S, Leggett AJ, Mun-ro WJ. A strict experimental test of macroscopic realism in a superconducting flux qubit. Nat Commun. 2016 Nov 4; 7:13253. doi: 10.1038/ncomms13253. PMID: 27811844; PMCID: PMC5097155.

Knill DC, Pouget A. The Bayesian brain: the role of uncertainty in neural coding and computation. Trends Neurosci. 2004 Dec;27(12):712-9. doi: 10.1016/j.tins.2004.10.007. PMID: 15541511.

Kogge. P, "The tops in flops," IEEE Spectrum, vol. 48, no. 2, pp. 48–54, Feb. 2011.

Kolb, B., & Milner, B. (1981). Performance of complex arm and facial movements after focal brain lesions. Neuropsychologia, 19:505-514.

Kompella V., Luciw M., Stollenga M., Pape L., Schmidhuber J. (2012). Autonomous learning of abstractions using curiosity-driven modular incremental slow feature analysis, in Proceedings of the Joint International Conference Development and

Learning and Epigenetic Robotics (ICDL-EPIROB-2012) (San Diego, CA:). 10.1109/DevLrn.2012.6400829.

Kong, Y.; Feng, M.; Li, X.; Lu, H.; Liu, X.; Yin, B. Spatial context-aware network for salient object detection. Pattern Recognit. 2021, 114, 107867.

Korn H, Faure P. Is there chaos in the brain? II. Experimental evidence and related models. CR Biol. 2003 Sep;326(9):787-840. doi: 10.1016/j.crvi.2003.09.011. PMID: 14694754.

Kowalczuk, Z, Czubenko, M. (2016). Computational Approaches to Modelling Artificial Emotion – An Overview of the Proposed Solutions. Frontiers in Robotics and AI. 3. 10.3389/frobt.2016.00021.

Kozima, H., Nakagawa, C., & Yano, H. (2004). Can a robot empathize with people? Artificial Life and Robotics, 8(1), 83–88.

Knudsen EI. Sensitive periods in the development of the brain and behaviour. J Cogn Neurosci. 2004 Oct;16(8):1412-25. doi: 10.1162/0898929042304796. PMID: 15509387.

Koculak M, Wierzchoń M. How much consciousness is there in complexity? Front Psychol. 2022 Sep 20; 13:983315. doi: 10.3389/fpsyg.2022.983315. PMID: 36204731; PMCID: PMC9530911.

Kolchinsky A, Wolpert David H. 2018 Semantic information, autonomous agency and non-equilibrium statistical physics Interface Focus.82018004120180041.

Kopp S, Krämer N. Revisiting Human-Agent Communication: The Importance of Joint Co-construction and Understanding Mental States. Front Psychol. 2021 Mar 23; 12:580955. doi: 10.3389/ fpsyg.2021.580955. PMID: 33833705; PMCID: PMC8021865.

Kosinski M.2023 Theory of Mind Might Have Spontaneously Emerged in Large Language Models. https://arxiv.org/pdf/2302.02083.pdf

Kruglanski, Shah, Fishbach, Friedman, Chun, & Sleeth-Keppler, 2002 f M.A. Conway, K. Meares, S. Standart Images & goals Memory, 12 (2004), pp. 525-531

KS Nagla, Moin Uddin, and Dilbag Singh International Journal of Robotics and Automation. Multi-sensor Data Fusion and Integration for Mobile Robots: A Review (IJRA) Vol. 3, No. 2, June 2014, pp. 131~138 ISSN: 2089-4856.

Kudithipudi D, et al. biological underpinnings for lifelong learning machines. Nat. Mach. In-tell. 2022; 4:196–210. doi: 10.1038/s42256-022-00452-0.

Kuffner J, "Cloud-enabled robots", Proc. IEEE-RAS Int. Conf. Humanoid Robot., 2010.

Kulke L, Feyerabend D, Schacht A. A Comparison of the Affectiva iMotions Facial Expression Analysis Software with EMG for Identifying Facial Expressions of Emotion. Front Psychol. 2020 Feb 28; 11:329. doi: 10.3389/fpsyg.2020.00329. PMID: 32184749; PMCID: PMC7058682.

Kümmerer, M, Theis, L, Bethge, M 2014. Deep Gaze I: Boosting Saliency Prediction with Feature Maps Trained on ImageNet.

Kurzweil, R. 2 https://www.bcbt.specs-lab.com/bcbt15/sites/default/files/event-files-public/BS4-DAC-DEF.pdf005. The singularity is near: When humans transcend biology. Penguin.

Kvanvig, J. The Value of Knowledge and the Pursuit of Understanding. NY: Cambridge University Press, 2003.

Kvanvig, J. "The Value of Understanding" In D. Pritchard, A. Haddock and A. Millar (eds.), Epis-temic Value. Oxford: Oxford University Press, 2009.

L

Lacin E, Muller A, Fernando M, Kleinfeld D, Slesinger PA. Construction of Cell-based Neuro-transmitter Fluorescent Engineered Reporters (CNiFERs) for Optical Detection of Neurotransmitters In Vivo. J Vis Exp. 2016 May 12;(111):53290. doi: 10.3791/53290. PMID: 27214050; PMCID: PMC4939270.

Lagioia A, Eliez S, Schneider M, Simons JS, Van der Linden M, Debbané M. Neural correlates of reality monitoring during adolescence. Neuroimage. 2011 Apr 1;55(3):1393-400. doi: 10.1016/j.neuroim-age.2010.12.058. Epub 2010 Dec 30. PMID: 21195192.

Lähdepuro, A., Savolainen, K., Lahti-Pulkkinen, M. et al. The Impact of Early Life Stress on Anxi-ety Symptoms in Late Adulthood. Sci Rep 9, 4395 (2019). https://doi.org/10.1038/s41598-019-40698-0

Lai CS, Fisher SE, Hurst JA, Vargha-Khadem F, Monaco AP. A forkhead-domain gene is mutated in a severe speech and language disorder. Nature. 2001 Oct 4;413(6855):519-23. doi: 10.1038/35097076. PMID: 11586359.

Laing RD. The Divided Self: An Existential Study in Sanity and Madnes. London, UK: Penguin Books; 1960

Laird JE (2012a) The soar cognitive architecture. MIT Press, Cambridge

Lakoff, G; Johnson, M (1980). The metaphorical structure of the human conceptual Cognitive Sci-ence. 4 (2): 195–208. doi:10.1207/s15516709cog0402_4.

Lancaster MA, Renner M, Martin CA, Wenzel D, Bicknell LS, Hurles ME, Homfray T, Penninger JM, Jackson AP, Knoblich JA. Cerebral organoids model human brain development and micro-cephaly. Nature. 2013 Sep 19;501(7467):373-9. doi: 10.1038/nature12517. Epub 2013 Aug 28. PMID: 23995685; PMCID: PMC3817409.

Franklin S, Strain S, Snaider j, McCall R Faghihi U.

Landauer, R. "Information is a physical entity." Physica A. 263, no. 1-4 (1999): 63-67.

Lanza M, Sebastian A, Lu WD, Le Gallo M, Chang MF, Akinwande D, Puglisi FM, Alshareef HN, Liu M, Roldan JB. Memristive technologies for data storage, computation, encryption, and radio-frequency communication. Science. 2022 Jun 3;376(6597): eabj9979. doi: 10.1126/science. abj9979. Epub 2022 Jun 3. PMID: 35653464.

Lashley KS, Mass action in cerebral function. Science 73, 245–254 (1931). doi: 10.1126/sci-ence.73.1888.245.

LaterA.,2017.Source. https://www.nytimes.com/2017/03/06/science/technology-addiction-irresisti-ble-by-adam-alter.html#:~:text=How%20do%20you%20define%20%E2%80%9Cad

Lau EY, Wong ML, Lau KN, Hui FW, Tseng CH. Rapid-Eye-Movement-Sleep (REM) Associated Enhancement of Working Memory Performance after a Daytime Nap. PLoS One. 2015 May 13;10(5):e0125752. doi: 10.1371/journal.pone.0125752. PMID: 25970511; PMCID: PMC4430242.

Lau H, Rosenthal D. Empirical support for higher-order theories of conscious awareness. Trends Cogn Sci. 2011 Aug;15(8):365-73. doi: 10.1016/j.tics.2011.05.009. Epub 2011 Jul 6. PMID: 21737339.

Laughlin SB, de Ruyter van Steveninck R, and Anderson J.C. The metabolic cost of neural infor-mation. NatureNeuroscience, 1(1):36–41, 1998.

Lawler, I. Understanding why, knowing why, and cognitive achievements. Synthese 196, 4583–4603 (2019). https://doi.org/10.1007/s11229-017-1672-9

LeDoux JE. Thoughtful feelings. Curr Biol. 2020 Jun 8;30(11): R619-R623. doi: 10.1016/j. cub.2020.04.012. PMID: 32516605.

Lebow MA, Chen A. Correction: Overshadowed by the amygdala: the bed nucleus of the stria ter-minalis emerges as key to psychiatric disorders. Mol Psychiatry. 2021 Aug;26(8):4561. doi: 10.1038/s41380-019-0568-0. Erratum for: Mol Psychiatry. 2016 Apr;21(4):450-63. PMID: 31700191; PMCID: PMC8550976.

LeCun, Yann & Bengio, Y. & Hinton, Geoffrey. (2015). Deep Learning. Nature. 521. 436-44. 10.1038/ nature14539.

Lee B, Cai W, Young CB, Yuan R, Ryman S, Kim J, Santini V, Henderson VW, Poston KL, Menon V. Latent brain state dynamics and cognitive flexibility in older adults. Prog Neurobiol. 2022 Jan; 208:102180. doi: 10.1016/j.pneurobio.2021.102180. Epub 2021 Oct 7. PMID: 34627994; PMCID: PMC9585912.

LeDoux JE. Emotion circuits in the brain. Annu Rev Neurosci. 2000; 23:155-84. doi: 10.1146/annurev. neuro.23.1.155. PMID: 10845062.

LeDoux J E. 2015. https://www.huffpost.com/entry/the-amygdala-is-not-the-b_b_7977278

Lee TS. The visual system's internal model of the world. Proc IEEE Inst Electr Electron Eng. 2015 Aug;103(8):1359-1378. doi: 10.1109/JPROC.2015.2434601. Epub 2015 Jul 6. PMID: 26566294; PMCID: PMC4638327.

Lee W, Kim D, Matsuhisa N, Nagase M, Sekino M, Malliaras GG, Yokota T, Someya T. Transpar-ent, conformable, active multielectrode array using organic electrochemical transistors. Proc Natl Acad Sci US A. 2017 Oct 3;114(40):10554-10559. doi: 10.1073/pnas.1703886114. Epub 2017 Sep 18. PMID: 28923928; PMCID: PMC5635873.

Lee, Y. & Ahn, J.-H. Biomimetic tactile sensors based on nanomaterials. ACS Nano14, 1220–1226 (2020).

Lee IS, Jung WM, Lee YS, Wallraven C, Chae Y. Operant and classical learning principles underly-ing mind-body interaction in pain modulation: a pilot fMRI study. Sci Rep. 2021;11(1):1663. doi:10.1038/ s41598-021-81134-6.

Legg, S & Hutter, M. (2007). A Collection of Definitions of Intelligence. Advances in Artificial General Intelligence: Concepts, Architectures and Algorithms. 157.

Legg S & Hutter M. A formal measure of machine intelligence. In Proc. 15th Annual Machine Learning Conference of Belgium and The Netherlands (Benelearn'06), pages 73–80, Ghent, 2006.

Lehman J, Stanley KO (2013) Correction: Evolvability Is Inevitable: Increasing Evolvability with-out the Pressure to Adapt. PLOS ONE 8(5): 10.1371/annotation/f4c5a0f3-cb53-4c05-a84c-f0aead483b77.

Lehnert BP, Santiago C, Huey EL, Emanuel AJ, Renauld S, Africawala N, Alkislar I, Zheng Y, Bai L, Koutsioumpa C, Hong JT, Magee AR, Harvey CD, Ginty DD. Mechanoreceptor synapses in the brain-stem shape the central representation of touch. Cell. 2021 Oct 28;184(22):5608-5621.e18. doi: 10.1016/j. cell.2021.09.023. Epub 2021 Oct 11. PMID: 34637701; PMCID: PMC8556359.

Leite, I., Pereira, A., Mascarenhas, S., Martinho, C., Prada, R., and Paiva, A. (2013). The influence of empathy in human-robot relations. Int. J. Hum. Comput. Stud. 71, 250–260. doi: 10.1016/j.ijhcs.2012.09.005

Lenart-Bugla M, Szcześniak D, Bugla B, Kowalski K, Niwa S, Rymaszewska J, Misiak B. The asso-ci-ation between allostatic load and brain: A systematic review. Psychoneuroendocrinology. 2022 Nov; 145:105917. doi: 10.1016/j. psyneuen.2022.105917. Epub 2022 Sep 7. PMID: 36113380.

Lesica, N.A., Mehta, N., Manjaly, J.G. et al. Harnessing the power of artificial intelligence to trans-form hearing healthcare and research. Nat Mach Intell 3, 840–849 (2021). https://doi.org/10.1038/s42256-021-00394-z.

Leslie, Alan. (1994). ToMM, ToBy, and Agency: Core architecture and domain specificity.

Lesort, T, Lomonaco, V, Stoian, A, Maltoni, D, Filliat, D, Díaz-Rodríguez, N. (2019). Continual Learning for Robotics: Definition, Framework, Learning Strategies, Opportunities and Challenges. Information Fusion. 58. 10.1016/j.inffus.2019.12.004

Levine J (1983), "Materialism and Qualia: The Explanatory Gap," Pacific Philosophical Quarterly, 64(4), October 1983, pp. 354–361.

Levita, L., Bois, C., Healey, A. et al. The Behavioural Inhibition System, anxiety and hippocampal volume in a non-clinical population. Biol Mood Anxiety Disord 4, 4 (2014). https://doi.org/10.1186/2045-5380-4-4.

Levy, David. (2007). Love + Sex with Robots: The Evolution of Human-Robot Relationships.

Lewis, P. Durrant S., Overlapping memory replay during sleep builds cognitive schemata,Trends in Cognitive Sciences,Volume 15, Issue 8,2011,Pages 343-351,ISSN 13646613, https://doi.org/10.1016/j.tics.2011.06.004.(https://www.sciencedirect.com/science/article/pii/S1364661311001094)

Lewis KM, Matsumoto C, Cardinale E, Jones EL, Gold AL, Stringaris A, Leibenluft E, Pine DS, Brotman MA. Self-Efficacy as a Target for Neuroscience Research on Moderators of Treatment Outcomes in Paedi-atric Anxiety. J Child Adolesc Psychopharmacol. 2020 May;30(4):205-214. doi: 10.1089/cap.2019.0130. Epub 2020 Mar 11. PMID: 32167803; PMCID: PMC7360109.

Levinson SC. 2006. On the human 'interaction engine'. In Roots of human sociality: culture, cogni-tion and interaction (eds Enfield NJ, Levinson SC), pp. 39-69. Oxford, UK: Berg.

Lieberman M 2005 (https://www.theguardian.com/science/2005/jun/30)

Lillicrap, T.P., Santoro, A., Marris, L. et al. Backpropagation and the brain. Nat Rev Neurosci 21, 335–346 (2020). https://doi.org/10.1038/s41583-020-0277-3.

Lin, JH., Vitter, J.S. A Theory for Memory-Based Learning. Machine Learning 17, 143–167 (1994). https://doi.org/10.1023/A:1022667616941

Lindell AK (2013) Continuities in emotion lateralization in human and nonhuman primates.

Front Hum Neurosci 7: 464.

Lindholm, T., Jönsson, F. U., and Tullio-Liuzza, M. (2018). Retrieval effort cues predict eyewit-ness accuracy. J. Exp. Psychol. 24, 534–542. doi: 10.1037/xap0000175.

Lindsay GW. Attention in Psychology, Neuroscience, and Machine Learning. Front Comput Neuro-sci. 2020 Apr 16; 14:29. doi: 10.3389/fncom.2020.00029. Erratum in: Front Comput Neurosci. 2021 May 26; 15:698574. PMID: 32372937; PMCID: PMC7177153.

Lipson, H., Pollack, J. Automatic design and manufacture of robotic lifeforms. Nature 406, Harpin V, Mazzone L, Raynaud JP, Kahle J, Hodgkins P. Long-Term Outcomes of ADHD: A Systematic Review of Self-Esteem and Social Function. J Atten Disord. 2016 Apr;20(4):295-305. doi: 10.1177/1087054713486516. Epub 2013 May 22. PMID: 23698916. 974–978 (2000). https://doi.org/10.1038/35023115.

Li R, Zhang S, Yin S, Ren W, He R, Li J. The fronto-insular cortex causally mediates the default-mode and central-executive networks to contribute to individual cognitive performance in healthy elderly. Hum Brain Mapp. 2018 Nov;39(11):4302-4311. doi: 10.1002/hbm.24247. Epub 2018 Jul 4. PMID: 29974584; PMCID: PMC6866622.

Litovsky VG, Dusek JB. Perceptions of child rearing and self-concept development during the early adolescent years. J Youth Adolesc. 1985 Mar;14(5):373-87. doi: 10.1007/BF02138833. PMID: 24301316.

Liu, S,Delbruck, T, Indiveri, G, Whatley, A, Douglas, R. (2015). Event-Based Neuromorphic Systems. 10.1002/9781118927601.ch2.

Liu,T Lyu, E. Wang and H. Meng M. Q., "Unified Intention Inference and Learning for Human–Robot Cooperative Assembly," in IEEE Transactions on Automation Science and Engineering, vol. 19, no. 3, pp. 2256-2266, July 2022, doi: 10.1109/TASE.2021.3077255.

Liu X, Rosendo A, Ikemoto S, Shimizu M, Hosoda K. Robotic investigation on effect of stretch reflex and crossed inhibitory response on bipedal hopping. JR Soc Interface. 2018 Mar;15(140):20180024. doi: 10.1098/rsif.2018.0024. PMID: 29593088; PMCID: PMC5908536.

Liu X, Ramirez S, Pang PT, Puryear CB, Govindarajan A, Deisseroth K, Tonegawa S. Optogenetic stimulation of a hippocampal engram activates fear memory recall. Nature. 2012 Mar 22;484(7394):381-5. doi: 10.1038/nature11028. PMID: 22441246; PMCID: PMC3331914.

Li, Y., Lopez-Huerta, V.G., Adiconis, X. et al. Distinct subnetworks of the thalamic reticular nucle-us. Nature 583, 819–824 (2020). https://doi.org/10.1038/s41586-020-2504-5.

Lloyd, S., Mohseni, M. and Rebentrost, P. Quantum algorithms for supervised and unsupervised machine learning. ArXiv:1307.0411 (2013).

Lopes, Manuel & Oudeyer, Pierre-Yves. (2010). Guest Editorial Active Learning and Intrinsically Motivated Exploration in Robots: Advances and Challenges. Autonomous Mental Development, IEEE Transactions on. 2. 65 – 69. 10.1109/TAMD.2010.2052419.

Low, Jason & Perner, Josef. (2012). Implicit and explicit theory of MD: State of the art. The British journal of developmental psychology. 30. 1-13. 10.1111/j.2044-835X.2011.02074. x.

Low SC, Vouloutsi V, Verschure P.F.M.J., 2021.Complementary interactions between classical and top-down driven inhibitory mechanisms of attention, Cognitive Systems Research, Volume 67, Pages 66-72, ISSN 1389-0417, https://doi.org/10.1016/j.cogsys.2020.12.003.

Loewenstein G, Lerner JS (2003) The role of affect in decision making. In: Davidson RJ, Scherer K, Hill HH (eds) Handbook of affective sciences. Oxford University Press, New York, pp 619–642.

Lukianoff, G. & Haidt, J. (2018). The Coddling of the American Mind: How good intentions and bad ideas are setting up a generation for failure. New York: Penguin.

Lungarella M, G. Metta, R. Pfeifer, and G. Sandini. Developmental robotics: a survey. Connection Science, 15(4):151–190, 2003.

Luo YH, Zhong HS, Erhard M, Wang XL, Peng LC, Krenn M, Jiang X, Li L, Liu NL, Lu CY, Zeilinger A, Pan JW. Quantum Teleportation in High Dimensions. Phys Rev Lett. 2019 Aug 16;123(7):070505. doi: 10.1103/PhysRevLett.123.070505. PMID: 31491117.

Lupien SJ, Maheu F, Tu M, Fiocco A, Schramek TE. The effects of stress and stress hormones on human cognition: Implications for the field of brain and cognition. Brain Cogn. 2007 Dec;65(3):209-37. doi: 10.1016/j.bandc.2007.02.007. Epub 2007 Apr 26. PMID: 17466428.

L. Wuensch, E.R. Pool, D. Sander. Individual differences in learning positive affective value Curr Opin Behav Sci, 39 (2021), pp. 19-26.

L. Renoult, M. D. Rugg Norman, 2010. Dissociating memory traces and scenario construction in mental time travel. Neuroscience and biobehavioural reviews, 60, 82-89.

Loftus EF, Klinger MR. Is the unconscious smart or dumb? American Psychologist. 1992; 47:761–765.

Long, J. (2012). Darwin's Devices: What Evolving Robots Can Teach us about the History of Life and the Future of Technology. Basic Books.

Longstreet, P., and Brooks, S. (2017). Life satisfaction: a key to managing internet & social media addiction. Technol. Soc. 50, 73–77. doi: 10.1016/j.techsoc.2017.05.003.

Lorenz-Spreen, P., Mønsted, B.M., Hövel, P. et al. Accelerating dynamics of collective attention. Nat Commun 10, 1759 (2019). https://doi.org/10.1038/s41467-019-09311-w

Lumer ED, Friston KJ, Rees G. Neural correlates of perceptual rivalry in the human brain. Science. 1998 Jun 19;280(5371):1930-4. doi: 10.1126/science.280.5371.1930. PMID: 9632390.

Lusher, J.M., C. Chandler, C. Ball D.Dopamine D4 receptor gene (DRD4) is associated with Novel-ty seeking (NS) and substance abuse: The saga continues Molecular Psychiatry, 6 (5) (2001), pp. 497-499

M

Macrae CN, Moran JM, Heatherton TF, Banfield JF, Kelley WM. Medial pre-frontal activity predicts memory for self. Cereb Cortex. 2004 Jun;14(6):647-54. doi: 10.1093/cercor/bhh025. Epub 2004 Apr 14. PMID: 15084488.

Magaña A, Vlaeyen M, Haitjema H, Bauer P, Schmucker B, Reinhart G. Viewpoint Planning for Range Sensors Using Feature Cluster Constrained Spaces for Robot Vision Systems. Sensors (Ba-sel). 2023 Sep 18;23(18):7964. doi: 10.3390/s23187964. PMID: 37766019; PMCID: PMC10537344.

Makin, TR & Krakauer, JW. Against Cortical Reorganisation. eLife; 21 Nov 2023; DOI: doi.org/10.7554/eLife.84716

Malfaz M, Salichs M A, 2004, A new architecture for autonomous robots based on emotions, IFAC Proceedings Volumes, Volume 37, Issue 8, 2004, Pages 805-809, ISSN 1474-6670, https://doi.org/10.1016/S14

Manninen T, Aćimović J, Linne ML. Analysis of Network Models with Neuron-Astrocyte Interac-tions. Neuroinformatics. 2023 Apr;21(2):375-406. doi:

10.1007/s12021-023-09622-w. Epub 2023 Mar 23. PMID: 36959372; PMCID: PMC10085960.

McAdams D. P. (1996). Personality, modernity, and the storied self: A contemporary framework for studying persons. Psychological Inquiry, 7, 295–321.

Markett S, Montag C, Reuter M. Network Neuroscience and Personality. Personal Neurosci. 2018 Aug 10;1:e14. doi: 10.1017/pen.2018.12. PMID: 32435733; PMCID: PMC7219685.

Markus H, Wurf E. The dynamic self-concept. A social psychological perspective. Annual Review of Psychology. 1987; 38:299–337.

Marr David. Vision: A computational approach, 1982.

Marr D, Poggio T. Cooperative computation of stereo disparity. Science. 1976 Oct 15;194(4262):283-7. doi: 10.1126/science.968482. PMID: 968482.

Martín F, Ginés J, Rodríguez-Lera FJ, Guerrero-Higueras AM, Matellán Olivera V. Client-Server Approach for Managing Visual Attention, Integrated in a Cognitive Architecture for a Social Robot. Front Neurorobot. 2021 Sep 9; 15:630386. doi: 10.3389/fnbot.2021.630386. PMID: 34566618; PMCID: PMC8458947.

Mateos-Aparicio P, Rodríguez-Moreno A. The Impact of Studying Brain Plasticity. Front Cell Neu-rosci. 2019; 13:66.

Matsuo Y, LeCun Y, Sahani M, Precup D, Silver D, Sugiyama M, Uchibe E, Morimoto J. Deep learning, reinforcement learning, and world models. Neural Netw. 2022 Aug; 152:267-275. doi: 10.1016/j.neu-net.2022.03.037. Epub 2022 Apr 19. PMID: 35569196.

McCloskey, Cohen, N J. (1989). Catastrophic Interference in Connectionist Networks: The Sequen-tial Learning Problem. Psychology of Learning and Motivation. Vol. 24. pp. 109–165. doi:10.1016/S0079-7421(08)60536-8. ISBN 978-0-12-543324-2

McFadden J. Integrating information in the brain's EM field: the cemi field theory of conscious-ness, Neuroscience of Consciousness (2020). DOI: 10.1093/nc/niaa016

Mckinsey Global Institute Report (2023) https://www.mckinsey.com/featured-insights/future-of-work/jobs-lost-jobs-gained-what-the-future-of-work-will-mean-for-jobs-skills-and-wages).

McLelland D, VanRullen R. Theta-Gamma Coding Meets Communication-through-Coherence: Neu-ronal Oscillatory Multiplexing Theories Reconciled. PLoS Comput Biol. 2016 Oct 14;12(10): e1005162. doi: 10.1371/journal.pcbi.1005162. PMID: 27741229; PMCID: PMC5065198.

McRae M., March 2ND 2023. For-the-first time-ever-physicists-see-molecules-form-through-quantum-tunneling. Source https://www.sciencealert.com/for-the-first-time-ever-physicists-see-molecules-form-through-quantum-tunneling

Malmberg, K.J., Raaijmakers, J.G.W. & Shiffrin, R.M. 50 years of research sparked by Atkinson and Shiffrin (1968). Mem Cogn 47, 561–574 (2019). https://doi.org/10.3758/s13421-019-00896-7

Ma W.J. Bayesian inference with probabilistic population codes. Nat. Neurosci. 2006; 9:1432–1438

Man J, Chen G, Chen J. Recent Progress of Biomimetic Tactile Sensing Technology Based on Mag-netic Sensors. Biosensors (Basel). 2022 Nov 21;12(11):1054. doi: 10.3390/bios12111054. PMID: 36421172; PMCID: PMC9688171.

Man, K., Damasio, A. Homoeostasis and soft robotics in the design of feeling machines. Nat Mach Intell 1, 446–452 (2019). https://doi.org/10.1038/s42256-019-0103-7.

Manigault AW, Handley IM, Whillock SR. Assessment of unconscious decision aids applied to complex patient-centered medical decisions. J Med Internet Res. 2015 Feb 5;17(2):e37. doi: 10.2196/jmir.3739. PMID: 25677337; PMCID: PMC4342682.

Mankowska ND, Marcinkowska AB, Waskow M, Sharma RI, Kot J, Winklewski PJ. Critical Flicker Fusion Frequency: A Narrative Review. Medicina (Kaunas). 2021 Oct 13;57(10):1096. doi: 10.3390/me-dicina57101096. PMID: 34684133; PMCID: PMC8537539.

Mansell W, Carey TA. A century of psychology and psychotherapy: is an understanding of 'con-trol' the missing link between theory, research, and practice? Psychol Psychother. 2009 Sep;82(Pt 3):337-53. doi: 10.1348/147608309X432526. Epub 2009 Apr 1. PMID: 19341533.

Mareschal, D, Blakeman S, 'Fast and Slow Learning in Human-Like Intelligence', in Stephen Mug-gleton, and Nicholas Chater (eds), Human-Like Machine Intelligence (Oxford, 2021; online edn, Oxford Academic, 19 Aug. 2021), https://doi.org/10.1093/oso/9780198862536.003.0016, accessed 15 Aug. 2023.

Markus, H. (1977). Self-schemata and processing information about the self. Journal of Personality and Social Psychology, 35(2), 63–78. https://doi.org/10.1037/0022-3514.35.2.63

Marr D (2010) Vision: a computational investigation into the human representation and processing of visual information. MIT Press, Cambridge.

Marvin, J., Borghuis, B., Tian, L. et al. An optimized fluorescent probe for visualizing glutamate neuro-transmission. Nat Methods 10, 162–170 (2013). https://doi.org/10.1038/nmeth.2333.

McCrae RR, John OP. An introduction to the five-factor model and its applications. J Pers. 1992 Jun;60(2):175-215. doi: 10.1111/j.1467-6494. 1992.tb00970. x. PMID: 1635039.

McCloskey M and Cohen N. J Catastrophic interference in connectionist networks: The sequential learning problem. 1989.

Marie Vandekerckhove, Jaak Panksepp,The flow of anoetic to noetic and autonoetic conscious-ness: A vision of unknowing (anoetic) and knowing (noetic) consciousness in the remembrance of things past and imagined futures,Consciousness and Cognition,Volume 18, Issue 4,2009,Pages 1018-1028,ISSN 10538100, https://doi.org/10.1016/j.concog.2009.08.002.

Marinier RP, Laird JE, Lewis RL (2009) A computational unification of cognitive behaviour and emotion. Cogn Syst Res 10(1):48–69.

H. Markram, "The human brain project," Sci. Amer., vol. 306, no. 6, pp. 50–55, May 2012

Marshall JB (2002b) Metacat: a self-watching cognitive architecture for analogy-making. In: Pro-ceedings of the 24th annual conference of the cognitive science society

Martignon, L., Deco, G., Laskey, K., Diamond, M., Freiwald, W., & Vaadia, E. (2000). Neural cod-ing: higher-order temporal patterns in the neurostatistics of cell assemblies. Neural Computa-tion, 12(11), 2621–2653.

Masse, Freedman, et al. "Circuit mechanisms for the maintenance and manipulation of information in working memory" Nature Neuroscience. June 10, 2019. DOI: 10.1038/s41593-019-0414-3.

Mateos-Aparicio P, Rodríguez-Moreno A. The Impact of Studying Brain Plasticity. Front Cell Neu-rosci. 2019; 13:66.

Matthews G, Lin J, Panganiban A et al. (2020) Individual differences in trust in autonomous ro-bots. IEEE Trans Hum Mach Syst 50:234–244

McAdams, D. P., & Pals, J. L. (2006). A new Big Five: Fundamental principles for an integrative science of personality. American Psychologist, 61, 204–217.

McAdams, D. P. (1985). Power, intimacy, and the life story: Personological inquiries into identity. Home-wood: Dorsey Press.

McDonnell, M., Ward, L. The benefits of noise in neural systems: bridging theory and experi-ment. Nat Rev Neurosci 12, 415–425 (2011). https://doi.org/10.1038/nrn3061

McEwen BS. Stress, adaptation, and disease: allostasis and allostatic load. Annals of the New York Academy of Sciences. 1998;840(1):33-44. doi:10.1111/j.1749-6632. 1998. tb09546.x

McGrew WC, Marchant LF. Laterality of hand use pays off in foraging success for wild chimpan-zees. Primates 40: 509–513, 1999. doi:10.1007/BF02557586.

Mearns, Jack. (2008). Social learning theory of Julian B. Rotter. Retrieved March 21, 2009, from http://psych.fullerton.edu/jmearns/rotter.htm.

Medland SE, Duffy DL, Wright MJ, Geffen GM, Hay DA, Levy F, van-Beijsterveldt CE, Willemsen G, Townsend GC, White V, Hewitt AW, Mackey DA, Bailey JM, Slutske WS, Nyholt DR, Treloar SA, Martin NG, Boomsma DI. Genetic influences on handedness: data from 25,732 Australian and Dutch twin families. Neuropsychologia. 2009 Jan;47(2):330-7. doi: 10.1016/j.neuropsychologia.2008.09.005. Epub 2008 Sep 9. PMID: 18824185; PMCID: PMC2755095.

Mehrabian, A Communicating without words, Psychol. Today. (1968).

Meltzoff AN. 1988. The human infant as Homo imitans. In Social learning: psychological and bio-logical perspectives (eds TR Zentall, BG Galef), pp. 319–341. Hillsdale, NJ: Erlbaum.

Merker B., Williford K., Rudrauf D. (2022). The integrated information theory of consciousness: A case of mistaken identity. Behav. Brain Sci. 45: e41. 10.1017/S0140525X21000881

Meyniel F, Dehaene S. Brain networks for confidence weighting and hierarchical inference during probabilistic learning. Proc Natl Acad Sci US A. 2017 May 9;114(19):E3859-E3868. doi: 10.1073/ pnas.1615773114. Epub 2017 Apr 24. PMID: 28439014; PMCID: PMC5441722.

Michael, Elizabeth & Santamaria, Lorena & Leong, Victoria & Kourtzi, Zoe. (2022). Learning at your brain's rhythm: individualized entrainment boosts learning for perceptual decisions. Cerebral Cortex. 33. 10.1093/cercor/bhac426.

Michlmayr M. Simulation theory versus theory theory: Theories concerning the ability to read minds. University of Innsbruck, 2002

Ming Huan C, Mutlu B 2016. https://intuitivecomputing.github.io/publications/2016-hri-huang.pdf -hri-huang.pdf.

Mnih V, Kavukcuoglu K, Silver D, Rusu AA, Veness J, Bellemare MG, Graves A, Riedmiller M, Fid-jeland AK, Ostrovski G, Petersen S, Beattie C, Sadik A, Antonoglou I, King H, Kumaran D, Wierstra D, Legg S, Hassabis D. Human-level control through deep reinforcement learning. Nature. 2015 Feb 26;518(7540):529-33. doi: 10.1038/nature14236. PMID: 25719670.

Modinos G, Ormel J, Aleman A. Activation of anterior insula during self-reflection. PLoS One. 2009;4(2): e4618. doi: 10.1371/journal.pone.0004618. Epub 2009 Feb 26. PMID: 19242539; PMCID: PMC2643476.

Molenberghs P,The neuroscience of in-group bias,Neuroscience & Biobehavioral Reviews, Volume 37, Issue 8,2013, Pages 1530-1536, ISSN 0149-7634, https://doi.org/10.1016/j.neubior-ev.2013.06.002. (https://www.sciencedirect.com/science/article/pii/S0149763413001498)

Montag C, Sindermann C, Lester D, Davis KL. Linking individual differences in satisfaction with each of Maslow's needs to the Big Five personality traits and Panksepp's primary emotional sys-tems. Heliyon. 2020 Jul 23;6(7): e04325. doi: 10.1016/j.heliyon. 2020. e04325. PMID: 32743084; PMCID: PMC7387820.

Moioli, R.C., Vargas, P.A., Husbands, P. (2010). The Dynamics of a Neural Network of Coupled Phase Oscillators with Synaptic Plasticity Controlling a Minimally Cognitive Agent. In: Diaman-taras, K., Duch, W., Iliadis, L.S. (eds) Artificial Neural Networks – ICANN 2010. ICANN 2010. Lecture Notes in Computer Science, vol 6353. Springer, Berlin, Heidelberg. https://doi.org/10.1007/978-3-642-15822-3_32

Moravec, H. Mind Children: The Future of Robot and Human Intelligence. (Harvard University Press, 1988).

Mouret, J.-B., and Doncieux, S. (2012). Encouraging behavioural diversity in evolutionary robotics: an empirical study. Evol. Comput. 20, 91–133. doi:10.1162/EVCO_a_00048.

Moutoussis K, Zeki S (March 1997). "A direct demonstration of perceptual asynchrony in vi-sion". Proceedings. Biological Sciences. 264 (1380): 393–9.

Muhammad Ilyas Azeem, Fabio Palomba, Lin Shi, Qing Wang,

Machine learning techniques for code smell detection: A systematic literature review and meta-analysis, Information and Software Technology, Volume 108,2019, Pages 115-138, ISSN 0950-5849, https://doi.org/10.1016/j.infsof.2018.12.009.

Malone J, Liu S, Vaillant G, Rentz D, Waldinger R. Midlife Eriksonian psychosocial development: Setting the stage for cognitive and emotional health in late life. Dev Psychol. 2016;52(3):496-508. doi:10.1037/ a0039875.

Marques-Villarroya S, Castillo JC, Gamboa-Montero JJ, Sevilla-Salcedo J, Salichs MA. A Bio-Inspired Endogenous Attention-Based Architecture for a Social Robot. Sensors (Basel). 2022 Jul 13;22(14):5248. doi: 10.3390/s22145248. PMID: 35890931; PMCID: PMC9323278.

Mehrabian, A., & Russell, J. A. (1974). An Approach to Environmental Psychology. Cambridge, MA: MIT Press.

Mell P and Grance T "The NIST definition of cloud computing", National Institute of Standards and Technology, 2009.

Merolla P, Arthur, J Akopyan F, Imam N, Manohar R, and Modha D, "A digital neurosynaptic core using embedded crossbar memory with 45 pJ per spike in 45 nm," in Proc. Custom Integr. Circuits Conf., 2011, DOI: 10.1109/CICC.2011.6055294.

Miller P. Dynamical systems, attractors, and neural circuits. F1000Res. 2016 May 24;5: F1000 Faculty Rev-992. doi: 10.12688/f1000research.7698.1. PMID: 27408709; PMCID: PMC4930057.

Mitchell R. Kinesthetic-Visual Matching and the self-concept as explanations of Mirror -Self -Recognition 27 December 2001 https://doi.org/10.1111/1468-5914.00024

Mitchell, R. W. (2002). Kinesthetic-visual matching, imitation, and self-recognition. In M. Bekoff, C. Allen, & G. M. Burghardt (Eds.), The cognitive animal: Empirical and theoretical perspectives on animal cognition (pp. 345–351). MIT Press.

Mitra A, Kraft A, Wright P, Acland B, Snyder AZ, Rosenthal Z, Czerniewski L, Bauer A, Snyder L, Culver J, Lee JM, Raichle ME. Spontaneous Infra-slow Brain Activity Has Unique Spatiotemporal Dynamics and Laminar Structure. Neuron. 2018 Apr 18;98(2):297-305.e6. doi: 10.1016/j.neuron.2018.03.015. Epub 2018 Mar 29. PMID: 29606579; PMCID: PMC5910292.

Miwa, H. et al. (2001). "Robot personality based on the equations of emotion defined in the 3D mental space". Proceedings of the International Conference on Robotics and Automation, pp.2602–2607

Modha D. S. Akopyan F, Andreopoulos A, Appuswamy, R Arthur, J. V, Cassidy, A. S Datta, P. DeBole, M. V. et al, 2023.Science 382, 329

Moioli R. C., Vargas P. A., Husbands P. (2009). A multiple hormone approach to the homoeostatic control of conflicting behaviours in an autonomous mobile robot, in 2009 IEEE Congress on Evo-lutionary Computation (Trondheim: IEEE), 47–54.

Moioli, R.C., Vargas, P.A., Von Zuben, F.J., Husbands, P. (2008). Evolving an Artificial Homeo-stat-ic System. In: Zaverucha, G., da Costa, A.L. (eds) Advances in Artificial Intelligence – SBIA 2008. SBIA 2008. Lecture Notes in Computer Science (), vol 5249. Springer, Berlin, Heidelberg. https://doi.org/10.1007/978-3-540-88190-2_33

Moneta, G. B., & Csikszentmihalyi, M. (1996). The effect of perceived challenges and skills on the quality of subjective experience. Journal of Personality, 64, 274–310. doi:10.1111/ j.1467-6494. 1996.tb00512.x

Moore GE 1965 Cramming more components onto integrated circuits Electronics 38 114–7

Morewedge CK, Preston J, Wegner DM. Timescale bias in the attribution of mind. J Pers Soc Psy-chol. 2007 Jul;93(1):1-11. doi: 10.1037/0022-3514.93.1.1. PMID: 17605584.

Morelli AM, Ravera S, Panfoli I. The aerobic mitochondrial ATP synthesis from a comprehensive point of view. Open Biol. 2020;10(10):200224. doi:10.1098/ rsob.200224.

Moran, R. J., Campo, P., Symmonds, M., Stephan, K. E., Dolan, R. J., and Friston, K. J. (2013). Free energy, precision and learning: the role of cholinergic neuromodulation. J. Neurosci. 33, 8227–8236. doi: 10.1523/JNEUROSCI.4255-12.2013

Moray, Neville (1959), "Attention in Dichotic Listening: Affective Cues and the Influence of In-struc-tions," Quarterly Journal of Experimental Psychology, 11 (1), 56–60.

Morin, A 2006, 'Levels of consciousness and self-awareness: a comparison and integration of various neurocognitive views', Consciousness and Cognition, vol. 15, no. 2, pp. 358–71.

Mohr C, Binkofski F, Erdmann C, Büchel C, Helmchen C. The anterior cingulate cortex contains distinct areas dissociating external from self-administered painful stimulation: a parametric fMRI study. Pain. 2005 Apr;114(3):347-357. doi: 10.1016/j. pain.2004.12.036. PMID: 15777860.

Muller A, Joseph V, Slesinger PA, Kleinfeld D. Cell-based reporters reveal in vivo dynamics of dopamine and nor-epinephrine release in murine cortex. Nat Methods. 2014 Dec;11(12):1245-52. doi: 10.1038/ nmeth.3151. Epub 2014 Oct 26. PMID: 25344639; PMCID: PMC4245316.

Murayama K, Matsumoto M, Izuma K, Matsumoto K. Neural basis of the undermining effect of monetary reward on intrinsic motivation. Proc Natl Acad Sci U S A. 2010 Dec 7;107(49):20911-6. doi: 10.1073/ pnas.1013305107. Epub 2010 Nov 15. PMID: 21078974; PMCID: PMC3000299.

Mukherjee A, Lam NH, Wimmer RD, and. Halassa M M Thalamic circuits for independent control of pre-frontal signal and noise.Nature, pp. 1–8, October 2021. ISSN 1476-4687. doi: 10.1038/s41586-021-04056-3

Murray RJ, Schaer M, Debbané M. Degrees of separation: a quantitative neuroimaging meta-analysis investigating self-specificity and shared neural activation between self- and other reflec-tion. Neurosci Biobehav Rev. 2012 Mar;36(3):1043-59. doi: 10.1016/j.neubiorev.2011.12.013. Epub 2012 Jan 4. PMID: 22230705. Copy

Myers CE, Interian A, Moustafa AA. A practical introduction to using the drift diffusion model of decision-making in cognitive psychology, neuroscience, and health sciences. Front Psychol. 2022 Dec 9; 13:1039172. doi: 10.3389/fpsyg.2022.1039172. PMID: 36571016; PMCID: PMC9784241.

N

Nair L, Chernova S. Feature Guided Search for Creative Problem Solving Through Tool Construc-tion. Front Robot AI. 2020 Dec 23; 7:592382. doi: 10.3389/frobt.2020.592382. PMID: 33501352; PMCID: PMC7806064.

Najmi S, Kuckertz JM, Amir N. Attentional impairment in anxiety: inefficiency in expanding the scope of attention. Depress Anxiety. 2012 Mar;29(3):243-9. doi: 10.1002/da.20900. Epub 2011 Sep 21. PMID: 21948383; PMCID: PMC3569031.

Nanay B. The Importance of Amodal Completion in Everyday Perception. Iperception. 2018 Jul 31;9(4):2041669518788887. doi: 10.1177/2041669518788887. PMID: 30109014; PMCID: PMC6083800.

Nangyeon Lim Cultural differences in emotion: differences in emotional arousal level between the East and the West. Integrative Medicine Research Volume 5, Issue 2, June 2016, Pages 105-109

Nangyeon Lim Cultural differences in emotion: differences in emotional arousal level between the East and the West. Integrative Medicine Research Volume 5, Issue 2, June 2016, Pages 105-109.

Nano Lett. 2010, 10, 4, 1297–1301 Publication Date:March 1, 2010 https://doi.org/10.1021/nl904092h

Narcross F. Artificial nervous systems-A new paradigm for artificial intelligence. Patterns (N Y). 2021 Jun 11;2(6):100265. doi: 10.1016/j.patter.2021.100265. PMID: 34179847; PMCID: PMC8212122.

Nayyer A, Ajmal M, Wei L,, Zulqarnain S Mubarak S. (2019). Video Description: A Survey of Methods, Datasets, and Evaluation Metrics. ACM Computing Surveys. 52. 1-37. 10.1145/3355390.

Neill James. (2006). What is locus of control? Retrieved March 21, 2009, from wilder-dom.com/psycholo-gy/loc/LocusOfControlWhatIs.html.

Neri F,2023 10 August https://www.bbc.co.uk/news/technology-66465230

Nielsen JA, Zielinski BA, Ferguson MA, Lainhart JE, Anderson JS. An evaluation of the left-brain vs. right-brain hypothesis with resting state functional connectivity magnetic resonance imaging. PLoS One. 2013 Aug 14;8(8): e71275. doi: 10.1371/journal.pone.0071275. PMID: 23967180; PMCID: PMC3743825.

Neisser U (1988) Five kinds of self-knowledge. Philos Psychol 1, 35–59.

NelsonB, ParnasJ, SassLA. Disturbance of minimal self (ipseity) in schizophrenia: clarification and current status. Schizophr Bull. 2014; 40:479–482

Nelson K, Fivush R. The emergence of autobiographical memory: a social cultural developmental theory. Psychol Rev. 2004 Apr;111(2):486-511. doi: 10.1037/0033-295X.111.2.486. PMID: 15065919.

Neumann P.R. Options and strategies for countering online radicalization in the United States. Stud. Confl. Terror. 2013; 36:431–459. doi: 10.1080/1057610X.2013.784568.

Nguyen KN, Watanabe T, Andersen GJ. Role of endogenous and exogenous attention in task-relevant visual perceptual learning. PLoS One. 2020 Aug 28;15(8): e0237912. doi: 10.1371/journal.pone.0237912. PMID: 32857813; PMCID: PMC7454975.

Nguyen, M., Vanderwal, T., and Hasson, U. (2019). Shared understanding of narratives is correlat-ed with shared neural responses. NeuroImage 184, 161–170. doi: 10.1016/j.neuroimage.2018.09.010

Nguyen, QT., Schroeder, L., Mank, M. et al. An in vivo biosensor for neurotransmitter release and in situ receptor activity. Nat Neurosci 13, 127–132 (2010). https://doi.org/10.1038/nn.2469

Nguyen T., Zhao Q., Yan S. (2018). Attentive systems: a survey. Int. J. Comput. Vis. 126, 86–110. 10.1007/s11263-017-1042-6

Ngugi, A. K., Bottomley, C., Kleinschmidt, I., Sander, J. W. & Newton, C. R. Estimation of the bur-den of active and life-time epilepsy: a meta-analytic approach. Epilepsia 51, 883–890 (2010).

NicholasB.Turk-Browne,Do-JoonYi,MarvinM.Chun,LinkingImplicitandExplicitMemory: Common Encoding Factors and Shared Representations, Neuron, Volume 49, Issue 6, 2006, Pages 917-927, ISSN 0896-6273, https://doi.org/10.1016/j.neuron.2006.01.030. (https://www.sciencedirect.com/science/article/pii/S0896627306000870) Abstract: Summary.

Nie, X, Cao, J. Existence and global stability of equilibrium point for delayed competitive neural networks with discontinuous activation functions. Int. J. Syst. Sci. 43(3), 459-474 (2012)

Nikolaidou M, Fraser DS, Hinvest N. Attentional bias in Internet users with problematic use of so-cial networking sites. J Behav Addict. 2019 Dec 1;8(4):733-742. doi: 10.1556/2006.8.2019.60. Epub 2019 Dec 2. PMID: 31786935; PMCID: PMC7044579.

Nomura, K. Kawakami, Relationships between robot's self-disclosures and human's anxiety to-ward robots, In: Proceedingsof the 2011 IEEE/WIC/ACM International

Conferences on Web Intel-ligence and Intelligent Agent Technology, IEEE Computer So-ciety, 2011, 66–69.

Nocentini, O, Fiorini, L, Acerbi, G,Sorrentino, A, Mancioppi, G,Cavallo, F. (2019). A Survey of Behav-ioural Models for Social Robots. 10.20944/preprints201905. 0251. v1.

Noppeney U, Friston KJ, Price CJ. Degenerate neuronal systems sustaining cognitive functions. J Anat. 2004 Dec;205(6):433-42. doi: 10.1111/j.0021-8782.2004.00343. x. PMID: 15610392; PMCID: PMC1571441.

"Noradrenergic modulation of rhythmic neural activity shapes selective attention" by Martin J. Dahl et al. Trends in Cognitive Sciences

Northoff G., Heinzel A., De Greck M., et al. (2006). Self-referential processing in our brain—a me-ta-analysis of imaging studies on the self. NeuroImage, 31, 440–57.

Northoff, G., Heinzel, A. First-Person Neuroscience: A new methodological approach for linking mental and neuronal states. Philos Ethics Humanit Med 1, 3 (2006). https://doi.org/10.1186/1747-5341-1-3

Northoff, George, 'Self-Specificity and Self-Perspectival Organization' Unlocking the Brain: Volume 2:Pages 278–327 Consciousness (New York, 2013; online edn, Oxford Academic, 16 Apr. 2014), https://doi.org/10.1093/acprof:oso/9780199826995.003.0012,

Northoff G, Bermpohl F, Schoeneich F, Boeker H. How does our brain constitute defence mecha-nisms? First-person neuroscience and psychoanalysis. Psychother Psychosom. 2007;76(3):141-53. doi: 10.1159/000099841. PMID: 17426413.

N., Sam M.S., "EXOGENOUS CUE," in PsychologyDictionary.org, April 7, 2013, https://psychologydic-tionary.org/exogenous-cue/

Nunez, Paul & Srinivasan, Ramesh. (2006). Electric fields of the brain.

Nvidia 2022) Souece: https://blogs.nvidia.com/blog/2022/07/29/what-is-a-qpu/

O

Oatley, K., Keltner, D., and Jenkins, J. M. (2012). Understanding Emotions, 2nd Edn. Malden, MA: Blackwell Publishing.

O'Connor, Peg 2014. "Searching for the Self, and Other Unicorns." Psychology Today Nov/Dec 50-1.

Oertel-Knöchel V, Linden DE. Cerebral asymmetry in schizophrenia. Neuroscientist. 2011 Oct;17(5):456-67. doi: 10.1177/1073858410386493. Epub 2011 Apr 25. PMID: 21518811.

Oizumi M., Albantakis L., Tononi G. (2014). From the phenomenology to the mechanisms of con-scious-ness: Integrated information theory 3.0. PLoS Comput. Biol. 10:e1003588. 10.1371/journal.pcbi.1003588

O. Kinouchi and M. Copelli, Nat. Phys. 2, 348 (2006).

Olshausen, Bruno & Field, David. (2004). What is the other 85% of V1 doing?4. July 22nd, 2004. What is the other 85% of V1 doing? Problems in Systems Neuroscience. T.J. Sejnowski, L.van Hemmen, eds. Oxford University Press. (in press) http://www.rctn.org/bruno/papers/V1-chapter.pdf

Olshausen, B. and Field, D. Emergence of simple-cell receptive field properties by learning a sparse code for natural images. Nature, 1996

O'Mahony, J. F. (1984). Knowing others through the self: Influence of self-perception on percep-tion of others: A review. Current Psychological Research & Reviews, 3(4), 48–62. https://doi.org/10.1007/BF02686558.

Ognibene D, Foulsham T, Marchegiani L, Farinella GM. Editorial: Active Vision and Perception in Human-Robot Collaboration. Front Neurorobot. 2022 Feb 8;16:848065. doi: 10.3389/fnbot.2022.848065. PMID: 35211002; PMCID: PMC8860825.

Onishi KH, Baillargeon R. Do 15-month-old infants understand false beliefs? Science. 2005 Apr 8;308(5719):255-8. doi: 10.1126/science.1107621. PMID: 15821091; PMCID: PMC3357322.

O'Reilly RC, Frank MJ. Making working memory work: a computational model of learning in the pre-frontal cortex and basal ganglia. Neural Comput. 2006 Feb;18(2):283-328. doi: 10.1162/089976606775093909. PMID: 16378516.

Orenstein G, Lewis L. Erikson's stages of psychosocial development. In: StatPearls [Internet]. Statpearls Publishing.

Orseau, L., & Armstrong, S. (2016). Safely Interruptible Agents. Conference on Uncertainty in Ar-tificial Intelligence.

Osborne, R., & Wittrock, M.C. (1983). Learning science: A generative process. Science Education, 67, 489-508.

Owen AM, Coleman MR, Boly M, Davis MH, Laureys S, Pickard JD. Detecting awareness in the vegetative state. Science. 2006 Sep 8;313(5792):1402. doi: 10.1126/science.1130197. PMID: 16959998.

Oztop E, Kawato M, Arbib MA. Mirror neurons: functions, mechanisms and models. Neurosci Lett. 2013 Apr 12; 540:43-55. doi: 10.1016/j.neulet.2012.10.005. Epub 2012 Oct 11. PMID: 23063951.

P

Padesky CA. Schema change processes in cognitive therapy. Clinical Psychology and Psychothera-py. 1994; 1:267–278. doi:10.1002/cpp.5640010502.

Pagkalos M, Chavlis S, Poirazi P. Introducing the Dendrify framework for incorporating dendrites to spiking neural networks. Nat Commun. 2023 Jan 10;14(1):131. doi: 10.1038/s41467-022-35747-8. PMID: 36627284; PMCID: PMC9832130.

Pal MM. Glutamate: The Master Neurotransmitter and Its Implications in Chronic Stress and Mood Disorders. Front Hum Neurosci. 2021 Oct 29; 15:722323. doi: 10.3389/fnhum.2021.722323. PMID: 34776901; PMCID: PMC8586693.

Palacios ER, Razi A, Parr T, Kirchhoff M, Friston K. On Markov blankets and hierarchical self-organisa-tion. J Theor Biol. 2020 Feb 7; 486:110089. doi: 10.1016/j.jtbi.2019.110089. Epub 2019 Nov 20. PMID: 31756340; PMCID: PMC7284313.

Palhano-Fontes F, Andrade KC, Tofoli LF, Santos AC, Crippa JA, Hallak JE, Ribeiro S, de Araujo DB. The psychedelic state induced by ayahuasca modulates the activity and connectivity of the de-fault mode network. PLoS One. 2015 Feb 18;10(2):e0118143. doi: 10.1371/journal.pone.0118143. PMID: 25693169; PMCID: PMC4334486.

Palmer, S., Rosch, E., and Chase, P. (1981). Canonical Perspective and the Perception of Objects. In J. Long, & A. Baddeley (Eds.), International Symposium on Attention and Performance (Atten-tion and performance IX) (pp. 135-151). Hillsdale, NJ: Lawrence Erlbaum Associates.

Paparo, G. D., Dunjko, V., Makmal, A., MartinDelgado, M. A. and Briegel, H. J. Quantum speedup for active learning agents. Phys. Rev. X 4, 031002 (2014). URL http://journals.aps.org/prx/abstract/ 10.1103/PhysRevX.4.031002.

Parisi, G. I., Weber, C., and Wermter, S. (2015). Self-organizing neural integration of pose-motion features for human action recognition. Front. Neurorobot. 9:3. doi: 10.3389/fnbot.2015.00003.

Parisi, G.I. (2020). Human Action Recognition and Assessment via Deep Neural Network Self-Organization. ArXiv, abs/2001.05837.

Park IM, Meister ML, Huk AC, Pillow JW. Encoding and decoding in parietal cortex during sen-sorimotor decision-making. Nat Neurosci. 2014 Oct;17(10):1395-403. doi: 10.1038/nn.3800. Epub 2014 Aug 31. PMID: 25174005; PMCID: PMC4176983.

Park S, Whang M. Empathy in Human-Robot Interaction: Designing for Social Robots. Int J Envi-ron Res Public Health. 2022 Feb 8;19(3):1889. doi: 10.3390/ijerph19031889. PMID: 35162909; PMCID: PMC8835506.

Parker, A. J. and Newsome, W. T. 1998. 'Sense and the Single Neuron: Probing the Physiology of Perception'. Annual Review of Neuroscience, 21: 227–277.

Parisi GI, Kemker R, Part JL, Kanan C, Wermter S. Continual lifelong learning with neural net-works: A review. Neural Netw. 2019 May; 113:54-71. doi: 10.1016/j.neunet.2019.01.012. Epub 2019 Feb 6. PMID: 30780045.

Park S, Whang M. Empathy in Human-Robot Interaction: Designing for Social Robots. Int J Envi-ron Res Public Health. 2022 Feb 8;19(3):1889. doi: 10.3390/ijerph19031889. PMID: 35162909; PMCID: PMC8835506.

Parnas J, Handest P. Phenomenology of anomalous self-experience in early schizophrenia. Compr Psychiatry. 2003 Mar-Apr;44(2):121-34. doi: 10.1053/comp.2003.50017. PMID: 12658621.

Parnia S, Spearpoint K, de Vos G, Fenwick P, Goldberg D, Yang J, Zhu J, Baker K,

Killingback H, McLean P, Wood M, Zafari AM, Dickert N, Beisteiner R, Sterz F, Berger M, Warlow C, Bullock S, Lovett S, McPara RM, Marti-Navarette S, Cushing P, Wills P, Harris K, Sutton J, Walmsley A, Dea-kin CD, Little P, Farber M, Greyson B, Schoenfeld ER. AWARE-AWAreness during REsuscitation-a prospective study. Resuscitation. 2014 Dec;85(12):1799-805. doi: 10.1016/j.resuscitation.2014.09.004. Epub 2014 Oct 7. PMID: 25301715.

Passingham, R., Stephan, K. Kötter, R. The anatomical basis of functional localization in the cor-tex. Nat Rev Neurosci 3, 606–616 (2002). https://doi.org/10.1038/nrn893.

Patacchiola M, Cangelosi A. 2016 A developmental Bayesian model of trust in artificial cognitive systems. In 2016 Joint IEEE Int. Conf. on Development and Learning and Epigenetic Robotics (ICDL-EpiRob), pp. 117– 123. New York, NY: IEEE.

Patle B. K, Ganesh Babu L, Anish Pandey, D.R.K. Parhi, A. Jagadeesh, A review: On path planning strategies for navigation of mobile robot,Defence Technology,Volume 15, Issue 4, 2019, Pages 582-606, ISSN 2214-9147, https://doi.org/10.1016/j.dt.2019.04.011.

Payne P, Crane-Godreau MA. The preparatory set: a novel approach to understanding stress, trau-ma, and the bodymind therapies. Front Hum Neurosci. 2015 Apr 1; 9:178. doi: 10.3389/fnhum.2015.00178. PMID: 25883565; PMCID: PMC4381623.

Paradiso E, Gazzola V, Keysers C. Neural mechanisms necessary for empathy-related phenomena across species. Curr Opin Neurobiol. 2021 Jun; 68:107-115. doi: 10.1016/j.conb.2021.02.005. Epub 2021 Mar 20. PMID: 33756399.

Patle, B & Parhi, Dayal & Anne, Jagadeesh & Kashyap, Sunil Kumar. (2016). World Journal of En-gi-neering Probabilistic fuzzy controller-based robotics path decision theory Article information: For Authors Probabilistic fuzzy controller-based robotics path decision theory. World Journal of Engi-neering. 13. 181-192. 10.1108/WJE-04-2016-024.

P. C. Fletcher, R. N. A. Henson, Frontal lobes and human memory: Insights from functional neu-roimag-ing, Brain, Volume 124, Issue 5, May 2001, Pages 849–881, https://doi.org/10.1093/brain/124.5.849

P. Piolino et al.Do school-age children remember or know the personal past? Consciousness and Cogni-tion(2007)

Pearson JM, Heilbronner SR, Barack DL, Hayden BY, Platt ML (2011) Posterior cingulate cortex: adapting behaviour to a changing world. Trends Cogn Sci 15:143–151.

Pedersen ML, Frank MJ, Biele G. The drift diffusion model as the choice rule in reinforcement learning. Psychon Bull Rev. 2017 Aug;24(4):1234-1251. doi: 10.3758/s13423-016-1199-y. PMID: 27966103; PMCID: PMC5487295.

Pedro A.M. Mediano, Fernando E. Rosas, Daniel Bor, Anil K. Seth, Adam B. Barrett, The strength of weak integrated information theory,Trends in Cognitive Sciences,Volume

26, Issue 8,2022,Pages 646-655,ISSN 1364-6613, https://doi.org/10.1016/j.tics.2022.04.008.

Peuzin-Jubert M., Polette A., Nozais D., Mari J.L., Pernot J.P. Survey on the View Planning Prob-lem for Reverse Engineering and Automated Control Applications. Comput.-Aided Des. 2021; 141:103094. doi: 10.1016/j.cad.2021.103094.

Perez C. Intuition Machine 2016 https://medium.com/intuitionmachine/the-end-of-monolithic-deep-learning-86937c86bc1f

Perner, A., & Zeilinger, H. (2011). Action primitives for bionics inspired action planning system: Abstraction layers for action planning based on psychoanalytical concepts. 2011 9th IEEE Interna-tional Conference on Industrial Informatics, 63-68.

Perner J et al.Episodic memory and autonoetic consciousness: Developmental evidence and a the-ory of childhood amnesia Journal of Experimental Child Psychology(1995)

Perrine. Experimental Research on Dreaming: State of the Art and Neuropsychoanalytic Perspec-tives. Frontiers in Psychology. Vol 2 2011 https://www.frontiersin.org/articles/10.3389/fpsyg.2011.00286 DOI=10.3389/fpsyg.2011.00286

Pesaran B, Vinck M, Einevoll GT, Sirota A, Fries P, Siegel M, Truccolo W, Schroeder CE, Sriniva-san R. Investigating large-scale brain dynamics using field potential recordings: analysis and inter-pretation. Nat Neurosci. 2018 Jul;21(7):903-919. doi: 10.1038/s41593-018-0171-8. Epub 2018 Jun 25. PMID: 29942039; PMCID: PMC7386068.

Peters, R.A., Hambuchen, K.A. & Bodenheimer, R.E. The sensory ego-sphere: a mediating inter-face between sensors and cognition. Auton Robot 26, 1–19 (2009). https://doi.org/10.1007/s10514-008-9098-3.

Petersen SE, Posner MI. The attention system of the human brain: 20 years after. Annu Rev Neu-rosci. 2012; 35:73-89. doi: 10.1146/annurev-neuro-062111-150525. Epub 2012 Apr 12. PMID: 22524787; PMCID: PMC3413263.

Pezzulo G, Zorzi M, Corbetta M. The secret life of predictive brains: what's spontaneous activity for? Trends Cogn Sci. 2021 Sep;25(9):730-743. doi: 10.1016/j.tics.2021.05.007. Epub 2021 Jun 16. PMID: 34144895; PMCID: PMC8363551.

Pfeifer JH, Lieberman MD, Dapretto M. "I know you are but what am I?!": neural bases of self- and social knowledge retrieval in children and adults. J Cogn Neurosci. 2007 Aug;19(8):1323-37. doi: 10.1162/ jocn.2007.19.8.1323. PMID: 17651006; PMCID: PMC3407805.

Pfisterer, U., Petukhov, V., Demharter, S. et al. Identification of epilepsy-associated neuronal sub-types and gene expression underlying epileptogenesis. Nat Commun 11, 5038 (2020). https://doi.org/10.1038/s41467-020-18752-7.

Phepls EA (2006) Emotion and cognition: insights from studies of the human amygdala. Annu Rev Psy-chol 57:27–53

Philippi C.L., Duff M.C., Denburg N.L., et al. (2012). Medial PFC damage abolishes the self-reference effect. Journal of Cognitive Neuroscience, 24, 475–81.

Piaget, J. (1952) The Origins of Intelligence in Children (Vol. 8), International Universities Press

Picard L, Mayor-Dubois C, Maeder P, Kalenzaga S, Abram M, Duval C, Eustache F, Roulet-Perez E, Piolino P. Functional independence within the self-memory system: new insights from two cases of developmental amnesia. Cortex. 2013 Jun;49(6):1463-81. doi: 10.1016/j.cortex.2012.10.003. Epub 2012 Nov 20. PMID: 23261550.

Picard, R. W. (1995). Affective Computing. M.I.T Media Laboratory Perceptual Computing Section Technical Report No. 321.

Picard, R. W Affective Computing. Cambridge, MA, USA: MIT Press, 1997.

Pierre JM. Faith or delusion? At the crossroads of religion and psychosis. J Psychiatr Pract. 2001 May;7(3):163-72. doi: 10.1097/00131746-200105000-00004. PMID: 15990520.

Pink DH. Drive: The Surprising Truth about What Motivates Us. Reprint, paperback ed. Riverhead Books; 2012.)

Pinto Y, van der Leij AR, Sligte IG, Lamme VA, Scholte HS. Bottom-up and top-down attention are independent. J Vis. 2013 Jul 17;13(3):16. doi: 10.1167/13.3.16. PMID: 23863334

Pipitone A, Chella A, Robot passes the mirror test by inner speech, Robotics and Autonomous Systems, Volume 144, 2021, 103838, ISSN 0921-8890, https://doi.org/10.1016/j.robot.2021.103838.

Pipitone A, Chella A. What robots want? Hearing the inner voice of a robot. iScience, 2021; 102371 DOI: 10.1016/j.isci.2021.102371

Plate, Tony (2003). Distributed representations. Cognitive Science. 1-15.

Plaut, D. C. and McClelland, J. L. 2010. 'Locating Object Knowledge in the Brain: A Critique of Bowers's Citation(2009) Attempt to Revive the Grandmother Cell Hypothesis'. Psychological Re-view, 117: 284–290.

Plenz D, Thiagarajan TC. The organizing principles of neuronal avalanches: cell assemblies in the cortex? Trends Neurosci. 2007 Mar;30(3):101-10. doi: 10.1016/j.tins.2007.01.005. Epub 2007 Feb 1. PMID: 17275102.

Plutchik, R. (2001). The nature of emotions. Am. Sci. 89, 344. doi:10.1511/2001.4.344

Poirazi P, Mel BW. 2001 Impact of active dendrites and structural plasticity on the memory capaci-ty of neural tissue. Neuron. Mar;29(3):779-96. doi: 10.1016/s0896-6273(01)00252-5. PMID: 11301036.

Posner MI, Petersen SE. The attention system of the human brain. Annu Rev Neurosci. 1990; 13:25-42. doi: 10.1146/annurev.ne.13.030190.000325. PMID: 2183676.

Posner, M. I. (2008). Measuring alertness. Ann. N. Y. Acad. Sci. 1129, 193–199. doi: 10.1196/an-nals.1417.011.

Potter, M.C., Wyble, B., Hagmann, C.E. et al. Detecting meaning in RSVP at 13 ms per pic-ture. Atten Percept Psychophys 76, 270–279 (2014). https://doi.org/10.3758/s13414-013-0605-z.

Prescott SL, Liberles SD. Internal senses of the vagus nerve. Neuron. 2022 Feb 16;110(4):579-599. doi: 10.1016/j.neuron.2021.12.020. Epub 2022 Jan 19. PMID: 35051375; PMCID: PMC8857038.

Prescott TJ, Wilson SP. Understanding brain functional architecture through robotics. Sci Robot. 2023 May 31;8(78): eadg6014. doi: 10.1126/scirobotics. adg6014. Epub 2023 May 31. PMID: 37256968.

Preskill, J. Quantum computing in the NISQ era and beyond. Quantum2, 79. 10.22331/q-2018-08-06-79 (2018).

Preston JL, Ritter RS, Hepler J. Neuroscience and the soul: competing explanations for the human experience. Cognition. 2013 Apr;127(1):31-7. doi: 10.1016/j.cognition.2012.12.003. Epub 2013 Jan 12. PMID: 23318352.

Priebe NJ. Mechanisms of Orientation Selectivity in the Primary Visual Cortex. Annu Rev Vis Sci. 2016 Oct 14; 2:85-107. doi: 10.1146/annurev-vision-111815-114456. PMID: 28532362.

Pritchard, D. (2010). Knowledge and understanding. In D. Pritchard, A. Millar, & A. Haddock (Eds.), The nature and value of knowledge: Three investigations (pp. 5–88). Oxford, UK: Oxford University Press.

Pronin E. Perception and misperception of bias in human judgment. Trends Cogn Sci. 2007 Jan;11(1):37-43. doi: 10.1016/j.tics.2006.11.001. Epub 2006 Nov 28. PMID: 17129749.

Pronin, E., Gilovich, T., & Ross, L. (2004). Objectivity in the eye of the beholder: Divergent per-ceptions of bias in self versus others. Psychological Review, 111, 781-799. Public Library of Science. "The Human Brain Is on The Edge Of Chaos." ScienceDai-ly.ScienceDaily, 23 March 2009. <www.sciencedaily.com/releases/2009/03/090319224532.htm>.

Puderbaugh M, Emmady PD. Neuroplasticity. [Updated 2022 May 8]. In: StatPearls [Internet]. Treasure Island (FL): StatPearls Publishing; 2022 Jan-.

Pulvermüller, F., Tomasello, R., Henningsen-Schomers, M. R. & Wennekers, T. (2021) Biological constraints on neural network models of cognitive function. Nat. Rev. Neurosci. 22, 488–502.

Pynadath DV, Rosenbloom PS, Marsella SC (2014) Reinforcement learning for adaptive theory of mind in the sigma cognitive architecture. In: Proceedings of the conference on artificial general in-telligence

Q

Qi, M., Yang, R., Wang, Z., Liu, Y., Zhang, Q., He, B., Li, K., Yang, Q., Wei, L., Pan, C., & Chen, M. (2023). Bioinspired Self-healing Soft Electronics. Advanced Functional

Materials, 33."Questioning the role of sparse coding in the brain" by Anton Spanne and Henrik Jörntell in Trends in Neurosciences. Published online June 17 2015 doi: 10.1016/j.tins.2015.05.005

Quesque F, Brass M. The Role of the Temporoparietal Junction in Self-Other Distinction. Brain Topogr. 2019 Nov;32(6):943-955. doi: 10.1007/s10548-019-00737-5. Epub 2019 Nov 1. PMID: 31676934.

R

R.A. Davis, A cognitive-behavioral model of pathological Internet use, Computers in Human Be-havior, Volume 17, Issue 2, 2001, Pages 187-195, ISSN 0747-5632, https://doi.org/10.1016/S0747-5632(00)00041-8.

R.D Rafal, P.A Calabresi, C.W Brennan, T.K Sciolto Saccade preparation inhibits reorienting to re-cently attended locations. J. Exp. Psychol. Hum. Percept. Perform., 15 (1989), pp. 673-685

Rafael Vieira Bretas, Yumiko Yamazaki, Atsushi Iriki, Phase transitions of brain evolution that produced human language and beyond, Neuroscience Research,Volume 161,2020,Pages 1-7,ISSN 0168-0102,)

Rao, A., Plank, P., Wild, A. et al. A Long Short-Term Memory for AI Applications in Spike-based Neuro-morphic Hardware. Nat Mach Intell 4, 467–479 (2022). https://doi.org/10.1038/s42256-022-00480-w

Rahman T., Resnick P.J., Harry B. Anders Breivik: Extreme Beliefs Mistaken for Psychosis. J. Am. Acad. Psychiatry Law. 2016; 44:28–35.

Rahman T. Extreme Overvalued Beliefs: How Violent Extremist Beliefs Become "Normalized". Be-hav Sci (Basel). 2018 Jan 12;8(1):10. doi: 10.3390/bs8010010. PMID: 29329259; PMCID: PMC5791028.

Rajasethupathy Priya, Anteromedial Thalamus Gates the Selection and Stabilization of Long-Term Memories, Cell (2023). DOI: 10.1016/j.cell.2023.02.024. www.cell.com/cell/fulltext/S0092-8674(23)00167-8

Ramiro-Cortés Y, Hobbiss AF, Israely I. Synaptic competition in structural plasticity and cognitive function. Philos Trans R Soc Lond B Biol Sci. 2013 Dec 2;369(1633):20130157. doi: 10.1098/rstb.2013.0157. PMID: 24298158; PMCID: PMC3843888.

Ratcliff R, Smith PL, Brown SD, McKoon G. Diffusion Decision Model: Current Issues and Histo-ry. Trends Cogn Sci. 2016 Apr;20(4):260-281. doi: 10.1016/j.tics.2016.01.007. Epub 2016 Mar 5. PMID: 26952739; PMCID: PMC4928591

Ratcliff R. A theory of memory retrieval. Psychol. Rev. 1978; 85:59–108.

Ratcliff R, McKoon G. The diffusion decision model: theory and data for two-choice decision tasks. Neural Comput. 2008 Apr;20(4):873-922. doi: 10.1162/neco.2008.12-06-420. PMID: 18085991; PMCID: PMC2474742.

Raymundo, C.R.; Johnson, C.G.; Vargas, P.A. An Architecture for Emotional and

ContextAware Associative Learning for Robot Companions. In Proceedings of the Proceedings of the 24th IEEE International Symposium on Robot and Human Interactive Communication; 2015; Vol. 19, pp. 8799–8804.Raymundo, C.R.; Johnson, C.G.; Vargas, P.A 2017 A Situation-Aware Fear Learning (SAFEL) model for robots, Neurocomputing, Volume 221, 2017, Pages 32-47, ISSN 0925-2312, https://doi.org/10.1016/j.neucom.2016.09.035.

Rebentrost, P., Mohseni, M, and Lloyd, S. Quantum Support Vector Machine for Big Data Classifi-cation. Phys. Rev. Lett. 113, 130503 (2014).

Reggia JA, Katz GE, Davis GP. Humanoid Cognitive Robots That Learn by Imitating: Implications for Consciousness Studies. Front Robot AI. 2018 Jan 26; 5:1. doi: 10.3389/frobt.2018.00001. PMID: 33500888; PMCID: PMC7806019.

Reggia JA. The rise of machine consciousness: studying consciousness with computational models. Neural Netw. 2013 Aug; 44:112-31. doi: 10.1016/j.neunet.2013.03.011. Epub 2013 Mar 26. PMID: 23597599.

Reiff, J. S., Hershfield, H. E., & Quoidbach, J. (2020). Identity Over Time: Perceived Similarity Between Selves Predicts Well-Being 10 Years Later. Social Psychological and Personality Sci-ence, 11(2).

Reyner-Parra D, Huguet G (2022) Phase-locking patterns underlying effective communication in exact firing rate models of neural networks. PLoS Comput Biol 18(5): e1009342. https://doi.org/10.1371/jour-nal.pcbi.1009342

Ricoeur P.Oneself as Another. Chicago, IL: The University of Chicago Press; 1992

Ridgeway, Cecilia L. and Chris Bourg. 2004. "Gender as Status: An Expectation States Theory Approach." Pp. 217–241 in The Psychology of Gender, edited by A. H. Eagly, A. E. Beall, and R. J. Sternberg, New York: Guilford.

Ringach DL. Spontaneous and driven cortical activity: implications for computation. Curr Opin Neu-robiol. 2009 Aug;19(4):439-44. doi: 10.1016/j.conb.2009.07.005. Epub 2009 Aug 3. PMID: 19647992; PMCID: PMC3319344.

Robinson HM, MacDonald BA, Broadbent E. 2014. The role of healthcare robots for older people at home: a review. Int. J. Soc. Robot. 6:575–91.

Rodríguez L, Ramos, F. 2012. Computational models of emotions for autonomous agents: major challenges. Artificial Intelligence Review. 43. 10.1007/s10462-012-9380-9.

Rogers, T. B., Kuiper, N. A., & Kirker, W. S. (1977). Self-reference and the encoding of personal information. Journal of Personality and Social Psychology, 35, 677-688.

Rommelse, N. N., Van der Stigchel, S. & Sergeant, J. A. A review on eye movement studies in childhood and adolescent psychiatry. Brain Cogn. 68(3), 391–414 (2008). Connolly, J., Goodale, M., Menon, R. et al. Human fMRI evidence for the neural correlates of pre-parato-ry set. Nat Neurosci 5, 1345–1352 (2002). https://doi.org/10.1038/nn969

Rothbart, M.K., Ahadi, S.A., Hershey, K.L. and Fisher, P. (2001) Investigations of

Temperament at Three to Seven Years: The Children's Behaviour Questionnaire. Child Development, 72, 1394-1408.

Ryan, R. M., & Deci, E. L. (2017). Self-determination theory: Basic psychological needs in motiva-tion, development, and wellness. The Guilford Press. https://doi.org/10.1521/978.14625/28806

Reitter D & F. Ritter F.E. (Eds.), Proceedings of the 14th International Conference on Cognitive Model-ling (ICCM 2016). University Park, PA: Penn State.

Renoult L, Irish M, Moscovitch M, Rugg MD. From Knowing to Remembering: The Semantic-Episodic Distinction. Trends Cogn Sci. 2019 Dec;23(12):1041-1057. doi: 10.1016/j.tics.2019.09.008. Epub 2019 Oct 28. PMID: 31672430.

Renoult L., Davidson P.S.R., Palombo D.J., et al. (2012). Personal semantics: at the crossroads of semantic and episodic memory. Trends in Cognitive Sciences, 16, 550–8.

Rensink RA. The dynamic representation of scenes. Vis. Cog. 2000;7:17–42.

Revonsuo, A. (2000a). The reinterpretation of dreams: An evolutionary hypothesis of the function of dreaming. Behavioural and Brain Sciences, 23, 877–901.

Ringo JL, Doty RW, Demeter S, Simard PY. Time is of the essence: a conjecture that hemispheric specialization arises from interhemispheric conduction delay. Cereb Cortex. 1994 Jul-Aug;4(4):331-43. doi: 10.1093/cercor/4.4.331. PMID: 7950307.

Robyn Fivush The Development of Autobiographical Memory. Annu. Rev. Psychol. 2011. 62:559–82)

Rogers LJ, Zucca P, Vallortigara G. Advantages of having a lateralized brain. Proc Biol Sci. 2004 Dec 7;271 Suppl 6(Suppl 6):S420-2. doi: 10.1098/rsbl.2004.0200. PMID: 15801592; PMCID: PMC1810119.

Rogers, T. B., Kuiper, N. A., and Kirker, W. S. (1977). Self-reference and the encoding of personal information. J. Pers. Soc. Psychol. 35, 677–688. doi:10.1037/0022-3514.35.9.677.

Roli, A, Jaeger, J, Kauffman, S. (2022). How Organisms Come to Know the World: Fundamental Limits on Artificial General Intelligence. Frontiers in Ecology and Evolution. 9. 10.3389/fevo.2021.806283.

Rolls ET. Attractor networks. Wiley Interdiscip Rev Cogn Sci. 2010 Jan;1(1):119-134. doi: 10.1002/wcs.1. Epub 2009 Dec 17. PMID: 26272845.

Roozendaal B, McGaugh JL. Memory modulation. Behav Neurosci. 2011 Dec;125(6):797-824. doi: 10.1037/a0026187. PMID: 22122145; PMCID: PMC3236701.

Ros, R, Nalin, M, Wood, R, Baxter, P,, Looije, R & Demiris, Y, Belpaeme, T, Giusti, A, Pozzi, C (2011). Child-robot interaction in the wild: Advice to the aspiring experimenter. ICMI'11 – Proceed-ings of the 2011 ACM International Conference on Multi-modal Interaction. 335-342. 10.1145/2070481.2070545.

Rotenberg KJ, Petrocchi S, Lecciso F, Marchetti A. 2015 The relation between children's trust be-liefs and theory of mind abilities. Infant Child Dev. 24, 206– 214. (doi:10.1002/icd. v24.2) D. D'Amico Rosenberg A, Patterson JS, Angelaki DE. A computational perspective on autism. Proc Natl Acad Sci US A. 2015 Jul 28;112(30):9158-65. doi: 10.1073/pnas.1510583112. Epub 2015 Jul 13. PMID: 26170299; PMCID: PMC4522787.

Roy M, Shohamy D, Wager TD. Ventromedial prefrontal-subcortical systems and the generation of affective meaning. Trends Cogn Sci 16: 147-156VL – 16 DO – 10.1016/j.tics.2012.01.005

Rubin EB, Bernat JL. Ethical aspects of disordered states of consciousness. Neurol Clin 2011; 29:1055– 1071.

Rubio-Fernández P, Geurts B. How to pass the false-belief task before your fourth birthday. Psy-chol Sci. 2013 Jan 1;24(1):27-33. doi: 10.1177/0956797612447819. Epub 2012 Nov 21. PMID: 23175524; PMCID: PMC4107839.

Rubio-Fernandez, P et al. "How do you know that? Automatic belief inferences in passing conver-sation." Cognition 193 (December 2019): 104011 © 2019 Elsevier B.V.

Rueda, Maria & Posner, Michael & Rothbart, Mary. (2005). The Development of Executive Atten-tion: Contributions to the Emergence of Self-Regulation. Developmental neuropsychology. 28. 573-94. 10.1207/s15326942dn2802_2.

Reuters 2021.Reuters.com/technology/exclusive-softbank-shrinks-robotics-business-stops-pepper-pro-duction-sources-2021-06-28/#:~:text=Production%20of%20the%20humanoid%20Pepper,two%20of%20the%20sources%20said.

Ruge H., Jamadar S., Zimmermann U., Karayanidis F. (2013). The many faces of preparatory con-trol in task switching reviewing a decade of fMRI research. Hum. Brain Mapp. 34, 12–35. 10.1002/hbm.21420.

Ruminjo A, Mekinulov B. A Case Report of Cotard's Syndrome. Psychiatry (Edgmont). 2008 Jun;5(6):28-9. PMID: 19727279; PMCID: PMC2695744.

Rumelhart, D.E., Hinton, G.E., & Williams, R.J. (1986). Learning representations by back-propagating errors. Nature, 323, 533-536.

Russell, James & Mehrabian, Albert. (1977). Evidence for a Three-Factor Theory of Emotions. Journal of Research in Personality. 11. 273-294. 10.1016/0092-6566(77)90037-X.

Ruz, M., & Lupiáñez, J. (2002). A review of attentional capture: On its automaticity and sensitivity to endogenous control. Psicológica, 23(2), 283–309.

Rush AJ, Trivedi MH, Wisniewski SR, Nierenberg AA, Stewart JW, Warden D, Niederehe G, Thase ME, Lavori PW, Lebowitz BD, McGrath PJ, Rosenbaum JF, Sackeim HA, Kupfer DJ, Luther J, Fava M. Acute and longer-term outcomes in depressed outpatients requiring one or several treatment steps: a STAR*D report. Am J Psychiatry. 2006 Nov;163(11):1905-17. doi: 10.1176/ajp.2006.163.11.1905. PMID: 17074942.

Russakovsky, O., Deng, J., Su, H., Krause, J., Satheesh, S., Ma, S., et al. (2015). Imagenet large scale visual recognition challenge. Int. J. Comput. Vis. 115, 211–252. doi: 10.1007/s11263-015-0816-y.

Russell AF, Mihalaş S, von der Heydt R, Niebur E, Etienne-Cummings R. A model of proto-object-based saliency. Vision Res. 2014 Jan; 94:1-15. doi: 10.1016/j.visres.2013.10.005. Epub 2013 Oct 31. PMID: 24184601; PMCID: PMC3902215.

Russell JA. Core affect and the psychological construction of emotion. Psychol Rev. 2003 Jan;110(1):145-72. doi: 10.1037/0033-295x.110.1.145. PMID: 12529060.

Russell, S., Dietterich, T., Horvitz, E., Selman, B., Rossi, F., Hassabis, D., Legg, S., Suleyman, M., George, D., & Phoenix, S. (2015). Letter to the Editor: Research Priorities for Robust and Benefi-cial Artificial Intelligence: An Open Letter. AI Magazine, 36(4), 3-4. https://doi.org/10.1609/aimag.v36i4.2621

S

Sabour, S.; Frosst, N.; Geoffrey, E. Hinton Dynamic routing between capsules. In Proceedings of the 31st International Conference on Neural Information Processing Systems (NIPS'17), Curran Associates Inc., Red Hook, NY, USA, 4–9 December 2017; pp. 3859–3869.

Saggio V, Asenbeck BE, Hamann A, Strömberg, T. P, Schiansky P, Dunjko, V. N. Friis N, Harris N C, Hochberg M, Englund D, Wölk S, Briegel H. J,Walther P. Experimental quantum speed-up in reinforcement learning agents. Nature, 2021; 591 (7849): 229 DOI: 10.1038/s41586-021-03242-7

Salvucci, Dario D., 'ACT-R and Beyond', in Susan E. F. Chipman (ed.), The Oxford Handbook of Cognitive Science, Oxford Handbooks (2017; online edn, Oxford Academic, 3 Nov. 2014), https://doi.org/10.1093/oxfordhb/9780199842193.013.7, a.

Sami Haddadin 2021: Why robots need reflexes. -interview 2021. https://www.eurekalert.org/news-releas-es/821065

Santamaria, L. et al: Emotional valence modulates the topology of the parent-infant inter-brain net-work. Neuroimage (2019). DOI: 10.1016/j.neuroimage.2019.116341

Sarasso S., Casali A. G., Casarotto S., Rosanova M., Sinigaglia C., Massimini M. (2021). Consciousness and complexity: A consilience of evidence. Neurosci. Conscious. niab023. 10.1093/nc/niab023.

Sarno JE. The Mindbody prescription: healing the body, healing the pain. New York, NY: Warner Books; 1998.

Saxe, R; Kanwisher, N (August 2003). "People thinking about thinking people: The role of the tem-poro-parietal junction in "theory of mind"". NeuroImage. 19 (4): 1835–42. doi:10.1016/S1053-8119(03)00230-1. PMID 12948738.

Schmitt LI, Wimmer RD, Nakajima M, Happ M, Mofakham S, Halassa MM. Thalamic amplification of cortical connectivity sustains attentional control. Nature. 2017 May

11;545(7653):219-223. doi: 10.1038/ nature22073. Epub 2017 May 3. PMID: 28467827; PMCID: PMC5570520.

Schubiner H, Betzold M. Unlearn your pain. 3rd ed. Pleasant Ridge, MI: Mind Body Publishing; 2016.

Searle J. (2004). Mind. Oxford, UK: Oxford University Press.

Seghier ML. The angular gyrus: multiple functions and multiple subdivisions. Neuroscientist. 2013 Feb;19(1):43-61. doi: 10.1177/1073858412440596. Epub 2012 Apr 30. PMID: 22547530; PMCID: PMC4107834.

Semendeferi K, Lu A, Schenker N, Damasio H. Humans and great apes share a large frontal cortex. Nat Neurosci. 2002 Mar;5(3):272-6. doi: 10.1038/nn814. PMID: 11850633.

Semon, R. W., Bella, D., and Vernon, L. (1923). Mnemic Psychology. London, UK: G. Allen & Unwin.

Seminara L, Dosen S, Mastrogiovanni F, Bianchi M, Watt S, Beckerle P, Nanayakkara T, Drewing K, Moscatelli A, Klatzky RL, Loeb GE. A hierarchical sensorimotor control framework for human-in-the-loop robotic hands. Sci Robot. 2023 May. 17;8(78): eadd5434. doi: 10.1126/scirobotics. add5434. Epub 2023 May 17. PMID: 37196072.

Seung Ju Kim, Sangbum Kim, Ho Won Jang, Competing memristors for brain-inspired compu-ting,iS-cience, Volume 24, Issue 1, 2021, 101889, ISSN 2589-0042, https://doi.org/10.1016/j.isci.2020.101889.(https://www.sciencedirect.com/science/article/pii/S2589004220310865)

Scalabrini A, Wolman A, Northoff G. The Self and Its Right Insula-Differential Topography and Dynamic of Right vs. Left Insula. Brain Sci. 2021 Oct 2;11(10):1312. doi: 10.3390/brainsci11101312. PMID: 34679377; PMCID: PMC8533814.

Scarantino, A, 'The Motivational Theory of Emotions', in Justin D'Arms, and Daniel Jacobson (eds), Moral Psychology and Human Agency: Philosophical Essays on the Science of Eth-ics (Oxford, 2014; onlineedn, Oxford Academic, 18 Dec. 2014), https://doi.org/10.1093/acprof:oso/9780198717812.003.0008, accessed 15 Nov. 2023.

Schacter DL, Addis DR, Buckner RL. Remembering the past to imagine the future: the prospective brain. Nat Rev Neurosci. 2007 Sep;8(9):657-61. doi: 10.1038/nrn2213. PMID: 17700624.

Schacter DL, Addis DR. The cognitive neuroscience of constructive memory: remembering the past and imagining the future. Philos Trans R Soc Lond B Biol Sci. 2007 May 29;362(1481):773-86. doi: 10.1098/ rstb.2007.2087. PMID: 17395575; PMCID: PMC2429996.)

Scherer, K. R. (1999). Appraisal theory. In T. Dalgleish & M. J. Power (Eds.), Handbook of cogni-tion and emotion (pp. 637–663). John Wiley & Sons Ltd. https://doi.org/10.1002/0470013494.ch30

Scherer, K. R., Schorr, A., & Johnstone, T. (Eds.). (2001). Appraisal processes in emotion: Theo-ry, methods, research. Oxford University Press.

Schmidhuber, J. Optimal Ordered Problem Solver. Machine Learning 54, 211–254 (2004). https://doi.org/10.1023/B:MACH.0000015880.99707.b2

Schmidhuber J. 2010: Formal Theory of Creativity, Fun, and Intrinsic Motivation (1990–2010)

Schmidhuber. J, Deep learning in neural networks: An overview, Neural Networks, Volume 61, 2015, Pages 85-117, ISSN 0893-6080, https://doi.org/10.1016/j.neunet.2014.09.003.(https://www.sciencedirect.com/science/article/pii/S0893608014002135)

Schmidhuber, J., 2020. Deep Learning: Our Miraculous Year 1990-1991)

Schretlen, D. J., Hulst, E. J., Pearlson, G. D. and Gordon, B. (2010). Journal of clinical and experi-mental neuropsychology. 32(10). 1068-1073.

Shrobe H. https://www.darpa.mil/program/machine-common-sense.

Schurz M, Radua J, Tholen MG, Maliske L, Margulies DS, Mars RB, Sallet J, Kanske P. Toward a hierarchical model of social cognition: A neuroimaging meta-analysis and integrative review of em-pathy and theory of mind. Psychol Bull. 2021 Mar;147(3):293-327. doi: 10.1037/bul0000303. Epub 2020 Nov 5. PMID: 33151703.

Seeber, M., Cantonas, LM., Hoevels, M. et al. Sub-cortical electrophysiological activity is detectable with high-density EEG source imaging. Nat Commun 10, 753 (2019). https://doi.org/10.1038/s41467-019-08725-w

Semendeferi K, Lu A, Schenker N, Damasio H. Humans and great apes share a large frontal cortex. Nat Neurosci. 2002 Mar;5(3):272-6. doi: 10.1038/nn814. PMID: 11850633.

Seth AK. Interoceptive inference, emotion, and the embodied self. Trends Cogn Sci. 2013 Nov;17(11):565-73. doi: 10.1016/j.tics.2013.09.007. Epub 2013 Oct 12. PMID: 24126130. Serrano-Gotarredona R., et al., "CAVIAR: A 45k Neuron, 5M Synapse, 12G Connects/s AER Hardware Sensory–Processing– Learning–Actuating System for High-Speed Visual Object Recogni-tion and Tracking," in IEEE Transactions on Neural Networks, vol. 20, no. 9, pp. 1417-1438, Sept. 2009, doi: 10.1109/TNN.2009.2023653.

Seung, H. Half a century of Hebb. Nat Neurosci 3 (Suppl 11), 1166 (2000). https://doi.org/10.1038/81430

Sheldon K (1995) Creativity Research Journal, Vol. 8, NO. 1,25-36.

Sheldon S, Farb N, Palombo DJ, Levine B. Intrinsic medial temporal lobe connectivity relates to individual differences in episodic autobiographical remembering. Cortex. 2016 Jan; 74:206-16. doi: 10.1016/j. cortex.2015.11.005. Epub 2015 Nov 19. PMID: 26691735.)

Schrodinger E 1944.What is Life?

Scherer, K. R., Banziger, T., and Roesch, E. (2010). A Blueprint for Affective Computing: A Sourcebook and Manual. New York, NY: Oxford University Press, Inc.

Schulkin J, McEwen BS, Gold PW. Allostasis, amygdala, and anticipatory angst. Neurosci Biobehav Rev. 1994 Fall;18(3):385-96. doi: 10.1016/0149-7634(94)90051-5. PMID: 7984356.

Schmidhuber, Juergen. (2020). Deep Learning: Our Miraculous Year 1990-1991.

Schroeder CE, Wilson DA, Radman T, Scharfman H, Lakatos P. Dynamics of Active Sensing and perceptual selection. Curr Opin Neurobiol. 2010 Apr;20(2):172-6. doi: 10.1016/j.conb.2010.02.010. Epub 2010 Mar 20. PMID: 20307966; PMCID: PMC2963579.

Sebanz N, Bekkering H, Knoblich G. Joint action: bodies and minds moving together. Trends Cogn Sci. 2006 Feb;10(2):70-6. doi: 10.1016/j.tics.2005.12.009. Epub 2006 Jan 10. PMID: 16406326.

Shah K, Jain SB, Wadhwa R. Capgras Syndrome. 2022 Jun 4. In: StatPearls [Internet]. Treasure Is-land (FL): StatPearls Publishing; 2023 Jan–. PMID: 34033319.

Shallice T, Warrington E K, The dissociation between short term retention of meaningful sounds and verbal material, Neuropsychologia, Volume 12, Issue 4, 1974, Pages 553-555, ISSN 0028-3932, https://doi.org/10.1016/0028-3932(74)90087-6.

Sheldon, K. M., & Elliot, A. J. (1998). Not all personal goals are personal: Comparing autonomous and controlled reasons for goals as predictors of effort and attainment. Personality and Social Psy-chology Bulletin, 24(5), 546–557. https://doi.org/10.1177/0146167298245010

Shimoda, Shingo & Jamone, Lorenzo & Ognibene, Dimitri & Nagai, Takayuki & Sciutti, Alessan-dra & Costa-Garcia, Alvaro & Oseki, Yohei & Taniguchi, Tadahiro. (2021). What is the role of the next generation of cognitive robotics? Advanced Robotics. 36. 1-14. 10.1080/01691864.2021.2011780.

Shipp, S. (2004). The brain circuitry of attention. Trends Cogn. Sci. 8, 223–230. doi: 10.1016/j. tics.2004.03.004

Shubhendu and Vijay (2013) "Applicability of Artificial Intelligence in Different Fields of Life.

Shulman GL (2001) A default mode of brain function. Proc Natl Acad Sci US A 98:676–682.

Shultz S, Dunbar RIM. 2007. The evolution of the social brain: Anthropoid primates' (those re-sembling human beings) brains contrasts with other vertebrates. Proc R Soc Lond 274B:24292436.

Siakaluk PD, Pexman PM, Aguilera L, Owen WJ, Sears CR. Evidence for the activation of sen-sori-motor information during visual word recognition: the body-object interaction effect. Cognition. 2008 Jan;106(1):433-43. doi: 10.1016/j.cognition.2006.12.011. Epub 2007 Jan 26. PMID: 17258186.

Silvia, Paul; Duval, T. Shelley (2001). "Objective self-awareness theory: Recent progress and en-during problems". Personality and Social Psychology Review. 5 (3): 230–241.

Silveri MC. Frontotemporal dementia to Alzheimer's disease. Dialogues Clin Neurosci. 2007;9(2):153– 160.

Simpson EH, Balsam PD. The Behavioural Neuroscience of Motivation: An Overview of Concepts, Measures, and Translational Applications. Curr Top Behav Neurosci. 2016; 27:1-12. doi: 10.1007/7854_2015_402. PMID: 26602246; PMCID: PMC4864984.

Singer W. Consciousness and the binding problem. Ann NY Acad Sci. 2001 Apr; 929:123-46. doi: 10.1111/j.1749-6632. 2001.tb05712. x. PMID: 11349422.

Sinitsyn DO, Poydasheva AG, Bakulin IS, Legostaeva LA, Iazeva EG, Sergeev DV, Sergeeva AN, Kremneva EI, Morozova SN, Lagoda DY, Casarotto S, Comanducci A, Ryabinkina YV, Suponeva NA, Piradov MA. Detecting the Potential for Consciousness in Unresponsive Patients Using the Perturbational Complexity Index. Brain Sciences. 2020; 10(12):917. https://doi.org/10.3390/brainsci10120917.

Sloman, A. and M. Croucher. "Why Robots will have Emotions," IJCAI 1981.

Smith J. A., Flowers P., Larkin M. (2009). Interpretative phenomenological analysis. Theory, method and research. Los Angeles, CA: Sage.

Smith, J, Zhang, D, Balram, K. (2023). Robotic Vectorial Field Alignment for Spin⊠Based Quantum Sensors. Advanced Science. 10.1002/advs.202304449.

Social Brain Hypothesis". (R.I.M. Dunbar (2009) The social brain hypothesis and its implications for social evolution, Annals of Human Biology, 36:5, 562-572, DOI: 10.1080/03014460902960289

Sofge D, Perzanowski D, Skubic, M Bugajska M, J Trafton G, Cassimatis N, Brock D, William Ad-ams W, Scbultz A Cognitive Tools for Humanoid Robots in Space, IFAC Proceedings Vol-umes,Volume 37, Issue 6,2004,Pages 351-356,ISSN 1474-6670, https://doi.org/10.1016/S1474-6670(17)32199-7.

SoftBankRoboticsPepperRobot:Characteristics.2018.[(accessedon8August2018)].Available online: https://www.softbankrobotics.com/emea/en/robots/pepper/find-out-more-about-pepper

Solms, M., and Turnbull, O. (2011). What is neuropsychoanalysis? Neuropsychoanalysis 13, 133–145. doi: 10.1080/15294145.2011.10773670

Soltoggo A, de Velde F. Front. Neurorobot., 27 October 2015 Volume 9 – 2015 https://doi.org/10.3389/fnbot.2015.00012.

Somerville L, Whalen P, Kelley W (2010). "Human Bed Nucleus of the Stria Terminalis Indexes Hyper-vigilant Threat Monitoring". Biol Psychiatry. 68 (5): 416–424. doi:10.1016/j.biopsych.2010.04.002. PMC 2921460. PMID 20497902.

Somerville, L. H., Jones, R. M., and Casey, B. J. (2009). A time of change: Behavioural and neural correlates of adolescent sensitivity to appetitive and aversive environmental cues. Brain Cogn.

Song HF, Kennedy H, Wang XJ. Spatial embedding of structural similarity in the cerebral cortex. Proc Natl Acad Sci U S A. 2014 Nov 18;111(46):16580-5. doi: 10.1073/pnas.1414153111. Epub 2014 Nov 3. PMID: 25368200; PMCID: PMC4246295.

Song Q, Li S, Yang J, Bai Q, Hu J, Zhang X, Zhang A. Intelligent Optimization Algorithm-Based Path Planning for a Mobile Robot. Comput Intell Neurosci. 2021 Sep 29; 2021:8025730. doi: 10.1155/2021/8025730. PMID: 34630554; PMCID: PMC8494556.

Song, Yisheng & Wang, Ting & Cai, Puyu & Mondal, Subrota & Sahoo, Jyoti. (2023). A Compre-hensive Survey of Few-shot Learning: Evolution, Applications, Challenges, and Opportunities. ACM Computing Surveys. 55. 10.1145/3582688.

Song Y, Lukasiewicz T, Xu Z, Bogacz R. Can the Brain Do Backpropagation? -Exact Implementa-tion of Backpropagation in Predictive Coding Networks. Adv Neural Inf Process Syst. 2020; 33:22566-22579. PMID: 33840988; PMCID: PMC7610561.

Sonuga-Barke EJS, Kennedy M, Kumsta R, Knights N, Golm D, Rutter M, Maughan B, Schlotz W, Kreppner J. Child-to-adult neurodevelopmental and mental health trajectories after early life depri-vation: the young adult follow-up of the longitudinal English and Romanian Adoptees study. Lan-cet. 2017 Apr 15;389(10078):1539-1548. doi: 10.1016/S0140-6736(17)30045-4. Epub 2017 Feb 23. PMID: 28237264.

Sorrentino A., Fiorini L., Fabbricotti I., Sancarlo D., Ciccone F., Cavallo F. Exploring Human atti-tude during Human-Robot Interaction; Proceedings of the 29th IEEE International Symposium on Robot and Human Interactive Communication; Naples, Italy. 31 August–4 September 2020.

Souza, A, Ribeiro S, Handbook of Behavioural Neuroscience, 2018.

Spalding KN, Jones SH, Duff MC, Tranel D, Warren DE. Investigating the Neural Correlates of Schemas: Ventromedial Pre-frontal Cortex Is Necessary for Normal Schematic Influence on Memory. J Neurosci. 2015 Nov 25;35(47):15746-51. doi: 10.1523/JNEUROSCI.2767-15.2015. PMID: 26609165; PMCID: PMC4659831.

Spanne A, Jörntell H. Questioning the role of sparse coding in the brain. Trends Neurosci. 2015 Jul;38(7):417-27. doi: 10.1016/j.tins.2015.05.005. Epub 2015 Jun 17. PMID: 26093844.

Spalla D, Cornacchia I M Treves A 2021.Continuous attractors for dynamic memories eLife 10: e69499.Spatola, N., Marchesi, S. & Wykowska, A. Cognitive load affects early processes involved in men-talizing robot behaviour. Sci Rep 12, 14924 (2022). https://doi.org/10.1038/s41598-022-19213-5

Sperry, Roger W. "Cerebral Organization and Behaviour." Science 133 (1961): 1749–57. http://people.uncw.edu/puente/sperry/sperrypapers/60s/85-1961.pdf (Access December 8, 2017).

Sproull, L., Subramani, M., Kiesler, S., Walker, J. H., & Waters, K. (1996). When the interface is a face. Human-Computer Interaction, 11(2), 97–124. https://doi.org/10.1207/s15327051hci1102_1 s: //www.bcbt.specs-lab.com/bcbt15/sites/default/files/event-files-public/BS4-DAC-DEF.pdf

Spruston N. Pyramidal neurons: dendritic structure and synaptic integration. Nature reviews. Neu-rosci-ence, 9:206–221, 2008. ISSN 1471-003X. doi:10.1038/nrn2286.

Sreelekha.S (2017). Statistical Vs Rule Based Machine Translation, A Case Study on Indian Lan-guage Perspective.

Srikantan N & Ranasinghe, K & Vossel, K. (2016). Brain Imaging with Magnetoencephalography During Rest and During Speech and Language Processing. 10.1016/B978-0-12-800105-9.00015-9.

Srivastava, N, Hinton, G & Krizhevsky, A, Sutskever, I, Salakhutdinov, Rc. (2014). Dropout: A Simple Way to Prevent Neural Networks from Over-fitting. Journal of Machine Learning Research. 15. 19291958.

Starns, J.J., Ma, Q. Response biases in simple decision making: Faster decision making, faster re-sponse execution, or both? Psychon Bull Rev 25, 1535–1541 (2018). https://doi.org/10.3758/s13423-017-1358-9.

Stewart TC, Tripp B, Eliasmith C. Python scripting in the nengo simulator. Front Neuroinform. 2009 Mar 24; 3:7. doi: 10.3389/neuro.11.007.2009. PMID: 19352442; PMCID: PMC2666197.

Strain S, Kugele S and Franklin, S "The learning intelligent distribution agent (LIDA) and medical agent X (MAX): Computational intelligence for medical diagnosis," 2014 IEEE Symposium on Computational Intelligence for Human-like Intelligence (CIHLI), Orlando, FL, USA, 2014, pp. 1-8, doi: 10.1109/CIH-LI.2014.7013390.

Steenbergen L, Maraver MJ, Actis-Grosso R, Ricciardelli P, Colzato LS. Recognizing emotions in bodies: Vagus nerve stimulation enhances recognition of anger while impairing sadness. Cogn Af-fect Behav Neurosci. 2021 Dec;21(6):1246-1261. doi: 10.3758/s13415-021-00928-3. Epub 2021 Jul 15. PMID: 34268714; PMCID: PMC8563521.

Stefano Sandrone, Marco Bacigaluppi, Marco R. Galloni, Stefano F. Cappa, Andrea Moro, Marco Catani, Massimo Filippi, Martin M. Monti, Daniela Perani, Gianvito Martino, Weighing brain activi-ty with the balance: Angelo Mosso's original manuscripts come to light, Brain, Volume 137, Issue 2, February 2014, Pages 621–633, https://doi.org/10.1093/brain/awt091

Stel M, van Dijk E, Olivier E. You want to know the truth? Then don't mimic! Psychol Sci. 2009 Jun;20(6):693-9. doi: 10.1111/j.1467-9280.2009.02350. x. Epub 2009 Apr 28. PMID: 19422628.

Stel M, Vonk R. Mimicry in social interaction: benefits for mimickers, mimickees, and their inter-action. Br J Psychol. 2010 May;101(Pt 2):311-23. doi: 10.1348/000712609X465424. Epub 2009 Jul 29. PMID: 19646328.

Stendardi D, Biscotto F, Bertossi E, Ciaramelli E. Present and future self in memory: the role of vmPFC in the self-reference effect. Soc Cogn Affect Neurosci. 2021 Dec 30;16(12):1205-1213. doi: 10.1093/scan/ nsab071. PMID: 34086968; PMCID: PMC8716844.

Stephan KE, Friston KJ. Analysing effective connectivity with functional magnetic resonance imag-ing. Wiley Interdiscip Rev Cogn Sci. 2010 May;1(3):446-459. doi: 10.1002/wcs.58. Epub 2010 Apr 2. PMID: 21209846; PMCID: PMC3013343.

Sterling, P. (2004). "Principles of allostasis: optimal design, predictive regulation, pathophysiology and rational therapeutics," in Allostasis, Homoeostasis, and the Costs of Physiological Adaptation, ed. J. Schulkin (Cambridge, UK: Cambridge University Press), 17–64.

Sterling P and Laughlin S (2015) Principles of Neural Design (Cambridge, MA: MIT Press).

Sternberg, R. J. (1985). Beyond IQ: A triarchic theory of human intelligence. New York: Cambridge University Press.

Sternberg, R. J. (1997). Successful intelligence. New York: Plume Books.

Steve Furber 2016 J. Neural Eng. 13 051001.

Stone VE, Gerrans P. What's domain-specific about theory of mind? Soc Neurosci. 2006;1(3-4):309-19. doi: 10.1080/17470910601029221. PMID: 18633796.

Strigo IA, Craig AD. Interoception, homoeostatic emotions and sympathovagal balance. Philos Trans R Soc Lond B Biol Sci. 2016 Nov 19;371(1708):20160010. doi: 10.1098/rstb.2016.0010. Epub 2016 Oct 10. PMID: 28080968; PMCID: PMC5062099.

Suddendorf T, Corballis MC. Mental time travel and the evolution of the human mind. Genet Soc Gen Psychol Monogr. 1997 May;123(2):133-67. PMID: 9204544.

Sun R (2004) desiderata for cognitive architectures. Philos Psychol 17(3):341–373.

Sun H, Tan Q, Fan G, Tsui Q. Different effects of rumination on depression: Key role of hope. Int J Ment Health Syst. 2014;8(1):53. doi:10.1186/1752-4458-8-53

Sun, H., Guyon, I. (2023). Modularity in Deep Learning: A Survey. In: Arai, K. (eds) Intelligent Computing. SAI 2023. Lecture Notes in Networks and Systems, vol 739. Springer, Cham. https://doi.org/10.1007/978-3-031-37963-5_40

Surian L, Caldi S, Sperber D. Attribution of beliefs by 13-month-old infants. Psychol Sci. 2007 Jul;18(7):580-6. doi: 10.1111/j.1467-9280.2007.01943. x. PMID: 17614865.

Susoy Vladislav et al. "Natural sensory context drives diverse brain-wide activity during C. elegans mating." Cell 184 (2021): 5122-5137.e17. Sutton RS, Barto AG (2018) Reinforcement learning: An introduction. MIT press, Cambridge, MA

Swaab D (2007). "Sexual differentiation of the brain and behaviour". Best Pract Res Clin Endo-crinol Metab. 21 (3): 431–44.

Symons CS, Johnson BT. The self-reference effect in memory: a meta-analysis. Psychol Bull. 1997 May;121(3):371-94. doi: 10.1037/0033-2909.121.3.371. PMID: 9136641.

T

Taghia, J., Cai, W., Ryali, S. et al. Uncovering hidden brain state dynamics that regulate perfor-mance and decision-making during cognition. Nat Commun 9, 2505 (2018). https://doi.org/10.1038/s41467-018-04723-6

Takahata K, Kato M. [Neural mechanism underlying autistic savant and acquired savant syn-drome]. Brain Nerve. 2008 Jul;60(7):861-9. Japanese. PMID: 18646626.

Tanaka F, Cicourel A. Movellan J. Socialization between toddlers and robots at an early childhood education centre. PNAS, 104(46):17954–17958, 2007.

Tang G., Polykretis I.E., Ivanov V.A., Shah A., Michnizos K.P. Introducing Astrocytes on a Neu-romor-phic Processor: Synchronization, Local Plasticity and Edge of Chaos. NICE '19: Neuro-Inspired Computational Elements Workshop. 2019

Tang G, Gudsnuk K, Kuo SH, Cotrina ML, Rosoklija G, Sosunov A, Sonders MS, Kanter E, Castagna C, Yamamoto A, Yue Z, Arancio O, Peterson BS, Champagne F, Dwork AJ, Goldman J, Sulzer D. Loss of mTOR-dependent macroautophagy causes autistic-like synaptic pruning deficits. Neuron. 2014 Sep 3;83(5):1131-43. doi: 10.1016/j.neuron.2014.07.040. Epub 2014 Aug 21. Erra-tum in: Neuron. 2014 Sep 17;83(6):1482. PMID: 25155956; PMCID: PMC4159743.

Tang, J., LeBel, A., Jain, S. et al. Semantic reconstruction of continuous language from non-invasive brain recordings. Nat Neurosci 26, 858–866 (2023). https://doi.org/10.1038/s41593-023-01304-9.

Tappen RM, Williams C, Fishman S, Touhy T. Persistence of self in advanced Alzheimer's disease. Image J Nurs Sch. 1999;31(2):121-5. doi: 10.1111/j.1547-5069. 1999. tb00445. x. PMID: 10380386; PMCID: PMC1958994.

Taymour Mostafa, Ghada El Khouly, Ashraf Hassan, Pheromones in sex and reproduction: Do they have a role in humans?,Journal of Advanced Research,Volume 3, Issue 1, 2012, Pages 1-9.

The New Breed: What our History with Animals Reveals About our Future with Robots by Kate Darling (October 24, 2022). Osgoode Hall Law Journal 59:3 (summer 2022), Available at SSRN: https://ssrn.com/abstract=4416625.

"The Internet Trust Bubble Global Values, Beliefs and Practice" (2104) https://www3.weforum.org/docs/WEF_InternetTrustBubble_Report2_2014.pdfTensors for Data Processing (2022) Christos Chatzichristos and Sabine Van Huffel.

Terasawa Y, Fukushima H, Umeda S. How does interoceptive awareness interact with the subjec-tive experience of emotion? An fMRI study. Hum Brain Mapp. 2013 Mar;34(3):598-612. doi: 10.1002/ hbm.21458. Epub 2011 Nov 18. PMID: 22102377; PMCID: PMC6870042.

Tesser, A.; Campbell, J. (1982). "Self-evaluation maintenance and the perception of friends and strangers". Journal of Personality. 50 (3): 261–279. doi:10.1111/j.1467-6494. 1982.tb00750.x

Thomaz, Andrea & Hoffman, Guy & Cakmak, Maya. (2016). Computational Human-Robot Inter-action. Foundations and Trends in Robotics. 4. 104-223. 10.1561/2300000049. Thomas Nagel "What Is It Like to Be a Bat?" (1974)

Thompson KG, Biscoe KL, Sato TR. Neuronal basis of covert spatial attention in the frontal eye field. J Neurosci. 2005 Oct 12;25(41):9479-87. doi: 10.1523/JNEUROSCI.0741-05.2005. PMID: 16221858; PMCID: PMC2804969.

Timmermann Slater CB, Bauer P, Gosseries O, et al., 2023, A neurophenomenological approach to non-ordinary states of consciousness: hypnosis, meditation, and psychedelics, Trends in Cognitive Sciences, Vol:27, ISSN:1364-6613, Pages:139-159.

Timmis, J. and Neal, M. (2004). "Once More Unto the Breach: Towards Artificial Homoeostasis", in L. N. de Castro and F.J. Von Zuben, Recent Developments in Biologically Inspired Computing, Idea Group Inc., Chapter XIV, pp. 340-366.

Tingley, D., McClain, K., Kaya, E. et al. A metabolic function of the hippocampal sharp wave-ripple. Nature 597, 82–86 (2021). https://doi.org/10.1038/s41586-021-03811-w.

Tinwell, A, Abdel N, John D, John C. (2013). Perception of psychopathy and the Uncanny Valley in virtual characters. Computers in Human Behaviour. 29. 1617-1625. 10.1016/j.chb.2013.01.008.

Tobin RM, Graziano WG, Vanman EJ, Tassinary LG. Personality, emotional experience, and ef-forts tocontrol emotions. J Pers Soc Psychol. 2000; 79(4): 656–69. PMID.

Tomasello, M., Constructing a language: a usage-based theory of language acquisition. Cambridge, MA: Harvard University Press, 2003. Pp. 388.

Tomasello M. (2010). Origins of human communication. Cambridge, MA: MIT Press.

Tombu M, Jolicoeur P. Testing the predictions of the central capacity sharing model. Journal of Ex-peri-mental Psychology: Human Perception and Performance. 2005; 31:790–802.

Tononi G, Edelman GM. Consciousness and complexity. Science. 1998 Dec 4;282(5395):1846-51. doi: 10.1126/science.282.5395.1846. PMID: 9836628.

Tononi G., Boly M., Massimini M., Koch C. (2016). Integrated information theory: From con-sciousness to its physical substrate. Nat. Rev. Neurosci. 17 450–461. 10.1038/nrn.2016.44

Tononi, G., Boly, M., Massimini, M. et al. Integrated information theory: from consciousness to its physical substrate. Nat Rev Neurosci 17, 450–461 (2016). https://doi.org/10.1038/nrn.2016.44.

Tononi G, Sporns O, Edelman GM. A measure for brain complexity: relating functional segregation and integration in the nervous system. Proc. Natl Acad. Sci. USA. 1994; 91:5033–5037.

Tooby, J., and Cosmides, L. (1990). The past explains the present: emotional adaptations and the structure of ancestral environments. Ethol. Sociobiol. 11, 375–424. doi:10.1016/0162-

TPUs (Tovar J 2023 Source (https://www.linkedin.com/pulse/gpus-vs-tpus-comprehensive-compari-son-neural-network-workloads-joel/)

Tovote, P., Fadok, J. & Lüthi, A. Neuronal circuits for fear and anxiety. Nat Rev Neurosci 16, 317–331 (2015). https://doi.org/10.1038/nrn3945.

Tracy B 2008. “Universal Laws of Success”, p.7, Jaico Publishing House

Trafton JG, Hiatt LM, Harrison AM, Tamborello P, Khemlani SS, Schultz AC (2013) ACT-R/E: an embodied cognitive architecture for human-robot interaction. J Hum Robot Interact 2(1):30–54

Trafton JG, Cassimatis NL, Bugajska MD, Brock DP, Mintz FE, Schultz AC (2005) Enabling effec-tive human robot interaction using perspective-taking in robots. IEEE Trans Syst Man Cybern Part A Syst Hum 35(4):460–470

Trimble M, Hesdorffer D. Music and the brain: the neuroscience of music and musical apprecia-tion. BJPsych Int. 2017 May 1;14(2):28-31. doi: 10.1192/s2056474000001720. PMID: 29093933; PMCID: PMC5618809.

Trimble, M. R. (2012) Why Humans Like to Cry. Tragedy, Evolution and the Brain. Oxford Uni-versity Press.

Troccaz J., Lavallee S., Hellion E. A passive arm with dynamic constraints: A solution to safety problems in medical robotics; Proceedings of the IEEE Systems Man and Cybernetics Confer-ence—SMC; Le Touquet, France. 17–20 October 1993; pp. 166–171.

Tsotsos J K, Culhane SW, Wai Y L, Davis N, Nuflo F. Modelling visual attention via selective tun-ing. Artificial Intelligence, 78 (1995), pp. 507-547

Tsotsos JK. Attention: The Messy Reality. Yale J Biol Med. 2019 Mar 25;92(1):127-137. PMID: 30923480; PMCID: PMC6430176.

Tsotsos JK (2011) A computational perspective on visual attention. MIT Press, Cambridge

Tsotsos, J.K., & Kruijne, W. (2014). Cognitive programs: software for attention's executive. Frontiers in Psychology, 5.

Tulving E. 1972. Episodic and semantic memory. Organization of Memory, eds Tulving E, Don-aldson W (Academic, New York), pp 381–402.

Turjman, O. (2016) On the role of mirror neurons in the sense of self, J. Conscious. Explor. Res., 7, pp. 288-302. 61.

U

Uejima T, Niebur E, Etienne-Cummings R. Proto-Object Based Saliency Model with Texture Detec-tion Channel. Front Comput Neurosci. 2020 Sep 24; 14:541581. doi: 10.3389/fncom.2020.541581. PMID: 33071766; PMCID: PMC7541834.

Uget, A, Habibah, B., & Jegak, U. (2007). The influence of causal elements of control on academic achievement satisfaction. Journal of Instructional Psychology, 34(2), 120-8.

UNESCO:Report of Comest on Robotics Ethics Paris UNESCO.2017 http://unesdoc.unesco.org/images/0025/002539/253952E.pdfReturn to ref 6 in article

University College London. “First Out-of-body Experience Induced In Laboratory Setting.” Sci-enceDaily. ScienceDaily, 24 August 2007. <www.sciencedaily.com/releases/2007/08/070823141057.htm>.

un JJ, Steinmetz NA, Siegle JH, Denman DJ, Bauza M, Barbarits B, Lee AK, Anastassiou CA, An-drei A, Aydın Ç, Barbic M, Blanche TJ, Bonin V, Couto J, Dutta B, Gratiy SL, Gutnisky DA, Häusser M, Karsh B, Ledochowitsch P, Lopez CM, Mitelut C, Musa S, Okun M, Pachitariu M, Putzeys J, Rich PD, Rossant C, Sun WL, Svoboda K, Carandini M, Harris KD, Koch C, O'Keefe J, Harris TD. Fully integrated silicon probes for high-density recording of neural activity. Nature. 2017 Nov 8;551(7679):232-236. doi: 10.1038/ nature24636. PMID: 29120427; PMCID: PMC5955206.

V

V.A. Cardenas, K. Samuelson, M. Lenoci, C. Studholme, T.C. Neylan, C.R. Marmar, N. Schuff, M.W. Weiner. Changes in brain anatomy during the course of posttraumatic stress disorder. Psychia-try Res, 193 (2011), pp. 93-100, 10.1016/j.pscychresns.2011.01.013.

van de Ven GM, Tuytelaars T, Tolias AS. Three types of incremental learning. Nat Mach Intell. 2022;4(12):1185-1197. doi: 10.1038/s42256-022-00568-3. Epub 2022 Dec 5. PMID: 36567959; PMCID: PMC9771807.

van de Ven, G.M., Siegelmann, H.T. & Tolias, A.S. Brain-inspired replay for continual learning with artificial neural networks. Nat Commun 11, 4069 (2020). https://doi.org/10.1038/s41467-020-17866-2

van de Weijer-Bergsma E, Wijnroks L, Jongmans MJ. Attention development in infants and pre-school children born preterm: a review. Infant Behav Dev. 2008 Sep;31(3):333-51. doi: 10.1016/j.inf-beh.2007.12.003. Epub 2008 Feb 21. PMID: 18294695.

Van Overwalle, F., and Baetens, K. (2009). Understanding others' actions and goals by mirror and mental-izing systems: a meta-analysis. Neuroimage 48, 564–584. doi: 10.1016/j.neuroimage.2009.06.00

Velasco, C.D., Santarella-Mellwig, R., Schorb, M. et al. Microtubule depolymerization contributes to spontaneous neurotransmitter release in vitro. Commun Biol 6, 488 (2023). https://doi.org/10.1038/s42003-023-04779-1

Valiant,L.G (2008) Knowledge Infusion:In Pursuit of Robustness in Artificial Intelligence. Founda-tions of Software Technology and Computer Science.

Valenzo D, Ciria A, Schillaci G, Lara B. Grounding Context in Embodied Cognitive Robotics. Front Neurorobot. 2022 Jun 15; 16:843108. doi: 10.3389/fnbot.2022.843108. PMID: 35812785; PMCID: PMC9262126.

van de Ven GM, Tuytelaars T, Tolias AS. Three types of incremental learning. Nat Mach Intell. 2022;4(12):1185-1197. doi: 10.1038/s42256-022-00568-3. Epub 2022 Dec 5. PMID: 36567959; PMCID: PMC9771807.

Vandekerckhove M, et al Bulnes LC, Panksepp J. The emergence of primary anoetic consciousness in episodic memory. Front Behav Neurosci. 2014 Jan 3; 7:210. doi: 10.3389/fnbeh.2013.00210.

Vandekerckhove M, Panksepp J. A neurocognitive theory of higher mental emergence: from anoet-ic affective experiences to noetic knowledge and autonoetic awareness. Neurosci Biobehav Rev. 2011 Oct;35(9):2017-25.

Vandekerckhove M, Panksepp J, The flow of anoetic to noetic and autonoetic consciousness: A vision of unknowing (anoetic) and knowing (noetic) consciousness in the remembrance of things past and imagined futures, Consciousness and Cognition,Volume 18, Issue 4, 2009, Pages 1018-1028, ISSN 1053-8100, https://doi.org/10.1016/j.concog.2009.08.002.

Vanderbilt KE, Liu D, Heyman GD. 2011 The development of distrust. Child Dev. 82, 1372– 1380. (doi:10.1111/j.1467-8624.2011. 01629.x)

Van Gulick, Robert, 'Consciousness, intrinsic intentionality, and self-understanding machines', in A. J. Marcel, and E. Bisiach (eds), Consciousness in Contemporary Science (Oxford, 1992; online edn, Oxford Academic, 22 Mar. 2012), https://doi.org/10.1093/acprof:oso/9780198522379.003.0004, accessed 27 July 2023.

van Kesteren MT, Ruiter DJ, Fernández G, Henson RN. How schema and novelty augment memory formation. Trends Neurosci. 2012 Apr;35(4):211-9. doi: 10.1016/j.tins.2012.02.001. Epub 2012 Mar 5. PMID: 22398180.

Van Overwalle, F., and Baetens, K. (2009). Understanding others' actions and goals by mirror and mental-izing systems: a meta-analysis. Neuroimage 48, 564–584. doi: 10.1016/j.neuroimage.2009.06.009

Vann SD, Aggleton JP, Maguire EA. What does the retrosplenial cortex do? Nat Rev Neurosci. 2009 Nov;10(11):792-802. doi: 10.1038/nrn2733. Epub 2009 Oct 8. PMID: 19812579.-

Vargas P., Moioli R., de Castro L. N., Timmis J., Neal M., Zuben F. J. V. (2005). Artificial homeo-static system: a novel approach, in Advances in Artificial Life (Berlin; Heidelberg: Springer Berlin Heidelberg;), 754–764.

Varpula, S., Annila, A., and Beck, C. (2013). Thoughts about thinking: cognition according to the second law of thermodynamics. Adv. Stud. Biol. 5, 135–149.

Vasilopoulos T. 2023. https://www.bbc.com/news/technology-66465230.

Véniat, T., Denoyer, L., Ranzato, M. (2020). Efficient Continual Learning with Modular Networks and Task-Driven Priors. ArXiv, abs/2012.12631.

Vernon, D, von Hofsten C. Fadiga L., 2016: Desiderata for Developmental Cognitive Architectures.

Verschure PFMJ. 2012 Distributed adaptive control: a theory of the mind, brain, body nexus. Biol. Inspired Cognit. Archit. 1, 55–72. (doi: 10.1016/j.bica.2012.04.005)

Verschure PFMJ., 'A chronology of Distributed Adaptive Control', in Tony J. Prescott, Nathan Le-pora, and Paul F.M.J Verschure (eds), Living machines: A handbook of research in biomimetics and biohybrid systems (Oxford, 2018; online edn, Oxford Academic, 21 June 2018),

Verschure, P. 2012. Distributed adaptive control: a theory of the mind, brain, body nexus. Biologically Inspired Cognitive Architectures. 1:55-72

Verschure PFMJ. 2016 Synthetic consciousness: the distributed adaptive control perspective Phil. Trans. R. Soc. B3712015044820150448.

Viard A, Chételat G, Lebreton K, Desgranges B, Landeau B, de La Sayette V, Eustache F, Piolino P. Mental time travel into the past and the future in healthy aged adults: an fMRI study. Brain Cogn. 2011 Feb;75(1):1-9. doi: 10.1016/j.bandc.2010.10.009. Epub 2010 Nov 19. PMID: 21093970.

Vig E, Dorr M, and Cox D. Large-scale optimization of hierarchical features for saliency prediction in natural images. In Computer Vision and Pattern Recognition, 2014. CVPR'14. IEEE Conference on. IEEE, 2014.

Villalobos P, Sevilla J, Besiroglu T, Heim L, Ho A, 2022. Hobbhahn M. Machine Learning Model Sizes and the Parameter Gap. https://arxiv.org/pdf/2207.02852.pdf

Vinanzi, S., Patacchiola, M., Chella, A., and Cangelosi, A. (2019). Would a robot trust you? Devel-op-mental robotics model of trust and theory of mind. Phil. Trans. R. Soc. B. 374:20180032. doi: 10.1098/ rstb.2018.0032

Vincenzo Lomonaco., Oct 24, 2019. Source https://www.numenta.com/blog/2019/10/24/machine-learning-guide-to-htm/

Vircikova, Maria & Magyar, Gergely & Sincak, Peter. (2015). The Affective Loop: A Tool for Au-tono-mous and Adaptive Emotional Human-Robot Interaction. Advances in Intelligent Systems and Computing. 345. 247-254, 10.1007/978-3-319-16841-8_23.

Vogt P.A.and Divina F.a b. social symbol grounding and language evolution. Interaction Studies, 8(1):31–52, 2007.

Volman V., Ben-Jacob E., Levine H. The astrocyte as a gatekeeper of synaptic information trans-fer. Neural Comput. 2007; 19:303–326. doi: 10.1162/neco.2007.19.2.303

Volterra A and Meldolesi. J.2005. Astrocytes, from brain glue to communication elements: the rev-olution continues. Nature Reviews Neuroscience 6, 8 (2005), 626.

Vouloutsi V., Lallée S., Verschure P. F. M. J. (2013). Modulating behaviours using allostatic control, in Biomimetic and Biohybrid Systems (Berlin; Heidelberg: Springer Berlin Heidelberg;), 287–298Vygotsky, L. (1962). Thought and language. (E. Hanfmann & G. Vakar, Eds.). MIT Press. https://doi.org/10.1037/11193-000

W

Walker, E., & Blumer, D. The localization of sex in the brain. In K.J. Zulch, O. Creutzfeldt, and G. Galbraith, eds. Cerebral Localization, Berlin and New York: Springer-Verlag, 1975.

Walther D, Itti L, Riesenhuber M, Poggio, T,Koch C.. Attentional Selection for Object Recognition – a Gentle Way. Lecture Notes in Computer Science, 2525(1):472–479, 2002.

Wan, C. et al. Artificial sensory memory. Adv. Mater.32, 1902434 (2019).17.

Wan, J. Tang, S Yan, H. Wang D. Li, S and Vasilakos A. V., "Cloud robotics: Current status and open issues," in IEEE Access, vol. 4, pp. 2797-2807, 2016, doi: 10.1109/ACCESS.2016.2574979.

Wan, C., Cai, P., Guo, X. et al. An artificial sensory neuron with visual-haptic fusion. Nat Com-mun 11, 4602 (2020). https://doi.org/10.1038/s41467-020-18375-y.

Wang H, Chen C, Li L, Peng C. (2023). Video Saliency Object Detection with Motion Quality Com-pen-sation. Electronics. 12. 1618. 10.3390/electronics12071618.

Wang Z, Solloway T, Shiffrin RM, Busemeyer JR. Context effects produced by question orders re-veal quantum nature of human judgments. Proc Natl Acad Sci US A. 2014 Jul 1;111(26):9431-6. doi: 10.1073/ pnas.1407756111. Epub 2014 Jun 16. PMID: 24979797; PMCID: PMC4084470.

Wang, QI and Ding, Yongmei and Wu, Shiqian, 2022. Human Implicit Intention Inference Based on Gaze Information. Available at https://ssrn.com/abstract=4086566 or http://dx.doi.org/10.2139/ssrn.4086566.

Wang R, Thakur CS, Cohen G, Hamilton TJ, Tapson J, van Schaik A. Neuromorphic Hardware Ar-chitec-ture Using the Neural Engineering Framework for Pattern Recognition. IEEE Trans Biomed Circuits Syst. 2017 Jun;11(3):574-584. doi: 10.1109/TBCAS.2017.2666883. Epub 2017 May 19. PMID: 28436888.

Wang, S., Chen, X., Zhao, C. et al. An organic electrochemical transistor for multi-modal sensing, memory and processing. Nat Electron 6, 281–291 (2023). https://doi.org/10.1038/s41928-023-00950-y.

Wang Y, Wang, Y, & Patel, S. Patel, D. (2006). A layered reference model of the brain (LRMB). Systems, Man, and Cybernetics, Part C: Applications and Reviews, IEEE Transactions on. 124 – 133. 10.1109/ TSMCC.2006.871126.

Wang, Y, Song, W, Tao, W, Liotta, A, Yang, D, Li, X, Gao, S, Sun, Y, Ge, W, Zhang, W. Zhang, W. (2022). A Systematic Review on Affective Computing: Emotion Models, Databases, and Recent Advances.

Wang, X., and Krumhuber, E. G. (2018). Mind perception of robots varies with their economic versus social function. Front. Psychol. 9:1230. doi: 10.3389/fpsyg.2018.01230

Wang XJ, Kennedy H. Brain structure and dynamics across scales: in search of rules. Curr Opin Neurobi-ol. 2016 Apr; 37:92-98. doi: 10.1016/j.conb.2015.12.010. Epub 2016 Feb 8. PMID: 26868043; PMCID: PMC5029120.

Wannig A, Stanisor L, Roelfsema PR. Automatic spread of attentional response modulation along Gestalt criteria in primary visual cortex. Nat Neurosci. 2011 Sep 18;14(10):1243-4. doi: 10.1038/nn.2910. PMID: 21926984.

Ward AF.Brain Drain: The Mere Presence of One's Own Smartphone Reduces Available Cognitive Capacity. Adrian F. Ward, Kristen Duke, Ayelet Gneezy, and Maarten W. Box. Journal of the Asso-ciation for Consumer Research (April 2017).

Ward AF Bronkhorst, A W The Cocktail Party Phenomenon: A Review on Speech Intelligibility in Multiple-Talker Conditions. Acta Acustica – Acustica 86 (2000): 117–128.

Waters AM, Henry J, Mogg K, Bradley BP, Pine DS. Attentional bias towards angry faces in child-hood anxiety disorders. J Behav Ther Exp Psychiatry. 2010 Jun;41(2):158-64. doi: 10.1016/j.jbtep.2009.12.001. Epub 2009 Dec 28. PMID: 20060097.

Watkins, C.J.C.H. (1989). Learning from delayed rewards. PhD Thesis, University of Cambridge, England.

Watkins,C.J.C.H., Dayan, P. Q-learning. Mach Learn 8, 279–292 (1992). https://doi.org/10.1007/BF00992698.

Waydo, S., Kraskov, A., Quian Quiroga, R., Fried, I. and Koch, C. 2006. 'Sparse Representation in the Human Medial Temporal Lobe'. Journal of Neuroscience, 26: 10232–10234.

Waytz A, Epley N, Cacioppo JT. Social Cognition Unbound: Insights into Anthropomorphism and Dehumanization. Curr Dir Psychol Sci. 2010 Feb;19(1):58-62. doi: 10.1177/0963721409359302. PMID: 24839358; PMCID: PMC4020342.

Wearing, Deborah (2006). Forever Today: A Memoir of Love and Amnesia (Paper-back) (Updated ed.). Corgi. ISBN 0-552-77169-4.

Wegner, D. M. (2002). The illusion of conscious will. Boston Re-view. https://doi.org/10.7551/mit-press/3650.001.0001

Waibel M et al., "RoboEarth," in IEEE Robotics & Automation Magazine, vol. 18, no. 2, pp. 69-82, June 2011, doi: 10.1109/MRA.2011.941632.

Weigang Pan, Congcong Liu, Qian Yang, Yan Gu, Shouhang Yin, Antao Chen, The neural basis of trait self-esteem revealed by the amplitude of low-frequency fluctuations and resting state function-al connectivity, Social Cognitive and Affective Neuroscience, Volume 11, Issue 3, March 2016, Pages 367–376, https://doi.org/10.1093/scan/nsv119.

Weingarten CP, Doraiswamy PM, Fisher MP. A New Spin on Neural Processing: Quantum Cogni-tion. Front Hum Neurosci. 2016 Oct 26;10:541. doi: 10.3389/fnhum.2016.00541. PMID: 27833543; PMCID: PMC5080346.

Weizenbaum, J. (1966). ELIZA—a computer program for the study of natural language communi-cation between man and machine. Communications of the ACM, 9, 36 – 45.

Welford, A. T. (1952). The 'psychological refractory period' and the timing of high-speed perfor-mance—a review and a theory. British Journal of Psychology, 43, 2–19.

Wells, A. (1995). Meta-cognition and worry: A cognitive model of generalized anxiety disor-der. Behav-ioural and Cognitive Psychotherapy, 23(3), 301–320.

Wells, G. L., Memon, A., and Penrod, S. D. (2006). Eyewitness evidence: improving its probative value. Psychol. Sci. Public Inter. 7, 45–75. doi: 10.1111/j.1529-1006.2006.00027.x

Wells, A (2008). Meta-cognitive therapy for anxiety and depression. New York: Guilford Press.

Wellsby, M, Pexman, P. (2014). The Influence of Bodily Experience on Children's Language Pro-cessing. Topics in Cognitive Science. 6. 10.1111/tops.12092.

Weng J, McClelland J, Pentland A, Sporns O, Stockman I, Sur M, Thelen E. Artificial intelligence. Autonomous mental development by robots and animals. Science. 2001 Jan 26;291(5504):599-600. doi: 10.1126/science.291.5504.599. PMID: 11229402.

Werner G. Viewing brain processes as Critical State Transitions across levels of organization: Neu-ral events in Cognition and Consciousness, and general principles. Biosystems. 2009 Apr;96(1):114-9. doi: 10.1016/j.biosystems.2008.11.011. Epub 2008 Dec 11. PMID: 19124060.

Wesley A., 1989 Analog VLSI and Neural Systems' 2 J. Neural Eng. 13 (2016) 051001 Topical Review.

Wheatley T, Wegner DM. Automaticity of action, psychology of. In: International Encyclopedia of the Social & Behavioural Sciences. Elsevier; 2001:991-993. doi:10.1016/B0-08-043076-7/01747-2

Whitford T J, Bradley N Jack N B. Daniel Pearson D Oren Griffiths O David Luque D, Anthony WF Harris AWF, Kevin M Spencer KM, Mike E Le Pelley ME,(2017) Neurophysiological evidence of efference copies to inner speecheLife 6:e28197

Whittington JCR, Bogacz R. Theories of Error Backpropagation in the Brain. Trends Cogn Sci. 2019 Mar;23(3):235-250. doi: 10.1016/j.tics.2018.12.005. Epub 2019 Jan 28. PMID: 30704969; PMCID: PMC6382460.

Whyte CJ, Smith R. The predictive global neuronal workspace: A formal active inference model of visual consciousness. Prog Neurobiol. 2021 Apr; 199:101918. doi: 10.1016/j.pneurobio.2020.101918. Epub 2020 Oct 8. PMID: 33039416.

William James. The Principles of Psychology (1890),

Willmore, K.E. An Introduction to Evolutionary Developmental Biology. Evo Edu Outreach 5, 181–183 (2012). https://doi.org/10.1007/s12052-012-0422-1.

Wilming, N., Betz, T., Kietzmann, T. C., & König, P. (2011). Measures and limits of models of fix-ation selection. PLoS ONE, 6(9), Article e24038. https://doi.org/10.1371/journal.pone.0024038

Wilson, E., & MacLeod, C. (2003). Contrasting two accounts of anxiety-linked attentional bias: selective attention to varying levels of stimulus threat intensity. Journal of abnormal Psycholo-gy, 112(2), 212.

Wimmer H., Perner J. (1983). Beliefs about beliefs: Representation and constraining function of wrong beliefs in young children's understanding of deception. Cognition, 13, 103– 128

Wittgenstein, Ludwig (1958). The Blue and Brown Books: Preliminary Studies for the 'Philosophi-cal Investigations'. Oxford, England: Harper & Row.

Witkowski, Tomasz (1 January 2010). "Thirty-Five Years of Research on Neuro-Linguistic Pro-gramming. NLP Research Data Base. State of the Art or Pseudoscientific Decoration?". Polish Psy-chological Bulletin. 41 (2). doi:10.2478/v10059-010-0008-0.

Witton C., Sergeyev S. V., Turitsyna E. G., Furlong P. L., Seri S., Brookes M., et al. (2019). Rogue bioel-ectrical waves in the brain: the Hurst exponent as a potential measure for presurgical mapping in epilepsy. J. Neural Eng. 16:56019. 10.1088/1741-2552/ab225e.

Wittrock, M. C. (1990). Generative processes of comprehension. Educational Psychologist, 24, 345–376.

Wixted, J. T., and Wells, G. L. (2017). The relationship between eyewitness confidence and identi-fication accuracy: a new synthesis. Psychol. Sci. Public Inter. 18, 10–65. doi: 10.1177/1529100616686966.

Wolfe, JM., 'Guided Search 4.0: Current Progress with a Model of Visual Search', in Wayne D. Gray (ed.), Integrated Models of Cognitive Systems (New York, 2007; online edn, Oxford Aca-demic, 22 Mar. 2012),

Wolfe, J.M. Guided Search 6.0: An updated model of visual search. Psychon Bull Rev 28, 1060–1092 (2021). https://doi.org/10.3758/s13423-020-01859-9

Wolff M, Vann SD. The Cognitive Thalamus as a Gateway to Mental Representations. J Neurosci. 2019 Jan 2;39(1):3-14. doi: 10.1523/JNEUROSCI.0479-18.2018. Epub 2018 Nov 2. PMID: 30389839; PM-CID: PMC6325267.

Worden R, Bennett MS, Neacsu V. The Thalamus as a Blackboard for Perception and Planning. Front Behav Neurosci. 2021 Mar 1; 15:633872. doi: 10.3389/fnbeh.2021.633872. PMID: 33732119; PMCID: PMC7956969.

Wu, H. Lin, H. Guan, Y Harada K. and Rojas J, "Robot introspection with Bayesian nonparametric vector autoregressive hidden Markov models," 2017 IEEE-RAS 17th International Conference on Humanoid Robotics (Humanoids), Birmingham, UK, 2017, pp. 882-888, doi: 10.1109/HUMANOIDS.2017.8246976.

Wu Z, Kwon M, Daptardar S, Schrater P, Pitkow X. Rational thoughts in neural codes. Proc Natl Acad Sci US A. 2020 Nov 24;117(47):29311-29320. doi: 10.1073/pnas.1912336117. PMID: 33229521; PMCID: PMC7703629.

Wutz A, Muschter E, van Koningsbruggen MG, Weisz N, Melcher D. Temporal Integration Windows in Neural Processing and Perception Aligned to Saccadic Eye Movements. Curr Biol. 2016 Jul 11;26(13):1659-1668. doi: 10.1016/j.cub.2016.04.070. Epub 2016 Jun 9. PMID: 27291050; PMCID: PMC4942674.

X

Xu, F., Liu, L. & Xiao, J. Analysis of the equilibrium points of background neural networks with uniform firing rate. Adv Differ Equ 2017, 314 (2017). https://doi.org/10.1186/s13662-017-1322-z

Y

Yale University. "Brain Energy Use Key To Understanding Consciousness." ScienceDaily. ScienceDaily, 16 June 2009. <www.sciencedaily.com/releases/2009/06/090615171517.htm>.

Yamazaki K, Vo-Ho VK, Bulsara D, Le N. Spiking Neural Networks and Their Applications: A Review. Brain Sci. 2022 Jun 30;12(7):863. doi: 10.3390/brainsci12070863. PMID: 35884670; PMCID: PMC9313413.

Yan, F., Iliyasu, A.M. & Venegas-Andraca, S.E. A survey of quantum image representa-tions. Quantum Inf Process 15, 1–35 (2016). https://doi.org/10.1007/s11128-015-1195-6

Yang, Xin-She. (2010). Firefly Algorithms for Multi-modal Optimization. 5792. 10.1007/978-3-642-04944-6_14.

Yang GZ, Bellingham J, Dupont PE, Fischer P, Floridi L, Full R, Jacobstein N, Kumar V, McNutt M, Merrifield R, Nelson BJ, Scassellati B, Taddeo M, Taylor R, Veloso M, Wang ZL, Wood R. The grand challenges of Science Robotics. Sci Robot. 2018 Jan 31;3(14): eaar7650. doi: 10.1126/scirobotics. aar7650. PMID: 33141701.

Ye, Haosheng & ZENG, Hong & Wendeng, Yang. (2019). Enactive cognition: Theoretical rationale and practical approach. Acta Psychologica Sinica. 51. 1270. 10.3724/SP.J.1041.2019.01270.

Ye P, Wang T, Wang FY. A Survey of Cognitive Architectures in the Past 20 Years. IEEE Trans Cybern. 2018 Dec;48(12):3280-3290. doi: 10.1109/TCYB.2018.2857704. Epub 2018 Aug 2. PMID: 30072355.

Yeh IL, Holst-Wolf J, Elangovan N, Cuppone AV, Lakshminarayan K, Cappello L, Masia L, Konczak J. Effects of a robot-aided somatosensory training on proprioception and motor function in stroke survivors. J Neuroeng Rehabil. 2021 May 10;18(1):77. doi: 10.1186/s12984-021-00871-x. Erratum in: J Neuroeng Rehabil. 2022 Jul 18;19(1):74. PMID: 33971912; PMCID: PMC8112068.

Yim, M., Duff, D.G., Roufas, K.D.. (2000). PolyBot: a modular reconfigurable robot. 1. 514 – 520 vol.1. 10.1109/ROBOT.2000.844106.

Yu Bai, Dorien Emmers, Ying Li, Lei Tang, 2022 Parental locus of control and early childhood de-velop-ment: Evidence on parent and grandparent caregivers in rural China, China Economic Review,Volume 74,2022,101814,ISSN 1043-951X, https://doi.org/10.1016/j.chieco.2022.101814.

Yu, C,E and Ngan, H.F.B. (2019), "The power of head tilts: gender and cultural differences of per-ceived human vs human-like robot smile in service", Tourism Review, Vol. 74 No. 3, pp. 428-442. https://doi.org/10.1108/TR-07-2018-0097

Yu J, Yi Z, Zhang L. Representations of continuous attractors of recurrent neural networks. IEEE Trans Neural Netw. 2009 Feb;20(2):368-72. doi: 10.1109/TNN.2008.2010771. Epub 2009 Jan 13. PMID: 19150791.

Yufik YM, Friston K. Life and Understanding: The Origins of "Understanding" in Self-Organizing Nervous Systems. Front Syst Neurosci. 2016 Dec 9; 10:98. doi: 10.3389/fnsys.2016.00098. PMID: 28018185; PMCID: PMC5145877.

Z

Zadeh, L.A. (1965), “Fuzzy sets”, Information and Control, Vol. 8 No. 3, pp. 338-353.

Zador AM. A critique of pure learning and what artificial neural networks can learn from animal brains. Nat Commun. 2019 Aug 21;10(1):3770. doi: 10.1038/s41467-019-11786-6. PMID: 31434893; PMCID: PMC6704116.

Zadra A, Desjardins S, Marcotte E. Evolutionary function of dreams: A test of the threat simulation theory in recurrent dreams. Conscious Cogn. 2006 Jun;15(2):450-63. doi: 10.1016/j.concog.2005.02.002. PMID: 16720254.

Zagzebski, L. “Recovering Understanding” In M. Steup (ed.), Knowledge, Truth and Obligation. Oxford: Oxford University Press, 2001.

Zalewski L. J., Sink C. A., Yachimowicz D.J. (1992) Using Cerebral Dominance for Educational Programs. The Journal of General Psychology 119 (1), 45-57.

Zandersen M, Parnas J, Identity Disturbance, Feelings of Emptiness, and the Boundaries of the Schizophrenia Spectrum, Schizophrenia Bulletin, Volume 45, Issue 1, January 2019, Pages 106–113, https://doi.org/10.1093/schbul/sbx183

Zhang H, Lee S. Robot Bionic Vision Technologies: A Review. Applied Sciences. 2022; 12(16):7970. https://doi.org/10.3390/app12167970

Zeevi, L., klein Selle, N., Kellmann, E.L. et al. Bio-behavioral synchrony is a potential mechanism for mate selection in humans. Sci Rep 12, 4786 (2022). https://doi.org/10.1038/s41598-022-08582-6.

Zellig S. Harris (1954) Distributional Structure, WORD, 10:2-3, 146-162, DOI: 10.1080/00437956.1954.11659520.

Zhao, S, Jia, G, Yang, J, Ding, G, Keutzer, K (2021). Emotion Recognition from Multiple Modali-ties: Fundamentals and Methodologies.

Zhaoyuan G, Zhenzhong J,,Howie C. (2019). Adversary A3C for Robust Reinforcement Learning.

Zidan, M.A., Strachan, J.P. & Lu, W.D. The future of electronics based on memristive systems. Nat Electron 1, 22–29 (2018). https://doi.org/10.1038/s41928-017-0006-8

Zinkivskay A, Nazir F, Smulders TV. What-where-when memory in magpies (Pica pica) Anim Cogn. 2009;12(1):119–125.

Zhou, Xuefeng & Wu, Hongmin & Rojas, Juan & Xu, Zhihao & Li, Shuai. (2020). Introduction to Robot Introspection. 10.1007/978-981-15-6263-1_1.

Zhou, C., Huang, B. & Fränti, P. A review of motion planning algorithms for intelligent robots. J Intell Manuf 33, 387–424 (2022). https://doi.org/10.1007/s10845-021-01867-z

Zylberberg A, Fetsch C Shadlen MN (2016) The influence of evidence volatility on choice, reaction time and confidence in a perceptual deci-sion eLife 5: e17688.

Made in the USA
Monee, IL
11 June 2025

19209046R00372